Proceedings of the 9th International Conference

PATH INTEGRALS

NEW TRENDS AND PERSPECTIVES

Proceedings of the 9th International Conference

PATH INTEGRALS

NEW TRENDS AND PERSPECTIVES

Dresden, Germany September 23 – 28, 2007

editors

Wolfhard Janke

Universität Leipzig, Germany

Axel Pelster

Universität Duisburg-Essen, Germany

World Scientific

NEW JERSEY • LONDON • SINGAPORE • BEIJING • SHANGHAI • HONG KONG • TAIPEI • CHENNAI

Published by

World Scientific Publishing Co. Pte. Ltd.
5 Toh Tuck Link, Singapore 596224
USA office: 27 Warren Street, Suite 401-402, Hackensack, NJ 07601
UK office: 57 Shelton Street, Covent Garden, London WC2H 9HE

British Library Cataloguing-in-Publication Data
A catalogue record for this book is available from the British Library.

PATH INTEGRALS — NEW TRENDS AND PERSPECTIVES
Proceedings of the 9th International Conference

ISBN-13 978-981-283-726-4
ISBN-10 981-283-726-4

Printed in Singapore by Mainland Press Pte Ltd

PREFACE

The 9th International Conference on *Path Integrals – New Trends and Perspectives* (PI07) was held at the Max Planck Institute for the Physics of Complex Systems (MPI PKS) in Dresden, Germany during the period September 23–28, 2007. The present volume contains the material of most of the talks delivered at this meeting and a selection from the poster presentations.

The general format and style of the conference followed the accepted and well-developed pattern of the series, focusing on the development, refinement, and important applications of the techniques of path integration. Since its founding in 1983, the series of international path-integral conferences has been coordinated by an International Advisory Committee (IAC) whose permanent members are listed on page vii. The previous conferences in Albany, NY (1983), Bielefeld (1985), Bangkok (1989), Tutzing (1992), Dubna (1996), Florence (1998), Antwerp (2002), and Prague (2005) have emphasized the broad range of path-integral applications in many fields of physics and chemistry. Thus, the series has an interdisciplinary role, bringing together scientists whose interest in the path integral spans many fields, allowing them to exchange opinions, discuss problems, and disseminate new ideas. In particular, the series aims at fostering the exchange of ideas and techniques among physicists applying many-body techniques in such diverse areas as nuclear and subnuclear physics, atomic and molecular physics, quantum chemistry, condensed matter physics, complex systems, quantum field theory, elementary particle physics, and astrophysics.

Continuing this tradition, the 9th International Conferencex under the motto *Path Integrals – New Trends and Perspectives* focused on recent developments and new directions of the path-integral approach in this transdisciplinary setting. This is reflected in the main topics in this Proceedings volume, which first cover in two parts more traditional fields such as general quantum physics and quantum field theory. Applications of generalized path integrals to quantum gravity, astrophysics, and cosmology are separately collected in the next part. This is followed by articles on statistical field theory and an overview of the many modern Monte Carlo techniques based on the path-integral approach. The next two parts deal with topical subjects such as atomic and molecular Bose-Einstein condensation and many other modern developments in condensed matter physics where path integrals have turned out to be a useful tool. Next, the part on spin models surveys phenomena that can be modeled by discrete lattice formulations. Finally, the concluding part reports on many other challenging applications, ranging from the study of thermal fluctuations of polymers and membranes in the realm of biophysics to the treatment of stochastic processes such as option pricing and risk management in econophysics.

The Max Planck Institute for the Physics of Complex Systems (MPI PKS) in Dresden provided a truly perfect and stimulating environment for such a meeting. A total of about 120 scientists from all over the world attended, representing more than 20 different countries. Plenary and parallel lectures comprised the core of the scientific program and were supplemented by a poster presentation on Monday evening. The special social event was an excursion on Wednesday afternoon with Martin Gutzwiller's "River Elbe boat talk" posing the provocative question *Quo vadis, physica?*. The day was concluded by the conference dinner at Fortress Königstein. On Thursday evening, a very active round-table discussion, chaired by Cécille DeWitt-Morette, continued the theme by providing controversial answers to the question *Quo vadis, path integrals?*. This was preceded by historical reflections on both the personal and the scientific relationship between Einstein and Feynman in the evening talk of Tilman Sauer. These three special events are also documented in the prologue of the present volume.

We gratefully acknowledge generous financial support for the conference from the MPI PKS, the German Science Foundation (DFG), and the Wilhelm and Else Heraeus Foundation. Furthermore, we cordially thank Marita Schneider for her efficient assistance with all organisational matters and Andreas Nußbaumer for his invaluable technical assistance in producing the abstract booklet for the conference. Finally, we also thank him along with Elmar Bitter, Aristeu Lima, and Moritz Schütte for their help during the conference.

Wolfhard Janke
Universität Leipzig, Germany

Axel Pelster
Universität Duisburg-Essen, Germany
July 2008

CONFERENCE COMMITTEES

INTERNATIONAL ADVISORY COMMITTEE

for the Series of International Conferences on Path Integrals

Jozef T. Devreese	– Universiteit Antwerpen, Belgium
Hermann Grabert	– Universität Freiburg, Germany
Martin C. Gutzwiller	– Yale University, New York, USA
Akira Inomata	– State University of New York at Albany, USA
John R. Klauder	– University of Florida at Gainesville, USA
Virulh Sa-yakanit	– Chulalongkorn University, Bangkok, Thailand
Lawrence S. Schulman	– Clarkson University at Potsdam, USA
Valerio Tognetti	– Università di Firenze, Italy
Ulrich Weiß	– Universität Stuttgart, Germany
Vladimir S. Yarunin	– Joint Institute for Nuclear Research, Dubna, Russia

LOCAL ORGANIZING COMMITTEE

9th International Conference on
Path Integrals – New Trends and Perspectives,
Max Planck Institute for the Physics of Complex Systems,
Dresden, Germany, September 23–28, 2007

Wolfhard Janke	– Universität Leipzig, Germany
Axel Pelster	– Universität Duisburg-Essen, Germany

Conference secretary:

Marita Schneider	– MPI PKS Dresden, Germany

Group photo of the participants of the 9th International Conference on *Path Integrals – New Trends and Perspectives* at the Max Planck Institute for the Physics of Complex Systems, Dresden, Germany, September 23–28, 2007 (see list of participants on page 595).

CONTENTS

Part III Quantum Field Theory 117

Part IV Quantum Gravity 179

Part IX Spin Models 485

Part X Biophysics and Stochastics 529

PART I

History and Perspectives

REMARKS ON THE ORIGIN OF PATH INTEGRATION: EINSTEIN AND FEYNMAN

T. SAUER

Einstein Papers Project,
California Institute of Technology 20-7,
Pasadena, CA 91125, USA
E-mail: tilman@einstein.caltech.edu

I offer some historical comments about the origins of Feynman's path-integral approach, as an alternative approach to standard quantum mechanics. Looking at the interaction between Einstein and Feynman, which was mediated by Feynman's thesis supervisor John Wheeler, it is argued that, contrary to what one might expect, the significance of the interaction between Einstein and Feynman pertained to a critique of classical field theory, rather than to a direct critique of quantum mechanics itself. Nevertheless, the critical perspective on classical field theory became a motivation and point of departure for Feynman's space-time approach to non-relativistic quantum mechanics.

Keywords: History of quantum mechanics; Einstein; Feynman.

1. Introduction

In this paper, I am interested in the genesis of Feynman's path-integral approach to non-relativistic quantum mechanics. I take Feynman's 1948 paper on "A Space-Time Approach to Quantum Mechanics"[1] as the point in time when the approach was fully formulated and published and made available to the community of physicists. I will take a look into the prehistory of Feynman's 1948 paper. I shall not attempt to give anything like a balanced, or even complete historical account of this prehistory. Instead, I will focus on a little footnote in Feynman's paper:

> The theory of electromagnetism described by J. A. Wheeler and R. P. Feynman, Rev. Mod. Phys. **17**, 157 (1945) can be expressed in a principle of least action involving the coordinates of particles alone. It was an attempt to quantize this theory, without reference to the fields, which led the author to study the formulation of quantum mechanics given here. The extension of the ideas to cover the

> case of more general action functions was developed in his Ph.D. thesis, "The principle of least action in quantum mechanics" submitted to Princeton University, 1942. (Ref. 1, p. 385)

My guide in organizing my remarks will be to look at what we know about any direct and indirect interaction between Feynman and Einstein. Let me briefly motivate this focus on Einstein and Feynman.

Feynman was born in New York in 1918, did his undergraduate studies at MIT, and took his Ph.D. with John A. Wheeler at Princeton University in 1942, before going to Los Alamos during the war years. After the war, he was first at Cornell and in 1951 he went to Caltech. In 1954, Feynman received the Einstein Award,[2] as a 36-year old man for his work on quantum electrodynamics that in 1965 would earn him the Nobel prize for physics.

The Einstein Award was a prestigious award, established in 1949 in Einstein's honor, but it seems that Einstein had not much to do with the awarding of the prize to Feynman.

At the time of Feynman's receiving the Einstein award, Einstein himself was a 76-year old world famous man. He had been living in Princeton since his emigration from Nazi-Germany in 1933 and was scientifically engaged in a search for a unified field theory of gravitation and electromagnetism.[3] But he also still thought about problems of the foundations of quantum mechanics. Among his extensive research notes and manuscripts with calculations along the unified field theory program, there is, e.g., a manuscript page from around 1954 with a concise formulation of Einstein's of the famous Einstein-Podolsky-Rosen incompleteness argument for standard quantum mechanics. Probably in reaction to David Bohm's reformulation of the original argument, Einstein here also formulates the incompleteness argument for spin observables.[4]

With both Feynman and Einstein being concerned with the foundations of quantum mechanics, one might hope that an interaction between the two physicists, if there was any, might give us some insight into the historical development of our understanding of the principles of quantum theory.

A similar question was also asked once by Wheeler. In 1989, after Feynman's death, he recalled:

> Visiting Einstein one day, I could not resist telling him about Feynman's new way to express quantum theory.[5]

After explaining the basic ideas of Feynman's path-integral approach to Einstein, Wheeler recalls to have asked:

> "Doesn't this marvelous discovery make you willing to accept quantum theory, Professor Einstein?" He replied in a serious voice, "I still cannot believe that God plays dice. But maybe," he smiled, "I have earned the right to make my mistakes."[5]

So is this the end of my story?

According to Feynman's own account, he himself met Einstein only twice. One of these encounters was at the occasion of Feynman's first technical talk, as a young graduate student, in the Princeton physics department. The occasion probably took place in late 1940. Wheeler had suggested that Feynman was to talk on their joint work, and Feynman recalls

> Professor Wigner was in charge of the colloquium, so after I said I would do it, he told me that he had heard from Wheeler about the work and he knew something about it. I think we had discussed it a little bit with him. And he thought it was important enough that he had taken the liberty to invite especially Professor Henry Norris Russell from the astronomy department, the great astronomer, you know, John von Neumann from the mathematics department, the world's great mathematician, and Professor Pauli, who was visiting from Zurich, would be there. And Professor Einstein had been especially invited—and although he never comes to the colloquia, he thinks he will come!
> So I went through fire on my first. I must have turned a yellowish-green or something [...].[a]

Feynman continues to recount details of this seminar, he relates how his excitement and anxiety abated once he started to talk about physics, and indicates how some members of his audience, including Einstein, reacted to his presentation in question time.

In the following, I will take this encounter between Feynman and Einstein as a point of departure for a historical argument: the interaction between Feynman and Einstein reminds us of a significant historical context of discovery of the path-integral method. This original context is still prominently visible in Feynman's 1942 thesis[9] but it is already reduced to a footnote in his 1948 publication.[1] For an appreciation of the path-integral method, even today, it may nevertheless still be useful to recall the historical circumstances of its discovery.

---

[a]Ref. 6, p. 133. I am quoting from the transcript of an oral history interview conducted by C. Weiner with Feynman in 1966. For slightly different versions of the episode, see also Ref. 7, pp. 64f, and Ref. 8, p. 66.

Specifically, I will address and discuss the following four questions:

(1) What is the Wheeler-Feynman theory that Feynman presented in his first seminar at Princeton?
(2) What does Einstein have to do with this?
(3) What does this have to do with path integrals?
(4) Why is this context of the origin of the path-integral approach only mentioned in a footnote in Feynman's 1948 paper?

Most of the information that the argument is based on can be found in Schweber's book on the history of quantum electrodynamics.[10] The significance of the Wheeler-Feynman theory for Feynman's subsequent development is also emphasized by Feynman himself in his Nobel lecture.[11]

2. The Wheeler-Feynman Absorber Theory

The results of their joint work that Feynman presented in the Princeton physics colloquium were not published at the time. Feynman gave his presentation again, shortly thereafter, at a meeting of the American Physical Society in Cambridge, Massachussetts, that took place on 21 and 22 February 1941. Of this talk, an abstract was published.[12] There exists also a typescript by Feynman giving an account of the theory,[13] dated to spring 1941 (Ref. 10, p. 383). The abstract identifies radiative damping as a problem in Lorentz's classical electron theory and in Dirac's theory of a point electron, and then summarizes the main points of Feynman's paper:

> We postulate (1) that an accelerated point charge in otherwise free space does *not* radiate energy; (2) that, in general, the fields which act on a given particle arise only from *other* particles; (3) that these fields are represented by one-half the retarded plus one-half the advanced Liénard-Wiechert solutions of Maxwell's equations. In a universe in which all light is eventually absorbed, the absorbing material scatters back to an accelerated charge a field, part of which is found to be independent of the properties of the material. This part is equivalent to one-half the retarded field *minus* one-half the advanced field generated by the charge. It produces radiative damping (Dirac's expression) and combines with the field of the source to give retarded effects alone.[12]

A detailed account of the Wheeler-Feynman theory was published after the war in two papers. The first paper appeared in 1945 under the title

'Interaction with the Absorber as the Mechanism of Radiation.'[14] As indicated in a first footnote to the title, this paper essentially gives an account of the theory that Feynman had presented in 1941.

In a long second footnote to the title of the paper, Wheeler then explains that this first paper was actually planned to be the third part of a projected series of five papers. Of the missing four parts, only the second one actually appeared, four years later, in 1949, under the title 'Classical Electrodynamics in Terms of Direct Interparticle Action'.[15]

The core of the Wheeler-Feynman theory thus concerned a special problem that arose out of a broader research program, laid out, in part, in the later 1949 paper. In qualitative terms, the broader research program concerned this.

Among the many difficulties with attempts to come to a quantum theory of electrodynamics in the late thirties, Wheeler and Feynman thought some had to do with difficulties that occur already at the level of classical electrodynamic field theory. As a radical response, Wheeler and Feynman questioned whether the notion of an electromagnetic field is, in fact, a useful one. They argued that one should in principle be able to express all electromagnetic phenomena in terms of direct interaction between point-like particles. Any notion of a field would be a derived concept. The primary notion would be a collection of point-like charges that interact with each other through Liénard-Wiechert retarded and advanced potentials.

They found that such a theory is expressible in terms of an action principle that involves a variation over the world-lines of charged electrons. They wrote the action principle as[15]

$$J = -\sum_a m_a c \int (-da_\mu da^\mu)^{\frac{1}{2}} + \sum_{a<b} (e_a e_b / c) \\ \times \iint \delta(ab_\mu ab^\mu)(da_\nu db^\nu) = \text{extremum}, \quad (1)$$

where the sums are over electrons of mass m_a and charge e_a, da denotes derivative with respect to the respective proper time, and $ab^\mu \equiv a^\mu - b^\mu$ is short for the four-vector of the separation between the particles, a somewhat unusual notation introduced in order to be able to make use of the Einstein summation convention. The attractive feature of this action is that "all of mechanics and electrodynamics is contained in this single variational principle" (Ref. 15, p. 425). Note that the single action principle incorporates both the Maxwell equations and the Lorentz force law. The idea and the action (1) were known before, they can be found, in more or less explicit terms, in older papers by Schwarzschild,[16] Tetrode,[17] and Fokker.[18]

The only problem with this formulation was the issue of radiative reaction. In classical theory, an accelerated electron radiates and loses energy to the field. To avoid the notion of a field, Wheeler and Feynman postulated that a single electron alone in the universe, if accelerated, would, in fact, *not* radiate. Instead, they succeeded to show that radiative reaction can arise in a universe with a surrounding material that absorbs all outgoing radiation. The electrons of the absorber interact with the electron at the source through advanced potentials, such that an accelerated electron feels a radiative force. This is the main point that Feynman was elaborating on in his Princeton seminar.

3. Einstein and the Electromagnetic Arrow of Time

Feynman recalled that immediately after his presentation, Pauli asked critical questions and then asked Einstein whether he would agree.

> Anyway, Professor Pauli got up immediately after the lecture. He was sitting next to Einstein. And he says, "I do not think this theory can be right because of this, that and the other thing—" it's too bad that I cannot remember what, because the theory is not right, and the gentleman may well have hit the nail on the bazeeto, but I don't know, unfortunately, what he said. I guess I was too nervous to listen, and didn't understand the objections. "Don't you agree, Professor Einstein?" Pauli said at the end of his criticism. "I don't believe this is right—don't you agree, Professor Einstein?"
> Einstein said, "No," in a soft German voice that sounded very pleasant to me, and said that he felt that the one idea, the one thing that seemed to him, was that the principles of action and distance which were involved here were inconsistent with the field views, the theory of gravitation, of general relativity. But after all general relativity is not so well established as electrodynamics, and with this prospect I would not use that as an argument against you, because maybe we can develop a different way of doing gravitational interaction, too.
> Very nice. Very interesting. I remember that.[b]

We also know, both from Feynman (Ref. 6, p. 133) and Wheeler (Ref. 19, p. 167) as well as, independently, from a letter by Wheeler to Einstein, that

[b]Ref. 6, p. 134, see also Ref. 7, p. 66, Ref. 8, pp. 67–68.

Feynman and Wheeler visited Einstein once in his house in Princeton and discussed the "interpretation of the force of radiation in terms of advanced and retarded action at a distance."[20] It is unclear when the meeting took place,[c] and I am not aware of any detailed account of the discussion that took place, but it seems that Einstein alerted Feynman and Wheeler to existing literature on the subject, including some in which he himself was involved. In a footnote to their 1945 paper, Wheeler and Feynman acknowledge Einstein's input:

> We are indebted to Professor Einstein for bringing to our attention the ideas of Tetrode and also of Ritz, [...] (Ref. 14, p. 10).

Somewhere else in the article, they

> recall an inconclusive but illuminating discussion carried on by Ritz and Einstein in 1909, in which "Ritz treats the limitation to retarded potentials as one of the foundations of the second law of thermodynamics while Einstein believes that the irreversibility of radiation depends exclusively on considerations of probability." (Ref. 14, p. 160)

The Einstein-Ritz controversy,[22–24] from which they quoted, was about the origin of irreversibility of electromagnetic radiation phenomena.[26] In the 1941 typescript, Feynman observed that their theory is in full agreement with Einstein's position against Ritz, that the fundamental electrodynamical equations are time-reversal invariant, and that the radiative irreversibility is a macroscopic, statistical phenomenon:

> The apparent irreversibility in a closed system, then, either from our point of view or the point of view of Lorentz is a purely macroscopic irreversibility. The present authors believe that all physical phenomena are microscopically reversible, and that, therefore, all apparently irreversible phenomena are solely macroscopically irreversible. (Ref. 13, p. 13.1; quoted in Ref. 10, p. 386)

Feynman here has a footnote saying

---

[c]It is even unclear whether the meeting in Einstein's house took place before or after Feynman's Princeton colloquium. In 1966, Feynman did not remember but was "pretty sure" that it was before the colloquium "because he knew me," see Ref. 6, pp. 133, 139, Wheeler (Ref. 19, p. 167) recalls that it was "while working on our second action-at-a-distance paper", but from his letter to Einstein, we know that it must have been before November 1943, see also Ref. 21, p. 118.

> That this and the following statement are true in the Lorentz theory was emphasized by Einstein in a discussion with Ritz. (Einstein and Ritz, Phys. Zeits. **10**, 323 (1909)). Our viewpoint on the matter discussed is essentially that of Einstein. (We should like to thank Prof. W. Pauli for calling our attention to this discussion.) (ibid.)

Although Pauli is credited here for alerting Feynman to the Ritz-Einstein controversy, we may assume that the point was also a topic when Feynman and Wheeler discussed their ideas with Einstein during their visit at his Princeton home. There is, in any case, an English translation, in Feynman's hand, of the Ritz-Einstein controversy[24] in the Feynman papers.[25]

4. Path Integrals for Actions with No Hamiltonian

In 1942, Feynman was recruited for the Los Alamos project. Before leaving for Los Alamos, Wheeler urged Feynman to write up his thesis.[27] Feynman's thesis[9] is not directly dealing with the Wheeler-Feynman absorber theory but it rather gives a discussion of the 'Principle of Least Action in Quantum Mechanics', and is, in fact, a direct forerunner of Feynman's 1948 paper. But the thesis is very explicit about its original motivation. The discussion of quantizing systems expressed in terms of a Lagrangian is given in the context of solving the general problem of finding a quantum version of the Wheeler-Feynman theory of action-at-a-distance. The main point here is that

> the theory of action at a distance finds its most natural expression in a principle of least action, which is of such a nature that no Hamiltonian may be derived from it. That is to say the equations of motion of the particles cannot be put into Hamiltonian form in a simple way. This is essentially because the motion of one particle at one time depends on what another particle is doing at some other time, since the interactions are not instantaneous.[28]

This is not just a remark made in (a draft version of) the preface to motivate the approach. An example that derives directly from the action (1) is discussed also in the body of the text. At some point, Feynman explains how to generalize the quantization procedure to more general actions, for example those involving time-displaced interactions:

> The obvious suggestion is, then, to replace this exponent by $\frac{i}{\hbar}$ times the more general action. The action must of course first be

expressed in an approximate way in terms of q_i, t_i in such a way that as the subdivision becomes finer and finer it more nearly approaches the action expressed as a functional of $q(t)$.
In order to get a clearer idea of what this will lead to, let us choose a simple action function to keep in mind, for which no Hamiltonian exists. We may take,

$$\mathcal{A} = \int_{-\infty}^{\infty} \left\{ \frac{m\dot{x}(t)^2}{2} - V(x(t)) + k^2 \dot{x}(t)\dot{x}(t+\tau) \right\} dt,$$

which is an approximate action function for a particle in a potential $V(x)$ and which also interacts with itself in a mirror by half advanced and half retarded waves, [...]. (Ref. 9, p. 41)

In the 1941 typescript Feynman comes close to showing how this simple action follows from the general action (1) by considering the special case of two charges at a distance apart in otherwise free space, neglecting their electrostatic interaction. Of course, the path-integral quantization of actions that are non-local in time is considerably more involved[29] and Feynman does not give an explicit discussion of his example. Nevertheless, it confirms his remark in the (actual) preface of the thesis which

is concerned with the problem of finding a quantum mechanical description applicable to systems which in their classical analogue are expressible by a principle of least action, and not necessarily by Hamiltonian equations of motion. (Ref. 9, p. 6)

5. The Demise of the Early Context of Path Integration

In 1949, even before the second of the Wheeler-Feynman papers appeared in print, Feynman himself submitted another one of his famous papers, entitled 'Space-Time Approach to Quantum Electrodynamics.'[30] In it, one finds this little footnote:

These considerations make it appear unlikely that the contention of J. A. Wheeler and R. P. Feynman, Rev. Mod. Phys. **17**, 157 (1945), that electrons do not act on themselves, will be a successful concept in quantum electrodynamics. (Ref. 30, p. 773)

Why did Feynman retract a basic assumption of his joint work with Wheeler, with explicit reference to their earlier paper? Two years later, Feynman wrote a letter to Wheeler asking him about his opinion about the status of their earlier work:

> I wanted to know what your opinion was about our old theory of action at a distance. It was based on two assumptions:
> (1) Electrons act only on other electrons;
> (2) They do so with the mean of retarded and advanced potentials.
> The second proposition may be correct but I wish to deny the correctness of the first. The evidence is two-fold. First there is the Lamb shift in hydrogen which is supposedly due to the self-action of the electron. [...]
> The second argument involves the idea that the positrons are electrons going backwards in time. [...]
> So I think we guessed wrong in 1941. Do you agree?[31]

I am not aware of an explicit response by Wheeler to this letter, but several remarks in his autobiography[19] indicate that he, too, eventually gave up his belief in an action-at-a-distance electrodynamics: "[...] until the early 1950s, I was in the grip of the idea that Everything is Particles." (Ref. 19, p. 63)

For Feynman, one of the two reasons for giving up the theory of action-at-a-distance was an experimental finding, the Lamb shift. Lamb had presented data from his experiments on the fine structure of hydrogen at the Shelter Island conference. This conference, devoted to problems of the quantum mechanics of the electron, took place in June 1947 and was an event of considerable impact in the history of post-war physics (Ref. 10, ch. 4). It brought together the leading theorists for the first time after the war for a meeting which helped to determine the course of American physics in the atomic age. 9 of the 23 participants ended up being awarded the Nobel prize, a significant fraction of the participants were of the young generation. It was at this conference that Feynman presented his 'space-time approach to quantum mechanics', essentially the work of his thesis, and soon after the conference he penned his classic 1948 paper.[1]

Incidentally, the Shelter Island conference could have provided an occasion for a third encounter between Feynman and Einstein: following a suggestion of Wheeler, who was present as well, Einstein was among the invitees but he declined, due to ill health (Ref. 10, pp. 169f). It is tempting to speculate how Einstein would have reacted to Feynman's presentation of his new approach to quantum mechanics at this meeting.

When Feynman wrote up his approach for publication, he decided to mention the original motivation for his work only in passing. Given the generality of the path-integral formulation, it may be seen a wise decision on Feynman's part to reduce the historical context of its genesis to a footnote.

References

1. R. P. Feynman, *Rev. Mod. Phys.* **20**, 367 (1948). Reprinted in Ref. 9, pp. 71–112.
2. *New York Times*, 14 March 1954.
3. T. Sauer, in *The Cambridge Companion to Einstein*, ed. M. Janssen and C. Lehner (Cambridge University Press, Cambridge, forthcoming); http://philsci-archive.pitt.edu/archive/00003293.
4. T. Sauer, *Stud. Hist. Phil. Mod. Phys.* **38**, 879 (2007).
5. J. A. Wheeler, *Phys. Today* **42**, 24 (1989).
6. C. Weiner, *Interview with Dr. Richard Feynman, March 4 to June 28, 1966*, Niels Bohr Library & Archives, Amer. Inst. of Physics, College Park, MD.
7. R. P. Feynman, *"Surely You're Joking, Mr. Feynman!" Adventures of a Curious Character* (Bantam, Toronto, 1986).
8. R. Leighton (ed.), *Classic Feynman. All the Adventures of a Curious Character* (Norton, New York and London, 2006).
9. L. M. Brown (ed.), *Feynman's Thesis. A New Approach to Quantum Theory* (World Scientific, Singapore, 2005).
10. S. S. Schweber, *QED and the Men Who Made It: Dyson, Feynman, Schwinger, and Tomonaga* (Princeton University Press, Princeton, 1994).
11. R. P. Feynman, *Science* **153**, 699 (1966).
12. R. P. Feynman and J. A. Wheeler, *Phys. Rev.* **59**, 683 (1941).
13. "The Interaction Theory of Radiation." Feynman Papers, Caltech, folder 6.1.
14. J. A. Wheeler and R. P. Feynman, *Rev. Mod. Phys.* **17**, 157 (1945).
15. J. A. Wheeler and R. P. Feynman, *Rev. Mod. Phys.* **21**, 425 (1949).
16. K. Schwarzschild, *Nachr. Königl. Ges. Wiss. Göttingen, Math.-Phys. Kl.* (1903) 126.
17. H. Tetrode, *Zs. Phys.* **10**, 317 (1922).
18. A. D. Fokker, *Zs. Phys.* **58**, 386 (1929).
19. J. A. Wheeler with K. Ford, *Geons, Black Holes, and Quantum Foam* (Norton, New York, 1998).
20. J. A. Wheeler to A. Einstein, 3 November 1943, Albert Einstein Archives, The Hebrew University of Jerusalem, call nr. 23-442.
21. J. Gleick, *Genius. The Life and Science of Richard Feynman* (Pantheon, New York, 1992).
22. A. Einstein, *Phys. Zs.* **10**, 185 (1909).
23. W. Ritz, *Phys. Zs.* **9**, 903 (1908); **10**, 224 (1908).
24. W. Ritz and A. Einstein, *Phys. Zs.* **10**, 323 (1909).
25. Feynman Papers, Caltech, folder 6.19.
26. J. Earman, "Sharpening the electromagnetic arrow of time," preprint, 2007.
27. J. A. Wheeler to R. P. Feynman, 26 March 1942, Feynman Papers, Caltech.
28. R. P. Feynman, Thesis draft, Feynman Papers, Caltech, folder 15.4, f. 32.
29. L. S. Schulman, *J. Math. Phys.* **36**, 2546 (1995).
30. R. P. Feynman, *Phys. Rev.* **76**, 769 (1949).
31. R. P. Feynman to J. A. Wheeler, 4 May 1951, Feynman Papers, Caltech, folder 3.10; quoted in full in Ref. 10, p. 462.

QUO VADIS, PHYSICA?

M. GUTZWILLER

IBM Research Emeritus and Yale University,
Department of Physics, New York, USA
E-mail: moongutz@aol.com

The history of physics is short, compared with mathematics and astronomy, starting in about 1800. After the first century of electrodynamics and thermodynamics, come 60 years of quantum, relativity, and nuclei. In the last 50 years, physicists have profited from an incredible boom in their research. That is threatened now, because the applications of physics have become very complicated engineering. At the other end of the spectrum, a lot of esoteric physics is going nowhere. It is not the job of the universities to produce specialists. Young people have to learn and get a chance to use their intelligence and imagination. Professors have to show them the way, and industry has to have confidence in them.

1. Introduction

"Quo Vadis?" is the title of a very famous novel by the Polish writer Hendryk Sienkiewicz who got the Nobel Prize of Literature in 1905, partially for this work. Hollywood then made probably more than one very successful movie, but I neither have read the book nor seen the movie. The title comes directly from the Gospel of St. John. The apostles and disciples were anxious and confused when they saw Christ again after his crucification, and were afraid to ask the simple question, "Where are you going?"

The title of this paper has been used already in a talk I was invited to present 4 years ago in Valencia, Spain, for a general audience of scientists. It was then translated into Catalan by a colleague, J. Navarro. I like the solemn character of the expanded question because I feel quite strongly that physics is at a critical moment of its development. Young physicists don't know which way to turn, nor are the older ones clear about the future.

Physics is in a difficult position. Its further rise in the 21th century is hard to imagine, because its success all through the 20th century will be hard to match. It has been reluctantly recognized as the foundation of most other sciences. But physics plays this role only in a general and abstract way, and the connection particularly with the life sciences is tenuous.

In order to find some views of the future, we are bound to look into three separate questions: 1) What is our position now among the other hard sciences? 2) When was physics generally recognized, and how did we get there? 3) Are we getting closer to the firm foundations of the sciences that we are seeking and claiming? To answer question 1) we need to worry about our successors, like students looking for a job, and scientific contacts outside physics. For 2) some knowledge of our history is necessary, and for 3) we have to look very critically at the important claims we love to make.

These are very far-ranging problems, and I will only try to present some of the viewpoints and answers on the basis of my special interests and my own work. Physicists generally are too busy to worry about the history and philosophy of their subject. But we would be reduced to the status of mere technicians if we do not have some perspective on our work. Among other things I study history and philosophy hoping to learn some physics.

2. Some Input from Philosophy

Descartes was a first-class mathematician and a very good physicist. In one of his more general treatises he says in the preface: "Philosophy is like a tree". There are the roots, the trunk, and then three main branches with lots of leaves. The analogy of Descartes specifies: the roots represent "metaphysics", the trunk stands for "physics", the three branches carry assignments from left to right, "mechanics", "morals", "medicine". The more detailed interpretation is my invention, but I think that many scientists would agree with my highly simplified picture.

"Mechanics" includes all of astronomy, physics as we see it today, chemistry, geology and their subfields, most importantly engineering of all sorts. "Medicine" contains the whole range of the life-sciences from botany and agriculture all the way to medicine and psychiatry. "Morals" is the center of the tree, and in my view there is the whole organization of our society, the laws, politics, education, social services including responsibility and solidarity.

"Metaphysics" is a word of the middle ages, and is applied to one of the volumes of Aristotle's work on logic and related topics. I include all classical and modern mathematics as well as computers. Now what about the trunk called "physics", and its role in this picture? Descartes believed in demonstrating the existence of God by starting in the roots. Few scientists would agree. But many physicists believe that the tools in "metaphysics" should suffice to go very far in the explanation of the world around us.

Are we really able to reach the "morals" branch starting with the ideas contained in the roots? Nevertheless, most of us believe that most of the sciences in the "medicine" branch can be reached. The fundamental laws and facts of today's physics go a long way in explaining biophysics, and then on to the stupendous construction of our living organs. And what about the brain? If anything, it is a very close call. Of course, we are still a long way from the complete understanding, but there is most interesting work to do.

The optimism of Descartes (contemporary of Galileo and Kepler), did not survive the progress in the work of Isaac Newton. The role of "physics" was not denied, but it was reduced in the opinion of Immanuel Kant. Whatever we observe and then explain, is not nature itself, but our picture and the ways we think about it. Our reason can only recognize in nature whatever we have first accepted in our own way of thinking. Our experiments serve only to force an approval by nature for what we have figured out. If we are lucky we get an affirmative answer. But we have to be very careful in basing further speculation on such a limited result. Every possible consequence of our thoughts should be tested as specifically as possible, and should never be considered final.

As an example, the standard model of particle physics does a credible job in explaining the very-high-energy experiments. But our present techniques allow us only to measure the probability of exceedingly complicated scattering phenomena. The standard model has as yet to explain the well known low-energy properties of simple nuclei. The magnetic moment of the proton is known for 70 years, and the shell-model for 60 years. The nuclear physicists, however, have had only empirical models and energy-level statistics in the last 50 years. The applicable theory is a generalization of the original quantum electro-dynamics (QED). Large-scale computations now are barely quite able to handle the mathematics, and they get more difficult at the low energies of ordinary nuclei.

These difficulties, including many ad-hoc remedies, have not held back many theoretical physicists from going way-out beyond what is known from feasible experiments. It goes under various names like string theory, quantizing general relativity, inflationary model of cosmology, and others. It is jumping to the Planck length, which is smaller than the radius of the proton by 18 factors 10! No experiment has even been proposed, although the ideas have been bantered around for some 40 years. Every person is free to entertain ideas about nature that have found no confirmation. Such research programs, however, should not be supported in the long run by public funds. They are the entertainment as well as the personal responsibility of the researchers themselves with few exceptions.

3. The Short History of Physics

The word 'physics' was introduced by Aristotle as a reference to nature in general. For 2000 years '*Physica*' was used in a very wide sense. Nowadays it should be translated as '*Natural Philosophy*'. Newton used it in the title of his "Principia Mathematica Philosophiae Naturalis". The publications of the academic societies that started in France, England, Germany in the second half of the 17th century used 'physics' in various ways. A lot of the works were put into the well-known categories, like mathematics, astronomy, chemistry, geography, and the remainder was called (sometimes 'general') 'physics'. Pieces from the life sciences were put together with special topics of the hard sciences. After long discussions the modern interpretation was adopted at the end of the 18th century.

As a further illustration let me present a list of famous physicists born in the 18th and working in the 19th century. As a starting honorary physicist there is Immanuel Kant (1724-1804), then C. A. Coulomb (1736-1806), Alessandro Volta (1745-1827), Thomas Young (1773-1829), Joseph Fourier (1768-1830), Sadi Carnot (1796-1832), A. M. Ampere (1775-1836), S. D. Poisson (1781-1840), John Dalton (1791-1876), J. L. Gay-Lussac (1778-1950), H. C. Oersted (1773-1829), Amadeo Avogadro (1776-1856), Michael Faraday (1791-1867). The choice is perhaps arbitrary. There are 6 French, 3 British, 2 Italian, and 1 Danish physicist, but no German unless we include the mathematician Gauss.

Can we set up a list of great physicists for the earlier periods in human history? The obvious heroes like Galileo, Huygens, and Newton did most of their remarkable work in astronomy. Galileo's reports on his experiments in mechanics are ambiguous; Huygens' discussion of the centripetal force found its first application in deriving Kepler's third law; Newton's superb experiments in optics did not receive a valid interpretation. We should compare that to the great Greek mathematicians and astronomers whose basic texts are still very fundamental and enjoyable reading. The chemists profited from many practical applications that were refined over the centuries and recognized as being in a special category. All the marvelous mechanics in the advanced civilizations, for building temples, bridges, ships, transport equipment, weapons, and so on, did not become objects of special scientific investigations. The French Academy of Science had a section of Mechanics, and the Royal Society of London classified papers as 'mechanical philosophy'. It was considered useful, but not very fundamental.

Around 1800, both the French Revolution and the British Industrial Revolution were promising a better life. A deeper technical understanding was required for all the many new applications, mechanical, optical, acoustical, even electrical and magnetic. Physics became a new branch of scientific activity, next to its much older companions. The 19th century was dominated by the advances in chemistry, particularly after many of the 'organic' processes inside living beings were found to be of the same chemical nature as the 'inorganic' processes. The fundamental role of atoms was recognized and accepted by most chemists after Dalton. But physicists resisted the idea and were looking for a deeper justification of such a strange foundation of nature.

4. Quantum Physics, Relativity, and the Nucleus until 1957

No doubt, the years from 1900 to about 1957 (!) are the most productive in the history of physics. The most fundamental principles of physics were discovered, and many of the basic applications were successfully worked out. The latter include superconductivity, nuclear power, semi-conducting devices, lasers, nuclear magnetic resonance. At the end of this epoch the USA were undoubtedly the leaders in physics, because the European countries were only recovering from their two World Wars. After the unusual effort in WWII, most research in physics, in the USA and in Europe, returned to its academic base, with few institutes that depended directly from the government. In 1945 I started my professional life as a student at the ETH (Swiss Federal Institute of Technology) in Zürich. My further comments on physics in the modern world will be mostly based on my own experiences and observations.

The ETH until 1928 had only one professor of physics; at that time it was Peter Debye. He was succeeded for the next 30 years by two full professors, Wolfgang Pauli and Paul Scherer. When I was a student, the Department of Mathematics and Physics had moreover 10 full professors in mathematics and 1 in astronomy. In spring 1949 I asked Pauli, whether he would supervise me for the required short (6 months) thesis. With the help of his assistant Felix Villars (later at MIT), I calculated the magnetic moment of the proton/neutron coupled to a vectorial, charged and massive field, using the tricks recently invented by Schwinger and Feynman. No course was taught in quantum mechanics, let alone in field theory. I got it directly from Sommerfeld, Dirac, and Wentzel.

After passing the oral exams I was ready in principle to start working on a PhD thesis. There was no assistantship or any other source of support available, and I decided to accept a job at Brown Boveri Co. (BBC). The necessary equipment for the telephone connection between Zürich and Geneva via microwaves was all invented and built at BBC, and I was helping the installation. I enjoyed doing it; but after a year and a half I wanted to continue in physics. I applied for a stipend to the USA much too late, but I got an offer. Wentzel from the University of Chicago encouraged me, and I came to the USA as a graduate student in 1951.

The little known University of Kansas had 4 full professors of physics, very active, including Max Dresden, a small group of junior professors, and graduate students; everybody congenial, receptive for ideas, free of ranking, reputation, and external pressures from funding agencies. After two years of a very pleasant life I had finished my PhD thesis on QED in deSitter space.

In spite of my immigration visum I had a hard time finding a job, because of the communist scare in the USA. The geophysical laboratory of Shell Oil Company in Houston had no trouble because it had no contract with the government. It was the best thing for me to happen. My thesis work in quantum field theory was in line with all my colleagues who had essentially the same esoteric background. They were working on completely different problems. I was first asked to look into the plastic flow of crystals, then into the magnetic behavior of sedimentary rocks, and finally into the generation and propagation of seismic waves from a disturbance on the surface.

5. The Boom in Physics from 1957 to 2007

Exactly 50 years ago, October 1957, the Russians sent the first ever artificial moon into space, called Sputnik. The US government reacted by making a great show of going to promote almost any science much more than after WWII. Meanwhile the Europeans had recovered sufficiently to start their own scientific enterprises, nationally or internationally like CERN, Centre Europeen de Recherche Nucleaire. Industry in all countries was setting up their own research centers, like IBM near Zürich where I found my third job in 1960. I had an easy start after learning about magnetism at Shell. I moved back with my family in 1963, because I had a chance to teach a course of solid-state physics in the engineering school of Columbia University in New York City while continuing to work for IBM.

Some years after Sputnik, scientific institutes and universities in the USA and Western Europe started a great push to increase their research programs. The financial means were provided for by local and national governments as well as by industry according to the field of interest. Suddenly there existed scholarships for getting a PhD, junior positions for the lower academic ranks, and promotions to the higher ranks. Government and industry were offering quasi-permanent employment, with certain restrictions concerning the type of work and the area of science. New universities were founded, and many small ones hired people to start research projects.

In 1960 the physics institute of the ETH was still the same as during my years as a student. When I came back for the second time in 1973 to teach a special course on electron correlations in metals, I found myself in a palatial campus of 6 buildings on the outskirts of Zürich. All 6 were dedicated to physics and provided with large auditoria, library, and a restaurant. By 2005 there were 25 full professors and more than 400 other paid people from PhD candidates on up the ladder with all the other work to be done.

Industry stopped hiring and started closing their research facilities already in the 1980's. The USA federal government was somewhat protected, but a number of its laboratories changed their orientation, with the exception, of course, of defense oriented places. The university laboratories, however, kept on struggling quite successfully for the support of the government, trying to beat at least the inflation. The senior teaching staff likes to stay beyond the former maximum of 70 years, because the law does not allow discrimination against age, gender, etc.. They can even claim their pensions after 70 in addition to their salary!

The departure of physicists from industry was protected by the same statute. But after 2 years of large deficit, IBM decided to get rid of the scientists whose merit was to maintain the scientific reputation of their laboratory. In spring of 1993 we were faced with a proposal that offered a gift with promises or a threat of a new work assignment. The gift as well as the promises have been kept so far, and at my age I felt quite satisfied.

6. Things to be Expected

The first three sections of this report tried to show that physics has a special role among the sciences. Originally it was "Natural Philosophy", whereas mathematics, astronomy, chemistry, geology, and mechanics had each a well understood domain. But with mathematics starting to play an essential role in astronomy, somewhat later in mechanics, and finally in electricity and magnetism, 'physics' took on a special task, like the trunk in Descartes' picture. Physics was destined to dig deeper to find the origins for its main topics. During the 19-th century it competed with the chemists who had gotten ahead of them about the atoms. But since 1900, physics took on itself a very fundamental role.

The physics boom fell into the time when a large number of simple, but fundamental applications had been discovered, e.g. for lasers, electromagnetic and semi-conducting devices. Also ideas for sophisticated machines in biophysics and medicine were proposed. They came from physics laboratories, and have since produced an avalanche of useful products. A vast engineering effort has effectively become almost independent of physics.

Similarly, the late 1950's and then 1960's coincided with great progress in the explanation of particle physics that lead eventually to the Standard Model. A number of new accelerators were planned, then built, and finally run for many years. As the energy necessary increased things became ever more expensive and complicated. The boundary of the field was pushed forward until only the proton collider of CERN is left. Everybody waits for it. In both cases the physicists had found interesting problems, and started working on them in increasing numbers. But after two or three decades, either the engineers took over or the problem grew too much for continuing.

As a rule during the last 50 years, the number of physicists grows much faster than the number and variety of new fields in physics. Each new area demanded ever more complicated apparatus and theory. As a result the physics community divided itself into many more specialties. This can be seen in the advertisements of Job Opportunities in Physics Today. The various specialists belong explicitly or implicitly to some organization of their specialty. That gets compounded by the competition among universities and laboratories aiming at establishing a general class system in physics. The periodic worldwide ranking by an Anglo-Saxon outfit comes out with a long list where the Physics Department of highest rank in Germany gets number 200!

Life in the physics departments of the universities will change. Many professors will have to concentrate on teaching, and on reorienting their field of research. Such a development is taking place already in other departments, as unfortunately also in the classic languages Latin and Greek. Students will have to choose carefully the area of their research, because their opportunities will depend heavily later on what experience they have gained. Moreover, they cannot expect to get a lot of guidance in their choice.

In the last 50 years, physics has often proclaimed its role as the science that will find the foundations for all of nature. The support of research from outside sources was centered on this idea, in addition to the promise of new products for industry, and of course the military requirements. But high-energy physics slowed down because its experiments were so expensive. At the same time, however, the theory of the standard model is beyond our mathematical skills, let alone all the fancy models. It is doubtful, whether we are able to understand nature on the smallest and on the largest scale. Was our role of providing Descartes' tree with a trunk too much to ask?

Mathematics and astronomy will continue what they have done so well over more than 2000 years. Physics has to become more down-to-earth in its literal meaning. It is already broken up into numerous areas with many associated engineering fields. The danger is that physicists bury themselves into one particular, esoteric specialty while ignoring most of the others. As human being a physicist should try to maintain many contacts in order to mix life and profession. Each existence might resemble the life of a skilled artisan. Many variations would allow each individual to show special talents and tastes, as well as methods and style.

ROUND-TABLE DISCUSSION

The round-table discussion on Thursday evening was held under the provocative title *Quo vadis, path integrals – new trends and perspectives?*. After a brief introduction by the chair woman Professsor Cécille DeWitt-Morette, the panel members presented their opinions in the form of a few concise "theses" written on a black board. The final outcome is recorded in the photograph on the next page. These initial thoughts by the panel members were then taken up by the other participants who very actively contributed with their own ideas on how the field should develop, so that a vivid discussion emerged.

Chair:

Cécille DeWitt-Morette – University of Texas at Austin, USA

Panel members:

Jozef T. Devreese – Universiteit Antwerpen, Belgium
Martin C. Gutzwiller – Yale University, New York, USA
Akira Inomata – State University of New York at Albany, USA
John R. Klauder – University of Florida at Gainesville, USA
Hagen Kleinert – Freie Universität Berlin, Germany
Lawrence S. Schulman – Clarkson University at Potsdam, USA
Jean Zinn-Justin – CEA-Saclay, France

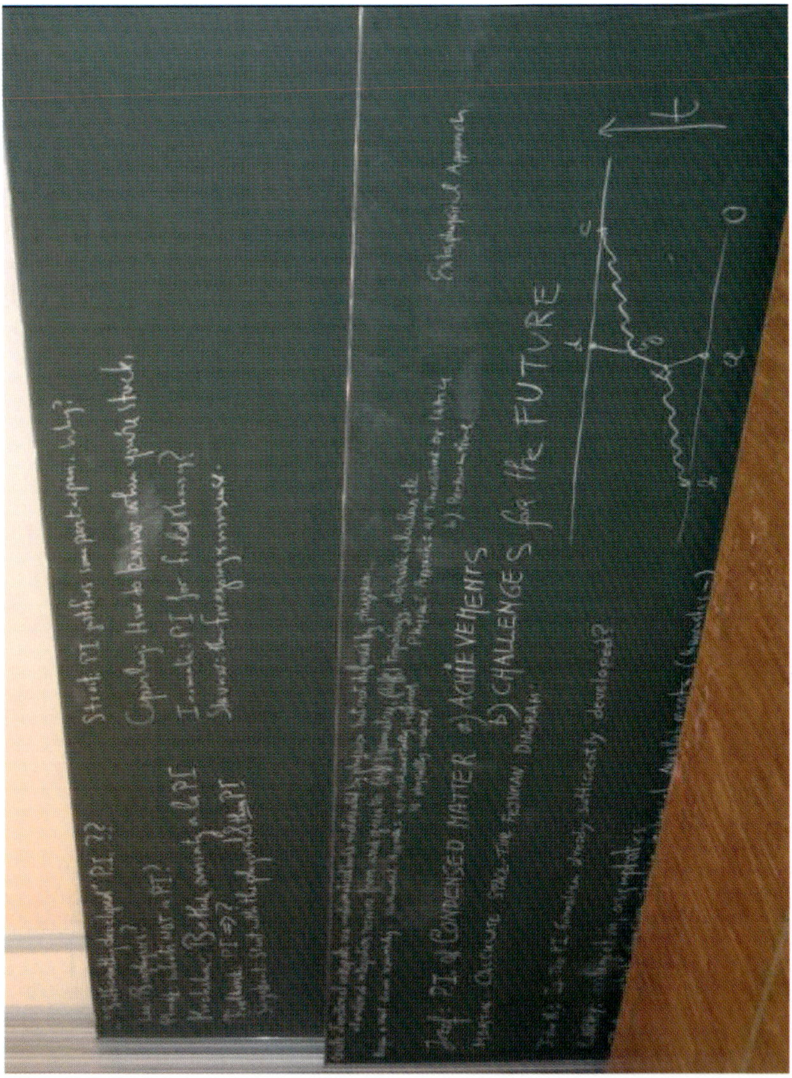

Fig. 1. Black board of the round-table discussion.

PART II

Quantum Physics

AN APPETIZER: A SAMPLER OF MAIN COURSES

C. DEWITT-MORETTE

Department of Physics – Center for Relativity, University of Texas at Austin, Austin, Texas 78712-1081, USA

E-mail: cdewitt@physics.utexas.edu, www.utexas.edu

This paper discusses two basic issues of functional integration: domains of integration and volume elements adapted to a given domain of integration. Two examples of domain of integration are given explicitly in Sections 2 and 3 respectively: the domain of integration is a space of contractible paths and the domain of integration is a space of Poisson paths. A property of volume element, presented in Section 3, namely the Koszul formula, valid on totally different geometries (riemannian, symplectic, grassman) can be used for some infinite dimensional geometries.

Keywords: Domain of integration; Volume element; Contractible space; Symmetry; Poisson path; Koszul formula.

1. Introduction

I chose two topics relevant to all aspects of functional integration, namely identifying domains of integration, say $\mathbb{X}$, and choosing volume elements appropriate for $\mathbb{X}$.

Examples of domain of integration are presented in the first two appetizers. A property (the Koszul formula) of volume elements valid in very different geometries, riemannian, symplectic, grassmann, and expected to be valid in function spaces is presented in the third appetizer.

The other issues in the general theory of functional integration are:

- characterizing functionals integrable with respect to the chosen volume element
- computing the integral (or an approximation), or using the information encoded in the integral. For instance, the solution of a PDE consists of a large class of functions together with a choice of initial conditions compatible with the PDE, whereas a functional integral is the solution of a PDE satisfying a given set of initial conditions.

See Ref. 1 and references therein.

2. When Symmetries Define a Path Integral; A Basic Theorem on Spaces of Pointed Paths

(i) A space of pointed paths is a space of paths x mapping $\mathbb{R}$ into a D-dimensional manifold $\mathbb{M}^D$

$$x : \mathbb{T} \subset \mathbb{R} \longrightarrow \mathbb{M}^D, \qquad x \epsilon \mathcal{P}_0 \mathbb{M}^D \tag{1}$$

such that for a certain value $t_0 \,\epsilon\, \mathbb{T}$ all the paths take the same value $\mathbf{x}_0$

$$x(t_0) = \mathbf{x}_0 \epsilon \mathbb{M}^D \text{ for every } x \epsilon \mathcal{P}_0 \mathbb{M}^D \tag{2}$$

A space of pointed path $\mathcal{P}_0\mathbb{M}^D$ is a very desirable domain of integration because it is contractible and can be parametrized by pointed paths $z \,\epsilon\, \mathcal{P}_0\mathbb{R}^D$ taking their values on $\mathbb{R}^D$.

$$x(t, z) \epsilon \mathbb{M}^D, \qquad z \epsilon \mathcal{P}_0 \mathbb{R}^D \tag{3}$$

An integral over $\mathcal{P}_0\mathbb{M}^D$ can be expressed as an integral over $\mathcal{P}_0\mathbb{R}^D$. A well known example is the Cartan development map that maps a path on a riemannian manifold $\mathbb{M}^D$ into a path on $\mathbb{R}^D$.

(ii) Consider now a system having symmetries, i.e. a system with invariances under a group of transformations $\{\sigma_\alpha\}_\alpha$ defined by dynamical vector fields $\{X_\alpha\}_\alpha$, generators of integral curves $\{\sigma_\alpha\}_\alpha$ on $\mathbb{M}^D$

$$\frac{d}{dr}(\mathbf{x}_o \cdot \sigma_{(\alpha)}(r)) = X_{(\alpha)}(\mathbf{x}_o \cdot \sigma_{(\alpha)}(r)) \tag{4}$$

$$X_{(\alpha)}(\mathbf{x}) = \frac{d}{dr}(\mathbf{x} \cdot \sigma_{(\alpha)}(r))_{r=0} \tag{5}$$

The map from $\mathcal{P}_0\mathbb{R}^D$ into $\mathcal{P}_0\mathbb{M}^D$ that parametrizes the paths on $\mathbb{M}^D$ is rarely known explicitly, but is usually known implicitly by the equations:

$$\mathrm{dx}(t, z) = X_{(\alpha)}(x(t, z))\mathrm{dz}^\alpha(t) \tag{6}$$

$$x(t, z) = \mathbf{x}_0 \cdot \Sigma(t, z) \tag{7}$$

where $\{\Sigma(t, z)\}$ is a group action on $\mathbb{M}^D$.

(iii) Given the following information:
a "reasonable" function on $\mathbb{M}^D$, $\phi : \mathbb{M}^D \to \mathbb{R}$
a space of pointed paths $x \in \mathcal{P}_0\mathbb{M}^D$
a set of dynamical vector fields $\{\mathbb{X}_{(\alpha)}\}_\alpha$ on $\mathbb{M}^D$
a quadratic form on $\mathbb{R}^D$

$$Q(z) := \int_{\mathbb{T}} \mathrm{dt}\ h_{\alpha\beta}\dot{z}^\alpha(t)\dot{z}^\beta(t), \tag{8}$$

then the function Ψ on $\mathbb{R} \times \mathbb{M}^D$ defined by the following functional integral on $\mathbb{R}^D$

$$\Psi(t, \mathbf{x}_o) := \int_{P_o\mathbb{R}^D} D_{s,Q}z \cdot \exp(-\frac{\pi}{s}Q(z))\Phi(\mathbf{x}_o \cdot \Sigma(t, z)) \tag{9}$$

is a solution of the parabolic equation on $\mathbb{M}^D$:

$$\frac{\partial \Psi}{\partial t} = \frac{s}{4\pi} h^{\alpha\beta} \mathcal{L}_{X_{(\alpha)}} \mathcal{L}_{X_{(\beta)}} \Psi \tag{10}$$

with

$$\Psi(t_o, \mathbf{x}) = \Phi(\mathbf{x}) \tag{11}$$

3. Solutions of PDE Other Than Parabolic; Spaces of Poisson Paths

In 1956, Mark Kac delivered a Colloquium Lecture at the Magnolia Petroleum Company on "Some stochastic problems in physics and mathematics." He used the Monte Carlo method based on Poisson processes for solving the telegraph equation. There are two equivalent characteristics of Poisson Processes:

(i) by waiting times between events $\{T_1, ...T_n\}$ with a probability law (see Fig. 1)

$$Pr(t_k \leq T_k \leq t_k + \mathrm{dt}) = a\ e^{-at_k}\mathrm{dt} \tag{12}$$

(ii) by n jumping times $\{t_1, ...t_n\}$

$$Pr(x = n) = e^{-\lambda}\lambda^n/n! \tag{13}$$

A path x can then be written $\delta_{t_1} + ... + \delta_{t_n}$. Mark Kac notes that the waiting-times characterization is a "ridiculous scheme for a Monte Carlo calculation of the telegraph equation," but that the characterization by jumping times is very powerful.

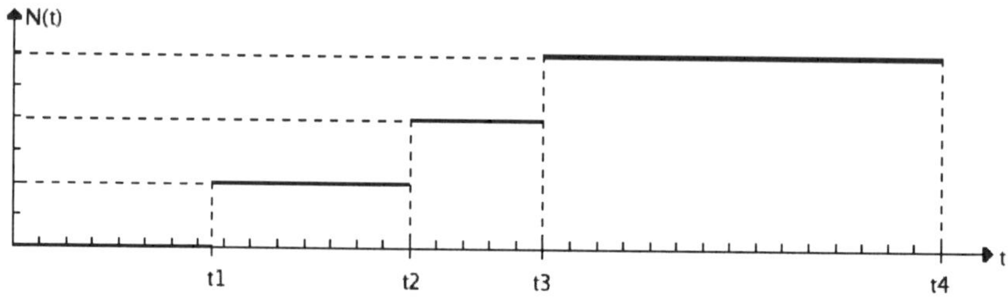

Fig. 1. A Poisson process can be stated in terms of waiting times between events or in terms of jumping times.

Interestingly, the waiting-times characterization cannot be adapted to quantum physics (i.e. to "imaginary time") but the jumping-times characterization makes it possible to define spaces of Poisson paths for functional integrals solutions of Klein and Dirac equations.

Let $\mathbb{X}_n$ be the space of Poisson paths jumping n times during the time interval $\mathbb{T} = [0, \mathrm{T}]$. The space $\mathbb{X}$ of Poisson paths is

$$\mathbb{X} = \cup_n \mathbb{X}_n \tag{14}$$

The following equation defines a volume element on $\mathbb{X}$:

$$\int_{\mathbb{X}} D_{a,\mathbb{T}} x e^{2\pi i < f,x >} := \exp \int_{\mathbb{T}} a \; \mathrm{d}t \; (e^{2\pi i f(t)} - 1) \tag{15}$$

where

$$\text{vol } \mathbb{X} = \exp \;\; \text{vol } \mathbb{T} = \exp a\mathrm{T} \tag{16}$$

An interesting example "A two-state system interacting with its environment" combines Poisson and Gaussian paths.

For other techniques, in particular solutions of elliptic equations in terms of first exit times, we refer to the Cartier/DeWitt book and references therein.

4. A General Property of Volume Elements; the Koszul Formula

Measures are not the appropriate concept for functional integrals; volume elements may be an easier concept than measures for use in infinite dimensional spaces.

On $\mathbb{M}^D$, a volume element is a D-form (a top form).

There are no top forms on infinite dimensional spaces.
There are not top forms in grassmann analysis.

On infinite dimensional spaces:

- Volume elements can be characterized by integrals (e.g. Gaussian volume elements and Poisson volume elements)
- The Koszul formula provides a differential characterization of volume elements in a great variety of cases

Let $\mathcal{L}_X$ be the Lie derivative defined by the vector field X.
Let ω be a volume element.
Let $D(x)$ be a (generalized) divergence of x: $\mathcal{L}_X\omega = D(x) \cdot \omega$.

The Koszul formula defines the divergence $D(x)$ as the change of the volume element ω under the group of transformations generated by X. It is valid in riemannian, symplectic, and grassmann spaces.

A Berezin integral is a derivation and does not readily generalize to functional integrals. However, Roepstorff has proposed a formalism for Berezin functional integrals and applied it to Euclidean Dirac Fields.

Acknowledgments

Thank you Peter Fulde, thank you Wolfhard Janke, thank you Axel Pelster, thank you Marita Schneider. Thanks to all of you we were treated to a magnificent conference superbly organized in a beautiful setting. The program was a very rich banquet; when asked to give an introduction to the incredibly varied topics listed in the program, I could only offer a few appetizers.

References

1. P. Cartier and C. DeWitt-Morette, *Functional Integration, Action and Symmetries* (Cambridge University Press, 2006).

WHAT DOES OPERATOR ORDERING HAVE TO DO WITH THE DENSITY OF PATHS?

L. S. SCHULMAN

Physics Department, Clarkson University,
Potsdam, NY 13699-5820, USA
E-mail: schulman@clarkson.edu

In joint work with M. Roncadelli, I have shown that the *quantum* Hamilton-Jacobi equation, which at one time attracted considerable attention but was abandoned as intractable, could be solved through its relation to the propagator. (Of course solving for the propagator is also far from trivial.) This article reviews that work and discusses applications. In one application we find that quantum operator ordering reveals the classical density of paths.

Keywords: Semiclassical propagator; Quantum Hamilton-Jacobi.

1. Introduction

This report is based on joint work with Marco Roncadelli.[1]

There is a famous article by Wigner[2] on the "Unreasonable effectiveness of mathematics." Sometimes I wonder about a narrower question: the unreasonable effectiveness of classical mechanics for quantum mechanics, and occasionally, vice-versa.

What I will tell you is how Roncadelli and I started from an old, and at the time unsuccessful, way to quantize, how we made this technique a bit more tractable, and how finally it forced us to consider that same "unreasonable effectiveness" to which I alluded. As you will see, our progress depended on recognizing the propagator when it appeared, and perhaps more important, exploiting our result brought us squarely into the world of path integrals, where the occasional solved propagator can lead to new results for this quantization scheme. In particular—and this is what the title of this article refers to—the solved quantum problem can come back and give subtle information about the related classical mechanics.

What you would like to do is start from the classical Hamilton-Jacobi equation and replace the "c-number" p's and q's by *operators*. The time dependence of these operators (in the Heisenberg picture) should be given by the usual rules for a canonical transformation. As for the classical problem, solving this system gives you the full time dependence. However, unlike the classical equation, which is itself far from trivial, you now encounter *operator ordering* problems. Because the canonical transformation mixes the "old" and "new" variables in a peculiar way, operator ordering issues arise even when the Hamiltonian does not by itself have these ambiguities. As a result, although one usually thinks of operator ordering issues as having $O(\hbar^2)$ consequences, the effects we shall see are $O(\hbar^1)$.

I will not dwell on the history of this problem except to say that quantum canonical transformations entered early into the fabric of quantum mechanics, especially in the works of Born, Heisenberg and Jordan,[3,4] Dirac[5] and London.[6] A later advocate was Schwinger,[7] especially in his work on a quantum action principal.[8]

In Sec. 2 I present the *operator* or *quantum* Hamilton-Jacobi equation. The next section describes the progress Roncadelli and I made towards its solution. Following that, in Sec. 4 I give actual and potential applications that we have found. The final section is a discussion.

2. The Quantum Hamilton-Jacobi Equation

As for classical canonical transformations,[9] one has a generating function that is a function of two (sets) out of the four (sets of) variables, $\hat{q}, \hat{p}, \hat{Q}, \hat{P}$, in the usual notation—except that now we place circumflexes over the variables to indicate their operator nature. We take the generating function to be a function of $\hat{q}$ and $\hat{Q}$. Calling that function $W(\hat{q}, \hat{Q}, t)$, with $\hat{q} = \{\hat{q}_k | k = 1, \ldots, N\}$, etc., the transformation is given by $(i = 1, \ldots, N)$

$$\hat{p}_i = \frac{\partial}{\partial \hat{q}_i} W(\hat{q}, \hat{Q}, t) , \tag{1}$$

$$\hat{P}_i = -\frac{\partial}{\partial \hat{Q}_i} W(\hat{q}, \hat{Q}, t) , \tag{2}$$

$$K(\hat{Q}, \hat{P}, t) = H(\hat{q}, \hat{p}, t) + \frac{\partial}{\partial t} W(\hat{q}, \hat{Q}, t) . \tag{3}$$

These expressions are meaningless until operator ordering is prescribed. We follow Jordan and Dirac and adopt what they call *well-ordering*: operators represented by capital letters go to the right of those labelled by lower case letters. Thus $W(\hat{q}, \hat{Q}, t)$ should have the structure

$$W(\hat{q}, \hat{Q}, t) = \sum_\alpha f_\alpha(\hat{q}, t)\, g_\alpha(\hat{Q}, t) , \tag{4}$$

for suitable functions $f_\alpha(\cdot)$ and $g_\alpha(\cdot)$. Throughout, we suppose that operator generating functions are *well-ordered.* With *well-ordering* a classical function $f(q,Q)$ is fully defined by the replacements $q \to \hat{q}$ and $Q \to \hat{Q}$.

Time evolution follows the classical method. One seeks a generating function that brings the canonical variables in the Heisenberg picture, $\hat{q}(t)$, $\hat{p}(t)$, to constant values at an initial time t_0, values that we call $\hat{Q}$ and $\hat{P}$. Such a transformation brings the transformed Hamiltonian to zero, so that, from Eq. (3), the generating function $W(\hat{q},\hat{Q},t)$ obeys the operator *quantum Hamilton-Jacobi equation*

$$H\left(\hat{q}, \frac{\partial}{\partial \hat{q}} W(\hat{q},\hat{Q},t), t\right) + \frac{\partial}{\partial t} W(\hat{q},\hat{Q},t) = 0\,. \tag{5}$$

The function $W(\hat{q},\hat{Q},t)$ should be a complete solution of Eq. (5), i.e., it should depend on N independent "integration constants" $\hat{Q}_i$.

This framework provides a complete formulation of quantum theory. But at the level of operator equations little progress was made in the many years since it was proposed. This is not surprising, given the difficulty of solving a nonlinear operator partial differential equation (PDE).

3. The Equivalent c-Number Problem

Our method is to find a c-number PDE whose solution can lead to a solution of Eq. (5). We do this for the general *Weyl-ordered* Hamiltonian

$$\begin{aligned} H(\hat{q},\hat{p},t) = {} & \tfrac{1}{2}a_{ij}(\hat{q})\hat{p}_i\hat{p}_j + \hat{p}_i a_{ij}(\hat{q})\hat{p}_j + \tfrac{1}{2}\hat{p}_i\hat{p}_j a_{ij}(\hat{q}) \\ & + b_i(\hat{q})\hat{p}_i + \hat{p}_i b_i(\hat{q}) + c(\hat{q})\,, \end{aligned} \tag{6}$$

where $a_{ij}(\cdot)$, $b_i(\cdot)$, and $c(\cdot)$ are functions of $\hat{q}_k$, and summation over repeated Latin indices is understood. Using the abbreviation $\hat{W}$ for $W(\hat{q},\hat{Q},t)$, Eq. (5) reads

$$\begin{aligned} & \tfrac{1}{2}a_{ij}(\hat{q})\frac{\partial \hat{W}}{\partial \hat{q}_i}\frac{\partial \hat{W}}{\partial \hat{q}_j} + \frac{\partial \hat{W}}{\partial \hat{q}_i}a_{ij}(\hat{q})\frac{\partial \hat{W}}{\partial \hat{q}_j} + \tfrac{1}{2}\frac{\partial \hat{W}}{\partial \hat{q}_i}\frac{\partial \hat{W}}{\partial \hat{q}_j}a_{ij}(\hat{q}) + b_i(\hat{q})\frac{\partial \hat{W}}{\partial \hat{q}_i} \\ & \quad + \frac{\partial \hat{W}}{\partial \hat{q}_i}b_i(\hat{q}) + c(\hat{q}) + \frac{\partial \hat{W}}{\partial t} = 0\,. \end{aligned} \tag{7}$$

The main simplifying step is to take the matrix element of $\hat{W}$ between states $\langle q|$ and $|Q\rangle$. This leads to the following c-number equation

$$\begin{aligned}
&\tfrac{1}{2}a_{ij}(q)\left\langle q\left|\frac{\partial\hat{W}}{\partial\hat{q}_i}\frac{\partial\hat{W}}{\partial\hat{q}_j}\right|Q\right\rangle + \left\langle q\left|\frac{\partial\hat{W}}{\partial\hat{q}_i}a_{ij}(\hat{q})\frac{\partial\hat{W}}{\partial\hat{q}_j}\right|Q\right\rangle \\
&+\tfrac{1}{2}\left\langle q\left|\frac{\partial\hat{W}}{\partial\hat{q}_i}\frac{\partial\hat{W}}{\partial\hat{q}_j}a_{ij}(\hat{q})\right|Q\right\rangle + b_i(q)\left\langle q\left|\frac{\partial\hat{W}}{\partial\hat{q}_i}\right|Q\right\rangle \\
&+\left\langle q\left|\frac{\partial\hat{W}}{\partial\hat{q}_i}b_i(\hat{q})\right|Q\right\rangle + c(q)\langle q|Q\rangle + \left\langle q\left|\frac{\partial\hat{W}}{\partial t}\right|Q\right\rangle = 0\,. \qquad (8)
\end{aligned}$$

Matrix elements of $\hat{W}$ alone would be no problem—it's assumed to be well-ordered. But the expressions in Eq. (8) are *not* well-ordered, and it is the unravelling of these terms to which we now turn our attention. In doing this our main tool is the canonical commutation relations. We use them together with the fact that for any function $G(\cdot)$

$$[G(\hat{q}),\hat{p}_i] = i\hbar\frac{\partial G(\hat{q})}{\partial\hat{q}_i}\,. \qquad (9)$$

Making use of $\hat{p}_i = \partial W(\hat{q},\hat{Q},t)/\partial\hat{q}_i$, Eq. (9) becomes

$$\frac{\partial\hat{W}}{\partial\hat{q}_i}G(\hat{q}) = G(\hat{q})\frac{\partial\hat{W}}{\partial\hat{q}_i} - i\hbar\frac{\partial G(\hat{q})}{\partial\hat{q}_i}\,. \qquad (10)$$

Now comes a lot of algebra. We take $G(\hat{q}) \equiv b_i(\hat{q})$, $G(\hat{q}) \equiv a_{ij}(\hat{q})$ and $G(\hat{q}) \equiv \partial a_{ij}(\hat{q})/\partial\hat{q}_j$. This allows us to move these functions, which may be inconveniently sandwiched to the right of Q operators, to the left, so that they can take as their arguments q's (since $\langle q|$ is on the left). We also make use of the expansion (4) in our intermediate steps. Defining the function $W(q,Q,t)$

$$\langle q|\hat{W}|Q\rangle = W(q,Q,t)\langle q|Q\rangle\,, \qquad (11)$$

after many steps (see Ref. 1) we arrive at

$$\begin{aligned}
&2a_{ij}(q)\left(\frac{\partial W(q,Q,t)}{\partial q_i}\frac{\partial W(q,Q,t)}{\partial q_j} - i\hbar\frac{\partial^2 W(q,Q,t)}{\partial q_i\partial q_j}\right) \\
&+2\left(b_i(q) - i\hbar\frac{\partial a_{ij}(q)}{\partial q_j}\right)\frac{\partial W(q,Q,t)}{\partial q_i} + c(q) - i\hbar\frac{\partial b_i(q)}{\partial q_i} \\
&\qquad -\frac{\hbar^2}{2}\frac{\partial^2 a_{ij}(q)}{\partial q_i\partial q_j} + \frac{\partial W(q,Q,t)}{\partial t} = 0\,. \qquad (12)
\end{aligned}$$

This equation should be familiar. It is the equation satisfied by W if

$$\psi(q,Q,t) \equiv \exp\{(i/\hbar)W(q,Q,t)\}\,, \qquad (13)$$

and ψ satisfies the time-dependent Schrödinger equation. One often sees this (perhaps for simpler Hamiltonians) when deriving the semiclassical approximation.[10]

What have we proved so far? If you know $\hat{W}$, then you can get something satisfying the Schrödinger equation. But to know *which* solution of the Schrödinger equation you have, you need to check the boundary conditions.

In terms of time, Schrödinger's equation is first order, so we only need establish the $t \to 0$ behavior of $W(q, Q, t)$. From the nature of the quantum Hamilton-Jacobi equation, $\hat{W}$ should become the identity transformation as $t \to 0$, but, as for the corresponding classical object, we run into a bit of trouble. For sufficiently small t, unless the potential is singular, the dynamics, whether classical or quantum, should be that of a free particle. So the classical solution is $F(q, Q) = m(Q - q)^2/2t$. Note that when one uses this in (the c-number version of) Eq. (1) or Eq. (2), the result is not well-defined for $t = 0$, but rather gives a limiting form.

As a way to guess the small-time solution of the *quantum* Hamilton-Jacobi equation we start with the classical version and well-order it. *This turns out to be insufficient*, so we will do our calculation on a slightly more general function than that given by the "F" just displayed, specifically we add a possible function of time. The candidate for a solution to Eq. (5) for small times is thus

$$\hat{W} = \frac{m}{2t}\left(\hat{Q}^2 - 2\hat{q}\hat{Q} + \hat{q}^2\right) + g(t) \,. \tag{14}$$

Substituting into Eq. (5), the squaring of $\hat{W}$ generates a term $-(\hat{q}\hat{Q} + \hat{Q}\hat{q})$, rather than $-2\hat{q}\hat{Q}$, so that satisfying Eq. (5) requires

$$0 = \frac{m}{2t^2}[\hat{q}, \hat{Q}] + \frac{\partial g(t)}{\partial t} \,. \tag{15}$$

To calculate the small-t commutator, one can again neglect the influence of the potential, and the commutator is deduced from the free particle Heisenberg picture relation $\hat{q} = \hat{Q} + \hat{P}t/m$. Equation (15) becomes $\partial g/\partial t = i\hbar/2t$ and we obtain

$$\hat{W} = \frac{m}{2t}\left(\hat{Q}^2 - 2\hat{q}\hat{Q} + \hat{q}^2\right) + \frac{i\hbar}{2}\ln t \,, \qquad \text{for } t \to 0 \,. \tag{16}$$

This in turn leads to

$$\psi(q, Q, t) = \text{const} \cdot \sqrt{\frac{1}{t}} \exp\left(\frac{i}{\hbar}\frac{m}{2t}\left(Q^2 - 2qQ - q^2\right)\right) \,, \qquad \text{for } t \to 0 \,, \tag{17}$$

with the constant arising from integrating $\partial g/\partial t = i\hbar/2t$.

The ψ of Eq. (17) is familiar: it is the free particle propagator and its $t \to 0$ limit is—up to a constant multiplier—a δ-function. This constant is not fixed by the properties of $\hat{W}$ and with an appropriate choice of constant we reach the following conclusion:

The solution of Schrödinger's equation in Eq. (13), for $W(q, Q, t) = \langle q|\hat{W}|Q\rangle$, is the propagator.

This is good news if you know $\hat{W}$ (which is seldom), but it's even better news if you know the propagator (which we'll call here $K(q, t; Q, t)$). That's because if you do know K you can construct $\hat{W}$ by taking $\log K$ and well-ordering. In this next section we exploit this feature.

4. Applications

The propagator of course is the central object of path-integral studies and there are a few known non-trivial explicit propagators.[11] In this article I will mention two examples. In one the quantum Hamilton-Jacobi equation reinforces the sentiments I expressed in opening this article: quantum mechanics seems to know about subtle properties of the classical mechanics. I'll also mention another example, but in less detail.

From Eq. (13), if you know the propagator, $K(q, Q, t)$ you can obtain $\hat{W}$ from

$$\hat{W} = -i\hbar \log K(\hat{q}, \hat{Q}, t)|_{\text{wo}}\,, \tag{18}$$

where the subscript "wo" indicates well-ordering. Thus one takes the c-number function K, substitutes the appropriate operators in the appropriate places, and well-orders.

4.1. *The semiclassical propagator*

Now consider a situation where the semiclassical approximation is valid and there is but one classical path between the initial and final points. As is well-known, in this approximation, $K(q, Q, t) = \text{const} \cdot \sqrt{\det \partial^2 S/\partial q \partial Q} \exp(iS(q, Q, t)/\hbar)$, with $S(q, Q, t)$ Hamilton's principal function (a solution of the *classical* Hamilton-Jacobi equation). It follows from Eq. (18) that $W(\hat{q}, \hat{Q}, t) = S(\hat{q}, \hat{Q}, t)|_{\text{wo}} - \frac{1}{2} i\hbar \log \det \partial^2 \hat{S}/\partial q \partial Q|_{\text{wo}}$. Now imagine that this expression is inserted in Eq. (5). If not for the well ordering, S alone would solve the equation. Therefore we conclude that the effect of the well-ordering is precisely to demand the presence of the additional term, $\frac{1}{2} i\hbar \log \det \partial^2 S/\partial q \partial Q$ (where "wo" has been dropped because there is already an $\hbar$ in the expression). But that additional term (famously) has a meaning of its own: it goes back to van Vleck and represents the density of paths along the classical path; it plays an essential role, for example, in the Gutzwiller trace formula. Our result says that this density of paths can be thought of as arising from the commutation operations necessary to bring S to well-ordered form.

This relation took us completely by surprise. To check it, we worked a simple but non-trivial example (our earlier demonstration comparing the boundary conditions for K and W already showed it to be true for the free particle). Let $H = p^2/2 + V$ with $V = V_0\Theta(a/2 - |x|)$ and x in one dimension. To lowest order in V the action is $S(x, y, t) = (x - y)^2/2t - V_0at/(x - y)$ for $y < -a/2$ and $x > a/2$. We checked our relation, with $x \to \hat{q}$ and $y \to \hat{Q}$ and with well-ordering implemented by $[1/(\hat{q} - \hat{Q})]_{\text{wo}} = \int_0^\infty du \exp(-u\hat{q}) \exp(u\hat{Q})$. Using the Baker-Campbell-Hausdorff formula and other techniques and keeping only lowest order in V and $\hbar$, indeed the relation checked out!

4.2. *The δ-function propagator*

For the Hamiltonian $H = p^2/2 + \lambda\delta(x)$, $x \in \mathbb{R}$, an integration by parts brings the known propagator[12,13] into the form

$$G(x, t; y) = g_0(x-y, t) - g_0(|x|+|y|, t) - \int_0^\infty e^{-\lambda u} g_0'(|x|+|y|+u, t) du \,, \quad (19)$$

where $g_0(\xi, t) \equiv \exp\left(i\xi^2/2\hbar t\right)/\sqrt{2\pi i\hbar t}$, and g_0' is the derivative with respect to the spatial argument. This holds for both positive and negative λ.

At present there is no sensible semiclassical theory for this Hamiltonian. With a barrier ($\lambda > 0$), classical mechanics predicts reflection, no matter how energetic the incoming particle. For a well ($\lambda < 0$), there is perfect transmission. Quantum mechanically, neither is true. By looking at the case $x > 0$, $y < 0$ one can use the known propagator to provide a complex classical mechanics, giving us some inkling of a more sensible limit. However, this work is still in preliminary form.

5. Discussion

In conclusion, I have shown how to construct solutions to the operator quantum Hamilton-Jacobi equation starting from the quantum propagator $K(q, Q, t)$ for the same Hamiltonian. Explicitly, once $K(q, Q, t)$ is known we get its "complex phase" $W(q, Q, t)$ via Eq. (13). Then, by demanding *well-ordering*, the replacement $q \to \hat{q}$, $Q \to \hat{Q}$ uniquely produces the operator $W(\hat{q}, \hat{Q}, t)$. Alternatively, by convolving $K(q, Q, t)$ with an arbitrary $\phi(Q)$ one can produce any solution of the Schrödinger equation. While this obviously works for exact propagators, it also enables one to find approximate operator solutions by exploiting approximate propagators. In particular we used the semiclassical approximation to the propagator to show that the commutation operations establishing well-ordering provide just what is needed to get the density of paths around the classical path. This density of paths satisfies a continuity equation which, as O'Raifeartaigh and Wipf[14] emphasize, is in a sense of order $\hbar$ (even though it involves classical quantities only and has no $\hbar$ in it!). Although our proof establishes this surprising relation, there remains the provocative question of understanding its intuitive basis.

References

1. M. Roncadelli and L. S. Schulman, *Phys. Rev. Lett.* **99**, 170406 (2007).
2. E. P. Wigner, *Comm. Pure Appl. Math.* **13**, 1 (1960); reprinted in Ref. 15.
3. M. Born, W. Heisenberg, and P. Jordan, *Zeit. Phys. A* **35**, 557 (1926).
4. P. Jordan, *Z. Phys.* **37**, 383 (1926), *ibid.* **38**, 513 (1926).
5. P. A. M. Dirac, *Physik. Zeits. Sowjetunion* **3**, 64 (1933), reprinted in Ref. 16.
6. F. London, *Z. Phys.* **37**, 915 (1926).
7. J. Schwinger, *Phys. Rev.* **82**, 914 (1951), *ibid.* **91**, 713 (1953).
8. J. Schwinger, *Quantum Kinematics and Dynamics* (Benjamin, New York, 1970).
9. H. Goldstein, *Classical Mechanics*, 2nd edition (Addison-Wesley, Reading, Massachusetts, 1980).
10. K. Gottfried, *Quantum Mechanics*, 1st edition (Benjamin, New York, 1966).
11. L. S. Schulman, *Techniques and Applications of Path Integration* (Dover, New York, 2005). With supplements. (Original publication, Wiley, New York, 1981).
12. See Ref. 11, pp. 381–382, Dover edition.
13. B. Gaveau and L. S. Schulman, *J. Phys. A* **19**, 1833 (1986).
14. L. O'Raifeartaigh and A. Wipf, *Found. Phys.* **18**, 307 (1987).
15. E. P. Wigner, *Symmetries and Reflections: Scientific Essays of Eugene P. Wigner* (Indiana Univ. Press, Bloomington, 1967). Eds. W. J. Moore and M. Scriven.
16. J. Schwinger, *Selected Papers on Quantum Electrodynamics* (Dover, New York, 1958).

NEAR ACTION-DEGENERATE PERIODIC-ORBIT BUNCHES: A SKELETON OF CHAOS

A. ALTLAND

Institut für Theoretische Physik, Zülpicher Str 77, D-50937 Köln, Germany

P. BRAUN

Fachbereich Physik, Universität Duisburg-Essen, D-47048 Duisburg, Germany
Institute of Physics, Saint-Petersburg University, 198504 Saint-Petersburg, Russia

F. HAAKE

Fachbereich Physik, Universität Duisburg-Essen, D-47048 Duisburg, Germany
E-mail: Fritz.Haake@uni-due.de

S. HEUSLER

Institut für Didaktik der Physik, Westfälische Wilhelms-Universität, D-48149 Münster, Germany

G. KNIEPER

Fakultät für Mathematik, Ruhr-Universität Bochum, D-44780 Bochum, Germany

S. MÜLLER

School of Mathematics, University of Bristol, University Walk, Bristol BS8 1TW, UK

Long periodic orbits of hyperbolic dynamics do not exist as independent individuals but rather come in closely packed bunches. Under weak resolution a bunch looks like a single orbit in configuration space, but close inspection reveals topological orbit-to-orbit differences. The construction principle of bunches involves close self-"encounters" of an orbit wherein two or more stretches stay close. A certain duality of encounters and the intervening "links" reveals an infinite hierarchical structure of orbit bunches. — The orbit-to-orbit action differences ΔS within a bunch can be arbitrarily small. Bunches with ΔS of the order of Planck's constant have constructively interfering Feynman amplitudes for quantum observables, and this is why the classical bunching phenomenon could yield the semiclassical explanation of universal fluctuations in quantum spectra and transport.

Keywords: Classical chaos; Orbit bunches; Encounters.

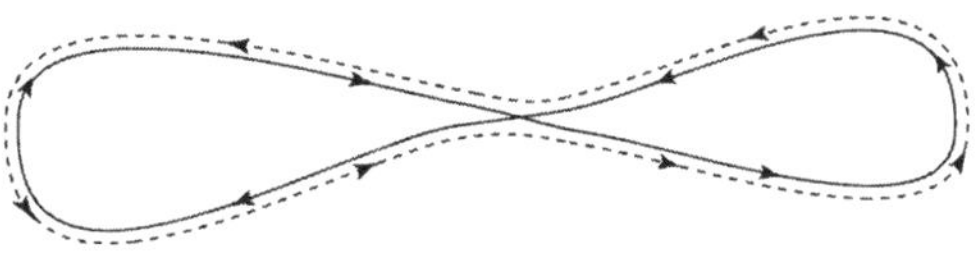

Fig. 1. Cartoon of simplest orbit bunch. One orbit has small-angle crossing which the other avoids. Difference between orbits grossly exaggerated.

1. Introduction

Extremely unstable motion, so sensitive to perturbation that long-term prediction is impossible, is the common notion of chaos, deterministic laws à la Newton notwithstanding. As a contrasting feature of chaos, within the continuum of unstable trajectories straying through the accessible space, there is a dense set of periodic orbits,[1] and these are robust against perturbations. Here we report that long periodic orbits associate to hierarchically structured bunches. — Orbits in a bunch are mutually close everywhere and yet topologically distinct. The construction principle for bunches is provided by close self-encounters where two or more stretches of an orbit run mutually close over a long (compared to the Lyapounov length) distance. The self-encounter stretches are linked by orbit pieces ("links") of any length. Different orbits in a bunch are hardly distinct geometrically along the links in between the self-encounters, but the links are differently connected in self-encounters. Under weak resolution a bunch looks like a single orbit, and the orbits in a bunch may have arbitrarily small action differences. — Orbit bunches are a new and largely unexplored topic in classical mechanics but also spell fascination by their strong influence on quantum phenomena: Bunches with orbit-to-orbit action differences smaller than Planck's constant are responsible for universal fluctuations in energy spectra,[2–6] as well as for universal features of transport through chaotic electronic devices.[7,8]

The simplest orbit bunch, exhibited in Fig. 1, is a pair of orbits differing in an encounter of two stretches (a "2-encounter").

Arrows on the two orbits indicate the sense of traversal. That elementary bunch was discovered by Sieber and Richter[2,3] who realized that each orbit with a small-angle crossing is "shadowed" by one with an avoided crossing.

Bunches may contain many orbits. Anticipating the discussion below we show a multi-orbit bunch in Fig. 2; besides a 2-encounter, two 3-encounters (each involving three stretches) are active; the various intra-encounter connections are resolved only in the inset blow-ups.

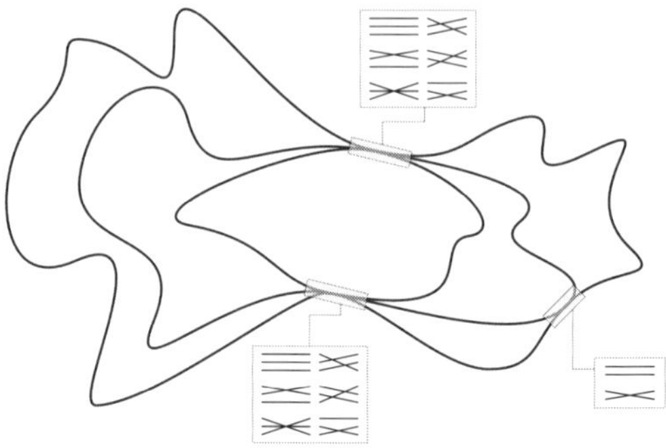

Fig. 2. Bunch of 72 orbits looking like a single orbit: Only blow-ups resolve intra-encounters connections distinguishing orbits; each inter-encounter link is 72-fold, with separations yet smaller than in encounters.

As an interesting phenomenon in bunches we meet "pseudo-orbits"; Figure 3 depicts the prototype where the replacement of a crossing by an avoided crossing entails decomposition of the original orbit into two shorter orbits; the latter are then said to form a pseudo-orbit.

The distinction of genuine periodic orbits and their pseudo-orbit partners is further illustrated in Fig. 4, for the bunch of Fig. 2.

We shall show how orbit bunches come about, illustrating our ideas for a particle moving in a two dimensional chaotic billiard, like the cardioid of Fig. 5. The particle moves on a straight line with constant velocity in between bounces and is specularly reflected at each bounce.

2. Unstable Initial Value Problem vs Stable Boundary Value Problem

To prepare for our explanation of orbit bunches it is useful to recall some basic facts about chaos. We consider hyperbolic dynamics where all trajectories, infinite or periodic, are unstable (i. e. have non-zero Lyapounov rate λ): Tiny changes of the initial data (coordinate and velocity) entail exponentially growing deflections both towards the future and the past.

Turning from the initial value problem to a boundary value problem we may specify initial and final positions (but no velocity) and ask for the connecting trajectory piece during a prescribed time span. No solution needs to exist, and if one exists it needs not be the only one. However, hyperbolicity forces a solution to be locally unique. Of foremost interest are time spans long compared to the inverse of the Lyapounov rate. Then, slightly shifted boundary points yield a trajectory piece approaching the original

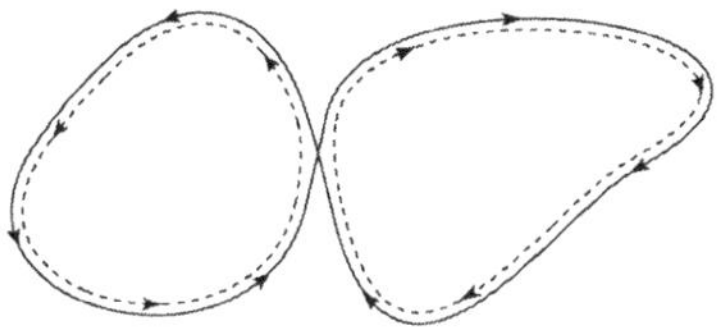

Fig. 3. Simplest pseudo-orbit, a partner of an orbit with a 2-encounter.

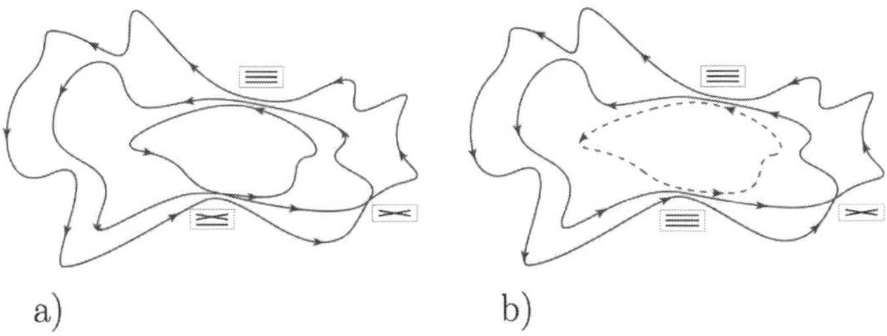

Fig. 4. Different intra-encounter connections of the bunch of Fig. 2: (a) One of the 24 genuine orbits, (b) one of the 48 pseudo-orbits.

one within intervals of duration $\sim 1/\lambda$ in the beginning and at the end, like ; towards the "inside" the distance between the perturbed and the original trajectory decays exponentially. (That fact is most easily comprehended by arguing in reverse: only an exponentially small transverse shift of position and velocity at some point deep inside the original trajectory piece can, if taken as initial data, result in but slightly shifted end points.)

When beginning and end points for the boundary value problem are merged, each solution in general has a cusp (i. e. different initial and final velocities) there. If the cusp angle is close to π one finds, by a small shift of the common beginning/end, a close by periodic orbit smoothing out the cusp and otherwise hardly distinguishable from the cusped loop, like in either loop of Fig. 3.

3. Self-Encounters

Figure 5 depicts a long periodic orbit. It appears to behave ergodically, i. e. it densely fills the available space. Figure 5 also indicates that a long orbit crosses itself many times. The smaller the crossing angle the longer

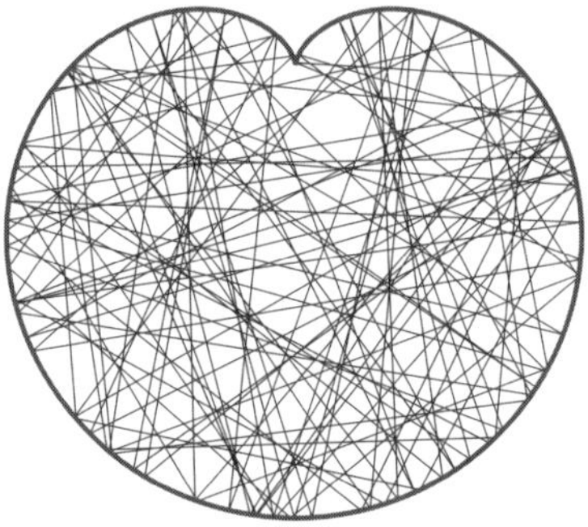

Fig. 5. Cardioid billiard with nearly ergodic orbit.

the two crossing stretches remain close; if the closeness persists through many bounces we speak of a 2-encounter.

A general close self-encounter of an orbit has two or more stretches close to one another. We speak of an l-encounter when l stretches are all mutually close throughout many bounces. For a precise definition one may pick one of the l encounter stretches as a reference and demand that none of the $(l-1)$ companions be further away than some distance d_{enc}; the latter "encounter width" must be chosen small compared to the billiard diameter D and such that the ensuing "encounter length" L_{enc} is much larger than D. To sum up, a close encounter is characterized by the following order of the various length scales,

$$d_{\text{enc}} \ll D \ll L_{\text{enc}} \ll L\,, \tag{1}$$

with L the orbit length.

4. Orbit Bunches

In the setting of Fig. 1 we can formulate an important insight into chaotic dynamics.[2,3] The equations of motion allowing for the self-crossing orbit also allow for a partner orbit which has the crossing replaced by a narrowly avoided crossing (see dashed line in Fig. 1). The existence of the partner follows from the shadowing theorem.[9] Arguing more explicitly, we invoke the exponential stability of the boundary value problem mentioned above. Namely, the two loops of the orbit with a crossing may be regarded as solutions of the boundary value problem with the beginning and end at the point of crossing. We may slightly shift apart beginning and end for each loop while retaining two junctions, as $\infty \to \infty$, with nearly no change for those loops away from the junctions. By tuning the shifts we can smooth out the cusps in the junctions, $\infty \to \infty$, and thus arrive at the partner orbit with an avoided crossing and reversed sense of traversal of one loop.

Throughout the links outside the encounter the two orbits are exponentially close. The length (and thus action) difference is the smaller the narrower the encounter:[2,3] it is quadratic in the crossing angle, $\Delta L \propto \epsilon^2$.

The foregoing mechanism for generating partner orbits works for l-encounters with any integer l. Each encounter serves as a "switch": Its l orbit stretches allow for $l!$ different connections of the nearly fixed links outside. For a given orbit a single l-encounter may and in a certain sense does give rise to $(l! - 1)$ partner orbits. The "certain sense" refers to the already mentioned fact that a partner so generated may be a pseudo-orbit, i. e. decompose into two or more shorter orbits (see Fig. 3).

A long orbit has many close self-encounters, some with $l = 2$, some with $l = 3$, etc. Every such encounter gives rise to partner (pseudo-)orbits whereupon many-orbit bunches come about; the number of orbits within a bunch acquires from each close self-encounter the pertinent factor $l!$. Figure 2 illustrates a multi-orbit bunch; among the $(3!)^2 2! = 72$ constituents there are 24 genuine periodic orbits and 48 pseudo-orbits. In the various (pseudo-)orbits of a bunch, links are traversed in different order, and even the sense of traversal of a link may change if time reversal invariance holds (see Fig. 1).

5. Hierarchies of Bunches

We would like to mention two ramifications of the concept of orbit bunches. At first, orbit bunches form hierarchical structures, due to the near indistinguishability of different orbits of a bunch within links. Every link of a bunch may thus be considered as an extremely close encounter of the participating (pseudo-)orbits, and reconnections therein produce new longer (pseudo-)orbits; the length (and action) of the new (pseudo-)orbit is approximately a multiple of that of the original one. (The Sieber-Richter pair of Fig. 1 makes for a pedagogical example: Considering, say, the two left links as stretches of an encounter we may switch these and thereby merge the two orbits.) This "process" of creating ever longer orbits by selecting a link of a bunch and treating the orbit links therein as inter-orbit encounters to switch stretches can be continued. Each step produces action differences between (pseudo-)orbits exponentially smaller than the previous step. A sequence of steps establishes an infinite hierarchical structure, and we may see a duality of encounters and links as its basis. — Second, when an orbit closely encounters an orbit from another bunch, the encounter stretches may be switched to merge the two orbits such that the original lengths are approximately added; clearly, the associated bunches then also unite.

6. Quantum Signatures of Bunches

The new perspective on classical chaos arose as a byproduct from work on discrete energy spectra of quantum dynamics with chaotic classical limits. As first discovered for atomic nuclei and later found for atomic, molecular, and many mesoscopic dynamics, the sequence of energy levels displays universal fluctuations on the scale of the mean level spacing. For instance, each such spectrum reveals universal statistical variants of repulsion of neighboring levels which depend on no other properties of the dynamics than presence or absence of certain symmetries, most notably time reversal invariance; correlation functions of the level density also fall in symmetry classes. A successful phenomenological description was provided by the Wigner/Dyson theory of random matrices (RMT); that theory employs averages over ensembles of Hermitian matrices modelling Hamiltonians, rather than dealing with any specific dynamical system.

Proving universal spectral fluctuations for individual chaotic dynamics was recognized as a challenge in the 1980's. Only quite recently it has become clear that orbit bunches generated by switching stretches of close self-encounters provide the clue, within the framework of Gutzwiller's periodic-orbit theory. The quantum mechanically relevant bunched orbits have, as already mentioned, action differences of the order of Planck's constant. Using such bunches the validity of RMT predictions for universal spectral fluctuations of individual chaotic dynamics has been demonstrated recently.[2–6]

Concurrently with the developing understanding of spectral fluctuations just sketched, it was realized that the role of encounters as switches is not restricted to periodic orbits but also arises for long entrance-to-exit trajectories between different leads of chaotic cavities.[7,8,10–12] Bunches of trajectories connecting entrance and exit leads could be invoked to explain universal conductance fluctuations of conductors. Like the semiclassical work on spectra, that explanation is a welcome step beyond RMT inasmuch as it applies to individual conductors rather than ensembles.

7. Conclusion

To conclude, bunches of periodic orbits are a hitherto unnoticed phenomenon in classical chaos, in close correspondence to universal quantum phenomena. System specific behavior in mesoscopic situations is also amenable to the new semiclassical methods[10–12] and we may expect further application there. A semiclassical theory of localization phenomena stands out as a challenge. — Classical applications of orbit bunches comprise action correlations among periodic orbits;[5,13] others could arise in a theory of the so-called Frobenius-Perron resonances which describe the approach of ergodic equilibrium for sets of trajectories. Similarly, orbit bunches can be expected to become relevant for the well known cycle expansions[14] of classical observables (where an important role of pseudo-orbits was first noticed long since). — The hierarchical structure of bunches, a beautiful phenomenon in its own right, deserves further study.

Acknowledgment

This work was supported by the DFG under the SFB/TR 12. We thank Thomas Dittrich for fruitful discussions and Stefan Thomae for art work.

References

1. H. Poincaré, *Methodes Nouvelles de la Méchanique Classique*, Vol. I, § 36 (reedit. Blanchard, Paris, 2003).
2. M. Sieber and K. Richter, *J. Phys. A* **T91**, 128 (2001).
3. M. Sieber, *J. Phys. A* **35**, L613 (2002).
4. S. Müller, S. Heusler, P. Braun, F. Haake, and A. Altland, *Phys. Rev. Lett.* **93**, 014103 (2004).
5. S. Müller, S. Heusler, P. Braun, F. Haake, and A. Altland, *Phys. Rev. E* **72**, 046207 (2005).
6. S. Heusler, S. Müller, A. Altland, P. Braun, and F. Haake, *Phys. Rev. Lett.* **98**, 044103 (2007).
7. K. Richter and M. Sieber, *Phys. Rev. Lett.* **89**, 206801 (2002).
8. S. Müller, S. Heusler, P. Braun, and F. Haake, *New. J. Phys.* **9**, 12 (2007).
9. A. Katok and B. Hasselblatt, *Introduction to the Modern Theory of Dynamical Systems* (Cambridge University Press, Cambridge, 1995).
10. S. Rahav and P. Brouwer, *Phys. Rev. Lett.* **95**, 056806 (2005).
11. S. Rahav and P. Brouwer, *Phys. Rev. Lett.* **96**, 196804 (2006).
12. R. S. Whitney and P. Jacquod, *Phys. Rev. Lett.* **96**, 206804 (2006).
13. N. Argaman, F.-M. Dittes, E. Doron, J. P. Keating, A. Yu. Kitaev, M. Sieber, and U. Smilansky, *Phys. Rev. Lett.* **71**, 4326 (1993).
14. P. Cvitanović, R. Artuso, P. Dahlqvist, R. Mainieri, G. Tanner, G. Vattay, N. Whelan, and A. Wirzba, *Chaos: Classical and Quantum*, online version 11 (2004), http://chaosbook.org/stable.

PATH INTEGRATION IN THE FIELD OF DISPIRATION

A. INOMATA

Department of Physics, State University of New York at Albany,
Albany, New York 12222, USA
E-mail: inomata@albany.edu

Path integration is carried out for the bound states of a particle in the combined field of a wedge disclination and a screw dislocation. The energy spectrum extracted from the Feynman kernel differs from that obtained by solving the Schrödinger equation.

Keywords: Linear defects; Dislocation; Disclination.

1. Introduction

In the present paper, we study path integration in the field of a dispiration – the combined structure of a screw dislocation and a wedge disclination – around which the squared line element is given by[1],

$$ds^2 = dr^2 + \sigma^2 r^2\, d\theta^2 + (dz + \beta d\theta)^2 \tag{1}$$

where β and σ are the parameters related to the Burgers vector of the dislocation and the Frank vector of the disclination, respectively. The medium characterized by (1) has a non-Euclidean structure with singular torsion and curvature along the dispiration line. The quantum behavior of a particle bound by the harmonic oscillator potential in such a medium has been studied by using Schrödinger's equation.[2]

Schrödinger's equation in curved space may be written as

$$\left\{-\frac{\hbar^2}{2m}\Delta + \eta\hbar^2 R + V(\mathbf{r})\right\}\psi(\mathbf{r},t) = i\hbar\frac{\partial}{\partial t}\psi(\mathbf{r},t) \tag{2}$$

where Δ is the Laplace-Beltrami operator, η a constant, and R the curvature scalar. While Podolsky[3] defined the Schrödinger equation without the curvature term ($\eta = 0$), DeWitt[4] proposed by comparing with Feynman's path integral that the curvature term is needed. More recently Kleinert[5] argued that there is no need of the curvature term. This controversy stems from the fact that the Hamiltonian cannot uniquely be converted into a differential operator because of the operator ordering problem, and has not fully been settled. Since both torsion and curvature of the present medium have a singularity at $r = 0$ making the medium topologically nontrivial, it is uncertain whether the curvature term has no role in Schrödinger's equation. Our purpose is not to resolve the controversy, but to report that the path integration results differ from the one obtained from Schrödinger's equation (2) with $\eta = 0$.

2. Gauge Approach to the Dispiration

The dispiration we consider is a combination of the screw dislocation along the z-axis and the wedge disclination about the z-axis. Although the gauge formulation of dislocation and disclination has been extensively discussed,[1,6] here we present very briefly the gauge approach to a dispiration.

A screw dislocation lying along the z-axis (with the Burgers vector pointing in the z-direction) is a translational deformation (obtained by gauging the z-translation group $T(1)$):

$$\mathbf{x}' = \mathbf{x} + \frac{b\theta}{2\pi}\mathbf{e}_z, \qquad \theta = \tan^{-1}(y/x) \in [0, 2\pi), \tag{3}$$

and the wedge disclination about the z-axis (with the Frank vector pointing in the z-direction) is a rotational deformation (obtained by gauging $SO(2)$),

$$\mathbf{x}' = \mathbf{D}(\theta)\mathbf{x} \tag{4}$$

with the rotation matrix

$$\mathbf{D}(\theta) = \begin{pmatrix} \cos(\gamma\theta/2\pi) & -\sin(\gamma\theta/2\pi) & 0 \\ \sin(\gamma\theta/2\pi) & \cos(\gamma\theta/2\pi) & 0 \\ 0 & 0 & 1 \end{pmatrix}. \tag{5}$$

The connection one-form for the dispiration is defined by

$$\omega = d\mathbf{x} + \Gamma^{(R)}\mathbf{x} + \Gamma^{(T)}. \tag{6}$$

Here the rotation connection has the form

$$\Gamma^{(R)} = \mathbf{D}\cdot d\mathbf{D}^{-1} = \frac{\gamma}{2\pi}\, d\theta\, \mathbf{m}, \qquad \mathbf{m} = \begin{pmatrix} 0 & -1 & 0 \\ 1 & 0 & 0 \\ 0 & 0 & 0 \end{pmatrix} \tag{7}$$

and the translation connection $\Gamma^{(T)} = (b/2\pi)\, d\theta\, \mathbf{e}_z$. Accordingly the connection one-form reads

$$\omega = \begin{pmatrix} dx - (\gamma/2\pi) y\, d\theta \\ dy + (\gamma/2\pi) x\, d\theta \\ dz + (b/2\pi) d\theta \end{pmatrix}. \tag{8}$$

Using this connection form as the coframe we get the line element in the vicinity of the dispiration,

$$ds^2 = \delta_{ab}\omega^a \otimes \omega^b = dr^2 + \sigma^2 r^2\, d\theta^2 + (dz + \beta d\theta)^2\,, \tag{9}$$

where $\sigma = 1 + \gamma/(2\pi)$ and $\beta = b/(2\pi)$.

The torsion two-form $\mathbf{T} = \mathbf{T}_{ij}\, dx^i \wedge dx^j$ for this system is given by,

$$\mathbf{T} = d\omega + \Gamma^{(R)} \wedge \omega = \frac{\gamma}{2\pi r^2}\mathbf{m} \cdot \mathbf{x}\, dx \wedge dy + b\delta^{(2)}(x,y)\, \mathbf{e}_z\, dx \wedge dy, \tag{10}$$

which has a singularity along the z-axis. The Burgers vector is calculated by

$$\mathbf{b} = \int_{\partial C} \mathbf{T} = b\, \mathbf{e}_z. \tag{11}$$

The parameter $b = 2\pi\beta$ in (3) is indeed the magnitude of the Burgers vector.

The curvature two-form $\mathbf{R} = \mathbf{R}_{ij}\, dx^i\, dx^j$ takes the form

$$\mathbf{R} = d\Gamma^{(R)} + \Gamma^{(R)} \wedge \Gamma^{(R)} = \frac{\gamma}{2\pi}\mathbf{m}\, d^2\theta = \gamma\delta^{(2)}(x,y)\, \mathbf{m}\, dx \wedge dy, \tag{12}$$

which has also a singularity along the z-axis and whose nonvanishing components are only $\mathbf{R}^{12} = -\mathbf{R}^{21}$. Thus the Frank vector defined by $\mathbf{f} = \{\Phi^{23}, \Phi^{31}, \Phi^{12}\}$ with the surface integral Φ of $\mathbf{R}$ becomes

$$\mathbf{f} = \Phi^{12}\, \mathbf{e}_z = \int_{\partial C} \mathbf{R}^{12} = \gamma\, \mathbf{e}_z. \tag{13}$$

The parameter γ appearing in (9) turns out to be the magnitude of the Frank vector, which is the deficit angle of the disclination. Apparently $\sigma > 1$ for the positive curvature and $\sigma < 1$ for the negative curvature.

3. The Propagator

The Lagrangian: Since the dispiration field is characterized by the line element (1), the Lagrangian for a particle of mass M bound in the field of the dispiration under the influence of a two-dimensional central potential $V(r)$ with $r^2 = x^2 + y^2$,

$$L = \frac{1}{2}M\left\{\dot{r}^2 + \sigma^2 r^2\dot{\theta}^2 + \left(\dot{z} + \beta\dot{\theta}\right)^2\right\} - V(r). \tag{14}$$

Here we assume the scalar potential of the form,

$$V(r) = \frac{1}{2}M\omega^2 r^2 + \frac{\kappa\hbar^2}{8M\sigma^2 r^2}. \tag{15}$$

Thus the Feynman kernel (or the propagator in quantum mechanics) for the system is

$$K\left(\mathbf{r}'', z''; \mathbf{r}', z'; \tau\right) = \int \exp\left[\frac{i}{\hbar}\int_{t'}^{t''}\left\{\frac{M}{2}\left(\dot{r}^2 + \sigma^2 r^2\dot{\theta}^2\right) + \frac{M}{2}\left(\dot{z} + \beta\dot{\theta}\right)^2 - V(r)\right\}dt\right]\mathcal{D}^2\mathbf{r}\,\mathcal{D}z \tag{16}$$

where $\mathbf{r}$ is a collective symbol for two variables (r, θ).

The z-integration: The z-integration can easily be carried out by changing the variable as $\zeta = z + \beta\theta$. The z-integration is nothing but the Gaussian path integral for a one-dimensional free particle, the result of integration being

$$\int \exp\left[\frac{i}{\hbar}\int_{t'}^{t''}\frac{M}{2}\dot{\zeta}^2 dt\right]\mathcal{D}\zeta = \sqrt{\frac{M}{2\pi i\hbar\tau}}\exp\left[\frac{iM\left(\zeta'' - \zeta'\right)^2}{2\hbar\tau}\right], \tag{17}$$

which we put into the form,

$$\sqrt{\frac{M}{2\pi i\hbar\tau}}\exp\left[\frac{iM\left(\zeta'' - \zeta'\right)^2}{2\hbar\tau}\right] = \frac{1}{2\pi}\int_{-\infty}^{\infty} e^{-i\hbar\tau k^2/2M}e^{i\left(\zeta'' - \zeta'\right)k}dk. \tag{18}$$

Noting that

$$\zeta'' - \zeta' = z'' - z' + \beta'\left(\theta'' - \theta'\right) = z'' - z' + \beta\int_{t'}^{t''}\dot{\theta}dt, \tag{19}$$

we obtain

$$K\left(\mathbf{x}'', \mathbf{x}'; \tau\right) = \frac{1}{2\pi}\int_{-\infty}^{\infty} dk e^{ik\left(z'' - z'\right)}e^{-i\hbar\tau k^2/2M}K^{(k)}\left(\mathbf{r}'', \mathbf{r}'; \tau\right) \tag{20}$$

with the two-dimensional propagator for a fixed k value,

$$K^{(k)}\left(\mathbf{r}'', \mathbf{r}'; \tau\right) = \int \exp\left[\frac{i}{\hbar}\int_{t'}^{t''}\left\{\frac{M}{2}\left(\dot{r}^2 + \sigma^2 r^2\dot{\theta}^2\right) + \beta k\hbar\dot{\theta} - V(r)\right\}dt\right]\mathcal{D}^2\mathbf{r}. \tag{21}$$

Discretized propagator: To evaluate the path integral for the k-propagator (21), we use the time-sliced approach:

$$K^{(k)}\left(\mathbf{r}'',\mathbf{r}';\tau\right)=\lim_{N\to\infty}\int_{r'=r(t')}^{r''=r(t'')}\prod_{j=1}^{N}K^{(k)}\left(\mathbf{r}_j,\mathbf{r}_{j-1};\epsilon\right)\prod_{j=1}^{N-1}d^2\mathbf{r}_j. \tag{22}$$

The short time propagator is given by

$$K^{(k)}\left(\mathbf{r}_j,\mathbf{r}_{j-1};\epsilon\right)=A_j\exp\left(\frac{i}{\hbar}S_j\right), \tag{23}$$

with the short time action

$$S_j=\int_{t_{j-1}}^{t_j}\left[\frac{M}{2}(\dot{r}^2+\sigma^2r^2\dot{\theta}^2)+\beta k\hbar\dot{\theta}-V(r)\right]dt. \tag{24}$$

The integration measure is chosen to be $A_j\,d^2\mathbf{r}_j=(M\sigma/2\pi i\hbar\epsilon)\,r_j\,dr_j\,d\theta_j$.

As is well-known, in path integration, $(\Delta\theta)^2\sim\epsilon$. Hence $(\Delta\theta)^4/\epsilon$ cannot be ignored. Thus, making the following replacement $\dot{\theta}^2\to 2[1-\cos(\Delta\theta)]/\epsilon$, we approximate the short time action by

$$S_j=\frac{M}{2\epsilon}\left\{(\Delta r_j)^2+2\sigma^2r_jr_{j-1}\left[1-\cos(\Delta\theta_j)\right]\right\}+\beta k\hbar\Delta\theta_j-V(r_j)\epsilon. \tag{25}$$

Moreover we use the approximation,[7]

$$\cos\left(\Delta\theta\right)-a\epsilon\Delta\theta\simeq\cos\left(\Delta\theta+a\epsilon\right)+\frac{1}{2}a^2\epsilon^2, \tag{26}$$

to write the short time action multiplied by $(i/\hbar)$ as

$$\begin{aligned}\frac{i}{\hbar}S_j&=\frac{iM}{2\epsilon}\left(r_j^2+r_{j-1}^2\right)+(\sigma^{-2}-1)\frac{M\sigma^2r_jr_{j-1}}{i\hbar\epsilon}\\&\quad+\frac{M\sigma^2r_jr_{j-1}}{i\hbar\epsilon}\cos\left(\Delta\theta_j+\frac{\beta k\hbar\epsilon}{M\sigma^2r_jr_{j-1}}\right)+\frac{(4\beta^2k^2+\kappa)\hbar\epsilon}{8Mi\sigma^2r_jr_{j-1}}-\frac{i\epsilon}{\hbar}\tilde{V}_j\end{aligned} \tag{27}$$

where $\tilde{V}_j=(M\omega^2/4)(r_j^2+r_{j-1}^2)$ is the contribution from the harmonic oscillator portion of the potential $V(r_j)$.

Angular integration: To separate the angular variable from the radial function, we employ the asymptotic recombination technique.[8] Basic to this technique is the conjecture that the Edwards-Gulyaev one-term asymptotic formula for the modified Bessel function,[9,10]

$$I_{|\nu|}\left(z\right)\sim\frac{1}{\sqrt{2\pi z}}\exp\left[z-\frac{1}{2z}\left(\nu^2-\frac{1}{4}\right)\right], \tag{28}$$

is valid for large z and for $\arg z\ <\pi/2$ when used in path integration. From (28) we can derive the following asymptotic relation,

$$I_{|\nu|}(az)\,e^{bz}\,e^{c/z}\sim\sqrt{\frac{a+b}{a}}\,I_{|\mu|}\left[(a+b)z\right]\,, \tag{29}$$

where a, b and c are real constants, and

$$\mu=\left[\frac{a+b}{a}\nu^2-\frac{b}{4a}-2(a+b)c\right]^{1/2}. \tag{30}$$

Combining this asymptotic relation with the Jacobi-Anger formula,

$$e^{z\cos\vartheta}=\sum_{m=-\infty}^{\infty}e^{im\vartheta}I_m(z) \tag{31}$$

leads to another asymptotic relation,[8]

$$\begin{aligned}\exp&\left\{bz+z\cos\left[\Delta\theta+i\frac{d}{z}\right]-\frac{(d^2-2c)}{2z}\right\}\\&\doteq\sqrt{1+b}\,\sum_{m=-\infty}^{\infty}e^{im\,\Delta\theta}\,I_{|\mu|}\left[(1+b)z\right]\,,\end{aligned} \tag{32}$$

where

$$\mu=\left[(1+b)\{(m+d)^2-2c\}-b/4\right]^{1/2}. \tag{33}$$

Letting $b=\sigma^{-2}-1$, $z=M\sigma^2 r_j r_{j-1}/(i\hbar\epsilon)$ and $d=-\beta k$, we utilize the asymptotic relation (32) to separate variables of the short time propagator (23) with the action (27) as

$$\begin{aligned}K^{(k)}\left(\mathbf{r}_j,\mathbf{r}_{j-1};\epsilon\right)&=A_j\exp\left(\frac{i}{\hbar}S_j\right)\\&=\frac{1}{2\pi}\sum_{m_j=-\infty}^{\infty}e^{im_j(\vartheta_j-\vartheta_{j-1})}R_{m_j}\left(r_j,r_{j-1};\epsilon\right)\end{aligned} \tag{34}$$

with the short time radial propagator,

$$R_{m_j}\left(r_j,r_{j-1};\epsilon\right)=\frac{M}{i\hbar\epsilon}\exp\left[\frac{iM}{2\hbar\epsilon}\left(r_j^2+r_{j-1}^2\right)-\frac{i\epsilon}{\hbar}\tilde{V}_j\right]I_{\mu(m_j)}\left(\frac{Mr_jr_{j-1}}{i\hbar\epsilon}\right) \tag{35}$$

where $\mu(m_j)=\left[4(m_j-\beta k)^2+\sigma^2-1+\kappa\right]^{1/2}/(2\sigma)$. Now the angular integration of (34) can easily be achieved, which results in $m_j=m_N\equiv m$.

Radial propagator: After angular integration the short time radial propagator (35) takes the form similar to that of the radial harmonic oscillator,

$$R_m(r_j, r_{j-1}; \epsilon) = \frac{M}{i\hbar\epsilon} \exp\left[\frac{iM}{2\hbar\epsilon}(r_j^2 + r_{j-1}^2) - \frac{iM\omega^2\epsilon}{4\hbar}(r_j^2 + r_{j-1}^2)\right] I_{\nu(m)}\left(\frac{Mr_j r_{j-1}}{i\hbar\epsilon}\right) \quad (36)$$

where

$$\mu(m) = \frac{1}{2\sigma}[4(m - \beta k)^2 + \sigma^2 - 1 + \kappa]^{1/2}. \quad (37)$$

The radial path integration for the harmonic oscillator has been calculated.[8,11] The finite time radial propagator for the present system can also be obtained by following the same procedure. Since the page number is limited, we just write down the result. Namely,

$$R_m(r'', r'; \tau) = \frac{M\omega}{i\hbar \sin\omega\tau} \exp\left[\frac{iM\omega}{2\hbar}(r'^2 + r''^2)\cot\omega\tau\right] I_{\mu(m)}\left(\frac{M\omega r' r''}{i\hbar \sin\omega\tau}\right) \quad (38)$$

where $\mu(m)$ is given by (37). This result can also be formally derived from the radial propagator for a free particle by applying the deAlfano-Fubini-Furlan-Jackiw transformation,[12] $r \to r\ \sec(\omega t)$ and $\omega t \to \tan(\omega t)$, and replacing the angular momentum quantum number m by $\mu(m)$ of (37).[13]

4. Energy Spectrum

The information concerning the energy spectrum and the radial wave functions is all contained in the radial propagator (38). In order to read off such information out of (38), we make use of the Hille-Hardy identity (GR: 8.976.1),[14]

$$I_\mu\left(\frac{2\sqrt{xyz}}{1-z}\right)\frac{(xyz)^{-\mu/2}}{1-z}\exp\left[-z\frac{x+y}{1-z}\right] = \sum_{n=0}^{\infty}\frac{n!}{\Gamma(n+\mu+1)} L_n^{(\mu)}(x)\, L_n^{(\mu)}(y) z^n\,, \quad (39)$$

where $L_n^{(\mu)}(x)$ is the Laguerre polynomial related to the confluent hypergeometric function as

$$L_n^{(\mu)}(x) = \frac{\Gamma(n+\mu+1)}{\Gamma(\mu+1)\, n!} F(-n, \mu+1; x). \quad (40)$$

Letting $x = (M\omega/\hbar)r'^2$, $y = (M\omega/\hbar)r''^2$, and $z = e^{-2i\omega\tau}$ in (39), we write (38) as

$$R_m(r'', r'; \tau) = \frac{2M\omega}{\hbar}\left(\frac{M\omega}{\hbar}r'r''\right)^{\mu/2} \exp\left[-\frac{iM\omega}{2\hbar}(r'^2+r''^2)\right] \times \sum_{n=0}^{\infty} \frac{n!\, e^{-i\omega(2n+\mu+1)\tau}}{\Gamma(n+\mu+1)} L_n^{(\mu)}\left(\frac{M\omega r'^2}{\hbar}\right) L_n^{(\mu)}\left(\frac{M\omega r''^2}{\hbar}\right). \quad (41)$$

From this radial propagator, we can read off the radial wave functions,

$$\chi_{m,n}(r) = N_{m,n}\, r^{\mu(m)}\, e^{-(M\omega/2\hbar)r^2}\, F\left(-n, \mu(m)+1; \frac{M\omega}{\hbar}r^2\right) \quad (42)$$

with the normalization factor

$$N_{m,n} = \frac{1}{\Gamma(\mu(m)+1)}\left[\frac{\Gamma(n+\mu(m)+1)}{n!}\right]^{1/2}\left(\frac{M\omega}{\hbar}\right)^{(\mu(m)+1)/2}, \quad (43)$$

and the energy spectrum,

$$E_{m,n} = \hbar\omega\left\{2n+1+\frac{1}{2\sigma}\sqrt{4(m-\beta k)^2+\sigma^2-1+\kappa}\right\}, \quad (44)$$

where we have used (37) for $\mu(m)$. If we let $\kappa = \sigma^2 - 1$, then the above spectrum reduces to the one obtained from Schrödinger's equation.[2]

5. Concluding Remarks

In the above we have discussed a partial wave expansion of the Feynman kernel for a particle bound in the field of a dispiration. The partial wave expansion can be converted into the winding number expansion classifying homotopically inequivalent paths. The expansion coefficients turns out to form a one-dimensional unitary representation of the fundamental group $\pi_1 \cong \mathbf{Z}$, being consistent with the Laidlaw-deWitt-Schulman theorem.[15] This result will be reported elsewhere. Next we wish to make the following remarks on the energy spectrum (44). (a) The screw dislocation plays a role similar to that of a magnetic flux tube for the Aharonov-Bohm effect and in Wilczek's anyon model.[16] (b) Although the Dirac fields are understood as giving rise to torsion in Einstein-Cartan theory,[17] the present spectrum shows that the torsion field due to the dislocation gives rise to spin. (c) The harmonic oscillator potential is introduced only for creating the bound states in two dimensions. (d) If the deficit angle is positive, then $\sigma^2 - 1 > 0$ and the short-range force is enhanced. The disclination with the positive singular curvature appears to generate a repulsive force. If the deficit angle is negative, then $\sigma^2 - 1 < 0$ and an attractive force is induced. The negative singular curvature could behave as a sink; if $m - \beta k$ is small and $\kappa = 0$ then there will be no bound states. (e) For reproducing the spectrum resulted from Schrödinger's equation with $\eta = 0$, the Lagrangian has to contain an inverse-square potential with $\kappa = -(\sigma^2 - 1)$ as a counter term.

References

1. R. A. Puntigam and H. H. Soleng, *Class. Quantum Grav.* **14**, 1129 (1997).
2. C. Furtado and F. Moraes, *J. Phys. A: Math. Gen.* **33**, 5513 (2000).
3. B. Podolsky, *Phys. Rev.* **32**, 812 (1928).
4. B. S. DeWitt, *Rev. Mod. Phys.* **29**, 337 (1957).
5. H. Kleinert, *Path Integrals in Quantum Mechanics, Statistics and Polymer Physics*, 2nd edition (World Scientific, Singapore, 1995).
6. H. Kleinert, *Gauge Fields in Condensed Matter* (World Scientific, Singapore, 1989), and references therein.
7. A. Inomata and V. A. Singh, J. Math. Phys. **19**, 23118 (1978).
8. A. Inomata, H. Kuratsuji, and C.C. Gerry, *Path Integrals and Coherent States of SU(2) and SU(1,1)* (World Scientific, Singapore, 1992).
9. S. F. Edwards and Y. V. Gulyaev, *Proc. Roy. Soc. London A* **279**, 229 (1964).
10. W. Langguth and A. Inomata, *J. Math. Phys.* **20**, 499 (1979).
11. D. Peak and A. Inomata, *J. Math. Phys.* **10**, 1422 (1969).
12. V. deAlfaro, S. Fubini, and G. Furlan, *Nuovo Cimento* **34**, 117 (1976); R. Jackiw, *Ann. Phys.* **129**, 183 (1980).
13. P. Y. Cai, A. Inomata, and P. Wang, *Phys. Lett. A* **91**, 331 (1982).
14. I. S. Gradshteyn and I. Ryzik, *Tables of Integrals, Series and Products* (Academic Press, New York, 1965).
15. M. G. G. Laidlaw and C. M. DeWitt, *Phys. Rev. D* **3**, 1375 (1971); L. S. Schulman, *J. Math. Phys.* **12**, 304 (1971).
16. F. Wilczek, *Phys. Rev. Lett.* **58**, 1144 (1982); D. Arovas, in *Geometric Phases in Physics*, eds. A. Shapere and F. Wilczek (World Scientific, Singapore, 1985), p. 284.
17. F. W. Hehl and B. K. Datta, *J. Math. Phys.* **12**, 1334 (1971); A. Inomata, *Phys. Rev. D* **18**, 3552 (1978).

REAL TIME PATH INTEGRALS IN STUDIES OF QUANTUM DOTS DYNAMICS: NON-MONOTONOUS DECAY RATE AND REAPPEARANCE OF RABI ROTATIONS

A. VAGOV*, M. D. CROITORU†, V. M. AXT‡, T. KUHN‡, and F. PEETERS§

*Physics Department, Lancaster University, LA1 4YB Lancaster, United Kingdom
†EMAT, Departement Fysica, Universiteit Antwerpen, Groenenborgerlaan 171, B-2020 Antwerpen, Belgium
‡Institut für Festkörpertheorie, Westfälische Wilhelms-Universität, D-48149 Münster, Germany
§CMT, Departement Fysica, Universiteit Antwerpen, Groenenborgerlaan 171, B-2020 Antwerpen, Belgium
*E-mail: a.vagov@lancaster.ac.uk

The dynamics of strongly confined laser driven semiconductor quantum dots coupled to phonons is studied theoretically by calculating the time evolution of the reduced density matrix using the path integral method. We explore the cases of long pulses, strong dot-phonon and dot-laser coupling and high temperatures, which up to now have been inaccessible. We find that the decay rate of the Rabi oscillations is a non-monotonic function of the laser field leading to the decay and reappearance of the Rabi oscillations in the field dependence of the dot exciton population.

Keywords: Quantum dynamics; Quantum dot; Path-integral approach.

1. Introduction

Coherent manipulation of exciton states in semiconductor nanosize quantum dots (QD) using Rabi oscillations is vital for their applications in quantum communication and information.[1] However, in the coherence of the QD carrier states is destroyed by coupling to the environment, leading to decay of the oscillations.[2–5] The phonon-dot coupling provides a basic dephasing mechanism which is present in any sample and all setups and thus marks a lower limit for the total decoherence.[6–9] This work focuses on the influence of the acoustic (LA) phonon modes on the dynamics of the QD exciton states.

A theoretical description of a strongly confined QD reduces can be done using a two-level independent boson model (TLIB) coupled to an external field.[9] A fully non-perturbative analysis of TLIB is possible only for ultra-short laser pulses.[6] A general situation of arbitrary excitation pulses requires approximations, e.g. perturbation analysis,[10] the correlation[7,11,12] and cumulant[8] expansions, non-interacting blip approximation[13] or a Gaussian approximation, where the phonon influence is reduced to simple damping.[5] These approaches are generally unsuitable for the strong phonon coupling and at large temperatures.

We present a solution of the TLIB obtained using the numerical real-time path integral calculations. Calculations of the real time path integrals usually run into the well known dynamical sign problem.[14] However, if the phonon memory time is finite this problem can be solved by reformulating the path integral calculation in the form of a pseudo-Markovian evolution of the so-called augmented density matrix.[15] We apply a modification of this formalism combined with the "on-fly" path selection algorithm[16–18] to study QD dynamics in the previously inaccessible regimes of long evolution times, strong phonon coupling and high temperatures.

2. Model and Numerical Path-Integral Calculation

The TLIB Hamiltonian coupled with the laser field is written (in Bohr units) in the rotation wave approximation as

$$H = \overline{\sigma}_z\Omega_0 + \overline{\sigma}_z \sum_q \gamma_q(b_q^+ + b_q) + \sum_q \omega_q b_q^+ b_q + ME(t)\sigma_+ + M^*E^*(t)\sigma_- \tag{1}$$

where $\overline{\sigma}_z = \sigma_z + 1/2$, $\sigma_\pm = \sigma_x \pm i\sigma_y$ are Pauli matrices acting on the QD states, Ω is the exciton energy, b_q are phonon operators, ω_q is phonon dispersion, γ_q is the phonon coupling, $M_\pm$ is the dot-field coupling constant and $E(t) = f(t)e^{i\omega t}$ is resonant laser field of a frequency $\omega = \Omega_0 - \sum_q(|\gamma_q|/\omega_q)^2$ with $f(t)$ being a real envelope function. As discussed, only LA phonons are accounted for in Eq. (1).[6,19] The dynamics of the density matrix is obtained by solving the Liouville-von Neumann equation with the initial condition that the system is in a temperature equilibrium. The dynamics of the exciton subsystem is obtained by taking the trace of the density matrix over the phonon states. For Eq. (1) the trace is calculated explicitly leading to the well known Feynman-Vernon functional.[20] A discretized version of the path integral for the time dependence of the reduced density matrix at time slices $t = N\Delta t$ assumes the following form

$$\rho_{\rho_{\sigma_N^+ \sigma_N^-}} = \sum_{\sigma_0^\pm \ldots \sigma_N^\pm = 0}^{1} R_N(\sigma_0^\pm \ldots \sigma_N^\pm) \rho_{\sigma_0^+ \sigma_0^-}, \quad (2)$$

where $\sigma_i^\pm = 0, 1$, $\rho_{\sigma_0^+ \sigma_0^-}$ is the initial exciton density matrix and

$$R_N(\sigma_0^\pm \ldots \sigma_N^\pm) = \prod_i M_{\sigma_{i+1}^+ \sigma_i^+} M_{\sigma_{i+1}^- \sigma_i^-} \prod_{j \le i} e^{S_{ij}}. \quad (3)$$

Here the rotation matrix is $M = \exp(i\sigma_x \int_{t_{i-1}}^{t_i} f(t)dt)$. The influence functional is $S_{ij} = -(\sigma_i^+ - \sigma_i^-)(K_{i-j}\sigma_j^+ - K_{i-j}^* \sigma_j^-)$, the memory kernel in which is

$$K_{i-j>0} = \int_{t_{i-1}}^{t_i} ds \int_{t_{j-1}}^{t_j} ds' \, \Gamma(s - s'), \; K_0 = \int_{t_{i-1}}^{t_i} ds \int_{t_{i-1}}^{t} ds' \, \Gamma(s - s'), \quad (4)$$

where

$$\Gamma(s) = \sum_q \gamma_q^2 \left\{ \coth\left(\frac{\omega}{2T}\right) \cos(\omega s) - i \sin(\omega s) \right\}. \quad (5)$$

In order to calculate Eq. (2) we utilize the fact that for the dot-phonon coupling the memory kernel decays rapidly with time so one uses $K_{i>N_c} = 0$. Then one writes the recurrence relation for the next time slice $R_{N+1} = T_{N+1} R_N$, where the "transfer" matrix T_{N+1} is

$$T_{N+1} = M_{\sigma_{N+1}^+ \sigma_N^+} M_{\sigma_{N+1}^- \sigma_N^-} \prod_{j=N+1-N_c}^{j_{N+1}} e^{S_{N+1-j}}, \quad (6)$$

which depends on N_c variables $\sigma_i^\pm$, $i = N + 1 - N_c ... N + 1$. Defining the augmented density matrix as $\overline{R}_N = \sum_{\sigma_0^\pm \ldots \sigma_{N-N_c}^\pm} R_N$ one obtains a new recurrence relation

$$\overline{R}_{N+1} = \sum_{\sigma_{N+1-N_c}^\pm} T_{N+1} \overline{R}_N. \quad (7)$$

Further calculations are done using a 3 step algorithm: 1) If all contributing configurations (paths) are known at time N, for which $\overline{R}_N > \epsilon$ (ϵ defines accuracy) then 2) multiply $\overline{R}_N$ by T_{N+1} as in Eq. (7) and find all new configurations with $\overline{R}_{N+1} > \epsilon$ and 3) sum over $\sigma_{N-N_c}^\pm$ in Eq. (7) and return to step 1). Then one finds $\overline{R}_N$ and thus ρ_d at arbitrary N. The described procedure is similar to the Monte-Carlo algorithms, where the paths are constructed recursively. The advantage of this procedure is that by calculating the augmented density matrix at the time ΔN one automatically obtains it for all $n < N$, i.e. for the entire time interval.

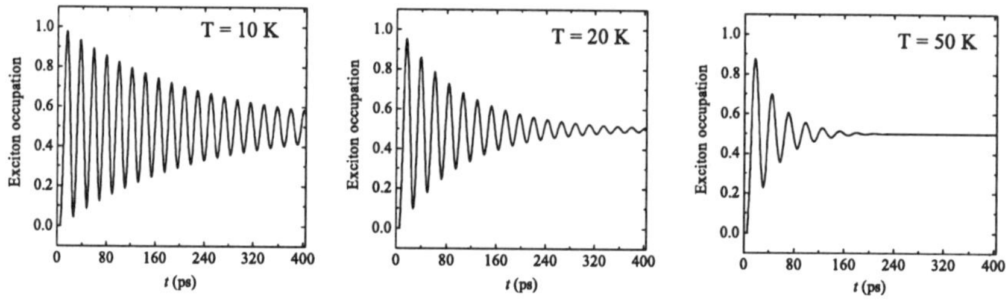

Fig. 1. Time evolution of the QD exciton population at $T = 10, 20, 50$ K and for the time integrated pulse area $f_a t_f = 80\pi$.

3. Results and Discussion

Numerical calculations are done for GaAs QD of a spherical shape with the diameter 5 nm. Details of the model and values of the material parameters for the dot-phonon coupling can be found elsewhere.[6] The memory kernel can be neglected at $t > 5$ ps which defines the cutoff as $K_{n>5\ \mathrm{ps}/\Delta} = 0$. For the sake of clarity we present results for the rectangular pulses $f_a(t) = const$, which allow for a transparent physical interpretation while being qualitatively general.

The dynamics of a QD exciton occupation, ρ_{11}, shown in Fig. 1 displays decaying Rabi oscillations, that approach a stationary limit 1/2 at large time. The dynamics can be approximated by a simple formula $\frac{1}{2}\left[1 - \cos(\Omega_R t)\exp(-t/T_d)\right]$, where T_d and Ω_R serve as the decay time and the Rabi frequency, respectively, and are found from the best fit of the numerical solution. Both Ω_R and T_d depend on T and f. As expected T_d decreases with rising T. However, $T_d(f)$ is a non-monotonous function as shown in Fig. 2. At smaller fields $T_d(f)$ is a decreasing function, reaching its minimum at $f_a \approx 1-2$ ps^{-1}, while at larger fields it increases with f.

This leads to the decay and reappearance of the Rabi oscillations in the field dependence of the exciton occupation, measured after the pulses of a fixed duration, as demonstrated in Fig. 3. Such definition of the Rabi oscillations (Rabi rotations) is frequent in experimental literature, because the full time evolution of system is harder to obtain. At larger T one obtains a more pronounced decay-reappearance pattern.

The non-monotonous fields dependence of the decay time can be qualitatively explained by invoking results of the perturbation theory.[10] The critical field with the maximal decay is equal to the inverse of the phonon kinetic time, $\tau_{ph} \approx 0.5$ ps. This fact is not accidental and follows from that

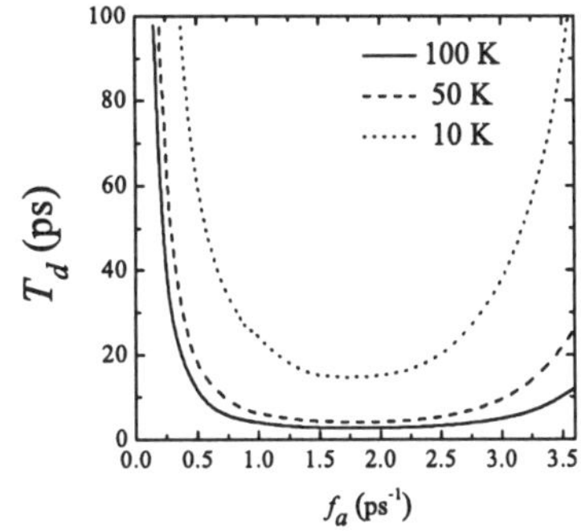

Fig. 2. Decay time T_d as a function of the field amplitude f and temperature T.

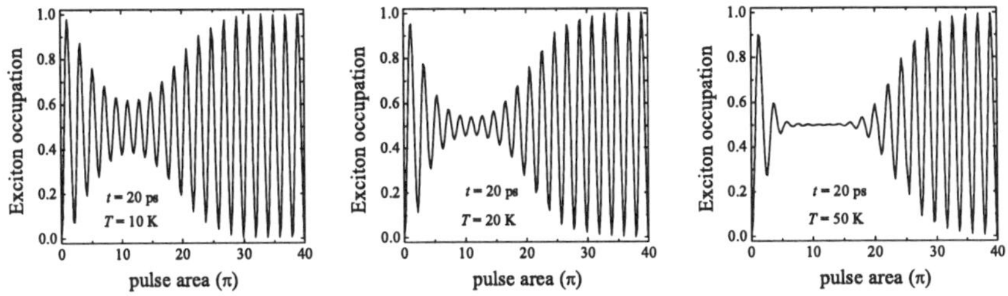

Fig. 3. Exciton occupation as a function of pulse area, calculated at $T = 10, 20, 50$ K for the pulse length $t = 20$ ps.

the Rabi rotations introduces an energy scale Ω_R into the system which has to be compared with the characteristic energy of the phonons. Exact numerical results confirms this and establishes the parameter range where the minimal value of T_d is reached.

4. Conclusions

We present numerical path integral calculations for the dynamics of a resonantly excited strongly confined QD. The long time evolution is obtained using the modified version of the augmented density matrix formalism. The decay time of the oscillations is a non-monotonous function of the field, having its minimum when the field is approximately equal to the characteristic frequency of the phonon coupling. This yields the decay and reappearance of the Rabi oscillations in the field dependence of the exciton occupation. We note that this should not be mixed with the collapse-revival of the oscillations in the Jaynes-Cummings model. The decay-reappearance phenomenon can be understood within a general physical picture, based on the commensurability of the Rabi oscillations and the phonon coupling characteristic energy scales, which indicates its wide applicability.

Acknowledgments

We acknowledge support from the EU NoE SANDiE and Marie-Curie excellence grant NEXT-CT-2005-023778 (Nanoelectrophotonics), and the Belgian Science Policy. M. D. C. is grateful to the Physics Department at Lancaster University for hospitality.

References

1. E. Biolatti, R. C. Iotti, P. Zanardi, and F. Rossi, *Phys. Rev. Lett.* **85**, 5647 (2000); F. Troiani, U. Hohenester, and E. Molinari *Phys. Rev. B* **62**, R2263 (2000).
2. D. DiVincenzo and B. Terhal, *Decoherence: the obstacle to quantum computation, Physics World*, March 1998, p. 53.
3. T. H. Stievater *et al. Phys. Rev. Lett.* **87**, 133603 (2001).
4. S. Stufler, P. Ester, and A. Zrenner, *Phys. Rev. B* **72**, 121301(R) (2005).
5. Q. Q. Wang *et al.*, *Phys. Rev. B* **72**, 035306 (2005).
6. A. Vagov, V. M. Axt, and T. Kuhn, *Phys. Rev. B* **66**, 165312 (2002).
7. J. Förstner, C. Weber, J. Danckwerts, and A. Knorr, *Phys. Rev. Lett.* **91**, 127401 (2003).
8. E. A. Muljarov, T. Takagahara, and R. Zimmermann, *Phys. Rev. Lett.* **95**, 177405 (2005).
9. A. Vagov, V. M. Axt, T. Kuhn, W. Langbein, P. Borri, and U. Woggon, *Phys. Rev. B* **70**, 201305 (2004).
10. P. Machnikowski and L. Jacak, *Phys. Rev. B* **69**, 193302 (2004).
11. V. M. Axt, P. Machnikowski, and T. Kuhn, *Phys. Rev. B* **71**, 155305 (2005).
12. U. Hohenester and G. Stadler, *Phys. Rev. Lett.* **92**, 196801 (2004).
13. A. J. Leggett *et al.*, *Rev. Mod. Phys.* **59**, 1 (1987).
14. *Quantum Monte Carlo Methods in Condensed Matter Physics*, edited by M. Suzuki (World Scientific, Singapore, 1993), and references therein.
15. N. Makri and D. Makarov, *J. Chem. Phys.* **102**, 4600 (1995), *ibid.* **102**, 4611 (1995).
16. E. Sim, *J. Chem. Phys.* **115**, 4450 (2001).
17. E. Sim and N. Makri, *Chem. Phys. Lett.* **249**, 224 (1996).
18. E. Sim and N. Makri, *Comp. Phys. Comm.* **99**, 335 (1997).
19. B. Krummheuer, V. M. Axt, and T. Kuhn, *Phys. Rev. B* **65**, 195313 (2002).
20. R. P. Feynman and F. L. Vernon, *Annals of Physics* **24**, 118 (1963).

THREE USEFUL BOUNDS IN QUANTUM MECHANICS – EASILY OBTAINED BY WIENER INTEGRATION

H. LESCHKE* and R. RUDER

Institut für Theoretische Physik, Universität Erlangen-Nürnberg,
D-91058 Erlangen, Germany
**E-mail: hajo.leschke@physik.uni-erlangen.de*
URL: www.theorie1.physik.uni-erlangen.de

In a reasonably self-contained and explicit presentation we illustrate the efficiency of the Feynman–Kac formula by the rigorous derivation of three inequalities of interest in non-relativistic quantum mechanics.

Keywords: Wiener integration; Feynman–Kac formula; Diamagnetic monotonicities; Integrated density of states.

1. Feynman–Kac Formula

In 1923 Norbert Wiener[1] introduced a certain probability distribution μ concentrated on the set of continuous paths $w : [0,\infty[\to \mathbb{R}^d, \tau \mapsto w(\tau)$ from the positive half-line $[0,\infty[\subset \mathbb{R}^1$ into d-dimensional Euclidean space $\mathbb{R}^d$, $d \in \{1,2,3,\dots\}$. This distribution is a mathematically well-defined positive measure in the sense of abstract measure theory[2] and therefore induces a corresponding concept of integration over paths, which we denote by $\int \mu(\mathrm{d}w)(\cdot)$. The Wiener measure μ is uniquely determined by requiring that it is Gaussian with normalization $\int \mu(\mathrm{d}w) = 1$, first and second moments

$$\int \mu(\mathrm{d}w)\, w_j(\tau) = 0, \qquad \int \mu(\mathrm{d}w)\, w_j(\tau) w_k(\sigma) = \delta_{jk} \min\{\tau, \sigma\} \tag{1}$$

for all $j,k \in \{1,\dots,d\}$ and all $\tau, \sigma \in [0,\infty[$. Here $w_j(\tau)$ denotes the j-th component of the path w evaluated at (time) parameter $\tau \geq 0$. The simple Wiener integrals (1) imply that the components of w are — in probabilistic language — centered, independent and identically distributed. Moreover, the Chebyshev inequality shows for $\sigma = \tau = 0$ and $k = j$ that for each j the variable $w_j(0)$ is in fact not random, but equal to its mean value μ-almost surely. In other words, all paths start with probability 1 at the origin of $\mathbb{R}^d$.

To Wiener integrals apply — in contrast to Feynman path integrals — all the rules and computational tools provided by abstract measure and integration theory, most notably Lebesgue's dominated-convergence theorem. As a consequence, Wiener integration often serves — via the Feynman–Kac formula — as an efficient technique for obtaining results in quantum mechanics with complete rigour. An impressive compilation of such results was given by Barry Simon already in 1979. Since then not much has changed which is reflected by the fact that the second edition of his book[3] differs from the first one only by an addition of bibliographic notes on some of the more recent developments. Still Wiener integration should be considered neither as a secret weapon nor as a panacea for obtaining rigorous results in quantum mechanics. In any case, the Feynman–Kac formula is more than just a poetic rewriting of a Lie–Trotter formula. Ironically Richard Feynman himself took advantage of that as early as 1955 in his celebrated paper on the polaron, in particular, by using the Jensen inequality.[4]

Now, what is the Feynman–Kac formula? Let us consider a spinless charged particle with configuration space $\mathbb{R}^d$ subjected to a scalar potential $v : \mathbb{R}^d \to \mathbb{R}^1, q \mapsto v(q), q = (q_1, \ldots, q_d)$ and a vector potential $a : \mathbb{R}^d \to \mathbb{R}^d, q \mapsto a(q), a = (a_1, \ldots, a_d)$. The latter generates a magnetic field (tensor) defined by $b_{jk} := \partial a_k/\partial q_j - \partial a_j/\partial q_k$. The corresponding (non-relativistic) quantum system is informally given by the Hamiltonian

$$H(a, v) := \big(P - a(Q)\big)^2/2 + v(Q)\,, \tag{2}$$

where $Q = (Q_1, \ldots, Q_d)$ and $P = (P_1, \ldots, P_d)$ denote the d-component operators of position and canonical momentum, respectively. They obey the usual commutation relations $Q_j P_k - P_k Q_j = \mathrm{i}\delta_{jk}\mathbb{1}$. Here $\mathrm{i} = \sqrt{-1}$ is the imaginary unit and we have chosen physical units where the mass and charge of the particle as well as Planck's constant (divided by 2π) are all equal to 1. Under rather weak assumptions[5] on v and a, $H(a, v)$ can be defined as a self-adjoint operator on the Hilbert space $\mathrm{L}^2(\mathbb{R}^d)$ of all (equivalence classes of) Lebesgue square-integrable complex-valued functions on $\mathbb{R}^d$. Furthermore its "Boltzmann–Gibbs operator" $\mathrm{e}^{-\beta H(a,v)}$ even possesses for each $\beta \in]0, \infty[$ an integral kernel $\langle q|\mathrm{e}^{-\beta H(a,v)}|q'\rangle$ (in other words, position representation or Euclidian propagator) which is jointly continuous in $q, q' \in \mathbb{R}^d$.

We are now prepared to recall the Feynman–Kac formula. Apart from mathematical subtleties its content is most concisely expressed by the following representation free version:

$$\mathrm{e}^{-\beta H(a,v)} = \int \mu(\mathrm{d}w)\, \mathrm{e}^{-\mathrm{i}w(\beta)\cdot P} \exp\Big\{ - \int_0^\beta \mathrm{d}\tau\, v(w(\tau)\mathbb{1} + Q)\Big\}$$

$$\times \exp\Big\{ \mathrm{i} \int_0^\beta \mathrm{d}\tau \dot{w}(\tau) \cdot a\big(w(\tau)\mathbb{1} + Q\big)\Big\}. \quad (3)$$

Here the dot "$\cdot$" denotes the usual scalar product of $\mathbb{R}^d$ and the integral containing the vector potential is a suggestive notation for a stochastic line integral in the sense of R. L. Stratonovich and D. L. Fisk (corresponding to a mid-point discretization). In (3) the Wiener integration serves to disentangle the non-commuting operators P and Q in $\mathrm{e}^{-\beta H(a,v)}$. By going informally to the position representation of (3) one gets the rigorously proven formula

$$\langle q|\mathrm{e}^{-\beta H(a,v)}|q'\rangle = \int \mu(\mathrm{d}w)\, \delta\big(w(\beta) + q' - q\big) \exp\Big\{ - \int_0^\beta \mathrm{d}\tau\, v\big(w(\tau) + q'\big)\Big\}$$

$$\times \exp\Big\{ \mathrm{i} \int_0^\beta \mathrm{d}\tau\, \dot{w}(\tau) \cdot a\big(w(\tau) + q'\big)\Big\}. \quad (4)$$

It even holds for a class of potentials v for which $H(a, v)$ is not bounded from below.[5] The Dirac delta in (4) indicates that all paths w to be integrated over arrive in $q - q' \in \mathbb{R}^d$ at time $\beta > 0$. In fact, they may be considered to end there, because μ is Markovian and the integrand in (4) does not depend on $w(\tau)$ for $\tau > \beta$. More precisely, the path integration may be performed with respect to the *Brownian bridge*.[3,5]

In the remaining three sections we are going to illustrate the usefulness of (4) by deriving three inequalities of interest in quantum mechanics.

2. Diamagnetic Inequality

Theorem 2.1.

$$\Big|\langle q|\mathrm{e}^{-\beta H(a,v)}|q'\rangle\Big| \leq \langle q|\mathrm{e}^{-\beta H(0,v)}|q'\rangle \quad (5)$$

holds for all $\beta > 0$ and all $q, q' \in \mathbb{R}^d$.

Proof. Inequality (5) is an immediate consequence of (4) by taking the absolute value and observing the "triangle inequality" $\big|\int \mu(\mathrm{d}w)\,(\cdot)\big| \leq \int \mu(\mathrm{d}w)\,|(\cdot)|$ and the elementary identity $\big|\mathrm{e}^{x+\mathrm{i}y}\big| = \mathrm{e}^x$ for $x, y \in \mathbb{R}^1$. □

Remarks:

(i) This elegant proof is due to Edward Nelson, see Ref. 6 (also for other historic aspects of (5) and related inequalities).

(ii) If $\int_{\mathbb{R}^d} \mathrm{d}q\, \mathrm{e}^{-\beta v(q)} < \infty$, the free energy $-\beta^{-1} \ln \int_{\mathbb{R}^d} \mathrm{d}q\, \langle q|\mathrm{e}^{-\beta H(0,v)}|q\rangle$ at inverse temperature $\beta > 0$ exists and (5) then implies that it cannot be lowered by turning on a magnetic field. Under weaker assumptions on v, for example for the hydrogen atom (that is, for $d = 3$ and $v(q) = -1/|q|$) (5) still implies in the limit $\beta \to \infty$ the same sort of stability for the ground-state energy. Altogether this explains the name *diamagnetic inequality.*

(iii) There are also diamagnetic inequalities in case the particle is confined to a bounded region in $\mathbb{R}^d$ with Dirichlet, Neumann or other boundary conditions.[7,8] Moreover, the proof of the diamagnetic inequality easily extends to the case of many (interacting) particles, provided there is no spin and no Fermi statistics involved.

An interesting question is what can be said if $a \neq 0$ is changed (pointwise) to another vector potential $a' \neq 0$. For a partial answer see Sec. 4 below.

3. Quasi-Classical Upper Bound on the Integrated Density of States in the Case of a Random Scalar Potential

In the single-particle theory of electronic properties of disordered or amorphous solids the scalar potential v in $H(a, v)$ is considered to be a *random* field on $\mathbb{R}^d$ which is distributed according to some probability measure ν. We denote by $\int \nu(\mathrm{d}v)\,(\cdot)$ the corresponding (functional) integration or averaging. One example is a *Gaussian* ν with vanishing first moments and second moments $\int \nu(\mathrm{d}v)\, v(q)v(q') = C(q - q')$ for all $q, q' \in \mathbb{R}^d$ with some (even) covariance function $C : \mathbb{R}^d \to \mathbb{R}^1$. The fact that the second moments only depend on the difference $q - q'$ reflects the assumed "homogeneity on average". We also assume that C is continuous, $C(q)$ tends to zero as $|q| \to \infty$ and the single-site variance obeys $0 < C(0) < \infty$. The $\mathbb{R}^d$-homogeneity together with the decay of the correlations of the fluctuations at different sites with increasing distance amounts to the $\mathbb{R}^d$-*ergodicity* of the (Gaussian) random potential.

A quantity of basic interest in the above-mentioned theory is the integrated density of states. It may be defined[5,9] as the non-decreasing function $N : \mathbb{R}^1 \to \mathbb{R}^1, E \mapsto N(E, a, q)$ where

$$N(E, a, q) := \int \nu(\mathrm{d}v)\, \langle q|\Theta(E\mathbb{1} - H(a, v)|q\rangle. \tag{6}$$

Here Θ denotes Heaviside's unit-step function and the (non-random) vector potential a as well as $q \in \mathbb{R}^d$ are considered as parameters. If the random potential (characterized by ν) and the magnetic field (generated by a)

are both homogeneous, then $N(E, a, q)$ actually does not depend on q. Of course, in the physically most relevant cases the random potential should be even ergodic, so that $N(E, a, 0)$ coincides in the infinite-volume limit for ν-almost all realizations v with the number of eigenvalues per volume of a finite-volume restriction of $H(a, v)$ below the energy $E \in \mathbb{R}^1$. Nevertheless the following estimate holds also for random potentials and magnetic fields which are not homogeneous.

Theorem 3.1. *If the probability measure ν of the random potential has the property*[a] *that $L_\beta := \operatorname{ess\,sup}_{r\in\mathbb{R}^d} \int \nu(\mathrm{d}v)\, \mathrm{e}^{-\beta v(r)} < \infty$ for all $\beta > 0$, then*

$$N(E, a, q) \leq (2\pi\beta)^{-d/2} L_\beta\, \mathrm{e}^{\beta E} \tag{7}$$

holds for all energies $E \in \mathbb{R}^1$ and all $\beta > 0$.

Proof.

$$N(E, a, q)\, \mathrm{e}^{-\beta E} \leq \int \nu(\mathrm{d}v)\, \langle q|\mathrm{e}^{-\beta H(a,v)}|q\rangle \leq \int \nu(\mathrm{d}v)\, \langle q|\mathrm{e}^{-\beta H(0,v)}|q\rangle \tag{8}$$

$$= \int \mu(\mathrm{d}w)\, \delta\big(w(\beta)\big) \int \nu(\mathrm{d}v)\, \exp\Big\{ - \int_0^\beta \mathrm{d}\tau\, v(w(\tau) + q)\Big\} \tag{9}$$

$$\leq \int_0^\beta \frac{\mathrm{d}\tau}{\beta} \int \mu(\mathrm{d}w)\, \delta\big(w(\beta)\big) \int \nu(\mathrm{d}v)\, \mathrm{e}^{-\beta v(w(\tau)+q)} \leq (2\pi\beta)^{-d/2}\, L_\beta. \tag{10}$$

Here (8) is due to the elementary inequality $\Theta(E\mathbb{1} - H(a, v)) \leq \mathrm{e}^{\beta(E\mathbb{1}-H(a,v))}$ and (5). Equation (4) then gives (9). The next inequality is Jensen's with the uniform average $\beta^{-1} \int_0^\beta \mathrm{d}\tau\, (\cdot)$. The claim now follows from the definition of L_β, Eq. (4) with $(a, v) = (0, 0)$ and $\int_0^\beta \mathrm{d}\tau = \beta$. The various interchanges of integrations can be justified by the Fubini-Tonelli theorem. □

Remarks:

(i) Theorem 3.1 is a slight extension of a result which goes back to Pastur, see Thm. 9.1 in Ref. 10. The right-hand side (RHS) of (7) is *quasi-classical* in the sense that it does not depend on a and does not take into account — due to the Jensen inequality in (10) — the non-commutativity of the kinetic and potential energy.

(ii) While the estimate (7) holds for rather general random potentials, the various inequalities in its proof are responsible for its roughness, even

[a] Given a function $f : \mathbb{R}^d \to \mathbb{R}^1$, then $\operatorname{ess\,sup}_{r\in\mathbb{R}^d} |f(r)|$ denotes the smallest $M \in [0, \infty]$ such that $|f(r)| \leq M$ holds for Lebesgue-almost all $r \in \mathbb{R}^d$.

when optimized with respect to $\beta > 0$. Nevertheless, it shows that $N(E)$ decreases to 0 at least exponentially fast as $E \to -\infty$. For a homogeneous Gaussian random potential (and a constant magnetic field) the optimized estimate even reflects the exact Gaussian decay[9,10] $\ln N(E) \sim -E^2/2C(0)$ as $E \to -\infty$. For a non-Gaussian random potential like a repulsive Poissonian one the leading low-energy decay of N, the *Lifshitz tail*, typically is of true quantum nature[9–11] and can therefore not be reflected by the RHS of (7). Although the (universal) leading high-energy growth $N(E) \sim (E/2\pi)^{d/2}/\Gamma(1+d/2)$ as $E \to \infty$, see Refs. 9,10 and references therein, is purely classical, it is overestimated by the RHS of (7) (due to the elementary inequality in (8)).

4. A Simple Diamagnetic Monotonicity

For a partial answer to the question raised at the end of Sec. 2 we only consider the planar case $\mathbb{R}^2$ with $v = 0$ and a perpendicular magnetic field not depending on the second co-ordinate q_2. We assume that $b := b_{12}$ is a continuously differentiable function of $q_1 \in \mathbb{R}^1$. One possible vector potential generating b, not depending on q_2 either, is given by $a^{(b)}(q) := \big(0, \int_0^{q_1} \mathrm{d}r\, b(r)\big)$. In the following theorem $H^{(b)}$ denotes any Hamiltonian on $\mathrm{L}^2(\mathbb{R}^2)$ which is gauge equivalent to $H(a^{(b)}, 0) = P_1^2/2 + \big(P_2 - a_2^{(b)}(Q_1)\big)^2/2$ for the given b. The assertion (11) is therefore gauge invariant. Nevertheless, in the proof we will use $H(a^{(b)}, 0)$ and see that one can dispense with the absolute value on the RHS of (11) in this particular gauge.

Theorem 4.1. *If b and B are two magnetic fields as just described and satisfy either $|b(r)| \leq B(r)$ or $|b(r)| \leq -B(r)$ for all $r \in \mathbb{R}^1$, then*

$$\left|\langle q|\mathrm{e}^{-\beta H^{(B)}}|q'\rangle\right| \leq \left|\langle (q_1, 0)|\mathrm{e}^{-\beta H^{(b)}}|(q_1', 0)\rangle\right| \mathrm{e}^{-(q_2 - q_2')^2/(2\beta)} \tag{11}$$

holds for all $\beta > 0$, all $q = (q_1, q_2) \in \mathbb{R}^2$ and all $q' = (q_1', q_2') \in \mathbb{R}^2$.

Proof. By (4) the LHS of (11) is invariant under a global sign change of B. Therefore it suffices to consider the case $B(r) \geq 0$. For notational transparency we write a and A for $a_2^{(b)}$ and $a_2^{(B)}$, respectively. For given $\beta > 0$, $q_1' \in \mathbb{R}^1$ and a one-dimensional Wiener path $w : [0, \infty[\to \mathbb{R}^1$ we introduce the notations

$$m_\beta(a, w, q_1') := \beta^{-1} \int_0^\beta \mathrm{d}\tau\, a(w(\tau) + q_1'), \tag{12}$$

$$s_\beta^2(a, w, q_1') := \beta^{-1} \int_0^\beta \mathrm{d}\tau\, \big(a(w(\tau) + q_1')\big)^2 - \big(m_\beta(a, w, q_1')\big)^2 \tag{13}$$

for the mean and variance of $a\big(w(\tau) + q_1'\big)$ with respect to the uniform average $\beta^{-1}\int_0^\beta \mathrm{d}\tau\,(\cdot)$, and similarly with A instead of a. Next we observe the following two "doubling identities"

$$2\beta^2\left[s_\beta^2(A,w,0) - s_\beta^2(a,w,0)\right] \tag{14}$$

$$= \int_0^\beta \mathrm{d}\tau \int_0^\beta \mathrm{d}\sigma \Big\{ \big[A(w(\tau)) - A(w(\sigma))\big]^2 - \big[a(w(\tau)) - a(w(\sigma))\big]^2 \Big\} \tag{15}$$

$$= \int_0^\beta \mathrm{d}\tau \int_0^\beta \mathrm{d}\sigma \big[a_+(w(\tau)) - a_+(w(\sigma))\big]\big[a_-(w(\tau)) - a_-(w(\sigma))\big]. \tag{16}$$

The last integrand is non-negative, because the two functions $r \mapsto a_\pm(r) := A(r) \pm a(r) = \int_0^r \mathrm{d}r'\big(B(r') \pm b(r')\big)$, $r \in \mathbb{R}^1$, are both non-decreasing since $B(r') \geq |b(r')| \geq \mp b(r')$ by assumption. The same arguments apply when the path w is replaced by the rigidly shifted one $w + q_1'$. To summarize, we have shown so far that

$$s_\beta^2(a,w,q_1') \leq s_\beta^2(A,w,q_1') \tag{17}$$

for all w, all q_1' and all $\beta > 0$.

Since $H^{(B)}$, in the particular gauge chosen, commutes with P_2, we may decompose it according to $H^{(B)} = \int_{\mathbb{R}^1} \mathrm{d}k H^{(B)}(k) \otimes |k\rangle\langle k|$, using an informal notation. Here the one-parameter family of effective Hamiltonians associated with the 1-direction

$$H^{(B)}(k) := P_1^2/2 + \big(k\mathbb{1} - A(Q_1)\big)^2/2, \quad k \in \mathbb{R}^1, \tag{18}$$

acts on the Hilbert space $\mathrm{L}^2(\mathbb{R}^1)$. By this decomposition we get

$$\langle q|\mathrm{e}^{-\beta H^{(B)}}|q'\rangle = (2\pi)^{-1}\int_{\mathbb{R}^1} \mathrm{d}k \langle q_1|\mathrm{e}^{-\beta H^{(B)}(k)}|q_1'\rangle\, \mathrm{e}^{\mathrm{i}k(q_2-q_2')} \tag{19}$$

$$\begin{aligned} &= (2\pi\beta)^{-1/2}\mathrm{e}^{-(q_2-q_2')^2/(2\beta)} \int \mu(\mathrm{d}w)\delta(w(\beta) + q_1' - q_1) \\ &\quad \times \exp\Big\{-\beta s_\beta^2(A,w,q_1')/2\Big\} \exp\Big\{\mathrm{i}(q_2 - q_2')m_\beta(A,w,q_1')\Big\}. \end{aligned} \tag{20}$$

Here we have used (4) with $d = 1$, $a = 0$ and $v = (k - A)^2/2$ and then performed the (Gaussian) integration with respect to k. By applying the "triangle inequality" to (20) and then using (17) we finally obtain

$$\begin{aligned} &\mathrm{e}^{(q_2-q_2')^2/(2\beta)} \left|\langle q|\mathrm{e}^{-\beta H^{(B)}}|q'\rangle\right| \\ &\leq (2\pi\beta)^{-1/2} \int \mu(\mathrm{d}w)\delta(w(\beta) + q_1' - q_1) \exp\Big\{-\beta s_\beta^2(a,w,q_1')/2\Big\} \end{aligned} \tag{21}$$

$$= \langle (q_1,0)|\mathrm{e}^{-\beta H^{(b)}}|(q_1',0)\rangle = \left|\langle (q_1,0)|\mathrm{e}^{-\beta H^{(b)}}|(q_1',0)\rangle\right|. \tag{22}$$

The last two equalities follow again from (20) with a instead of A. □

Remarks:

(i) To our knowledge, Theorem 4.1 first appeared in Ref. 12. It complements some of the results obtained by Loss, Thaller and Erdős.[13,14] For a survey of results of this genre see Sec. 9 in Ref. 15.

(ii) For a given sign-definite B the RHS of (11) becomes explicit by choosing for b the globally constant field $B_0 := \inf_{r\in\mathbb{R}} |B(r)|$, so that

$$\left|\langle q|\mathrm{e}^{-\beta H^{(B)}}|\langle q'\rangle\right| \leq \frac{B_0/4\pi}{\sinh(\beta B_0/2)} \exp\left\{-\frac{B_0}{4}\frac{(q_1-q_1')^2}{\tanh(\beta B_0/2)}-\frac{(q_2-q_2')^2}{2\beta}\right\}.$$

If $B_0 \neq 0$, the Gaussian decay on the RHS is faster along the 1- than along the 2-direction. Such an anisotropy has been found also for the almost-sure transport properties in the case that B is an ergodic random field with non-zero mean.[16]

References

1. N. Wiener, *J. Math. Phys. Sci.* **2**, 131 (1923).
2. W. Rudin, *Real and Complex Analysis* (McGraw-Hill, 1987, 3rd ed.).
3. B. Simon, *Functional Integration and Quantum Physics*, 2nd edition (AMS Chelsea Publishing, 2005).
4. R. P. Feynman, *Phys. Rev.* **97**, 660 (1955).
5. K. Broderix, H. Leschke, and P. Müller, *J. Func. Anal.* **212**, 287 (2004).
6. B. Simon, *Ed Nelson's work in quantum theory*, in *Diffusion, Quantum Theory, and Radically Elementary Mathematics*, ed. W. G. Faris, Mathematical Notes, Vol. 47 (Princeton University Press, 2006), pp. 75–93.
7. T. Hupfer, H. Leschke, P. Müller, and S. Warzel, *Commun. Math. Phys.* **221**, 229 (2001).
8. D. Hundertmark and B. Simon, *J. Reine Angew. Math.* **571**, 107 (2004).
9. H. Leschke, P. Müller, and S. Warzel, *Mark. Proc. Rel. Fields* **9**, 729 (2003).
10. L. A. Pastur and A. Figotin, *Spectra of Random and Almost-Periodic Operators* (Springer, Berlin, 1992).
11. H. Leschke and S. Warzel, *Phys. Rev. Lett.* **92**, 086402 (2004).
12. H. Leschke, R. Ruder, and S. Warzel, *J. Phys. A* **35**, 5701 (2002).
13. M. Loss and B. Thaller, *Commun. Math. Phys.* **186**, 95 (1997).
14. L. Erdős, *J. Math. Phys.* **38**, 1289 (1997).
15. G. Rozenblum and M. Melgaard, *Schrödinger operators with singular potentials*, in *Handbook of Differential Equations: Stationary Partial Differential Equations*, Vol. 2, eds. M. Chipot and P. Quittner (Elsevier, North Holland, 2005), pp. 407–517.
16. H. Leschke, S. Warzel, and A. Weichlein, *Ann. Henri Poincaré* **7**, 335 (2006).

PATH-INTEGRAL DERIVATION OF LIFSHITZ TAILS

V. SA-YAKANIT

Center of Excellence in Forum for Theoretical Science, Chulalongkorn University, Bangkok 10330, Thailand
E-mail: vsayakanit@gmail.com

The behavior of the electron density of states (DOS) for the Lifshitz tail states is studied in the limit of low energy using the Feynman path-integral method. This method was used to study the heavily doped semiconductors for the case of a Gaussian random potential. The main results obtained are that the tail states behave as DOS$\sim \exp\left(-B\left(E\right)\right)$, with $B\left(E\right)=E^{n}, n=\frac{1}{2}$ for short-range interaction and $n=2$ for long-range interaction. In this study it is shown that without the Gaussian approximation, the behavior of the Lifshitz tails for the Poisson distribution is obtained as DOS$\sim \exp\left(-B\left(E\right)\right)$ with $B\left(E\right)=E^{n}, n=-\frac{3}{2}$. As in the case of heavily doped semiconductor, the method can be easily generalized to long-range interactions. A comparison with the method developed by Friedberg and Luttinger based on the reformulation of the problem in terms of Brownian motion is given.

Keywords: Density of states; Lifshitz tails.

1. Introduction

In a previous paper[1] we presented a method for calculating the density of states (DOS) of an electron moving in a random potential containing high number of impurities or scatterers with weak scattering potential. This model is equivalent to treating the random system as Gaussian random potential. By formulating the random system using Feynman path integrals we have shown that the behavior of the DOS in the Gaussian random system in three dimensions deep in the tail goes roughly as $\exp(-B\left(E\right))$ where $B\left(E\right)$ is a function of energy and is proportional to E^{n} with n varying from $\frac{1}{2}$ for small autocorrelation length to 2 for large autocorrelation length. For general statistics without the Gaussian approximation the behavior of the DOS is drastically different from the Gaussian approximation. It was Lifshitz[2] who has first presented the intuitive arguments and showed that the DOS should behave as $\sim \exp\left(-E^{-\frac{3}{2}}\right)$. The Lifshitz ideas are as follows.

The probability of finding a large region of volume V being free of impurities is proportional to $\exp(-\rho V.)$. Then the main contribution to the probability of finding a low-level E for a system will be proportional to the probability of finding the region V whose energy level is E. The lowest-level E in the empty volume of radius R_0 with the boundary conditions that the wave function vanishes on its surface is given by $E = \frac{\pi^2}{2R_0^2}$; or $R_0 = \sqrt{\frac{\pi^2}{2E}}$ with $m = \hbar = 1$. The probability of such a region existing is proportional to $\exp\left(-\rho\left(\frac{4\pi}{3}\right)R_0^3\right)$ so that the DOS of the low-lying level will be given by

$$D(E) \underset{E\to 0}{\sim} \exp\left(-cE^{\left(-\frac{3}{2}\right)}\right) \quad \text{with} \quad c = \frac{1}{3}4\pi\left(\frac{\pi^2}{2}\right)^{\frac{3}{2}}\rho. \tag{1}$$

The first attempt to derive the Lifshitz conjecture is given by Friedberg and Luttinger[3] based on the reformulation of the problem in terms of Brownian motion. They showed that the solution of this problem is equivalent to the knowledge of $D(E)$ for small energy. A variational method is used for solving this problem and the solution is reduced to solving the nonlinear differential equation. This equation is solved in the long-time limit and obtained the Lifshitz conjecture. In this paper we derive the Lifshitz conjecture using the Feynman path-integral formulation developed in our previous paper for handling the heavily doped semiconductors. Without the Gaussian approximation we derive the average propagator for general statistics or Poisson distribution. After performing the random average the system becomes translation invariant and therefore it is reasonable to model the system with a nonlocal harmonic oscillator trial action. A variational principle is used to obtain the Lifshitz conjecture for the large-time limit. In the second section we derive the model action. The DOS is obtained in the third section and a comparison between the Friedberg and Luttinger result is given. The final section is devoted to the discussion.

2. Model System

Following the method given in Ref. 1, we consider an electron moving among a set of N rigid impurities or scatterers, confined within a volume V, and having a density $\rho = N/V$. Such a system is described by the Hamiltonian

$$H = -\frac{\hbar^2}{2m}\nabla^2 + \sum_i v\left(\vec{x} - \vec{R}_i\right), \tag{2}$$

where $v\left(\vec{x}-\vec{R}_i\right)$ represents the potential of a single scatterer at position $\vec{R}_i$. The propagator $K\left(\vec{x}_2,\vec{x}_1;\{v\}\right)$ of such a system can be expressed as

$$K\left(\vec{x}_2,\vec{x}_1;t,\{v\}\right) =$$
$$\int D\left(\vec{x}\left(\tau\right)\right)\exp\left[\frac{i}{\hbar}\int_0^t d\tau\left(\frac{m}{2}\dot{\vec{x}}^2\left(\tau\right)-\sum_i v\left(\vec{x}-\vec{R}_i\right)\right)\right], \qquad (3)$$

where $D\left(\vec{x}\left(\tau\right)\right)$ denotes the path integral to be carried out with the boundary conditions $\vec{x}\left(0\right)=\vec{x}_1$, $\vec{x}\left(t\right)=\vec{x}_2$. The probability distribution of the scattering potential is assumed to be $P\left[\vec{R}\right]d\left(\left[\vec{R}\right]\right)=\prod_{N,V\to\infty}\frac{dR_1dR_2...dR_N}{V^N}$. Performing the configuration average we obtain

$$G\left(\vec{x}_2,\vec{x}_1;t\right)=\lim_{N,V\to\infty}\int P\left[\vec{R}\right]d\left(\left[\vec{R}\right]\right)K\left(\vec{x}_2,\vec{x}_1;t,[v]\right) \qquad (4)$$
$$=\int D\left(\vec{x}\left(\tau\right)\right)\exp\left[\frac{i}{\hbar}S\right]$$

with the action

$$S=\int_0^t d\tau\left\{\frac{m}{2}\dot{\vec{x}}^2\left(\tau\right)-\rho\int d\vec{R}\left[\exp\left(-\frac{i}{\hbar}\int_0^t d\tau v\left(\vec{x}\left(\tau\right)-\vec{R}\right)\right)-1\right]\right\}. \qquad (5)$$

After the configuration average, the system becomes translation invariant, therefore it is reasonable to introduce the nonlocal harmonic oscillator trial action. This action was used to be applied to the band tail problems in heavily doped semiconductors. The model trial action is

$$S_0=\int_0^t d\tau\left(\frac{m}{2}\dot{\vec{x}}^2\left(\tau\right)-\frac{\omega^2}{2t}\int_0^t d\tau\left|\vec{x}_2-\vec{x}_1\right|^2\right), \qquad (6)$$

where ω is an unknown parameter to be determined. Such a translational invariant action proves to be important to obtain the prefactor in the DOS. The propagator associated with the nonlocal harmonic trial action is[3]

$$G\left(\vec{x}_2,\vec{x}_1;t,\omega\right)=$$
$$\int D\left(\vec{x}\left(\tau\right)\right)\exp\left[\frac{i}{\hbar}\int_0^t d\tau\left(\frac{m}{2}\dot{\vec{x}}^2\left(\tau\right)-\frac{\omega^2}{2t}\left|\vec{x}_2-\vec{x}_1\right|^2\right)\right]. \qquad (7)$$

3. Density of States

The DOS per unit volume is given by the Fourier transform of the diagonal part of the configurationally average propagator as

$$D\left(E\right)=\frac{1}{2\pi\hbar}\int_{-\infty}^{\infty}dtG\left(0,0;t\right)\exp\left(\frac{i}{\hbar}Et\right). \qquad (8)$$

Therefore the problem of the DOS calculation within the framework of the path-integral approach consists in evaluating the exact propagator by integrating over all possible paths using S. Rewriting the diagonal propagator

$$G(0,0;t) = G_0(0,0;t,\omega)\left\langle \exp\frac{i}{\hbar}(S-S_0)\right\rangle_{S_0}, \tag{9}$$

where $G_0(0,0;t,\omega) = \left(\frac{m}{2\pi i\hbar t}\right)^{\frac{3}{2}}\left[\frac{\omega t}{2\sin\left(\frac{1}{2}\omega t\right)}\right]^3$ is the diagonal propagator of the non-local trial action propagator given in Ref. 4. Hear the symbol $\langle O\rangle_{S_0}$ denotes the path-integral average with respect to the trial action S_0. To prove the Lifshitz conjecture we approximate the diagonal propagator $G(0,0;t)$ by keeping only the first cumulant, Eq. (7). To obtain $G_1(0,0;t,\omega)$ we have to find the average $\langle(S-S_0)\rangle_{S_0}$. Since the kinetic energy terms in the actions S and S_0 are always canceled out with each other, we shall denote $\langle S_0\rangle_{S_0}$ and $\langle S\rangle_{S_0}$ as the average without the kinetic terms. We first consider the average $\langle S\rangle_{S_0}$:

$$\langle S\rangle_{S_0} = \rho\int d\vec{R}\left[\left\langle \exp\left(-\frac{i}{\hbar}\int_0^t d\tau v\left(\vec{x}(\tau)-\vec{R}\right)\right)\right\rangle_{S_0} - 1\right]. \tag{10}$$

The average over the scattering potential can be conveniently evaluated by making the Fourier decomposition of the potential v. We obtain

$$\langle S\rangle_{S_0} = \rho\int d\vec{R}\left[\begin{array}{c}\exp\left(-\frac{i}{\hbar}\int_0^t d\tau\int\frac{d\vec{k}}{(2\pi)^3}v_{\vec{k}}\exp\left(-i\vec{k}\cdot\vec{R}\right)\right)\\ \times\exp\left(-\frac{1}{2}k^2\frac{1}{3}\langle x^2\rangle_{S_0}\right)-1\end{array}\right]. \tag{11}$$

Next we consider the average $\langle S_0\rangle_{S_0}$ which can be written as

$$\langle S_0\rangle_{S_0} = -\frac{\omega^2}{2t}\int_0^t d\tau\int_0^t d\sigma\left\langle\left|\vec{x}_2-\vec{x}_1\right|^2\right\rangle_{S_0}. \tag{12}$$

From the above Eqs. (11) and (12) we can see that the average $\langle(S-S_0)\rangle_{S_0}$ can be expressed solely in terms of the following averages: $\langle\vec{x}(\tau)\rangle_{S_0}, \langle\vec{x}(\tau)\cdot\vec{x}(\sigma)\rangle_{S_0}$. To obtain the Lifshitz tails, we assume that the scattering potential is a Gaussian function with finite interacting length. For short length the scattering potential reduces to the white noise delta function potential. This limit can be compared with the Friedberg and Luttinger model of using a square well with finite width and depth. We now take the scattering potential as a Gaussian function with finite length l and u, where u is the parameter introduced to take care of the dimension of the system. This parameter u is essential when one takes the white noise limit.

The scattering potential is

$$v\left(\overrightarrow{x}(\tau)-\overrightarrow{R}\right)=u\left(\pi l^2\right)^{-\frac{3}{2}}e^{\frac{\left(\overrightarrow{x}(\tau)-\overrightarrow{R}\right)^2}{l^2}}. \tag{13}$$

Taking the Fourier transform and performing the $\overrightarrow{k}$ integration, we have

$$\langle S\rangle_{S_0}=\rho\int d\overrightarrow{R}\left[\exp\left(\begin{array}{c}-\frac{i}{\hbar}\int_0^t d\tau u\left[4\pi\left(\frac{l^2}{4}+\frac{1}{6}\left\langle\overrightarrow{x}^2(\tau)\right\rangle_{S_0}\right)\right]^{-\frac{3}{2}}\\ \times e^{-\frac{\overrightarrow{R}^2}{4\left(\frac{l^2}{4}+\frac{1}{6}\left\langle\overrightarrow{x}^2(\tau)\right\rangle_{S_0}\right)}}\end{array}\right)-1\right]. \tag{14}$$

To obtain the ground states, we take the limit t→ ∞; then $\left\langle\overrightarrow{x}^2(\tau)\right\rangle_{S_0}\underset{t\to\infty}{=}\frac{\hbar}{2m\omega}$, so that $\left(\frac{l^2}{4}+\frac{\hbar}{2m\omega}\right)=\frac{l^2}{4}\left(1+\frac{4\hbar^2}{2ml^2\hbar\omega}\right)=\frac{l^2}{4}\left(1+\frac{4E_l}{E_\omega}\right)$, and we have defined $E_l=\frac{\hbar^2}{2ml^2}$ and $E_\omega=\hbar\omega$. Finally for large t, we obtain $\langle S_0\rangle_{S_0}\underset{t\to\infty}{=}\frac{3}{4}\omega t$ and

$$G_1(0,0;t,\omega)\underset{t\to\infty}{=} \tag{15}$$

$$\left(\frac{m}{2\pi i\hbar t}\right)^{\frac{3}{2}}\frac{\omega t}{2}e^{\frac{i}{4}\omega t}\exp\left\{\rho\int d\overrightarrow{R}\left[\exp\left(\begin{array}{c}-\frac{i}{\hbar}tu\left[\pi l^2\left(1+\frac{4E_l}{E_\omega}\right)\right]^{-\frac{3}{2}}\\ \times\exp\left(-\frac{\overrightarrow{R}^2}{l^2\left(1+\frac{4E_l}{E_\omega}\right)}\right)\end{array}\right)-1\right]\right\}.$$

Scaling the R integration, $\overrightarrow{R'}^2=\overrightarrow{R}^2l^{-2}\left(1+\frac{4E_l}{E_\omega}\right)^{-1}$, we then have

$$G_1(0,0;t,\omega)\underset{t\to\infty}{=}\left(\frac{m}{2\pi i\hbar t}\right)^{\frac{3}{2}}\left(\frac{i}{\hbar}E_\omega t\right)^3e^{\frac{i}{4}\omega t} \tag{16}$$

$$\times\exp\left[\rho l^3\left(1+\frac{4E_l}{E_\omega}\right)^{\frac{3}{2}}\int d\overrightarrow{R'}\left[\exp\left(-\frac{i}{\hbar}t\xi_l e^{-\overrightarrow{R'}^2}\right)-1\right]\right],$$

where we have defined $\xi_l=u\left[\pi l^2\left(1+\frac{4E_l}{E_\omega}\right)\right]^{-\frac{3}{2}}$. For white noise, $l\to 0$, corresponding to $\frac{4E_l}{E_\omega}\gg 1$, we get

$$G_1(0,0;t,\omega)\underset{t\to\infty}{=}$$

$$\left(\frac{m}{2\pi i\hbar t}\right)^{\frac{3}{2}}\exp\left[3\ln\left(\frac{i}{\hbar}E_\omega t\right)-\frac{3}{4}\frac{i}{\hbar}E_\omega t-\rho l^3\left(\frac{4E_l}{E_\omega}\right)^{\frac{3}{2}}f(t,\omega)\right], \tag{17}$$

where $f(t,\omega)=\int d\overrightarrow{R'}\left[\exp\left(-\frac{i}{\hbar}t\xi_l\exp\left(-\overrightarrow{R'}^2\right)\right)-1\right]$.

To consider the leading contribution, we assume that $\frac{i}{\hbar}t\xi_l=1$. Then the function becomes a constant, $f(t,\omega)=f$. Minimizing the exponent, and for large t we can neglect the $\ln\left(\frac{i}{\hbar}E_\omega t\right)$ term as compared to $E_\omega t$. Then we have $E_\omega = (it)^{-\frac{2}{5}}\left[\rho l^3 2\hbar\,(4E_l)^{\frac{3}{2}} f\right]^{\frac{2}{5}}$. The function f is

$$f = \int d\overrightarrow{R'}\left[\exp\left(-\exp\left(-\overrightarrow{R}'^2\right)\right) - 1\right] \approx \int_0^\infty 4\pi r^2 dr\left[\exp\left(-r^2\right)\right] = 2\pi^{\frac{3}{2}}, \tag{18}$$

and substituting the variational parameter E_ω in Eq. (18) we have

$$G_1(0,0;t) \underset{t\to\infty}{=} \left(\frac{m}{2\pi i\hbar t}\right)^{\frac{3}{2}}\left(\frac{i}{\hbar}E_\omega t\right)^3 \exp\left[-\frac{5}{4\hbar}(it)^{\frac{3}{5}}\left[\rho l^3 2\hbar\,(4E_l)^{\frac{3}{2}} f\right]^{\frac{2}{5}}\right]. \tag{19}$$

The density of states is

$$D(E) = \frac{1}{2\pi\hbar}\int_{-\infty}^{\infty} dt\left(\frac{m}{2\pi i\hbar t}\right)^{\frac{3}{2}}\left(\frac{i}{\hbar}E_\omega t\right)^3 \exp\left[-\alpha\,(it)^{\frac{3}{5}} + \frac{i}{\hbar}Et\right] \tag{20}$$

$$= \frac{5}{4\hbar}\left[\rho l^3 2\hbar\,(4E_l)^{\frac{3}{2}} f\right]^{\frac{2}{5}}.$$

Minimizing the exponent, we have $\frac{d}{dit}\left(-\alpha\,(it)^{\frac{3}{5}} + \frac{i}{\hbar}Et\right) = 0$ and performing the saddle-point calculation we obtain

$$D(E) \approx \exp\left[-\left(\frac{5}{4\hbar}\right)^{\frac{5}{2}}\left(\frac{5}{3\hbar}\right)^{-\frac{3}{2}}\left[l^3 2\hbar\,(4E_l)^{\frac{3}{2}} f\right]\rho E^{\left(-\frac{3}{2}\right)}\right]. \tag{21}$$

Finally we obtain the Lifshitz tail

$$D(E) \approx \exp\left[-\left(\frac{3}{4}\right)^{\frac{5}{2}}\left(\frac{5}{3\hbar}\right)\left[2\hbar^4\left(\frac{2}{m}\right)^{\frac{3}{2}} 4\pi\frac{1}{2}\sqrt{\pi}\right]\rho E^{\left(-\frac{3}{2}\right)}\right]. \tag{22}$$

In order to compare with the Friedberg-Luttinger result, we set $m = 1$ and $\hbar = 1$. We have $D(E) \approx \exp\left[-\frac{15\pi}{2}\sqrt{\frac{3\pi}{2}}\rho E^{-\frac{3}{2}}\right]$, while the Liftshitz result is $D(E) \underset{E\to 0}{\sim} \exp\left(-cE^{-\frac{3}{2}}\right)$ with $c = \frac{4}{3}\pi\left(\frac{\pi^2}{2}\right)^{\frac{3}{2}}\rho$ and the Friedberg-Luttinger method yields $D(E) \underset{E\to 0}{\sim} \exp\left(-cE^{-\frac{3}{2}}\right)$ with $c = \frac{4}{3}\pi\left(\frac{\pi^2}{2}\right)^{\frac{3}{2}}\rho$.

4. Discussion

In this paper we applied the Feynman path-integral method developed in our previous paper for handling the heavily doped semiconductors to the Lifshitz tail problem. The statistics used in this problem is the Poisson distribution. Instead of using a square well interaction potential as used by

Friedberg and Luttinger we use a Gaussian interaction with finite interaction length. This finite interaction allows us to discuss the short range interaction or the white noise problem. In order to be able to carry out the calculation analytically we introduce the quadratic model trial action. The variational principle is used to obtain the variational parameter and a saddle point calculation is used to obtain the Lifshitz tail behavior. It is noted that the factor in front of the Lifshitz tail is different from that of Friedberg and Luttinger. The difference is due to the different potential used in handling the problem. In the Friedberg and Luttinger approach the squared potential with finite depth and finite width is used as a variational parameter. In this approach a harmonic potential is used. However, both approaches lead to the Lifshitz tails behavior. The advantage of this approach is that it is easily generalized to finite interaction length. For very long range, this approach gives the behavior of the density of states as energy power of minus one. This approach can be generalized to study the Lifshitz tails in the case of strong magnetic field or the quantum Hall problem.

References

1. V. Sa-yakanit, *Phys. Rev.* **19**, 2266 (1979).
2. I. M. Lifshitz, *Vsp. Fiz. Nauk* **83**, 617 (1964).
3. R. Friedberg and J. M. Luttinger, *Phys. Rev. B* **12**, 4460 (1975).
4. V. Sa-yakanit, *J. Phys. C* **7**, 2849 (1974).
5. R. P. Feynman and A. R. Hibbs, *Quantum Mechanics and Path Integrals* (Mc Graw-Hill, New York, 1965).

FEYNMAN INTEGRALS AS GENERALIZED FUNCTIONS ON PATH SPACE: THINGS DONE AND OPEN PROBLEMS

L. STREIT

BiBoS - Universität Bielefeld, Germany
CCM - Universidade da Madeira, Portugal
E-mail: streit@physik.uni-bielefeld.de

Observing that distribution theory offers an extension of the concept of integration, we review a framework in which the Feynman integral becomes mathematically meaningful for large classes of interaction potentials. We present some examples and open problems.

Keywords: Path-integral methods; White noise theory; Schrödinger equation; Feynman integrals.

1. Introduction: The Feynman "Integral"

It must be said that many of the innumerable successes of Feynman quantization present themselves to the mathematician as just one more ill defined or open problem. Since the times of Gelfand-Yaglom[1] and of Cameron[2] we all know that the Feynman sum over histories is not an integral in the mathematical sense of the word. But something has to be right, even mathematically, about such an immensely successful concept: we are reminded of Dirac's delta function which, initially, the mathematicians had great doubts about.

Example 1.1.

$$I = \int_{-\infty}^{\infty} dx \exp(-\frac{x^2}{2})$$

is an integral.

$$J(f) = \int_{-\infty}^{\infty} dx \delta'(x) f(x) = -f'(0)$$

is NOT an integral:

$$J : f \rightarrow J(f) = -f'(0)$$

$$f \in S(R) \subset L^2(R, dx) \subset S^*(R) \ni J \tag{1}$$

is a "generalized function", a continuous linear map from the test functions to the real (or complex) numbers.

Example 1.2.

$$I = \int d\mu_{B.M.}(x) \exp\left(-\int_0^t V(x(s))\, ds\right) f(x(t))$$

is an integral (Feynman-Kac).

$$J = \int d^\infty x \exp\left(i \int_0^t (T(\dot{x}) - V(x(s)))\, ds\right) f(x(t))$$

is NOT an integral.

Is

$$J = J(f) = \int d^\infty x \exp(iS[x])\, f[x],$$

if not an integral, a continuous map from test functions (on path space) to complex numbers, in analogy to the improper "integrals" of distribution theory? Trouble is, there is no reasonable flat measure $d^\infty x$. The next best choice is an infinite-dimensional Gaussian one, characterized by its Fourier transform

$$\int_{S^*(R)} d\mu(\omega) \exp\left(i \int \omega(s) f(s) ds\right) = \exp\left(-\frac{1}{2}\int f^2(s) ds\right).$$

With this "white noise" measure we are given an L^2 space

$$(L^2) = L^2(S^*(R), d\mu).$$

Smooth and generalized functions can then be constructed in analogy to the finite-dimensional case:[3–6]

$$(S) \subset (L^2) \subset (S)^*.$$

White noise ω is a model for (Brownian motion) velocity:

$$\omega(t) = \frac{dB}{dt}, \qquad B(t) = \int_0^t \omega(s) ds.$$

2. How to Recognize a Generalized Function When You See One

Among white noise test functions $\varphi = \varphi(\omega) \in (S)$, a prominent example is $\varphi_f(\omega) \equiv \exp\left(i\int \omega(t)f(t)dt\right)$ with $f \in S(R)$.

Example 2.1. Generalized functions $\Phi = \Phi(\omega) \in (S)^*$:

(1) $\Phi_K(\omega) \equiv N\exp\left(\frac{1}{2}(\omega, K\omega)\right)$ if $1 \notin$ *spectrum*(K) "Gauss kernel"
(2) $\Phi_a(\omega) \equiv \delta(B(t) - a)$ "Donsker's δ-function".

It is not hard to calculate[3] the action of these distributions Φ on the exponential test functions $\varphi_f(\omega) = e^{i\int \omega(t)f(t)dt}$:

$$\langle \Phi_K, \varphi_f\rangle = \exp\left(-\frac{1}{2}\left(f, (1-K)^{-1}f\right)\right),$$

$$\langle \Phi_a, \varphi_f\rangle = \frac{1}{\sqrt{2\pi t}}\exp\left(-\frac{\left(\int_0^t if(s)ds - a\right)^2}{2t}\right)\exp\left(-\frac{1}{2}(f,f)\right).$$

Setting $\langle \Phi, \varphi_f\rangle \equiv T\Phi(f)$ we note that in both examples $T\Phi(zf_1 + f_2)$ is analytic in z for any f_1, f_2, and $|T\Phi(zf)| < ae^{b|zf|_p^2}$. Remarkably, these two conditions are sufficient to characterize generalized white noise functions $\Phi \in (S)^*$.

Theorem 2.1.[3] *A functional $G(f)$, $f \in S(R)$, is the T-transform of a unique "Hida distribution" $\Phi \in (S)^*$ iff for all $f_i \in S(R)$, $G(zf_1 + f_2)$ is analytic in the whole complex z-plane and of second order exponential growth.*

A bigger distribution space is obtained if we admit T-transforms which are analytic and bounded near zero only ("Kondratiev distributions"[7]).

3. Construction

We describe the paths by a Brownian ansatz

$$x(t_0 + \tau) = x_0 + \left(\frac{\hbar}{m}\right)^{1/2} B(\tau). \tag{2}$$

Informally, the Feynman integrand yielding the propagator should be

$$I = I_0 \cdot \exp\left(-i\int_{t_0}^t V(x(\tau))\,d\tau\right) \tag{3}$$

with

$$I_0\left(x,t \mid x_0,t_0\right)(\omega) = N \exp\left(\frac{i+1}{2}\int_{\mathbf{R}} \omega^2(\tau)\,\mathrm{d}\tau\right)\delta\left(x(t)-x\right). \quad (4)$$

Remark 3.1. $\frac{i}{2}\int_{\mathbf{R}} \omega^2(\tau)\,\mathrm{d}\tau$ in the expression for I_0 is just the free action. The extra real part compensates the Gaussian "non-flat" nature of the white noise measure. The mathematical validation amounts to a verification that the heuristic expression

$$\langle I, \varphi_f \rangle \text{ ``=''} \int d\mu(\omega)\, I(\omega) \exp\left(i \int \omega(s) f(s) ds\right)$$

is indeed the T-transform of a well defined generalized function $I \in (S)^{-1}$, via the characterization theorem above.

4. Classes of Admissible Interactions

As with any mathematical formulation of Feynman integrals, the crucial question to ask of it is: "What interactions can you handle?" For this see the recent Ref. 8 and references therein.

A rather surprising class is given by potentials which are Laplace transforms of measures:[9]

$$V(x) = \int_{\mathbf{R}^d} e^{\alpha\cdot x}\mathrm{d}m(\alpha)\,, \quad (5)$$

where m is any complex measure with

$$\int_{\mathbf{R}^d} e^{C|\alpha|}\,\mathrm{d}|m|(\alpha) < \infty, \quad \forall C > 0. \quad (6)$$

Example 4.1. $V(x) = g\,e^{ax}$. Likewise, one obtains potentials such as, e.g., $\sinh(ax)$, $\cosh(ax)$, and the *Morse potential*

$$V(x) = g(e^{-2ax} - 2\gamma e^{-ax}) \text{ with } g, a, x \in \mathbf{R} \text{ and } \gamma > 0.$$

Example 4.2. *A Gaussian measure m gives the anharmonic oscillator potentials $V(x) = ge^{bx^2}$. Entire functions of arbitrary high order of growth are also in this class.*

Remarkably, in each case the construction of the Feynman integrand is *perturbative*,

$$I = I_0 \cdot \exp\left(-i \int_{t_0}^{t} V(x(\tau))\, \mathrm{d}\tau\right) \tag{7}$$

$$= \sum_{n=0}^{\infty} \frac{(-i)^n}{n!} \int_{[t_0,t]^n} \int_{\mathbf{R}^{dn}} I_0 \cdot \prod_{j=1}^{n} e^{\alpha_j \cdot x(\tau_j)} \prod_{j=1}^{n} \mathrm{d}m(\alpha_j)\, \mathrm{d}^n\tau \tag{8}$$

Because of its interest in applications we take a close look at one of the above examples, the Morse potential. Its Hamilton operator is

$$H = -\frac{m}{2}\Delta_x + g\left(e^{-2ax} - 2\gamma e^{-ax}\right) \tag{9}$$

Remark 4.1. *H* is essentially self-adjoint for $g \geq 0$ and it is not essentially self-adjoint for $g < 0$. The Green function, the eigenvectors and the discrete eigenvalues are **not analytic** in g near zero.

On the other hand it follows from the above series that *the propagator admits a convergent perturbation series*! In fact, quite unexpectedly, the propagators for all the potentials (5) admit convergent perturbation series *for their propagators* while this is not the case for the corresponding "Euclidean" *heat equations*!

The potentials considered above are singular at infinity but smooth locally. This however is not a crucial limitation, and can be relaxed.[8]

5. Boundary Value Problems

Quantum dynamics on a restricted manifold $M \subset \mathbf{R}^d$ will in general require boundary conditions on ∂M.[10]

A simple case is the restriction of position space to the positive half-line:

$$M = \mathbf{R}_+ \subset \mathbf{R}^1.$$

Boundary conditions are required at $x = 0$. To make the Hamiltonian $H = H_0 + V$ self-adjoint and time development unitary, all wave functions in the domain of $H = H_{\alpha,\beta}$ should obey

$$\lim_{x \to +0} \left(\alpha\psi(x) - \beta\psi'(x)\right) = 0$$

for a fixed pair of real α, β. Such general boundary conditions are fulfilled for

$$K_V(x, t \mid x_0, t_0) = K_{\hat{V}}(x, t \mid x_0, t_0) + K_{\hat{V}}(x, t \mid -x_0, t_0)$$

if we put

$$\widehat{V}(x) = \gamma\delta(x) + V\left(|x|\right), \quad \gamma = \alpha/\beta.$$

This kind of potential is admissible as stated in Ref. 8.

6. Beyond Perturbation Theory

All the above examples admit a perturbation expansion for the Green function. This restriction, however, is not an essential one. Up to trivial factors, the T-transform of the Feynman integrand solves the Schrödinger equation

$$\left(i\frac{\partial}{\partial t} + \frac{1}{2}\triangle_d - V\left(x\right) - x \cdot f'\left(t\right)\right) K^{(f)}\left(x, t \mid x_0, t_0\right) = 0 \qquad (10)$$

and admissibility hinges on the existence of a perturbation series only in the Kato-small auxiliary term $x \cdot f'\left(t\right)$.

For quantum dynamics in a bounded domain this perturbation is in fact even bounded and convergence of the corresponding series, hence the required holomorphy of the T-transform can be shown easily.[11] For $M = \mathbf{R}^d$ an operator theoretic proof is complicated by the fact that the perturbation is time-dependent; a path space argument may be more effective.

As pointed out by Doss[12] there is a solution ψ on $[0, T] \times \mathbb{R}^d$ for the Schrödinger equation, obtainable through complex scaling of Brownian motion in the Feynman-Kac formula for the heat equation

$$\psi(t, x) = E_{B.M.}\left(\exp\left(-i\int_0^t \tilde{V}(x + \sqrt{i}B_r)dr \right) \tilde{f}(x + \sqrt{i}B_t) \right)$$

for wave functions f and potentials V which admit suitable analytic continuations, such as e.g. the non-perturbative interactions

$$V(x) = gx^{4n+2}.$$

Clearly the expression inside the expectation needs to be integrable in this approach; for the existence of the T-transform in the sense of the characterization theorem we only need to include the extra factor

$$\varphi_f\left(\omega\right) \equiv \exp\left(i \int \omega(t) f(t) dt \right), \qquad f \in S(R),$$

which can easily be dominated.[13]

7. A Glimpse at (E)QFT

Is there a generalized "vacuum density" $\rho \in (S)^*$, such that for quantum fields φ – canonical, and/or Euclidean – and hopefully non-trivial,

$$\langle \Omega \,|F(\varphi)|\, \Omega\rangle = < \rho, F > \qquad \text{with } \rho \in (S)^*?$$

In other words: can we make sense of this elusive quantity, the "vacuum density"[14] ρ, as a *generalized* function of white noise? The answer turned out to be in the affirmative: physical vacuum densities (Euclidean and Minkowski, $P(\varphi)_2$, Sine-Gordon, etc.) are positive Hida distributions.[15–17] Proofs of this use the "Froehlich bounds". As a result, the Schwinger functions admit a convergent generating function

$$\sigma(f) = \sum \frac{i^n}{n!} \langle \Omega \,|\varphi^n_{eucl}(f)|\, \Omega\rangle = \sum \frac{i^n}{n!} \int d^n x f(x_1)\ldots f(x_n)\sigma(x_1,\ldots,x_n).$$

Nothing is known about the convergence after continuation from Schwinger's σ to time-ordered functions τ with the proper Minkowski space arguments, except for the trivial case of free fields where the series continues to converge. This would imply a representation

$$\tau(f) = \int_{S^*(R)} e^{i\langle \omega, f\rangle} \rho\,(\omega)\, d\mu(\omega),$$

again with a generalized white noise function $\rho\,(\omega)$, but in this case not giving rise to a measure.

8. Things to be Done

- More on non-perturbative Feynman integrals: relax integrability.
- Time ordered expectation values of quantum fields: Is $\left(\Omega, T e^{i\varphi(f)}\Omega\right)$ holomorphic in f?
- A new look at the change of variables in the Feynman integral.

References

1. I. M. Gelfand and A. M. Yaglom, *J. Math. Phys.* **1**, 48 (1960).
2. H. Cameron, *J. Math. and Phys.* **39**, 126 (1960).
3. T. Hida, H. H. Kuo, J. Potthoff, and L. Streit, *White Noise. An Infinite Dimensional Calculus* (Kluwer Academic, Dordrecht, 1993).
4. Z.-Y. Huang and J.-A. Yan, *Introduction to Infinite Dimensional Stochastic Analysis* (Kluwer, Dordrecht, 2000).
5. H. H. Kuo, *White Noise Distribution Theory* (CRC Press, Boca Raton, 1996).
6. N. Obata, *White Noise Calculus and Fock Space*, Lect. Notes Math. **1577** (Springer, Berlin/Heidelberg, 1994).

7. Yu. G. Kondratiev and L. Streit, *Reports on Math. Phys.* **33**, 341 (1993).
8. M. de Faria, M.-J. Oliveira, and L. Streit, *J. Math. Phys.* **46**, 063505 (2005).
9. T. Kuna, L. Streit, and W. Westerkamp, *J. Math. Phys.* **39**, 4476 (1998).
10. C. Bernido, V. Bernido, and L. Streit, in preparation.
11. M. de Faria, J. Potthoff, and L. Streit, *J. Math. Phys.* **32**, 2123 (1991).
12. H. Doss, *Comm. Math. Phys.* **73**, 247 (1980).
13. M. Grothaus, L. Streit, and A. Vogel, in preparation.
14. F. Coester and R. Haag, *Phys. Rev.* **117**, 1137 (1960).
15. S. Albeverio, T. Hida, J. Potthoff, and L. Streit, *Phys. Lett. B* **217**, 511 (1989).
16. S. Albeverio, T. Hida, J. Potthoff, M. Röckner, and L. Streit, *Rev. Math. Phys.* **1**, 291 and 313 (1990).
17. J. Potthoff and L. Streit, *J. Funct. Anal.* **111**, 295 (1993).

ACCELERATED PATH-INTEGRAL CALCULATIONS VIA EFFECTIVE ACTIONS

A. BALAŽ*, I. VIDANOVIĆ, A. BOGOJEVIĆ, and A. BELIĆ

Scientific Computing Laboratory, Institute of Physics Belgrade, Pregrevica 118, 11080 Belgrade, Serbia
** E-mail: antun@phy.bg.ac.yu*
http://scl.phy.bg.ac.yu/

We give an overview of a recently developed method which systematically improves the convergence of generic path integrals for transition amplitudes, partition functions, expectation values, and energy spectra. This was achieved by analytically constructing a hierarchy of discretized effective actions indexed by a natural number p and converging to the continuum limit as $1/N^p$. We analyze and compare the ensuing increase in efficiency of several orders of magnitude, and perform series of Monte Carlo simulations to verify the results.

Keywords: Effective action; Many-body system; Monte Carlo simulation.

1. Introduction

Path integral formalism offers a general framework for treatment of quantum theories.[1] Functional integrals provide easy way for generalization and extension of quantization methods to more complex physical systems, including systems with no classical counterparts. Originally introduced in quantum mechanics and later most widely used in high energy theory and condensed matter, path integrals can today be found in almost all areas of physics, ranging from atomic, molecular and nuclear physics, to the physics of polymers, biophysics, and chemistry. Moreover, path integrals are starting to play important roles in several areas of mathematics, even in modern finance. An up to date overview of the path integral formalism and its various applications can be found in Kleinert's book.[2]

The definition of path integrals as a limit of multiple integrals over a discretized theory makes their numerical evaluation quite natural. However, path integrals remain notoriously demanding of computing. Considerable research effort has been devoted to the development of approaches that enable faster numerical convergence to the continuum. Efficient implementation of path integral Monte Carlo algorithms, coupled with various model-related approximations, has enabled the application of path integrals to real-world problems.[3] For a long time the state of the art result was $1/N^4$ convergence of discretized partition functions.[4,5] This was achieved using a generalized form[6] of the Trotter formula. However, it is only the integral over all the diagonal amplitudes (i.e. cyclicity of the trace) that has the $O(1/N^4)$ behavior, so such approach cannot be applied to the calculation of general amplitudes or associated (non-thermal) expectation values.

A recently developed method for analytical construction of improved discretized effective actions,[7–9] based on the study of relationship between discretizations of different coarseness,[10,11] has led to substantial speedup of numerical path-integral calculations of several orders of magnitude. Until now, the method has been limited to one-particle one-dimensional systems. Here we present the generalization of this formalism to generic non-relativistic many-particle quantum systems in arbitrary dimensions.

2. Improved Discretized Effective Actions

The presented method is applicable to all quantum theories. For simplicity, we illustrate the details of the derivation on the case of a non-relativistic quantum system consisting of M distinguishable particles in d spatial dimensions, interacting through the potential V. The imaginary time amplitude $A(a, b; T)$ for a transition from initial state $|a\rangle$ to final state $|b\rangle$ in time T is given as the $N \to \infty$ limit of the discretized amplitude

$$A_N(a, b; T) = \frac{1}{(2\pi\varepsilon)^{\frac{MNd}{2}}} \int dq_1 \cdots dq_{N-1}\, e^{-S_N}\,. \tag{1}$$

In this expression, N is the discretization coarseness (number of time slices), while S_N is the naively discretized action,

$$S_N = \sum_{n=0}^{N-1} \varepsilon \left(\sum_{i=1}^{M} \frac{1}{2} \left(\frac{\delta_{n,i}}{\varepsilon} \right)^2 + V(\bar{q}_n) \right) . \tag{2}$$

Here the time step is $\varepsilon = T/N$, while discretized velocities are defined as $\delta_{n,i} = q_{n+1,i} - q_{n,i}$, and mid-point coordinates $\bar{q}_n = (q_n + q_{n+1})/2$. The index n goes over N time steps, and index i goes over M particles.

The above definition of the path integral requires the transition from continuum to discretized theory, i.e. the introduction of coarseness N. Such expressions converge to the continuum very slowly, typically as $O(1/N)$.

One of the key features of definition (1) is that the discretization is not unique. In fact, the choice of discretization strongly affects convergence of discretized amplitudes to the continuum. In a recent paper[10] we have shown that for a general theory there exists an ideal discretization (equivalently, an ideal discretized action S^*), giving the exact (continuum limit) result for *any* discretization coarseness N

$$A_N^*(a, b; T) = A(a, b; T) .$$

This is easily seen if we recall that the defining relation for path integrals as the continuum limit of discretized amplitudes follows from the completeness relation

$$A(a, b; T) = \int \mathrm{d}q_1 \cdots \mathrm{d}q_{N-1} \; A(a, q_1; \varepsilon) \cdots A(q_{N-1}, b; \varepsilon) , \qquad (3)$$

through the substitution of short-time amplitudes $A(q_n, q_{n+1}; \varepsilon)$ calculated to first order in time step ε. A faster converging result may be obtained by evaluating the amplitudes under the integral in Eq. (3) to higher orders in ε. From the above relation we directly see that the ideal discretized action S^* leads to exact propagation, and is given in terms of the exact amplitude,

$$A(q_n, q_{n+1}; \varepsilon) = (2\pi\varepsilon)^{-\frac{Md}{2}} \; e^{-S_n^*} . \qquad (4)$$

The ideal discretized action S^* is simply the sum of expressions S_n^*. We will use Eq. (4) to calculate the ideal discretized action as a power series in ε, which starts from the naive action (2) as the zeroth order term. The details of this expansion have been inspired by an analogous derivation given in Kleinert's book.[2] The outlined approach makes possible the systematic improvement of numerical convergence of path integral calculations, and the construction of a hierarchy of discretized actions $S_N^{(p)}$, denoted by level number p, giving improved convergence

$$A_N^{(p)}(a, b; T) = A(a, b; T) + O(1/N^p). \qquad (5)$$

In order to calculate the short-time amplitude to the desired order in ε, we shift integration variable $q = \xi + x$ about a fixed referent trajectory ξ, and the time to $s \in [-\varepsilon/2, \varepsilon/2]$, so that the short-time amplitude becomes

$$A(q_n, q_{n+1}; \varepsilon) = e^{-S_n[\xi]} \int_{x(-\varepsilon/2)=0}^{x(\varepsilon/2)=0} [\mathrm{d}x] \; e^{-\int_{-\varepsilon/2}^{\varepsilon/2} \mathrm{d}s \left(\frac{1}{2}\dot{x}^2 + U(x;\xi)\right)} . \qquad (6)$$

The referent trajectory ξ satisfies the same boundary conditions as q, which implies that x vanishes at the boundaries. The action $S_n[\xi]$ is defined as

$$S_n[\xi] = \int_{-\varepsilon/2}^{\varepsilon/2} \mathrm{d}s \left(\frac{1}{2}\dot{\xi}^2 + V(\xi) \right) , \qquad (7)$$

and $U(x;\xi) = V(\xi + x) - V(\xi) - x\ddot{\xi}$. The amplitude may now be written as

$$A(q_n, q_{n+1}; \varepsilon) = \frac{e^{-S_n[\xi]}}{(2\pi\varepsilon)^{\frac{Md}{2}}} \left\langle e^{-\int_{-\varepsilon/2}^{\varepsilon/2} ds\, U(x;\xi)} \right\rangle, \tag{8}$$

where $\langle ... \rangle$ denotes expectation values with respect to free massless theory. The above expression holds for any choice of referent trajectory ξ.

Expansion in powers of U gives

$$\left\langle e^{-\int ds\, U(x;\xi)} \right\rangle = 1 - \int ds\, \langle U(x;\xi) \rangle + \frac{1}{2} \iint ds ds' \, \langle U(x;\xi) U(x';\xi') \rangle + \ldots,$$

where $x' = x(s'), \xi' = \xi(s')$, and $U(x;\xi)$ is further expanded around the referent trajectory ξ. The expectation values of products $\langle x_i(s) \ldots x_j(s') \rangle$ is calculated through the use of the massless free theory generating functional. Note that the generating functional (and as a result the expectation values) do not depend on the choice of ξ. However, different choices of ξ are related to different approximation techniques: the choice of classical trajectory for ξ corresponds to the semiclassical expansion, while the choice of a linear referent trajectory is the simplest way to obtain short-time expansion.

In order to perform the remaining integrations over s in Eq. (8), due to the explicit dependence of the referent trajectory on s, we first expand the potential U and all its derivatives around some reference point. For example, in the mid-point prescription, we choose $\bar{q}_n$ as that reference point. Once one chooses the trajectory $\xi(s)$, all expectation values in Eq. (8) are given in terms of quadratures. In this way we obtain a double expansion for S^* in ε and in $\delta n, i^2$. In order to retain only the terms that contribute up to a certain order in ε, we further use the fact that the short time propagation of the considered class of theories satisfies, to leading order, the diffusion relation $\delta_{n,i}^2 \propto \varepsilon$.

The explicit analytical expressions for the many-particle discretized effective action have been so far derived for $p \leq 12$. The derived expressions become algebraically more complex, and such calculations require the use of some of the available packages for symbolic calculus. Note that, in principle, there are no obstacles in going to as high values of p as desired. The derived higher level effective actions can be found on our web site.[12] We stress that in would also be quite interesting to attack the problem of solving Eq. (4) through the use of other approximation schemes, particularly those that are non-perturbative in ε, e.g. the Feynman-Kleinert variational approach.[13–15]

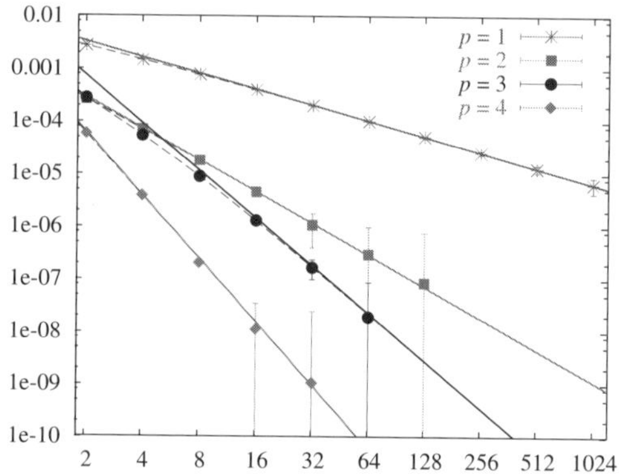

Fig. 1. Deviations of amplitudes from the continuum limit vs. N for two-particle system (9) in two dimensions. Solid lines give the leading $1/N^p$ behavior, while dashed lines correspond to the fitted polynomial functions.

3. Numerical Results

In order to verify the analytically derived speedup in convergence of discretized path integrals, we have performed a series of path integral Monte Carlo simulations of transition amplitudes for a two-dimensional system of two particles interacting through the potential

$$V(\vec{r}_1, \vec{r}_2) = \frac{1}{2}(\vec{r}_1 - \vec{r}_2)^2 + \frac{g_1}{24}(\vec{r}_1 - \vec{r}_2)^4 + \frac{g_2}{2}(\vec{r}_1 + \vec{r}_2)^2 . \qquad (9)$$

All numerical simulations were done using the latest version of our SPEEDUP[12] program that has been extended so as to include multi-particle multi-dimensional systems. The simulations have been performed for different values of couplings g_1 and g_2 and for a variety of initial and final states. The associated continuum limit amplitudes $A^{(p)}$ have been estimated by fitting polynomials in $1/N$ to the discretized values $A_N^{(p)}$, according to the analytically derived relation (5). For all values of p the fitted continuum values $A^{(p)}$ agree within the error bars. Figure 1 gives the plot of the deviations of discretized amplitudes from the continuum limit for two-particle system (9) in two dimensions, with $g_1 = 10$, $g_2 = 0$, $T = 1$, and initial and final states $a = (0, 0; 0.2, 0.5)$, $b = (1, 1; 0.3, 0.6)$. The number of MC samples was from 10^6 for $p = 1$ to 10^{10} for $p = 4$. The increase of level p leads to an ever faster approach to the continuum. The obtained $1/N^p$ dependence gives explicit verification of the analytically derived increase in convergence. As a result of the newly presented method, the usual simulations proceed much faster than by using standard calculation schemes. Note that even the $p = 4$ curve corresponds to a precision of four decimal places in the case of an extremely coarse discretization such as $N = 2$.

4. Conclusions

We have presented a derivation of discretized effective actions leading to substantial, systematical speedup of numerical calculation of path integrals of a generic many-particle non-relativistic theory. The derived speedup holds for all path integrals - for transition amplitudes, partition functions, expectation values, energy levels. The newly calculated discretized effective actions agree with previous approaches. The obtained analytical results have been numerically verified through simulations of path integrals for an anharmonic oscillator with quartic coupling for two particles in two spatial dimensions. The two principle advantages of the new method are: simpler derivation and straightforward generalization to more complex systems.

Acknowledgments

This work was supported in part by the Ministry of Science of the Republic of Serbia, under project No. OI141035, and the EC under EU Centre of Excellence grant CX-CMCS. The numerical results were obtained on the AEGIS e-Infrastructure, which is supported in part by FP6 projects EGEE-II and SEE-GRID-2.

References

1. R. P. Feynman, *Rev. Mod. Phys.* **20**, 367 (1948).
2. H. Kleinert, *Path Integrals in Quantum Mechanics, Statistics, Polymer Physics, and Financial Markets*, 4th Ed., World Scientific, Singapore, 2006.
3. D. M. Ceperley, *Rev. Mod. Phys.* **67**, 279 (1995).
4. M. Takahashi and M. Imada, *J. Phys. Soc. Jpn.* **53**, 3765 (1984).
5. X. P. Li and J. Q. Broughton, *J. Chem. Phys.* **86**, 5094 (1987).
6. H. De Raedt, and B. De Raedt, *Phys. Rev. A* **28**, 3575 (1983).
7. A. Bogojević, A. Balaž, and A. Belić, *Phys. Rev. Lett.* **94**, 180403 (2005).
8. A. Bogojević, A. Balaž, and A. Belić, *Phys. Rev. B* **72**, 064302 (2005).
9. A. Bogojević, A. Balaž, and A. Belić, *Phys. Lett. A* **344**, 84 (2005).
10. A. Bogojević, A. Balaž, and A. Belić, *Phys. Rev. E* **72**, 036128 (2005).
11. A. Bogojević, A. Balaž, and A. Belić, *Phys. Lett. A* **345**, 258 (2005).
12. `http://scl.phy.bg.ac.yu/speedup/`
13. R. P. Feynman and H. Kleinert, *Phys. Rev. A* **34**, 5080 (1986).
14. H. Kleinert, *Phys. Lett. B* **280**, 251 (1992).
15. A. Pelster, H. Kleinert, and M. Schanz, *Phys. Rev. E* **67**, 016604 (2003).

SYSTEMATIC SPEEDUP OF ENERGY SPECTRA CALCULATIONS FOR MANY-BODY SYSTEMS

I. VIDANOVIĆ*, A. BALAŽ, A. BOGOJEVIĆ, and A. BELIĆ

Scientific Computing Laboratory, Institute of Physics Belgrade, Pregrevica 118, 11080 Belgrade, Serbia

** E-mail: ivanavi@phy.bg.ac.yu*

URL: http://scl.phy.bg.ac.yu/

We present an application of a recently developed method for accelerated Monte Carlo computations of path integrals to the problem of energy spectra calculation of generic many-particle systems. We calculate the energy spectra of a two-particle two dimensional system in a quartic potential using the hierarchy of discretized effective actions, and demonstrate agreement with analytical results governing the increase in efficiency of the new method.

Keywords: Path integral; Effective action; Many-body system; Energy spectra.

In addition to their key position in the analytical approach to quantum theory, path integrals also play an important role in the computational simulations of realistic many-body systems. The starting point in these calculations is the time-sliced expression for the general quantum-mechanical amplitude,[1]

$$A_N(a,b;T) = \frac{1}{(2\pi\varepsilon)^{\frac{MNd}{2}}} \int dq_1 \cdots dq_{N-1}\, e^{-S_N}, \tag{1}$$

where N is the number of time slices $\varepsilon = T/N$ and S_N is the naively discretized action for a system of M nonrelativistic particles in d spatial dimensions. The $N \to \infty$ limit of the above discretized amplitude gives the continuum amplitude $A(a,b;T)$.

The performance of the numerical algorithms for the calculation of path integrals is directly determined by the efficiency of the applied integration technique, as well as by the speed of convergence of the discretized amplitudes to the continuum. The first problem has been essentially solved through the development of modern efficient integration techniques. Improvement of performance now only depends on improved convergence of discretized expressions to the continuum.

In a recent series of papers[2–5] we focused on the systematic analytical construction of effective actions $S_N^{(p)}$ that improve convergence of discretized transition amplitudes, partition functions and expectation values to the continuum limit for one-particle one-dimensional systems. More recently[6] we have extended this to general many-particle non-relativistic systems, with the convergence of discretized amplitudes

$$A_N^{(p)}(a, b; T) = A(a, b; T) + O(1/N^p), \tag{2}$$

with $p \leq 12$, where $p = 1$ corresponds to naive discretization. This improved convergence translates directly into significant speedup of numerical calculations.

In this paper we focus on the benefits of the above improved convergence of amplitudes for the evaluation of energy spectra of many-body systems. The speedup in the calculation of amplitudes leads to the same improvement in the convergence of partition functions, owing to the relation

$$Z_N(\beta) = \int dq A_N(q, q; \beta) \ , \tag{3}$$

where β is the inverse temperature. The partition function is the central object for obtaining information about various thermodynamical quantities.

The numerical results of a series of Monte Carlo (MC) simulations performed clearly demonstrate the analytically derived speedup. Figure 1 shows increase in convergence of discretized free energy $F_N(\beta) = -\log Z_N(\beta)/\beta$ as the level p of the used effective action $S_N^{(p)}$ increases. This is illustrated on the case of a two-dimensional system of two distinguishable particles interacting through a quartic potential

$$V(\vec{r}_1, \vec{r}_2) = \frac{1}{2}(\vec{r}_1 - \vec{r}_2)^2 + \frac{g_1}{24}(\vec{r}_1 - \vec{r}_2)^4 + \frac{g_2}{2}(\vec{r}_1 + \vec{r}_2)^2 . \tag{4}$$

The convergence of discretized free energies, presented by solid lines in Fig. 1, is governed by a $1/N^p$ term, corresponding to the analog of the equation Eq. (2) for F_N.

The partition function offers a straightforward way for extracting information about the energy spectrum,[7] since

$$Z(\beta) = \sum_{n=0}^{\infty} d_n e^{-\beta E_n} \, , \tag{5}$$

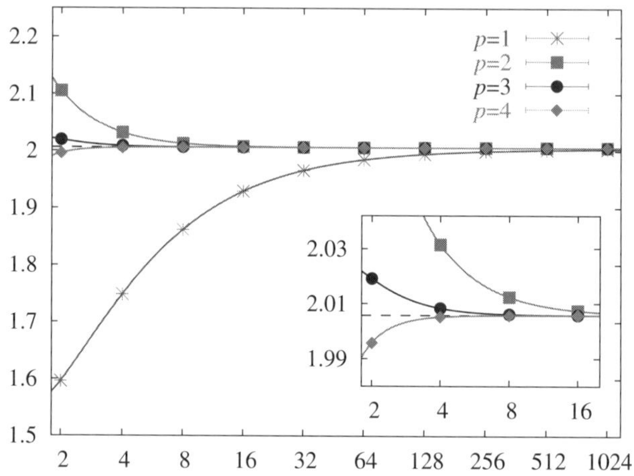

Fig. 1. Convergence of discretized free energies $F_N(\beta)$ to the continuum as a function of N for the quartic potential ($g_1 = 1$, $g_2 = 1$, $\beta = 1$). Number of MC samples is 10^7.

where E_n and d_n denote corresponding energy levels and degeneracies. Similarly to the free energy, we can introduce a set of auxiliary functions

$$f_n(\beta) = -\frac{1}{\beta} \ln \frac{Z(\beta) - \sum_{i=0}^{n-1} d_i e^{-\beta E_i}}{d_n} \,, \tag{6}$$

which can be fitted for large β to

$$f_n(\beta) \approx E_n - \frac{1}{\beta} \ln(1 + a e^{-\beta b}) \,, \tag{7}$$

and which tend to the corresponding energy levels E_n. In this way, by studying the large β behavior of the functions f_n, one can obtain the energy spectrum of the model.

In numerical simulations we are inevitably limited to the finite range of inverse temperatures. In addition, the above procedure for the construction of the auxiliary functions f_n is recursive, i.e. in order to construct f_n we need to know all the energy levels below E_n. This leads to the accumulation of errors as n increases, and practically limits the number of energy levels that can be calculated. Note that the orders of magnitude increase in precision of the presented method reduces the magnitude of the accumulated error and thus allows us to extract viable information about a larger number of energy levels.

Table 1. Energy levels of the quartic potential, $g_2 = 1/9$. The corresponding degeneracies of the calculated energy levels are found to be $d_0 = 1$, $d_1 = 2$, $d_2 = 3$, $d_3 = 6$.

g_1	E_0	$E_0^{\rm pert}$	E_1	E_2	E_3
0	1.8857(1)	1.88562	2.3571(6)	2.83(1)	3.3(2)
0.1	1.9019(2)	1.90187	2.374(2)	2.82(1)	—
1	2.0228(2)	2.03384	2.497(3)	2.94(3)	—
10	2.6327(6)	—	3.098(4)	3.57(3)	—

Table 1 gives the calculated energy levels for quartic coupling from $g_1 = 0$ (free theory) to $g_1 = 10$ (strongly interacting theory), obtained using the $p = 5$ effective action. The number of MC samples was 10^9 with discretization coarseness $N = 64$, chosen so that $F_N(\beta) = F(\beta)$ within the statistical error over the range $1 \leq \beta \leq 11$, yielding four decimal places precision for the ground energy level. The employment of higher level effective actions makes it possible to use much coarser discretization, thus substantially reducing the CPU time. For comparison, the table also contains values of the ground state energy $E_0^{\rm pert}$ obtained using perturbation expansion up to third order in g_1.

To conclude, we have applied the recently derived many-particle discretized effective actions to the calculation of low-lying energy spectra. Numerical results confirm the analytically derived increase in convergence of discretized expressions, resulting in substantially more efficient simulations.

Acknowledgments

This work was supported in part by the Ministry of Science of the Republic of Serbia, under project No. OI141035, and the EC under EU Centre of Excellence grant CX-CMCS. The numerical results were obtained on the AEGIS e-Infrastructure, which is supported in part by FP6 projects EGEE-II and SEE-GRID-2.

References

1. R. P. Feynman and A. R. Hibbs, *Quantum Mechanics and Path Integrals* (McGraw-Hill, New York, 1965).
2. A. Bogojević, A. Balaž, and A. Belić, *Phys. Rev. Lett.* **94**, 180403 (2005).
3. A. Bogojević, A. Balaž, and A. Belić, *Phys. Rev. B* **72**, 064302 (2005).
4. A. Bogojević, A. Balaž, and A. Belić, *Phys. Lett. A* **344**, 84 (2005).
5. A. Bogojević, A. Balaž, and A. Belić, *Phys. Rev. E* **72**, 036128 (2005).
6. A. Balaž, I. Vidanović, A. Bogojević, and A. Belić, talk at the PI07 Conference, Dresden, Germany (2007) [in this proceedings].
7. D. Stojiljković, A. Bogojević, and A. Balaž, *Phys. Lett. A* **360**, 217 (2006).

GEOMETRIC PHASE AND CHIRAL ANOMALY IN PATH INTEGRAL FORMULATION

K. FUJIKAWA

Institute of Quantum Science, College of Science and Technology, Nihon University, Chiyoda-ku, Tokyo, Japan
Email: fujikawa@phys.cst.nihon-u.ac.jp

All the geometric phases, adiabatic and non-adiabatic, are formulated in a unified manner in the second quantized path integral formulation. The exact hidden local symmetry inherent in the Schrödinger equation defines the holonomy. All the geometric phases are shown to be topologically trivial. The geometric phases are briefly compared to the chiral anomaly which is naturally formulated in the path integral.

Keywords: Geometric phase; Chiral anomaly; Holonomy.

1. Second Quantization

To analyze various geometric phases in a unified manner,[1-8] we start with an *arbitrary* complete basis set $\int d^3x v_n^\star(t,\vec{x}) v_m(t,\vec{x}) = \delta_{nm}$ and expand the field variable as $\psi(t,\vec{x}) = \sum_n b_n(t) v(t,\vec{x})$. The action

$$S = \int_0^T dt d^3x [\psi^\star(t,\vec{x}) i\hbar \frac{\partial}{\partial t} \psi(t,\vec{x}) - \psi^\star(t,\vec{x}) \hat{H}(\hat{\vec{p}}, \hat{\vec{x}}, X(t)) \psi(t,\vec{x})] \quad (1)$$

with background variables $X(t) = (X_1(t), X_2(t), ..)$ then becomes $S = \int_0^T dt \{\sum_n b_n^\star(t) i\hbar\partial_t b_n(t) - H_{\rm eff}\}$ with the effective Hamiltonian in the second quantized version

$$\hat{H}_{\rm eff}(t) = \sum_{n,m} \hat{b}_n^\dagger(t) [\int d^3x v_n^\star(t,\vec{x}) \hat{H}(\hat{\vec{p}}, \hat{\vec{x}}, X(t)) v_m(t,\vec{x}) - \int d^3x v_n^\star(t,\vec{x}) i\hbar \frac{\partial}{\partial t} v_m(t,\vec{x})] \hat{b}_m(t), \quad (2)$$

and $[\hat{b}_n(t), \hat{b}_m^\dagger(t)]_\mp = \delta_{n,m}$, but statistics is not important in our application. We use the fermions (Grassmann variables) in the path integral.

The Schrödinger probability amplitude with $\psi_n(0,\vec{x}) = v_n(0,\vec{x})$ is defined by[11]

$$\begin{aligned}\psi_n(t,\vec{x}) &= \langle 0|\hat{\psi}(t,\vec{x})\hat{b}_n^\dagger(0)|0\rangle \\ &= \sum_m v_m(t,\vec{x})\langle m|T^\star \exp\{-\frac{i}{\hbar}\int_0^t \hat{\mathcal{H}}_{\rm eff}(t)dt\}|n\rangle\,, \end{aligned} \tag{3}$$

where the Schrödinger picture $\hat{\mathcal{H}}_{\rm eff}(t)$ is defined by replacing $\hat{b}_n(t)$ by $\hat{b}_n(0)$ in $\hat{H}_{\rm eff}(t)$ with time ordering symbol $T^\star$ and $|n\rangle = \hat{b}_n^\dagger(0)|0\rangle$. The path integral representation is given by

$$\begin{aligned}\langle m|T^\star \exp\{-\frac{i}{\hbar}\int_0^t \hat{\mathcal{H}}_{\rm eff}(t)dt\}|n\rangle &= \int\prod_n \mathcal{D}b_n^\star\mathcal{D}b_n\phi_m^\star(b_n^\star(t)) \\ &\times \exp\{\frac{i}{\hbar}\int_0^t dt[b_n^\star(t)i\hbar\partial_t b_n(t) - H_{\rm eff}(t)\}\phi_n(b_n^\star(0))\end{aligned} \tag{4}$$

with suitable wave functions $\phi_m^\star(b_n^\star(t))$ and $\phi_n(b_n^\star(0))$ in the holomorphic representation.[9] The general geometric terms automatically appear in the second term of the *exact* $H_{\rm eff}(t)$ of Eq. (2) and thus the naive holomorphic wave functions are sufficient. This means that the analysis of geometric phases is reduced to a simple functional analysis in the second quantized path integral.

If one uses a specific basis

$$\hat{H}(\hat{\vec{p}},\hat{\vec{x}},X(t))v_n(\vec{x};X(t)) = \mathcal{E}_n(X(t))v_n(\vec{x};X(t))$$

and assumes "diagonal dominance" in the effective Hamiltonian in (4), we have the adiabatic formula

$$\psi_n(t,\vec{x}) \simeq v_n(\vec{x};X(t))\exp\{-\frac{i}{\hbar}\int_0^t [\mathcal{E}_n(X(t)) - v_n^\star i\hbar\frac{\partial}{\partial t}v_n dt\}\,, \tag{5}$$

which shows that the *adiabatic approximation* is equivalent to the *approximate diagonalization* of $H_{\rm eff}$, and thus the geometric phases are *dynamical*.[10,11]

2. Hidden Local Gauge Symmetry

Since $\psi(t,\vec{x}) = \sum_n b_n(t)v_n(t,\vec{x})$, we have an exact local symmetry[11]

$$v_n(t,\vec{x}) \to v_n'(t,\vec{x}) = e^{i\alpha_n(t)}v_n(t,\vec{x}),\quad b_n(t)\to b_n'(t) = e^{-i\alpha_n(t)}b_n(t)$$

for $n = 1,2,\ldots$. This symmetry means an arbitrariness in the choice of the coordinates in the functional space. The exact Schrödinger amplitude $\psi_n(t,\vec{x}) = \langle 0|\hat{\psi}(t,\vec{x})\hat{b}_n^\dagger(0)|0\rangle$ is transformed under this substitution rule as

$\psi'_n(t,\vec{x}) = e^{i\alpha_n(0)}\psi_n(t,\vec{x})$ for any t. Namely, we have the ray representation with a constant phase change, and we have an enormous hitherto unrecognized exact hidden symmetry behind the ray representation.

The combination $\psi_n(0,\vec{x})^\star\psi_n(T,\vec{x})$ thus becomes manifestly gauge invariant. By this way, one can identify the unique gauge invariant phase, for example, in the adiabatic approximation (5).[11]

Parallel transport and holonomy: The parallel transport of $v_n(t,\vec{x})$ is defined by $\int d^3x v_n^\dagger(t,\vec{x})\frac{\partial}{\partial t}v_n(t,\vec{x}) = 0$. By using the hidden local gauge $\bar{v}_n(t,\vec{x}) = e^{i\alpha_n(t)}v_n(t,\vec{x})$ for general $v_n(t,\vec{x})$, one may impose the condition $\int d^3x\bar{v}_n^\dagger(t,\vec{x})\frac{\partial}{\partial t}\bar{v}_n(t,\vec{x}) = 0$ which gives

$$\bar{v}_n(t,\vec{x}) = \exp[i\int_0^t dt' \int d^3x v_n^\dagger(t',\vec{x}) i\partial_{t'} v_n(t',\vec{x})] v_n(t,\vec{x}). \quad (6)$$

The *holonomy* for a cyclic motion is then defined by

$$\bar{v}_n^\dagger(0,\vec{x})\bar{v}_n(T,\vec{x}) = v_n^\dagger(0,\vec{x})v_n(T,\vec{x})\exp[i\int_0^T dt' \int d^3x v_n^\dagger(t',\vec{x}) i\partial_{t'} v_n(t',\vec{x})].$$

This holonomy of *basis vectors*, not Schrödinger amplitude, determines *all* the geometric phases in the second quantized formulation.

3. Non-Adiabatic Phase

3.1. *Cyclic evolution*

The cyclic evolution is defined by $\psi(T,\vec{x}) = e^{i\phi}\psi(0,\vec{x})$ or by

$$\psi(t,\vec{x}) = e^{i\phi(t)}\tilde{\psi}(t,\vec{x}), \quad \tilde{\psi}(T,\vec{x}) = \tilde{\psi}(0,\vec{x}) \quad (7)$$

with $\phi(T) = \phi$, $\phi(0) = 0$. If one chooses the first element of the arbitrary basis set $\{v_n(t,\vec{x})\}$ such that $v_1(t,\vec{x}) = \tilde{\psi}(t,\vec{x})$, one has diagonal $\hat{H}_{\rm eff}(t)$ and

$$\psi(t,\vec{x}) = v_1(t,\vec{x})\exp\{-\frac{i}{\hbar}[\int_0^t dt \int d^3x v_1^\star(t,\vec{x})\hat{H}v_1(t,\vec{x}) - \int_0^t dt \int d^3x v_1^\star(t,\vec{x}) i\hbar\partial_t v_1(t,\vec{x})]\} \quad (8)$$

in (3), and the factor[12]

$$\beta = \oint dt \int d^3x v_1^\star(t,\vec{x}) i\frac{\partial}{\partial t} v_1(t,\vec{x}) \quad (9)$$

gives the *non-adiabatic phase*.[5]

Note that the so-called "projective Hilbert space" and the transformation $\psi(t,\vec{x}) \rightarrow e^{i\omega(t)}\psi(t,\vec{x})$, which is not the symmetry of the Schrödinger equation, is not used in our formulation. Also, the holonomy of the basis vector, not the Schrödinger amplitude, determines the non-adiabatic phase. Our derivation of the non-adiabatic phase (9), which works in the path integral (4) also, is quite different from that in Ref. 5.

3.2. *Non-cyclic evolution*

It is shown that *any* exact solution of the Schrödinger equation is written in the form,[13]

$$\psi_k(\vec{x},t) = v_k(\vec{x},t)\exp\{-\frac{i}{\hbar}\int_0^t \int d^3x[v_k^\dagger(\vec{x},t)\hat{H}(t)v_k(\vec{x},t) \\ - v_k^\dagger(\vec{x},t)i\hbar\frac{\partial}{\partial t}v_k(\vec{x},t)]\}\,, \tag{10}$$

if one suitably chooses the basis set $\{v_k(\vec{x},t)\}$, though the periodicity is generally lost, $v_k(\vec{x},0) \neq v_k(\vec{x},T)$.

By choosing a suitable hidden symmetry $v_k(t,\vec{x}) \rightarrow e^{i\alpha_k(t)}v_k(t,\vec{x})$, one can identify the gauge invariant non-cyclic and non-adiabatic phase[6] as the second term in Eq. (10).[13] The present definition also works in the path integral (4).

It is shown that geometric phases for mixed states[7,8] are similarly formulated in the second quantized formulation.[13]

4. Exactly Solvable Example

We study the model

$$\hat{H} = -\mu\hbar\vec{B}(t)\vec{\sigma}, \quad \vec{B}(t) = B(\sin\theta\cos\varphi(t), \sin\theta\sin\varphi(t), \cos\theta) \tag{11}$$

with $\varphi(t) = \omega t$ and constant ω, B and θ.

One can *diagonalize* the second quantized H_{eff} by defining the constant α by $2\mu\hbar B\sin\alpha = \hbar\omega\sin(\theta-\alpha)$ and the basis vectors

$$w_+(t) = \begin{pmatrix} \cos\frac{1}{2}(\theta-\alpha)e^{-i\varphi(t)} \\ \sin\frac{1}{2}(\theta-\alpha) \end{pmatrix}, \quad w_-(t) = \begin{pmatrix} \sin\frac{1}{2}(\theta-\alpha)e^{-i\varphi(t)} \\ -\cos\frac{1}{2}(\theta-\alpha) \end{pmatrix}$$

which satisfy $w_\pm(0) = w_\pm(T)$ with $T = \frac{2\pi}{\omega}$, and

$$w_\pm^\dagger(t)\hat{H}w_\pm(t) = \mp\mu\hbar B\cos\alpha, \quad w_\pm^\dagger(t)i\hbar\partial_t w_\pm(t) = \frac{\hbar\omega}{2}(1\pm\cos(\theta-\alpha)).$$

The effective Hamiltonian $H_{\rm eff}$ is now diagonal, and the *exact* solution of the Schrödinger equation $i\hbar\partial_t\psi(t) = \hat{H}\psi(t)$ is given by[13]

$$\psi_\pm(t) = w_\pm(t)\exp\{-\frac{i}{\hbar}\int_0^t dt'[w_\pm^\dagger(t')\hat{H}w_\pm(t') - w_\pm^\dagger(t')i\hbar\partial_{t'}w_\pm(t')]\}. \quad (12)$$

This exact solution is regarded either as an exact version of the adiabatic phase or as the cyclic non-adiabatic phase in our formulation.

One can examine some limiting cases of the exact solution:

(i) For *adiabatic limit* $\hbar\omega/(\hbar\mu B) \ll 1$, we have the parameter $\alpha \simeq [\hbar\omega/2\hbar\mu B]\sin\theta$, and one recovers the Berry's phase by setting $\alpha = 0$

$$\psi_\pm(T) \simeq \exp\{i\pi(1\pm\cos\theta)\}\exp\{\pm\frac{i}{\hbar}\int_0^T dt\mu\hbar B\}w_\pm(T). \quad (13)$$

(ii) For *non-adiabatic* limit $\hbar\mu B/(\hbar\omega) \ll 1$, we have the parameter $\theta - \alpha \simeq [2\hbar\mu B/\hbar\omega]\sin\theta$, and one obtains the trivial geometric phase by setting $\alpha = \theta$

$$\psi_\pm(T) \simeq w_\pm(T)\exp\{\pm\frac{i}{\hbar}\int_0^T dt[\mu\hbar B\cos\theta]\}.$$

This shows that the "monopole-like" phase in (13) is smoothly connected to a trivial phase in the exact solution (12), and thus the geometric phase is *topologically trivial.*[10]

5. Chiral Anomaly

It is known that all the anomalies in gauge field theory[14,15] are understood in the path integral as arising from the non-trivial Jacobians under symmetry transformations.[16,17] For example, in the fermionic path integral

$$\int \mathcal{D}\bar{\psi}\mathcal{D}\psi\exp\{i\int d^4x[\bar{\psi}i\gamma^\mu(\partial_\mu - igA_\mu)\psi]\} \quad (14)$$

and for infinitesimal chiral transformation $\psi(x) \to e^{i\omega(x)\gamma_5}\psi(x)$, $\bar{\psi}(x) \to \bar{\psi}(x)e^{i\omega(x)\gamma_5}$, we have

$$\mathcal{D}\bar{\psi}\mathcal{D}\psi \to \exp\{-i\int d^4x\omega(x)\frac{g^2}{16\pi^2}\epsilon^{\mu\nu\alpha\beta}F_{\mu\nu}F_{\alpha\beta}\}\mathcal{D}\bar{\psi}\mathcal{D}\psi. \quad (15)$$

The anomaly is integrated for a finite transformation, and it gives rise to the so-called Wess-Zumino term.[18]

Based on this observation, one recognizes the following differences between the geometric phases and chiral anomaly:[19]

1. The Wess-Zumino term is added to the classical action in path integral,

whereas the geometric term appears *inside* the classical action sandwiched by field variables as in (2). Geometric phases are thus state-dependent.
2. The topology of chiral anomaly, which is provided by gauge fields, is exact, whereas the topology of the adiabatic geometric phase, which is valid only in the adiabatic limit, is trivial.

References

1. C. Mead and D. Truhlar, *J. Chem. Phys.* **70**, 2284 (1978).
2. M. V. Berry, *Proc. Roy. Soc. A* **392**, 45 (1984).
3. B. Simon, *Phys. Rev. Lett.* **51**, 2167 (1983).
4. H. Kuratsuji and S. Iida, *Prog. Theor. Phys.* **74**, 439 (1985).
5. Y. Aharonov and J. Anandan, *Phys. Rev. Lett.* **58**, 1593 (1987).
6. J. Samuel and R. Bhandari, *Phys. Rev. Lett.* **60**, 2339 (1988).
7. E. Sjöqvist *et al.*, *Phys. Rev. Lett.* **85**, 2845 (2000).
8. K. Singh, D. M. Tong, K. Basu, J. L. Cheng, and J. F. Du, *Phys. Rev. A* **67**, 032106 (2003).
9. L. D. Faddeev, *Introduction to Functional Methods*, in *Methods in Field Theory* (Les Houches Lectures), edited by R. Balian and J. Zinn-Justin (North Holland, 1976), pp. 1.
10. K. Fujikawa, *Mod. Phys. Lett. A* **20**, 335 (2005); S. Deguchi and K. Fujikawa, *Phys. Rev. A* **72**, 012111 (2005).
11. K. Fujikawa, *Phys. Rev. D* **72**, 025009 (2005).
12. K. Fujikawa, *Int. J. Mod. Phys. A* **21**, 5333 (2006).
13. K. Fujikawa, *Ann. of Phys.* **322**, 1500 (2007).
14. J. S. Bell and R. Jackiw, *Nuovo Cim. A* **60**, 47 (1969).
15. S. L. Adler, *Phys. Rev.* **177**, 2426 (1969).
16. K. Fujikawa, *Phys. Rev. Lett.* **42**, 1195 (1979); *Phys. Rev. D* **21**, 2848 (1980); *Phys. Rev. Lett.* **44**, 1733 (1980).
17. K. Fujikawa and H. Suzuki, *Path Integrals and Quantum Anomalies* (Oxford University Press, Oxford, 2004).
18. J. Wess and B. Zumino, *Phys. Lett. B* **37**, 95 (1971).
19. K. Fujikawa, *Phys. Rev. D* **73**, 025017 (2006).

PHASE SPACE PATH INTEGRALS AND THEIR SEMICLASSICAL APPROXIMATIONS

N. KUMANO-GO*

Mathematics, Kogakuin University,
1-24-2 Nishishinjuku, Shinjuku-ku, Tokyo 163-8677, Japan
E-mail: ft24343@ns.kogakuin.ac.jp

D. FUJIWARA†

Department of Mathematics, Gakushuin University,
1-5-1 Mejiro, Toshima-ku, Tokyo 171-8588, Japan
E-mail: daisuke.fujiwara@gakushuin.ac.jp

This is a survey of our work in Ref. 1. We give a fairly general class of functionals for which the phase space Feynman path integrals have a mathematically rigorous meaning. More precisely, for any functional belonging to our class, the time slicing approximation of the phase space path integral converges uniformly on compact subsets of the phase space. Our class of functionals is rich because it is closed under addition and multiplication. The interchange of the order with the Riemann integrals, the interchange of the order with a limit and the perturbation expansion formula hold in the phase space path integrals. The use of piecewise bicharacteristic paths naturally leads us to the semiclassical approximation on the phase space.

Keywords: Path integrals; Fourier integral operators; Semiclassical approximation.

1. Introduction

Let $u(T)$ be the solution for the Schrödinger equation such that

$$(i\hbar\partial_T - H(T,x,\frac{\hbar}{i}\partial_x))u(T) = 0\,, \quad u(0) = v\,, \tag{1}$$

*This work was supported by MEXT. KAKENHI 18740077 and at GFMUL in Portugal by POCTI/MAT/34924.

†This work was supported by JSPS. KAKENHI(C) 17540170.

where $0 < \hbar < 1$ is the Planck parameter. In the theory of Fourier integral operators (cf. H. Kitada and H. Kumano-go[2]), we treat $u(T)$ as

$$\begin{aligned} u(T) &= \left(\frac{1}{2\pi\hbar}\right)^d \int_{\mathbf{R}^d} K(T,x,\xi_0)\hat{v}(\xi_0)d\xi_0 \\ &= \left(\frac{1}{2\pi\hbar}\right)^d \int_{\mathbf{R}^d}\int_{\mathbf{R}^d} K(T,x,\xi_0)e^{-ix_0\cdot\xi_0}v(x_0)dx_0d\xi_0\,. \end{aligned} \tag{2}$$

Using the phase space Feynman path integral,[3] we formally write

$$K(T,x,\xi_0)e^{-ix_0\cdot\xi_0} = \int e^{\frac{i}{\hbar}\phi[q,p]}\mathcal{D}[q,p]\,. \tag{3}$$

Here $(q,p) : [0,T] \to \mathbf{R}^{2d}$ is the path in the phase space with $q(T) = x$, $q(0) = x_0$ and $p(0) = \xi_0$, and $\phi[q,p]$ is the action defined by

$$\phi[q,p] = \int_{[0,T)} p(t)\cdot dq(t) - \int_{[0,T)} H(t,q(t),p(t))dt\,, \tag{4}$$

and the phase space path integral $\int \sim \mathcal{D}[q,p]$ is a sum over all the paths (q,p). Feynman[4] explained his original configuration space path integral as a limit of a finite dimensional integral, which is now called the time slicing approximation. Furthermore, Feynman considered the configuration space path integrals with general functional as integrand (cf. L. S. Schulman[5]). In some papers of physics, the phase space path integral has some interpretations of the paths (q,p) (cf. Schulman[5]).

In this paper, using the time slicing approximation via piecewise bicharacteristic paths, we give a fairly general class $\mathcal{F}$ of functionals $F[q,p]$ so that the phase space path integrals with functionals $F[q,p]$

$$\int e^{\frac{i}{\hbar}\phi[q,p]}F[q,p]\mathcal{D}[q,p]\,, \tag{5}$$

have a mathematically rigorous meaning. More precisely, for any $F[q,p] \in \mathcal{F}$, the time slicing approximation of (5) converges uniformly on compact subsets of $\mathbf{R}^{3d}$ with respect to the endpoints (x,ξ_0,x_0).

For other mathematical rigorous definitions of phase space path integrals, see I. Daubechies–J. R. Klauder[6] , S. Albeverio–G. Guatteri–S. Mazzucchi[7] and O. G. Smolyanov–A. G. Tokarev–A. Truman.[8]

2. Main Results

Assumption 2.1. *The Hamiltonian function $H(t,x,\xi)$ of (1) is a real-valued function of $(t,x,\xi) \in \mathbf{R}\times\mathbf{R}^d\times\mathbf{R}^d$. For any multi-indices α, β,*

$\partial_x^\alpha \partial_\xi^\beta H(t,x,\xi)$ is continuous in $\mathbf{R} \times \mathbf{R}^d \times \mathbf{R}^d$, and there exists a positive constant $\kappa_{\alpha,\beta}$ such that

$$|\partial_x^\alpha \partial_\xi^\beta H(t,x,\xi)| \le \kappa_{\alpha,\beta}(1+|x|+|\xi|)^{\max(2-|\alpha+\beta|,0)} . \tag{6}$$

Let $\Delta_{T,0} = (T_{J+1}, T_J, \ldots, T_1, T_0)$ be an arbitrary division of the interval $[0,T]$ into subintervals, i.e.,

$$\Delta_{T,0} : T = T_{J+1} > T_J > \cdots > T_1 > T_0 = 0 . \tag{7}$$

Set $t_j = T_j - T_{j-1}$ for $j = 1,2,\ldots,J,J+1$. Let $|\Delta_{T,0}| = \max_{1\le j\le J+1} t_j$. Set $x_{J+1} = x$. Let $x_j \in \mathbf{R}^d$ and $\xi_j \in \mathbf{R}^d$ for $j = 1,2,\ldots,J$.

Lemma 2.1. *Let $j = 1,2,\ldots,J,J+1$ and $\kappa_2 d(T_j - T_{j-1}) < 1/2$. Then, for any $(x_j, \xi_{j-1}) \in \mathbf{R}^d \times \mathbf{R}^d$, there exists a unique solution*

$$\bar{q}_{T_j,T_{j-1}}(t) = \bar{q}_{T_j,T_{j-1}}(t, x_j, \xi_{j-1}) , \ \bar{p}_{T_j,T_{j-1}}(t) = \bar{p}_{T_j,T_{j-1}}(t, x_j, \xi_{j-1}) , \tag{8}$$

of the system of equations

$$\begin{aligned}
&\partial_t \bar{q}_{T_j,T_{j-1}}(t) = (\partial_\xi H)(t, \bar{q}_{T_j,T_{j-1}}(t), \bar{p}_{T_j,T_{j-1}}(t)) , \quad T_{j-1} \le t \le T_j , \\
&\partial_t \bar{p}_{T_j,T_{j-1}}(t) = -(\partial_x H)(t, \bar{q}_{T_j,T_{j-1}}(t), \bar{p}_{T_j,T_{j-1}}(t)) , \quad T_{j-1} \le t \le T_j , \\
&\bar{q}_{T_j,T_{j-1}}(T_j) = x_j , \quad \bar{p}_{T_j,T_{j-1}}(T_{j-1}) = \xi_{j-1} .
\end{aligned} \tag{9}$$

Using $\bar{q}_{T_j,T_{j-1}}(t)$, $\bar{p}_{T_j,T_{j-1}}(t)$, we define the piecewise bicharacteristic path

$$\begin{aligned}
q_{\Delta_{T,0}}(t) &= q_{\Delta_{T,0}}(t, x_{J+1}, \xi_J, x_J, \xi_{J-1}, \ldots, x_1, \xi_0, x_0) , \\
p_{\Delta_{T,0}}(t) &= p_{\Delta_{T,0}}(t, x_{J+1}, \xi_J, x_J, \xi_{J-1}, \ldots, x_1, \xi_0) ,
\end{aligned} \tag{10}$$

by

$$\begin{aligned}
&q_{\Delta_{T,0}}(t) = \bar{q}_{T_j,T_{j-1}}(t, x_j, \xi_{j-1}) , \ T_{j-1} < t \le T_j , \quad q_{\Delta_{T,0}}(0) = x_0 , \\
&p_{\Delta_{T,0}}(t) = \bar{p}_{T_j,T_{j-1}}(t, x_j, \xi_{j-1}) , \ T_{j-1} \le t < T_j ,
\end{aligned} \tag{11}$$

for $j = 1,2,\ldots,J,J+1$ (see Fig. 1).

Remark 2.1. As a simple case of (10) when $T = T_1 > T_0 = 0$, $q_{T,0}(t) = q_{T,0}(t,x,\xi_0,x_0)$, $p_{T,0}(t) = p_{T,0}(t,x,\xi_0)$ of Theorem 2.4 satisfy

$$\begin{aligned}
&q_{T,0}(t) = \bar{q}_{T,0}(t,x,\xi_0) , \quad 0 < t \le T , \quad q_{T,0}(0) = x_0 , \\
&p_{T,0}(t) = \bar{p}_{T,0}(t,x,\xi_0) , \quad 0 \le t < T .
\end{aligned} \tag{12}$$

Then $\phi[q_{\Delta_{T,0}}, p_{\Delta_{T,0}}]$, $F[q_{\Delta_{T,0}}, p_{\Delta_{T,0}}]$ are functions of $x_{J+1}, \xi_J, x_J, \ldots, \xi_0, x_0$.

$$\begin{aligned}
\phi[q_{\Delta_{T,0}}, p_{\Delta_{T,0}}] &= \phi_{\Delta_{T,0}}(x_{J+1}, \xi_J, x_J, \ldots, \xi_1, x_1, \xi_0, x_0) , \\
F[q_{\Delta_{T,0}}, p_{\Delta_{T,0}}] &= F_{\Delta_{T,0}}(x_{J+1}, \xi_J, x_J, \ldots, \xi_1, x_1, \xi_0, x_0) .
\end{aligned} \tag{13}$$

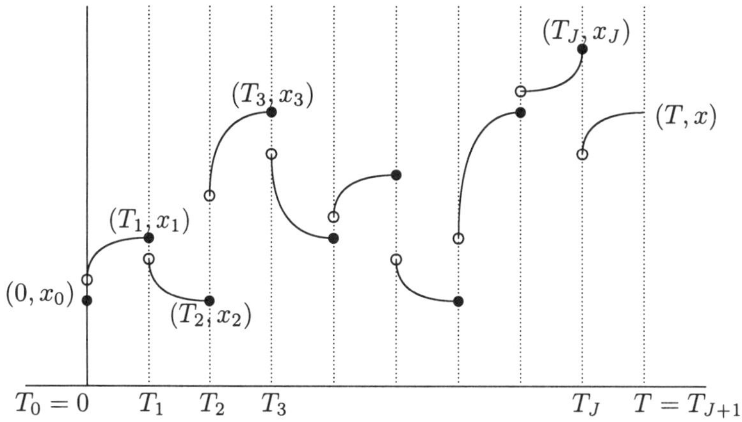

Fig. 1. The graphical explanation of $q_{\Delta_{T,0}}$.

We define the phase space Feynman path integral (5) by

$$\int e^{\frac{i}{\hbar}\phi[q,p]} F[q,p] \mathcal{D}[q,p] \tag{14}$$

$$= \lim_{|\Delta_{T,0}|\to 0} \left(\frac{1}{2\pi\hbar}\right)^{dJ} \int_{\mathbf{R}^{2dJ}} e^{\frac{i}{\hbar}\phi[q_{\Delta_{T,0}}, p_{\Delta_{T,0}}]} F[q_{\Delta_{T,0}}, p_{\Delta_{T,0}}] \prod_{j=1}^{J} d\xi_j dx_j\,,$$

if the limit of the right-hand side exists. Even when $F[q,p] \equiv 1$, the integrals of the right-hand side of (14) do not converge absolutely. We treat the multiple integral of (14) as an oscillatory integral.

Definition 2.1 (The class $\mathcal{F}$). *Let $F[q,p]$ be a functional whose domain contains all the piecewise bicharacteristic paths $q_{\Delta_{T,0}}(t)$, $p_{\Delta_{T,0}}(t)$ of (10). We say that $F[q,p] \in \mathcal{F}$ if $F[q,p]$ satisfies Assumption 2.2.*

Assumption 2.2. *Let m be a non-negative integer. Let U be a non-negative constant and let u_j, $j = 1, 2, \ldots, J, J+1$ be non-negative parameters depending on the division $\Delta_{T,0}$ such that $\sum_{j=1}^{J+1} u_j \le U < \infty$. For any non-negative integer M, there exist positive constants A_M, X_M such that for any division $\Delta_{T,0}$, any multi-indices α_j, β_{j-1} with $|\alpha_j|$, $|\beta_{j-1}| \le M$, $j = 1, 2, \ldots, J, J+1$ and any integer k with $1 \le k \le J$,*

$$|(\prod_{j=1}^{J+1} \partial_{x_j}^{\alpha_j} \partial_{\xi_{j-1}}^{\beta_{j-1}}) F_{\Delta_{T,0}}(x_{J+1}, \xi_J, \ldots, x_1, \xi_0, x_0)| \tag{15}$$

$$\le A_M (X_M)^{J+1} (\prod_{j=1}^{J+1} (t_j)^{\min(|\beta_{j-1}|,1)})(1 + \sum_{j=1}^{J+1}(|x_j| + |\xi_{j-1}|) + |x_0|)^m\,,$$

$$
\begin{aligned}
&|(\prod_{j=1}^{J+1} \partial_{x_j}^{\alpha_j}\partial_{\xi_{j-1}}^{\beta_{j-1}})\partial_{x_k} F_{\Delta_{T,0}}(x_{J+1},\xi_J,\dots,x_1,\xi_0,x_0)| \qquad (16)\\
&\le A_M(X_M)^{J+1}u_k(\prod_{j\neq k}(t_j)^{\min(|\beta_{j-1}|,1)})(1+\sum_{j=1}^{J+1}(|x_j|+|\xi_{j-1}|)+|x_0|)^m .
\end{aligned}
$$

Theorem 2.1 (Existence of phase space path integrals). *Let T be sufficiently small. Then, for any $F[q,p]\in\mathcal{F}$, the right-hand side of (14) converges on compact subsets of $(x,\xi_0,x_0)\in\mathbf{R}^d\times\mathbf{R}^d\times\mathbf{R}^d$.*

Theorem 2.2 (Algebra). *If $F[q,p]\in\mathcal{F}$ and $G[q,p]\in\mathcal{F}$, then $F[q,p]+G[q,p]\in\mathcal{F}$ and $F[q,p]G[q,p]\in\mathcal{F}$.*

Remark 2.2. Roughly speaking, $\mathcal{F}$ contains the following examples:

(1) The evaluation functionals with respect to (t,q) independent of the momentum p, $F[q]=B(t,q(t))\in\mathcal{F}$ if $|\partial_x^\alpha B(t,x)|\le C_\alpha(1+|x|)^m$. In particular, $F[q,p]\equiv 1\in\mathcal{F}$.
(2) The Riemann integrals $F[q,p] = \int_{T'}^{T''} B(t,q(t),p(t))dt \in \mathcal{F}$ if $|\partial_x^\alpha\partial_\xi^\beta B(t,x,\xi)|\le C_{\alpha,\beta}(1+|x|+|\xi|)^m$.
(3) If $|\partial_x^\alpha\partial_\xi^\beta B(t,x,\xi)|\le C_{\alpha,\beta}$, then $F[q,p]=e^{\int_{T'}^{T''} B(t,q(t),p(t))dt}\in\mathcal{F}$.

For the details of the conditions of B, see our paper.[1] Applying Theorem 2.2 to these examples, we can produce many functionals $F[q,p]\in\mathcal{F}$.

Lemma 2.2. *Let $4\kappa_2 dT<1/2$. Then, for any $(x_{J+1},\xi_0)\in\mathbf{R}^d\times\mathbf{R}^d$, there exists the solution $(\xi_J^*,x_J^*,\dots,\xi_1^*,x_1^*)$ such that*

$$
(\partial_{(\xi_J,x_J,\dots,\xi_1,x_1)}\phi_{\Delta_{T,0}})(x_{J+1},\xi_J^*,x_J^*,\dots,\xi_1^*,x_1^*,\xi_0)=0 . \qquad (17)
$$

We define $D_{\Delta_{T,0}}(x_{J+1},\xi_0)$ by

$$
\begin{aligned}
&D_{\Delta_{T,0}}(x_{J+1},\xi_0) \qquad (18)\\
&=(-1)^{dJ}\det(\partial^2_{(\xi_J,x_J,\dots,\xi_1,x_1)}\phi_{\Delta_{T,0}})(x_{J+1},\xi_J^*,x_J^*,\dots,\xi_1^*,x_1^*,\xi_0) .
\end{aligned}
$$

Theorem 2.3. *For any multi-indices α, β, there exists a positive constant $C_{\alpha,\beta}$ independent of $\Delta_{T,0}$ such that*

$$
|\partial_x^\alpha\partial_{\xi_0}^\beta(D_{\Delta_{T,0}}(x,\xi_0)-D(T,x,\xi_0))|\le C_{\alpha,\beta}|\Delta_{T,0}|T , \qquad (19)
$$

with a function $D(T,x,\xi_0)$. We use this limit function $D(T,x,\xi_0)$ as a Hamiltonian version of the Morette Van Vleck determinant.[9]

Theorem 2.4 (Semiclassical approximation as $\hbar \to 0$). *Let T be sufficiently small. Then, for any $F[q,p] \in \mathcal{F}$, we can write*

$$\int e^{\frac{i}{\hbar}\phi[q,p]} F[q,p]\mathcal{D}[q,p] \tag{20}$$
$$= e^{\frac{i}{\hbar}\phi[q_{T,0},p_{T,0}]}\left(D(T,x,\xi_0)^{-1/2}F[q_{T,0},p_{T,0}] + \hbar\Upsilon(T,\hbar,x,\xi_0,x_0)\right).$$

Furthermore, for any non-negative integer M, there exist a positive constant C_M and a positive integer M' independent of $0 < \hbar < 1$ such that

$$|\partial_x^\alpha \partial_{\xi_0}^\beta \Upsilon(T,\hbar,x,\xi_0,x_0)| \le C_M A_{M'} T(U+T)(1+|x|+|\xi_0|+|x_0|)^m, \tag{21}$$

for any multi-indices α, β with $|\alpha|$, $|\beta| \le M$.

Remark 2.3. As a simple case of (13) when $T = T_1 > T_0 = 0$, we have

$$\phi[q_{T,0},p_{T,0}] = \phi_{T,0}(x,\xi_0,x_0), \quad F[q_{T,0},p_{T,0}] = F_{T,0}(x,\xi_0,x_0). \tag{22}$$

Then $\phi_{T,0}(x,\xi_0,x_0) = (x-x_0)\cdot\xi_0 + \omega_{T,0}(x,\xi_0)$ and

$$|\partial_x^\alpha \partial_{\xi_0}^\beta \omega_{T,0}(x,\xi_0)| \le C_{\alpha,\beta} T(1+|x|+|\xi_0|)^{\max(2-|\alpha+\beta|,0)}. \tag{23}$$

By (15), $F_{T,0}(x,\xi_0,x_0)$ satisfies

$$|\partial_x^\alpha \partial_{\xi_0}^\beta F_{T,0}(x,\xi_0,x_0)| \le A_M X_M(t_1)^{\min(|\beta_0|,1)}(1+|x|+|\xi_0|+|x_0|)^m. \tag{24}$$

References

1. N. Kumano-go and D. Fujiwara, *Phase space Feynman path integrals via piecewise bicharacteristic paths and their semiclassical approximations*, *Bull. Sci. Math.* (in print).
2. H. Kitada and H. Kumano-go, *A family of Fourier integral operators and the fundamental solution for a Schrödinger equation*, *Osaka J. Math.* **18**, 291–360 (1981).
3. R. P. Feynman, *An operator calculus having applications in quantum electrodynamics*, *Phys. Rev.* **84**, 108–236 (1951), App. B.
4. R. P. Feynman, *Space-time approach to non-relativistic quantum mechanics*, *Rev. Mod. Phys.* **20**, 367–387 (1948).
5. L. S. Schulman, *Techniques and Applications of Path Integration* (Wiley-Interscience, New York, 1981).
6. I. Daubechies and J. R. Klauder, *Quantum mechanical path integrals with Wiener measure for all polynomial Hamiltonians*, *J. Math. Phys.* **26**, 2239–2256 (1985).
7. S. Albeverio, G. Guatteri, and S. Mazzucchi, *Phase space Feynman path integrals*, *J. Math. Phys.* **43**, 2847–2857 (2002).
8. O. G. Smolyanov, A. G. Tokarev, and A. Truman, *Hamiltonian Feynman path integrals via Chernoff formula*, *J. Math. Phys.* **43**, 5161–5171 (2002).
9. C. Morette, *On the definition and approximation of Feynman path integral*, *Phys. Rev.* **81**, 848–852 (1951).

GELFAND-YAGLOM TYPE EQUATION FOR WIENER UNCONDITIONAL MEASURE FUNCTIONAL INTEGRAL WITH x^4 TERM IN POTENTIAL

J. BOHÁČIK

Institute of Physics, Slovak Academy of Sciences, Dúbravská cesta 9, 811 45 Bratislava, Slovakia
Email: fyzibohj@savba.sk

P. PREŠNAJDER

Department of Theoretical Physics and Physics Education, Faculty of Mathematics, Physics and Informatics, Comenius University, Mlynská dolina F2, 842 48 Bratislava, Slovakia

We propose a non-perturbative method for the evaluation of the functional integral with fourth order term in the action. We found the result in the form of an asymptotic series.

Keywords: Non-perturbative methods.

1. Evaluation of x^4 Wiener Functional Integral

The simplest non-gaussian functional integral is the Wiener functional integral with x^4 term in the action. In the Euclidean sector of the theory we have to evaluate the continuum Wiener functional integral:

$$\mathcal{Z} = \int [\mathcal{D}\varphi(x)] \exp(-\mathcal{S}) \,. \tag{1}$$

In this case the action possesses the fourth order term:

$$\mathcal{S} = \int_0^\beta d\tau \left[c/2 \left(\frac{\partial \varphi(\tau)}{\partial \tau} \right)^2 + b\varphi(\tau)^2 + a\varphi(\tau)^4 \right] \,.$$

The continuum Wiener functional integral is defined by a formal limit $\mathcal{Z} = \lim_{N\to\infty} \mathcal{Z}_N$. The finite dimensional integral $\mathcal{Z}_N$ is defined by the time-slicing method:

$$\mathcal{Z}_N = \int\limits_{-\infty}^{+\infty} \prod_{i=1}^{N} \left(\frac{d\varphi_i}{\sqrt{\frac{2\pi\triangle}{c}}} \right) \exp\left\{ -\sum_{i=1}^{N} \triangle \left[\frac{c}{2} \left(\frac{\varphi_i - \varphi_{i-1}}{\triangle} \right)^2 + b\varphi_i^2 + a\varphi_i^4 \right] \right\} \quad (2)$$

where $\triangle = \beta/N$, and a, b, c are the parameters of the model. The quantity Z_N represents the unconditional propagation from $\varphi_0 = 0$ to any $\varphi = \varphi_N$ (Eq. (2) contains an integration over φ_N).

1.1. *One-dimensional integral*

An important task is to calculate the one dimensional integral

$$I_1 = \int\limits_{-\infty}^{+\infty} dx \, \exp\{-(Ax^4 + Bx^2 + Cx)\} \, ,$$

where $Re\, A > 0$. The standard perturbative procedure rely on Taylor's decomposition of $\exp(-Ax^4)$ term with consecutive replacements of the integration and summation order. The integrals can be calculated, but the sum is divergent.

Instead we propose the power expansion in C:

$$I_1 = \sum_{n=0}^{\infty} \frac{(-C)^n}{n!} \int\limits_{-\infty}^{+\infty} dx \, x^n \exp\{-(Ax^4 + Bx^2)\} \, .$$

The integrals in the above relation can be expressed by the parabolic cylinder functions $D_{-\nu-1/2}(z)$. Then, the integral I_1 reads:

$$I_1 = \frac{\Gamma(1/2)}{\sqrt{B}} \sum_{m=0}^{\infty} \frac{\xi^m}{m!} \mathcal{D}_{-m-1/2}(z) \, , \; \xi = \frac{C^2}{4B} \, , \; z = \frac{B}{\sqrt{2A}} \, , \quad (3)$$

where we used the abbreviation:

$$\mathcal{D}_{-m-1/2}(z) = z^{m+1/2} \, e^{\frac{z^2}{4}} \, D_{-m-1/2}(z) \, .$$

It was shown, that the sum in the Eq. (3) is convergent and for finite values z this sum converges uniformly in ξ.

Applying this idea of integration to the N dimensional integral (2) we proved[1] the exact formula:

$$\mathcal{Z}_N = \left[\prod_{i=0}^{N} 2(1 + b\triangle^2/c)\omega_i \right]^{-\frac{1}{2}} \mathcal{S}_N$$

with

$$\mathcal{S}_N = \sum_{k_1,\cdots,k_{N-1}=0}^{\infty} \prod_{i=0}^{N} \left[\frac{(\rho)^{2k_i}}{(2k_i)!} \Gamma(k_{i-1}+k_i+1/2) \sqrt{\omega_i}\, \mathcal{D}_{-k_{i-1}-k_i-1/2}(z) \right],$$

where $k_0 = k_N = 0$, $\rho = (1+b\triangle^2/c)^{-1}$, $z = c(1+b\triangle^2/c)/\sqrt{2a\triangle^3}$, $\omega_i = 1 - A^2/\omega_{i-1}$, $\omega_0 = 1/2 + Ab\triangle^2/c$, $A = 1/2(1+b\triangle^2/c)$.

To evaluate S_N, we must solve the problem how to sum up the product of two parabolic cylinder functions. The parabolic cylinder functions are related to the representation of the group of the upper triangular matrices, so we implicitly expect a simplification of their product. This problem is not solved completely yet. We adopt less complex method of summation, namely we use the asymptotic expansion of one of them, then, exchanging the order of summations we can sum over k_i. The result is degraded to an asymptotic expansion, but we have an analytical solution of the problem. This procedure was widely discussed in detail in Ref. 1, here we remind the result:

$$S_N = \sum_{\mu=0}^{\mathcal{J}} \frac{(-1)^\mu}{\mu!\,(2z^2\triangle^3)^\mu} \triangle^{3\mu} \left\{C^{2\mu}(N)\right\}_{2\mu,0} \tag{4}$$

The evaluation of the symbols $\left\{C^{2\mu}(N)\right\}_{2\mu,0}$ is described in Ref. 2.

1.2. *Gel'fand–Yaglom equation*

Gel'fand and Yaglom in their article[3] for the harmonic oscillator proved that nontrivial continuum limit of the finite dimensional integral approximation to the functional integral should be evaluated from N-dimensional integral results by a recurrent procedure. For harmonic oscillator Gel'fand and Yaglom derived for the unconditional measure integral $Z(\beta)$ the equation:

$$\frac{\partial^2}{\partial \tau^2} y(\tau) = \frac{2b}{c}\, y(\tau), \tag{5}$$

where $Z(\beta) = \lim_{N\to\infty} \mathcal{Z}_N = y(\beta)^{-1/2}$

Following the idea of Gel'fand and Yaglom we found for an-harmonic oscillator the generalized Gel'fand–Yaglom equation. We define the unconditional measure functional integral $Z(\beta)$ by relation:

$$Z(\beta) = \lim_{N\to\infty} \mathcal{Z}_N = \frac{1}{\sqrt{F(\beta)}}\,, \quad F(\beta) = \frac{y(\beta)}{S(\beta)^2}.$$

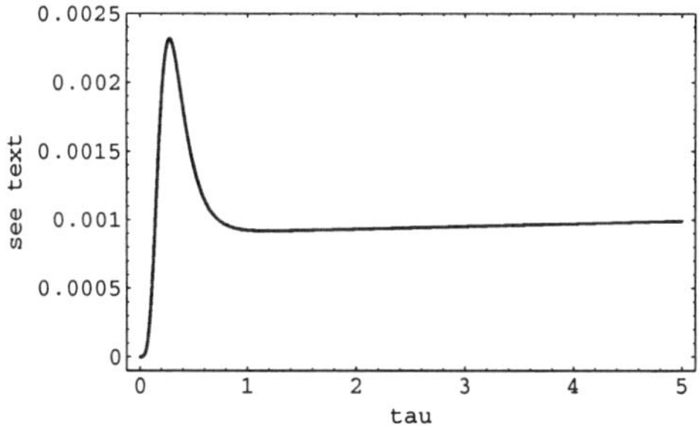

Fig. 1. A typical dependence of the functions $4\left(\frac{\partial}{\partial\tau}\ln S(\tau)\right)^2$ on τ for positive fixed values of parameters a, b, c is shown on the figure.

The function $S(\tau)$ is given as the continuum limit of Eq. (4). The generalized Gel'fand–Yaglom equation read:

$$\frac{\partial^2}{\partial\tau^2}y(\tau) = y(\tau)\left[\frac{2b}{c} + 4\left(\frac{\partial}{\partial\tau}\ln S(\tau)\right)^2\right] . \tag{6}$$

Initial conditions: $y(0) = S(0)^2$, and $\left.\frac{\partial y(\tau)}{\partial\tau}\right|_{\tau=0} = \left.\frac{\partial}{\partial\tau}S(\tau)^2\right|_{\tau=0}$.

For $S(\tau)$ one can use a perturbative expansion in coupling constant a and then solve Eq. (6). This procedure gives a non-perturbative approximation of the functional integral (1), see Fig. (1).

2. Conclusions

We find for the functional integral of an an-harmonic oscillator the non-perturbative equation. Solving Eq. (6) we in principle can find the analytical solution of the an-harmonic oscillator problems as energy levels, measurable quantities, etc. Our opinion is that the theory of the differential equations is more elaborated and flexible than approaches based on naive perturbative theory and it can give more reliable results than the perturbative theory.

Acknowledgments

This work was supported by VEGA projects No. 2/6074/26.

References

1. J. Boháčik and P. Prešnajder, eprint: arXiv:hep-th/0503235.
2. J. Boháčik and P. Prešnajder, to appear.
3. I. M. Gelfand and A. M. Yaglom, *J. Math. Phys.* **1**, 48 (1960).

COHERENT STATES FOR A QUANTUM PARTICLE ON A MÖBIUS STRIP

D. J. CIRILO-LOMBARDO

Bogoliubov Laboratory of Theoretical Physics Joint Institute for Nuclear Research, 141980, Dubna, Russian Federation
E-mail: diego@theor.jinr.ru

The coherent states for a quantum particle on a Möbius strip are constructed and their relation with the natural phase space for fermionic fields is shown. An explicit comparison of the obtained states with previous works, where the cylinder quantization was used and the spin 1/2 was introduced by hand, is worked out.

Keywords: Coherent states; Möbius strip.

1. Introduction

Coherent States (CS) have attracted much attention in many branches of physics.[1] In spite of their importance, the theory of CS, when the configuration space has nontrivial topology, is far from complete. CS for a quantum particle on a circle[2] and a sphere have been introduced very recently, and also the case of the torus has been treated. Although in all these works the different constructions of the CS for the boson case are practically straightforward, the simple addition by hand of 1/2 to the angular momentum operator J for the fermionic case into the corresponding CS remains obscure and non-natural. The question that naturally arises is: does here exist any geometry for the phase space in which the CS construction leads precisely to a fermionic quantization condition? The purpose of this paper is to demonstrate the positive answer to this question showing that the CS for a quantum particle on the Möbius strip geometry is the natural candidate to describe fermions exactly as the cylinder geometry for bosons.

2. Abstract Coherent States

The position of a point into the Möbius strip geometry can be parameterized as $P_0 = (X_0, Y_0, Z_0)$ and $P_0 = (X_0 + X_1, Y_0 + Y_1, Z_0 + Z_1)$. The coordinates of P_0 describe the central cylinder (generated by the invariant fiber of the middle of the weight of the strip): $Z_0 = l$, $X_0 = R\cos\varphi$, $Y_0 = R\sin\varphi$. We use the standard spherical coordinates: R, θ, φ with $d\Omega = R^2(d\theta^2 + \sin^2\theta d\varphi^2)$ and r is the secondary radius of the torus.

The coordinates of P_1 (the boundaries of the Möbius band) are of P_0 plus $Z_1 = r\cos\theta$, $X_1 = r\sin\theta\cos\varphi$, $Y_1 = r\sin\theta\sin\varphi$. The weight of the band is obviously $2r$, then our space of phase is embedded into of the Torus: $X = R\cos\varphi + r\sin\theta\cos\varphi$, $Y = R\sin\varphi + r\sin\theta\sin\varphi$, $Z = l + r\cos\theta$. The important point is that the angles are not independent in the case of the Möbius band and are related by the following constraint: $\theta = (\varphi + \pi)/2$.

In order to introduce the coherent states for a quantum particle on the Möbius strip geometry we follow the Barut-Girardello construction and we seek the CS as the solution of the eigenvalue equation $X|\xi\rangle = \xi|\xi\rangle$ with complex ξ. Taking $R = 1$ and inserting the constraint into the parameterization of the torus we obtain the parameterization of the band: $X = \cos\varphi + r\cos(\varphi/2)\cos\varphi$, $Y = \sin\varphi + r\cos(\varphi/2)\sin\varphi$, $Z = l + r\sin(\varphi/2)$.

Taking into account the initial condition and the transformation $X' = e^{-Z}X$, $Y' = e^{-Z}Y$, $Z' = Z$, we finally obtain

$$\xi = e^{-(l+r\sin(\varphi/2))+i\varphi}\left(1 + r\cos(\varphi/2)\right).$$

Inserting the above expression in the expansion of the coherent state in the j basis we obtain the CS in explicit form

$$|\xi\rangle = \sum_{j=-\infty}^{\infty} \xi^{-j} e^{-\frac{j^2}{2}} |j\rangle = \sum_{j=-\infty}^{\infty} e^{l'j - i\varphi j} e^{-\frac{j^2}{2}} |j\rangle,$$

where $l' \equiv (l + r\sin(\varphi/2)) - \ln(1 + r\cos(\varphi/2)) - i\varphi$. From the above expression, the fiducial vector is $|1\rangle = \sum_{j=-\infty}^{\infty} e^{-\frac{j^2}{2}}|j\rangle$, then

$$|\xi\rangle = \sum_{j=-\infty}^{\infty} e^{-(\ln\xi)\widehat{J}}|1\rangle \tag{1}$$

As is easily seen the fiducial vector is $|1\rangle = |0,0\rangle_{r=0}$ in the (l,φ) parameterization. This fact permit us to rewrite expression (1) as

$$|l,\varphi\rangle = e^{[(l+r\sin(\varphi/2))-\ln(1+r\cos(\varphi/2))-i\varphi]j}|0,0\rangle_{r=0}.$$

The overlapping or non-orthogonality formulas are explicitly derived*:

*The normalization as a function of Θ_3: $\langle\xi|\xi\rangle = \Theta_3\left(\frac{i}{\pi}\ln|\xi| \mid \frac{i}{\pi}\right)$ or $\langle l,\varphi|l,\varphi\rangle = \Theta_3\left(\frac{il'}{\pi} \mid \frac{i}{\pi}\right)$.

$$\langle \xi \,|\eta \rangle = \sum_{j=-\infty}^{\infty} \left(\xi^{*}\eta\right)^{-j} e^{-j^{2}} = \Theta_{3}\left(\frac{i}{2\pi}\ln\left(\xi^{*}\eta\right) \mid \frac{i}{\pi}\right)$$

$$\langle l,\varphi \,|h,\psi \rangle = \Theta_{3}\left(\frac{i}{2\pi}\left(\varphi-\psi\right) - \frac{l'+h'}{2}\frac{i}{\pi} \mid \frac{i}{\pi}\right).$$

3. Physical Phase Space and Natural Quantization

From the expressions obtained in the previous section and $\widehat{J}\,|j\rangle = j\,|j\rangle$ we notice that the normalization for the cylinder[2] (boson case) which doesn't depend on φ, depends now on φ through $l' \equiv (l + r\sin(\varphi/2)) - \ln(1 + r\cos(\varphi/2))$. Also $\widehat{J}\,|l,\varphi\rangle = j\,|l,\varphi\rangle$, then

$$\frac{\langle \xi |\, \widehat{J}\, |\xi\rangle}{\langle \xi\, |\xi\rangle} = l' + \sum_{n=1}^{\infty} \frac{2\pi \sin\left(2l'\pi\right) e^{-\pi^{2}(2n-1)}}{\left(1 + e^{-\pi^{2}(2n-1)} e^{2i\pi l'}\right)\left(1 + e^{-\pi^{2}(2n-1)} e^{-2i\pi l'}\right)},$$

where the well known identities for theta functions was introduced. Notice the important result coming from the above expression: the fourth condition required for the CS,[3] namely $\left\langle \widehat{J} \right\rangle = l$ for the CS, demands not only l to be integer or semi-integer (as the case for the circle quantization) but also that $\varphi = (2k+1)\,\pi$ which leads to a natural quantization similar to the charge quantization of the Dirac monopole. Precisely this condition over the angle leads the position of the particle in the internal or the external border of the Möbius band, that for $r = 1/2$ is $s = \pm 1/2$ as it is requested to be.

In order to compare our case with the CS constructed in Ref. 2 we consider the existence of the unitary operator $U \equiv e^{i\varphi}$, that $[J, U] = U$ then $U\,|j\rangle = |j+1\rangle$ such that the same average as in the previous case for the $\widehat{J}$ operator is:

$$\frac{\langle \xi |\, U\, |\xi\rangle}{\langle \xi\, |\xi\rangle} = e^{-\frac{1}{4}} e^{i\varphi} \frac{\Theta_{2}\left(\frac{il'}{\pi} \mid \frac{i}{\pi}\right)}{\Theta_{3}\left(\frac{il'}{\pi} \mid \frac{i}{\pi}\right)} = e^{-\frac{1}{4}} e^{i\varphi} \frac{\Theta_{3}\left(l' + 1/2 \mid i\pi\right)}{\Theta_{3}\left(l' \mid i\pi\right)}, \qquad (2)$$

where in the last equality the relation $\Theta_{2}(\nu) = e^{i\pi\left(\frac{1}{4}\tau+\nu\right)}\Theta_{3}(\nu + \tau/2)$ was introduced. Also, as in Ref. 2, we can make the relative average for the operator U in order to eliminate the factor $e^{-\frac{1}{4}}$ then at the first order expression (2) coincides with the unitary circle. It is clear that the denominator in the quotient (2), average with respect to the fiducial CS state, plays the role to centralize the expression of the numerator. However, the claim that U is the best candidate for the position operator is still obscure and requires a special analysis that we will be given elsewhere.[4]

4. Dynamics

In order to study the dynamics in this non-trivial geometry, we construct the non-relativistic Lagrangian and the corresponding Hamiltonian:

$$H = \frac{1}{2}\left\{\dot{\varphi}^2\left[(1 + r\cos(\varphi/2)\,)^2 - \frac{r^2}{4}\cos\varphi\right] + L_0^2\right\}$$

With $\widehat{H}\,|E\rangle = E\,|E\rangle$ if $|E\rangle = |j\rangle$, imposing the fourth CS requirement[3] we have $\varphi = (2k+1)\pi$ and the expression for the energy takes the form: $E = 2j^2/(4+r^2) + L_0^2/2$.

From the dynamical expressions given above, it is not difficult to make the following remarks:

1) the Hamiltonian is not a priori T invariant. The H_{MS} is T invariant iff $TL_0 = -L_0$: the variable conjugate to the external momenta l changes under T as J manifesting with this symmetry the full inversion of the motion of the particle on a Möbius strip (evidently this is not the case for the particle motion on the circle).

2) the distribution of energies is Gaussian: from the Bargmann representation follows $\phi_j\left(\xi^*\right) \equiv \langle\xi|\,E\rangle = \left(\xi^*\right)^{-j} e^{-\frac{j^2}{2}}$, and by using the approximate relation from the definition of the Theta function†, the expression for the distribution of energies can be written as

$$\frac{\left|\langle j\,|\xi\rangle\right|^2}{\langle\xi\,|\xi\rangle} \approx \frac{1}{\sqrt{\pi}}e^{-(j-l')^2}.$$

It is useful to remark here that when $\varphi = (2k+1)\pi$, $l = l'$ and the above equation coincides exactly in form with the boson case given in Ref. 2 but l now is semi-integer valuated.

References

1. J. R. Klauder and B. S. Skagerstam, *Coherent States: Applications in Physics and Mathemetical Physics* (World Scientific, Singapore, 1985).
2. K. Kowalski, J. Rembielinski, and L. C. Papaloucas, *J. Phys. A* **29**, 4149 (1996).
3. J. P. Gazeau and J. R. Klauder, *J. Phys. A* **32**, 123 (1999).
4. D. J. Cirilo-Lombardo, in preparation.

†$\Theta_3\left(\frac{il'}{\pi} \mid \frac{i}{\pi}\right) = e^{(l')^2}\sqrt{\pi}\left(1 + 2\sum_{n=1}^{\infty} e^{-\pi^2 n^2}\cos(2l'\pi n)\right) \approx e^{(l')^2}\sqrt{\pi}.$

PART III

Quantum Field Theory

CHALLENGES TO PATH-INTEGRAL FORMULATIONS OF QUANTUM THEORIES

R. JACKIW

Center for Theoretical Physics,
MIT, Cambridge, MA 02139-4307, USA
E-mail: ab_jackiw@mit.edu

The functional integral has many triumphs in elucidating quantum theory. But incorporating charge fractionalization into that formalism remains a challenge.

Keywords: Topological defects; Polyacetylene; Graphene.

This conference celebrates the achievements of path and functional integration in quantum physics. But it is good to remember that some of these achievements were hard to attain, because they required resolving unexpected subtleties of the functional formalism. Two historical examples will illustrate my point, and then I shall posit a new challenge.

For the first example, we look at the description of rotationally symmetric motion in a potential $V(r)$. The classical effective Hamiltonian reads $H = \frac{p_r^2}{2m} + \frac{L^2}{2mr^2} + V(r)$, where the centrifugal term represents $r^2\,\dot{\theta}^2/2m$ and vanishes for radially symmetric motion. But in the quantum description of this motion, when it is confined to a plane, the classical centrifugal term is replaced by $\hbar^2(M^2 - 1/4)/2mr^2$, where M is any integer. In particular a residual attraction remains for s-waves, $M = 0$. This quantal attraction has the important physical consequence that in planar physics bound states exist, no matter how weakly attractive might be the potential V. (In this way, planar bound states follow the behavior of one dimensional bound states, rather than three dimensional, which require a minimum strength to achieve binding.)

The functional integral involves integration over c-number functions. A direct change of variables from Cartesian to circular coordinates reproduces the classical centrifugal barrier, but misses the residual, attractive potential $-\hbar^2/8mr^2$. This is particularly vexing, since coordinate changes are point canonical transformations, which are allowed in quantum mechanics.

The resolution of this problem was given by Edwards and Gulyaev.[1] But it required returning to the discretized formulation of the path integral and realizing that the angular step $\Delta\theta$, is not of order of the temporal step $\Delta\tau$, but rather $0(\sqrt{\Delta\tau})$. [This issue reappeared in the collective coordinate quantization of solitons. Investigations based on functional integrals missed $0(\hbar^2)$ terms that were found in an operator approach.[2]]

For my second example, I turn to the anomaly phenomenon: the classical action can possess an invariance against a transformation that is not an invariance after quantization. Where is this effect in the functional integral, which involves "functional integration" over the exponentiated classical action? The answer, which was found by Fujikawa (after anomalies were discovered by conventional methods), located the effect in the functional measure. Evidently in the anomalous situation it is not invariant; a fact that is established after the measure is discretized.[3]

In both instances we see that the validity of formal changes of variables in a functional integral must be assessed by a return to the discretized functional sum, in fact, by a return to conventional quantum theory.

Now I shall present another peculiar quantum effect, which as far as I know has not had a functional integral description, even though a conventional quantal argument establishes it rather easily. I have in mind the phenomenon of fermion charge fractionalization in the presence of a topological defect.[4]

It is well known and also guaranteed by various index theorems, that the Dirac equation in the presence of a topological defect (kink in one spatial dimension, vortex in two, monopole in three etc.; we call these "solitons") possesses an isolated, normalized zero-energy, mid-gap bound state. Of course the Dirac equation also possesses positive energy solutions and negative energy solutions. The conventional instruction, given by Dirac, is to define the vacuum by filling the negative energy states and leaving the positive states empty; with this definition the vacuum charge vanishes. But what should one do with the mid-gap, zero-energy bound states, if it is present? Dirac is silent on this; he did not know about mid-gap bound states. The answer is that there is double degeneracy in energy, since filling the mid-gap state costs no energy. Moreover, the empty mid-gap state carries fermion charge $-1/2$, and the filled one, $+1/2$. Remarkably, this effect has been observed in polyacetylene — a 1-dimensional lineal material — and it has been proposed for 2-dimensional planar graphene. (Note: the Dirac equation is relevant to these condensed matter materials not because of relativistic considerations; rather a well defined linearization of the energy dispersion near the Fermi surface gives rise to a linear matrix equation, which is of the Dirac type.)

Fermion number fractionalization is established by the following argument. Consider the Dirac Hamiltonian $H(\varphi)$ in a background field φ, which can be topologically trivial or non-trivial. In the trivial case φ takes a homogenous value φ_v, which gives rise to a mass gap in the Dirac spectrum, between the positive and negative continuum energy eigenstates, which we call "vacuum states" and denote them by ψ_E,

$$H(\varphi_v)\,\psi_E = E\,\psi_E, \qquad E \gtrless 0. \tag{1}$$

With the topologically non trivial background φ takes a soliton profile φ_S and the Dirac continuum eigenfunctions are called "soliton states," denoted by Ψ_E. Also there is an isolated normalized state at zero energy, Ψ_0,

$$\begin{aligned} H(\varphi_S)\Psi_E &= E\Psi_E, \quad E \gtrless 0, \\ H(\varphi_S)\Psi_0 &= \quad 0. \end{aligned} \tag{2}$$

We wish to compute the charge density in the presence of the soliton. This is defined relative to the charge density in the vacuum, where the background field is topologically trivial,

$$\rho(\mathbf{r}) = \int\limits_{-\infty}^{0} dE \Big(\Psi_E^* \left(\mathbf{r}\right) \Psi \left(r\right) - \psi_E^* \left(\mathbf{r}\right) \psi_E \left(r\right) \Big). \tag{3}$$

The further argument proceeds in its simplest form if the Dirac Hamiltonian possesses an energy reflection symmetry; *viz.* if there exists a unitary matrix R, which anti-commutes with the Dirac Hamiltonian $HR + RH = 0$. Then R acting on negative energy states produces positive energy states and vice-versa, while the zero energy state is an eigenstate of R,

$$\begin{aligned} R\,\Psi_{\pm E} = \Psi_{\mp E},\, R\,\psi_{\pm E} &= \psi_{\mp E}, \\ R\,\Psi_0 &= \pm\Psi_0. \end{aligned} \tag{4}$$

(The effective Dirac Hamiltonians for polyacetelyne and graphene possess this property.) As a consequence the charge density at energy E is an even function of E, both in the vacuum and soliton sectors: $\psi_E^*\,\psi_E = \psi_{-E}^*\,\psi_{-E}$, $\Psi_E^*\,\Psi_E = \Psi_{-E}^*\,\Psi_{-E}, \Rightarrow \rho_E = \rho_{-E}$; and (3) may be presented as

$$\rho = \frac{1}{2} \int\limits_{-\infty}^{\infty} dE \;\left(\Psi_E^*\,\Psi_E - \psi_E^*\,\psi_E\right). \tag{5}$$

The vacuum continuum wave functions ψ_E in (5) are complete; the solitonic vave functions Ψ_E are one short of being complete because the zero energy state Ψ_0 is absent. Therefore

$$\rho(\mathbf{r}) = -\frac{1}{2}\,\Psi_0^*(\mathbf{r})\,\Psi_0(\mathbf{r}) \tag{6}$$

and the charge Q is

$$Q = \int d\mathbf{r}\,\rho(\mathbf{r}) = -\frac{1}{2}\,. \tag{7}$$

Although this simple argument does not show it, one can prove that the fraction is an eigenvalue without fluctuations, rather than an expectation value with fluctuations — the latter would not be interesting.

One may easily contemplate Hamiltonians that do not possess energy reflection. For example, take a Hamiltonian with that property, and append to it a term proportional to the previously described matrix R. The new Hamiltonian no longer anti-commutes with R, and calculation of the fractional charge becomes much more involved.

Two methods have been developed for dealing with this more difficult situation. One can take a field theoretic approach and calculate in perturbation theory the induced vacuum current in the presence of a soliton $<\bar{\psi}\,\gamma^\mu\,\psi>$. Here $\bar{\psi}\,\gamma^\mu\,\psi$ is the field theoretic current operator, and $<|>$ signifies the field theoretic vacuum in an external background. This determines the induced charge density and therefore the charge. [The induced current approach is available only for problems without energy reflection symmetry — the charge in the symmetric case can be obtained in the limit of vanishing asymmetric effects.[5]] Alternatively one can show that the charge density is related to the spectral asymmetry in the soliton sector,[6]

$$\rho(\mathbf{r}) = -\frac{1}{2}\int_{-\infty}^{\infty} dE\ (\operatorname{sign} E)\ \Psi_E^*(\mathbf{r})\,\Psi_E(\mathbf{r}) - \frac{1}{2}\,\Psi_0^*(\mathbf{r})\,\Psi_0(\mathbf{r}). \tag{8}$$

The calculated vacuum charge in the absence of energy reflection can become an irrational quantity, approaching $-\frac{1}{2}$ when the strength ϵ of the term violating energy reflection vanishes. (But later I shall describe a more involved scenario.) Note that the spectral asymmetry expression (8) immediately yields $Q = -1/2$ when energy reflection symmetry is present, because in that case the integral vanishes.

All these are fascinating quantum effects, but it remains a challenge to find them in the functional integral.

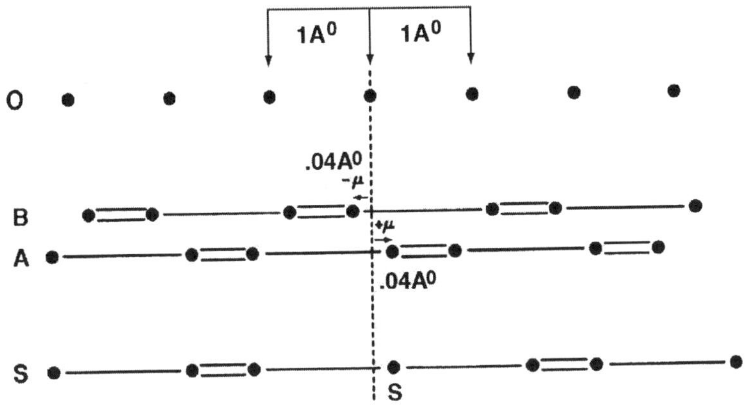

Fig. 1. The equally spaced configuration of carbon atoms in polyacetylene (O) possesses a left-right symmetry, which however is energetically unstable. Rather in the ground states the carbon atoms shift a distance μ to the left or right, breaking the symmetry and producing two degenerate vacua (A, B). A soliton (S) is a defect in the alteration pattern; it provides a domain wall between configurations (A) and (B).

Next I shall turn away from elaborate formalism, and give a simple counting argument that establishes the fractionalization effect in one dimension, in the linear polymer of polyacetylene.

To begin a description of polyacetylene, we imagine a rigid array of carbon atoms, about 1Å apart, exhibiting a left-right symmetry. One might expect that thermal and quantal fluctuations lead to oscillations about the equidistant equilibrium positions. But in fact something more dramatic happens. The energetics of the system force the atoms to shift by about 0.04 Å to the left or to the right — both are allowed due to the left-right symmetry. This is a consequence of an instability of the rigid lattice, identified by Peierls. Therefore the polyacetylene chain presents two equivalent vacua A and B (see Fig. 1). The potential energy of the distortion field (phonon) shows a familiar double well shape: the left-right symmetric point at the origin is unstable; stable configurations at the two minima break the left-right symmetry (see Fig. 2).

By now it is well known that a double well potential, like in Fig. 2, supports also a kink configuration that interpolates between the two vacua. Physically this represents a defect in the bonding pattern. These profiles are shown in Fig. 3.

Let us now consider a sample of the A configuration, with two solitons — two defects — inserted. Let us count the kinks in the A configuration

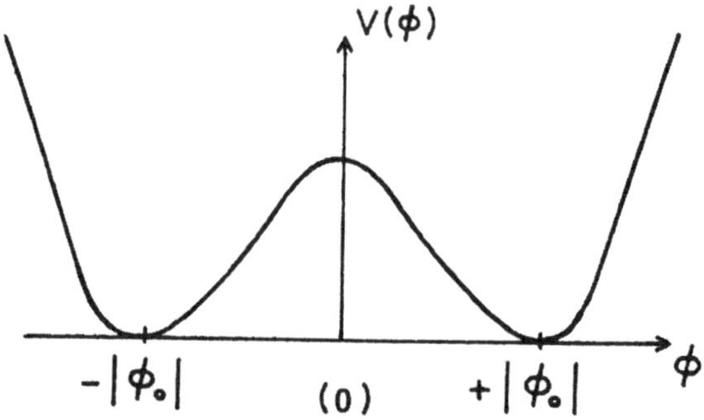

Fig. 2. Energy density $V(\phi)$, as a function of a constant phonon field ϕ. The symmetric stationary point, $\phi = 0$, is unstable. Stable vacua are at $\phi = +|\phi_0|$, (A) and $\phi = -|\phi_0|$, (B).

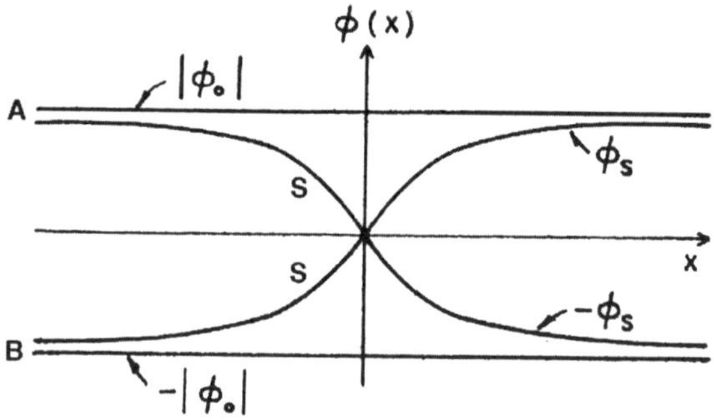

Fig. 3. The two constant fields, $\pm \mid \phi_0 \mid$, correspond to the two vacua (A and B). The two kink fields, $\pm\phi_s$, interpolate between the vacua and represent domain walls.

without and with two solitons. These numbers need be considered only in the region between the solitons, because elsewhere the patterns are identical (see Fig. 4).

One observes that in presence of the two solitons, there is one fewer kink. Imagine now separating the solitons to great distance, so that each acts independently. We conclude that the 1-kink deficit must be equally divided, producing a fermion state with number $-\frac{1}{2}$.

In fact this has been indirectly observed. However, the theory has to be elaborated before it confronts experiment. Our argument has ignored spin. Since electrons have spin $\pm\frac{1}{2}$ and two fit in each level, all our results are doubled. The charge defect in the presence of a soliton is -1, but there is no net spin since all other electrons are pairwise aligned. Inserting one elec-

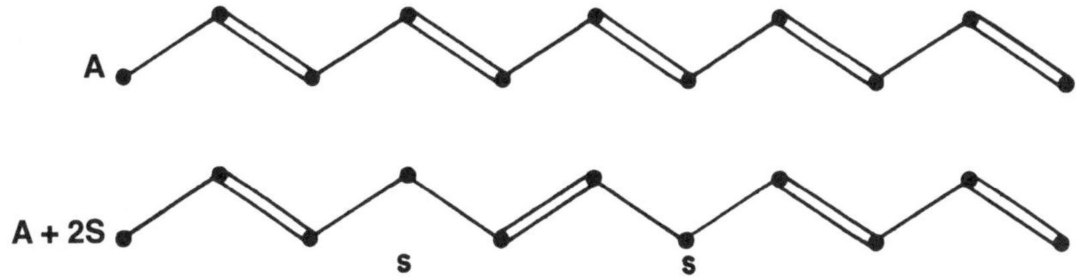

Fig. 4. Two soliton state carries one fewer link relative to the no-soliton vacuum A.

tron erases the defect, but produces a spin $\frac{1}{2}$ excitation with zero charge. This charge-spin separation, which has been observed experimentally, gives a physical realization to the phenomenon of charge fractionalization in polyacetylene.[7]

The Dirac Hamiltonian matrix, relevant to this 1-dimensional problem, is a 2×2 matrix acting on a 2-component "spinor", thereby producing a matrix equation in one spatial dimension: $H(\varphi) = \alpha p + \beta\varphi$. The kinetic term $\alpha p\,(\alpha \equiv \sigma^3, p \equiv -i\,\partial_x)$ comes from linearizing a quadratic dispersion law around the two points, called "Dirac points," where it intersects the Fermi surface, thereby giving rise to a 2-component structure. The interaction with the phonon field $\beta\varphi$ $(\beta \equiv \sigma^2)$ produces a gap with homogenous φ and a zero mode with a kink profile for φ. The σ^1 matrix anti-commutes with $H(\varphi)$ and acts as the R matrix that implements the energy reflection symmetry, which is lost when $\epsilon\sigma^1$ is added to the Hamiltonian, thereby producing an irrational charge.

More recently there appeared a 2-dimensional material, for which a similar analysis has been performed. This is graphene, which is described by a hexagonal lattice of carbon atoms that is presented as a superposition of two triangular sublattices, see Fig. 5. Each sublattice supports two Dirac points, hence a 4-dimensional matrix equation on the two dimensional plane becomes relevant. The background field is taken to be a combination of scalar φ and vector $\mathbf{A}$.[8] The Hamiltonian governing electron motion reads

$$H(\varphi, \mathbf{A}) = \boldsymbol{\alpha} \cdot [\mathbf{p} - \gamma_5 \mathbf{A}] + \beta[\varphi^r - i\,\gamma_5\,\varphi^i].$$

$$\text{Here } \boldsymbol{\alpha} = \begin{pmatrix} \boldsymbol{\sigma} & 0 \\ 0 & -\boldsymbol{\sigma} \end{pmatrix}, \; \boldsymbol{\sigma} = (\sigma^1, \sigma^2), \; \gamma_5 = \begin{pmatrix} I & 0 \\ 0 & -I \end{pmatrix},$$

$$\beta = \begin{pmatrix} 0 & I \\ I & 0 \end{pmatrix}, \; \mathbf{p} = \frac{1}{i}\,(\partial_x, \partial_y),$$

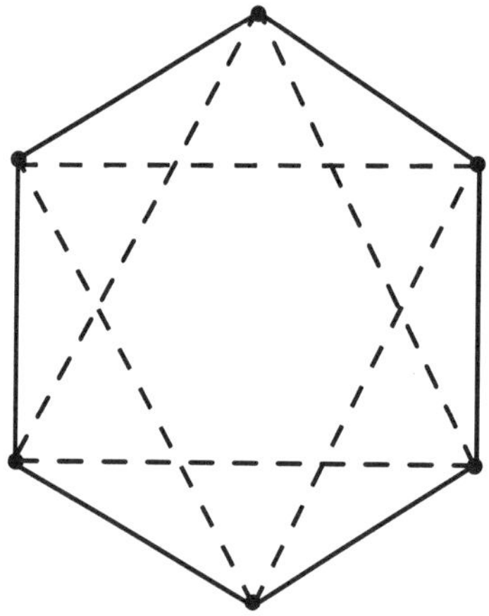

Fig. 5. The hexagonal graphene lattice, taken as a superposition of two triangular sublattices.

where φ^r, φ^i form the real and imaginary parts of a complex scalar field $\varphi = \varphi^r + i\varphi^i$. The matrix that effects energy reflection is $R = \alpha^3 = \begin{pmatrix} \sigma^3 & 0 \\ 0 & -\sigma^3 \end{pmatrix}$.

The topologically trivial background consist of constant φ and vanishing **A**. This gives rise to a gap in the energy spectrum. For the topologically non-trivial background we take for φ and **A** a vortex profile. A zero mode ensues, and charge becomes $-\frac{1}{2}$ (with a single vortex). The mid-gap state is bound with just a vortex configuration for φ and vanishing **A**. **A** is not needed for binding the zero energy state. Its presence does affect the profile of the zero-energy wave function, but not the vanishing of the eigenvalue. However, a vector potential is needed to give the vortex finite energy: the scalar field and vector field vortex configurations separately carry infinite energy, which is rendered finite in the combination. In other words a pure φ vortex cannot "move" because it is infinitely heavy. Evidently this infinity is screened away by the contribution from **A**.

When $\varepsilon R = \varepsilon\alpha^3$ is appended to H, the energy reflection symmetry is lost, and the induced charge structure becomes interestingly complicated. For a pure φ vortex, without a vector potential, the charge becomes irrational — as expected and as it happens in the 1-dimensional case. However, including a vector potential in its vortex configuration changes the charge back to $-\frac{1}{2}$ — its value in the energy reflection symmetric case. Just as for energy, the vector potential screens away a contribution from the scalar field.

The above 2-dimensional story is very interesting in its elegant intricacies. But it must be stated that its validity as a theoretical description of actually realized physical graphene remains to be established. Also as yet there is no experimental verification of the charge fractionalization phenomenon in graphene. Nevertheless it would be most interesting to develop a functional integral analysis of this phenomenon.

This work was supported in part by funds provided by the U.S. Department of Energy under cooperative research agreement #DE-FG02-05ER41360.

References

1. S. F. Edwards and Y. V. Gulyaev, *Proc. Roy. Soc. A* **279**, 229 (1964).
2. J. L. Gervais and A. Jevicki, *Nucl. Phys. B* **110**, 93 (1976).
3. K. Fujikawa and H. Suzuki, *Path Integrals and Quantum Anomalies* (Oxford, Oxford UK, 2004).
4. R. Jackiw and C. Rebbi, *Phys. Rev. D* **13**, 3398 (1976).
5. J. Goldstone and F. Wilczek, *Phys. Rev. Lett.* **47**, 987 (1981).
6. A. Niemi and G. Semenoff, *Phys. Rep.* **135**, 100 (1986).
7. W.-P. Su, J. Schrieffer, and A. Heeger, *Phys. Rev. Lett.* **42**, 1698 (1979).
8. C. Y. Hou, C. Chamon, and C. Mudry, *Phys. Rev. Lett.* **98**, 186809 (2007); R. Jackiw and S.-Y. Pi, *Phys. Rev. Lett.* **98**, 266402 (2007).

MULTIVALUED FIELDS AND THIRD QUANTIZATION

H. KLEINERT

Institute for Theoretical Physics, Freie Universität Berlin,
Arnimallee 14, D-14195 Berlin, Germany
E-mail: kleinert@physik.fu-berlin.de

Changes of field variables may lead to multivalued fields which do not satisfy the Schwarz integrability conditions. Their quantum field theory needs special care as is illustrated here in applications to superfluid and superconducting phase transitions. Extending the notions that first qantization governs fluctuating orbits while second quantization deals with fluctuating field, the theory of multivalued fields may be considered as a theory of third quantization. The lecture is an introduction to my new book on this subject.

Keywords: Path integrals; Vortices; Defects; Gauge theories; Superfluids; Superconductors; Phase transition; Distributions; Third quantization.

1. Coordinate Transformations in Path Integrals

Changes of coordinates or field variables must not change the physical content of a theory. This trivial requirement is automatically guaranteed in quantum field theories. As a simple example consider the path integral of a harmonic oscillator

$$Z_\omega = \int \mathcal{D}x\, e^{-\mathcal{A}_\omega[x]} = \exp\left[-\frac{D}{2}\mathrm{Tr}\log(-\partial^2+\omega^2)\right] \equiv e^{-\beta F_\omega} \qquad (1)$$

with an action

$$\mathcal{A}_\omega[x] = \frac{1}{2}\int_0^\beta d\tau\, \left[\dot{x}^2(\tau) + \omega^2 x^2(\tau)\right] \qquad (2)$$

and a free energy $F = \beta^{-1}\ln[2\sinh(\omega\beta/2)]$. Let us subject this path integral to a simple coordinate transformation such as

$$x = x_\eta(q) = q - \eta q^3/3, \qquad (3)$$

where η is some expansion parameter. The tranformed path integral

$$Z = \int \mathcal{D}q(\tau)\, e^{-\mathcal{A}_\omega[q]-\mathcal{A}^{\mathrm{int}}[q]-\mathcal{A}_J[q]} \equiv e^{-\beta F} \qquad (4)$$

has an interaction $\mathcal{A}_{\text{int}}[q] = \mathcal{A}^{\text{ah}}[q] + \mathcal{A}_J[q]$, consisting of the anharmonic part of the transformed action

$$\mathcal{A}^{\text{ah}}[q] = \int_0^\beta d\tau \left\{ -\eta \left[q^2(\tau)\dot{q}^2(\tau) + \frac{\omega^2}{3} q^4(\tau) \right] \right. \\ \left. + \eta^2 \left[\frac{1}{2} q^4(\tau)\dot{q}^2(\tau) + \frac{\omega^2}{18} q^6(\tau) \right] + \mathcal{O}(\eta^3) \right\}, \tag{5}$$

and an anharmonic part due to the Jacobian $\mathcal{D}x/\mathcal{D}q = \exp[\delta(0) \log \partial x(q)/\partial q]$:

$$\mathcal{A}_J[q] = -\delta(0) \int d\tau \, \log \frac{\partial x_\eta(q)}{\partial q} \\ = -\delta(0) \int_0^\beta d\tau \left[-\eta q^2(\tau) - \frac{\eta^2}{2} q^4(\tau) + \cdots + \mathcal{O}(\eta^3) \right]. \tag{6}$$

The transformed path integral (4) can no longer be solved exactly but only perturbatively as an expansion in powers of the parameter η:

$$\beta F = \beta F_\omega + \langle \mathcal{A}_{\text{int}} \rangle_c - \frac{1}{2!} \langle \mathcal{A}_{\text{int}}^2 \rangle_c + \ldots \; = \beta F_\omega + \beta \sum_{n=1}^{\infty} \eta^n F_n. \tag{7}$$

In order to guarantee coordinate invariance, all coefficients F_n have to vanish.

The Feynman diagrams contributing to F_n consist of vertices and three kinds of lines representing the one-dimensional versions of the correlation functions

$$G^{(2)}_{\mu\nu}(\tau,\tau') \equiv \langle q_\mu(\tau) q_\nu(\tau') \rangle = \text{———}, \tag{8}$$

$$\partial_\tau G^{(2)}_{\mu\nu}(\tau,\tau') \equiv \langle \dot{q}_\mu(\tau) q_\nu(\tau') \rangle = \text{——}\cdots\cdot, \tag{9}$$

$$\partial_{\tau'} G^{(2)}_{\mu\nu}(\tau,\tau') \equiv \langle q_\mu(\tau) \dot{q}_\nu(\tau') \rangle = \cdots\text{——}, \tag{10}$$

$$\partial_\tau \partial_{\tau'} G^{(2)}_{\mu\nu}(\tau,\tau') \equiv \langle \dot{q}_\mu(\tau) \dot{q}_\nu(\tau') \rangle = \cdots\cdots\cdot. \tag{11}$$

These contain distributions $\Theta(\tau - \tau')$ and $\delta(\tau - \tau')$ (see Fig. 1), so that the Feynman integrals run over products of distributions which in the standard theory of generalized functions are undefined. Recently, however, it has been shown that there is a way of defining products of distributions in such a way that all F_n vanish, i.e., that coordinate invariance can be maintained.[1]

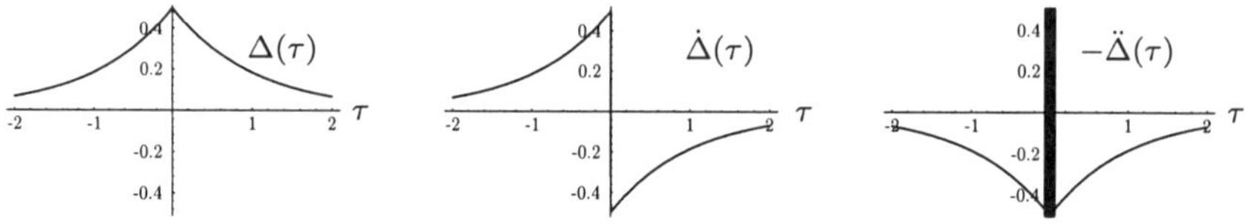

Fig. 1. Green functions for perturbation expansions in curvilinear coordinates in natural units with $\omega = 1$. The second contains a Heaviside function $\Theta(\tau)$, the third a Dirac δ-function at the origin.

2. Multivalued Fields

The situation becomes much more involved if the transformed coordinates $q(\tau)$ are *multivalued* fields in D dimensions. This happens, for instance, if a complex field $\psi(\mathbf{x})$ in a functional integral is replaced by its radial and azimuthal parts of $\rho(\mathbf{x}) \equiv |\psi(\mathbf{x})|$ and $\theta(\mathbf{x}) \equiv \arctan[\mathrm{Im}\psi(\mathbf{x})/\mathrm{Re}\psi(\mathbf{x})]$. A good example is the Landau-Pitaevski energy density of superfluid helium near the critical point:

$$\mathcal{H}_{\mathrm{He}}(\psi, \boldsymbol{\nabla}\psi) = \frac{1}{2}\left\{|\boldsymbol{\nabla}\psi|^2 + \tau|\psi|^2 + \frac{g}{2}|\psi|^4\right\}. \tag{12}$$

The parameter $\tau \equiv T/T_c^{\mathrm{MF}} - 1$ is a reduced temperature measuring the distance from the characteristic temperature T_c^{MF} at which the $|\psi|^2$-term changes sign. Under a field transformation $\psi(x) \to \rho(x)e^{i\theta(x)}$, the energy density *cannot* be simply replaced by

$$\mathcal{H}_1 = \frac{\rho^2}{2}(\boldsymbol{\nabla}\theta)^2 + \frac{1}{2}(\boldsymbol{\nabla}\rho)^2 + \frac{\tau}{2}\rho^2 + \frac{g}{4}\rho^4, \tag{13}$$

as we might be tempted to do following the naive Leibniz rule

$$D\psi = (i\rho\boldsymbol{\nabla}\theta + \boldsymbol{\nabla}\rho)e^{i\theta}. \tag{14}$$

This rule is no longer valid. Since $\theta(\mathbf{x})$ and $\theta(\mathbf{x}) + 2\pi$ correspond to the same complex field $\psi(\mathbf{x})$, the corrected Leibniz rule reads

$$D\psi = [i\rho(\boldsymbol{\nabla}\theta - 2\pi\ {}^{v}) + \boldsymbol{\nabla}\rho]e^{i\theta}. \tag{15}$$

The cyclic nature of the scalar field $\theta(\mathbf{x})$ requires the presence of a vector field $\ {}^{v}(\mathbf{x})$ called *vortex gauge field*. This field is a sum of δ-functions on Volterra surfaces across which $\theta(\mathbf{x})$ has jumps by 2π. The boundary lines of the surfaces are vortex lines. They are found from the vortex gauge field $\ {}^{v}(\mathbf{x})$ by forming the curl

$$\boldsymbol{\nabla} \times\ {}^{v}(\mathbf{x}) = \mathbf{j}^{v}(\mathbf{x}), \tag{16}$$

where $\mathbf{j}^v(\mathbf{x})$ is the *vortex density*, a sum over δ-functions $(L;\mathbf{x}) \equiv \int_L d\bar{\mathbf{x}}\,\delta(\mathbf{x}-\bar{\mathbf{x}})$ along the vortex lines L.

Vortex gauge transformations correspond to deformations of the surfaces at fixed boundary lines which add to $^v(\mathbf{x})$ pure gradients of the form $\boldsymbol{\nabla}\delta(V;\mathbf{x})$, where $\delta(V;\mathbf{x}) \equiv \int_V d^3\bar{x}\,\delta(\mathbf{x}-\bar{\mathbf{x}})$ are δ-functions on the volumes V over which the surfaces have swept. The theory of these fields has been developed in the textbook[2] and the Cambridge lectures.[3] Being a gauge field, $^v(\mathbf{x})$ may be modified by a further gradient of a smooth function to make it purely transverse, $\boldsymbol{\nabla}\cdot\ {}^v_T(\mathbf{x}) = 0$, as indicated by the subscript T.

Since the vortex gauge field is not a gradient, it cannot be absorbed into the vector potential by a gauge transformation. Hence it survives in the last term in Eq. (13), and the correct partition function is

$$Z_{\text{He}} \approx \int \mathcal{D}\ {}^v_T \int \mathcal{D}\rho\rho \exp\left[-\frac{\rho^2}{2}(\boldsymbol{\nabla}\theta)^2 - \frac{1}{2}(\boldsymbol{\nabla}\rho)^2 - \frac{\tau}{2}\rho^2 - \frac{g}{4}\rho^4 - \frac{4\pi^2\rho^2}{2}\ {}^{v\,2}_T\right]. \tag{17}$$

The symbol $\int \mathcal{D}\ {}^v_T$ does not denote an ordinary functional integral. It is defined as a sum over any number and all shapes of Volterra surfaces S in ${}^v_T(\mathbf{x})$, across which the phase jumps by 2π.[3]

The important observation is now that due to the fluctuations of the vortex gauge field ${}^v_T(\mathbf{x})$, the partition function (17) possesses a second-order phase transition, the famous λ-transition observed in superfluid helium at 2.18 K. The critical exponents of this transition are in the same universality class as those of the so-called XY model, which describes only interacting phase angles $\theta(\mathbf{x}) \in (0, 2\pi)$ on a lattice.

At the mean-field level, the λ-transition of (17) takes place if τ drops below zero where the pair field $\psi(x)$ acquires the nonzero expectation value $\langle\psi(x)\rangle = \rho_0 = \sqrt{-\tau/g}$, the order parameter of the system. The ρ-fluctuations around this value have a *coherence length* $\xi = 1/\sqrt{-2\tau}$.

3. Apparent First Order of Superconducting Transition

For a long time it has been a debate whether this transition persists if a fluctuating vector potential $\mathbf{A}(\mathbf{x})$ is coupled minimally to the field $\psi(\mathbf{x})$ in (12), which then becomes the Ginzburg-Landau Hamiltonian density of superconductivity,

$$\mathcal{H}_{\text{sc}}(\psi, \boldsymbol{\nabla}\psi, \mathbf{A}, \boldsymbol{\nabla}\mathbf{A}) = \frac{1}{2}\left\{[(\boldsymbol{\nabla} - iq\mathbf{A})\psi]^2 + \tau|\psi|^2 + \frac{g}{2}|\psi|^4\right\} + \frac{1}{2}(\boldsymbol{\nabla}\times\mathbf{A})^2 . \tag{18}$$

Now $\psi(\mathbf{x})$ is the field describing Cooper pairs of charge $q = 2e$. The theory needs gauge fixing, which may be done by absorbing the gradient of the phase $\theta(\mathbf{x})$ of the field $\psi(\mathbf{x})$ in the vector potential, so that we can replace $\psi(\mathbf{x}) \to \rho(\mathbf{x})$. The transverse vortex gauge field ${}^{v}_{T}(\mathbf{x})$, however, cannot be absorbed and it interacts with the vector potential $A(\mathbf{x})$. This has a partial partition function

$$Z_A[\rho] \equiv \int \mathcal{D}\, {}^{v}_{T} \mathcal{D}\mathbf{A} \exp\left\{-\frac{1}{2}\int d^3x(\boldsymbol{\nabla}\times\mathbf{A})^2 - \frac{1}{2}\int d^3x\,\rho^2(e\mathbf{A} - 2\pi\, {}^{v}_{T})^2\right\}. \tag{19}$$

Without the vortex gauge field ${}^{v}_{T}(\mathbf{x})$, the partition function (19) describes free bosons of space-dependent mass $\rho^2(\mathbf{x})$. If we ignore ${}^{v}_{T}(\mathbf{x})$ and $\mathbf{A}(\mathbf{x})$, the total partition function has the same form as in (17) and describes a second-order phase transition.

Let us now admit the vector partition function (19), but still ignore vortices by setting ${}^{v}_{T}(\mathbf{x}) \equiv 0$, and ignoring the space dependence of $\rho(\mathbf{x})$. Then the second term in (19) in the condensed phase with $\rho_0 \neq 0$ generates a Meissner-Higgs mass term. This gives rise to a finite *penetration depth* of the magnetic field $\lambda = 1/m_A = 1/\rho_0 q$. The ratio of the two length scales $\kappa \equiv \lambda/\sqrt{2}\xi$ (which for historic reasons carries a factor $\sqrt{2}$) is the Ginzburg parameter whose mean field value is $\kappa_{\rm MF} \equiv \sqrt{g/q^2}$. Type I superconductors have small values of κ, type-II superconductors have large values. At the mean-field level, the dividing line lies at $\kappa = 1/\sqrt{2}$.

Let us now allow for $\mathbf{A}(\mathbf{x})$-fluctuations while still ignoring the vortex gauge field ${}^{v}_{T}(\mathbf{x})$. At very smooth $\rho(\mathbf{x})$, they can be integrated out in (19) which becomes

$$Z_A^0[\rho] = \exp\left[\int d^3x \frac{e^3\rho^3}{6\pi}\right]. \tag{20}$$

This adds to the energy density (13) a cubic term $-e^3\rho^3/6\pi$. Such a term makes the transition first order. The free energy has now a minimum at

$$\tilde{\rho}_0 = \frac{c}{2g}\left(1 + \sqrt{1 - \frac{4\tau g}{c^2}}\right). \tag{21}$$

If τ decreases below

$$\tau_1 = 2c^2/9g, \tag{22}$$

the new minimum lies *lower* than the one at the origin (see Fig. 2), so that the order parameter jumps from zero to

$$\rho_1 = 2c/3g \tag{23}$$

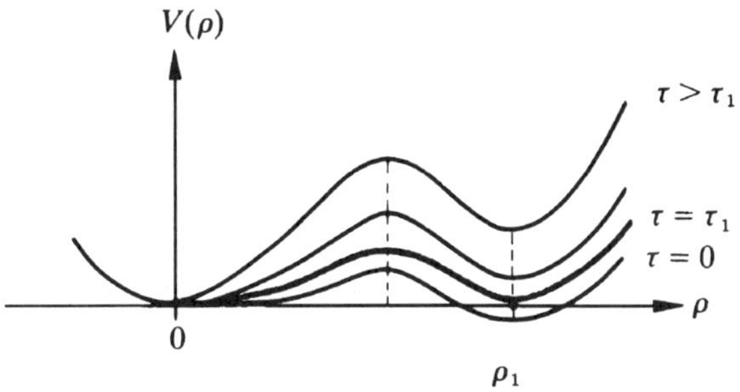

Fig. 2. Potential for the order parameter ρ with cubic term. At τ_1, the order parameter jumps from $\rho = 0$ to ρ_1, corresponding to a phase transition of first order.

in a phase transition. At this point, the coherence length of the ρ-fluctuations $\xi = 1/\sqrt{\tau + 3g\rho^2 - 2c\rho}$ has the finite value

$$\xi_1 = \frac{3}{c}\sqrt{\frac{g}{2}}, \tag{24}$$

this being the same as at $\rho = 0$. The jump from $\rho = 0$ to ρ_1 implies a phase transition of first order.[7]

4. Second-Order Transition

However, this result is reliable only under the assumption of a smooth $\rho(\mathbf{x})$. It is applicable only in the type-I regime where vortex loops are strongly suppressed. In the type-II regime, such lines can more easily be excited thermally and we can no longer ignore the vortex gauge field $^{v}_{T}(\mathbf{x})$. This invalidates the above conclusion and gives rise to a second-order transition (of the same XY universality class as in the superfluid) if the Ginzburg parameter κ is sufficiently large. In fact, if we now integrating out the **A**-field, we obtain

$$Z_A[\rho] = \exp\left[\int d^3x \frac{e^3\rho^3}{6\pi}\right] \times \int \mathcal{D}\, ^{v}_{T} \exp\left[\frac{4\pi^2\rho^2}{2}\int d^3x \left(\frac{1}{2}\, ^{v}_{T}{}^{2} - \, ^{v}_{T} \frac{\rho^2 q^2}{-\boldsymbol{\nabla}^2 + \rho^2 q^2}\, ^{v}_{T}\right)\right], \tag{25}$$

rather than (20). The second integral can be simplified to

$$\frac{4\pi^2\rho^2}{2}\int d^3x \left(\, ^{v}_{T} \frac{-\boldsymbol{\nabla}^2}{-\boldsymbol{\nabla}^2 + \rho^2 q^2}\, ^{v}_{T}\right). \tag{26}$$

Integrating this by parts, and replacing $\nabla_i \overset{v}{T} \nabla_i \overset{v}{T}$ by $(\boldsymbol{\nabla} \times \overset{v}{T})^2 = \mathbf{j}^{v\,2}$, since $\boldsymbol{\nabla} \cdot \overset{v}{T} = 0$, the partition function (25) takes the form

$$Z_A[\rho]=\exp\left[\int d^3x \frac{e^3\rho^3}{6\pi}\right]\int \mathcal{D}\overset{v}{T}\exp\left[-\frac{4\pi^2\rho^2}{2}\int d^3x \left(\mathbf{j}^v \frac{1}{-\boldsymbol{\nabla}^2+\rho^2q^2}\mathbf{j}^v\right)\right].$$

This is the partition function of a grand-canonical ensemble of closed fluctuating vortex lines L described by the δ-functions over lines in $\mathbf{j}^v(\mathbf{x})$. The interaction between them has a finite range equal to the penetration depth $\lambda = 1/\rho q$. It is well-known how to compute pair and magnetic fields of the Ginzburg-Landau theory for a single straight vortex line from the extrema of the energy density (18). In an external magnetic field, there exist triangular and various other regular arrays of vortex lattices and various phase transitions. In the core of each vortex line, the pair field ρ goes to zero over a distance ξ. If we want to sum over grand-canonical ensemble of fluctuating vortex lines of any shape in the partition function (17), the space dependence of ρ causes complications. These can be avoided by an approximation, in which the system is placed on a simple-cubic lattice of spacing $a = \alpha\,\xi$, with α of the order of unity, and replacing the variable $\rho(\mathbf{x})$ by a *fixed* $\rho = \tilde{\rho}_0$ given by Eq. (21). Thus we replace the partial partition function (27) approximately by

$$Z_2[\tilde{\rho}_0]=\sum_{\{\mathbf{l};\boldsymbol{\nabla}\cdot\mathbf{l}=0\}}\exp\left[-\frac{4\pi^2\tilde{\rho}_0^2 a}{2}\sum_{\mathbf{x}}\mathbf{l}(\mathbf{x})v_{\tilde{\rho}_0 e}(\mathbf{x}-\mathbf{x}')\mathbf{l}(\mathbf{x}')\right]. \tag{27}$$

The sum runs over the discrete versions of the vortex density in (16). These are integer-valued vectors $\mathbf{l}(\mathbf{x}) = (l_1(\mathbf{x}), l_2(\mathbf{x}), l_3(\mathbf{x}))$ which satisfy $\boldsymbol{\nabla}\cdot\mathbf{l}(\mathbf{x}) = 0$, where $\boldsymbol{\nabla}$ denotes the lattice derivative. This condition restricts the sum over all $\mathbf{l}(\mathbf{x})$-configurations in (27) to all non-self-backtracking integer-valued closed loops. The function

$$\begin{aligned} v_m(\mathbf{x}) &= \prod_{i=1}^{3}\int \frac{d^3(ak_i)}{(2\pi)^3}\frac{e^{i(k_1x_1+k_2x_2+k_3x_3)}}{2\sum_{i=1}^{3}(1-\cos ak_i)+a^2m^2} \\ &= \int_0^\infty ds e^{-(6+m^2)s} I_{x_1}(2s) I_{x_2}(2s) I_{x_3}(2s) \end{aligned} \tag{28}$$

is the lattice Yukawa potential.[4]

The lattice partition function (27) is known to have a second-oder phase transition in the universality class of the XY model. This can be seen by a comparison with the Villain approximation[5] to the XY model, whose partition function is a lattice version of

$$Z_V[\rho] = \int \mathcal{D}\theta \int \mathcal{D}\ {}^{v}_{T} \exp\left[-\frac{b}{2}\int d^3x\ (\boldsymbol{\nabla}\theta - {}^{v}_{T})^2\right]. \tag{29}$$

After integrating out $\theta(\mathbf{x})$, this becomes

$$Z_V[\rho] = \mathrm{Det}^{-1/2}(-\boldsymbol{\nabla}^2)\int \mathcal{D}\ {}^{v}_{T} \exp\left(-\frac{b}{2}\int d^3x\ {}^{v}_{T}{}^{2}\right), \tag{30}$$

and we can replace ${}^{v}_{T}{}^{2}$ by $(\boldsymbol{\nabla}\times\ {}^{v}_{T})(-\boldsymbol{\nabla}^2)^{-1}(\boldsymbol{\nabla}\times\ {}^{v}_{T}) = \mathbf{j}^v(-\boldsymbol{\nabla}^2)^{-1}\mathbf{j}^v$. By taking this expression to a simple-cubic lattice we obtain the partition function (27), but with $\tilde{\rho}_0^2 a$ replaced by $\beta_V \equiv ba$, and the Yukawa potential $v_{\tilde{\rho}_0 e}(\mathbf{x})$ replaced by the Coulomb potential $v_0(\mathbf{x})$.

The partition function (27) has the same transition at roughly the same place as its local approximation

$$Z_2[\tilde{\rho}_0] \approx \sum_{\{\mathbf{l};\boldsymbol{\nabla}\cdot\mathbf{l}=0\}} \exp\left[-\frac{4\pi^2\tilde{\rho}_0^2 a}{2} v_{\tilde{\rho}_0 e}(\mathbf{0})\sum_{\mathbf{x}}\mathbf{l}^2(\mathbf{x})\right]. \tag{31}$$

A similar approximation holds for the Villain model with $v_0(\mathbf{x})$ instead of $v_{\tilde{\rho}_0 e}(\mathbf{x})$, and $\tilde{\rho}_0^2 a$ replaced by $\beta_V \equiv ba$.

The Villain model is known to undergo a second-order phase transition of the XY model type at $\beta_V = r/3$ with $r \approx 1$, where the vortex lines become infinitely long.[5,8] Thus we conclude that also the partition function (31) has a second-order phase transition of the XY model type at $\tilde{\rho}^2 v_{\tilde{\rho}_0 e}(\mathbf{0})a \approx v_0(\mathbf{0})/3$. The potential (28) at the origin has the hopping expansion[9]

$$v_m(\mathbf{0}) = \sum_{n=0,2,4} \frac{H_n}{(a^2m^2+6)^{n+1}},\ H_0 = 1, H_2 = 6, \dots . \tag{32}$$

To lowest order, this yields the ratio $v_m(\mathbf{0})/v_0(\mathbf{0}) \equiv 1/(m^2/6+1)$. A more accurate numerical fit to the ratio $v_m(\mathbf{0})/v_0(\mathbf{0})$ which is good up to $m^2 \approx 10$ (thus comprising all interesting κ-values since m^2 is of the order of $3/\kappa^2$) is $1/(\sigma\, m^2/6+1)$ with $\sigma \approx 1.38$. Hence the transition takes place at

$$\frac{\tilde{\rho}_0^2 a}{(\sigma\, a^2\tilde{\rho}_0^2 q^2/6+1)} \approx \frac{r}{3} \text{ or } \tilde{\rho}_0 \approx \frac{1}{\sqrt{3a}}\sqrt{\frac{r}{1-\sigma raq^2/18}}. \tag{33}$$

The important point is now that this transition can occur only until $\tilde{\rho}_0$ reaches the value $\rho_1 = 2c/3g$ of Eq. (23). From there on, the transition will no longer be of the XY model type but occur discontinuously as a first-order transition.

Replacing in (33) a by $\alpha\xi_1$ of Eq. (24), and $\tilde{\rho}_0$ by ρ_1, we find the equation for the mean-field Ginzburg parameter $\kappa_{\rm MF} = \sqrt{g/q^2}$:

$$\kappa_{\rm MF}^3 + \alpha^2\sigma\frac{\kappa_{\rm MF}}{3} - \frac{\sqrt{2}\alpha}{\pi r} = 0. \tag{34}$$

Inserting $\sigma \approx 1.38$ and choosing $\alpha \approx r \approx 1$, the solution of this equation yields the tricritical value

$$\kappa_{\rm MF}^{\rm tric} \approx 0.81/\sqrt{2}. \tag{35}$$

In spite of the roughness of the approximations, this result is very close to the value $0.8/\sqrt{2}$ derived from the dual disorder field theory.[6] The approximation has three uncertainties. First, the identification of the effective lattice spacing $a = \alpha\xi$ with $\alpha \approx 1$; second the associated neglect of the $\mathbf{x}$-dependence of ρ and its fluctuations, and third the localization of the critical point of the XY model type transition in Eq. (33).

Our goal has been achieved: We have shown the existence of a tricritical point in a superconductor directly within the fluctuating Ginzburg-Landau theory, by taking the vortex fluctuations into account. This became possible after correcting the covariant derivative (14) of $\psi = \rho e^{i\theta}$ to (15). For $\kappa > 0.81/\sqrt{2}$, vortex fluctuations give rise to an XY model type second-order transition before the cubic term becomes relevant. This happens for $\kappa < 0.81/\sqrt{2}$ where the cubic term causes a discontinuous transition.

5. Crystal Defects and the Melting Transition

These examples show that the subtleties of functional integration over multivalued fields are crucial for understanding important physical phenomena such as phase transitions. Similar considerations are necessary in the context of elasticity theory where the energy is usually expressed in terms of the strain $u_{ij} = \partial_i u_j(\mathbf{x})$ of the displacement field $u_i(\mathbf{x})$ of the atoms from their rest position. Such a description is also false since the displacement field $u_i(\mathbf{x})$ is a multivalued field. It is defined only up to multiples of the lattice vectors. This multivaluedness must be taken into account with the help of a *defect gauge field* similar to $^{v}(\mathbf{x})$, one for each lattice direction. Its fluctuations give rise to the melting transition, as has been shown in the textbook.[10]

6. Crystal Defects and Geometry

The defects in a crystal form a Riemann-Cartan space with curvature and torsion, and it is possible to rederive all properties of the Riemann-

Cartan geometry by starting from a multivalued crystal world and converting the variables to metric and affine connection. For more details see the textbook.[11]

References

1. H. Kleinert, *Path Integrals in Quantum Mechanics, Statistics, Polymer Physics, and Financial Markets*, 3rd edition (World Scientific, Singapore, 2006) [`kl/b5`, where `kl` is short for `http://www.physik.fu-berlin.de/~kleinert`].
2. H. Kleinert, *Gauge Fields in Condensed Matter*, Vol. I: Superflow and Vortex Lines (World Scientific, Singapore, 1989) [`kl/re.html#B1`].
3. H. Kleinert, *Theory of fluctuating nonholonomic fields and applications: statistical mechanics of vortices and defects and new physical laws in spaces with curvature and torsion*, in Proceedings of NATO Advanced Study Institute on *Formation and Interaction of Topological Defects* (Plenum, New York, 1995), pp. 201–232 [`kl/227`].
4. See Eq. (6.110) of Ref. 2 [`kl/b1/gifs/v1-165s.html`].
5. J. Villain, *J. Phys. (Paris)* **36**, 581 (1977). See also `kl/b1/gifs/ v1-489s.html`.
6. H. Kleinert, *Lett. Nuovo Cimento* **37**, 295 (1982) [`kl/108`]. The tricritical value $\kappa \approx 0.8/\sqrt{2}$ derived in this paper was confirmed only recently by Monte Carlo simulations: J. Hove, S. Mo, and A. Sudbø, *Phys. Rev. B* **66**, 64524 (2002).
7. See Eq. (3.111) of Ref. 10 [`kl/b1/gifs/v1-337s.html`].
8. See the figure on p. 501 of Ref. 2 [`kl/b1/gifs/v1-501s.html`].
9. See Eq. (6.122) of Ref. 2 [`kl/b1/gifs/v1-168s.html`].
10. H. Kleinert, *Gauge Fields in Condensed Matter*, Vol. II: Stresses and Defects (World Scientific, Singapore, 1989) [`kl/re.html#B2`].
11. H. Kleinert, *Multivalued Fields in Condensed Matter, Electromagnetism, and Gravitation* (World Scientific, Singapore, 2008) [`kl/b11`].

GAUSSIAN EQUIVALENT REPRESENTATION OF PATH INTEGRALS OVER A GAUSSIAN MEASURE

G. V. EFIMOV

Laboratory of Theoretical Physics, Joint Institute for Nuclear Research, Dubna, Russia
Email: efimovg@thsun1.jinr.ru

A method named the Gaussian equivalent representation and developed to calculate path integrals over a Gaussian measure, is presented. As an example partition functions for simple liquids and proton plasma are calculated by this method. Free energy and a pair correlation function in the lowest and next approximations are obtained.

Keywords: Gaussian equivalent representation; Gaussian measure; Screening; Coulomb potential; Plasma.

1. Introduction

At present solutions of many problems from quite different branches of modern physics (quantum mechanics and quantum field theory, statistics, plasma and polymer physics and so on) can be represented in the form of functional or path integrals over a Gaussian measure. However, the methods of calculating these integrals is not developed in a suitable form. Particularly, it also concerns the complex path integrals. In this talk I want to formulate a method, called *the Gaussian equivalent representation* to calculate path integrals over a Gaussian measure. For demonstration of efficiency of the proposed method the screened Coulomb or Debye-Hückel potential in plasma and the next corrections to it will be calculated directly from the canonical partition function.

We shall consider the path integrals of the type:

$$I(g) = \int \frac{D\phi}{C_A} e^{-\frac{1}{2}(\phi A^{-1}\phi)+gW[\phi]}, \quad I(0) = 1, \quad x \in \Gamma \subset \mathbf{R}^d. \qquad (1)$$

The kernel of the Gaussian measure and its Green function satisfying appropriate boundary conditions are defined as

$$A^{-1} = K\,(\partial_x)\,\delta(x-x'), \qquad \int\limits_{\Gamma} dy\; A^{-1}(x,y)A(y,x') = \delta(x-x')\,.$$

The interaction functionals are supposed to be represented in the form of the Fourier integrals so that the perturbation expansion over the coupling constant g does exist. In particular,

$$\int \frac{D\phi}{C_A} e^{-\frac{1}{2}(\phi A^{-1}\phi)} gW[\phi] = g \int d\mu_\eta e^{-\frac{1}{2}(\eta A \eta)}\,.$$

If in Eq. (1) the measure is positive and the interaction functional is real one can use variational calculations based on the Jensen inequality. We can introduce a new kernel B

$$\begin{aligned} I(g) &= \int \frac{D\phi}{C_A} e^{-\frac{1}{2}(\phi A^{-1}\phi)} \cdot e^{gW[\phi]} \\ &= \frac{C_B}{C_A} \int \frac{D\phi}{C_B} e^{-\frac{1}{2}(\phi B^{-1}\phi)} \cdot e^{-\frac{1}{2}(\phi[A^{-1}-B^{-1}]\phi)+gW[\phi]}\,. \end{aligned}$$

The variational estimation looks like

$$I(g) \geq \frac{C_B}{C_A} e^{\int \frac{D\phi}{C_B} e^{-\frac{1}{2}(\phi B^{-1}\phi)}\left[-\frac{1}{2}(\phi[A^{-1}-B^{-1}]\phi)+gW[\phi]\right]} = e^{E(g)}\,,$$

$$E(g) = \max_B \left[\frac{1}{2}\ln\left(\frac{\det(B)}{\det(A)}\right) - \frac{1}{2}([A^{-1}-B^{-1}]B) + g\int d\mu_\eta e^{-(\eta B \eta)}\right].$$

We formulate two problems:

- Is it possible to calculate the next corrections to the variational estimation?
- Is it possible to extend this technique to complex functional?

Our answer – *YES.*

2. Gaussian Equivalent Representation

Our idea comes from quantum field theory (QFT) in which vacuum loops, so-called tadpoles, give the main contributions to value of Eq. (1). These contributions in QFT are divergent and can be removed by going to *the normal product* of field operators. Thus we should find an appropriate representation of the functional integral (1) in which these contributions would be taken into account automatically, so that the lowest approximation would coincide with the variational estimation. We proceed in the following way.

We introduce the conception *normal form with respect to a given Gaussian measure with a given kernel A*. The normal form $e^{i(b\phi)}$ with respect to a given Gaussian measure with the kernel A is called the product

$$\vdots e^{i(b\phi)} \vdots_A \equiv e^{i(b\phi)+\frac{1}{2}(bAb)}.$$

According to this definition, we have

$$\int d\sigma_{\phi,A} \vdots e^{i(b\phi)} \vdots_A = \int \frac{D\phi}{C_A} e^{-\frac{1}{2}(\phi A^{-1}\phi)} \cdot e^{i(b\phi)+\frac{1}{2}(bAb)} \equiv 1 .$$

The following formula is important for consequent calculations

$$\vdots e^{i(b\phi)} \vdots_A = e^{\frac{1}{2}(b[A-B]b)} \vdots e^{i(b\phi)} \vdots_B .$$

The interaction functional can be written in the normal form

$$\begin{aligned} W[\phi] &= \int d\mu_\eta e^{i\eta\phi)} = \int d\mu_\eta \; e^{-\frac{1}{2}(bAb)} \cdot \vdots e^{i(b\phi)} \vdots_A \\ &= W_0 + i(W_1\phi) - \frac{1}{2} \vdots (\phi W_2 \phi) \vdots_A + \vdots W_I[\phi] \vdots_A, \\ & \quad W_0 = \int d\mu_\eta \; e^{-\frac{1}{2}(\eta A \eta)}, \\ & \quad i(W_1\phi) = \int d\mu_\eta \; e^{-\frac{1}{2}(\eta A\eta)} i(\eta\phi) = O(\phi), \\ & \quad \frac{1}{2} \vdots (\phi W_2 \phi) \vdots_A = \frac{1}{2} \int d\mu_\eta \; e^{-\frac{1}{2}(\eta A \eta)} \cdot \left(\vdots (\eta\phi)(\eta\phi) \vdots_A \right) = O(\vdots \phi^2 \vdots_A), \\ & \quad \vdots W_I[\phi] \vdots_A = \int d\mu_\eta e^{-\frac{1}{2}(\eta A\eta)} \cdot \vdots e_2^{i(\eta\phi)} \vdots_A = O(\vdots \phi^3 \vdots_A), \\ & \qquad\qquad e_2^z \equiv e^z - 1 - z - \frac{z^2}{2}. \end{aligned}$$

Now we can formulate *the Gaussian equivalent representation*. We have the integral

$$I(g) = \int \frac{D\phi}{C_A} e^{-\frac{1}{2}(\phi A^{-1}\phi)} \cdot e^{gW[\phi]} = \int d\sigma_{\phi,A} e^{gW[\phi]} .$$

Let us produce the displacement of the field ϕ and introduce a new kernel B^{-1}:

$$(1)\ \phi(x) \longrightarrow \phi(x) + \xi(x), \qquad (2)\ A^{-1} \longrightarrow B^{-1}.$$

The interaction functional in the integrand should be rewritten in the normal form with respect to the Gaussian measure with the kernel B. We get

$$
\begin{aligned}
I(g) &= \frac{C_B}{C_A}\int d\sigma_{\phi,B}\cdot e^{F[\phi]},\\
F[\phi] &= -\frac{1}{2}(\xi A^{-1}\xi) - \frac{1}{2}([A^{-1}-B^{-1}]B)\\
&\quad + \left\{-(\boldsymbol{\phi} A^{-1}\xi) - \frac{1}{2}\vdots(\boldsymbol{\phi}[A^{-1}-B^{-1}]\boldsymbol{\phi})\vdots_B\right\}\\
&\quad + gW_0 + \left\{ig(W_1\boldsymbol{\phi}) - \frac{g}{2}\vdots(\boldsymbol{\phi} W_2\boldsymbol{\phi})\vdots_B\right\} + g\vdots W_I[\phi]\vdots_B .
\end{aligned}
$$

We consider the main contribution coming from the Gaussian measure with the kernel B^{-1} so that the terms with boldface $\boldsymbol{\phi}$ inside curly brackets should cancel each other. This requirement leads to the equations:

$$
\begin{cases}
-(\boldsymbol{\phi} A^{-1}\xi) + ig(W_1\boldsymbol{\phi}) = 0,\\
-\frac{1}{2}\vdots(\boldsymbol{\phi}[A^{-1}-B^{-1}]\boldsymbol{\phi})\vdots_B - \frac{g}{2}\vdots(\boldsymbol{\phi} W_2\boldsymbol{\phi})\vdots_B = 0.
\end{cases}
$$

Solution of these equations gives B and ξ.

Finally we get *the Gaussian equivalent representation*:

$$
I(g) = \int d\sigma_{\phi,A} e^{gW[\phi]} = e^{E_0}\int d\sigma_{\phi,B} e^{g\vdots W_I[\phi]\vdots_B}\,, \tag{2}
$$

where

$$
E_0 = \frac{1}{2}\ln\left(\frac{C_B}{C_A}\right) - \frac{1}{2}([A^{-1}-B^{-1}]B) - \frac{1}{2}(\xi A^{-1}\xi) + g\int d\mu_\eta\; e^{i(\eta\xi)-\frac{1}{2}(\eta A\eta)}
$$

and the interaction functional is

$$
\vdots W_I[\phi]\vdots_B = \int d\mu_\eta\; e^{i(\eta\xi)-\frac{1}{2}(\eta B\eta)}\cdot \vdots e_2^{i(\eta\phi)}\vdots_B \;= O(\vdots\phi^3\vdots_B)\,.
$$

The value of E_0 is nothing else but the variational estimation of the initial integral (1).

The representation (2) permits us to find the next corrections to E_0 by standard perturbation calculations. One can get

$$
I(g) = e^{E_0}\int d\sigma_{\phi,B} e^{g\vdots W_I[\phi]\vdots_B} = e^{E_0+g^2W_2+\cdots}.
$$

The accuracy of the zeroth approximation can be evaluated by

$$
\delta = \frac{g^2|W_2|}{g|W_0|} =?\,. \tag{3}
$$

3. Screening of the Coulomb Potential in Plasma

It is known (see, for example, Ref. 4) that the effective electrostatic potential between two charged particles in the plasma differs from the Coulomb one owing to collective electrostatic interaction with other particles. The famous Debye-Hückel approach supposes that the pair distribution function $n_2(r)$ defines the charge density of particles in the plasma and is defined by an effective potential

$$n_2(r) = n\ e^{-\beta V_{\text{eff}}(r)}\,.$$

The effective electrostatic potential satisfies the Poisson equation

$$\begin{aligned} -\nabla^2 V_{\text{eff}}(r) = 4\pi\rho(r) = 4\pi e^2 n\left[e^{-\beta V_{\text{eff}}(r)} - 1\right] \ \Rightarrow -4\pi e^2\beta V_{\text{eff}}(r), \\ \left[-\nabla^2 + 4\pi\beta e^2 n\right] V_{\text{eff}}(r) = 0. \end{aligned} \tag{4}$$

The solution of the last equation represents the screened Debye-Hückel potential:

$$V_{\text{eff}}(r) = \frac{e^2}{r} e^{-\kappa_D r}, \qquad \kappa_D = \sqrt{4\pi\beta e^2 n},$$

where $\lambda_D = 1/\kappa_D$ is the Debye radius. All modifications and improvements of the Debye-Hückel approach are reduced to changing the equation (4) (see, for example, Ref. 3).

Our approach permits us to calculate the screened Coulomb potential directly from the canonical ensemble describing plasma. The canonical partition function looks like

$$Z_\Lambda = \int\limits_\Lambda \frac{dx_1}{|\Lambda|} \cdots \int\limits_\Lambda \frac{dx_N}{|\Lambda|} e^{-\beta \sum\limits_{i<j} \frac{e^2}{|x_i - x_j|}} \ \Rightarrow\ e^{-|\Lambda| F}, \qquad n = \frac{N}{|\Lambda|}\,.$$

The pair radial distribution function is defined as

$$n_2(x_1 - x_2) = \frac{1}{Z_\Lambda} \int\limits_\Lambda \frac{dx_3}{|\Lambda|} \cdots \int\limits_\Lambda \frac{dx_N}{|\Lambda|} e^{-\sum\limits_{1\le i<j\le N} \frac{\beta e^2}{|x_i - x_j|}} \ \Longrightarrow\ e^{-\beta V_{\text{eff}}(x_1 - x_2)}\,.$$

Using the path integral representation one can get

$$\begin{aligned} Z_\Lambda &= \frac{1}{|\Lambda|^N} \int \frac{D\phi}{\sqrt{\det V}} e^{-\frac{1}{2}(\phi V^{-1}\phi)} \left[\int\limits_\Lambda dx\ \vdots e^{i\sqrt{\beta}\phi(x)} \vdots_V\right]^N \\ &= \frac{N!}{|\Lambda|^N} \frac{1}{2\pi i} \oint \frac{dz}{z^{1+N}} \int d\sigma_{\phi,V} e^{z\int\limits_\Lambda dx\ \vdots e^{i\sqrt{\beta}\phi(x)} \vdots_V}\,. \end{aligned} \tag{5}$$

The kernel of the Gaussian measure

$$V^{-1}(x-y) = -\frac{1}{4\pi e^2}\triangle\delta(x-y)$$

contains the so-called zeroth mode, so that in the path integral we should integrate over functions satisfying the condition $\int dx\ \phi(x) = 0$. Let us write down the Gaussian equivalent representation for the path integral in Eq. (5). In this case, it is sufficient to introduce a new kernel

$$V^{-1}(x-y) \Longrightarrow D^{-1}(x-y).$$

We get

$$\int d\sigma_{\phi,V}\ e^{z\int\limits_{\Lambda} dx\ \vdots e^{i\sqrt{\beta}\phi(x)}\vdots_V} = e^{|\Lambda|F(z)}\cdot\int d\sigma_{\phi,D}\cdot e^{W_I[\phi]},$$

$$|\Lambda|F(z) = \frac{1}{2}\ln\frac{\det D}{\det V} - \frac{1}{2}(D[V^{-1}-D^{-1}]) + |\Lambda|ze^{\frac{\beta}{2}[V(0)-D(0)]}.$$

The interaction functional acquires the form

$$W_I[\phi] = c\int\limits_{\Lambda} dx \vdots e_2^{i\sqrt{\beta}\phi(x)}\vdots_D, \qquad c = z\beta e^{\frac{\beta}{2}[V(0)-D(0)]}.$$

The kernel D is defined by the equation

$$-\frac{1}{2}\vdots(\phi[V^{-1}-D^{-1}]\phi)\vdots_D - c\int\limits_{\Lambda} dx\ \vdots\phi^2(x)\vdots_D = 0,$$

so that

$$\tilde{D}(k) = \frac{\tilde{V}(k)}{1+c\tilde{V}(k)} = \frac{4\pi e^2}{k^2 + c\ 4\pi e^2}.$$

The next step is to calculate by the saddle-point method the complex contour integral over dz for $|\Lambda|\to\infty$. The saddle point is realized for $z = n\beta$ and

$$I = \frac{1}{2\pi i}\oint\frac{dz}{z^{1+N}}e^{|\Lambda|F(c(z))} \Longrightarrow e^{|\Lambda|[-n\ln n + F(n\beta)]}.$$

Finally we get

$$Z_\Lambda = e^{|\Lambda|F_0(n,\beta)}\cdot\int d\sigma_{\phi,D}e^{W_I[\phi]},$$

$$W_I[\phi] = n\int\limits_{\Lambda} dx\ \vdots e_2^{i\sqrt{\beta}\phi(x)}\vdots_D,$$

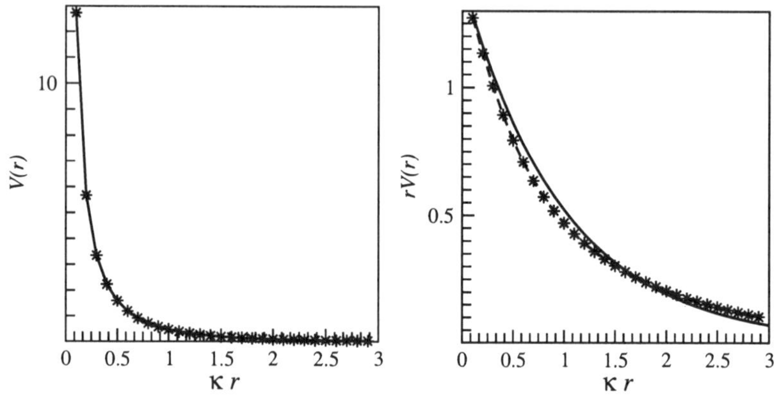

Fig. 1. LHS plot: Debye-Hückel potential (solid line), Debye-Hückel potential plus the first correction (stars). RHS plot is for the potential multiplied by r.

$$\beta D(x) = \frac{\beta e^2}{r} e^{-r\beta e^2} = \frac{A}{\kappa r} e^{-\kappa r} .$$

Here $\kappa = 1/\lambda_D = \sqrt{4\pi n \beta e^2}$ is the Debye parameter and

$$A = \frac{1}{4\pi} \frac{\kappa^3}{n} \sim \left(\frac{\ell_{\text{mean}}}{\lambda_D}\right)^3$$

is the plasma parameter where $\lambda_{\text{mean}} = n^{-1/3}$ is an average distance between particles in plasma.

Let us calculate the next corrections to free energy and pair distribution function, and evaluate the accuracy of our approach. Free energy looks like

$$\begin{aligned} F(n,\beta) &\approx F_0(n,\beta) + F_2(n,\beta) \\ &= -\frac{n}{3}\left\{A - \frac{3}{2}\left[1 - \frac{A}{4} - \frac{1}{A}\int_0^\infty ds\ s^2 \left[1 - e^{-\frac{A}{s}e^{-s}}\right]\right]\right\} . \end{aligned}$$

Accuracy of the Gaussian equivalent representation can be evaluated by the ratio

$$\delta(A) = \frac{|F_2(A)|}{|F_0(A)|} \approx \frac{3}{8} \frac{A}{(1 + A^{\frac{3}{4}})^{\frac{4}{3}}} < \frac{3}{8}, \qquad \forall A > 0 .$$

One can see that the accuracy is acceptable practically for all A. The standard point of view is that the Debey-Hückel approach is valid for a small plasma parameter $A \ll 1$.

Now let us calculate the screened Coulomb potential. The pair distribution function is defined as

$$n_2(\mathbf{x}-\mathbf{x}') = \frac{1}{\mathcal{Z}} \cdot \int \frac{D\phi}{\sqrt{\det D}} e^{-\frac{1}{2}(\phi D^{-1}\phi)} \vdots e^{i\sqrt{\beta}\phi(\mathbf{x})} \vdots_D \vdots e^{i\sqrt{\beta}\phi(\mathbf{x}')} \vdots_D e^{W_I[\phi]}$$
$$= e^{-\beta V_{\text{eff}}(r)}, \qquad r = |\mathbf{x}-\mathbf{x}'|.$$

The effective screened potential in the lowest and next approximation is

$$\beta\, V_{\text{eff}}(r) \approx \frac{\beta e^2}{r} e^{-\kappa r} + \beta V_2(r),$$
$$\beta V_2(r) = n \int d\mathbf{y} \left\{ \beta^2 D(\mathbf{r}-\mathbf{y}) D(\mathbf{y}) - \left(1 - e^{-\beta D(\mathbf{y})}\right)\left(1 - e^{-\beta D(\mathbf{r}-\mathbf{y})}\right) \right\}.$$

These equations lead to the screened Coulomb potential for proton plasma at the center of our Sun shown in Fig. 1. For the calculation we used $kT_\odot = 1.3 \cdot 10^3\ eV$, $\rho_\odot = 100\ \frac{\text{g}}{\text{cm}^3}$. The resulting Debye and plasma parameters are

$$\kappa_\odot = \sqrt{4\pi n_\odot \beta_\odot e^2} = 1.02 \cdot 10^9\ \text{cm}^{-1}, \qquad \lambda_\odot = 0.98 \cdot 10^{-9}\ \text{cm},$$
$$A_\odot = \frac{1}{4\pi} \frac{\kappa_\odot^3}{n_\odot} = 1.42.$$

The accuracy according to (3) is

$$\delta(A_\odot) = \frac{|F_2(A_\odot)|}{|F_0(A_\odot)|} = 0.2.$$

One can see that the Debye-Hückel approximation is indeed a very good approximation for a true effective potential.

In summery one can conclude that

- The Gaussian equivalent representation is an effective method of approximate calculation of path integrals over a Gaussian measure.
- In the case of plasma, the corrections to the Debye-Hückel potential are small for any plasma parameter A.

References

1. R. P. Feynman and A. R. Hibbs, *Quantum Mechanics and Path Integrals* (McGraw-Hill Book Company, N.Y., 1965).
2. V. Dineykhan, G. V. Efimov, G. Ganbold, and S. N. Nedelko, *Oscillator Representation in Quantun Physics*, Lecture Notes in Physics New Series M, Vol. **26** (Springer-Verlag, Berlin, 1995).
3. L. M. Varela, M. Garcia, and V. Mosquera, *Phys. Rep.* **382**, 1 (2003).
4. R. Balescu, *Equilibrium and Nonequilibrium Statistical Mechanics* (Wiley-Interscience Publication, N.Y., 1975).

PATH INTEGRAL INSPIRED GAUGE INVARIANT INFRARED REGULARIZATION

A. A. SLAVNOV

Steklov Mathematical Institute,
Moscow 119991, Russia
E-mail: slavnov@mi.ras.ru

On the basis of the path-integral formulation of the Yang-Mills theory a gauge invariant infrared regularization is introduced. The regularized model includes higher derivatives, but in the limit when the regularization is removed, unphysical excitations decouple.

Keywords: Gauge invariance; Infrared regularization; Higher derivatives.

1. Introduction

Gauge invariance is one of the most essential ingredients of modern field theory. It is a common wisdom that the scattering matrix of the Yang-Mills theory does not depend on gauge fixing. However this statement in fact makes no sense as in the Yang-Mills theory a scattering matrix connecting the free asymptotic states, which include massless particles, does not exist. Some gauge invariant infrared regularization allowing to make sense of formal manipulations with the S-matrix generating functional certainly would be welcome. Moreover in nondiagonal gauges even the Green functions are plagued with infrared divergences which may lead to certain problems in studying Slavnov-Taylor identities in such theories.

In this talk I discuss an infrared regularization for the Yang-Mills theory, which may be described by a local gauge invariant Lagrangian. This Lagrangian contains higher derivatives and hence the regularized theory includes nonpositive norm states. However, in the limit when the regularization is removed the nonphysical states decouple.

In the next section a path-integral formulation of such a regularization will be presented. The third section deals with the field theoretical realization of this construction.

2. A Path-Integral Regularization

To illustrate the main idea I present a heuristic derivation of the regularized action for the $SU(2)$ gauge theory. Generalization to other groups is straightforward.

The following formal equality obviously holds:

$$\int \exp\{i \int [L_{YM} + m^2 \varphi^* \varphi] dx\} d\mu =$$
$$\int \exp\{i \int [L_{YM} + m^{-2}(D^2\varphi')^*(D^2\varphi') - d^* D^2 b - b^* D^2 d] dx\} d\mu' \quad (1)$$

Here L_{YM} is the usual Yang-Mills Lagrangian

$$L_{YM} = -\frac{1}{4} F^j_{\mu\nu} F^j_{\mu\nu} \quad (2)$$

and the complex scalar fields φ form the $SU(2)$ doublet, which may be conveniently written as follows:

$$\varphi_1 = \frac{iB^1 + B^2}{\sqrt{2}}; \quad \varphi_2 = \frac{B^0 + iB^3}{\sqrt{2}}. \quad (3)$$

Anticommuting complex scalar fields b, d form the similar doublets. The measure $d\mu$ includes differentials of all fields as well as gauge fixing factors. The measure $d\mu'$ differs by the presence of the differentials of the fields b, b^*, d, d^*. The operator D^2 denotes the sum $\sum_\mu D_\mu D_\mu$, where D_μ is the covariant derivative

$$D_\mu \varphi = (\partial_\mu + \frac{ig\tau^j}{2} A^j_\mu)\varphi. \quad (4)$$

Equation (1) is formal, as neither l.h.s. nor r.h.s. exist because of infrared divergences. We define the infrared regularized theory in the following way. Let us add to the action in the r.h.s. the gauge invariant term

$$\int \{\alpha (D_\mu \varphi')^* (D_\mu \varphi') - \alpha m^2 (d^* b + b^* d)\} dx. \quad (5)$$

The integral on the r.h.s. of Eq. (1) is still infrared divergent. However if we make the shift

$$\varphi' \to \varphi' + \hat{a}, \quad \hat{a}_1 = 0, \quad \hat{a}_2 = a, \quad (6)$$

the regularized action acquires a form

$$A_R = \int \{ -\frac{1}{4} F^j_{\mu\nu} F^j_{\mu\nu} - m^{-2}(D^2\varphi)^*(D^2\varphi) - (D_\mu d)^* D_\mu b - (D_\mu b)^* D_\mu d$$
$$-\frac{a^2 g^2}{4m^2}(\partial_\mu A_\mu)^2 + \frac{ag}{\sqrt{2}m^2} \partial^2 B^j \partial_\mu A^j_\mu + \alpha (D_\mu \varphi)^* (D_\mu \varphi)$$

$$+\frac{\alpha g^2 a^2}{4} A_\mu^2 - \frac{\alpha g a}{2\sqrt{2}} B^j \partial_\mu A_\mu^j - \alpha m^2 (d^* b + b^* d) + \dots \tag{7}$$

Here ... denote the interaction terms which arise due to shift (6).

One sees that in complete analogy with the Higgs model the shift (6) generates the mass term for the vector field, the term $(\partial_\mu A_\mu)^2$, the mixing of B^j with $\partial_\mu A_\mu^j$ and additional interaction terms. For simplicity in the following we choose $g^2 a^2 = 2m^2$. Then the mass of the Yang-Mills field is $\sqrt{\alpha} m$.

The theory described by the action (7) is free of infrared singularities. At the same time the action is local and invariant with respect to the gauge transformations. This invariance allows to use in the corresponding path integral any admissible gauge condition. Particularly convenient is the Lorentz gauge $\partial_\mu A_\mu = 0$. In this gauge the mixing between A_μ^j and B^j is absent and renormalizability is manifest.

One has to understand that the transformation (6) is not a simple change of variables. It changes the boundary conditions in the path integral. Rather it is a definition of the infrared regularized Yang-Mills theory. More precisely

$$\int \exp\{i \int L_{YM} dx\} d\mu|_{reg} = \int \exp\{i A_R\} d\mu'. \tag{8}$$

Equation (8) gives a definition of the infrared regularized scattering matrix for the Yang-Mills theory as a path integral of the exponent of a local gauge invariant action. It also allows to give a sensible definition of the correlation functions as in the regularized theory one can perform the Wick rotation in all Feynman integrals making the transition $\alpha \to 0$ legitimate.

In the next section we shall show that this path-integral regularization admits an elegant field theoretical realization, similar to the BRST quantization of gauge invariant models.

3. Canonical Quantization and Unitarity of Regularized Theory

It was shown in our papers[1,2] that a change of variables in a path integral which introduces higher derivatives may be interpreted as a transition to a field theory model including unphysical ghost fields. This theory possesses a (super)symmetry which leads via Noether theorem to existence of a conserved nilpotent charge Q. Existence of such a charge allows to separate the physical states by imposing the condition

$$Q|\psi\rangle_{\text{phys}} = 0. \tag{9}$$

These states have nonnegative norms and the scattering matrix is unitary in the subspace (9).

Below we shall show that a similar construction may be done in the present model. A peculiar feature of our model is related to the fact that contrary to the cases considered before the conserved charge Q is not nilpotent. Nilpotency is recovered only in the limit $\alpha \to 0$, and this limit, when it exists, determines the Yang-Mills theory. The limit $\alpha \to 0$ for the on-shell S-matrix does not exist due to infrared divergences, but the formal expression for the matrix elements in the limit when the regularization is removed coincides with the S-matrix elements of the original Yang-Mills theory.

Our starting point is the regularized action

$$A_R = \int \{-\frac{1}{4}F^j_{\mu\nu}F^j_{\mu\nu} - m^{-2}(D^2(\varphi+\hat{a}))^* D^2(\varphi+\hat{a}) + (D_\mu d)^* D_\mu b \\ + (D_\mu b)^* D_\mu d + \alpha[(D_\mu(\varphi+\hat{a}))^* D_\mu(\varphi+\hat{a}) - m^2(d^* b + b^* d)]\} dx. \quad (10)$$

This action is invariant with respect to the gauge transformations and the supersymmetry transformations

$$\delta\varphi = \varepsilon b, \\ \delta d = m^{-2} D^2(\varphi+\hat{a})\varepsilon, \quad (11)$$

where ε is an anti-Hermitean parameter anticommuting with b, d. Note that these transformations are not nilpotent: $\delta^2 d \neq 0$. The nilpotency is restored only in the limit $\alpha = 0$.

The action (10) is invariant both with respect to the gauge transformations and the transformations (11). The supersymmetry transformations do not change the fields A_μ, so it is convenient to choose for quantization a manifestly supersymmetric and renormalizable gauge $\partial_\mu A_\mu = 0$.

In this gauge the Lagrangian may be written in terms of the components B^a, B^0, and the similar components for the fields b, d,

$$b_1 = \frac{ib^1 + b^2}{\sqrt{2}}, \quad b_2 = \frac{b^0 + ib^3}{\sqrt{2}}, \\ d_1 = \frac{d^1 - id^2}{\sqrt{2}}, \quad d_2 = \frac{-id^0 + d^3}{\sqrt{2}}, \quad (12)$$

as follows

$$L_R = -\frac{m^{-2}}{2}\partial^2 B^\rho \partial^2 B^\rho - i\partial_\mu b^\rho \partial_\mu d^\rho - \frac{1}{4}(\partial_\mu A^a_\nu - \partial_\nu A^a_\mu)^2 \\ + \alpha[\partial_\mu B^\rho \partial_\mu B^\rho - im^2 b^\rho d^\rho] + \dots \quad (13)$$

where $\rho = 0, 1, 2, 3$ and $\dots$ denote the interaction terms.

The quantization of the Yang-Mills fields A^j_μ is performed in a standard way and requires the introduction of the Faddeev-Popov ghosts $\bar{c}, c$. The scalar fields b, d also make no problems.

The fields B^ρ are described by the higher derivative Lagrangian and for their quantization we use the Ostrogradsky canonical formalism.[1–3]

Introducing the creation and annihilation operators for the excitations corresponding to the B-fields one can write the free Hamiltonian in the form

$$H^0 = \int [\omega_1(k) q_1^{\rho+}(k) q_1^{\rho-}(k) - \omega_2(k) q_2^{\rho+}(k) q_2^{\rho-}(k)] dk. \qquad (14)$$

In these equations $\omega_1 = \sqrt{k^2}$, $\omega_2 = \sqrt{k^2 + \alpha m^2}$, and the operators $q^{\rho\pm}_{1,2}$ satisfy the commutation relations

$$[q_1^{\rho-}(k), q_1^{\sigma+}(k')] = \delta^{\rho\sigma}\delta(k-k'), \quad [q_2^{\rho-}(k), q_2^{\sigma+}(k')] = -\delta^{\rho\sigma}\delta(k-k'). \quad (15)$$

One sees that the operators $q_2^{\rho+}$ create negative norm states.

The creation and annihilation operators for the supersymmetry ghosts may be introduced in a standard way. They are given by the equations:

$$d^{\rho\pm} = \frac{d^\rho \omega_2 \pm p_b^\rho}{\sqrt{2\omega_2}}, \quad b^{\rho\pm} = \frac{b^\rho \omega_2 \mp p_d^\rho}{\sqrt{2\omega_2}}. \qquad (16)$$

They satisfy the anticommutation relations

$$[b^{\rho-}(k), d^{\sigma+}(k')]_+ = -i\delta^{\rho\sigma}\delta(k-k'), \quad [d^{\rho-}(k), b^{\sigma+}(k')]_+ = i\delta^{\rho\sigma}\delta(k-k'). \qquad (17)$$

The space of states includes many unphysical excitations, like the supersymmetry ghost states, states corresponding to the fields B^ρ, unphysical components of A_μ and Faddeev-Popov ghosts. The real physical states including only transversal components of the Yang-Mills field may be separated by imposing on the asymptotic states the conditions

$$Q_0|\psi\rangle_{\text{phys}} = 0, \qquad (18)$$

$$Q_0^{\text{BRST}}|\psi\rangle_{\text{phys}} = 0, \qquad (19)$$

and taking the limit $\alpha \to 0$. Here Q_0 is the free charge associated with the supersymmetry transformations (11), and Q_0^{BRST} is the free BRST charge.

The invariance of the action (10) with respect to the supersymmetry transformations (11) generates via Noether's theorem the conserved current, and the corresponding conserved asymptotic charge may be written as follows

$$Q_0 = \frac{1}{\sqrt{2\omega_2}} \int \{ b^{\rho+}(P_1^\rho + i\omega_2 P_2^\rho - \alpha X_2^\rho) + (P_1^\rho - i\omega_2 P_2^\rho - \alpha X_2^\rho) b^{\rho-} \} dk$$

$$\sim \int \{b^{\rho+}(k)\frac{q_1^{\rho-}(k)+q_2^{\rho-}(k)}{2}+\frac{q_1^{\rho+}(k)+q_2^{\rho+}(k)}{2}b^{\rho-}(k)\}dk+O(\alpha). \quad (20)$$

One sees that although for a finite α the charge Q_0 is not nilpotent, in the limit $\alpha \to 0$ the nilpotency is recovered as the operators $b^{\rho+}, b^{\rho-}$ and $q_+^{\rho\pm} = q_1^{\rho\pm} + q_2^{\rho\pm}$ are mutually (anti)commuting.

Any vector annihilated by Q_0 may be presented in the form

$$|\varphi\rangle = |\varphi\rangle_A + Q_0|\chi\rangle + O(\alpha). \quad (21)$$

Here $|\varphi\rangle_A$ is a vector which does not include the excitations, corresponding to the ghost fields $q^\rho_{1,2}$ and b^ρ, c^ρ. This vector depends only on the Yang-Mills field excitations and the Faddeev–Popov ghosts. Imposing on it the condition (19), which is compatible with the condition (18), we conclude that the vectors $|\psi\rangle_{\mathrm{phys}}$ have a form

$$|\psi\rangle_{\mathrm{phys}} = |\psi\rangle_{\mathrm{tr}} + |N\rangle + O(\alpha) \quad (22)$$

where $|\psi\rangle_{\mathrm{tr}}$ depends only on transversal polarizations of the Yang-Mills field, and $|N\rangle$ is a zero norm vector. Hence in the limit $\alpha \to 0$ we recover the usual Yang-Mills theory. It completes the proof.

Acknowledgments

I thank the organizers of the conference for a generous support. This work was also partially supported by RBRF, under grant 050100541 and by the RAS program "Theoretical problems of mathematics".

References

1. A. A. Slavnov, *Phys. Lett. B* **258**, 391 (1991).
2. A. A. Slavnov, *Phys. Lett. B* **620**, 97 (2005).
3. A. A. Slavnov, *Nucl. Phys. B* **31**, 301 (1971).

CAUSAL SIGNAL TRANSMISSION BY INTERACTING QUANTUM FIELDS

L. I. PLIMAK* and W. P. SCHLEICH

Abteilung Quantenphysik, Universität Ulm, D-89069 Ulm, Germany
**E-mail: lev.plimak@uni-ulm.de*

S. STENHOLM

Physics Department, Royal Institute of Technology, KTH, Stockholm, Sweden
Laboratory of Computational Engineering, HUT, Espoo, Finland

General response properties of interacting bosonic fields are investigated.

Keywords: Quantum field theory; Nonlinear quantum response problem; Phase-space methods.

Dedicated to the memory of A. N. Vasil'ev

1. Response of the Harmonic Oscillator

We assume that the reader is familiar with standard facts from the theory of the driven harmonic oscillator.[1] We shall use without definition or proof the concepts of normal ordering, time ordering, double-time ordering on the Schwinger-Perel-Keldysh C-contour, and the related Wick's theorems.[2–4] For omitted technical details see the extended version of these notes in the Annals of Physics.[5,6]

We consider a driven quantum harmonic oscillator, with the Hamiltonian in the Schrödinger picture

$$\hat{\mathcal{H}}_{\mathrm{S}} = \hat{\mathcal{H}}_{0\mathrm{S}} - j(t)\hat{q}, \quad \hat{\mathcal{H}}_{0\mathrm{S}} = \frac{\hat{p}^2}{2m} + \frac{m\omega_0^2\hat{q}^2}{2} = \hbar\omega_0\left(\hat{a}^\dagger\hat{a} + \frac{1}{2}\right), \tag{1}$$

where $\hat{q}$ and $\hat{p}$ are the position and momentum operators, $[\hat{q},\hat{p}] = i\hbar$, and $\hat{a}^\dagger, \hat{a}$ are the creation and annihilation operators, $\left[\hat{a},\hat{a}^\dagger\right] = 1$. The Heisenberg position operator associated with $\hat{\mathcal{H}}_S$ reads,

$$\hat{q}_j(t) = \hat{q}(t) + q_j(t), \tag{2}$$

$$\hat{q}(t) = \sqrt{\hbar/2m\omega_0}\left(\hat{a}e^{-i\omega_0 t} + \hat{a}^\dagger e^{i\omega_0 t}\right), \tag{3}$$

$$q_j(t) = \int dt' D_R(t-t')j(t'), \tag{4}$$

where $\hat{q}(t)$ is the "interaction-picture" operator, and $q_j(t)$ is the c-number amplitude created by the c-number source $j(t)$. All integrals in these notes are from minus to plus infinity. In (4), D_R is the retarded Green's function of the classical oscillator,

$$\ddot{D}_R(t) + \omega_0^2 D_R(t) = \frac{\delta(t)}{m}, \quad D_R(t) = \theta(t)\frac{\sin\omega_0 t}{m\omega_0}. \tag{5}$$

With future extension to arbitrary interacting nonlinear nonequilibrium quantum-statistical systems in mind, we describe the oscillator within Schwinger's closed-time-loop formalism.[3,4] We construct the generating functional of Keldysh-style Green's functions,

$$\Phi(\eta_-,\eta_+;j) = \left\langle T_-\exp\left[-i\int dt\, \eta_-(t)\hat{q}_j(t)\right] T_+\exp\left[i\int dt\, \eta_+(t)\hat{q}_j(t)\right]\right\rangle, \tag{6}$$

where the functional arguments $\eta_\pm(t)$ are arbitrary smooth c-number functions, and the quantum averaging is with respect to the initial (Heisenberg) state of the oscillator, $\langle\cdots\rangle = \mathrm{Tr}\rho_0(\cdots)$. A systematic introduction to functional techniques may be found in Ref. 7.

Applying, firstly, Eq. (2), and, secondly, Wick's theorem for the double-time ordering on the C-contour, we split $\Phi(\eta_-,\eta_+;j)$ in three factors,

$$\Phi(\eta_-,\eta_+;j) = \Phi_{cl}(\eta;j)\Phi_{in}(\eta)\Phi_{vac}(\eta_-,\eta_+), \tag{7}$$

describing contributions of the external source, initial state and vacuum fluctuations, respectively:

$$\eta(t) = -i\left[\eta_+(t) - \eta_-(t)\right], \tag{8}$$

$$\Phi_{cl}(\eta;j) = \exp\int dt\, \eta(t)q_j(t), \tag{9}$$

$$\Phi_{in}(\eta) = \left\langle :\exp\int dt\, \eta(t)\hat{q}(t): \right\rangle, \tag{10}$$

$$\Phi_{vac}(\eta_-,\eta_+) = \exp\left\{i\hbar\int dtdt'\left[\frac{1}{2}\eta_+(t)\eta_+(t')D_F(t-t') - \frac{1}{2}\eta_-(t)\eta_-(t')D_F^*(t-t') - \eta_-(t)\eta_+(t')D(t-t')\right]\right\}. \tag{11}$$

The functional Φ_{vac} is "supplied" by Wick's theorem and is expressed by contractions,

$$-i\hbar D_F(t-t') = \langle 0\,|T_+\hat{q}(t)\hat{q}(t')|\,0\rangle\,, \tag{12}$$

$$i\hbar D_F^*(t-t') = \langle 0\,|T_-\hat{q}(t)\hat{q}(t')|\,0\rangle\,, \tag{13}$$

$$-i\hbar D(t-t') = \langle 0\,|\hat{q}(t)\hat{q}(t')|\,0\rangle\,. \tag{14}$$

Its interpretation as "vacuum fluctuations" follows from the fact that it coincides with $\Phi(\eta_-,\eta_+;j)$ for $j=0$ and $\rho_0 = |0\rangle\langle 0|$.

Of the three factors in Eq. (7), Φ_{cl} affords a fully classical interpretation, Φ_{in} may be regarded classical or quantum depending on whether the oscillator is in a classical or quantum state, while, it would seem, the quantum nature of Φ_{vac} is beyond doubt, because replacing $\hat{a},\hat{a}^\dagger$ by c-number amplitudes turns it into unity. However, belief in the "fully quantum" nature of Φ_{vac} is shuttered if we note that by a mere change of functional variables we express it by the "fully classical" Φ_{vac}:

$$\Phi_{\text{vac}}(\eta_-,\eta_+) = \Phi_{\text{cl}}(\eta;\sigma), \tag{15}$$

$$\eta(t) = -i\big[\eta_+(t)-\eta_-(t)\big],\quad \sigma(t) = \hbar\left[\eta_+^{(+)}(t)+\eta_-^{(-)}(t)\right], \tag{16}$$

$$\eta_+(t) = i\eta^{(-)}(t)+\frac{1}{\hbar}\sigma(t),\quad \eta_-(t) = -i\eta^{(+)}(t)+\frac{1}{\hbar}\sigma(t). \tag{17}$$

The change of variables $\eta_-,\eta_+ \leftrightarrow \eta,\sigma$ employs separating of the frequency-positive and negative parts,

$$g^{(\pm)}(t) = \int\frac{d\omega}{2\pi}\,\mathrm{e}^{-i\omega t}\theta(\pm\omega)g_\omega,\quad g_\omega = \int dt\,\mathrm{e}^{i\omega t}g(t). \tag{18}$$

Equation (15) is a consequence of the relations

$$D(t) = D_{\mathrm{R}}^{(+)}(t) - D_{\mathrm{R}}^{(-)}(-t),\quad D_{\mathrm{F}}(t) = D_{\mathrm{R}}^{(+)}(t) + D_{\mathrm{R}}^{(+)}(-t). \tag{19}$$

For the oscillator, they can be verified directly.

Equation (15) shows that, for the harmonic oscillator, "vacuum fluctuations" are a (rather tangled) way of expressing response properties of the system. This is also the case for an arbitrary free bosonic field, see Ref. 5.

2. Response of Interacting Bosons

Here we consider the generic case of a nonlinear quantum oscillator interacting with an environment. For the general case of an interacting bosonic field see Ref. 6. Formally, we generalise the treatment of the harmonic oscillator by assuming that H_{OS} in Eq. (1) is arbitrary. The Heisenberg operator of coordinate associated with H_{S} is denoted as $\hat{Q}_j(t)$. The system is characterised by the Keldysh-style Green's functions of $\hat{\mathcal{Q}}_j(t)$ through their generating functional $\Phi(\eta_-, \eta_+; j)$, given by Eq. (6) with $\hat{q}_j(t) \to \hat{\mathcal{Q}}_j(t)$. Applying the substitution (16) defines a new functional $\Phi_{\mathrm{R}}(\eta; j)$,

$$\Phi(\eta_-, \eta_+; j) = \Phi_{\mathrm{R}}(\eta; j + \sigma). \tag{20}$$

The critical feature is that the functional variable $\sigma(t)$ and the external source $j(t)$ enter in Eq. (20) as a sum. This is a general property of an arbitrary interacting bosonic field, see Ref. 6.

Interpretation of $\Phi_{\mathrm{R}}(\eta; j)$ follows the analogy with the harmonic oscillator. For the latter, using Eqs. (7) and (15) we find

$$\Phi_{\mathrm{R}}(\eta; j) = \left\langle :\exp \int dt\, \eta(t) \hat{q}_j(t): \right\rangle = \exp \int dt dt' \eta(t) D_{\mathrm{R}}(t - t') j(t'). \tag{21}$$

For interacting bosons, we postulate that, regarded as a functonal of $\eta(t)$, $\Phi_{\mathrm{R}}(\eta; j)$ generates the *time-normal averages*[1,8] of $\hat{\mathcal{Q}}_j(t)$,

$$\Phi_{\mathrm{R}}(\eta; j) = \left\langle \mathcal{T}\!:\exp \int dt\, \eta(t) \hat{\mathcal{Q}}_j(t): \right\rangle. \tag{22}$$

As a functional of both $\eta(t)$ and $j(t)$, $\Phi_{\mathrm{R}}(\eta; j)$ generates *quantum-statistical response functions*,

$$\begin{aligned} \mathcal{D}_{\mathrm{R}}^{(m;n)}(t_1, \cdots, t_m; t'_1, \cdots, t'_n) &= \left. \frac{\delta^n \langle \mathcal{T}\!:\hat{\mathcal{Q}}_j(t_1) \cdots \hat{\mathcal{Q}}_j(t_m):\rangle}{\delta j(t'_1) \cdots \delta j(t'_n)} \right|_{j=0} \\ &= \left. \frac{\delta^{m+n} \Phi_{\mathrm{R}}(\eta; j)}{\delta \eta(t_1) \cdots \delta \eta(t_m) \delta j(t'_1) \cdots \delta j(t'_n)} \right|_{\eta=j=0}. \end{aligned} \tag{23}$$

Implications of Eqs. (22) and (23) are discussed in detail in Ref. 6. Here we only mention, without a proof, the most important points.

- Glauber-Kelley-Kleiner's definition of the time-normal ordering[8] is recovered from (22) in the approximation of slowly varying amplitudes.
- Equation (23) for $m = n = 1$ produces Kubo's famous formula for the linear response function,[9]

$$\mathcal{D}_{\mathrm{R}}^{(1;1)}(t; t') = \frac{i}{\hbar} \theta(t - t') \langle [\hat{\mathcal{Q}}_j(t), \hat{\mathcal{Q}}_j(t')] \rangle.$$

- Explicit causality we see in Kubo's formula is a general property of the response functions, namely,

$$\begin{aligned} &\mathcal{D}_{\mathrm{R}}^{(m;n)}(t_1, \cdots, t_m; t'_1, \cdots, t'_n) = 0, \ \max\{t\} < \max\{t'\}, \\ &\mathcal{D}_{\mathrm{R}}^{(0;n)}(; t'_1, \cdots, t'_n) = 0. \end{aligned}$$

- The functional $\Phi_{\mathrm{R}}(\eta;\sigma)$ emerges if applying the substitution (17) to $\Phi(\eta_-,\eta_+) = \Phi(\eta_-,\eta_+;j)|_{j=0}$. This means that full information about response properties of the system is contained in the field operator $\hat{Q}(t) = \hat{Q}_j(t)|_{j=0}$. This fact has profound consequences for the quantum measurement theory.
- A close inspection of explicit formulae for the response functions reveals a fundamental connection between response and noncommutativity of operators. In particular, neglecting formally noncommutativity of the field operators does not make the system classical, as could be expected, but turns it into an "external quantum source." Hence the classical limit of quantum mechanics can only be taken by analysing physics, and not by formal means like setting $\hbar$ to zero or disregarding noncommutativity.

References

1. L. Mandel and E. Wolf, *Optical Coherence and Quantum Optics* (Cambridge University Press, 1995).
2. S. S. Schweber, *An Introduction to Relativistic Quantum Field Theory* (Dover, 2005).
3. J. S. Schwinger, *J. Math. Phys.* **2**, 407 (1961).
4. O. V. Konstantinov and V. I. Perel, *Zh. Eksp. Theor. Phys.* **39**, 197 (1960) [*Sov. Phys. JETP* **12**, 142 (1961)]; L. V. Keldysh, *ibid.* **47**, 1515 (1964).
5. L. I. Plimak and S. Stenholm, *Annals of Physics*, in print (DOI 10.1016/j.aop.2007.11.013, 2008).
6. L. I. Plimak and S. Stenholm, *Annals of Physics*, in print (DOI 10.1016/j.aop.2007.11.014, 2008).
7. A. N. Vasil'ev, *Functional Methods in Quantum Field Theory and Statistical Physics* (CRC Press, 1998).
8. P. L. Kelley and W. H. Kleiner, *Phys. Rev.* **136**, A316 (1964); R. J. Glauber, in *Quantum Optics and Electronics*, Les Houches Summer School of Theoretical Physics (Gordon and Breach, New York, 1965).
9. R. Kubo, *Statistical Mechanics* (North-Holland, Amsterdam, 1965).

RIGOROUS FUNCTIONAL INTEGRATION WITH APPLICATIONS IN QFT

J. LÖRINCZI

School of Mathematics, Loughborough University,
Loughborough LE11 3TU, United Kingdom
E-mail: J.Lorinczi@lboro.ac.uk

Recent developments in rigorous functional integration applied to the Nelson and Pauli-Fierz models are presented.

Keywords: Functional integration; Gibbs measure; Nelson model; Pauli-Fierz model.

1. Introduction

The purpose of this note is to illustrate through Nelson's and the Pauli-Fierz models how functional integration can be applied in a rigorous way to solve problems of quantum field theory. The mathematical approach has specific benefits. Although for technical reasons particular proofs may involve restrictions on the reach of statements, the conclusions are reliable against a clear background of hypotheses. Secondly, this approach does allow to derive genuinely new results or understanding. Finally, the concepts and tools of modern stochastic analysis have a far reaching potential of cross-fertilizing the methods developed by physicists. For the details and proofs of the summary presentation below we refer to the (severely selected) bibliography.

2. Functional Integration for Nelson's Model

The *Nelson model* describes an electrically charged spinless quantum particle coupled to a scalar boson field. The Hamiltonian is given by

$$H_{\mathrm{N}} = H_{\mathrm{p}} + H_{\mathrm{f}} + H_{\mathrm{i}} \quad \text{on } L^2(\mathbb{R}^d, dx) \otimes \mathcal{F}$$

with free particle Hamiltonian $H_{\mathrm{p}} = (-(1/2)\Delta + V(x)) \otimes 1$, free field Hamiltonian $H_{\mathrm{f}} = 1 \otimes \int_{\mathbb{R}^d} \omega(k) a^*(k) a(k) dk$, and interaction $H_{\mathrm{i}} =$

$\int_{\mathbb{R}^d} (2\omega(k))^{-1/2} \left(\widehat{\varrho}(k)e^{ik\cdot x} \otimes a(k) + \text{h.c.}\right)$. Here $\mathcal{F}$ stands for Fock space, a and a^* are the boson field operators, ω is the dispersion relation (equal to $|k|$ in the interesting case of massless bosons), and ϱ is the charge distribution function (whose Fourier transform appears in the formula), imposing a UV cutoff. The external potential V acting on the particle is assumed to be Kato-decomposable, see Ref. 1; this class is reasonably large to accommodate semibounded potentials having local singularities, for instance, Coulomb potential.

Under the mild conditions $|\widehat{\varrho}|\omega^{-1/2}, |\widehat{\varrho}|\omega^{-1} \in L^2$ Nelson's Hamiltonian is self-adjoint on $D(H_\mathrm{p}) \cap D(H_\mathrm{f})$, the joint domain of the free Hamiltonians. Moreover, H_N has a unique, strictly positive ground state $\Psi \in L^2 \otimes \mathcal{F}$ as soon as $|\widehat{\varrho}|\omega^{-3/2} \in L^2$.[2] The latter condition is interpreted to be an IR cutoff.

Traditional physics offers as an alternative a path-integrals framework for studying the features of quantum mechanical models. A similar reformulation can be made for the Nelson model as well. By writing down the Euclidean actions $S_\mathrm{p}, S_\mathrm{f}, S_\mathrm{i}$ associated with the three operators above, and denoting by X_t and ξ_t the particle resp. field processes (see more below), the path measure is formally written as

$$\prod_{t\in\mathbb{R}} dX_t e^{-S_\mathrm{p}(\{X_t\})} \prod_{(t,x)\in\mathbb{R}^{d+1}} d\xi_t(x)\, e^{-S_\mathrm{f}(\{\xi_t\})}\, e^{-S_\mathrm{i}(\{X_t,\xi_t\})}.$$

This is a mathematically unappealing almost surely $0 \times \infty$ situation, but happily we can prove that cancellations do occur and can make sense of a path measure for the full time line $\mathbb{R}$ through a limiting procedure.

A joint use of the so called *ground state transform* and *Wiener-Itô transform* allows mapping the space $L^2(\mathbb{R}^d, dx) \otimes \mathcal{F}$ into a suitable function space. Through this procedure it is found that e^{-tH_p} gives rise to a $P(\phi)_1$*-process* $\mathbb{R} \ni t \mapsto X_t \in \mathbb{R}^d$ whose path measure can be written as $d\mathcal{N}^0_T(X) = e^{-\int_{-T}^{T} V(X_t)dt} d\mathcal{W}_T(X)$ by using the Feynman-Kac formula for Schrödinger operators. Here $\mathcal{W}_T$ denotes Wiener measure for $\mathbb{R}^d$-valued Brownian motion running in the time interval $[-T, T] \subset \mathbb{R}$. The boson field is mapped into the space of tempered distributions $\mathcal{S}'(\mathbb{R}^d)$ over Schwartz space, and e^{-tH_f} gives rise to the *Ornstein-Uhlenbeck process* $\mathbb{R} \ni t \mapsto \xi_t \in \mathcal{S}'(\mathbb{R}^d)$ with Gaussian path measure $\mathcal{G}$ characterized by $\mathbb{E}_\mathcal{G}[\xi_t(f)] = 0$ and $\mathbb{E}_\mathcal{G}[\xi_t(f)\xi_s(g)] = \int_{\mathbb{R}^d} \widehat{f}(k)\overline{\widehat{g}(k)}(2\omega(k))^{-1}e^{-\omega(k)|t-s|}dk$. To operate with continuous paths of the Ornstein-Uhlenbeck process we prefer to replace $\mathcal{S}'(\mathbb{R}^d)$ with a full-measure subset $\mathcal{B}$ which we do not give here explicitly. Under the joint map the interaction Hamiltonian H_i turns into the multiplication operator $(\xi_t * \varrho)(x)$ (where the star denotes convolution). The *Euclidean Hamiltonian* $\tilde{H}_{\mathrm{N}}$ can then be identified as the sum of the images of all the three above Hamiltonians apart, which at the same time are the generators of the stochastic processes we have just discussed.

The path measure for bounded intervals $[-T, T]$ is then given by a suitable extension of the Feynman-Kac formula.

Theorem 2.1 (Nelson). *For square integrable F, G we have*

$$(F, e^{-2T\tilde{H}_{\mathrm{N}}} G)_{L^2} = \int \overline{F(X_{-T}, \xi_{-T})} G(X_T, \xi_T) d\mathcal{P}_T$$

where

$$d\mathcal{P}_T = \frac{1}{Z_T} e^{-\int_{-T}^{T} (\xi_t * \varrho)(X_t) dt} d\mathcal{P}_T^0$$

is the path measure describing the interacting system, and $\mathcal{P}_T^0 = \mathcal{N}_T^0 \times \mathcal{G}$ is that of the non-interacting system serving here for reference measure.

As the field variables ξ_t appear linearly in the exponent, the Gaussian integrals can be made explicitly. Hence for sets E and F in the Borel σ-fields (algebras of events) of $C(\mathbb{R}, \mathbb{R}^d)$ resp. $C(\mathbb{R}, \mathcal{B})$, by the identity $\mathcal{P}_T(E \times F) = \int_E \mathcal{P}_T(F|X = Y) d\mathcal{N}_T(Y)$ we get

$$d\mathcal{N}_T(X) = \frac{1}{Z_T} e^{-\int_{-T}^{T} \int_{-T}^{T} W(X_s - X_t, s-t) ds dt} d\mathcal{N}_T^0(X),$$

$$W(x, t) = -\frac{1}{2} \int_{\mathbb{R}^d} \frac{\widehat{\varrho}(k)}{2\omega(k)} \cos(k \cdot x) e^{-\omega(k)|t|} dk,$$

$$Z_T = \int e^{-\int_{-T}^{T} (\xi_t * \varrho)(X_t) dt} d(\mathcal{N}_T^0 \times \mathcal{G}) = \int e^{-\int_{-T}^{T} \int_{-T}^{T} W(X_s - X_t, s-t) ds dt} d\mathcal{N}_T^0.$$

It can be checked that $\mathcal{N}_T$ is a *Gibbs measure* (i.e., satisfies the Dobrushin-Lanford-Ruelle consistency conditions) on path space $C(\mathbb{R}, \mathbb{R}^d)$ with respect to the infinite range pair interaction potential W and reference measure $\mathcal{N}_T^0$. The last formula above gives the partition function (normalizing constant) for $\mathcal{N}_T$ turning it, and thus the full path measure $\mathcal{P}_T$, into a probability measure, for all $T > 0$. In lack of space here we can only briefly discuss what results can be derived by using the path measure $\mathcal{P}_T$, and more closely, the Gibbs measure $\mathcal{N}_T$.

Infinite time limit and ground state: Provided that a weak limit of $\mathcal{P}_T$ exists as $T \to \infty$, it can be shown that the ground state Φ of $\tilde{H}_{\mathrm{N}}$ (by the above constructions unitary equivalent with Ψ) can be written as

$$\Phi^2 = \frac{d\mathsf{P}}{d\mathsf{P}^0}$$

whenever P is absolutely continuous with respect to P^0, where P is the stationary distribution of the $T \to \infty$ limit of $\mathcal{P}_T$, and P^0 is that of the limit of $\mathcal{P}_T^0$. On the other hand, since it can be shown that the conditional probability $\mathcal{P}_T(F|X = Y)$ appearing in the identity quoted above is Gaussian, as this presents no particular difficulty the key to getting the weak limit of the path measure $\mathcal{P}_T$ lies in controlling $\mathcal{N}_T$. In Ref. 3 we constructed weak limits of Gibbs measures including $\mathcal{N}_T$ by using a cluster expansion and have furthermore obtained results on uniqueness of this limit, and properties of the limit measure such as mixing, typical path behaviour, and more.

Ground state properties: By using the formula above we are able to express ground state expectations (for operators A on the original Hilbert space) in terms of Gibbs averages (for associated functions f_A on its image space), which turn out to be easier to calculate or at least meaningfully to estimate than directly the scalar products. In particular, we can derive superexponential decay of boson sectors (a long standing open question), establish Coulomb's law for the average field and derive the field fluctuations, and show exponential localization of the particle in space.[4]

Infrared behaviour: Having the path measure at hand, we can define an independent way of quantization. As seen thus far, Fock space quantization is performed by taking the Hilbert space $\mathcal{H}^0 = L^2(\mathbb{R}^d \times \mathcal{B}, d\mathsf{P}^0)$ with Hamiltonian $H_{\rm N}$ unitary equivalent with $\widetilde{H}_{\rm N}$. In Euclidean quantization we can take $\mathcal{H} = L^2(\mathbb{R}^d \times \mathcal{B}, d\mathsf{P})$ instead. Then a semigroup T_t associated with the time reversible Markov stochastic process (X_t, ξ_t) can be defined through $(F, T_t G)_{\mathcal{H}} = \mathbb{E}_{\mathsf{P}}[F(X_0, \xi_0)G(X_t, \xi_t)]$, $\forall t > 0$. Since this is symmetric and contractive, there is a self-adjoint semibounded operator $H_{\rm euc}$ such that $T_t = e^{-tH_{\rm euc}}$. We will view this by definition as the Euclidean Hamiltonian. The function $1 \in \mathcal{H}$ is the unique ground state of $H_{\rm euc}$. In Ref. 5 we prove that in 3D there is a genuine IR divergence problem: if the IR cutoff is not applied, $\widetilde{H}_{\rm N}$ has no ground state in $\mathcal{H}^0$ (and hence $H_{\rm N}$ in Fock space). In particular, $H_{\rm N}$ and $H_{\rm euc}$ are not unitary equivalent, disproving the existing belief that Fock and Euclidean quantizations are equivalent schemes. In higher dimensions the problem disappears; both Hamiltonians and Hilbert spaces are unitary equivalent. 3D infrared divergence occurs due to infinitely many soft bosons in the ground state which cannot be accommodated in Fock space; this is ultimately due to too slow decay of Coulomb potential in this case. Even then, however, a physical ground state can be identified, which just fails to be a Fock space vector. By using the path measure we can make an IR-renormalization in 3D, i.e., compute a Hamiltonian that is unitary equivalent with $H_{\rm euc}$.[6]

Ultraviolet behaviour: Nelson has shown that, roughly speaking, by subtracting from H_{N} a constant logarithmically divergent in the point charge limit (i.e., when $\varrho \to \delta$), the so obtained UV renormalized Hamiltonian is well defined as a self-adjoint operator. In Ref. 7 we have shown that UV renormalization can be done on the level of the Gibbs measure and Nelson's energy renormalization scheme is a simple consequence of the Itô formula. We are able to show that the so obtained Hamiltonian is unitary equivalent with Nelson's and, going beyond his results, that furthermore it has a ground state.

3. Functional Integration for the Pauli-Fierz Model

The *Pauli-Fierz model* describes a charged particle coupled to a quantized Maxwell field. With similar notations as before, the Hamiltonian in 3D is

$$H_{\mathrm{PF}} = \frac{1}{2}(\sigma \cdot (-i\nabla \otimes 1 - eA))^2 + V \otimes 1 + 1 \otimes H_{\mathrm{f}}$$

with coupling constant $e = \int \varrho(x)dx$, vector potential

$$A_\mu(x) = \frac{1}{\sqrt{2}} \sum_{j=\pm 1} \int e_\mu(k,j) \left(\frac{\widehat{\varrho}(k)}{\sqrt{\omega(k)}} e^{ik\cdot x} \otimes a(k,j) + \text{h.c.} \right) dk,$$

polarization vectors satisfying $\sum_{j=\pm 1} e_\mu(k,j)e_\nu(k,j) = \delta_{\mu\nu} - (1/|k|^2)k_\mu k_\nu$, and Pauli matrices $\sigma_1, \sigma_2, \sigma_3$ accounting for spin. The bottom of the spectrum of the Pauli-Fierz Hamiltonian is contained in the absolutely continuous spectrum, no matter how weak the coupling is. Nonetheless, a ground state exists in Fock space for arbitrary $e \neq 0$, with no infrared cutoff.[8,9]

By a procedure in its essentials similar to the one described above, we can derive a path measure to the Pauli-Fierz model. In case spin is disregarded, on integrating out the boson field we arrive at the same effective pair potential as above, however, the Gibbs measure formally becomes

$$d\mathcal{N}_T(X) = \frac{1}{Z_T} e^{-\int_{-T}^{T}\int_{-T}^{T} W(X_s - X_t, s-t)dX_s dX_t} d\mathcal{N}_T^0(X),$$

i.e., the double Riemann integrals in the exponential densities turn into double stochastic integrals. To make sense of Gibbs measures in this case we have introduced the novel framework of Brownian currents combining rough paths analysis and cluster expansion.[10] When the spin is applied, we have a more complicated Feynman-Kac-type formula.[11] A corollary to this is the energy comparison inequality

$$\max_{\pi \in \Pi_3} E\left(0, 0, \sqrt{B_{\pi(1)}^2 + B_{\pi(2)}^2}, 0, B_{\pi(3)}\right) \le E(P, A, B_1, B_2, B_3),$$

where $E(P, A, B_1, B_2, B_3)$ is the ground state energy at fixed momentum P, $(B_1, B_2, B_3) = \nabla \times A$, and Π_3 is the permutation group of order 3. Moreover, in the spinless case we also prove for the ground state Φ that provided the external potential is sufficiently confining, there is a $b^* > 0$ such that $\Phi \in D(e^{bN})$ for all $b < b^*$, where N is the boson number operator, i.e., the boson sectors decay exponentially fast.[12] We also refer to the book (Ref. 13) covering these methods and results in great detail.

Acknowledgment

It is a pleasure to thank Max Planck Institute, Dresden, for kind hospitality.

References

1. V. Betz and J. Lőrinczi, *Uniqueness of Gibbs measures relative to Brownian motion*, *Ann. I. H. Poincaré* **39**, 877–889 (2003).
2. H. Spohn, *Ground state of a quantum particle coupled to a scalar boson field*, *Lett. Math. Phys.* **44**, 9–16 (1998).
3. J. Lőrinczi and R. A. Minlos, *Gibbs measures for Brownian paths under the effect of an external and a small pair potential*, *J. Stat. Phys.* **105**, 605–647 (2001).
4. V. Betz, F. Hiroshima, J. Lőrinczi, R. A. Minlos, and H. Spohn, *Ground state properties of the Nelson Hamiltonian – A Gibbs measure-based approach*, *Rev. Math. Phys.* **14**, 173–198 (2002).
5. J. Lőrinczi, R. A. Minlos, and H. Spohn, *The infrared behaviour in Nelson's model of a quantum particle coupled to a massless scalar field*, *Ann. Henri Poincaré* **3**, 1–28 (2002).
6. J. Lőrinczi, R. A. Minlos, and H. Spohn, *Infrared regular representation of the three dimensional massless Nelson model*, *Lett. Math. Phys.* **59**, 189–198 (2002).
7. M. Gubinelli and J. Lőrinczi, *Gibbs measures on Brownian currents*, arXiv:0704.3237 (math-ph), to appear in *Commun. Pure Appl. Math.* (2008).
8. V. Bach, J. Fröhlich, and I. M. Sigal, *Spectral analysis for systems of atoms and molecules coupled to the quantized radiation field*, *Commun. Math. Phys.* **207**, 249–290 (1999).
9. M. Griesemer, E. Lieb, and M. Loss, *Ground states in non-relativistic quantum electrodynamics*, *Invent. Math.* **145**, 557–595 (2001).
10. M. Gubinelli and J. Lőrinczi, *Ultraviolet renormalization of Nelson's model through functional integration*, submitted for publication (2007).
11. F. Hiroshima and J. Lőrinczi, *Functional integral representation of the Pauli-Fierz model with spin* 1/2, *J. Funct. Anal.* **254**, 2127–2185 (2008).
12. F. Hiroshima and J. Lőrinczi, *Exponential localization of the ground state in the Pauli-Fierz model for weak couplings*, preprint (2007).
13. J. Lőrinczi, F. Hiroshima, and V. Betz, *Feynman-Kac-Type Theorems and Gibbs Measures on Path Space. With Applications to Rigorous Quantum Field Theory* (Walter de Gruyter, Berlin-New York, 2008) (in print).

STABLE EXTENDED STRING-VORTEX SOLITONS

I. L. BOGOLUBSKY* and A. A. BOGOLUBSKAYA†

Laboratory of Information Technologies, Joint Institute for Nuclear Research, Joliot-Courie 6, 141980 Dubna, Moscow region, Russia
**E-mail: bogolubs@jinr.ru, †E-mail: abogol@jinr.ru*

A. I. BOGOLUBSKY

Department of Mathematics and Mechanics, Moscow State University, Leninskiye Gory 1, 119899 Moscow, Russia
E-mail: bogolubs@gmail.com

S. L. SKOROKHODOV

Dorodnicyn Computing Centre RAS, Vavilov st. 40, 119991 Moscow, Russia
E-mail: skor@ccas.ru

We discuss the properties of classical string-like topological solitons in the $A3M$ model. This model describes gauge-invariant interaction of Maxwell and "easy-axis" 3-component spin field

Keywords: Topological solitons; Strings; Hedgehog; Vortices.

1. Introduction

The investigation of field-theoretical models with extended string-like solutions opens possibilities for considering nonperturbative phenomena in quantum field theory, condensed matter physics (e.g. high-temperature superconductivity) and cosmology (the "cosmic strings" hypothesis). Such studies should include two main steps: (i) analytical or numerical search for and investigation of extended stable 2-dimensional solutions in physically motivated and aesthetically appealing models and (ii) their quantization using known methods based on the notion of *the path integral.*

2. 2D Stable Solitons in the $A3M$ Model

In the present contribution we focus on string-like localised solutions in a gauge-invariant S^2 sigma-model in (D+1)-dimensional ($D = 2$) spacetime. We discuss properties of two-dimensional (2D) stationary topological solitons in the A3M model,[1] which describes minimal interaction of the scalar Heisenberg 3-component unit isovector field $s_a(x)$, $s_a s_a = 1$, $a = 1, 2, 3$, with the Maxwell field.

The Lagrangian of the A3M model is:

$$\begin{aligned} \mathcal{L} = \eta^2 \left(\bar{\mathcal{D}}_\mu s_- \mathcal{D}^\mu s_+ + \partial_\mu s_3 \partial^\mu s_3\right) - V(s_a) - \frac{1}{4} F_{\mu\nu}^2, \\ \bar{\mathcal{D}}_\mu = \partial_\mu + igA_\mu, \quad \mathcal{D}_\mu = \partial_\mu - igA_\mu, \quad F_{\mu\nu} = \partial_\mu A_\nu - \partial_\nu A_\mu, \\ s_+ = s_1 + is_2, \quad s_- = s_1 - is_2, \quad V(s_a) = \beta^2(1 - s_3^2). \end{aligned} \tag{1}$$

where β^2, η^2 are constants, $\left[\eta^2\right] = L^{(1-D)}$, $\left[\beta^2\right] = L^{-(1+D)}$, g is a coupling constant, $[g^2] = L^{(D-3)}$, $\mu, \nu = 0, 1, ..., D$, and summation over repeated indices μ, ν is meant. The A3M model is a gauge-invariant extension of the classical Heisenberg antiferromagnet model with the *"easy-axis"* anisotropy;[2] it possesses global $Z(2)$ and local $U(1)$ symmetries; note that $Z(2)$ symmetry is spontaneously broken both on vacuum and soliton solutions.

The Euler-Lagrange equations of the model in dimensionless form are obtained by rescaling $x_\mu \to g^{-1}\eta^{-1}x_\mu$, $A_\mu \to \eta^{-1}A_\mu$, (we denote $p = \beta^2 g^{-2}\eta^{-4}$) and take the simplest form if the Lorentz gauge, $\partial_\mu A^\mu = 0$,

$$\begin{aligned} &\partial_\mu \partial^\mu s_i + [\partial_\mu s_a \partial^\mu s_a + 2A_\mu j^\mu + p(s_3^2 - \delta_{i3}) \\ &\quad + A_\mu A^\mu (s_1^2 + s_2^2 - \delta_{1i} - \delta_{2i})] s_i - 2A_\mu(\delta_{2i}\partial^\mu s_1 - \delta_{1i}\partial^\mu s_2) = 0, \\ &j_\mu = s_2 \partial_\mu s_1 - s_1 \partial_\mu s_2, \quad \partial_\mu \partial^\mu A_\nu + 2j_\nu + 2(s_1^2 + s_2^2)A_\nu = 0, \\ &\qquad \mu, \nu = 0, 1, ..., D, \quad i = 1, 2, 3. \end{aligned} \tag{2}$$

In angular variables, $s_1 = \sin\theta\cos\phi$, $s_2 = \sin\theta\sin\phi$, $s_3 = \cos\theta$, the Lagrangian density reads:

$$g^{-2}\eta^{-4}\mathcal{L} = \partial_\mu\theta\partial^\mu\theta + \sin^2\theta\left[\partial_\mu\phi\partial^\mu\phi - 2A_\mu\partial^\mu\phi + A_\mu A^\mu - p\right] - \frac{1}{4}F_{\mu\nu}^2 \tag{3}$$

and the Euler-Lagrange equations become:

$$\begin{aligned} \partial_\mu\partial^\mu\theta + \frac{1}{2}\sin 2\theta\left[p - \partial_\mu\phi\partial^\mu\phi + 2A_\mu\partial^\mu\phi - A_\mu A^\mu\right] = 0, \\ \partial_\mu\left[\sin^2\theta(\partial^\mu\phi - A^\mu)\right] = 0, \\ \partial_\mu\partial^\mu A_\nu + 2j_\nu + 2A_\nu\sin^2\theta = 0, \quad j_\nu = -\sin^2\theta\partial_\nu\phi. \end{aligned} \tag{4}$$

Time-independent soliton solutions $\phi(x) = \phi(\mathbf{x})$, $A_0 = 0$, $A_k(x) = A_k(\mathbf{x})$, $\theta(x) = \theta(\mathbf{x})$, $k = 1, ..., D$, obey equations:

$$\partial_k^2\theta - \frac{1}{2}\sin 2\theta\left[p + (\partial_k\phi - A_k)^2\right] = 0,$$

$$\partial_k \left[\sin^2\theta(\partial_k\phi - A_k)\right] = 0,$$
$$\partial_k^2 A_m + 2\sin^2\theta(\partial_m\phi - A_m) = 0. \tag{5}$$

($k, m = 1, \dots, D$, summation over repeated k is meant). To find $D = 2$ soliton solutions we use the "hedgehog" ansatz for the unit isovector field $s_i(\mathbf{x})$, $i = 1, 2, 3$,

$$s_1 = \cos m\chi \sin\theta(r), \quad s_2 = \sin m\chi \sin\theta(r), \quad s_3 = \cos\theta(r),$$
$$\sin\chi = \frac{y}{r}, \quad \cos\chi = \frac{x}{r}, \quad r^2 = x^2 + y^2, \tag{6}$$

where m is an integer number, and the "vortex" ansatz for the Maxwell field $A_\mu(\mathbf{x})$,

$$A_0 = 0, \quad A_1 = A_x = -m\alpha(r)\frac{y}{r^2}, \quad A_2 = A_y = m\alpha(r)\frac{x}{r^2}. \tag{7}$$

As a result we obtain equations for $\theta(r)$ and $\alpha(r)$,

$$\frac{d^2\theta}{dr^2} + \frac{1}{r}\frac{d\theta}{dr} - \sin\theta\cos\theta\left[\frac{m^2(\alpha-1)^2}{r^2} + p\right] = 0, \tag{8}$$

$$\frac{d^2\alpha}{dr^2} - \frac{1}{r}\frac{d\alpha}{dr} + 2(1-\alpha)\sin^2\theta = 0 \tag{9}$$

with boundary conditions

$$\theta(0) = \pi, \quad \theta(\infty) = 0, \tag{10}$$

$$\alpha(0) = 0, \quad \frac{d\alpha}{dr}(\infty) = 0. \tag{11}$$

Localised field configurations $s_i(\mathbf{x})$ given by Eqs. (6), (10) correspond to maps from R^2_{comp} to S^2 with integer homotopic indices ("winding numbers"),[2,3] and we shall refer to solutions of the problem (8)–(11) as topological solitons with the "topological charges" $Q_t = m$.

Detailed numerical and analytical investigation shows[1,5–7] that for positive p values, $0 < p < p_{\text{cr}}$ there exists a unique solution to the problem (8)–(11). Note that the asymptotic value $\alpha_\infty = \alpha(\infty)$ decreases monotonically as p is increased, with $\alpha_\infty \to 1$ when $p \to 0$.

The dependence of the soliton energy $E = 2\pi \int \mathcal{H}(r) r dr$ on p, where

$$\mathcal{H}(r) = \left(\frac{d\theta}{dr}\right)^2 + \sin^2\theta\left[p + \frac{m^2(\alpha-1)^2}{r^2}\right] + \frac{m^2}{2}\left(\frac{1}{r}\frac{d\alpha}{dr}\right)^2, \tag{12}$$

is depicted in Fig. 1. It is important to note that $E(p) < 8\pi$ for $p < p_{\text{cr}} \approx 0.4088$ (recall that 8π is the energy value of the Belavin-Polyakov localized solutions in the 2D isotropic Heisenberg ferromagnet[4]). It means that for $0 < p < p_{\text{cr}}$ the string-like solutions (8)–(11) of the A3M model describe

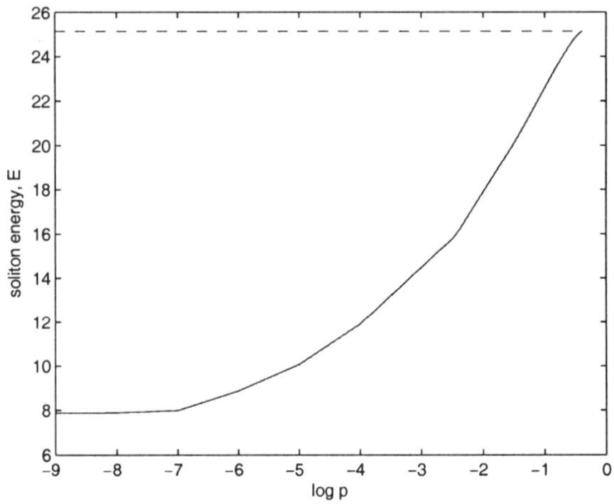

Fig. 1. Soliton energy versus p (solid line), dashed line represents $E_{\mathrm{BP}} = 8\pi$.

spatially localized *bound* states of the spin and Maxwell fields, and hence it is natural to conjecture these $2D$ solitons to be stable for $p < p_{\mathrm{cr}}$. Stability of these solitons for $0 < p < p_{\mathrm{cr}}$ has been confirmed by computer simulation.

Acknowledgments

The work of A. I. B. and S. L. S. was supported by the RFBR (project 07-01-00295), and by the Program No. 3 of the RAS.

References

1. I. L. Bogolubsky and A. A. Bogolubskaya, *Phys. Lett. B* **395**, 269 (1997).
2. A. M. Kosevich, B. A. Ivanov, A. S. Kovalev, *Nonlinear magnetisation waves. Dynamical and topological solitons* (Naukova dumka, Kiev, 1983) (in Russian).
3. A. S. Schwartz, *Quantum Field Theory and Topology* (Springer, Berlin, 1993).
4. A. A. Belavin and A. M. Polyakov, *Pis'ma v Zh.E.T.F.* **22**, 503 (1975) [*JETP Lett.* **22**, 245 (1975)].
5. I. L. Bogolubsky, B. Piette, and W. J. Zakrzewski, *Phys. Lett. B* **432**, 151 (1998).
6. S. Ai and X. Chen, *J. Diff. Eqs.* **153**, 61 (1999).
7. A. I. Bogolubsky and S. L. Skorokhodov, *Programming and Computer Software* **30**, 95 (2004).

RECENT RESULTS FOR YANG-MILLS THEORY RESTRICTED TO THE GRIBOV REGION

J. A. GRACEY

Theoretical Physics Division, Department of Mathematical Sciences, University of Liverpool, P.O. Box 147, Liverpool, L69 3BX, United Kingdom
E-mail: gracey@liv.ac.uk

We summarize recent results for the Gribov-Zwanziger Lagrangian which includes the effect of restricting the path integral to the first Gribov region. These include the two loop $\overline{\text{MS}}$ and one loop MOM gap equations for the Gribov mass.

Keywords: Gribov problem; Renormalization.

1. Introduction

The generalization of quantum electrodynamics to include non-abelian gauge fields produces the asymptotically free gauge theory called quantum chromodynamics (QCD) which describes the strong interactions. The natural forum to construct the properly gauge fixed (renormalizable) Lagrangian with which to perform calculations, is provided by the path integral machinery. For instance in the Landau gauge, which we concentrate on here, the Faddeev-Popov ghosts naturally emerge as a consequence of the non-gauge invariance of the path integral measure. Whilst the resulting Lagrangian more than adequately describes the ultraviolet structure of asymptotically free quarks and gluons, the infrared behaviour has not been fully established. For instance, it is evident that, as a result of confinement, gluons and quarks cannot have propagators of a fundamental type. Over the last few years there has been intense activity into measuring gluon and ghost form factors using lattice methods and the Dyson Schwinger formalism. Denoting these respectively by $D_A(p^2)$ and $D_c(p^2)$ a general picture emerges in that there is gluon suppression with $D_A(0) = 0$ and ghost enhancement where $D_c(p^2) \sim 1/(p^2)^\lambda$ as $p^2 \to 0$ with $\lambda > 0$. Such behaviour is not inconsistent with general considerations from confinement criteria.[1–9] Ideally given that these properties are now accepted, it is important that they can be explained from general field theory considerations. This was the approach of Zwanziger[4,5,7,8] in treating the Gribov problem from the path integral point of view. Therefore we will briefly review the construction of the Gribov-Zwanziger Lagrangian before giving a summary of recent results of using it in the Landau gauge.

2. Gribov-Zwanziger Lagrangian

Gribov pointed out[1] that in non-abelian gauge theories it is not possible to uniquely fix the gauge globally due to the existence of copies of the gauge field. To handle this the path integral was restricted to the first Gribov region, Ω, where $\partial\Omega$ is defined by the place where the Faddeev-Popov operator $\mathcal{M} = -\partial^\mu D_\mu$ first vanishes. Within the first Gribov region $\mathcal{M}$ is always positive and in the Landau gauge it is hermitian. Moreover Ω is convex and bounded[3] and all gauge copies transit[3] Ω. Any copy in the subsequent regions defined by the other zeroes of $\mathcal{M}$ can be mapped into Ω. Whilst the path integral is constrained to Ω, within Ω there is a region, Λ, known as the fundamental modular region where there are no gauge copies and the gauge is properly fixed. Although Λ is difficult to define, for practical purposes expectation values over Λ or Ω give the same values.[10] Consequently the gluon form factor is modified to $D_A(p^2) = (p^2)^2/[(p^2)^2 + C_A\gamma^4]$ where γ is the Gribov mass, whence suppression emerges.[1] The parameter γ is not independent and satisfies a gap equation. The theory can only be interpreted as a gauge theory when γ takes the value defined in the gap equation. Thence computing the one loop ghost propagator, it is enhanced precisely when the gap equation is satisfied.[1]

Gribov's revolutionary analysis was based on a semi-classical approach and then Zwanziger[4,5] extended it to a path integral construction by modifying the measure to restrict the integration region to Ω via the defining criterion known as the horizon condition,

$$\int A^a_\mu(x) \frac{1}{\partial^\nu D_\nu} A^{a\,\mu}(x) = \frac{dN_A}{C_A g^2} , \tag{1}$$

where d is the dimension of spacetime and N_A is the adjoint representation dimension.[5] For the Landau gauge the convexity and ellipsoidal properties of Ω allow one to modify the usual Yang-Mills Lagrangian to include the horizon condition, (1), producing the non-local Yang-Mills Lagrangian[4,5]

$$L^\gamma = -\frac{1}{4} G^a_{\mu\nu} G^{a\,\mu\nu} + \frac{C_A\gamma^4}{2} A^a_\mu \frac{1}{\partial^\nu D_\nu} A^{a\,\mu} - \frac{dN_A\gamma^4}{2g^2} . \tag{2}$$

Again (2) only has meaning when γ satisfies (1) which is equivalent to the Gribov gap equation. Finally the non-locality can be handled by using localizing fields to produce the Gribov-Zwanziger Lagrangian[5]

$$\begin{aligned} L^Z = L^{QCD} &+ \bar{\phi}^{ab\,\mu}\partial^\nu \left(D_\nu\phi_\mu\right)^{ab} - \bar{\omega}^{ab\,\mu}\partial^\nu \left(D_\nu\omega_\mu\right)^{ab} \\ &- g f^{abc}\partial^\nu \bar{\omega}^{ae}_\mu \left(D_\nu c\right)^b \phi^{ec\,\mu} \\ &+ \frac{\gamma^2}{\sqrt{2}}\left(f^{abc}A^{a\,\mu}\phi^{bc}_\mu + f^{abc}A^{a\,\mu}\bar{\phi}^{bc}_\mu\right) - \frac{dN_A\gamma^4}{2g^2}, \end{aligned} \tag{3}$$

where ϕ^{ab}_μ and ω^{ab}_μ are localizing ghost fields with the latter anti-commuting. This Lagrangian is renormalizable[7,11,12] and reproduces Gribov's one loop gap equation and ghost enhancement.[8] For (3) the horizon condition equates to

$$f^{abc}\langle A^{a\,\mu}(x)\phi^{bc}_\mu(x)\rangle = \frac{dN_A\gamma^2}{\sqrt{2}g^2}. \tag{4}$$

3. Calculations

As the Zwanziger construction has produced a renormalizable Lagrangian with extra fields incorporating infrared features without upsetting ultraviolet properties, such as asymptotic freedom, it is possible to extend the earlier one loop analysis.[1,8] For instance in $\overline{\text{MS}}$ the two loop gap equation results from (4) after computing 17 vacuum bubble graphs, giving,[13]

$$\begin{aligned} 1 = C_A &\left[\frac{5}{8} - \frac{3}{8}\ln\left(\frac{C_A\gamma^4}{\mu^4}\right)\right] a \\ &+ \left[C_A^2\left(\frac{2017}{768} - \frac{11097}{2048}s_2 + \frac{95}{256}\zeta(2) - \frac{65}{48}\ln\left(\frac{C_A\gamma^4}{\mu^4}\right)\right.\right. \\ &\qquad \left. + \frac{35}{128}\left(\ln\left(\frac{C_A\gamma^4}{\mu^4}\right)\right)^2 + \frac{1137}{2560}\sqrt{5}\zeta(2) - \frac{205\pi^2}{512}\right) \\ &\quad + C_A T_F N_f\left(-\frac{25}{24} - \zeta(2) + \frac{7}{12}\ln\left(\frac{C_A\gamma^4}{\mu^4}\right)\right. \\ &\qquad \left.\left. - \frac{1}{8}\left(\ln\left(\frac{C_A\gamma^4}{\mu^4}\right)\right)^2 + \frac{\pi^2}{8}\right)\right] a^2 + O(a^3), \end{aligned} \tag{5}$$

where $s_2 = (2\sqrt{3}/9)\mathrm{Cl}_2(2\pi/3)$ with $\mathrm{Cl}_2(x)$ the Clausen function, $\zeta(n)$ is the Riemann zeta function, and $a = \alpha_S/(4\pi)$. To appreciate the non-perturbative nature of γ one can formally solve for it with the ansatz

$$\frac{C_A\gamma^4}{\mu^4} = c_0[1 + c_1 C_A\alpha_S]\exp\left[-\frac{b_0}{C_A\alpha_S}\right] \tag{6}$$

giving

$$b_0 = \frac{32\pi\left[3C_A - \sqrt{79C_A^2 - 32C_AT_FN_f}\right]}{[35C_A - 16T_FN_f]} , \quad (7)$$

$$c_0 = \exp\left[\frac{1}{[105C_A - 48T_FN_f]}\left[260C_A - 112T_FN_f - \frac{[255C_A - 96T_FN_f]C_A}{\sqrt{79C_A^2 - 32C_AT_FN_f}}\right]\right] \quad (8)$$

and

$$\begin{aligned} c_1 = \Big[& 8940981420\sqrt{5}C_A^4\zeta(2) - 11330632512\sqrt{5}C_A^3N_fT_F\zeta(2) \\ & + 4778237952\sqrt{5}C_A^2N_f^2T_F^2\zeta(2) - 670629888\sqrt{5}C_AN_f^3T_F^3\zeta(2) \\ & - 8060251500\pi^2C_A^4 - 109078793775s_2C_A^4 \\ & + 7470477000C_A^4\zeta(2) + 19529637400C_A^4 \\ & + 12730881600\pi^2C_A^3N_fT_F + 138232221840s_2C_A^3N_fT_F \\ & - 29598076800C_A^3N_fT_F\zeta(2) - 32025280640C_A^3N_fT_F \\ & - 7496478720\pi^2C_A^2N_f^2T_F^2 - 58293872640s_2C_A^2N_f^2T_F^2 \\ & + 29503733760C_A^2N_f^2T_F^2\zeta(2) + 19655024640C_A^2N_f^2T_F^2 \\ & + 1949368320\pi^2C_AN_f^3T_F^3 + 8181596160s_2C_AN_f^3T_F^3 \\ & - 11318722560C_AN_f^3T_F^3\zeta(2) - 5351014400C_AN_f^3T_F^3 \\ & - 188743680\pi^2N_f^4T_F^4 + 1509949440N_f^4T_F^4\zeta(2) + 545259520N_f^4T_F^4\Big] \\ & \times \frac{1}{46080\pi[79C_A - 32T_FN_f]^{5/2}[35C_A - 16T_FN_f]\sqrt{C_A}} . \end{aligned} \quad (9)$$

So, in principle, one could now compute with a gluon propagator which includes renormalon type singularities. Further, with (5) there is two loop ghost enhancement with the Kugo-Ojima confinement criterion[9] precisely fulfilled at this order consistent with Zwanziger's all orders proof.[7] Also at one loop it has been shown[14] that $D_A(0) = 0$. The final quantity of interest is the renormalization group invariant effective coupling constant $\alpha_S^{\mathrm{eff}}(p^2) = \alpha_S(\mu)D_A(p^2)\left(D_c(p^2)\right)^2$ which is believed to freeze at zero momentum. From the $\overline{\mathrm{MS}}$ one loop form factors it was shown[14] that $\alpha_S^{\mathrm{eff}}(0) = 50/3\pi C_A$.

Whilst the previous expressions have all been in the $\overline{\mathrm{MS}}$ scheme it is worth considering other renormalization schemes such as MOM. Given that one loop calculations[14] produced exact form factors, the derivation of the one loop MOM gap equation is straightforward, giving

$$1 = \left[\frac{5}{8} + \frac{3}{8} \ln \left(\frac{C_A \gamma^4}{[C_A \gamma^4 + \mu^4]} \right) - \frac{C_A \gamma^4}{8\mu^4} \ln \left(\frac{C_A \gamma^4}{[C_A \gamma^4 + \mu^4]} \right) - \frac{3\pi \sqrt{C_A} \gamma^2}{8\mu^2} \right. \\ \left. + \left[\frac{3\sqrt{C_A}\gamma^2}{4\mu^2} - \frac{\mu^2}{4\sqrt{C_A}\gamma^2} \right] \tan^{-1} \left[\frac{\sqrt{C_A}\gamma^2}{\mu^2} \right] \right] C_A a + O(a^2) \; . \qquad (10)$$

For later we formally define this as $1 = \text{gap}(\gamma, \mu, \text{MOM}) C_A a + O(a^2)$. Central to deriving this was the preservation of the Slavnov-Taylor identities in MOM. For instance defining Z_A and Z_c from the respective gluon and ghost 2-point functions in MOM, then the coupling constant and γ renormalization constants are already fixed and these must be used in computing the horizon function. Given (10) we have reproduced the one loop ghost enhancement in MOM and the *same* freezing value for $\alpha_S^{\text{eff}}(0)$. Since the numerical structure is different from the $\overline{\text{MS}}$ calculation we record that the analogous[14] computation is

$$\alpha_S^{\text{eff}}(0) \;=\; \lim_{p^2 \to 0} \left[\frac{\alpha_S(\mu) \left[1 - C_A \left(\text{gap}(\gamma, \mu, \text{MOM}) + \frac{5}{8} - \frac{265}{384} \right) a \right] (p^2)^2}{C_A \gamma^4 \left[1 - C_A \left(\text{gap}(\gamma, \mu, \text{MOM}) - \frac{\pi p^2}{8\sqrt{C_A}\gamma^2} \right) a \right]^2} \right] , \qquad (11)$$

whence $\alpha_S^{\text{eff}}(0) = 50/3\pi C_A$.

4. Discussion

To conclude we note that we have reviewed the path integral construction of Zwanziger's localised renormalizable Lagrangian for the Landau gauge which incorporates the restriction of gauge configurations to the first Gribov region. A picture emerges of the infrared structure which is consistent with the gluon being confined. Crucial to the analysis was the geometry of the Gribov region. This can be appreciated from another point of view given recent work in trying to extend the path integral construction to other gauges.[15–17] For linear covariant gauges other than Landau the Faddeev-Popov operator is not hermitian[15] and convexity of the Gribov region is only valid when the covariant gauge fixing parameter is small.[15] Moreover, given that the Faddeev-Popov operator in this instance would involve the transverse part of the gauge field, then the non-local operator of (2) would itself contain a non-locality in the covariant derivative.[15] Another example is the construction of a Gribov-Zwanziger type Lagrangian for $SU(2)$ Yang-Mills fixed in the maximal abelian gauge.[16,17] Whilst a localised renormalizable

Lagrangian analogous to (3) can be constructed, the algebraic renormalization analysis demonstrates that there is an additional free parameter which has no analogue in the Landau gauge.[17] Given these recent considerations it would seem therefore that in the Gribov context the Landau gauge is peculiarly special.

Acknowledgments

The author thanks Prof. S. Sorella, Prof. D. Zwanziger and Dr. D. Dudal for useful discussions concerning the Gribov problem.

References

1. V. N. Gribov, *Nucl. Phys. B* **139**, 1 (1978).
2. G. Dell'Antonio and D. Zwanziger, *Commun. Math. Phys.* **138**, 291 (1991).
3. D. Zwanziger, *Nucl. Phys. B* **209**, 336 (1982).
4. D. Zwanziger, *Nucl. Phys. B* **321**, 591 (1989).
5. D. Zwanziger, *Nucl. Phys. B* **323**, 513 (1989).
6. G. Dell'Antonio and D. Zwanziger, *Nucl. Phys. B* **326**, 333 (1989).
7. D. Zwanziger, *Nucl. Phys. B* **399**, 477 (1993).
8. D. Zwanziger, *Nucl. Phys. B* **378**, 525 (1992).
9. T. Kugo, and I. Ojima, *Prog. Theor. Phys. Suppl.* **66**, 1 (1979); T. Kugo and I. Ojima, *Prog. Theor. Phys. Suppl.* **77**, 1121 (1984).
10. D. Zwanziger, *Phys. Rev. D* **69**, 016002 (2004).
11. N. Maggiore and M. Schaden, *Phys. Rev. D* **50**, 6616 (1994).
12. D. Dudal, R. F. Sobreiro, S. P. Sorella, and H. Verschelde, *Phys. Rev. D* **72**, 014016 (2005).
13. J. A. Gracey, *Phys. Lett. B* **632**, 282 (2006).
14. J. A. Gracey, *JHEP* **05**, 052 (2006).
15. R. F. Sobreiro and S. P. Sorella, *JHEP* **06**, 054 (2005).
16. M. A. L. Capri, V. E. R. Lemes, R. F. Sobreiro, S. P. Sorella, and R. Thibes, *Phys. Rev. D* **72**, 085021 (2005).
17. M. A. L. Capri, V. E. R. Lemes, R. F. Sobreiro, S. P. Sorella, and R. Thibes, *Phys. Rev. D* **74**, 105007 (2006).

MASS SPECTRA OF THE LIGHT AND HEAVY MESONS AND THE GLUEBALL

G. GANBOLD

Bogoliubov Laboratory of Theoretical Physics, Joint Institute for Nuclear Research, Dubna, 141980 Russia
E-mail: ganbold@thsun1.jinr.ru

It is shown that the conventional mesons and the lowest glueball state can be reasonably described within a simple relativistic quantum-field model of interacting quarks and gluons under the analytic confinement by using a path-integral approach. The ladder Bethe-Salpeter equation is solved for the meson ($\bar{q}q'$) and glueball (gg) spectra. A minimal set of parameters (the quark masses m_f, the coupling constant α_s and the confinement scale Λ) is used to fit the latest experimental data. In spite of the simplicity, the model provides a reasonable framework to estimate the decay constants f_π and f_K as well as the non-exotic meson and glueball masses in a wide range of energy up to 10 GeV.

Keywords: Confinement; Quark; Gluon; Meson; Glueball; Bethe-Salpeter equation; Decay constant.

1. Introduction

The color confinement of the QCD is achieved by taking into account non-perturbative and nonlinear gluon interaction. In the hadron distance the coupling becomes stronger and the correct summation of the higher-order contributions becomes a problem. The structure of the QCD vacuum and the explicit quark and gluon propagator at the confinement scale is not well established yet. However one may suppose that the confinement is not obligatory connected with the strong-coupling regime in QCD. There exists a conception of the *analytic confinement* based on the assumption that the QCD vacuum is realized by a nontrivial homogeneous vacuum gluon field[1] which is the classical solution of the Yang-Mills equation. The quark and gluon propagators in the gluon background $\breve{B}_\mu(x)$ with constant strength Λ have been calculated[2] and found entire analytic functions on the complex p^2-plan. However, a direct use of these propagators to the hadron amplitudes is not convenient and leads to long, cumbersome, and complicated formulae. In our earlier investigations[3] we used simple Gaussian exponents for the propagators and obtained the mass spectrum of the two-particle bound states in a self-consistent form. Particularly, qualitative Regge behaviors of the excitations[4] were obtained.

Below we extend the consideration by taking into account the spin, color, and flavor degrees of freedom and develop a simple relativistic quantum field model of interacting quarks and gluons.

2. The Model

Assume that the analytic confinement takes place and the quark and gluon propagators S and D are entire analytic functions. The effective QCD Lagrangian may be written as follows:[5]

$$\mathcal{L}=(\bar{q}S^{-1}q)+\frac{1}{2}(\mathcal{A}D^{-1}\mathcal{A})+g\left(\bar{q}^a_\alpha\left[\Gamma^C_\mu\right]^{\alpha\beta}_{ab}q^b_\beta\mathcal{A}^C_\mu\right)+\frac{g}{2}f^{ABC}(\mathcal{A}^A_\alpha\mathcal{A}^B_\beta F^C_{\alpha\beta}),$$

where g is the coupling strength, q and $\mathcal{A}$ are the quark and gluon fields, $\left[\Gamma^C_\mu\right]^{\alpha\beta}_{ab}=i\gamma^{\alpha\beta}_\mu t^C_{ab}$, $F^C_{\mu\nu}=\partial_\mu\mathcal{A}^C_\nu-\partial_\nu\mathcal{A}^C_\mu$. Consider the path integral representation for the generating functional

$$Z=\int\delta q\delta\bar{q}\int\delta\mathcal{A}\ \exp\{-\mathcal{L}\}. \tag{1}$$

Within the ladder approximation the following terms correspond to the two-quark and two-gluon bound states:

$$\begin{aligned}\mathcal{L}_M &= \frac{g^2}{2}\int\delta\mathcal{A}\ e^{-\frac{1}{2}(\mathcal{A}D^{-1}\mathcal{A})}\left(\bar{q}\Gamma qD\bar{q}\Gamma q\right),\\ \mathcal{L}_G &= \frac{g^2}{8}\int\delta\mathcal{A}\ e^{-\frac{1}{2}(\mathcal{A}D^{-1}\mathcal{A})}\left(\mathcal{A}\mathcal{A}Ff\right)^2.\end{aligned}$$

The matrix elements of hadron processes at large distance are integrated characteristics of the quark and gluon propagators and the solution of the Bethe-Salpeter equation in a variational approach[5,6] should not be too sensitive on the details of propagators. Taking into account the correct global symmetry properties and their breaking by introducing additional physical parameters may be more important than the working out in detail (e.g., Ref. 7). We consider the effective quark and gluon propagators:

$$\begin{aligned}\tilde{S}^{ab}_{\alpha\beta}(\hat{p}) &= \delta^{ab}\frac{\{i\hat{p}+m_f\,[1+\gamma_5\omega(m_f)]\}_{\alpha\beta}}{m_f^2}\exp\left\{-\frac{p^2+m^2}{2\Lambda^2}\right\},\\ \tilde{D}^{AB}_{\alpha\beta}(p) &= \delta_{\alpha\beta}\delta^{AB}\ \frac{1}{\Lambda^2}\ \exp\left\{-\frac{p^2}{\Lambda^2}\right\},\end{aligned} \tag{2}$$

where $\omega(m_f) = 1/(1+m_f^2/4)$. These entire analytic functions in Euclidean metric extend our earlier investigations[3,4] and represent reasonable approximations to the explicit propagators.[2] In doing so, we use a minimal set of physical parameters, the effective coupling constant $\alpha_s = g^2/4\pi$, the scale of confinement Λ, and the quark masses $\{m_{ud}, m_s, m_c, m_b\}$.

3. Meson Spectrum

We consider the meson ground states, the pseudoscalar ($P:\ 0^{-+}$) and vector ($V:\ 1^{--}$) mesons. In the following we describe shortly the important steps of our approach.[3,6] First, we allocate one-gluon exchange between quark currents and go to the relative co-ordinates in the center of mass system. Then, perform Fierz transformations to obtain colorless bilocal quark currents and diagonalize $\mathcal{L}_M$ on ortonormalized basis functions $\{U_Q(x)\}$. Introduce a Gaussian path-integral representation for the quark currents (in exponentials) by using auxiliary meson fields $B_\mathcal{N}$ and rewrite the generating function in a new variables:

$$Z_\mathcal{N} = \int DB_\mathcal{N} \exp\left\{-\frac{1}{2}(B_\mathcal{N}[1+g^2\mathbf{Tr}(V_\mathcal{N}SV_\mathcal{N}S)]B_\mathcal{N}) + W_I[B_\mathcal{N}]\right\}, \quad (3)$$

where $V_\mathcal{N} = \Gamma_J \int dy\ U(y)\sqrt{D(y)}\exp\{y\mu_f \overleftrightarrow{\partial}\}$ is a vertex function and $\mu_f = m_f/(m_1+m_2)$ is the relative mass of the f-flavor of quark. The residue interaction between mesons is described by $W_I[B_\mathcal{N}] = O[B_\mathcal{N}^3]$. Apply the *Hadronization Ansatz* to identify $B_\mathcal{N}(x)$ with meson fields, where $\mathcal{N} = \{J, Q, f_1, f_2\}$ and J-the spin, $Q = n, l, \mu$,-the radial, orbital, magnetic quantum numbers.

The diagonalization of the quadratic form in $\mathcal{L}_M$ on $\{U_Q\}$ is nothing else but the solution of the Bethe-Salpeter equation:

$$1+\lambda_\mathcal{N}(-p^2) = \iint dxdy\ U_Q(x)\left\{1+g^2\sqrt{D(x)}\int\frac{d^4k}{(2\pi)^4}\,e^{-ik(x-y)}\right.$$
$$\left.\cdot\mathbf{Tr}\left[\Gamma_J\tilde{S}\left(\hat{k}+\mu_1\hat{p}\right)\Gamma_J\tilde{S}\left(\hat{k}-\mu_2\hat{p}\right)\right]\sqrt{D(y)}\right\}U_Q(y)\,, \quad (4)$$

where $\Gamma_P = i\gamma_5$ and $\Gamma_V = i\gamma_\mu$. The meson mass is derived from

$$1+\lambda_\mathcal{N}(M_\mathcal{N}^2) = 0\,. \quad (5)$$

Our estimates for the pseudoscalar and vector meson masses compared with experimental data (see Fig. 1) show that the relative error does not exceed $1 \div 3$ percent. The optimal values of model parameters read:

$$\alpha_s = 0.0764\,,\qquad \Lambda = 464\ \text{MeV}\,,\qquad m_{ud} = 124\ \text{MeV}\,,$$

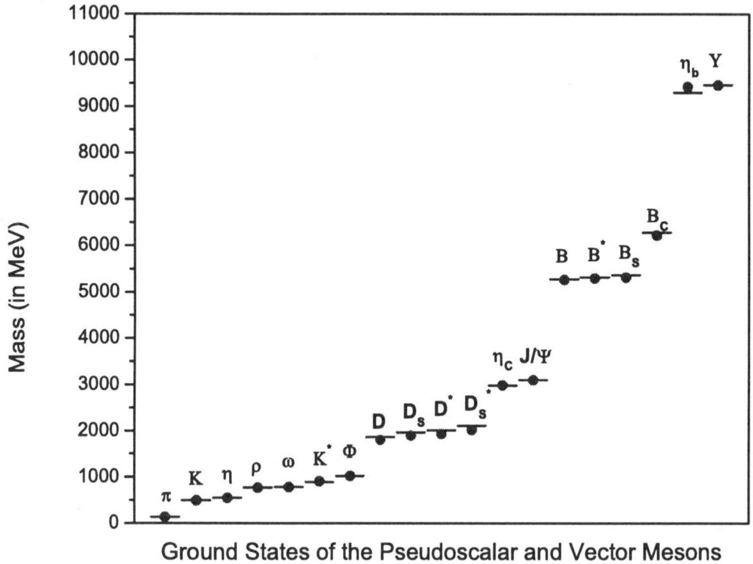

Fig. 1. The estimated meson masses (dots) compared with experimental data from PDG-2006 (dashes).

$$m_s = 156 \text{ MeV}, \qquad m_c = 1007 \text{ MeV}, \qquad m_b = 4500 \text{ MeV}. \tag{6}$$

4. Lowest Glueball State

Due to the self-interaction of gluons the QCD predicts the existence of glueballs. The lightest glueball is expected to be a scalar particle with mass around $1.2 \div 1.8$ GeV.[8–11] The experimental basis for the glueball parameters is still rather weak and there are predictions expecting the lightest glueball in the mass range $\sim 1.5 \div 1.8$ MeV.[12,13]

The Gaussian character of the gluon propagator (2) allows us to calculate explicitly the equation for the lowest glueball mass:

$$e^{-\frac{M_G^2}{2\Lambda^2}} = \frac{\alpha_s}{96\pi} \max_{0<b<1} \left[\frac{b\,(4-3b)}{2-b}\right]^2 \left(\frac{19}{8} + \frac{b}{2} + \frac{3b^2}{32} + \frac{7-b}{16}\,\frac{M_G^2}{\Lambda^2} + \frac{1}{64}\,\frac{M_G^4}{\Lambda^4}\right). \tag{7}$$

Evolution of the lowest-state glueball scaled mass M_G/Λ with α_s is plotted in Fig. 2. Note, the glueball mass (7) depends on α_s in a nonperturbative way. Particularly, for the parameters $\alpha_s = 0.0764$, $\Lambda = 464$ MeV we estimate $M_G = 1592$ MeV.

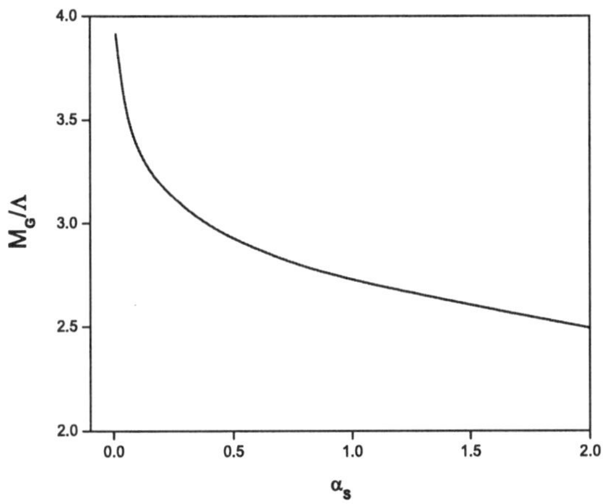

Fig. 2. Evolution of the M_G (scaled to Λ) with the coupling constant α_s.

5. Light Meson Decay Constants

An important quantity in the meson physics is the decay constant f_P defined as follows

$$if_P \, p_\mu = \langle 0 | J_\mu(0) | U_R(p) \rangle \,, \tag{8}$$

where J_μ is the axial vector part of the weak current and $U_R(p)$ is the normalized state vector. Particularly, the decay constant f_{π^+} for π^+-meson is determined from the combined rate for $\pi^+ \to \mu^+\nu_\mu + \mu^+\nu_\mu\gamma$. With the recent experimental data one obtains:[14] $f_\pi = 130.7 \pm 0.5$MeV and $f_K = 159.8 \pm 1.9$MeV. We estimate the decay constants by using a path-integral form of (8) as follows:

$$\begin{aligned} if_P \, p_\mu = \frac{g}{6}\sqrt{-\dot{\lambda}_P(M_P^2)} \int \frac{dk}{(2\pi)^4} \int dx \; e^{-ikx} \; U_P(x)\sqrt{D(x)} \\ \cdot \mathrm{Tr}\Big\{ i\gamma_5 \tilde{S}\left(\hat{k} + \mu_1 \hat{p}\right) i\gamma_5\gamma_\mu \tilde{S}\left(\hat{k} - \mu_2 \hat{p}\right) \Big\} \,. \end{aligned} \tag{9}$$

The evolution of the pseudoscalar meson decay constants with Λ is given in Fig. 3. Particularly, for values of $\{m_u, m_s, \alpha_s\}$ defined in (6) and for $\Lambda = 460 \div 470$ MeV we obtain

$$f_\pi = 105 \div 142 \text{ MeV}\,, \qquad f_K = 155 \div 225 \text{ MeV}\,.$$

Comparing these estimates with the latest data from Refs. 14,15 we see that our results are in satisfactorily agreement with the experiment despite the simplicity of our model.

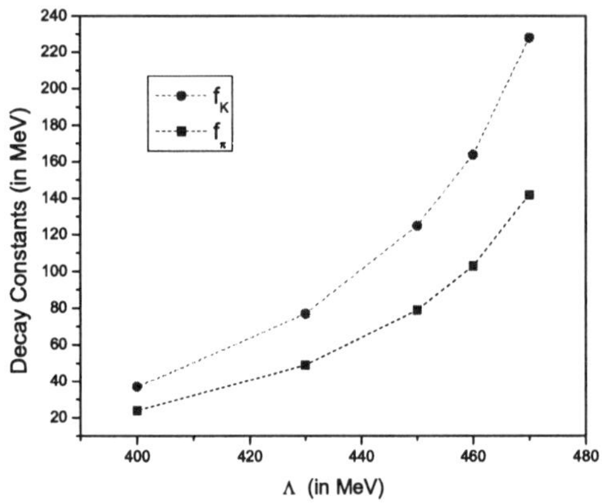

Fig. 3. The light meson decay constants f_π and f_K versus the confinement scale Λ.

In conclusion, we have considered a relativistic quantum field model of interacting quarks and gluons under the analytic confinement and estimated the meson ground state spectrum as well as the lowest glueball mass. The merit of our approach is that we describe simultaneously two different sectors of hadron physics by using a minimal set of physical parameters.

References

1. H. Leutwyler, *Nucl. Phys. B* **179**, 129 (1981).
2. Ja.V. Burdanov *et al.*, *Phys. Rev. D* **54**, 4483 (1996).
3. G. V. Efimov and G. Ganbold, *Phys. Rev. D* **65**, 054012 (2002).
4. G. Ganbold, *AIP Conf. Proc.* **717**, 285 (2004).
5. G. Ganbold, arXiv:hep-ph/0610399.
6. G. Ganbold, *AIP Conf. Proc.* **796**, 127 (2005).
7. T. Feldman, *Int. J. Mod. Phys. A* **15**, 159 (2000).
8. C. Morningstar and M. Peardon, *Phys. Rev. D* **60**, 034509 (1999).
9. S. Narison, *Nucl. Phys. A* **675**, 54 (2000).
10. H. Forkel, *Phys. Rev. D* **71**, 054008 (2005).
11. Y. Chen *et al.*, *Phys. Rev. D* **73**, 014516 (2006).
12. C. Amsler, N. A. Tornqvist, *Phys. Rep.* **389**, 61 (2004).
13. D. V. Bugg, *Phys. Lett. C* **397**, 257 (2004).
14. W.-M. Yao *et al.*, *J. Phys. G* **33**, 1 (2006).
15. C. Bernard *et al.*, *PoS LAT 2005*, 025 (2005); arXiv:hep-lat/0509137.

PART IV
Quantum Gravity

PATH INTEGRALS IN QUANTUM GRAVITY: GENERAL CONCEPTS AND RECENT DEVELOPMENTS

C. KIEFER

Institute for Theoretical Physics, University of Cologne,
Zülpicher Straße 77, D-50937 Köln, Germany
E-mail: kiefer@thp.uni-koeln.de
www.thp.uni-koeln.de/gravitation

I give a brief introduction into quantum-gravitational path integrals and discuss some recent developments and applications.

Keywords: Quantum gravity.

1. Introduction

Path integrals have been very successful in all branches of quantum theory. It is therefore natural to apply them also to quantum gravity. Such a theory aims at a consistent accommodation of the gravitational interaction into the quantum framework. It does not yet exist as a complete theory, but many promising approaches exist.[1] The simplest is the direct quantization of Einstein's theory of general relativity, with which I shall mainly deal with. Since there the central quantity is the four-dimensional metric, a path-integral quantization would entail a functional integral over the allowed class (to be specified) of four-metrics. In addition, one would envisage a sum over all four-dimensional topologies.

To my knowledge, the first paper on the quantum-gravitational path integral was written by Charles Misner, who was then a PhD student of John Wheeler in Princeton. At the end of his paper he writes:[2]

> Above all, we thank Professor J. A. Wheeler. He originally suggested the problem of formulating the Feynman quantization of general relativity. We regret that our approach to this problem has not the conciseness of his solution to it, which was simply to write
>
> $$\int \exp\{(i/\hbar)(\text{Einstein action})\}\, d(\text{field histories})$$

As he already alludes to here, it is not a simple task to specify how the exact sum has to be taken, that is, how the measure has to be chosen and how the path integral actually has to be evaluated. In my contribution to these Proceedings, I shall give a brief introduction into quantum gravitational path integrals and a brief overview of some recent developments; for more details I refer to Ref. 1 and the references therein.

2. Euclidean versus Lorentzian Path Integrals

Euclidean path integrals are widely applied in quantum field theory. The reason is that the convergence properties are much better and that there exist mathematical theorems which guarantee the equivalence of the results (the correlation functions) with the results from the original Lorentzian path integral.

In the quantization of general relativity, the Euclidean path integral has the form

$$Z[h_{ab}, \Phi, \Sigma] = \sum_{\mathcal{M}} \nu(\mathcal{M}) \int_{\mathcal{M}} \mathcal{D}g\mathcal{D}\Phi \, \mathrm{e}^{-S_{\mathrm{E}}[g_{\mu\nu}, \Phi]} \,, \tag{1}$$

where g is a short-hand for the four-metric, Φ stands for all non-gravitational fields, and the additional sum indicates a sum over all four-manifolds $\mathcal{M}$ with an unspecified weight $\nu(\mathcal{M})$; S_{E} is the Euclidean Einstein–Hilbert action. The sum over all topologies can at best be considered formal because four-manifolds are not classifiable.

Euclidean path integrals in quantum gravity have first been studied extensively in order to study black-hole thermodynamics in which the topological properties of the Euclidean Schwarzschild solution seem to play a crucial role. As Stephen Hawking put it in his Cargèse lectures:[3]

> I am going to describe an approach to quantum gravity using path integrals in the Euclidean regime. The motivation for this is the belief that the topological properties of the gravitational fields play an essential role in quantum theory.

Twenty-five years later, in his contribution in honour of Hawking's sixtieth birthday, Gary Gibbons remarked the following on Euclidean quantum gravity:[4]

> ...from today's standpoint, *Euclidean quantum gravity* should not be (and never should have been) viewed as a 'fundamental theory of physics' but rather an efficient and elegant means of extracting *non-perturbative* information about quantum gravity.

This is perhaps the prevailing viewpoint about Euclidean quantum gravity: one can extract from it much useful information, but one should not interpret it as *the* fundamental theory of quantum gravity.

Among the useful information are results about the thermodynamic behaviour of black holes. One calculates the partition sum via a Euclidean path integral with the Euclidean Einstein–Hilbert action

$$S_{\mathrm{E}} = -\frac{1}{16\pi G}\int \mathrm{d}^4x\ \sqrt{g}R + \frac{1}{8\pi G}\int \mathrm{d}^3x\ \sqrt{h}(K-K^0)\ , \qquad (2)$$

in which the second term refers to a three-dimensional boundary of the four-dimensional spacetime region under consideration; h denotes the determinant of the three-metric, K the trace of the second fundamental form, and K^0 the trace of the second fundamental form for a corresponding three-hypersurface in Minkowski spacetime, which has to be subtracted for reasons of regularization. The speed of light has been set equal to one.

The full path integral cannot be calculated exactly with such a complicated action. One thus resorts to a semiclassical approximation. The path integral is then evaluated by finding the appropriate saddle point, that is, the dominating classical solution. For the Euclidean Schwarzschild solution one finds for the partition sum with $\beta\hbar$ as 'Euclidean time':[3,5]

$$Z = \int_{\mathcal{M}} \mathcal{D}g\ \exp\left(-S_{\mathrm{E}}/\hbar\right) \approx \exp\left(-S_{\mathrm{E}}^{\mathrm{cl}}/\hbar\right) = \exp\left(-\frac{\beta^2\hbar}{16\pi G}\right)\ . \qquad (3)$$

Since the Ricci tensor vanishes for the Schwarzschild solution (because it is a vacuum solution), only the boundary term contributes. Once the partition function $Z \equiv \exp(-\beta F)$ is known, one can derive from it the thermodynamical quantities of interest:

- $E = -\frac{\partial \ln Z}{\partial \beta} = \frac{\hbar\beta}{8\pi G} = M$: energy;
- $T_{\mathrm{BH}} = \frac{1}{8\pi k_{\mathrm{B}} G M} \approx 6.17 \times 10^{-8}\left(\frac{M_\odot}{M}\right)$ K: Hawking temperature;
- $S_{\mathrm{BH}} = -\frac{\partial F}{\partial T} = \frac{\hbar\beta^2}{16\pi G} = \frac{k_{\mathrm{B}}A}{4G\hbar}$: Bekenstein–Hawking entropy.

The Euclidean path integral in gravity has various novel features. The main reason is the background independence of general relativity: whereas in non-gravitational situations one has a fixed non-dynamical background spacetime, in general relativity spacetime itself becomes dynamical and there is no fixed background. For this reason one cannot just 'Wick rotate' the various metrics because Wick rotation is not a diffeomorphism-invariant procedure; a real Lorentzian metric does not, in general, possess a real Euclidean version. This is, however, not considered a problem. The idea is to take the path integral (1) as the starting point and to integrate over all Euclidean metrics *per se*. The result is then interpreted as the relevant partition function or wave function. Moreover, in many applications one even envisages an integration over all complex four-metrics in order to improve the convergence properties of the path integral. In this sense, the difference between the Lorentzian and the Euclidean approach becomes fuzzy.

A more serious obstacle is given by the 'conformal-factor problem': in general relativity, the Euclidean action can be made arbitrarily negative. Starting from (2), a conformal transformation $g_{\mu\nu} \to \tilde{g}_{\mu\nu} = \Omega^2 g_{\mu\nu}$ leads to the action

$$\begin{aligned} S_{\mathrm{E}}[\tilde{g}] = &-\frac{1}{16\pi G}\int_{\mathcal{M}} \mathrm{d}^4x\ \sqrt{g}(\Omega^2 R + 6\Omega_{;\mu}\Omega_{;\nu}g^{\mu\nu} - 2\Lambda\Omega^4) \\ &-\frac{1}{8\pi G}\int_{\partial\mathcal{M}} \mathrm{d}^3x\ \sqrt{h}\Omega^2 K\ , \end{aligned} \tag{4}$$

in which we have also considered the cosmological constant, Λ. Upon choosing a highly varying factor Ω one can make the action arbitrarily negative; since such action functions contribute to the path integral (1), its convergence is far from obvious.

How can this problem be dealt with? A possible solution may be found through a careful investigation of the integration measure in the path integral. As in ordinary gauge theories, the measure can be determined by the Faddeev–Popov procedure and contains gauge-fixing terms as well as the Faddeev–Popov determinant; a description can be found, for example, in Ref. 1. The detailed discussion presented in Ref. 6 has exhibited, at least on the formal level, that the divergence from the conformal modes of the metric is cancelled non-perturbatively by the Faddeev–Popov determinant in the measure. (This has confirmed an earlier result derived on the perturbative level.[7])

Due to the complexity of the path integral, it is not surprising that much effort has been spent into attempts for a numerical evaluation. In particular, one distinguishes between[1]

(1) *Regge calculus*: this is an application to the Euclidean path integral; the central idea is to decompose four-dimensional space into a set of simplices and treat the edge lengths as dynamical entities.
(2) *Dynamical triangulation*: this is an application to the Lorentzian path integral; in contrast to Regge calculus, all edge lengths are kept fixed, and the sum in the path integral is instead taken over all possible manifold-gluings of equilateral simplices. The evaluation is thus reduced to a combinatorical problem.

Some recent results are obtained in the framework of dynamical triangulation:[8]

- The resulting space possesses a Hausdorff dimension $H = 3.10 \pm 0.15$, which is compatible with the expected three-dimensionality of space;
- spacetime appears to be two-dimensional on smallest scales (a similar result is found in the asymptotic-safety approach, see below);
- a positive cosmological constant is needed, although its precise value remains unspecified;
- in a cosmological situation, an effective action for the scale factor of the universe can be found by integrating out in the full quantum theory all other degrees of freedom;
- the question of the continuum limit remains, however, open.

More details can also be found in the contribution of Jan Ambjørn to these Proceedings.

3. Perturbation Theory and Asymptotic Safety

The quantum-gravitational path integral is also a useful tool for the development of perturbation theory. In the 'background-field method',[9] the starting point is the expansion of the four-metric around a classical background $\bar{g}_{\mu\nu}$,

$$g_{\mu\nu} = \bar{g}_{\mu\nu} + \sqrt{32\pi G} f_{\mu\nu} \ , \tag{5}$$

where $f_{\mu\nu}$ is the perturbation to be quantized. After the application of the Faddeev–Popov procedure, the path integral assumes the form

$$Z = \int \mathcal{D}f \mathcal{D}\eta^{\alpha} \mathcal{D}\eta^{*\alpha} \ \mathrm{e}^{\mathrm{i}S_{\mathrm{tot}}[f,\eta,\bar{g}]/\hbar} \ , \tag{6}$$

where the action can be written as a sum of the original gravitational action, a gauge-fixing and a ghost contribution,[1]

$$\begin{aligned} S_{\mathrm{tot}}[f,\eta,\bar{g}] &= S[f,\bar{g}] - \frac{1}{4\xi} \int \mathrm{d}^4x \ G_{\alpha}[f,\bar{g}] G^{\alpha}[f,\bar{g}] \\ &+ \int \mathrm{d}^4x \ \eta^{*\alpha}(x) A_{\alpha\beta}[f,\bar{g}](x) \eta^{\beta}(x) \equiv \int \mathrm{d}^4x \ (\mathcal{L}_{\mathrm{g}} + \mathcal{L}_{\mathrm{gf}} + \mathcal{L}_{\mathrm{ghost}}) \ . \end{aligned} \tag{7}$$

This is the starting point for the derivation of Feynman rules. A major difference from the Standard Model of particle physics is the non-renormalizability of the quantum-gravitational perturbation theory. In spite of this problem, one can calculate unambiguous predictions at low energy from the path integral (6); this is similar to chiral perturbation theory in quantum chromodynamics (limit of small pion mass). One can, for example, calculate the quantum gravitational correction to the Newtonian potential between two masses m_1 and m_2.[10] The result is

$$V(r) = -\frac{Gm_1m_2}{r}\left(1 + \underbrace{3\frac{G(m_1+m_2)}{rc^2}}_{\text{GR-correction}} + \underbrace{\frac{41}{10\pi}\frac{G\hbar}{r^2c^3}}_{\text{QG-correction}}\right) . \tag{8}$$

While the first correction term is a known correction term from general relativity, the second term is a genuine quantum-gravity prediction (I have re-inserted here c for illustration). Albeit unmeasurably small in the laboratory, the second term is nevertheless a concrete prediction from quantum gravity.

Other recent results using path-integral methods have been found in the context of asymptotic safety. A theory is called *asymptotically safe* if all essential coupling parameters g_i of the theory approach a non-trivial fixed point for momentum $k \to \infty$.[11] Explicit and detailed discussions of the corresponding renormalization-group equations lead to the following preliminary results, see Ref. 12 and the references therein:

- The effective (scale-dependent) gravitational constant vanishes for $k \to \infty$, leading to an asymptotically free theory;
- the effective gravitational constant increases with distance and could thus simulate the existence of dark matter;
- a small positive cosmological constant emerges as an infrared effect and could thus simulate dark energy;
- spacetime appears two-dimensional on smallest scales, similar to the dynamical-triangulation approach.

Although preliminary, it seems that one could understand from this rather straightforward approach both dark matter and dark energy as two aspects of the same theory of quantum gravity.

4. Path Integrals and Canonical Quantum Gravity

The quantization of general relativity can also be described in the canonical formalism.[1,13] In this section I want to describe briefly the connection with the path integral.

Starting point for canonical gravity is the foliation of spacetime into three-dimensional hypersurfaces; a necessary assumption for this to work is that spacetime be globally hyperbolic. The canonical variables are then the three-dimensional metric, h_{ab}, of a hypersurface and the momenta, π^{ab}, which are a linear combination of the extrinsic curvature. Alternatively, one can choose variables that have the interpretation of holonomies and fluxes, which leads to the framework of loop quantum gravity; for definiteness, I restrict myself to the original geometrodynamical formulation. Spacetime can then be interpreted as a 'trajectory of spaces'.

The Einstein equations then fall into two classes: six evolution equations for h_{ab} and π^{ab}, and four constraints. The constraints read

$$H[h,\pi] = 16\pi G\, G_{ab\,cd}\pi^{ab}\pi^{cd} - (16\pi G)^{-1}\sqrt{h}({}^{(3)}R - 2\Lambda) \approx 0\,, \tag{9}$$

$$D^a[h,\pi] = -2\nabla_b\pi^{ab} \approx 0\,, \tag{10}$$

with

$$G_{ab\,cd} = \frac{1}{2\sqrt{h}}(h_{ac}h_{bd} + h_{ad}h_{bc} - h_{ab}h_{cd})$$

as the 'DeWitt-metric', which plays the role of a metric on the configuration space of all three-metrics. Furthermore, h is the determinant of the three-metric, and ${}^{(3)}R$ the three-dimensional Ricci scalar. For simplicity, restriction is here made to the vacuum case. The constraint (9) is called Hamiltonian constraint, the constraints (10) are called diffeomorphism (or momentum) constraints.

In the quantum theory, the three-metric and the momenta obey non-trivial commutation rules. Therefore, spacetime is absent in quantum gravity because the three-metric and the extrinsic curvature cannot be specified simultaneously; this is analogous to the disappearance of trajectories in quantum mechanics. Consequently, only the constraints remain. If one quantizes them in the sense of Dirac, they become restriction on physically allowed wave functionals $\Psi[h_{ab}]$. One gets[1]

$$\hat{H}\Psi \equiv \left(-16\pi G\hbar^2 G_{abcd}\frac{\delta^2}{\delta h_{ab}\delta h_{cd}} - (16\pi G)^{-1}\sqrt{h}\left({}^{(3)}R - 2\Lambda\right)\right)\Psi = 0 \tag{11}$$

from (9) and

$$\hat{D}^a\Psi \equiv -2\nabla_b\frac{\hbar}{\mathrm{i}}\frac{\delta\Psi}{\delta h_{ab}} = 0 \tag{12}$$

from (10). The first equation is referred to as the *Wheeler–DeWitt equation* and the second as the *quantum diffeomorphism (momentum) constraints.*

We do not discuss here the many subtleties which are connected with these equations.[1]

What, then, is the connection with the path-integral approach? As for the quantum-mechanical path integral, one knows that it obeys the Schrödinger equation. Similarly, one would expect that the quantum-gravitational path integral obeys the quantum constraints (11) and (12). On a formal level, it is rather straightforward to check that this is indeed the case, see, for example, Ref. 1. With some effort, one can explicitly check that the quantum-gravitational path integral satisfies the constraints in the one-loop approximation.[14]

Path integrals are used in quantum cosmology in connection with boundary conditions. By restricting the class of geometries to be summed over in the path integral, one can arrive at a restricted class of solutions to the Wheeler–DeWitt equation and thus to a selected class of 'wave functions of the universe'. The most popular proposal is the no-boundary condition of Hartle and Hawking.[15] The idea there is to take the Euclidean path integral (1) as the starting point and to sum only over four-geometries with no initial boundary. The original idea was to find in this way a unique solution for the wave function. As it turned out, this is not the case, but the no-boundary proposal may still serve as a heuristic guide to find solutions of the Wheeler–DeWitt equation.[1] Its main application is in the semiclassical approximation where the path integral can be evaluated through the selection of extrema for the classical action.

In quantum cosmology, restriction is usually being made to homogeneous metrics. Then, the path integral can be written as an ordinary path integral *plus* the integration over the time separation, cf. Ref. 16,

$$\int \mathrm{d}T \; \langle q'', T | q', 0 \rangle \equiv G(q'', q'; E)|_{E=0} \; , \tag{13}$$

where

$$G(q'', q'; E) = \int \mathrm{d}T \; \mathrm{e}^{\mathrm{i}ET/\hbar} \langle q'', T | q', 0 \rangle$$

is an 'energy Green function'. The quantum-gravitational path integral thus resembles more an energy Green function with energy zero than a propagator. The reason is, of course, the absence of an external time parameter t in quantum gravity. Consequently, there is also no composition law with respect to time.

In simple models, the path integrals can even be evaluated exactly. For a Friedmann universe with a conformally-coupled scalar field, the Wheeler–DeWitt equation can be written as the equation for an 'indefinite harmonic oscillator',

$$\hat{H}\psi(a,\chi) \equiv \left(\frac{\partial^2}{\partial a^2} - \frac{\partial^2}{\partial \chi^2} - a^2 + \chi^2\right)\psi = 0 \; . \tag{14}$$

Normalizable solutions to this equation, which describe the evolution of wave packets, have been constructed in Ref. 17. Unfortunately, none of these solutions can be found from the quantum-gravitational path integral.[16] This includes the solutions from the no-boundary proposal, which are all non-normalizable. Therefore, either one has to embark on a new interpretation of the wave functions obtained from the path integral, or these solutions have to be rejected as unphysical.

5. Conclusion

In order to summarize the above discussion one can state that

- the properties of the quantum-gravitational path integral are understood on the formal level;
- they are useful for perturbation theory and effective field theory and allow even the calculation of effects;
- numerical methods such as dynamical triangulation may shed light on the spacetime structure at smallest scales;
- these path integrals find widespread applications to black holes and quantum cosmology, although the interpretation of the quantum-cosmological wave functions found in this way remains unclear.

Path integrals are also an important tool in string theory;[18] I have omitted a discussion here for lack of space. It is thus expected that path integrals will play an important role in the final formulation of quantum gravity.

Acknowledgments

I thank the organizers of this conferences for inviting me to a very interesting and stimulating meeting.

References

1. C. Kiefer, *Quantum Gravity*, 2nd edition (Oxford University Press, Oxford, 2007).
2. C. W. Misner, *Rev. Mod. Phys.* **29**, 497 (1957).
3. S. W. Hawking, in *Recent Developments in Gravitation – Cargèse 1978: Proceedings*, edited by M. Levy and S. Deser (Plenum Press, New York, 1978).

4. G. W. Gibbons, in *The Future of Theoretical Physics and Cosmology*, edited by G. W. Gibbons, E. P. S. Shellard, and S. J. Rankin (Cambridge University Press, Cambridge, 2003).
5. S. W. Hawking and R. Penrose, *The Nature of Space and Time* (Cambridge University Press, Cambridge, 1996).
6. A. Dasgupta and R. Loll, *Nucl. Phys. B* **606**, 357 (2001).
7. P. O. Mazur and E. Mottola, *Nucl. Phys. B* **341**, 187 (1990).
8. R. Loll, J. Ambjørn, and J. Jurkiewicz, *Contemp. Phys.* **47**, 103 (2006).
9. B. S. DeWitt, *The Global Approach to Quantum Field Theory*, Vols. I and II (Clarendon Press, Oxford, 2003).
10. N. E. J. Bjerrum-Bohr, J. F. Donoghue, and B. R. Holstein, *Phys. Rev. D* **67**, 084033 (2003).
11. S. Weinberg, in *General Relativity*, edited by S. W. Hawking and W. Israel (Cambridge University Press, Cambridge, 1979).
12. M. Reuter and F. Saueressig, arXiv:0708.1317 (hep-th).
13. D. Giulini and C. Kiefer, in *Theoretical Approaches to Fundamental Physics – an Assessment of Current Theoretical Ideas*, edited by E. Seiler and I.-O. Stamatescu (Springer, Berlin, 2007).
14. A. O. Barvinsky, *Nucl. Phys. B* **520**, 533 (1998).
15. J. B. Hartle and S. W. Hawking, *Phys. Rev. D* **28**, 2960 (1983).
16. C. Kiefer, *Ann. Phys. (NY)* **207**, 53 (1991); C. Kiefer, in *Fluctuating Paths and Fields*, edited by W. Janke, A. Pelster, H.-J. Schmidt, and M. Bachmann (World Scientific, Singapore, 2001).
17. C. Kiefer, *Nucl. Phys. B* **341**, 273 (1990).
18. J. Polchinski, *String Theory*, Vols. I and II (Cambridge University Press, Cambridge, 1998).

THE EMERGENCE OF (EUCLIDEAN) DE SITTER SPACE-TIME

J. AMBJØRN[†,*]

[†] *The Niels Bohr Institute, Copenhagen University,*
Blegdamsvej 17, DK-2100 Copenhagen Ø, Denmark
E-mail: ambjorn@nbi.dk

A. GÖRLICH AND J. JURKIEWICZ

Institute of Physics, Jagellonian University,
Reymonta 4, PL 30-059 Krakow, Poland
Email: jurkiewi@thrisc.if.uj.edu.pl, atg@th.if.uj.edu.pl

R. LOLL

[*] *Institute for Theoretical Physics, Utrecht University,*
Leuvenlaan 4, NL-3584 CE Utrecht, The Netherlands
Email: R.Loll@phys.uu.nl

We show how the four-sphere appears as a background geometry with superimposed quantum fluctuations in the background independent approach to quantum gravity which uses causal dynamical triangulations as a regularization.

Keywords: Quantum gravity; Dynamical triangulations.

1. Introduction

The theory of four-dimensional quantum gravity called Lorentzian simplicial quantum gravity is based on the concept of causal dynamical triangulations (CDT). While we refer to Refs. 1–3 for details, stated shortly it defines in a non-perturbative way the summation over four geometries by constructing the four-geometries from "building blocks" glued together such that only causal space-time histories are included. In order to perform the actual summation over space-time histories one rotates to space-times of Euclidean signature. The "building blocks", four-simplices, are characterized by a cut-off a, the side length of the simplices. The continuum limit of the regularized path integral should correspond to the limit $a \to 0$, possibly accompanied by a readjustment of the bare coupling constants such that the physics stays invariant.

The challenge in a theory of quantum gravity is to find a theory which behaves in this way. The challenge is three-fold: 1) to find a suitable tentative non-perturbative formulation of such a theory, 2) to find observables which can be used to test the physics of the model and 3) to actually show that one can adjust the coupling constants of the theory such that the physics stays invariant when the lattice spacing is changed. Here we focus on 1).

As mentioned above we have proposed a non-perturbative formulation of four-dimensional quantum gravity. How do we judge if it should be taken seriously as a quantum field theory of gravity? From "observations" we know that the following picture should emerge: a background geometry around which we have small quantum fluctuations. The challenge is to obtain such a picture from a background independent formalism where no background geometry is put in by hand. We have in earlier articles provided indirect evidence for such a scenario[4,5] based on computer simulations. We here discuss new simulations which confirm the above mentioned picture in a much more direct way.

2. The Macroscopic S^4 Universe

By construction we have a foliation in proper time t (here taken Euclidean by the analytic continuation mentioned above). Each geometry is constructed by gluing together four-simplices in a way respecting this foliation. All spatial slices have the same topology which we take to be that of S^3. We take the time direction to be periodic of some fixed length. This length is so large that the periodicity is of no importance, just a convenience from a simulation point of view. The action used in the Monte Carlo simulations is the Einstein-Hilbert action in the form of the so-called Regge action which is a geometric realization of the Einstein-Hilbert action on piecewise linear geometries.

The simulations are for computer technical reasons performed for a (almost) fixed four volume. For a certain range of bare coupling constants we observe a four-dimensional universe of well defined time and well defined space extension.[4] As discussed elsewhere[1] this is actually a non-trivial result and there are choices of bare coupling constants where this is not the case.

A Monte-Carlo simulation of the universe will generate a sequence of four-dimensional geometries, space-time histories represented by triangulations. We have by construction a time foliation, but nothing except the Regge version of the Einstein-Hilbert action as an input. A plot of the output, represented as a plot of $V_3(t)$, leads to a "blob" of time extension $T \sim V_4^{1/4}$. This blob is a configuration in the path integral and thus not "physical", in the same way as a path in the path integral of the particle is not an observable and not physical. We obtain the average trajectory of the particle by taking the average of the paths in the path integral (with the weight dictated by the classical action) and this trajectory agrees with the classical trajectory of the particle up to corrections of order $\hbar$. In the process of performing an average, for instance by Monte Carlo simulations, one obtains not only information about the average trajectory but also about the size of the quantum fluctuations. These will also in general be of order $\hbar$.

We want to do the same in the case of our universe: we have the quantum universe. It is fluctuating. Each configuration we "observe" in the computer is different. By taking the average we produce the background geometry around which the quantum universe fluctuates. We have chosen a total time T_{tot} much larger than the observed time extension T of our "blob". The center of mass of the "blob" (the universe) will perform a random walk along the time axis. In order to compare different universes we fix this center of mass and in this way we can obtain an "average" universe. The results of the measurements of the average spatial size of the "universe" at various times t are shown in Fig. 1 and can be summarized in the following formula:

$$N_3^{cl}(i) = \frac{3}{4}\frac{N_4}{s_0 N_4^{1/4}} \cos^3\left(\frac{i}{s_0 N_4^{1/4}}\right), \quad s_0 = 0.72. \tag{1}$$

Here $N_3(i)$ denotes the number of three-simplices making up the spatial slice at discretized time i and N_4 the total number of four-simplices making up the total universe.

We have verified (1) for N_4 ranging from 20.000 to 160.000. The collapse of the various curves to one, parameterized as in Eq. (1), is a beautiful example of finite size scaling. Eq. (1) shows that the spatial volume increases as $N_4^{3/4}$ and the time extension of the universe as $N_4^{1/4}$ as one would indeed expect for a genuine four-dimensional object. It is natural to write Eq. (1) in continuum notation:

$$V_3^{cl}(t) = V_4 \, \frac{3}{4B} \cos^3\left(\frac{t}{B}\right), \tag{2}$$

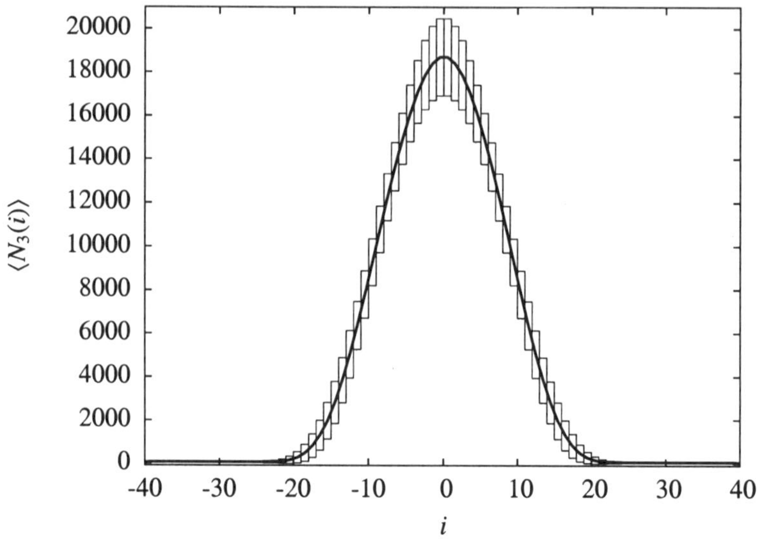

Fig. 1. The background geometry from measurement and the best fit, see Eq. (1). One cannot distinguish the two curves. In addition we have shown the average error bars for an individual configuration, i.e. the typical size of fluctuations.

where we have made the identifications:

$$V_4 \propto a^4 N_4, \qquad V_3 \propto a^3 N_3, \qquad t_i \propto ai, \tag{3}$$

a denoting the lattice spacing and t_i being proportional to proper time. If we write $V_4 = 8\pi^2 R^4/3$, Eq. (2) can be viewed as a parameterization of S^4 by the time t related to proper time τ by $\tau = t\, R/B$.

The background geometry of S^4 in the parameterization (2) is the classical solution of the following action (which is just the Einstein action if one assumes spatial homogeneity and isotropy):

$$S = \frac{1}{G}\int \mathrm{d}t \left(\frac{\left(\dot{V}_3(t)\right)^2}{V_3(t)} + k_2 V_3^{1/3} - \lambda V_3(t) \right), \tag{4}$$

where $k_2 = 9(3\pi^2 R^4/B^4)^{2/3}$ and λ in (4) is a Lagrange multiplier which is fixed by the requirement that the total four-volume is fixed to V_4: $\int \mathrm{d}t\; V_3(t) = V_4$.

Let us note for future use that if we start out with a discretized, dimensionless version of (4) and take the continuum limit using the identifications

(3), we obtain:

$$S_{discrete} = k_1 \sum_i \left(\frac{(N_3(i+1) - N_3(i))^2}{N_3(i)} + k_2 N_3^{1/3} \right) \tag{5}$$

$$\rightarrow \frac{k_1}{a^2} \int \mathrm{d}t \left(\frac{\dot{V}_3^2(t)}{V_3(t)} + k_2 V_3^{1/3} \right). \tag{6}$$

Thus the naive continuum limit of $S_{discrete}$ corresponds to a "gravitational" coupling constant:

$$G \sim \frac{a^2 2}{k_1} \sim \frac{1}{k_1} \sqrt{\frac{V_4}{N_4}}. \tag{7}$$

If k_1 is a constant, we conclude that the discretized action $S_{discrete}$ corresponds to a "gravitational" coupling constant G of order a^2, a being the lattice spacing, or stated differently: *the lattice spacing is the Planck scale.*

While the classical solution (2) will not provide any information about the constant G in front of (4), the study of the fluctuations will. A simple saddle point calculation shows that the fluctuations around V_3^{cl} will be of the order:

$$\langle (\delta V_3)^2 \rangle \sim G\, V_4. \tag{8}$$

We now turn to the study of these fluctuations.

3. Fluctuations Around the Four-Sphere

The measurement of the average three-volumes $V_3^{cl}(t) = \langle V_3(t) \rangle$ shows that a background geometry compatible with S^4 is formed dynamically from the background independent CDT approach. Having established this, we can now measure the fluctuations around the background geometry. The fluctuations are defined by the correlator

$$C(t,t') = \langle (V_3(t) - V_3^{cl}(t))(V_3(t') - V_3^{cl}(t')) \rangle. \tag{9}$$

While the effective action (4) gives a perfect description of the measured $V_3^{cl}(t)$, it is not clear to which extent it also describes the fluctuations. We will show that it does, and we will, as mentioned above, obtain information about the coupling constant G (recall that G is not put in by hand but is an effective coupling constant obtained after integrating out all modes except the scale factor).

The first observation is that the discretized version of (9) scales as follows with the total four-volume:

$$C_{N_4}(i,i') = \langle (N_3(i) - N_3^{cl}(i))(N_3(i) - N_3^{cl}(i')) \rangle = N_4\, F\left(\frac{i}{N_4^{1/4}}, \frac{i'}{N_4^{1/4}} \right), \tag{10}$$

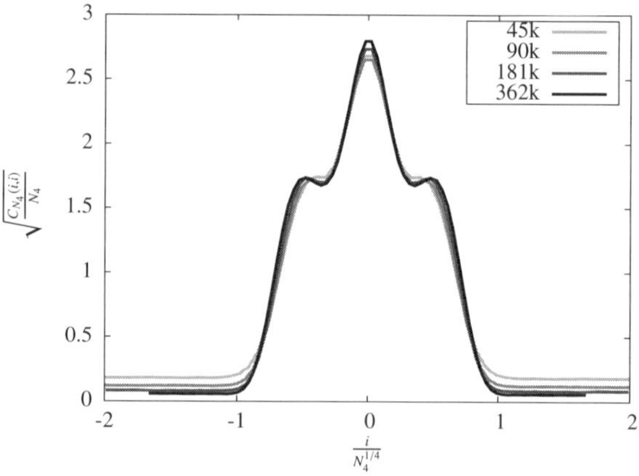

Fig. 2. The plot of the scaling function $\sqrt{F(t/B,t/B)}$ from (10) for $N_4 = 20\,000$, $40\,000$, $80\,000$, and $160\,000$.

where $F(i,j)$ is of order one. This is shown in Fig. 2 for the diagonal elements of C_{N_4}. The square roots of the diagonal elements, $\sqrt{C_{N_4}(i,i)}$, are precisely the fluctuations shown in Fig. 1.

It is easy to show that *if* the fluctuations arise from the discretized action (5) they have the scaling (10) only if k_1 in (5) is independent of N_4, i.e. if we have the scenario where the lattice spacing is of Planck length. On the other hand, with this interpretation of G we have from (7) and (10) that the continuum version of (10) reads:

$$C(t,t') \sim a_s^6\, N_4\, F(t/B,t'/B) \sim G\, V_4\, F(t/B,t'/B), \tag{11}$$

in agreement with the estimate (8). We will now check to which degree $F(t,t')$ is actually described by the effective action (4).

Let us expand the action (4) around the classical solution: $V_3(t) = V_3^{cl}(t) + x(t)$. The quadratic fluctuations around this solution are given by:

$$\langle x(t)x(t')\rangle \sim \int \mathcal{D}x(s)\; x(t)x(t')\; e^{-\int\int ds ds' x(s) M(s,s') x(s')} \sim M^{-1}(t,t'), \tag{12}$$

where the quadratic form $M(t,t')$ is determined by expanding the effective action S to second order in the fluctuation $x(t)$:

$$S(V_3) = S(V_3^{cl}) + \frac{4Bk_1}{3V_4} \int \mathrm{d}t\; x(t)\hat{H}x(t) + \xi \left(\int \mathrm{d}t\; x(t)\right)^2. \tag{13}$$

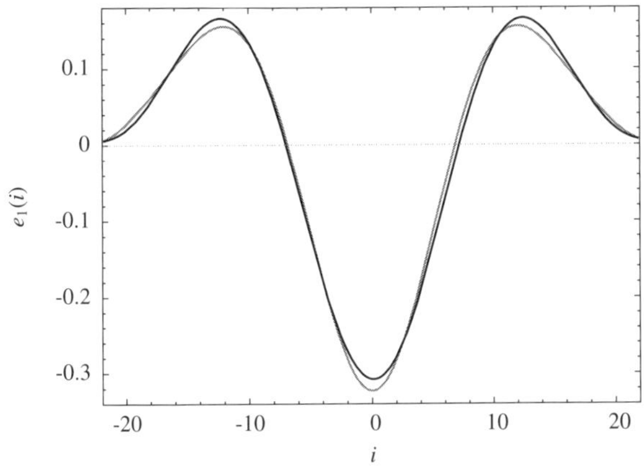

Fig. 3. The lowest eigenvector as calculated from the data, i.e. from $C(t,t')$ (grey curve), and from $M^{-1}(t,t')$ (black curve).

$\hat{H}$ denotes the Hermitian operator

$$\hat{H} = -\frac{\mathrm{d}}{\mathrm{d}t}\frac{1}{\cos^3(t/B)}\frac{\mathrm{d}}{\mathrm{d}t} - \frac{4}{B^2\cos^5(t/B)} \tag{14}$$

and the last term in (13) is added to enforce the constraint $\int \mathrm{d}t\; x(t) = 0$. It is understood that we have to take $\xi \to \infty$ in Eq. (13) to enforce this.

We ask to which extent $M^{-1}(t,t')$ agrees with $C(t,t')$. Since time in the computer simulations is discrete and typically has a range $r \sim 2\,N^{1/4}$, the matrix $C(t,t')$ will be a $r \times r$ matrix. In order to compare $C(t,t')$ and $M^{-1}(t,t')$ we discretize the integro-differential like operator $M(t,t')$ such that it also becomes an $r \times r$ matrix, of which we can readily find eigenvalues and eigenvectors. Fig. 3 shows the lowest eigenfunction calculated from the data, i.e. from the matrix $C(t,t')$ and calculated from the effective action, i.e. from the matrix $M(t,t')$. The agreement is very good, in particular taking into account that no parameter is adjusted in the action (we simply take the constant B in (2) and (13) to be $s_0 N_4^{1/4}$, i.e. 14.37 for $N_4 =$ 160.000).

A full analysis of the data will be published elsewhere,[6] here it can be summarized by saying that within the accuracy of the measurements the fluctuations of the spatial volume are described by the mini-superspace action (4).

4. Discussion

The CDT model of quantum gravity is extremely simple, but still we obtain the emergence of a classical background geometry around which there are small quantum fluctuations, which is precisely the goal of a theory of quantum gravity.

Since the relation $G \sim a^2$ we know that the spatial and temporal extensions of our computer generated universes are around $N_4^{1/4}$ Planck lengths, i.e. for the largest universes we have simulated so far around 20 Planck lengths. It is interesting that even for universes so small at least some aspects of the universes behave semiclassically.

Is it possible to penetrate beyond the Planck scale using CDT? It requires that we can adjust the bare coupling constants which enter in the model in such a way that k_1 can go to zero. In that case we can change N_4 and k_1 simultaneously such that $V_4 = N_4 a^4$ and $G = a^2/k_1$ stay constant while the lattice spacing a goes to zero. If such an adjustment was possible one would be able to address sub-Planckian physics. This possibility is presently under investigation.

Acknowledgment

All authors acknowledge support by ENRAGE (European Network on Random Geometry), a Marie Curie Research Training Network in the European Community's Sixth Framework Programme, network contract MRTN-CT-2004-005616. A.G and J.J. acknowledge a support by COCOS (Correlations in Complex Systems), a Marie Curie Transfer of Knowledge Project in the E.C. Sixth Framework Programme, contract MTKD-CT-2004-517186.

References

1. J. Ambjorn, J. Jurkiewicz, and R. Loll, *Phys. Rev. D* **72**, 064014 (2005) [arXiv:hep-th/0505154].
2. J. Ambjorn, J. Jurkiewicz, and R. Loll, *Nucl. Phys. B* **610**, 347 (2001) [arXiv:hep-th/0105267].
3. J. Ambjorn and R. Loll, *Nucl. Phys. B* **536**, 407 (1998) [arXiv:hep-th/9805108].
4. J. Ambjorn, J. Jurkiewicz, and R. Loll, *Phys. Rev. Lett.* **93**, 131301 (2004) [arXiv:hep-th/0404156].
5. J. Ambjorn, J. Jurkiewicz, and R. Loll, *Phys. Lett. B* **607**, 205 (2005) [arXiv:hep-th/0411152].
6. J. Ambjorn, A. Goerlich, J. Jurkiewicz, and R. Loll, *Phys. Rev. Lett.* **100**, 091304 (2008) [arXiv:0712.2485].

FUNCTIONAL INTEGRALS IN AFFINE QUANTUM GRAVITY

J. R. KLAUDER

Department of Physics and Department of Mathematics,
University of Florida, Gainesville, FL 32611, USA
E-mail: klauder@phys.ufl.edu

A sketch of a recent approach to quantum gravity is presented which involves several unconventional aspects. The basic ingredients include: (1) Affine kinematical variables; (2) Affine coherent states; (3) Projection operator approach for quantum constraints; (4) Continuous-time regularized functional integral representation without/with constraints; and (5) Hard core picture of nonrenormalizability. Emphasis is given to the functional integral expressions.

Keywords: Quantum gravity; Affine variables; Quantum constraints; Functional integral.

1. Introduction

This paper offers an introduction to the program of Affine Quantum Gravity (AQG) and its use of functional integrals. It is important at the outset to remark that this program is not string theory nor is it loop quantum gravity, the two most commonly studied approaches to quantum gravity at the present time. Although many aspects of this approach are still to be developed, AQG seems to the author to be more natural than most traditional views, and, moreover, it lies closer to classical (Einstein) gravity as well. Some general references for this paper are Refs. 1–3.

2. Affine Kinematical Variables

Metric positivity

A fundamental requirement of AQG is the strict positivity of the spatial metric. For the classical metric, this property means that for any nonvanishing set $\{u^a\}$ of real numbers and any nonvanishing, nonnegative test function, $f(x) \geq 0$, that

$$\int f(x) u^a g_{ab}(x) u^b d^3x > 0 \,, \tag{1}$$

where $1 \leq a, b \leq 3$. We also insist that this inequality holds when the classical metric field $g_{ab}(x)$ is replaced with the 3×3 operator metric field $\hat{g}_{ab}(x)$.

Affine commutation relations

Since the canonical commutation relations are in conflict with the requirement of metric positivity, our initial step involves replacing the classical Arnowitt-Deser-Misner (ADM) canonical momentum $\pi^{ab}(x)$ with the classical mixed-index momentum $\pi^a_b(x) \equiv \pi^{ac}(x) g_{cb}(x)$. We refer to $\pi^a_b(x)$ as the "momentric" tensor being a combination of the canonical *momen*tum and the canonical me*tric*. Besides the metric being promoted to an operator $\hat{g}_{ab}(x)$, we also promote the classical momentric tensor to an operator field $\hat{\pi}^a_b(x)$; this pair of operators form the basic kinematical affine operator fields, and all operators of interest are given as functions of this fundamental pair. The basic kinematical operators are chosen so that they satisfy the following set of *affine commutation relations* (in units where $\hbar = 1$, which are normally used throughout):

$$\begin{aligned}
[\hat{\pi}^a_b(x), \, \hat{\pi}^c_d(y)] &= \tfrac{1}{2} i [\delta^c_b \hat{\pi}^a_d(x) - \delta^a_d \hat{\pi}^c_b(x)] \, \delta(x,y) \,, \\
[\hat{g}_{ab}(x), \, \hat{\pi}^c_d(y)] &= \tfrac{1}{2} i [\delta^c_a \hat{g}_{bd}(x) + \delta^c_b \hat{g}_{ad}(x)] \, \delta(x,y) \,, \\
[\hat{g}_{ab}(x), \, \hat{g}_{cd}(y)] &= 0 \,.
\end{aligned} \tag{2}$$

These commutation relations arise as the transcription into operators of equivalent Poisson brackets for the corresponding classical fields, namely, the spatial metric $g_{ab}(x)$ and the momentric field $\pi^c_d(x) \equiv \pi^{cb}(x) g_{bd}(x)$, along with the usual Poisson brackets between the canonical metric field $g_{ab}(x)$ and the canonical momentum field $\pi^{cd}(x)$.

The virtue of the affine variables and their associated commutation relations is evident in the relation

$$e^{i\int \gamma^a_b(y) \hat{\pi}^b_a(y) \, d^3y} \, \hat{g}_{cd}(x) \, e^{-i\int \gamma^a_b(y) \hat{\pi}^b_a(y) \, d^3y} = \{e^{\gamma(x)/2}\}^e_c \, \hat{g}_{ef}(x) \, \{e^{\gamma(x)/2}\}^f_d \,. \tag{3}$$

This algebraic relation confirms that suitable transformations by the momentric field preserve metric positivity.

3. Affine Coherent States

It is noteworthy that the algebra generated by $\hat{g}_{ab}$ and $\hat{\pi}^a_b$ closes. These operators form the generators of the *affine group* whose elements may be defined by

$$U[\pi,\gamma] \equiv e^{i\int \pi^{ab}(y)\hat{g}_{ab}(y)\,d^3y}\, e^{-i\int \gamma^a_b(y)\hat{\pi}^b_a(y)\,d^3y} \,, \tag{4}$$

e.g., for all real, smooth c-number functions π^{ab} and γ^a_b of compact support. Since we assume that the smeared $\hat{g}_{ab}$ and $\hat{\pi}^a_b$ fields are self-adjoint operators, it follows that $U[\pi,\gamma]$ are unitary operators for all π and γ, and moreover, these unitary operators are strongly continuous in the label fields π and γ.

To define a representation of the basic operators it suffices to choose a fiducial vector and thereby to introduce a set of affine coherent states, i.e., coherent states formed with the help of the affine group. We choose $|\eta\rangle$ as a normalized fiducial vector in the original Hilbert space $\mathfrak{H}$, and we consider a set of unit vectors each of which is given by

$$|\pi,\gamma\rangle \equiv e^{i\int \pi^{ab}(x)\,\hat{g}_{ab}(x)\,d^3x}\, e^{-i\int \gamma^d_c(x)\,\hat{\pi}^c_d(x)\,d^3x}\,|\eta\rangle \,. \tag{5}$$

As π and γ range over the space of smooth functions of compact support, such vectors form the desired set of coherent states. The specific representation of the kinematical operators is fixed once the vector $|\eta\rangle$ has been chosen. As minimum requirements on $|\eta\rangle$ we impose

$$\langle\eta|\hat{\pi}^a_b(x)|\eta\rangle = 0 \,, \tag{6}$$

$$\langle\eta|\hat{g}_{ab}(x)|\eta\rangle = \tilde{g}_{ab}(x) \,, \tag{7}$$

where $\tilde{g}_{ab}(x)$ is a metric that determines the topology of the underlying space-like surface. As algebraic consequences of these conditions, it follows that

$$\langle\pi,\gamma|\hat{g}_{ab}(x)|\pi,\gamma\rangle = \{e^{\gamma(x)/2}\}^c_a\, \tilde{g}_{cd}(x)\, \{e^{\gamma(x)/2}\}^d_b \equiv g_{ab}(x) \,, \tag{8}$$

$$\langle\pi,\gamma|\hat{\pi}^a_c(x)|\pi,\gamma\rangle = \pi^{ab}(x)g_{bc}(x) \equiv \pi^a_c(x) \,. \tag{9}$$

These expectations are not gauge invariant since they are taken in the original Hilbert space where the constraints are not fulfilled.

By definition, the coherent states span the original, or kinematical, Hilbert space $\mathfrak{H}$, and thus we can characterize the coherent states themselves by giving their overlap with an arbitrary coherent state. In so doing, we choose the fiducial vector $|\eta\rangle$ so that the overlap is given by

$$\langle\pi'',\gamma''|\pi',\gamma'\rangle = \exp\bigg[-2\int b(x)\,d^3x \\ \times \ln\left(\frac{\det\{\frac{1}{2}[g''^{ab}(x)+g'^{ab}(x)] + \frac{1}{2}ib(x)^{-1}[\pi''^{ab}(x)-\pi'^{ab}(x)]\}}{\{\det[g''^{ab}(x)]\,\det[g'^{ab}(x)]\}^{1/2}}\right)\bigg] \tag{10}$$

where $b(x)$, $0 < b(x) < \infty$, is a scalar density which is discussed below.

Additionally, we observe that γ'' and γ' do *not* appear in the explicit functional form given in (10). In particular, the smooth matrix γ has been replaced by the smooth matrix g which is defined at every point by

$$g(x) \equiv e^{\gamma(x)/2}\, \tilde{g}(x)\, e^{\gamma(x)^T/2} \equiv \{g_{ab}(x)\}\ , \tag{11}$$

where T denotes transpose, and the matrix $\tilde{g}(x) \equiv \{\tilde{g}_{ab}(x)\}$ is given by (7). The map $\gamma \to g$ is clearly many-to-one since γ has nine independent variables at each point while g, which is symmetric, has only six. In view of this functional dependence we may denote the given functional in (10) by $\langle \pi'', g''|\pi', g'\rangle$, and henceforth we adopt this notation. In particular, we note that (8) and (9) become

$$\langle \pi, g|\hat{g}_{ab}(x)|\pi, g\rangle \equiv g_{ab}(x)\ , \tag{12}$$

$$\langle \pi, g|\hat{\pi}^a_c(x)|\pi, g\rangle = \pi^{ab}(x) g_{bc}(x) \equiv \pi^a_c(x)\ , \tag{13}$$

which show that the meaning of the labels π and g is that of *mean* values rather than sharp eigenvalues.

In addition, we observe that the coherent state overlap function (10) is a continuous function that can serve as a reproducing kernel for a reproducing kernel Hilbert space which provides a representation of the original Hilbert space $\mathfrak{H}$ by continuous functions of π and g. For details of such spaces, see Ref. 4.

4. Projection Operator Approach for Quantum Constraints

Classically, constraints are either: (i) first class, for which the Lagrange multipliers are undetermined and must be chosen to find a solution; or (ii) second class, for which the Lagrange multipliers are fixed by the equations of motion.

The Dirac approach to the quantization of constraints requires quantization before reduction. Thus the constraints are first promoted to self-adjoint operators,

$$\phi_\alpha(p, q) \to \Phi_\alpha(P, Q)\ , \tag{14}$$

for all α, and then the physical Hilbert space $\mathfrak{H}_{\rm phys}$ is defined by those vectors $|\psi\rangle_{\rm phys}$ for which

$$\Phi_\alpha(P, Q)|\psi\rangle_{\rm phys} = 0 \tag{15}$$

for all α. This procedure works for a limited set of classical first-class constraint systems, but it does not work in general and especially not for second-class constraints.

The projection operator approach to quantum constraints involves a slight relaxation of the Dirac procedure. Instead of insisting that (15) holds exactly, we introduce a projection operator $\mathbb{E}$ defined by

$$\mathbb{E} = \mathbb{E}(\Sigma_\alpha \Phi_\alpha^2 \leq \delta(\hbar)^2) , \tag{16}$$

where $\delta(\hbar)$ is a positive *regularization parameter* and we have assumed that $\Sigma_\alpha \Phi_\alpha^2$ is self adjoint. This relation means that $\mathbb{E}$ projects onto the spectral range of the self-adjoint operator $\Sigma_\alpha \Phi_\alpha^2$ in the interval $[0, \delta(\hbar)^2]$, and then $\mathfrak{H}_{\rm phys} = \mathbb{E}\mathfrak{H}$. As a final step, the parameter $\delta(\hbar)$ is reduced as much as required, and, in particular, when some second-class constraints are involved, $\delta(\hbar)$ ultimately remains strictly positive. This general procedure treats all constraints simultaneously and treats them all on an equal basis; for details see Ref. 5.

A few examples illustrate how the projection operator method works. If $\Sigma_\alpha \Phi_\alpha^2 = J_1^2 + J_2^2 + J_3^2$, the Casimir operator of $su(2)$, then $0 \leq \delta(\hbar)^2 < 3\hbar^2/4$ works for this first-class example. If $\Sigma_\alpha \Phi_\alpha^2 = P^2 + Q^2$, where $[Q, P] = i\hbar 1\!\!1$, then $\hbar \leq \delta(\hbar)^2 < 3\hbar$ covers this second-class example. If the single constraint $\Phi = Q$, an operator whose zero lies in the continuous spectrum, then it is convenient to take an appropriate form limit of the projection operator as $\delta \to 0$; see Ref. 5. The projection operator scheme can also deal with irregular constraints such as $\Phi = Q^3$, and even mixed examples with regular and irregular constraints such as $\Phi = Q^3(1 - Q)$, etc.; see Ref. 6.

It is also of interest that the desired projection operator has a general, time-ordered integral representation (see Ref. 7) given by

$$\mathbb{E} = \mathbb{E}(\Sigma_\alpha \Phi_\alpha^2 \leq \delta(\hbar)^2) = \int \mathsf{T} e^{-i\int \lambda^\alpha(t)\Phi_\alpha\, dt}\, \mathcal{D}R(\lambda) . \tag{17}$$

The weak measure R depends on the number of Lagrange multipliers, the time interval, and the regularization parameter $\delta(\hbar)^2$. The measure R does *not* depend on the constraint operators, and thus this relation is an operator identity, holding for any set of operators $\{\Phi_\alpha\}$. The time-ordered integral representation for $\mathbb{E}$ given in (17) can be used in path-integral representations as will become clear below.

5. Continuous-Time Regularized Functional Integral Representation without/with Constraints

It is pedagogically useful to reexpress the coherent-state overlap function by means of a functional integral. This process can be aided by the fact that the expression (10) is analytic in the variable $g''^{ab}(x) + ib(x)^{-1}\pi''^{ab}(x)$ up to a factor. As a consequence, the coherent-state overlap function satisfies a complex polarization condition, which leads to a second-order differential operator that annihilates it. This fact can be used to generate a functional integral representation of the form

$$\begin{aligned}\langle\pi'',g''|\pi',g'\rangle &= \exp\bigg[-2\int b(x)\,d^3x \\ &\quad\times\ln\bigg(\frac{\det\{\frac{1}{2}[g''^{ab}(x)+g'^{ab}(x)]+\frac{1}{2}ib(x)^{-1}[\pi''^{ab}(x)-\pi'^{ab}(x)]\}}{\{\det[g''^{ab}(x)]\,\det[g'^{ab}(x)]\}^{1/2}}\bigg)\bigg] \\ &= \lim_{\nu\to\infty}\overline{\mathcal{N}}_\nu\int\exp[-i\textstyle\int g_{ab}\dot{\pi}^{ab}\,d^3x\,dt] \\ &\quad\times\exp\{-(1/2\nu)\textstyle\int[b(x)^{-1}g_{ab}g_{cd}\dot{\pi}^{bc}\dot{\pi}^{da}+b(x)g^{ab}g^{cd}\dot{g}_{bc}\dot{g}_{da}]\,d^3x\,dt\} \\ &\quad\times[\Pi_{x,t}\Pi_{a\le b}\,d\pi^{ab}(x,t)\,dg_{ab}(x,t)]\,. \end{aligned} \tag{18}$$

Because of the way the new independent variable t appears in the right-hand term of this equation, it is natural to interpret t, $0\le t\le T$, $T>0$, as coordinate "time". The fields on the right-hand side all depend on space and time, i.e., $g_{ab}=g_{ab}(x,t)$, $\dot{g}_{ab}=\partial g_{ab}(x,t)/\partial t$, etc., and, importantly, the integration domain of the formal measure is strictly limited to the domain where $\{g_{ab}(x,t)\}$ is a positive-definite matrix for all x and t. For the boundary conditions, we have $\pi'^{ab}(x)\equiv\pi^{ab}(x,0)$, $g'_{ab}(x)\equiv g_{ab}(x,0)$, as well as $\pi''^{ab}(x)\equiv\pi^{ab}(x,T)$, $g''_{ab}(x)\equiv g_{ab}(x,T)$ for all x. Observe that the right-hand term holds for any T, $0<T<\infty$, while the left-hand and middle terms are independent of T altogether.

In like manner, we can incorporate the constraints into a functional integral by using an appropriate form of the integral representation (17). The resultant expression has a functional integral representation given by

$$\begin{aligned}\langle\pi'',g''|\mathbb{E}|\pi',g'\rangle &= \int\langle\pi'',g''|\mathbf{T}\,e^{-i\int[N^a\mathcal{H}_a+N\mathcal{H}]\,d^3x\,dt}|\pi',g'\rangle\,\mathcal{D}R(N^a,N) \\ &= \lim_{\nu\to\infty}\overline{\mathcal{N}}_\nu\int e^{-i\int[g_{ab}\dot{\pi}^{ab}+N^aH_a+NH]\,d^3x\,dt} \\ &\quad\times\exp\{-(1/2\nu)\textstyle\int[b(x)^{-1}g_{ab}g_{cd}\dot{\pi}^{bc}\dot{\pi}^{da}+b(x)g^{ab}g^{cd}\dot{g}_{bc}\dot{g}_{da}]\,d^3x\,dt\} \\ &\quad\times[\Pi_{x,t}\Pi_{a\le b}\,d\pi^{ab}(x,t)\,dg_{ab}(x,t)]\,\mathcal{D}R(N^a,N)\,. \end{aligned} \tag{19}$$

Despite the general appearance of (19), we emphasize once again that this representation has been based on the affine commutation relations and *not* on any canonical commutation relations.

The expression $\langle\pi'',g''|\mathbb{E}|\pi',g'\rangle$ denotes the coherent-state matrix elements of the projection operator $\mathbb{E}$ which projects onto a subspace of the original Hilbert space on which the quantum constraints are fulfilled in a regularized fashion. Furthermore, the expression $\langle\pi'', g''|\mathbb{E}|\pi', g'\rangle$ is another continuous functional that can be used as a reproducing kernel and thus used directly to generate the reproducing kernel physical Hilbert space on which the quantum constraints are fulfilled in a regularized manner. Observe that N^a and N denote Lagrange multiplier fields (classically interpreted as the shift and lapse), while H_a and H denote phase-space symbols (since $\hbar \neq 0$) associated with the quantum diffeomorphism and Hamiltonian constraint field operators, respectively. Up to a surface term, therefore, the phase factor in the functional integral represents the canonical action for general relativity.

6. Hard-Core Picture of Nonrenormalizability

Nonrenormalizable quantum field theories involve an infinite number of distinct counterterms when approached by a regularized, renormalized perturbation analysis. Focusing on scalar field theories, a qualitative Euclidean functional integral formulation is given by

$$S_\lambda(h) = \mathcal{N}_\lambda \int e^{\int h\phi\, d^nx - W_o(\phi) - \lambda V(\phi)}\, \mathcal{D}\phi\,, \tag{20}$$

where $W_o(\phi) \geq 0$ denotes the free action and $V(\phi) \geq 0$ the interaction term. If $\lambda = 0$, the support of the integral is determined by $W_o(\phi)$; when $\lambda > 0$, the support is determined by $W_o(\phi) + \lambda V(\phi)$. Formally, as $\lambda \to 0$, $S_\lambda(h) \to S_0(h)$, the functional integral for the free theory. However, it may happen that

$$\lim_{\lambda \to 0} S_\lambda(h) = S_0'(h) \neq S_0(h)\,, \tag{21}$$

where $S_0'(h)$ defines a so-called *pseudofree* theory. Such behavior arises formally if $V(\phi)$ acts partially as a *hard core*, projecting out certain fields that are not restored to the support of the free theory as $\lambda \to 0$.[8]

It is noteworthy that there exist highly idealized nonrenormalizable model quantum field theories with exactly the behavior described; see Ref. 9. Such examples involve counterterms not suggested by a renormalized perturbation analysis. It is our belief that these soluble models strongly suggest that nonrenormalizable φ_n^4, $n \geq 5$, models can be understood by the same mechanism, and that they too can be properly formulated by the incorporation of a limited number of counterterms distinct from those suggested by a perturbation treatment. Although technically more complicated, we see no fundamental obstacle in dealing with quantum gravity on the basis of an analogous hard-core interpretation.

References

1. J. R. Klauder, *Noncanonical quantization of gravity. I. Foundations of affine quantum gravity*, *J. Math. Phys.* **40**, 5860 (1999).
2. J. R. Klauder, *Noncanonical quantization of gravity. II. Constraints and the physical Hilbert space*, *J. Math. Phys.* **42**, 4440 (2001).
3. J. R. Klauder, *Ultralocal fields and their relevance for reparametrization invariant quantum field theory*, *J. Phys. A: Math. Gen.* **34**, 3277 (2001).
4. N. Aronszajn, *Théorie générale de noyaux reproduisants – première partie*, *Proc. Cambridge Phil. Soc.* **39**, 133 (1943); N. Aronszajn, *Theory of reproducing kernels*, *Trans. Am. Math. Soc.* **68**, 337 (1950); H. Meschkowski, *Hilbertsche Räume mit Kernfunktion* (Springer, Berlin, 1962).
5. J. R. Klauder, *Coherent state quantization of constraint systems*, *Ann. Phys.* **254**, 419 (1997); J. R. Klauder, *Quantization of constrained systems*, *Lect. Notes Phys.* **572**, 143 (2001) [arXiv:hep-th/0003297].
6. J. R. Klauder and J. S. Little, *Highly irregular quantum constraints*, *Class. Quant. Grav.* **23**, 3641 (2006).
7. J. R. Klauder, *Universal procedure for enforcing quantum constraints*, *Nucl. Phys. B* **547**, 397 (1999).
8. J. R. Klauder, *Field structure through model studies: Aspects of nonrenormalizable theories*, *Acta Physica Austriaca, Suppl.* **XI**, 341 (1973); J. R. Klauder, *On the meaning of a nonrenormalizable theory of gravitation*, *Gen. Rel. Grav.* **6**, 13 (1975); J. R. Klauder, *Continuous and discontinuous perturbations*, *Science* **199**, 735 (1978).
9. J. R. Klauder, *Beyond Conventional Quantization* (Cambridge University Press, Cambridge, 2000 & 2005).

THE ROLE OF THOMAS-FERMI APPROACH IN NEUTRON STAR MATTER

R. RUFFINI

Department of Physics and ICRA, Sapienza, Rome, 00185, Italy
ICRANet, p.le della Repubblica 10, Pescara, 65122, Italy
ICRANet, Sophia Antipolis, Nice, 06103, France
E-mail: ruffini@icra.it

The role of the Thomas-Fermi approach in Neutron Star matter cores is presented and discussed with special attention to solutions which are globally neutral and do not fulfil the traditional condition of local charge neutrality. A new stable and energetically favorable configuration is found. This new solution can be of relevance in understanding unsolved issues of gravitational collapse processes and their energetics.

Keywords: Degenerate Fermi gases; Neutron stars; Thomas-Fermi model.

1. Introduction

We first recall how certainly one of the greatest successes in human understanding of the Universe has been the research activity started in 1054 by Chinese, Korean, and Japanese astronomers by the observations of a "Guest Star"(see e.g. Shklovsky[1]), followed by the discovery of the Pulsar $NPO532$ in the Crab Nebula in 1967, (see e.g. Manchester and Taylor[2]), still presenting challenges in the yet not identified physical process originating the expulsion of the remnant in the Supernova explosion (see e.g. Mezzacappa and Fuller[3] and Fig. 1(a)). We are currently exploring the neutron star equilibrium configuration for a missing process which may lead to the solution of the above mentioned astrophysical puzzle.

We also recall an additional astrophysical observation which is currently capturing the attention of Astrophysicists worldwide: the Gamma ray Bursts or for short GRBs. Their discovery was accidental and triggered by a very unconventional idea proposed by Yacov Borisovich Zel'dovich (see e.g. Ref. 4). It is likely that this idea served as an additional motivation for the United States of America to put a set of four Vela Satellites into

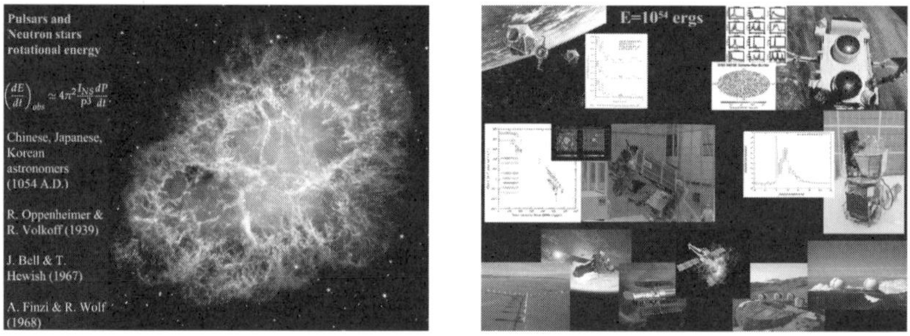

(a) (b)

Fig. 1. (a) The expanding shell of the remnant of the Crab Nebulae as observed by the Hubble Space Telescope. Reproduced from Hubble Telescope web site with their kind permission (News Release Number: STScl-2005-37). (b) On the upper left the Vela 5A and 5B satellites and a typical event as recorded by three of the Vela satellites; on the upper right the Compton satellite and the first evidence of the isotropy of distribution of GRB in the sky; on the center left the Beppo Sax satellite and the discovery of the after glow; on the center right a GRB from Integral satellite; in the lower part the Socorro very large array radiotelescope, the Hubble, the Chandra, and the XMM telescopes, as well as the VLT of Chile and KECK observatory in Hawaii. All these instruments are operating for the observations of GRBs.[6]

orbit, 150 000 miles above the Earth. They were top-secret omnidirectional detectors using atomic clocks to precisely record the arrival times of both X-rays and γ-rays (see Fig. 1(b)). When they were made operational they immediately produced results (see Fig. 1(b)). It was thought at first that the signals originated from nuclear bomb explosions on the earth but they were much too frequent, one per day! A systematic analysis showed that they had not originated on the earth, nor even in the solar system. These Vela satellites had discovered GRBs! The first public announcement of this came at the AAAS meeting in San Francisco in a special session on neutron stars, black holes, and binary X-ray sources, organized by Herb Gursky and myself.[5] A few months later, Thibault Damour and myself published a theoretical framework for GRBs based on the vacuum polarization process in the field of a Kerr–Newman black hole.[7] We showed how the pair creation predicted by the Heisenberg–Euler–Schwinger theory[8,9] would lead to a transformation of the black hole, asymptotically close to reversibility. The electron–positron pairs created by this process were generated by what we now call the blackholic energy.[4] In that paper we concluded that this "naturally leads to a very simple model for the explanation of the recently discovered GRBs". Our theory had two very clear signatures. It could only operate for black holes with mass M_{BH} in the range of 3.2 10^6 $M_{\odot}$ and the energy released had a characteristic value of

$$E = 1.8 \times 10^{54} M_{BH}/M_{\odot} \, \text{ergs} \,. \tag{1}$$

Since nothing was then known about the location and the energetics of these sources we stopped working in the field, waiting for a clarification of the astrophysical scenario.

The situation changed drastically with the discovery of the "afterglow" of GRBs[10] by the joint Italian-Dutch satellite BeppoSAX (see Fig. 1(b)). This X-ray emission lasted for months after the "prompt" emission of a few seconds duration and allowed the GRB sources to be identified much more accurately. This then led to the optical identification of the GRBs by the largest telescopes in the world, including the Hubble Space Telescope, the KECK telescope in Hawaii, and the VLT in Chile (see Fig. 1(b)). Also, the very large array in Socorro made the radio identification of GRBs possible. The optical identification of GRBs made the determination of their distances possible. The first distance measurement for a GRB was made in 1997 for GRB970228 and the truly enormous of isotropical energy of this was determined to be 10^{54} ergs per burst. This proved the existence of a single astrophysical system emitting as much energy during its short lifetime as that emitted in the same time by all other stars of all galaxies in the Universe![a] It is interesting that this "quantum" of astrophysical energy coincided with the one Thibault Damour and I had already predicted, see Eq. (1). Much more has been learned on GRBs in recent years confirming this basic result (see e.g. Ref. 11). The critical new important step now is to understand the physical process leading to the critical fields needed for the pair creation process during the gravitational collapse process from a Neutron Star to a Black Hole.

As third example, we recall the galactic 'X-ray bursters' as well as some observed X-ray emission precursor of supernovae events.[12] It is our opinion that the solution of: **a)** the problem of explaining the energetics of the emission of the remnant during the collapse to a Neutron Star, **b)** the problem of formation of the supercritical fields during the collapse to a Black Hole, **c)** the less energetics of galactic 'X-ray bursters' and of the precursor of the supernovae explosion event, will find their natural explanation from a yet unexplored field: the electrodynamical structure of a neutron star. We will outline a few crucial ideas of how a Thomas-Fermi approach to a

[a]Luminosity of average star = 10^{33} erg/s, Stars per galaxy = 10^{12}, Number of galaxies = 10^9. Finally, $33 + 12 + 9 = 54$!

neutron star can, indeed, represent an important step in identifying this crucial new feature.

2. Thomas-Fermi Model

We first recall the basic Thomas-Fermi non relativistic Equations (see e.g. Landau and Lifshitz[13]). They describe a degenerate Fermi gas of N_{el} electrons in the field of a point-like nucleus of charge Ze. The Coulomb potential $V(r)$ satisfies the Poisson equation

$$\nabla^2 V(r) = 4\pi e n, \tag{2}$$

where the electron number density $n(r)$ is related to the Fermi momentum p_F by $n = p_F^3/(3\pi^2\hbar^3)$. The equilibrium condition for an electron of mass m inside the atom is expressed by $\frac{p_F^2}{2m} - eV = E_F$. To put Eq. (2) in dimensionless form, we introduce a function ϕ, related to Coulomb potential by $\phi(r) = V(r) + \frac{E_F}{e} = Ze\frac{\chi(r)}{r}$. Assuming $r = bx$ with $b = \frac{(3\pi)^{3/2}}{2^{7/3}}\frac{1}{Z^{1/3}}\frac{\hbar^2}{me^2}$, we then have the universal equation[14,15]

$$\frac{d^2\chi(x)}{dx^2} = \frac{\chi(x)^{3/2}}{x^{1/2}}. \tag{3}$$

The first boundary condition for this equation follows from the request that approaching the nucleus one gets the ordinary Coulomb potential, therefore we have $\chi(0) = 1$. The second boundary condition comes from the fact that the number of electrons N_{el} is $1 - \frac{N_{el}}{Z} = \chi(x_0) - x_0\chi'(x_0)$.

3. White Dwarfs and Neutron Stars as Thomas-Fermi Systems

It was at the 1972 Les Houches organized by Bryce and Cecile de Witt summer School (see Fig. 2(a) and Ref. 16) that, generalizing a splendid paper by Landau,[17] I introduced a Thomas-Fermi description of both White Dwarfs and Neutron Stars within a Newtonian gravitational theory and describing the microphysical quantities by a relativistic treatment. The equilibrium condition for a self-gravitating system of fermions in relativistic regime is $c\sqrt{p_F^2 + m_n^2c^2} - m_nc^2 - m_nV = -m_nV_0$, where p_F is the Fermi momentum of a particle of mass m_n, related to the particle density n by $n = \frac{1}{3\pi^2\hbar^3}p_F^3$. Here $V(r)$ is the gravitational potential at a point at distance r from the center of the configuration and V_0 is the value of the potential at the boundary R_c of the configuration $V_0 = \frac{GNm_n}{R_c}$. N is the total number of particles. The Poisson equation is $\nabla^2 V = -4\pi G m_n n$. Assuming $V - V_0 = GNm_n\frac{\chi(r)}{r}$

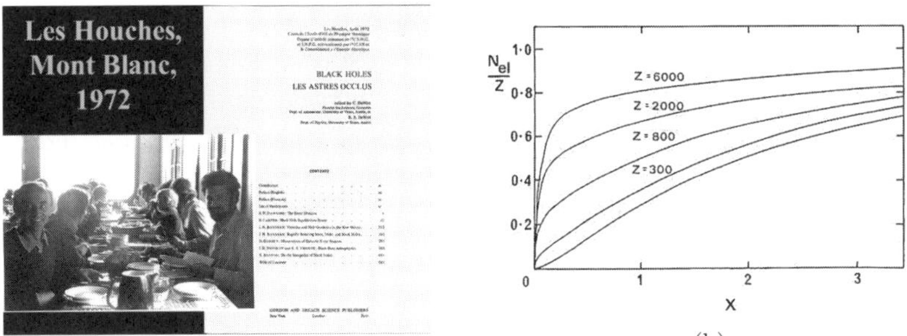

(a) (b)

Fig. 2. (a) Lunch at Les Houches summer school on 'Black Holes'. In front, face to face, Igor Novikov and the author; in the right the title of the book in English and in French. It is interesting that in that occasion Cecile de Witt founded the French translation of the word 'Back Hole' in 'Trou Noir' objectionable and she introduced instead the even more objectionable term 'Astres Occlus'. The French nevertheless happily adopted in the following years the literally translated word 'Trou Noir' for the astrophysical concept I introduced in 1971 with J. A. Wheeler.[18] (b) The number of electrons contained within a distance x of the origin, as a function of the total number Z for a neutral atom. The lowest curve is that given by the solution of the non-relativistic Thomas-Fermi equation.

and $r = bx$ with $b = \frac{(3\pi)^{2/3}}{2^{7/3}} \frac{1}{N^{1/3}} \left(\frac{\hbar}{m_n c}\right) \left(\frac{m_{Planck}}{m_n}\right)^2$ we obtain the gravitational Thomas-Fermi equation

$$\frac{d^2\chi}{dx^2} = -\frac{\chi^{3/2}}{\sqrt{x}} \left[1 + \left(\frac{N}{N^*}\right)^{4/3} \frac{\chi}{x}\right]^{3/2}, \tag{4}$$

where $N^* = \left(\frac{3\pi}{4}\right)^{1/2} \left(\frac{m_{Planck}}{m_n}\right)^3$. Equation (4) has to be integrated with the boundary conditions $\chi(0) = 0$ and $-x_b \left(\frac{d\chi}{dx}\right)_{x=x_b} = 1$. Equation (4) can be applied as well to the case of white dwarfs. It is sufficient to assume $b = \frac{(3\pi)^{2/3}}{2^{7/3}} \frac{1}{N^{1/3}} \left(\frac{\hbar}{m_e c}\right) \left(\frac{m_{Planck}}{\mu m_n}\right)^2$, $N^* = \left(\frac{3\pi}{4}\right)^{1/2} \left(\frac{m_{Planck}}{\mu m_n}\right)^3$, $M = \int_0^{R_c} 4\pi r^2 n_e(r) \mu m_n dr$. For the equilibrium condition we have $c\sqrt{p_F^2 + m^2c^2} - mc^2 - \mu m_n V = -\mu m_n V_0$, in order to obtain for the critical mass the value $M_{crit} \approx 5.7 M_{sun} \mu_e^{-2} \approx 1.5 M_{sun}$.

4. Relativistic Thomas-Fermi Equation

In the intervening years my attention was dedicated to an apparently academic problem: the solution of a relativistic Thomas-Fermi Equation and extrapolating the Thomas-Fermi solution to large atomic numbers of

$Z \approx 10^4 - 10^6$. Three new features were outlined: **a)** the necessity of introducing a physical size for the nucleus, **b)** the penetration of the electrons in the nucleus, **c)** the definition of an effective nuclear charge.[19,20] The electrostatic potential is given by $\nabla^2 V(r) = 4\pi e n$, where the number density of electrons is related to the Fermi momentum p_F by $n = \frac{p_F^3}{3\pi^2\hbar^3}$. In order to have equilibrium we have $c\sqrt{p_F^2 + m^2c^2} - mc^2 - eV(r) = E_F$. Assuming $\phi(r) = V(r) + \frac{E_F}{e} = Ze\frac{\chi(r)}{r}$, $Z_c = \left(\frac{3\pi}{4}\right)^{1/2}\left(\frac{\hbar c}{e^2}\right)^{3/2}$, and $r = bx$ with $b = \frac{(3\pi)^{3/2}}{2^{7/3}}\frac{1}{Z^{1/3}}\frac{\hbar^2}{me^2}$, Eq. (3) becomes

$$\frac{d^2\chi(x)}{dx^2} = \frac{\chi(x)^{3/2}}{x^{1/2}}\left[1 + \left(\frac{Z}{Z_c}\right)^{4/3}\frac{\chi(x)}{x}\right]^{3/2}. \tag{5}$$

5. Essential Role of Non-Pointlike Nucleus

The point-like assumption for the nucleus leads, in the relativistic case, to a non-integrable expression for the electron density near the origin. We assumed a uniformly charged nucleus with a radius r_{nuc} and a mass number A given by the following semi-empirical formulae

$$r_{nuc} = r_0 A^{1/3}, \quad r_0 \approx 1.5 \times 10^{-13}\ \text{cm}, \quad Z \simeq \left[\frac{2}{A} + \frac{3}{200}\frac{1}{A^{1/3}}\right]^{-1}, \tag{6}$$

Equation (5) then becomes

$$\frac{d^2\chi(x)}{dx^2} = \frac{\chi(x)^{3/2}}{x^{1/2}}\left[1 + \left(\frac{Z}{Z_c}\right)^{4/3}\frac{\chi(x)}{x}\right]^{3/2} - \frac{3x}{x_{nuc}^3}\theta(x_{nuc} - x), \tag{7}$$

where $\theta = 1$ for $r < r_{nuc}$, $\theta = 0$ for $r > r_{nuc}$, $\chi(0) = 0$, $\chi(\infty) = 0$. Equation (7) has been integrated numerically for selected values of Z (see Fig. 2(b) and Refs. 19,20). Similar results had been obtained by Greiner and his school and by Popov and his school with special emphasis on the existence of a critical electric field at the surface of heavy nuclei. Their work was mainly interested in the study of the possibility of having processes of vacuum polarization at the surface of heavy nuclei to be possibly achieved by heavy nuclei collisions (see for a review Ref. 21). Paradoxically at the time we were not interested in this very important aspect and we did not compute the strength of the field in our relativistic Thomas-Fermi model which is, indeed, of the order of the critical field $E_c = m^2c^3/e\hbar$.

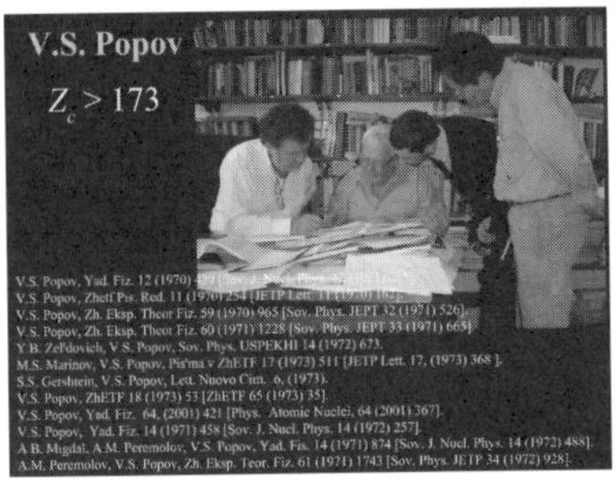

(a)

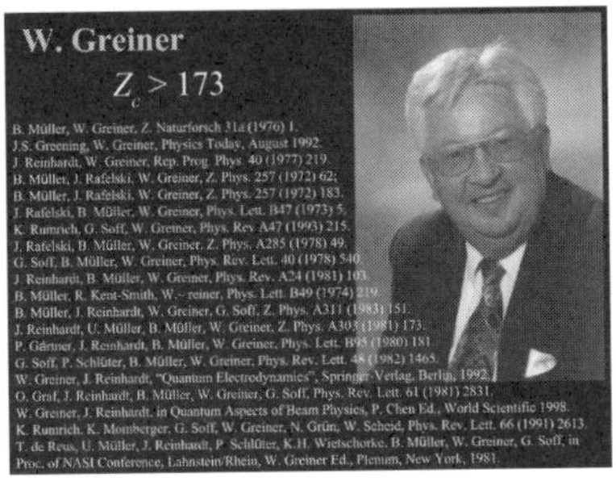

(b)

Fig. 3. (a) Vladimir Popov discussing with the author and Professors She Sheng Xue and Gregory Vereshchagin (Roma 2007). Also quoted the classical contributions of Popov and his school. (b) Walter Greiner and the citation of classical papers by him and his school.

6. Nuclear Matter in Bulk: $A \approx 300$ or $A \approx (m_{Planck}/m_n)^3$

The situation clearly changed with the discovery of GRBs and the understanding that the process of vacuum polarization unsuccessfully sought in earthbound experiments could, indeed, be observed in the process of formation of a Black Hole from the gravitational collapse of a neutron star.[21] The concept of a Dyadosphere[22,23] was introduced around an already formed Black Hole and it became clear that this concept was of paramount importance in understanding the energy source for GRBs. It soon became clear that the initial conditions for such a process had to be found in the electrodynamical properties of neutron stars. Similarly manifest came the crucial factor which had hampered the analysis of the true electrodynamical properties of a neutron star; the unjustified imposition of local charge neutrality as opposed to the global charge neutrality of the system. We have therefore proceeded to make a model of a nuclear matter core of $A \approx (m_{Planck}/m_n)^3$ nucleons.[24] We generalized to this more general case the concept introduced in their important work by W. Greiner and V. Popov (see Fig. 3(a) and Fig. 3(b)) as follows. I have assumed that proton's number-density is constant inside core $r \leq R_c$ and vanishes outside the core $r > R_c$: $n_p = \frac{1}{3\pi^2\hbar^3}(P_p^F)^3 = \frac{3N_p}{4\pi R_c^3}\theta(R_c - r)$, $\quad R_c = \Delta\frac{\hbar}{m_\pi c}N_p^{1/3}$, where P_p^F is the Fermi momentum of protons, $\theta(R_c - r)$ is the step-function, and Δ is a parameter. The proton Fermi energy is

$$\mathcal{E}_p(P_p^F) = [(P_p^F c)^2 + m_p^2 c^4]^{1/2} - m_p c^2 + eV, \tag{8}$$

where e is the proton charge and V is the Coulomb potential. Based on the Gauss law, $V(r)$ obeys the Poisson equation $\nabla^2 V(r) = -4\pi e\,[n_p(r) - n_e(r)]$

and the boundary conditions $V(\infty) = 0$, $V(0) =$ finite, where the electron number density $n_e(r)$ is given by

$$n_e(r) = \frac{1}{3\pi^2\hbar^3}(P_e^F)^3 \tag{9}$$

with P_e^F being the electron Fermi momentum. The electron Fermi energy is

$$\mathcal{E}_e(P_e^F) = [(P_e^F c)^2 + m^2c^4]^{1/2} - mc^2 - eV. \tag{10}$$

The energetic equation for an electrodynamic equilibrium of electrons in the Coulomb potential $V(r)$ is $\mathcal{E}_e(P_e^F) = 0$, hence the Fermi momentum and the electron number density can be written as $n_e(r) = \frac{1}{3\pi^2\hbar^3c^3}\left[e^2V^2(r) + 2mc^2eV(r)\right]^{3/2}$. Introducing the new variable $x = r/(\hbar/m_\pi c)$, where the radial coordinate is given in units of pion Compton length $(\hbar/m_\pi c)$, with $x_c = x(r = R_c)$ I have obtained the following relativistic Thomas-Fermi equation:[25,26]

$$\frac{1}{3x}\frac{d^2\chi(x)}{dx^2} = -\alpha\left\{\frac{1}{\Delta^3}\theta(x_c - x) - \frac{4}{9\pi}\left[\frac{\chi^2(x)}{x^2} + 2\frac{m}{m_\pi}\frac{\chi}{x}\right]^{3/2}\right\}, \tag{11}$$

where χ is a dimensionless function defined by $\frac{\chi}{r} = \frac{eV}{c\hbar}$ and α is the fine structure constant $\alpha = e^2/(\hbar c)$. The boundary conditions of the function $\chi(x)$ are $\chi(0) = 0$, $\chi(\infty) = 0$, and $N_e = \int_0^\infty 4\pi r^2 dr n_e(r)$. Instead of using the phenomenological relation between Z and A, given by Eq. (6), we determine directly the relation between A and Z by requiring the β-equilibrium

$$\mathcal{E}_n = \mathcal{E}_p + \mathcal{E}_e\,. \tag{12}$$

The number-density of degenerate neutrons is given by $n_n(r) = \frac{1}{3\pi^2\hbar^3}(P_n^F)^3$, where P_n^F is the Fermi momentum of neutrons. The Fermi energy of degenerate neutrons is

$$\mathcal{E}_n(P_n^F) = [(P_n^F c)^2 + m_n^2c^4]^{1/2} - m_nc^2, \tag{13}$$

where m_n is the neutron mass. Substituting Eqs. (8), (10), (13) into Eq. (12), we obtain $[(P_n^F c)^2 + m_n^2c^4]^{1/2} - m_nc^2 = [(P_p^F c)^2 + m_p^2c^4]^{1/2} - m_pc^2 + eV$. These equations and boundary conditions form a closed set of nonlinear boundary value problem with a unique solution for Coulomb potential $V(r)$ and electron distribution (9) as functions of the parameter Δ, i.e., the proton number-density n_p. The solution is given in Fig. 4(a). A relevant quantity for exploring the physical significance of the solution is given by the number of electrons within a given radius r, $N_e(r) = \int_0^r 4\pi(r')^2 n_e(r')dr'$. This allows to determine, for selected values of the $A = N_p + N_n$ parameter,

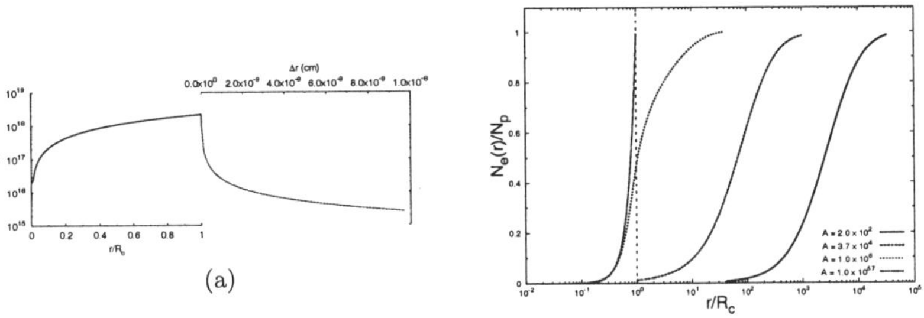

Fig. 4. (a) The solution χ of the relativistic Thomas-Fermi equation for $A = 10^{57}$ and core radius $R_c = 10$ km, is plotted as a function of radial coordinate. The left solid line corresponds to the internal solution and it is plotted as a function of radial coordinate in unit of R_c in logarithmic scale. The right dotted line corresponds to the solution external to the core and is plotted as function of the distance Δr from the surface in the logarithmic scale in centimeter. (b) The electron number in the unit of the total proton number N_p, for selected values of A, is given as a function of radial distance in the unit of the core radius R_c, again in logarithmic scale. It is clear how, by increasing the value of A, the penetration of electrons inside the core increases.

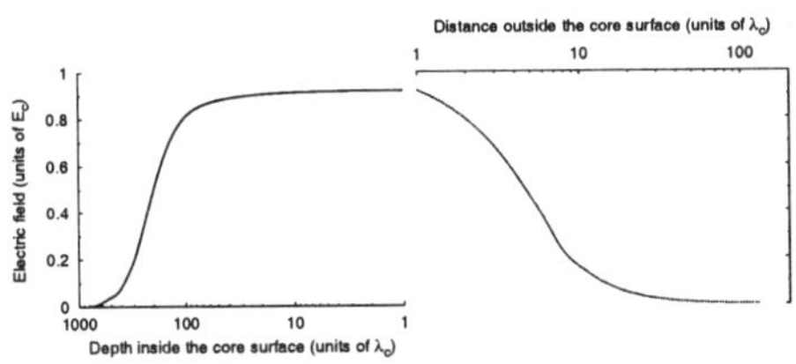

Fig. 5. The electric field in the unit of the critical field E_c is plotted around the core radius R_c. The left (right) solid (dotted) diagram refers to the region just inside (outside) the core radius plotted logarithmically. By increasing the density of the star the field approaches the critical field.

the distribution of the electrons within and outside the core and follow the progressive penetration of the electrons in the core at increasing values of A (see Fig. 4(b)). We can then evaluate, generalizing the results in Refs. 19,20, the net charge inside the core $N_{\rm net} = N_p - N_e(R_c) < N_p$, and consequently determine of the electric field at the core surface, as well as within and outside the core (see Fig. 5).

7. Energetically Favorable Configurations

Introducing the new function ϕ defined by $\phi = \Delta \left[\frac{4}{9\pi}\right]^{1/3} \frac{\chi}{x}$, and putting $\hat{x} = \Delta^{-1}\sqrt{\alpha}\,(12/\pi)^{1/6}\,x$, $\xi = \hat{x} - \hat{x}_c$ the ultra-relativistic Thomas-Fermi equation can be written as

$$\frac{d^2\hat{\phi}(\xi)}{d\xi^2} = -\theta(-\xi) + \hat{\phi}(\xi)^3, \tag{14}$$

where $\hat{\phi}(\xi) = \phi(\xi + \hat{x}_c)$. The boundary conditions on $\hat{\phi}$ are: $\hat{\phi}(\xi) \to 1$ as $\xi \to -\hat{x}_c \ll 0$ (at massive core center) and $\hat{\phi}(\xi) \to 0$ as $\xi \to \infty$. We must also have the continuity of the function $\hat{\phi}$ and the continuity of its first derivative $\hat{\phi}'$ at the surface of massive core $\xi = 0$.
Equation (14) admits an exact solution

$$\hat{\phi}(\xi) = \begin{cases} 1 - 3\left[1 + 2^{-1/2}\sinh(a - \sqrt{3}\xi)\right]^{-1}, & \xi < 0, \\ \frac{\sqrt{2}}{(\xi+b)}, & \xi > 0, \end{cases} \tag{15}$$

where integration constants a and b are: $\sinh a = 11\sqrt{2}$, $a = 3.439$; $b = (4/3)\sqrt{2}$.
We then have for the Coulomb potential energy, in terms of the variable ξ, $eV(\xi) = \left(\frac{1}{\Delta^3}\frac{9\pi}{4}\right)^{1/3} m_\pi c^2 \hat{\phi}(\xi)$, and at the center of massive core $eV(0) = \hbar c(3\pi^2 n_p)^{1/3} = \left(\frac{1}{\Delta^3}\frac{9\pi}{4}\right)^{1/3} m_\pi c^2$, which plays a fundamental role in order to determine the stability of the configuration.
It is possible to compare energetic properties of different configurations satisfying the different neutrality conditions $n_e = n_p$ and $N_e = N_p$, with the same core radius R_c and total nucleon number A. The total energy in the case $n_e = n_p$ is $\mathcal{E}^{\rm loc}_{\rm tot} = \sum_{i=e,p,n} \mathcal{E}^i_{\rm loc}$, where

$$\mathcal{E}^i_{\rm loc} = 2\int \frac{d^3r d^3p}{(2\pi\hbar)^3}\epsilon^i_{\rm loc}(p) == \frac{cV_c}{8\pi^2\hbar^3}\Big\{\bar{P}^F_i[2(\bar{P}^F_i)^2 + (m_i c)^2] \times[(\bar{P}^F_i)^2 + (m_i c)^2]^{1/2} - (m_i c)^4 \mathrm{Arsh}\left(\frac{\bar{P}^F_i}{m_i c}\right)\Big\}. \tag{16}$$

The total energy in the case $N_e = N_p$ is $\mathcal{E}^{\rm glob}_{\rm tot} = \mathcal{E}_{\rm elec} + \mathcal{E}_{\rm binding} + \sum_{i=e,p,n} \mathcal{E}^i_{\rm glob}$, where

$$\mathcal{E}_{\rm elec} = \int \frac{E^2}{8\pi} d^3r \approx \frac{3^{3/2}\pi^{1/2}}{4}\frac{N_p^{2/3}}{\sqrt{\alpha}\Delta c} m_\pi \int_{-\kappa R_c}^{+\infty} dx\, [\phi'(x)]^2\,, \tag{17}$$

$$\mathcal{E}_{\rm binding} = -2\int \frac{d^3r d^3p}{(2\pi\bar{h})^3} eV(r) \approx -\frac{V_c}{3\pi^2\hbar^3}(P^F_e)^3 eV(0)\,, \tag{18}$$

$$\mathcal{E}^i_{\rm glob} = 2\int \frac{d^3r d^3p}{(2\pi\bar{h})^3}\epsilon^i_{\rm glob}(p) = \frac{cV_c}{8\pi^2\hbar^3}\Big\{P^F_i[2(P^F_i)^2 + (m_i c)^2] \times[(P^F_i)^2 + (m_i c)^2]^{1/2} - (m_i c)^4 \mathrm{Arsh}\left(\frac{P^F_i}{m_i c}\right)\Big\}. \tag{19}$$

We have indicated with $\bar{P}_i^F$ $(i = n, e, p)$ the Fermi momentum in the case of local charge neutrality $(V = 0)$ and with P_i^F $(i = n, e, p)$ the Fermi momentum in the case of global charge neutrality $(V \neq 0)$. The energetic difference between local neutrality and global neutrality configurations is positive, $\Delta\mathcal{E} = \mathcal{E}_{tot}^{loc} - \mathcal{E}_{\rm tot}^{\rm glob} > 0$, so configurations which obey to the condition of global charge neutrality are energetically favorable with respect to one which obey to the condition of local charge neutrality.[25,27] For a core of 10 km the difference in binding energy reaches 10^{49} ergs which gives an upper limit to the energy emittable by a neutron star, reaching its electrodynamical ground state.
The current work is three fold: **a)** generalize our results considering the heavy nuclei as special limiting cases of macroscopic nuclear matter cores,[26] **b)** describe a macroscopic nuclear matter core within the realm of General Relativity fulfilling the generalized Tolman, Oppenheimer, Volkoff equation,[28] **c)** generalyze the concept of a Dyadosphere to a Kerr-Newman Geometry.[29]

8. Conclusions

It is clear that any neutron star has two very different components: the core with pressure dominated by a baryonic component and the outer crust with pressure dominated by a leptonic component and density dominated by the nuclear species. The considerations that we have presented above apply to the first component where the baryonic pressure dominates. It is clear that when the density increases and baryons become ultra-relativistic is this baryonic component which undergoes the process of gravitational collapse and its dynamics is completely dominated by the electrodynamical process which we have presented in this talk.

Acknowledgments

I thank Michael Rotondo and Carlo Luciano Bianco for the preparation of this manuscript.

References

1. I. S. Shlkvoskii, *Supernovae* (Wiley, New York, 1968).
2. R. N. Manchester and J. H. Taylor, *Pulsars* (Freeman and Co., San Francisco, CA, 1977).
3. A. Mezzacappa and G. M. Fuller, *Open Issues in Core Collapse Supernovae Theory* (World Scientific, Singapore, 2005).

4. R. Ruffini, *From the ergosphere to the dyadosphere* in *Festschrift in Honor of Roy Kerr*, D. Wildshire (ed.) (Cambridge University Press, 2008).
5. H. Gursky and R. Ruffini, *Neutron Stars, Black Holes, and Binary X-Ray Sources* (Springer, Berlin, 1975).
6. R. Ruffini, M. G. Bernardini, C. L. Bianco, L. Caito, P. Chardonnet, M. G. Dainotti, F. Fraschetti, R. Guida, M. Rotondo, G. Vareshchagin, L. Vitagliano, and S. S. Xue, in *Cosmology and Gravitation: XIIth Brazilian School of Cosmology and Gravitation*, M. Novello and S. E. Perez Bergliaffa (eds.), AIP Conf. Proc. **910**, 55 (2007).
7. T. Damour and R. Ruffini, *Phys. Rev. Lett.* **35**, 463 (1975).
8. W. Heisenberg and H. Euler, *Z. Phys.* **98**, 714 (1936).
9. J. Schwinger, *Phys. Rev.* **82**, 664 (1951).
10. E. Costa *et al.*, *Nature* **387**, 783 (1997).
11. R. Ruffini, in *Proceedings of the eleventh Marcel Grossman meeting*, R. Jantzen, H. Kleinert, and R. Ruffini (eds.) (World Scientific, Singapore, in press).
12. A. M. Soderberg *et al.*, `arXiv:0802.1712v1`.
13. L. D. Landau and E. M. Lifshitz, *Non-Relativistic Quantum Mechanics* (Pergamon Press, Oxford, 1975).
14. L. H. Thomas, *Proc. Cambridge Phil. Society* **23**, 542 (1927).
15. E. Fermi, *Rend. Accad. Lincei* **6**, 602 (1927).
16. R. Ruffini, *On the Energetics of Black Holes* in *Black Holes – Les astres occlus*, C. and B.S. De Witt (eds.) (Gordon and Breach, New York, 1972).
17. L. D. Landau, *Phys. Z. Sowjetunion* **1**, 285 (1932).
18. R. Ruffini and J. A. Wheeler, *Phys. Today* **24**, 30 (1971).
19. J. Ferreirinho, R. Ruffini, and L. Stella, *Phys. Lett. B* **91**, 314 (1980).
20. R. Ruffini and L. Stella, *Phys. Lett. B* **102**, 442 (1981).
21. R. Ruffini, G. Vereshchagin, and S. S. Xue, Progress Report, in preparation (2008).
22. R. Ruffini, in *Black Holes and High Energy Astrophysics*, H. Sato and N. Sugiyama (eds.), 167 (Universal Academic Press, Tokyo, 1998).
23. G. Preparata, R. Ruffini, and S. S. Xue, *Astron. Astroph.* **338**, L87 (1998).
24. R. Ruffini , M. Rotondo, and S. S. Xue, *Int. J. Mod. Phys. D* **16**, 1 (2007).
25. M. Rotondo, R. Ruffini and S. S. Xue, in *Relativistic Astrophysics: 4th Italian-Sino Workshop*, C. L. Bianco and S. S. Xue (eds.), AIP Conf. Proc. **966**, 147 (2007).
26. B. Patricelli, M. Rotondo, and R. Ruffini, in *Relativistic Astrophysics: 4th Italian-Sino Workshop*, C. L. Bianco and S. S. Xue (eds.), AIP Conf. Proc. **966**, 143 (2007).
27. R. Ruffini, M. Rotondo, and S. S. Xue, in preparation (2008).
28. J. Rueda, M. Rotondo, and R. Ruffini, in preparation (2008).
29. C. Cherubini, A. Geralico, J. A. H. Rueda, and R. Ruffini, in *Relativistic Astrophysics: 4th Italian-Sino Workshop*, C. L. Bianco and S. S. Xue (eds.), AIP Conf. Proc. **966**, 123 (2007).

A VARIATIONAL APPROACH TO THE COMPUTATION OF THE COSMOLOGICAL CONSTANT IN A MODIFIED GRAVITY THEORY

R. GARATTINI

Università degli Studi di Bergamo, Facoltà di Ingegneria,
Viale Marconi 5, 24044 Dalmine (Bergamo) Italy
INFN - sezione di Milano, Via Celoria 16, Milan, Italy
E-mail: remo.garattini@unibg.it

We discuss how to extract information about the cosmological constant from the Wheeler-DeWitt equation, considered as an eigenvalue of a Sturm-Liouville problem. A generalization to a $f(R)$ theory is taken under examination. The equation is approximated to one loop with the help of a variational approach with Gaussian trial wave functionals. We use a zeta function regularization to handle with divergences. A renormalization procedure is introduced to remove the infinities together with a renormalization group equation.

Keywords: Quantum cosmology; Cosmological constant; Renormalization; Modified gravity.

1. Introduction

It is well known that there exists a huge discrepancy between the observed[1] and the computed value of the cosmological constant. It amounts approximately to a factor of 120 orders of magnitude: this is the *cosmological constant problem.* One possible approach to such a problem is given by the Wheeler-DeWitt equation (WDW).[2] The WDW equation can be extracted from the Einstein's field equations with and without matter fields in a very simple way. If we introduce a time-like unit vector u^μ such that $u \cdot u = -1$, then after a little rearrangement, we get[3]

$$\mathcal{H} = (2\kappa)\, G_{ijkl}\pi^{ij}\pi^{kl} - \frac{\sqrt{g}}{2\kappa}\left({}^3R - 2\Lambda_c\right) = 0. \tag{1}$$

Here ${}^3R$ is the scalar curvature in three dimensions. This is the time-time component of the Einstein Field Equations in vacuum. It represents a constraint at the classical level and the invariance under *time* reparametrization. Its quantum counterpart $\mathcal{H}\Psi = 0$ is the WDW equation. This one can be cast into the form of an eigenvalue equation

$$\hat{\Lambda}_\Sigma \Psi [g_{ij}] = \Lambda (x) \Psi [g_{ij}], \tag{2}$$

where

$$\hat{\Lambda}_\Sigma = (2\kappa) G_{ijkl} \pi^{ij} \pi^{kl} - \frac{\sqrt{g}}{2\kappa} {}^3R, \qquad \Lambda (x) = -\sqrt{g} \frac{\Lambda_c}{\kappa}. \tag{3}$$

If we multiply Eq. (2) by $\Psi^* [g_{ij}]$ and we functionally integrate over the three spatial metric g_{ij}, we get

$$\int \mathcal{D} [g_{ij}] \Psi^* [g_{ij}] \hat{\Lambda}_\Sigma \Psi [g_{ij}] = \int \mathcal{D} [g_{ij}] \Lambda (x) \Psi^* [g_{ij}] \Psi [g_{ij}]. \tag{4}$$

After having integrated over the hypersurface Σ, one can formally re-write the modified WDW equation as

$$\frac{1}{V} \frac{\int \mathcal{D} [g_{ij}] \Psi^* [g_{ij}] \int_\Sigma d^3x \hat{\Lambda}_\Sigma \Psi [g_{ij}]}{\int \mathcal{D} [g_{ij}] \Psi^* [g_{ij}] \Psi [g_{ij}]} = \frac{1}{V} \frac{\left\langle \Psi \left| \int_\Sigma d^3x \hat{\Lambda}_\Sigma \right| \Psi \right\rangle}{\langle \Psi | \Psi \rangle} = -\frac{\Lambda_c}{\kappa}, \tag{5}$$

where the explicit expression of $\Lambda (x)$ has been used and where we have defined the volume of the hypersurface Σ as $V = \int_\Sigma d^3x \sqrt{g}$. We can gain more information considering a separation of the spatial part of the metric into a background term $\bar{g}_{ij}$ and a quantum fluctuation h_{ij} so that $g_{ij} = \bar{g}_{ij} + h_{ij}$. Equation (5) represents the Sturm-Liouville problem associated with the cosmological constant. The related boundary conditions are dictated by the choice of the trial wavefunctionals which, in our case, are of the Gaussian type. Different types of wavefunctionals correspond to different boundary conditions. Extracting the TT tensor contribution from Eq. (5) approximated to second order in perturbation of the spatial part of the metric into a background term $\bar{g}_{ij}$ and a perturbation h_{ij} we get

$$\hat{\Lambda}_\Sigma^\perp = \frac{1}{4V} \int_\Sigma d^3x \sqrt{\bar{g}} G^{ijkl} \left[(2\kappa) K^{-1\perp} (x,x)_{ijkl} + \frac{1}{(2\kappa)} (\triangle_2)_j^a K^\perp (x,x)_{iakl} \right]. \tag{6}$$

Here G^{ijkl} represents the inverse DeWitt metric and all indices run from one to three. The propagator $K^\perp (x,x)_{iakl}$ can be represented as

$$K^\perp (\overrightarrow{x}, \overrightarrow{y})_{iakl} := \sum_\tau \frac{h_{ia}^{(\tau)\perp} (\overrightarrow{x}) h_{kl}^{(\tau)\perp} (\overrightarrow{y})}{2\lambda (\tau)}, \tag{7}$$

where $h_{ia}^{(\tau)\perp} (\overrightarrow{x})$ are the eigenfunctions of $\triangle_2$, whose explicit expression for the massive case will be shown in the next section. Furthermore, τ denotes a complete set of indices and $\lambda (\tau)$ are a set of variational parameters to be determined by the minimization of Eq. (6). The expectation value of $\hat{\Lambda}^{\perp}_{\Sigma}$ is easily obtained by inserting the form of the propagator into Eq. (6) and minimizing with respect to the variational function $\lambda(\tau)$. Thus the total one loop energy density for TT tensors becomes

$$\frac{\Lambda}{8\pi G} = -\frac{1}{4}\sum_{\tau}\left[\sqrt{\omega_1^2(\tau)} + \sqrt{\omega_2^2(\tau)}\right]. \tag{8}$$

The above expression makes sense only for $\omega_i^2(\tau) > 0$, where ω_i are the eigenvalues of $\triangle_2$.

2. One Loop Energy Regularization and Renormalization for a $f(R) = R$ Theory

The Spin-two operator for the Schwarzschild metric in the Regge and Wheeler representation,[4] leads to the following system of equations ($r \equiv r(x)$)

$$\left[-\frac{d^2}{dx^2} + \frac{l(l+1)}{r^2} + m_i^2(r)\right] f_i(x) = \omega_{i,l}^2 f_i(x) \qquad i = 1, 2. \tag{9}$$

In Eq. (9), reduced fields have been used and the proper geodesic distance from the *throat* of the bridge has been considered. Close to the throat, the effective masses are $m_1^2(r) \simeq -m_0^2(M)$ and $m_2^2(r) \simeq m_0^2(M)$, where we have defined a parameter $r_0 > 2MG$ and $m_0^2(M) = 3MG/r_0^3$. The main reason for introducing a new parameter resides in the fluctuation of the horizon that forbids any kind of approach. It is now possible to explicitly evaluate Eq. (8) in terms of the effective mass. To further proceed we use the W.K.B. method used by 't Hooft in the brick wall problem[5] and we count the number of modes with frequency less than ω_i, $i = 1, 2$. Thus the one loop total energy for TT tensors becomes

$$\frac{\Lambda}{8\pi G} = \rho_1 + \rho_2 = -\frac{1}{16\pi^2}\sum_{i=1}^{2}\int_{\sqrt{m_i^2(r)}}^{+\infty}\omega_i^2\sqrt{\omega_i^2 - m_i^2(r)}\,d\omega_i, \tag{10}$$

where we have included an additional 4π coming from the angular integration. Here, we use the zeta function regularization method to compute the energy densities ρ_1 and ρ_2. Note that this procedure is completely equivalent to the subtraction procedure of the Casimir energy computation where the zero point energy (ZPE) in different backgrounds with the same asymptotic properties is involved. To this purpose, we introduce the additional mass parameter μ in order to restore the correct dimension for the regularized quantities. Such an arbitrary mass scale emerges unavoidably in any regularization scheme. One gets

$$\rho_i(\varepsilon) = \frac{m_i^4(r)}{256\pi^2}\left[\frac{1}{\varepsilon} + \ln\left(\frac{\mu^2}{m_i^2(r)}\right) + 2\ln 2 - \frac{1}{2}\right] \tag{11}$$

for $i = 1, 2$. The renormalization is performed via the absorption of the divergent part into the re-definition of the bare classical constant $\Lambda = \Lambda_0 + \Lambda^{\text{div}}$. The remaining finite value for the cosmological constant depends on the arbitrary mass scale μ. It is appropriate to use the renormalization group equation to eliminate such a dependence. To this aim, we impose[6]

$$\frac{1}{8\pi G}\mu\frac{\partial\Lambda_0^{TT}(\mu)}{\partial\mu} = \mu\frac{d}{d\mu}\rho_{\text{eff}}^{TT}(\mu, r). \tag{12}$$

Solving it we find that the renormalized constant Λ_0 should be treated as a running one in the sense that it varies provided that the scale μ is changing

$$\Lambda_0(\mu, r) = \Lambda_0(\mu_0, r) + \frac{G}{16\pi}\left(m_1^4(r) + m_2^4(r)\right)\ln\frac{\mu}{\mu_0}. \tag{13}$$

Finally, we find that

$$\frac{\Lambda_0(\mu_0, M)}{8\pi G} = -\frac{1}{128\pi^2}\left\{m_0^4(M)\left[\ln\left(\frac{m_0^2(M)}{4\mu_0^2}\right) + \frac{1}{2}\right]\right\}. \tag{14}$$

Eq. (14) has a maximum when[a]

$$\frac{1}{e} = \frac{m_0^2(M)}{4\mu_0^2} \qquad \Longrightarrow \qquad \Lambda_0(\mu_0, \bar{x}) = \frac{G m_0^4(M)}{32\pi} = \frac{G\mu_0^4}{2\pi e^2}. \tag{15}$$

Nothing prevents us to consider a more general situation where the scalar curvature R is replaced by a generic function of R. Therefore, we will consider the Sturm-Liouville problem of Eq. (5) in the context of a $f(R)$ theory[b].

3. One Loop Energy Regularization and Renormalization for a Generic $f(R)$ Theory in Hamiltonian Formulation

In this section, we would like to connect a $f(R)$ theory with the Sturm-Liouville problem of Eq. (5).[9] Let us consider now the Lagrangian density

[a]Note that in any case, the maximum of Λ corresponds to the minimum of the energy density.

[b]A recent review on the problem of $f(R)$ theories can be found in Ref. 7. A more general discussion on modified gravities of the type $f(R), f(G)$ and $f(R, G)$ where G is the Gauss-Bonnet invariant, can be found in Ref. 8.

describing a generic $f(R)$ theory of gravity

$$\mathcal{L} = \sqrt{-g}\left(f\left(R\right) - 2\Lambda\right) \qquad \text{with} \qquad f'' \neq 0, \tag{16}$$

where $f\left(R\right)$ is an arbitrary smooth function of the scalar curvature and primes denote differentiation with respect to the scalar curvature. A cosmological term is added also in this case for the sake of generality. Obviously $f'' = 0$ corresponds to GR. Equation (5) can be generalized to give[10]

$$\frac{1}{V}\frac{\left\langle \Psi \left| \int_\Sigma d^3x \left[\hat{\Lambda}^{(2)}_{\Sigma,f(R)}\right] \right| \Psi \right\rangle}{\langle \Psi | \Psi \rangle} + \frac{\kappa}{V}\frac{\left(f'\left(R\right) - 1\right)}{f'\left(R\right)}\frac{\left\langle \Psi \left| \int_\Sigma d^3x \left[\pi^2\right] \right| \Psi \right\rangle}{\langle \Psi | \Psi \rangle}$$

$$+\frac{1}{V}\frac{\left\langle \Psi \left| \int_\Sigma d^3x V(\mathcal{P}) / \left(2\kappa f'\left(R\right)\right) \right| \Psi \right\rangle}{\langle \Psi | \Psi \rangle} = -\frac{\Lambda_c}{\kappa}, \tag{17}$$

where

$$V(\mathcal{P}) = \sqrt{g}\left[Rf'\left(R\right) - f\left(R\right)\right]. \tag{18}$$

When $f\left(R\right) = R$, $V(\mathcal{P}) = 0$ as it should be. In Eq. (17), we have defined a "*modified*" $\hat{\Lambda}^{(2)}_{\Sigma}$ operator which includes $f'\left(R\right)$ in the following way

$$\hat{\Lambda}^{(2)}_{\Sigma,f(R)} = (2\kappa)\, h\left(R\right) G_{ijkl}\pi^{ij}\pi^{kl} - \frac{\sqrt{g}}{2\kappa}\ {}^3R^{\text{lin}}, \tag{19}$$

with

$$h\left(R\right) = 1 + \frac{2\left[f'\left(R\right) - 1\right]}{f'\left(R\right)} \tag{20}$$

and where ${}^3R^{\text{lin}}$ is the linearized scalar curvature. Note that when $f\left(R\right) = R$, consistently it is $h\left(R\right) = 1$. From Eq. (17), we redefine Λ_c

$$\Lambda'_c = \Lambda_c + \frac{1}{2V}\frac{\left\langle \Psi \left| \int_\Sigma d^3x \frac{V(\mathcal{P})}{f'(R)} \right| \Psi \right\rangle}{\langle \Psi | \Psi \rangle} = \Lambda_c + \frac{1}{2V}\int_\Sigma d^3x \sqrt{g}\frac{Rf'\left(R\right) - f\left(R\right)}{f'\left(R\right)}, \tag{21}$$

where we have explicitly used the definition of $V(\mathcal{P})$. Thus, in Eq. (14) we have to replace $\Lambda_0\left(\mu_0, r\right)$ with $\Lambda'_0\left(\mu_0, r\right)$. Adopting the same procedure of the $f\left(R\right) = R$ case, we find a maximum for $\Lambda'_0\left(\mu_0, x\right)$ which is related to $\Lambda_0\left(\mu_0, \bar{x}\right)$ by the following relation

$$\frac{1}{\sqrt{h\left(R\right)}}\left[\Lambda_0\left(\mu_0, \bar{x}\right) + \frac{1}{2V}\int_\Sigma d^3x \sqrt{g}\frac{Rf'\left(R\right) - f\left(R\right)}{f'\left(R\right)}\right] = \frac{G\mu_0^4}{2\pi e^2}. \tag{22}$$

Isolating $\Lambda_0\left(\mu_0, \bar{x}\right)$, we get

$$\Lambda_0\left(\mu_0, \bar{x}\right) = \sqrt{h\left(R\right)}\frac{G\mu_0^4}{2\pi e^2} - \frac{1}{2V}\int_\Sigma d^3x \sqrt{g}\frac{Rf'\left(R\right) - f\left(R\right)}{f'\left(R\right)}. \tag{23}$$

Note that $\Lambda_0(\mu_0, \bar{x})$ can be set to zero when

$$\sqrt{h(R)}\frac{G\mu_0^4}{2\pi e^2} = \frac{1}{2V}\int_\Sigma d^3x\sqrt{g}\frac{Rf'(R) - f(R)}{f'(R)}. \tag{24}$$

Let us see what happens when $f(R) = \exp(-\alpha R)$. This choice is simply suggested by the regularity of the function at every scale. In this case, Eq. (24)becomes

$$\sqrt{\frac{3\alpha\exp(-\alpha R) + 2}{\alpha\exp(-\alpha R)}}\frac{G\mu_0^4}{\pi e^2} = \frac{1}{\alpha V}\int_\Sigma d^3x\sqrt{g}(1 + \alpha R). \tag{25}$$

For Schwarzschild, it is $R = 0$, then by setting $\alpha = G$, we have the relation

$$\mu_0^4 = \frac{\pi e^2}{G}\sqrt{\frac{1}{(3G + 2)G}}. \tag{26}$$

References

1. For a pioneering review on this problem see S. Weinberg, *Rev. Mod. Phys.* **61**, 1 (1989). For more recent and detailed reviews see V. Sahni and A. Starobinsky, *Int. J. Mod. Phys. D* **9**, 373 (2000) [arXiv:astro-ph/9904398]; N. Straumann, arXiv:gr-qc/0208027; T. Padmanabhan, *Phys. Rep.* **380**, 235 (2003) [arXiv:hep-th/0212290].
2. B. S. DeWitt, *Phys. Rev.* **160**, 1113 (1967).
3. R. Garattini, *J. Phys. A* **39**, 6393 (2006) [arXiv:gr-qc/0510061]; R. Garattini, *J. Phys. Conf. Ser.* **33**, 215 (2006)[arXiv:gr-qc/0510062].
4. T. Regge and J. A. Wheeler, *Phys. Rev.* **108**, 1063 (1957).
5. G. 't Hooft, *Nucl. Phys. B* **256**, 727 (1985).
6. J. Perez-Mercader and S. D. Odintsov, *Int. J. Mod. Phys. D* **1**, 401 (1992); I. O. Cherednikov, *Acta Phys, Slov.* **52**, 221 (2002); I. O. Cherednikov, *Acta Phys. Polon. B* **35**, 1607 (2004); M. Bordag, U. Mohideen, and V. M. Mostepanenko, *Phys. Rep.* **353**, 1 (2001); R. Garattini, *TSPU Vestnik* **44 N7**, 72 (2004) [arXiv:gr-qc/0409016]. Inclusion of non-perturbative effects, namely beyond one-loop, in de Sitter Quantum Gravity have been discussed in S. Falkenberg and S. D. Odintsov, *Int. J. Mod. Phys. A* **13**, 607 (1998) [arXiv:hep-th 9612019].
7. S. Capozziello and M. Francaviglia, arXiv:0706.1146.
8. S. Nojiri and S. D. Odintsov, *Int. J. Geom. Meth. Mod. Phys* **4**, 115 (2007) [arXiv:hep-th/0601213].
9. S. Capozziello and R. Garattini, *Class. Quant. Grav.* **24**, 1627 (2007) [arXiv:gr-qc/0702075].
10. L. Querella, *Variational Principles and Cosmological Models in Higher-Order Gravity*, Ph.D. Thesis (1999) [arXiv:gr-qc/9902044].

ON NATURE OF THE COSMOLOGICAL CONSTANT

L. V. PROKHOROV
V. A. Fock Institute of Physics, Sankt-Petersburg State University, St. Petersburg, Russia
E-mail: lev.prokhorov@pobox.spbu.ru

It is shown that the problem of the cosmological constant is connected with the problem of emergence of quantum mechanics. Both of them are principal aspects of physics at the Planck scale. Probability amplitudes describe evolution of the harmonic oscillator non-equilibrium distributions in a thermal bath. The Planck constant $\hbar$, the Fock space and the Schrödinger equation appear in the natural way. For massless fields it leads, in particular, to appearance of masses, and, consequently, of the cosmological constant in the gravitational equations. The path integral for the relativistic particle propagator is presented.

Keywords: Planck scale; Quantum mechanics; Cosmological constant.

Among the hot spots of the modern physics problems of dark matter[1–3] and dark energy[4,5] are the most actual. It is clear that their solution presumes understanding physics at the Planck scale. Solution of another fundamental problem — the emergence of quantum mechanics (description of dynamical systems by probability amplitudes) evidently, is also rooted at the Planck scale. It is remarkable that there is simple classical model (set of harmonic oscillators in a thermal bath) where the probability amplitudes appear together with the cosmological constant. Wave functions describe evolution of non-equilibrium states of the system, and the latter are specified by some relaxation time t_r. As a result, all the massless fields become massive. In the gravitation theory it leads to the appearance of the cosmological constant $\Lambda = \alpha^2/2$, $\alpha^{-1} = t_r$. In particular, it gives correct recipe for resolving the singularities in the retarded and causal propagators. It also agrees with the path integral presentation of propagators for a relativistic particle.

The Hibbs distribution for a harmonic oscillator with the Hamiltonian

$$H = \frac{\omega}{2}(p^2 + q^2) = \frac{\omega}{2}(\bar{z}z + z\bar{z}), \quad z = \frac{q + ip}{\sqrt{2}}, \tag{1}$$

defines the following measure in the phase space

$$d\mu(\bar{z},z) = \frac{d\bar{z}\wedge dz}{2\pi i\hbar}e^{-\beta\omega\bar{z}z}, \quad \beta = \frac{1}{k_B T}, \quad \hbar = \frac{1}{\beta\omega}, \tag{2}$$

where T is the temperature, and k_B — the Boltzmann constant; $h = 2\pi\hbar$ has the dimension of action and will be identified with the Planck constant, $\int d\mu(\bar{z},z) = 1$. Any other distribution $d\mu_p(\bar{z},z) = p(\bar{z},z)d\mu(\bar{z},z)$ with $p(\bar{z},z) \geq 0$ describes a non-equilibrium state.[6] The complex canonical variables $\bar{z}, z$ with the Poisson bracket

$$\{f,g\} = i\frac{\partial(f,g)}{\partial(\bar{z},z)} = i\left(\frac{\partial f}{\partial\bar{z}}\frac{\partial g}{\partial z} - \frac{\partial f}{\partial z}\frac{\partial g}{\partial\bar{z}}\right) \tag{3}$$

play the principal role in the present approach. In this case one can use only the first of two Hamiltonian equations $\dot{z} = \{f,H\} = -i\omega z$, while the second one can be obtained from it by complex conjugation.

The non-equilibrium distribution appears after deformation of the Hibbs distribution (2) via $z \to z + c$:

$$d\mu_f(\bar{z},z) = |f_c(z)|^2 d\mu(\bar{z},z), \quad f_c(z) = e^{-\bar{c}z/\hbar}e^{-\bar{c}c/2\hbar}, \quad \int d\mu_f = 1. \tag{4}$$

Suppose that the relaxation time of this state is large, $t_r \gg \omega^{-1}$. Then its evolution with time for $t \ll t_r$ can be described by only one equation

$$\dot{f}_c = -i\omega z\frac{df_c}{dz}. \tag{5}$$

This construction in fact defines the Fock space of analytic functions $f(z)$ of order $\rho \leq 2$ with the scalar product[7]

$$(g,f) = \int d\mu(\bar{z},z)\bar{g}(z)f(z). \tag{6}$$

Thus, it allows to introduce operators: $\hat{z}f(z) = zf(z)$, $\hat{\bar{z}}f(z) = \hbar df(z)/dz$. They can be identified with the standard creation and annihilation operators $\hat{a}^+, \hat{a}$, $[\hat{a},\hat{a}^+] = \hbar$ (it can be proved using integration by parts in (7)). Then the Hamiltonian (1) should be taken in the form

$$\hat{H} = \frac{\omega}{2}(\hat{\bar{z}}\hat{z} + \hat{z}\hat{\bar{z}}) = \hbar\omega\left(z\frac{d}{dz} + \frac{1}{2}\right). \tag{7}$$

Multiplying the classical Eq. (6) by $i\hbar$ one obtains in fact the Schrödinger equation, though without the "vacuum energy" $E_0 = \hbar\omega/2$. Taking into consideration the commutation relation of $\hat{\bar{z}}, \hat{z}$ one obtains the correct expression (8) for $\hat{H}$ and correct Schrödinger's equation

$$i\hbar\dot{f} = \hbar\omega\left(z\frac{d}{dz} + \frac{1}{2}\right)f, \tag{8}$$

if we identify $\hbar$ in (2) with the Planck constant.

One has to take into consideration the damping of oscillations. To do it one may modify the Poisson bracket[8]

$$\{f,g\} = \omega^{ij} f_{,i} g_{,j} \to \{f,g\}_\alpha = \{f,g\} + \{f,g\}_+, \quad \omega^{ij} = -\omega^{ji}. \tag{9}$$

We denoted $f_{,i} = \partial f/\partial x^i$, $H = \sum h^{ij} x^i x^j/2$, $h_{ij} = \omega\delta_{ij}$ and $x(q^1, ..., q^n, p_{n+1}, ..., p_{2n})$. We define

$$\{f,g\}_+ = -\alpha h^{ij} f_{,i} g_{,j}, \quad h^{ij} = \frac{1}{\omega}\delta^{ij}, \quad \alpha > 1. \tag{10}$$

The new bracket (10) modifies the Hamiltonian equations: $\dot{f} = \{f,H\}_\alpha$, i.e., for a single oscillator ($n = 1$) one has $\dot{q} = \omega p - \alpha q$, $\dot{p} = -\omega q - \alpha p$, or $D_t q = \omega p$, $D_t p = -\omega q$, $D_t = \partial_t + \alpha$. Thus, $D_t^2 q = -\omega^2 q$, and $\ddot{q} + 2\alpha\dot{q} + (\omega^2 + \alpha^2)q = 0$, $q(t) = e^{-\alpha t}(c_1 \sin\omega t + c_2 \cos\omega t)$. In the Lagrangian formalism the terms $\sim \alpha^2$ may seem inessential. But the Hamiltonian mechanics is superior (in transformation of the first-order equations into the second-order ones one we lose some information), and the parameter α enters there linearly.

The importance of change $\partial_t \to D_t$ lies in the fact that all the fields are ordered sets of harmonic oscillators. Then, this substitution modifies the equations of motion, e.g.

$$(\Box - m^2)\varphi = 0 \to [\Box - 2\alpha\partial_t - (m^2 + \alpha^2)]\varphi_\alpha \equiv [\Box_\alpha - m^2]\varphi_\alpha = 0, \tag{11}$$

where $\Box = -\partial_t^2 + \Delta$ and $\varphi_\alpha = e^{-\alpha t}\varphi$. Thus, massless fields acquire masses, and the cosmological term appears in the gravitational equations.[8]

It solves the problem of dark energy. The Newton potential energy $V(r) = -Gm_1m_2/r$ now becomes the Yukawa one $V_\alpha(r) = -e^{-\alpha r}Gm_1m_2/r$, so at large distances ($r > \alpha^{-1}$) the attraction becomes negligible, and it imitates "acceleration" of the Universe. The astrophysical data[9] gives $\alpha^{-1} \sim 8 \cdot 10^9$ yr.

According to quantum field theory the matter in the Universe is nothing but excitation of fields. However, as we have seen all the excitations should disappear with time. So, matter disappears, and according to the present model instead of thermal death of XIX century we should expect the "quantum death" of the Universe.

It turns out that all this leads to correct prescriptions for resolving the retarded and causal propagator singularities. The scalar field causal propagator is defined by the sum of two vacuum expectations $G_c(\mathbf{x} - \mathbf{x}', t - t') = \theta(t - t')\langle\hat{\varphi}(x)\hat{\varphi}(x')\rangle_0 + \theta(t' - t)\langle\hat{\varphi}(x')\hat{\varphi}(x)\rangle_0$. All the fields acquire the factor $e^{-\alpha t}$, so we should replace $G_c \to e^{-\alpha(t-t')}G_c \equiv G_{\alpha c}$. As a result one finds the following representation for $G_{\alpha c}$

$$G_{\alpha c}(x,x') = \tag{12}$$
$$\int \frac{d^4p}{(2\pi)^4 i2E_p} e^{-i(p_0-i\alpha)\Delta t + i\mathbf{p}\Delta\mathbf{x}} \left(\frac{1}{-p_0+E_p-i\epsilon} + \frac{1}{p_0+E_p-i\epsilon} \right),$$

where $E_p^2 = \mathbf{p}^2 + m^2$. Moving the p_0 integration contour up in the complex plane ($p_0 \to p_0'$, Im $p_0' = i\alpha$) and taking into account the pole at $p_0 = -E_p + i\epsilon$ in the second term of the sum in (14) ($\alpha > \epsilon$) one obtains

$$G_{\alpha c}(x-x') = \int \frac{d^4p}{(2\pi)^4} \frac{ie^{-ip(x-x')}}{p^2 - m^2 + 2ip_0\alpha - \alpha^2} + e^{-\alpha(t-t')} \int \frac{d^3p}{(2\pi)^3 2E_p} e^{-ipx}, \tag{13}$$

and $G_{\alpha c}(x-x') \to G_c(x-x')$ in the limit $\alpha \to 0$, while the first term in Eq. (15) gives $G_{ret}(x-x')$. The modified propagator $G_{\alpha c}(x-x')$ satisfies the equation $(\Box_\alpha - m^2)G_{\alpha c}(x) = i\delta^{(4)}(x)$.

The path-integral representation of the relativistic particle propagator meets considerable difficulties. First, the classical Lagrangian contains the square root: $L = -m\sqrt{\dot{x}^2}$, $x^2 = x_0^2 - \mathbf{x}^2$. Second, the corresponding Hamiltonian is zero:[10] $H = \dot{x}_0(p_0 + \operatorname{sgn} \dot{x}_0 E_p) = 0$ with $\dot{x} = dx/d\tau$, where τ is some invariant parameter ("time"). Consider the kernel of the evolution operator

$$U_\omega(x,x') = \langle x|e^{-i\omega'\hat{H}}|x'\rangle = \int \frac{d^4p}{(2\pi)^4} e^{i[p\Delta x - \omega(p_0+E_p)]}, \quad \omega \to 0 \tag{14}$$

(H is given by (16); we demand $\dot{x}_0 > 0$, define $\omega = \omega'\dot{x}_0 > 0$, $\Delta x = x - x'$ and use the formula $\langle x|p\rangle = (2\pi)^{-2}\exp(ipx)$). Integrating in (17) over p_0, one obtains

$$\begin{aligned} U_\omega(x,x') &= \int \frac{d^3p}{(2\pi)^3} e^{i(\mathbf{p}\Delta\mathbf{x} - \omega E_p)} \delta(x_0 - x_0' - \omega) \\ &= \int \frac{d^3p}{(2\pi)^3} e^{i(\mathbf{p}\Delta\mathbf{x} - \Delta x_0 E_p)} \delta(\Delta x_0 - \omega). \end{aligned} \tag{15}$$

Substituting the kernel (18) into the formula $\tilde{\psi}(x,\tau+\omega) = \int d^4x' U_\omega(x,x')\tilde{\psi}(x,\tau)$ and integrating in it over x'^0 one gets

$$\psi(x^0,\mathbf{x}) = \int \frac{d^3pd^3x'}{(2\pi)^3} e^{i(p_i\Delta x^i - \Delta x^0 E_p)} \psi(x^0 - \omega\tau, \mathbf{x}'), \quad \Delta x^0 = x^0 - x'^0 = \omega \tag{16}$$

(we define $\psi(x^0,\mathbf{x}) = \tilde{\psi}(x^0,\mathbf{x},\tau)$, i.e. omit τ because according to (19) $\Delta\tau = \Delta x^0$). In the relativistic quantum mechanics the scalar product of state vectors is defined as

$$(\psi_1, \psi_2) = i \int d^3x \psi_1^* \overleftrightarrow{\partial_0} \psi_2 = i \int d^3x (\psi_1^* \partial_t \psi_2 - \partial_t \psi_1^* \psi_2),$$

so we have to take it into consideration in writing the path integral, for example, in the product $\hat{U}_\omega ... \hat{U}_\omega$. To this end we introduce the unity $(2E_p)^{-1} 2E_p$ under the integral and substitute $2E_p \to i \overleftrightarrow{\partial_0}$. Then the operator $\hat{U}_\omega$ acts like

$$\psi(x^0 + \omega, \mathbf{x}) = \hat{U}_\omega \psi(x^0, \mathbf{x}) = i \int d^3x U_\omega(\mathbf{x}, \mathbf{x}') \overleftrightarrow{\partial_0} \psi(x^0, \mathbf{x}'), \quad \partial_t = \partial_{x^0}, \tag{17}$$

where

$$U_\omega(\mathbf{x}, \mathbf{x}') = \int \frac{d^3p}{(2\pi)^3 2E_p} e^{-i(E_p \omega - \mathbf{p}\Delta\mathbf{x}')}. \tag{18}$$

Thus, we obtain the following path integral for the kernel $U_{t-t'}$:

$$U_{t-t'}(\mathbf{x}, \mathbf{x}') = \int \prod_{\tau=t'}^{t} \frac{d^3p(\tau) d^3x(\tau)}{(2\pi)^3} \frac{1}{2E_p(t')} e^{i \int_{t'}^{t} d\tau [\mathbf{p}\dot{\mathbf{x}} - E_{p(\tau)}]} \delta(\mathbf{x}(t') - \mathbf{x}'). \tag{19}$$

Integrating in (22) first over $\mathbf{x}(\tau)$ and then over $\mathbf{p}(\tau)$, one obtains $U_{t-t'}(\mathbf{x}, \mathbf{x}') = \langle \hat{\varphi}(x) \hat{\varphi}(x') \rangle_0$,

$$U_{t-t'}(\mathbf{x}, \mathbf{x}') = \int \frac{d^3p}{(2\pi)^3 2E_p} e^{-ip(x-x')} = \Delta_+(x - x'), \tag{20}$$

and in the limit $\alpha \to 0$, Eqs. (14), (22), and (23) give the path integral for the causal propagator.

References

1. F. Zwicky, *Helv. Phys. Acta.* **6**, 110 (1933).
2. J. Oort, *Bull. Astron. Inst. Neth.* **6**, 249 (1932); **15**, 45 (1960).
3. *Dark Matter in the Universe*, eds. G. R. Knapp and J. Kormendy (Reidel, Dordrecht, 1987).
4. A. G. Riess *et al.*, *Astron. J.* **116**, 1009 (1998).
5. S. Perlmutter *et al.*, *Astrophys. J.* **517**, 565 (1999).
6. Ya. G. Sinai, *Introduction into Ergodic Theory* (FASIS, Moscow, 1996) [in Russian].
7. V. Bergmann, *Comm. Pure Appl. Math.* **14**, 187 (1961).
8. L. V. Prokhorov, *Phys. Part. Nucl.* **38**, 364 (2007).
9. O. Lahav and A. R. Liddle, *J. Phys. G: Nucl. Part. Phys.* **33**, 224 (2006).
10. L. V. Prokhorov and S. V. Shabanov, *Hamiltonian Mechanics of Gauge Systems*, 2nd edition (KomKniga, 2006) [in Russian].

DIMENSIONAL REDUCTION NEAR THE HORIZON

Z. HABA

Institute of Theoretical Physics, University of Wroclaw, Poland
E-mail: zhab@ift.uni.wroc.pl

In the Euclidean formulation of functional integration we discuss a dimensional reduction of quantum field theory near the horizon in terms of Green functions. We show that a massless scalar quantum field in D dimensions can be approximated near the bifurcate Killing horizon by a massless two-dimensional conformal field .

Keywords: Quantum black hole; Dimensional reduction.

1. Introduction

There are well-known difficulties concerning a description of gravity in quantum theory. Some of these difficulties can be resolved only in a complete theory of matter and gravity. Nevertheless, there are some interesting phenomena appearing when gravity is treated as a classical background (pseudoRiemannian) metric. One could study such problems either in the framework of quantum mechanics with the wave function representing a vector in a state space or in the framework of quantum field theory (QFT) on a classical gravitational background. In the latter case the conventional quantization procedure relying on classical solutions of wave equations is not unique. In this paper we impose a requirement that the pseudoRiemannian manifold is one of the real sections of a complex manifold.[1] Another real section is the Riemannian manifold. The Riemannian manifold which can be continued to the pseudoRiemannian one is distinguished by a reflection symmetry.[2] The analytic continuation is to be performed across a reflection invariant curve.

We apply the Euclidean version to massless scalar QFT on a manifold with a bifurcate Killing horizon.[3] We show that, if a surface defining the bifurcate Killing horizon is compact, then at large distances such a field theory can be approximated by the two dimensional conformal invariant Euclidean field theory defined on R^2.

2. Reflection Positivity and Euclidean Functional Integration

We consider a D-dimensional Riemannian manifold $\mathcal{N}$ with a reflection symmetry θ such that

$$\mathcal{N} = \mathcal{N}_+ \cup \mathcal{N}_- \cup \mathcal{N}_0 ,$$

where for $x \in \mathcal{N}_+$ we have $\theta x \in \mathcal{N}_-$, whereas if $x \in \mathcal{N}_0$ then $\theta x = x$.

Let $\mathcal{G}$ be a two-point function on $\mathcal{N} \times \mathcal{N}$. We say that $\mathcal{G}$ is reflection positive if

$$\int dx dx' \overline{f}(x) \mathcal{G}(\theta x, x') f(x') \geq 0 \tag{1}$$

for any f with its support in $\mathcal{N}_+$. As an example, let $\mathcal{N} = R \times \mathcal{M}$, where $\mathcal{M}$ is a $D-1$ dimensional manifold, and $\mathcal{G} = (-\partial_0^2 + H^2)^{-1}$, where H is a positive operator defined on $\mathcal{M}$, then

$$\mathcal{G}(x_0, y; x_0', y') = \Big(\frac{1}{2H} \exp(-|x_0 - x_0'|H) \Big)(y, y') .$$

In such a case if $\theta x_0 = -x_0$, $x_0 \geq 0$ and $x_0' \geq 0$ then

$$\mathcal{G}(\theta x, x') = \exp(-x_0 H)(2H)^{-1} \exp(-x_0' H) \tag{2}$$

(in the sense of the composition of kernels). The reflection positivity is a simple consequence of the positive definiteness of H^{-1} and the formula (2). If $\mathcal{G}$ is a non-negative function and a non-negative bilinear form then we can define Euclidean free fields ϕ by means of their expectation values (for this purpose it is sufficient that H^2 in Eq. (2) is a second order differential operator)

$$\int d\mu(\phi) \phi(x) \phi(x') = \mathcal{G}(x, x') . \tag{3}$$

Equation (3) defines the Gaussian measure μ of the free field ϕ. A local interaction splits as

$$\int_N V(\phi) = \int_{N_+} V(\phi) + \int_{N_-} V(\phi) .$$

Hence, if we define $\Theta\phi(x) = \phi(\theta x)$ then for F being a function of $\phi(f)$, where f have their supports in $\mathcal{N}_+$, we have

$$\int d\mu(\phi) \overline{\Theta F} F \exp\Big(- \int V \Big) \geq 0 . \tag{4}$$

The formula (4) defines the reflection positivity for Euclidean field theory with an interaction V. It can be applied to define the physical Hilbert space. For stationary Riemannian metrics it also allows to define an analytic continuation of Euclidean Green functions to Wightman functions.[2]

3. An Approximation at the Horizon

We consider a D dimensional pseudoRiemannian manifold $\mathcal{N}$ with bifurcate Killing horizon.[3] This means that there is a Killing vector ξ^A orthogonal to a (past oriented) $D-1$ dimensional hypersurface $\mathcal{H}_A$ and a (future oriented) hypersurface $\mathcal{H}_B$ and $\xi^A\xi_A = 0$ on an intersection of $\mathcal{H}_A$ and $\mathcal{H}_B$ defining a $D-2$ dimensional surface $\mathcal{M}$. The metric splits into a block form

$$ds^2 = \sum_{a,b=0,1} g_{ab}dx^a dx^b + \sum_{jk>1} g_{jk}dx^j dx^k \,.$$

In the adapted coordinates such that $\xi^A = \partial_0$ we have $g_{10} = 0$ and $g_{00}(y = 0, \mathbf{x}) = 0$. The Taylor expansion of g_{00} starts with y^2. Hence, in the Euclidean version we have

$$ds^2 \equiv g_{AB}dx^A dx^B = y^2 dx_0^2 + dy^2 + ds_{D-2}^2 \,. \tag{5}$$

This means that $\mathcal{N} = \mathcal{R}_2 \times \mathcal{M}$, where $\mathcal{R}_2$ is the two-dimensional Rindler space.

Example: the four dimensional Schwarzschild black hole is approximated by $\mathcal{N} \simeq \mathcal{R}_2 \times S^2$.

We consider the Laplace-Beltrami operator on $\mathcal{N}$

$$\triangle_N = \frac{1}{\sqrt{g}}\partial_A g^{AB}\sqrt{g}\partial_B \,. \tag{6}$$

The massless Green function is defined as a solution of the equation

$$-\triangle_N \mathcal{G}_N = \frac{1}{\sqrt{g}}\delta \,. \tag{7}$$

The Green function can be expressed by the fundamental solution of the diffusion equation

$$\partial_\tau P = \frac{1}{2}\triangle_N P \,.$$

Then we obtain

$$\mathcal{G} = \frac{1}{2}\int_0^\infty d\tau P_\tau \,. \tag{8}$$

Now we have the decomposition

$$\triangle_N = y^{-2}\partial_0^2 + y^{-1}\partial_y y \partial_y + \triangle_M \,,$$

where

$$\triangle_M = \frac{1}{\sqrt{g_M}} \partial_j g^{jk} \sqrt{g_M} \partial_k$$

with $g_M = \det(g_{jk})$ being the Laplace-Beltrami operator on $\mathcal{M}$.

In the coordinates

$$y = b \exp u \tag{9}$$

Eq. (7) reads

$$\Big[- \partial_0^2 - \partial_u^2 - b^2 \exp(2u) \triangle_M \Big] \mathcal{G} = g_M^{-\frac{1}{2}} \delta(x_0 - x_0') \delta(u - u') \delta(\mathbf{x} - \mathbf{x}') \,. \tag{10}$$

4. Green Functions Near the Horizon

We define

$$\hat{\mathcal{G}} = (yy')^{-\frac{D-1}{2}} \mathcal{G}$$

and

$$\hat{\triangle}_N = y \partial_y y \partial_y - \tfrac{(D-2)^2}{4} + y^2 \triangle_M \,. \tag{11}$$

Let $\hat{P}$ be the fundamental solution of the heat equation

$$\partial_\tau \hat{P}_\tau = \frac{1}{2} \hat{\triangle}_N \hat{P}_\tau \,. \tag{12}$$

Then, the Green function can be expressed by the heat kernel

$$\hat{\mathcal{G}}_N(y, \mathbf{x}; y', \mathbf{x}') = \int_0^\infty d\tau (2\pi\tau)^{-\frac{1}{2}} \exp \Big[- \frac{1}{2\tau} (x_0 - x_0')^2 + \frac{\tau}{8} (D-2)^2 \Big] \hat{P}_\tau (y, \mathbf{x}; y', \mathbf{x}') \,. \tag{13}$$

We have estimated[4] the heat kernel and the Green functions in Eq. (13) with the conclusion that there is no dimensional reduction if the spectrum of the Laplace-Beltrami operator on $\mathcal{M}$ is continuous. We show that the behaviour of the Green function is different if the spectrum of the Laplace-Beltrami operator is discrete.

If $\mathcal{M}$ is a compact manifold without a boundary then the spectrum of the Laplace-Beltrami operator $\triangle_M$ is discrete:

$$-\triangle_M u_k = \epsilon_k u_k \,. \tag{14}$$

A constant is an eigenfunction with the lowest eigenvalue 0 as

$$\triangle_M 1 = 0 \,.$$

We expand the Green function into the Kaluza-Klein modes u_k

$$\mathcal{G}_N(x_0, x_1, \mathbf{x}; x_0', x_1', \mathbf{x}') = \sum_k g_k(x_0, x_1; x_0', x_1')\overline{u}_k(\mathbf{x})u_k(\mathbf{x}') . \tag{15}$$

Then, g_k can be expanded in a complete basis of eigenfunctions

$$\phi_k^E = \exp(ip_0x_0)\phi_k^{p_1}(x_1) , \tag{16}$$

where

$$(-\partial_1^2 + b^2\epsilon_k \exp(2x_1))\phi_k^{p_1} = p_1^2\phi_k^{p_1} . \tag{17}$$

In terms of ϕ^E we obtain

$$g_k(x_0, x_1; x_0', x_1') = \int dp_0 dp_1 (p_0^2 + p_1^2)^{-1}\overline{\phi}_k^E(x_0, x_1)\phi_k^E(x_0', x_1') . \tag{18}$$

The solution, which behaves like a plane wave with momentum p_1 for $x_1 \to -\infty$ and decays exponentially for $x_1 \to +\infty$, reads

$$\phi_k^{p_1} = N_{p_1} K_{ip_1}(b\sqrt{\epsilon_k}\exp(x_1)) , \tag{19}$$

where K_ν is the modified Bessel function of the third kind.

$\mathcal{G}_N$ contains the zero mode $u_0 = 1$. When we subtract the zero mode part G of $\mathcal{G}_N$, then we obtain

$$\begin{aligned} &\mathcal{G}_N(x_0, x_1, \mathbf{x}; x_0', x_1', \mathbf{x}') - G(x_0, x_1; x_0', x_1') \\ &= \frac{4}{\pi^2}\int_0^\infty dp_1 \sinh(\pi p_1)\exp(-p_1|x_0 - x_0'|) \\ &\quad \times \sum_{k\neq 0} K_{ip_1}(b\sqrt{\epsilon_k}\exp(x_1))K_{ip_1}(b\sqrt{\epsilon_k}\exp(x_1'))\overline{u}_k(\mathbf{x})u_k(\mathbf{x}') , \end{aligned} \tag{20}$$

where

$$G(x_0, x_1; x_0', x_1') = -\frac{1}{4\pi}\ln\left((x_0 - x_0')^2 + (x_1 - x_1')^2\right) \tag{21}$$

is the Green function for the two-dimensional free field on the plane.

Each term on the rhs of Eq. (20) is decaying exponentially. We wish to estimate the series. We estimate the rhs of Eq. (20) for large x_1 and x_1' by means of a simplified argument applicable when $\mathbf{x} = \mathbf{x}'$ and $x_1 = x_1'$. Let the lowest non-zero eigenvalue satisfy $\epsilon_1 \geq \delta$. We apply the Weyl approximation for large eigenvalues in order to bound the rhs of Eq. (20) as follows:

$$\begin{aligned} &|\mathcal{G}_N(x_0, x_1, \mathbf{x}; x_0', x_1, \mathbf{x}) - G(x_0, x_1; x_0', x_1)| \\ &\leq C^2\frac{4}{\pi^2}\int_0^\infty dp_1 \sinh(\pi p_1)\exp(-p_1|x_0 - x_0'|) \sum_{\delta\leq\epsilon_k\leq n} |K_{ip_1}(b\sqrt{\epsilon_k}\exp(x_1))|^2 \end{aligned} \tag{22}$$

$$+R\int_0^\infty dp_1 \sinh(\pi p_1)\exp(-p_1|x_0 - x_0'|)\int_{|\mathbf{k}|\geq\sqrt{n}} d\mathbf{k}|K_{ip_1}(b|\mathbf{k}|\exp(x_1))|^2 .$$

The finite sum as well as the integral on the rhs are decaying exponentially.

5. Euclidean Free Fields Near the Horizon

We introduce a free Euclidean field as a random field with the correlation function equal to the Green function

$$\begin{aligned}\Phi(x_0,x_1,\mathbf{x}) &= \int dp_0 dp_1 \sum_k a_k(p_0,p_1)\phi_k^{p_1}(x_1)u_k(\mathbf{x})\exp(ip_0x_0)\\ &= \Phi_0(x_0,x_1) + \sum_{k>0}\Phi_k(x_0,x_1,\mathbf{x}), \end{aligned} \tag{23}$$

where

$$\langle \overline{a}_k(p_0,p_1)a_{k'}(p_0',p_1')\rangle = \delta_{kk'}\delta(p_0-p_0')\delta(p_1-p_1')(p_0^2+p_1^2)^{-1} . \tag{24}$$

Here Φ_0 is a two-dimensional conformal invariant free field with the correlation function

$$\begin{aligned}\langle \Phi_0(x_0,x_1)\Phi_0(x_0',x_1')\rangle &= G(x_0,x_1;x_0',x_1')\\ &= -\frac{1}{4\pi}\ln\left((x_0-x_0')^2+(x_1-x_1')^2\right). \end{aligned} \tag{25}$$

The correlation functions of Φ_k are decaying exponentially.

The analytic continuation of free Euclidean fields $x_0 \to ix_0$ follows from the reflection positivity. The quantum free fields can be represented in the Fock space according to

$$\begin{aligned}\Phi(x_0,x_1,\mathbf{x}) &= \int dp_1 \sum_k a_k(p_1)\phi_k^{p_1}(x_1)u_k(\mathbf{x})\exp(-i|p_1|x_0)\\ &\quad + \int dp_1 \sum_k a_k^+(p_1)\phi_k^{p_1}(x_1)\overline{u_k}(\mathbf{x})\exp(i|p_1|x_0), \end{aligned} \tag{26}$$

where a and a^+ are annihilation and creation operators satisfying the commutation relations

$$[a_k(p_1), a_l^+(p_1')] = \delta_{kl}\delta(p_1-p_1') . \tag{27}$$

References

1. P. T. Chrusciel, *Acta Phys. Polon. B* **36**, 17 (2005).
2. A. Jaffe and G. Ritter, *Commun. Math. Phys.* **270**, 545 (2007).
3. R. M. Wald, *Quantum Field Theory in Curvd Spacetime and Black Hole Thermodynamics* (The University of Chicago Press, Chicago, 1994).
4. Z. Haba, arXiv:0708.1742.

PATH INTEGRAL FOR HALF-BINDING POTENTIALS AS QUANTUM MECHANICAL ANALOG FOR BLACK HOLE PARTITION FUNCTIONS

D. GRUMILLER

Center for Theoretical Physics, Massachusetts Institute of Technology, 77 Massachusetts Ave., Cambridge, MA 02139, USA
E-mail: grumil@lns.mit.edu

The semi-classical approximation to black hole partition functions is not well-defined, because the classical action is unbounded and the first variation of the uncorrected action does not vanish for all variations preserving the boundary conditions. Both problems can be solved by adding a Hamilton-Jacobi counterterm. I show that the same problem and solution arises in quantum mechanics for half-binding potentials.

Keywords: Hamilton-Jacobi counterterm; Black hole analog.

1. Introduction and Statement of the Problem

Path integrals have illuminated many aspects of quantum mechanics and quantum field theory,[1] but there remain some challenges to path integral formulations of quantum theories.[2] In this proceedings contribution I describe a problem arising for quantum mechanical potentials that are 'half-binding' (the definition of this term will be given below). I shall demonstrate that the naive semi-classical approximation to the path integral breaks down for two reasons: the leading contribution to the partition function is singular and the first variation of the action does not vanish for all variations preserving the boundary conditions. I discuss how both issues can be resolved by adding an appropriate (Hamilton-Jacobi) counterterm as boundary term to the action. Moreover, I shall point out formal similarities to black hole (BH) partition functions, so in that sense these quantum mechanical systems may serve as toy models to elucidate certain aspects of BH physics. For sake of clarity I focus on a specific Hamiltonian,[3]

$$H(q,p) = \frac{p^2}{2} + V(q)\,, \qquad V(q) = \frac{1}{q^2}\,, \tag{1}$$

where q is restricted to positive values. If q is small the Hamiltonian rises without bound, like for a binding potential. If q is large the potential is negligible, and the asymptotic dynamics is dominated by free propagation. I refer to a potential with these properties as 'half-binding'. [The conformal properties[3] of (1) will not play any role in this discussion.]

Consistency of the variational principle based on the Lagrangian action,

$$I[q] = \int_{t_i}^{t_f} dt \left(\frac{\dot{q}^2}{2} - \frac{1}{q^2}\right), \tag{2}$$

requires to fix the initial and final value of q at t_i and t_f, respectively. I am interested here mostly in the limit $t_f \to \infty$, which implies that $q|^{t_f} = \infty$ is the appropriate asymptotic boundary condition. The initial time is set to zero, $t_i = 0$, without loss of generality. The Lagrangian path integral,

$$\mathcal{Z} = \int \mathcal{D}q \, \exp\Big(-\frac{1}{\hbar}\, I[q]\Big), \tag{3}$$

consists of a coherent sum over all field configurations consistent with the boundary data. Even though (2) is exactly soluble, it is illustrative to consider the semi-classical expansion of the action,

$$I[q_{\rm cl} + \delta q] = I[q_{\rm cl}] + \delta I\big|_{\rm EOM} + \mathcal{O}(\delta q^2), \tag{4}$$

and of the partition function

$$\mathcal{Z} = \exp\Big(-\frac{1}{\hbar}\, I[q_{\rm cl}]\Big) \int \mathcal{D}\delta q \, \exp\Big(-\frac{1}{\hbar}\, \mathcal{O}(\delta q^2)\Big). \tag{5}$$

The semi-classical approximation (5) is well-defined only if the on-shell action is bounded, $|I[q_{\rm cl}]| < \infty$, and only if the first variation of the action vanishes on-shell, $\delta I|_{\rm EOM} = 0$, for all field configurations preserving the boundary conditions. I demonstrate now that neither is the case for the example (2).

The on-shell action diverges because asymptotically the propagation is essentially free, and because of the assumption $t_f \to \infty$. This is an idealization of situations where boundary conditions are imposed at late times, $t_f \sim 1/\epsilon$, with $\epsilon \ll 1$. In that case also $q_f \sim 1/\epsilon$ classically. However, the path integral does not only take into account classical contributions, but also samples nearby field configurations whose asymptotic behavior is $q \sim q_f[1 + \epsilon\, \Delta q + \mathcal{O}(\epsilon^2)]$, where Δq is finite. Therefore, the first variation of the action, evaluated on-shell, is given by the boundary term

$$\dot{q}\, \delta q|^{t_f} - \dot{q}\, \delta q|^{t_i=0} = \dot{q}\, \delta q|^{t_f} \sim \lim_{\epsilon \to 0} [\dot{q}\, \Delta q + \mathcal{O}(\epsilon)]\big|^{t_f} \neq 0\,. \tag{6}$$

The inequality emerges, because arbitrary finite variations $\delta q|^{t_f}$ certainly preserve the boundary condition $q|^{t_f} = \infty$. The two problems described here spoil the semi-classical approximation (5) to the partition function.

2. Hamilton-Jacobi Counterterm for Half-Binding Potentials

Both problems can be solved by considering an improved action[a]

$$\Gamma[q] = \int\limits_0^{t_f} dt \left(\frac{\dot{q}^2}{2} - \frac{1}{q^2} \right) - S(q,t)\Big|_0^{t_f} , \tag{7}$$

which differs from (2) by a boundary counterterm depending solely on quantities that are kept fixed at the boundary. The variation of (7),

$$\delta\Gamma|_{\mathrm{EOM}} = \left(\dot{q} - \frac{\partial S}{\partial q} \right) \delta q \Big|_0^{t_f} = \left(\dot{q} - \frac{\partial S}{\partial q} \right) \delta q \Big|^{t_f} , \tag{8}$$

does not necessarily suffer from the second problem if $\partial S/\partial q$ asymptotically behaves like $\dot{q}$, i.e., like the momentum p.

The method[8,9] that I am going to review does not involve the subtraction of the action evaluated on a specific field configuration, but rather is intrinsic. Moreover, the amount of guesswork is minimal: Hamilton's principal function is a well-known function of the boundary data with the property $\partial S/\partial q = p$. Therefore it is natural to postulate that S in (7) solves the Hamilton-Jacobi equation,

$$H\left(q, \frac{\partial S}{\partial q} \right) + \frac{\partial S}{\partial t} = 0 . \tag{9}$$

The complete integral[10]

$$S(q,t) = c_0 - Et + \sqrt{2(Eq^2 - 1)} + \sqrt{2} \arctan \frac{1}{\sqrt{Eq^2 - 1}} \tag{10}$$

allows to construct the enveloping solution

$$S(q,t) = \frac{q^2}{2t} \left(\sqrt{4\Delta_+ - 8t^2/q^4} - \Delta_+ \right) + \sqrt{2} \arctan \frac{1}{\sqrt{q^4 \Delta_+/(2t^2) - 1}} , \tag{11}$$

[a]There exists a variety of subtraction methods in quantum mechanics,[4] in General Relativity (and generalizations thereof)[5] and in holographic renormalization within the context of AdS/CFT.[6] Many of them have ad-hoc elements and require the subtraction of the action evaluated on a specific field configuration (like the ground state solution); in some cases there are several "natural" candidates, in others there is none, and in at least one example the "natural" guess even turned out to be wrong.[7]

where $\Delta_+ := \frac{1}{2}(1+\sqrt{1-8t^2/q^4})$.[b] The asymptotic expansion $S = q^2/(2t) + \mathcal{O}(t/q^2)$ is consistent with the intuitive idea that the asymptotically free propagation is the source of all subtleties. But the expression (11) contains a great deal of additional (non-asymptotic) information, which can be physically relevant, as mentioned in the next section.

Let me now come back to the two problems. Since asymptotically $\dot{q}|_{\mathrm{EOM}} = v = \text{const.}$, the on-shell action

$$\Gamma\big|_{\mathrm{EOM}} = \frac{v^2}{2}\int\limits_0^{t_f} dt - \frac{v^2}{2}t^f + \mathcal{O}(1) = \mathcal{O}(1) \tag{12}$$

evidently is finite. The terms of order of unity entail the information about the potential $V(q)$. The first variation

$$\delta\Gamma\big|_{\mathrm{EOM}} = \Big(\underbrace{\dot{q} - \frac{q}{t}}_{\mathcal{O}(1/t)} + \mathcal{O}(1/t^2)\Big)\delta q\Big|^{t_f} = \mathcal{O}(1/t)\underbrace{\delta q}_{\text{finite}}\Big|^{t_f} = 0 \tag{13}$$

vanishes for all variations preserving the boundary conditions. The two problems mentioned in the previous section indeed are resolved by the improved action (7) with (11).

The considerations above apply in the same way to the Hamiltonian (1) with a more general class of half-binding potentials $V(q)$. In particular, the (manifestly positive) potential $V(q)$ is required to be monotonically decreasing, and to vanish faster than $1/q$ for large q. Going through the same steps as above is straightforward. Other generalizations, e.g. to non-monotonic potentials or potentials with Coulomb-like behavior, may involve technical refinements, but the general procedure is always the same: one has to solve the Hamilton-Jacobi equation (9) to obtain the correct counterterm S in (7).

3. Comparison with Black Hole Partition Functions

The same issues as in the previous section arise when evaluating BH partition functions. Probably the simplest non-trivial model is 2-dimensional dilaton gravity (cf., e.g., Ref. 11 for recent reviews),

[b] One is forced to take the enveloping solution, since S is part of the definition of the improved action and therefore cannot depend on constants of motion. The energy E is eliminated from (10) by solving $\partial S/\partial E = 0$ for E. The other constant, c_0, is set to zero by hand, but other choices are possible. Such an ambiguity always remains in this (and any other) approach. It reflects the freedom to shift the free energy of the ground state.

$$I[g,X] = -\frac{1}{16\pi G_2}\int_{\mathcal{M}} d^2x\,\sqrt{g}\Big(X\,R - U(X)\,(\nabla X)^2 - 2\,V(X)\Big) \\ - \frac{1}{8\pi G_2}\int_{\partial\mathcal{M}} dx\,\sqrt{\gamma}\,X\,K\,. \quad (14)$$

An explanation of the notation can be found in.[9] The boundary term in (14) is the dilaton gravity analog of the Gibbons-Hawking-York boundary term. The latter arises in quantum mechanics if one converts the action $I = \int dt[-q\dot{p} - H(q,p)]$ into standard form, but it is *not* related to the Hamilton-Jacobi counterterm. It was shown first (second) in the second[9] (first[12]) order formulation that the improved action is given by

$$\Gamma[g,X] = I[g,X] + \frac{1}{8\pi G_2}\int_{\partial\mathcal{M}} dx\,\sqrt{\gamma}\,S(X)\,, \quad (15)$$

with the solution of the Hamilton-Jacobi equation ($V(X) \leq 0$)

$$S(X) = \Big(-2e^{-\int^X dy\,U(y)}\int^X dy\,V(y)\,e^{\int^y dz\,U(z)}\Big)^{1/2}. \quad (16)$$

The lower integration constant in the integrals over the function U is always the same and therefore cancels; the lower integration constant in the remaining integral represents the ambiguity mentioned in footnote b.

The BH partition function based upon the improved action (15),

$$\mathcal{Z} = \int \mathcal{D}g\,\mathcal{D}X\,\exp\Big(-\frac{1}{\hbar}\Gamma[g,X]\Big) \approx \exp\Big(-\frac{1}{\hbar}\Gamma[g_{\rm cl},X_{\rm cl}]\Big)\,, \quad (17)$$

by standard methods establishes the BH free energy. The asymptotic part of the counterterm (16) leads to the correct asymptotic charges for BHs with (essentially) arbitrary asymptotic behavior, and to consistency with the first law of thermodynamics (which is non-trivial[7]). The finite part of the counterterm (16) allows a quasi-local description of BH thermodynamics.[9]

Perhaps one might exploit the formal analogy between BH partition functions and quantum mechanical partition functions described in this work to construct interesting condensed matter analogs[13] mimicking thermodynamical aspects of BHs.

Acknowledgments

I am grateful to Roman Jackiw and Robert McNees for enjoyable discussions. I thank Luzi Bergamin and René Meyer for reading the manuscript and Wolfhard Janke for valuable administrative help.

This work is supported in part by funds provided by the U.S. Department of Energy (DoE) under the cooperative research agreement DEFG02-05ER41360 and by the project MC-OIF 021421 of the European Commission under the Sixth EU Framework Programme (FP6).

References

1. W. Janke and A. Pelster (Editors), *Path Integrals – New Trends and Perspectives* (World Scientific, Singapore, 2008).
2. R. Jackiw, arXiv:0711.1514.
3. V. de Alfaro, S. Fubini, and G. Furlan, *Nuovo Cim. A* **34**, 569 (1976).
4. H. Kleinert, *Path Integrals in Quantum Mechanics, Statistics, Polymer Physics, and Financial Markets*, Fourth Edition (World Scientific, Singapore, 2006); cf., e.g., the last paragraph in Sect. 5.4.
5. T. Regge and C. Teitelboim, *Ann. Phys.* **88**, 286 (1974) 286; G. W. Gibbons and S. W. Hawking, *Phys. Rev. D* **15**, 2752 (1977); H. Liebl, D. V. Vassilevich, and S. Alexandrov, *Class. Quant. Grav.* **14**, 889 (1997); R.-G. Cai and N. Ohta, *Phys. Rev. D* **62**, 024006 (2000); R. B. Mann and D. Marolf, *Class. Quant. Grav.* **23**, 2927 (2006); R. B. Mann, D. Marolf, and A. Virmani, *Class. Quant. Grav.* **23**, 6357 (2006).
6. M. Henningson and K. Skenderis, *JHEP* **7**, 23 (1998); V. Balasubramanian and P. Kraus, *Commun. Math. Phys.* **208**, 413 (1999); P. Kraus, F. Larsen, and R. Siebelink, *Nucl. Phys. B* **563**, 259 (1999); R. Emparan, C. V. Johnson, and R. C. Myers, *Phys. Rev. D* **60**, 104001 (1999) 104001; S. de Haro, S. N. Solodukhin, and K. Skenderis, *Commun. Math. Phys.* **217**, 595 (2001); I. Papadimitriou and K. Skenderis, *JHEP* **8**, 4 (2005); R. McNees, arXiv:hep-th/0512297.
7. Here it is shown where such a problem arises and how it can be resolved: J. L. Davis and R. McNees, *JHEP* **9**, 72 (2005).
8. J. de Boer, E. P. Verlinde, and H. L. Verlinde, *JHEP* **8**, 3 (2000); D. Martelli and W. Muck, *Nucl. Phys. B* **654**, 248 (2003); F. Larsen and R. McNees, *JHEP* **7**, 62 (2004); A. Batrachenko, J. T. Liu, R. McNees, W. A. Sabra, and W. Y. Wen, *JHEP* **5**, 34 (2005).
9. D. Grumiller and R. McNees, *JHEP* **4**, 74 (2007).
10. E. Kamke, *I. Differentialgleichungen: Lösungsmethoden und Lösungen, II. Partielle Differentialgleichungen Erster Ordnung für eine gesuchte Funktion* (Akademische Verlagsgesellschaft Geest & Portig, Leipzig, 1965).
11. D. Grumiller, W. Kummer, and D. V. Vassilevich, *Phys. Rep.* **369**, 327 (2002); D. Grumiller and R. Meyer, *Turk. J. Phys.* **30**, 349 (2006).
12. L. Bergamin, D. Grumiller, R. McNees, and R. Meyer, arXiv:0710.4140.
13. M. Novello, M. Visser, and G. Volovik (Editors), *Artificial Black Holes* (World Scientific, River Edge, 2002); C. Barcelo, S. Liberati, and M. Visser, *Living Rev. Rel.* **8**, 12 (2005), and references therein.

EFFECTIVE LAGRANGIANS FOR NONCOMMUTATIVE MECHANICS

C. S. ACATRINEI

Smoluchowski Institute of Physics, Jagellonian University,
Reymonta 4, 30-059, Cracow, Poland
and
National Institute for Nuclear Physics and Engineering,
Bucharest, MG-077125, Romania
E-mail: acatrine@th.if.uj.edu.pl

By using path integral methods we obtain effective Lagrangians for noncommutative Quantum Mechanics. The starting point is a relatively simple modification of standard phase-space path integrals, which leads in configuration space to Lagrangians depending also on the accelerations. We comment on the subtleties involved.

Keywords: Effective Lagrangian; Noncommutative quantum mechanics; Phase-space path integral.

1. Introduction

Phase-space path integrals usually take the form[1]

$$\int D\vec{q}D\vec{p}e^{i\int_0^T dt[\vec{p}\cdot\dot{\vec{q}}-H(\vec{p},\vec{q})]}, \tag{1}$$

with boundary conditions enforced by the type of quantum mechanical amplitude to be evaluated. Such integrals (or their Lagrangian counterparts) suffice for most of physical applications, provided the symplectic structure is canonical, i.e. $\omega_0 = \sum_i dp_i \wedge dq^i$.

We would like to discuss here the following modified path integral

$$\int D\vec{q}D\vec{p}e^{i\int_0^T dt[\vec{p}\cdot\dot{\vec{q}}-H(\vec{p},\vec{q})+\theta/2(p_1\dot{p}_2-p_2\dot{p}_1)]}, \tag{2}$$

where θ is a constant of dimension length-squared. For simplicity we work in two space dimensions and with all indices down, $\vec{q}=(q_1,q_2)$, $\vec{p}=(p_1,p_2)$. Standard notation is employed for velocity $v_i = \dot{q}_i \equiv \frac{dq_i}{dt}$, acceleration $a_i = \ddot{q}_i \equiv \frac{d^2q_i}{dt^2}$ and mass (m). The Planck constant is set to one throughout.

The apparently innocuous modification (2) actually amounts to a change in the symplectic structure, $\omega_0 \to \omega = \sum_{i=1}^{2}(dp_i \wedge dq^i + \frac{\theta}{2}dp_i \wedge dp_j)$ and has important consequences discussed below.

In fact, the path integral with modified symplectic structure (2) describes transition amplitudes in noncommutative quantum mechanics, a subject first introduced in Ref. 2 and intensively studied in the last decade, see Refs. 3–9. More precisely (2) describes quantum mechanics with an additional nonvanishing commutator between coordinates, $[q_1, q_2] = i\theta$. This theory admits a first principles path integral formulation only in phase space, as detailed in Ref. 8. At the classical level, the extended symplectic structure features an additional nonzero Poisson bracket, $\{q_1, q_2\} = \theta \neq 0$, and the resulting equations of motion do not admit a standard Lagrangian formulation.[9]

Nevertheless one can enforce an effective Lagrangian formulation in configuration space[10] by integrating over the momenta in the path integral (2). We describe this process, and conclude with a discussion of the result.

2. Path Integral

We path-integrate over the momenta in (2), to obtain the effective Lagrangian. Starting from the partition function

$$\int Dq_1 Dq_2 Dp_1 Dp_2 e^{iS} \tag{3}$$

with action

$$S = \int_0^T dt \left[p_1\dot{q}_1 + p_2\dot{q}_2 + \frac{\theta}{2}(p_1\dot{p}_2 - p_2\dot{p}_1) - \frac{p_1^2}{2m} - \frac{p_2^2}{2m} - V(q)\right], \tag{4}$$

we wish to integrate over the momenta p_1, p_2. The potential part $V(q)$ depends only on q_1 and q_2 and plays no role in what follows (the method is valid for any $V(q)$, more precisely for any Hamiltonian with separate quadratic dependence upon momenta). We divide the time interval T in n subintervals $\epsilon = \frac{T}{n}$ ($n \to \infty$ achieves the continuum limit), and choose for simplicity the discrete derivative $v^{(k)} \equiv \dot{x}^{(k)} \equiv \frac{x^{(k+1)} - x^{(k)}}{\epsilon}$; no issues requiring symmetric operations of any kind appear in the following. The relevant part of the discretized action (excluding $V(q)$ for now) becomes

$$\tilde{S} = \sum_{k=0}^{n} \left[\epsilon p_1^{(k)} v_1^{(k)} + \epsilon p_2^{(k)} v_2^{(k)} + \frac{\theta}{2}\left(p_1^{(k)} p_2^{(k+1)} - p_2^{(k)} p_1^{(k+1)}\right) \right.$$
$$\left. -\epsilon \frac{(p_1^{(k)})^2 + (p_2^{(k)})^2}{2m}\right]. \tag{5}$$

The clearest way to proceed with the coupled Gaussian integrals is to introduce matrix notation. Define the column vectors

$$V \equiv \epsilon(v_1^{(0)}, v_1^{(1)}, \ldots, v_1^{(n)} \ldots, v_2^{(0)}, v_2^{(1)}, \ldots, v_2^{(n)} \ldots)^T \tag{6}$$

$$P \equiv (p_1^{(0)}, p_1^{(1)}, \ldots, p_1^{(n)} \ldots, p_2^{(0)}, p_2^{(1)}, \ldots, p_2^{(n)} \ldots)^T \tag{7}$$

and the matrix

$$J = -a \begin{pmatrix} 1 & 0 & 0 & \cdot\cdot & 0 & b & 0\cdot\cdot \\ 0 & 1 & 0 & \cdot\cdot & 0 & 0 & b\cdot\cdot \\ \cdot & \cdot & \cdot & \cdot & \cdot & \cdot & \cdot\,\cdot \\ 0 & -b & 0 & \cdot\cdot & 1 & 0 & 0\cdot\cdot \\ 0 & 0 & -b & \cdot\cdot & 0 & 1 & 0\cdot\cdot \\ \cdot & \cdot & \cdot & \cdot & \cdot & \cdot & \cdot\,\cdot \end{pmatrix}.$$

where $a = \frac{\epsilon}{2m}$, $b = \frac{m\theta}{\epsilon}$. Its inverse J^{-1} has the same form as above, but with different entries a', b', namely $a' = 1/a$ and $b' = -b$ (the off diagonal part changes sign and the overall factor is reversed). In matrix notation the discrete action becomes

$$\tilde{S} = P^T V + P^T J P. \tag{8}$$

The coordinate transformation

$$\bar{P} \equiv P + \frac{1}{2} J^{-1} V \tag{9}$$

does not change the path integral measure ($D\bar{P} = DP$), and leads to

$$\tilde{S} = \bar{P}^T J \bar{P} - \frac{1}{4} V^T J^{-1} V. \tag{10}$$

The first term is now integrated out - and no more dependency upon momenta appears, whereas the second term leads to an exponent of the form (modulo a factor of i)

$$-\frac{1}{4} V^T J^{-1} V = \sum_{k=0}^{n} [\epsilon \frac{m}{2} (v_1^{(k)})^2 + \epsilon \frac{m}{2} (v_2^{(k)})^2 - \frac{\theta m^2}{2} (v_1^{(k)} v_2^{(k+1)} - v_2^{(k)} v_1^{(k+1)})]. \tag{11}$$

Upon taking the continuum limit $\epsilon \to 0$ our main result follows:

$$\int Dq_1 Dq_2 Dp_1 Dp_2 e^{iS} = N \int Dq_1 Dq_2 e^{i \int_0^T dt L_{eff}(q_i, v_i, a_i)} \tag{12}$$

with

$$L_{eff} = \frac{m}{2} (\dot{q}_1^2 + \dot{q}_2^2) - \frac{\theta m^2}{2} (\dot{q}_1 \ddot{q}_2 - \dot{q}_2 \ddot{q}_1) - V(q_1, q_2) \tag{13}$$

and N is a constant not depending on the q's. We have reintroduced the potential term, which passed unscathed through Eqs. (4) – (13). The second term in (13) is the correction due to noncommutativity; it depends on velocities *and* accelerations, and has an universal character. Its relative simplicity is striking and somehow unexpected. One is reconforted to find that the Lagrangian (13) was studied by Lukierski *et al.*[4] and shown to engender a noncommutative structure. A more detailed discussion follows.

3. Discussion

As already mentioned, see Refs. 8,9, the resulting effective Lagrangian could not be a standard one, depending only on coordinates and velocities. Given the complications introduced by noncommutativity, one may have expected *a priori* an involved function, perhaps nonlocal or potential-dependent. Remarkably, the effective Lagrangian turned out to be the usual one, plus an universal correction depending also on the particle accelerations,

$$\Delta L = -\frac{1}{2}\theta m^2 (v_1 a_2 - v_2 a_1), \tag{14}$$

θ denoting the noncommutative scale, m, v_i, a_i the mass, velocity, respectively acceleration along the i-axis, of a given particle

Exactly the term (14) was previously studied in detail in Ref. 4, although its appearance can be traced back to earlier developments (see Refs. 5–7). Lukierski *et al.*[4] started from considerations of Galilean invariance in (2+1)-dimensions, and added (14) to a free Lagrangian $\frac{m}{2}\vec{v}^2$, to provide a dynamical realization for a free particle Galilean algebra with one extra central charge. Upon constrained quantization of this higher order action (which thus circumvents the no-go theorem of Ref. 9) noncommutative dynamics was shown to emerge for appropriate choices of canonical variables. Two negative-energy "internal modes" were proved harmless since they decoupled from the four relevant degrees of freedom. Interactions were subsequently introduced in a constrained way in order to keep the ghosts harmless, and were described by potentials depending on noncommutative coordinates.

We briefly review the derivation in Ref. 4 from our perspective, putting more emphasis on the Faddeev-Jackiw approach.[11] Let us start from

$$L_{LSZ} = \frac{m}{2}(\dot{q}_1^2 + \dot{q}_2^2) - \frac{\theta m^2}{2}(\dot{q}_1\ddot{q}_2 - \dot{q}_2\ddot{q}_1) - v(q_1, q_2, \dot{q}_1, \dot{q}_2), \tag{15}$$

with q_1, q_2 commuting, and a more general potential considered (Lukierski *et al.* took initially $v = 0$). We pass to the Hamiltonian form using the

Ostrogradski formalism

$$y_i \equiv \dot{q}_i, \quad \tilde{p}_i \equiv \frac{\partial L}{\partial \ddot{q}_i} = k\epsilon_{ij}y_j, \quad k \equiv \frac{\theta m^2}{2}, \tag{16}$$

$$x_i \equiv q_i, \quad p_i \equiv \frac{\partial L}{\partial \dot{q}_i} - \frac{d}{dt}\frac{\partial L}{\partial \ddot{q}_i} = my_i - 2k\epsilon_{ij}\dot{y}_j, \tag{17}$$

$$L - p_i\dot{x}_i + \tilde{p}_i\dot{y}_i = H(x, y, p, \tilde{p}) = \tilde{p}_i\frac{\epsilon_{ij}}{k}p_j - \frac{m\tilde{p}_i^2}{2k^2}. \tag{18}$$

The action $\int dtL$ is already in Faddeev-Jackiw form, except for the constraint $ky_i + \epsilon_{ij}\tilde{p}_j = 0$, which is easily solved

$$\int dtL = \int (p_i dx_i - \tilde{p}_i\frac{\epsilon_{ij}}{k}d\tilde{p}_j) - \int dt\left(\tilde{p}_i\frac{\epsilon_{ij}}{k}dp_j - \frac{m\tilde{p}_i^2}{2k^2}\right). \tag{19}$$

The commutation relations are read out from the (inverse of) the symplectic form, and read $\{x_i, p_j\} = \delta_{ij}$, $\{\tilde{p}_1, \tilde{p}_2\} = k/2$.

We must now identify the 'true' coordinates X_i of the system. A Noether symmetry analysis immediately unveils that in the $v = 0$ case[4] the Galilean boosts are given by $G_i = p_i t - mx_i + 2\tilde{p}_i \equiv p_i t - K_i$. We have $\dot{G}_i = 0$; $\dot{K}_i = p_i$. Since no time appears in X_i, only the K_i part of G_i matters for the definition of X_i, $X_i \equiv \frac{K_i}{m} = x_i - 2\tilde{p}_i/m$. Momenta keep the same form, $P_i \equiv p_i$, and the extraneous pair of them is required to commute with X_i, P_i, leading to $\tilde{P}_i \equiv kp_i/m + \epsilon_{ij}\tilde{p}_j$.

The Hamiltonian reads in the new variables $X_i, P_i, \tilde{P}_i, i = 1, 2$,

$$H = \frac{P_i^2}{2m} - \frac{m\tilde{P}_i^2}{2k^2} + v\left(q_i = X_i + \frac{2\epsilon_{ij}}{m}(\frac{kp_j}{m} - \tilde{P}_j), \dot{q}_i = \frac{kp_i/m - \tilde{P}_i}{k}\right) \tag{20}$$

whereas the commutators are

$$\{X_1, X_2\} = \frac{2k}{m^2} = \theta, \quad \{X_i, P_j\} = \delta_{ij}, \quad \{\tilde{P}_1, \tilde{P}_2\} = k/2. \tag{21}$$

The last two equations define a noncommutative theory. The second term in the Hamiltonian is however negative. To keep it decoupled from the X, P variables one has to impose that no $\tilde{P}_i$ appear in v. Thus v must depend on the linear combination $q_i - \frac{2k\epsilon_{ij}}{m}\dot{q}_i$ of its variables, which results exactly in the noncommuting coordinates X_i.

No obvious reciprocal of the canonical analysis of Ref. 4 is known to us at present; actually a classical canonical approach leads to second (not third!) order equations of motion.[9]

We provided in Ref. 10 for the first time a univoque path integral description. Interestingly enough, it reconstructed the *constrained* higher-order Lagrangian of Ref. 4 up to coefficients. The correction (14) turned out to be the *only* possibility available for noncommutative systems of Heisenberg type and Hamiltonians of the form $H = \frac{1}{2m}(p_1^2 + p_2^2) + V(q_1, q_2)$.

Acknowledgments

The author is grateful to the organizers of *Path Integral 07*, Dresden, for providing a stimulating and pleasant atmosphere, and to the MPI-PKS in Dresden for covering a significant part of the conference costs. Financial support through the EU Marie Curie Host Fellowships for Transfer of Knowledge Project COCOS (contract MTKD-CT-2004-517186) and the NATO Grant PST.EAP.RIG.981202 is acknowledged.

References

1. L. S. Schulman, *Techniques and Applications of Path Integration* (Wiley, N.Y., 1981).
2. G. Dunne and R. Jackiw, *Nucl. Phys. Proc. Suppl. C* **33**, 114 (1993).
3. V. P. Nair and A. P. Polychronakos, *Phys. Lett. B* **505**, 267 (2001), and quotations thereof. We do not attempt to appropriately quote the whole noncommutative mechanics literature here.
4. J. Lukierski, P. C. Stichel, and W. J. Zakrzewski, *Ann. Phys.* **260**, 224 (1997).
5. C. Duval and P. A. Horvathy, *Phys. Lett. B* **479**, 284 (2000).
6. R. Jackiw and V. P. Nair, *Phys. Lett. B* **480**, 237 (2000).
7. P. A. Horvathy, *Acta Phys. Pol. B* **34**, 2611 (2003).
8. C. S. Acatrinei, *JHEP* **0109**, 7 (2001).
9. C. S. Acatrinei, *J. Phys. A* **37**, 1225 (2004); *Rom. J. Phys.* **52**, 3 (2007).
10. C. S. Acatrinei, *J. Phys. A* **40**, F929 (2007).
11. L. D. Faddeev and R. Jackiw, *Phys. Rev. Lett.* **60**, 1692 (1988). For a beautiful review, see R. Jackiw, in *Constraint Theory and Quantization Methods*, eds. F. Colomo *et al.* (World Scientific, Singapore, 1994), p. 163 [arXiv:hep-th/9306075].

PART V

Statistical Field Theory

FUNCTIONAL INTEGRALS IN PHYSICS: THE MAIN ACHIEVEMENTS

J. ZINN-JUSTIN

IRFU, DSM-CEA/Saclay,
F-91191 Gif-sur-Yvette cédex, France
and
Institut de Mathématiques de Jussieu–Chevaleret,
Université de Paris VII, France
E-mail: zinn@dsm-mail.saclay.cea.fr

I shall briefly review what I consider as some of the important physics results in which the use of path or field integrals has played an essential role. This is by now a long history.[1,2]

Keywords: Variational principle; Relativistic quantum field theory; Non-Abelian gauge theories; Non-linear σ-model; Statistical physics; Critical phenomena; Non-perturbative methods.

1. The Mystery of the Classical Variational Principle

1.1. *Euler–Lagrange equations*

Around 1750, Euler and Lagrange develop the variational calculus. Lagrange (1788) then shows that the equations of motion of Newtonian mechanics can be derived from a variational principle. He constructs a mathematical quantity, the action integral of a Lagrangian,

$$\mathcal{A}(q) = \int_{t'}^{t''} \mathrm{d}t\, \mathcal{L}\big(\mathbf{q}(t), \dot{\mathbf{q}}(t); t\big),$$

and recovers the equation of the classical motion by expressing the stationarity of the action with respect to variations of the trajectory $\mathbf{q}(t)$:

$$\delta\mathcal{A} = 0 \;\Rightarrow\; \frac{\mathrm{d}}{\mathrm{d}t}\frac{\partial\mathcal{L}}{\partial\dot{q}_i} = \frac{\partial\mathcal{L}}{\partial q_i}\,,$$

a form called Euler–Lagrange equations.

The property extends to the Hamiltonian formalism, the action defined on phase space variables (q, p) then takes the form

$$\mathcal{A}(p,q) = \int \mathrm{d}t \, \left[\mathbf{p}(t) \cdot \mathbf{q}(t) - H\big(\mathbf{p}(t), \mathbf{q}(t); t\big)\right].$$

1.2. *The particle in a magnetic field*

The equation of motion of a particle in a magnetic field $\mathbf{B}$ takes the form

$$m\ddot{\mathbf{q}} = e\mathbf{B}(\mathbf{q}) \times \dot{\mathbf{q}} \quad \text{where} \quad \nabla \cdot \mathbf{B}(\mathbf{q}) = 0 .$$

Quite remarkably, it can also be derived from an action principle, *provided an additional mathematical quantity is introduced*, the *vector potential*:

$$\mathbf{B}(\mathbf{q}) = \nabla \times \mathbf{A}(\mathbf{q}).$$

The Lagrangian can be written as

$$\mathcal{L}(\mathbf{q}, \dot{\mathbf{q}}) = \tfrac{1}{2} m\dot{\mathbf{q}}^2 - e\mathbf{A}(\mathbf{q}) \cdot \dot{\mathbf{q}} .$$

The vector potential is not considered physical since it is defined only up to a gradient term. Equivalent vector potentials are related by a *gauge transformation*:

$$\mathbf{A}(\mathbf{q}) \mapsto \mathbf{A}(\mathbf{q}) + \nabla\Omega(\mathbf{q}).$$

1.3. *Electromagnetism and Maxwell's equations*

Maxwell's equations (in the vacuum) can be written as

$$\nabla \cdot \mathbf{E} = \rho , \quad \nabla \times \mathbf{B} - \frac{\partial \mathbf{E}}{\partial t} = \mathbf{J} ,$$
$$\nabla \cdot \mathbf{B} = 0 , \quad \nabla \times \mathbf{E} + \frac{\partial \mathbf{B}}{\partial t} = 0 ,$$

where $\mathbf{E}$ and $\mathbf{B}$ are the electric and magnetic fields, resp., ρ and $\mathbf{J}$ the charge and current densities, resp..

In quadri-covariant notation where $(i, j = 1, 2, 3)$

$$t \equiv x_0 , \ F_{i0} = E_i , \ F_{ij} = \sum_k \epsilon_{ijk} B_k , \ J_0 = \rho ,$$

they can be written as

$$\partial^\mu F_{\mu\nu} = J_\nu \ \Rightarrow \ \partial^\nu J_\nu = 0 \quad \text{and} \quad \partial_\mu F_{\nu\rho} + \partial_\rho F_{\mu\nu} + \partial_\nu F_{\rho\mu} = 0 .$$

The second equation implies that the tensor $F_{\mu\nu}$ can be expressed in terms of a vector potential, or *gauge field*, $A_\mu(x)$ under the form

$$F_{\mu\nu} = \partial_\mu A_\nu - \partial_\nu A_\mu .$$

The gauge field is defined only up to an *Abelian gauge transformation*

$$A_\mu(x) \mapsto A_\mu(x) + \partial_\mu \Omega(x).$$

Then again, remarkably enough, with the introduction of this mathematical quantity, Maxwell's equations can be derived from an action principle. The gauge invariant action,

$$\mathcal{A} = \int \mathrm{d}^4 x\, \mathcal{L}(\mathbf{A}, \dot{\mathbf{A}}),$$

is the integral of the Lagrangian density

$$\mathcal{L}(\mathbf{A}, \dot{\mathbf{A}}) = -\frac{1}{4} F^{\mu\nu} F_{\mu\nu} - J^\mu A_\mu \ .$$

1.4. *General relativity*

In *Einstein's relativistic theory of gravitation* (or General Relativity), the equations of motion can also be derived from an action principle. For example, in the absence of matter, in terms of the metric tensor $g_{ij}(x)$ they read

$$R_{ij}\big(\mathbf{g}(x)\big) - \tfrac{1}{2} R\big(\mathbf{g}(x)\big) g_{ij} = 0\,,$$

where R is the scalar curvature and R_{ij} the Ricci tensor.

These equations can be derived from Einstein–Hilbert's action,

$$\mathcal{A}(\mathbf{g}) = \int \mathrm{d}^4 x \left(-g(x)\right)^{1/2} R\big(\mathbf{g}(x)\big),$$

where $g(x)$ is the determinant of the metric tensor. This property generalizes to a cosmological term (as seems to be required by recent observations). A question then arises: *why can all classical fundamental dynamical equations be derived from the stationarity of a local action?*

1.5. *Path integral in quantum mechanics: The stationary phase method*

At first sight, quantum mechanics in its Hamiltonian formulation does not provide a direct answer to this question. It should thus be considered as a first *major success of quantum mechanics in the path integral formulation*, quantum field theory in the field integral formulation, that it provides a very simple explanation to this property. Quantum evolution is given by an integral of the form

$$\langle q''|\, U(t'', t')\, |q'\rangle = \int_{q(t')=q'}^{q(t'')=q''} [\mathrm{d}q(t)]\, \mathrm{e}^{i\mathcal{A}(q)/\hbar},$$

where $\mathcal{A}(q)$ is the classical action. In the classical limit, for $\hbar \to 0$, the path integral can be calculated by the *stationary phase method* and thus is dominated by paths that leave the action stationary: the *classical paths.* This property generalizes to relativistic quantum field theory.

2. Covariance of the Relativistic Quantum Field Theory

The standard Hamiltonian formulation of relativistic quantum field theory is not explicitly covariant. By contrast as first noticed by Dirac, *the matrix elements of the evolution operator for infinitesimal times involve the Lagrangian, which is explicitly covariant.*

The path integral formulation allows to generalize this property to finite times. Indeed, quite generally the evolution operator is given by an integral over phase space variables p and q,

$$\langle q'' | U(t'', t') | q' \rangle = \int [\mathrm{d}p(t)\mathrm{d}q(t)] \exp\left(\frac{i}{\hbar}\mathcal{A}(p,q)\right),$$

where $\mathcal{A}(p,q)$ is the classical action in the Hamiltonian formalism:

$$\mathcal{A}(p,q) = \int \mathrm{d}t \, \left[p(t)\dot{q}(t) - H\left(p(t), q(t); t\right)\right] .$$

When H is a quadratic form in p, like $p^2/2m + V(q)$, the integral over p is Gaussian and can be performed explicitly. This amounts to replacing p by the solution of the classical equation and thus generates the Lagrangian.

3. Quantization of Non-Abelian Gauge Theories

Unlike QED, non-Abelian gauge theories, even without matter fields, cannot be quantized covariantly by using simple heuristic arguments.

3.1. *Non-Abelian gauge theories*

The gauge field $\mathbf{A}_\mu(x)$ is associated to the Lie algebra of a group G and transforms under the adjoint representation. In a gauge transformation with space-dependent group elements $\mathbf{g}(x)$, it transforms like

$$\mathbf{A}_\mu(x) \mapsto \mathbf{g}(x)\mathbf{A}_\mu(x)\mathbf{g}^{-1}(x) + \mathbf{g}(x)\partial_\mu \mathbf{g}^{-1}(x).$$

For matter fields, in a globally G invariant theory, gauge invariance is enforced by replacing derivatives by covariant derivatives: $\mathbf{1}\,\partial_\mu \mapsto \mathbf{D}_\mu = \mathbf{1}\,\partial_\mu + \mathbf{A}_\mu$.

The curvature associated to the gauge field,

$$\mathbf{F}_{\mu\nu}(x) = [\mathbf{D}_\mu, \mathbf{D}_\nu] = \partial_\mu \mathbf{A}_\nu - \partial_\nu \mathbf{A}_\mu + [\mathbf{A}_\mu, \mathbf{A}_\nu]\,,$$

is a tensor for gauge transformations:

$$\mathbf{F}_{\mu\nu}(x) \mapsto \mathbf{g}(x)\mathbf{F}_{\mu\nu}(x)\mathbf{g}^{-1}(x).$$

The local gauge action

$$\mathcal{A}(\mathbf{A}) = \frac{1}{4g^2}\int \mathrm{d}^4x \,\operatorname{tr} \mathbf{F}_{\mu\nu}(x)\mathbf{F}^{\mu\nu}(x),$$

thus is gauge-invariant.

However, due to gauge invariance, not all components of the gauge field are dynamical and a simple canonical quantization is impossible. Ideas that worked for QED, like the direct elimination of the auxiliary components in the Coulomb gauge fail. All concepts required to quantize covariantly non-Abelian gauge theories, like the so-called *Faddeev–Popov trick and ghosts*, are based on field integrals. BRS symmetry has emerged from this formalism.

3.2. *Faddeev–Popov trick*

The goal is to factorize the integration over gauge transformations. One starts from a *non-gauge invariant equation* for the space-dependent group element $\mathbf{g}(x)$, for example,

$$F(\mathbf{A}^{\mathbf{g}}_\mu) \equiv \partial_\mu \mathbf{A}^{\mathbf{g}}_\mu(x) - \nu(x) = 0\,,$$

where $\mathbf{A}^{\mathbf{g}}_\mu$ is the gauge transform of $\mathbf{A}_\mu$ by $\mathbf{g}$ and $\nu(x)$ an arbitrary field.

The variation of the equation with respect to $\mathbf{g}$: $\delta\mathbf{g}(x) = \omega(x)\mathbf{g}(x)$, $\omega(x)$ belonging to the Lie algebra, has the form

$$\delta F(\mathbf{A}^{\mathbf{g}}_\mu) = [\mathbf{M}(\mathbf{A}^{\mathbf{g}}_\mu)\omega](x), \quad \mathbf{M} = \partial_\mu \mathbf{D}_\mu\,.$$

One then introduces spinless fermions $\bar{\mathbf{C}}$ and $\mathbf{C}$, the Faddeev–Popov 'ghosts', and an auxiliary boson field λ all transforming under the adjoint representation.

One verifies the identity

$$1 = \int [\mathrm{d}\mathbf{g}\, \mathrm{d}\bar{\mathbf{C}}\, \mathrm{d}\mathbf{C}\, \mathrm{d}\lambda] \exp\left[-\mathcal{S}_{\text{gauge}}(\mathbf{A}^{\mathbf{g}}_\mu, \bar{\mathbf{C}}, \mathbf{C}, \lambda, \nu)\right]$$

with

$$\mathcal{S}_{\text{gauge}} = \int \mathrm{d}^d x \,\operatorname{tr}\left\{\lambda(x)\left[F(\mathbf{A}_\mu)(x) - \nu(x)\right] + \mathbf{C}(x)\mathbf{M}(\mathbf{A})\bar{\mathbf{C}}(x)\right\}.$$

This requires the introduction of *integration over Grassmann fields.*

Introducing the identity in the formal representation of the partition function, one obtains (in now Euclidean notation)

$$\mathcal{Z} = \int [\mathrm{d}\mathbf{g}\,\mathrm{d}\bar{\mathbf{C}}\,\mathrm{d}\mathbf{C}\,\mathrm{d}\lambda\,\mathrm{d}\mathbf{A}_\mu] \times \exp\left[\frac{1}{4g^2}\int \mathrm{d}^d x\ \mathrm{tr}\,\mathbf{F}^2_{\mu\nu}(x) - \mathcal{S}_{\text{gauge}}(\mathbf{A}^{\mathbf{g}}_\mu, \bar{\mathbf{C}}, \mathbf{C}, \lambda, \nu)\right].$$

After the change of variables $\mathbf{A}^{\mathbf{g}}_\mu \mapsto \mathbf{A}_\mu$, the integration over $\mathbf{g}(x)$ factorizes and yields an infinite multiplicative constant.

After a few additional simple manipulations and a Gaussian average over $\nu(x)$, one obtains the quantized partition function

$$\mathcal{Z} = \int \left[\mathrm{d}\mathbf{A}_\mu\,\mathrm{d}\bar{\mathbf{C}}\,\mathrm{d}\mathbf{C}\,\mathrm{d}\lambda\right] \exp\left[-\mathcal{S}(\mathbf{A}_\mu, \bar{\mathbf{C}}, \mathbf{C}, \lambda)\right],$$

where $\mathcal{S}$ is the *local action*:

$$\mathcal{S}(\mathbf{A}_\mu, \bar{\mathbf{C}}, \mathbf{C}, \lambda) = \int \mathrm{d}^d x\ \mathrm{tr}\left[-\frac{1}{4e^2}\mathbf{F}^2_{\mu\nu} + \frac{\xi e^2}{2}\lambda^2(x) + \lambda(x)\partial_\mu \mathbf{A}_\mu(x) + \mathbf{C}(x)\partial_\mu \mathbf{D}_\mu \bar{\mathbf{C}}(x)\right].$$

It was later noticed that this quantized action has a *fermion-like symmetry, the BRS symmetry.* Its generalization is *supersymmetry.*

4. Quantization of the Non-Linear σ-Model

The non-linear σ-model is a model with a global $O(N)$ symmetry acting on an N-component scalar field $\phi(x)$ that lives on the sphere S_{N-1}:

$$\phi^2(x) = 1\,.$$

In terms of ϕ, the action takes the form of a free field action,

$$\mathcal{S}(\phi) = \frac{1}{2g}\int \mathrm{d}^d x\ \left[\partial_\mu \phi(x)\right]^2,$$

but the constraint generates interactions. Within an expansion in powers of the coupling g, the $O(N)$ symmetry is realized in the phase of spontaneous symmetry breaking and the dynamical fields correspond to Goldstone (massless) modes. First leading order calculations seemed to indicate that the $O(N)$ symmetry was explicitly broken by perturbative corrections, the fields becoming massive. Within the canonical formulation, a complicated calculation showed that the breaking term actually cancelled.

By contrast, the field integral representation gave directly both the correct quantized form to all orders and the geometric explanation of the problem as due to *omitting the $O(N)$ invariant measure* (Meetz and Honerkamp)

$$\mathcal{Z} = \int [\mathrm{d}\phi] \prod_x \delta(\phi^2(x) - 1) \exp\left[-\mathcal{S}(\phi)\right].$$

Moreover, to give a meaning to the model beyond perturbation theory, one can introduce a *lattice regularization* and this yields an $O(N)$ lattice spin model. In this way, one can establish a *connection between the non-linear σ-model and the $(\phi^2)^2$ statistical field theory.*

5. Critical Phenomena

Following Wilson, it was realized that universal critical properties of a large class of statistical models could be described by an Euclidean quantum or statistical field theory. The main ingredient is the appearance of a *divergent correlation length* that allows passing to the continuum limit.

For example, the critical properties of the Ising model

$$\mathcal{Z} = \sum_{S_i = \pm 1} \exp\left[J \sum_{\text{n.n.}} S_i S_j\right]$$

are described by the ϕ^4 field theory,

$$\mathcal{Z} = \int [\mathrm{d}\phi] \exp\left[-\mathcal{S}(\phi)\right] \text{ with } \mathcal{S}(\phi) = \int \mathrm{d}^d x \left[\tfrac{1}{2}\left(\partial_\mu \phi\right)^2 + \tfrac{1}{2}\phi^2 + \frac{1}{4!} g\phi^4\right].$$

This generalizes to spin models with $O(N)$ symmetry. In addition path integral techniques allow proving directly that the $N = 0$ limit describes the *statistical properties of polymers* (or self-avoiding walk).

6. Classical and Quantum Statistical Physics

For a scalar field ϕ in *d space dimensions*, the *quantum partition function* reads

$$\mathcal{Z} = \int [\mathrm{d}\phi] \exp\left[-\int_0^\beta \mathrm{d}t \int \mathrm{d}^d x\, \mathcal{S}(\phi)\right]$$

with (t is the Euclidean or imaginary time)

$$\phi(0, x) = \phi(\beta, x).$$

This field integral representation immediately shows that the same partition function has also the interpretation of a *classical partition function in* $(d+1)$

space dimensions with periodic boundary conditions in the direction of finite size β.

This remark plays an important role in the *theory of continuous phase transitions*, relating *classical transitions in* $(d+1)$ *dimensions* and *quantum transitions at zero temperature in d dimensions.*

6.1. *Finite temperature QFT, finite-size effects and dimensional reduction*

The relation between classical and quantum statistical physics maps finite temperature quantum effects onto finite-size effects in the classical theory. This is most useful from the renormalization group viewpoint. Also, in this framework, high temperature is associated to dimensional reduction. Technically, one expands the periodic field over Fourier (Matsubara) modes

$$\phi(t,x)=\sum_{\nu\in\mathbb{Z}}\mathrm{e}^{i2\pi\nu t/\beta}\,\phi_\nu(x),$$

integrates perturbatively over non-zero modes, generating an effective action for the zero mode,

$$\mathcal{Z}=\int[\mathrm{d}\phi_0]\,\mathrm{e}^{-\mathcal{S}_{\mathrm{eff}}(\phi)}\ \text{ with }\ \mathrm{e}^{-\mathcal{S}_{\mathrm{eff.}}(\phi)}=\int\prod_{\nu\neq 0}[\mathrm{d}\phi_\nu]\,\mathrm{e}^{-\mathcal{S}(\phi)}\,.$$

As an example, the technique has been applied to the dilute (thus weakly interacting) Bose gas. The initial field integral over complex fields ψ^*,ψ *periodic in Euclidean time* reads

$$\mathcal{Z}=\int[\mathrm{d}\psi(t,x)\mathrm{d}\psi^*(t,x)]\,\mathrm{e}^{-\mathcal{S}(\psi^*,\psi)/\hbar}\,.$$

When one is interested only in long wavelength phenomena, the effective Euclidean action of the system may be written as (μ is the chemical potential)

$$\mathcal{S}(\psi^*,\psi)=-\int_0^\beta\mathrm{d}t\int\mathrm{d}^3x\left[\psi^*(t,x)\left(\hbar\frac{\partial}{\partial t}+\frac{\hbar^2}{2m}\nabla_x^2+\mu\right)\psi(t,x)\right.$$
$$\left.+\frac{2\pi\hbar^2 a}{m}\left(\psi^*(t,x)\psi(t,x)\right)^2\right].$$

The reduced partition function, at leading order (which means setting $\phi_\nu=0$ for $\nu\neq 0$) , takes the form of the field integral

$$\mathcal{Z}=\int[\mathrm{d}\phi(x)]\exp\left[-\mathcal{S}(\phi)\right]$$

with

$$\mathcal{S}(\phi) = \int \mathrm{d}^d x \left\{ \frac{1}{2} \left[\partial_\mu \phi(x)\right]^2 + \frac{1}{2} r \phi^2(x) + \frac{u}{4!} \left[\phi^2(x)\right]^2 \right\},$$

where $(\psi, \psi^*) \mapsto (\phi_1, \phi_2)$, $r = -2mT\mu$ and, for $d = 3$, $u = 96\pi^2 a/\lambda^2$.

The Euclidean action reduces to the ordinary $O(2)$ symmetric $(\phi^2)^2$ field theory, which also describes the *universal critical properties of the superfluid Helium transition.*

6.2. *Numerical simulations in quantum field theory*

The relation between classical and quantum partition function has led to the *application of statistical methods to the non-perturbative study of quantum field theories*, most notably QCD. The idea is to replace the continuum field integral by a lattice regularized form. It leads to a *non-perturbative definition of field theory.* Moreover, *non-perturbative numerical techniques* then become available, like strong coupling expansions or Monte Carlo type simulations. For pure QCD, they are based on Wilson's plaquette partition function (i, j, k, l are lattice sites):

$$\mathcal{Z} = \int \prod_{\text{links } (i,j)} \mathrm{d}\mathbf{U}_{ij}\, \mathrm{e}^{-\beta_p \mathcal{S}(\mathbf{U})}, \quad \mathcal{S}(\mathbf{U}) = -\sum_{\text{plaquettes}} \operatorname{tr} \mathbf{U}_{ij}\mathbf{U}_{jk}\mathbf{U}_{kl}\mathbf{U}_{li}\,,$$

where $\mathbf{U}_{ij}$ is a unitary matrix, group element associated to the link (i, j) and $\mathcal{S}$ the *plaquette action.*

7. Non-Perturbative Methods

7.1. *Large-N techniques*

In quantum field theories with $O(N)$ or $U(N)$ symmetries and fields in the vector representation, physical quantities can be calculated in the large-N limit, leading to non-perturbative results. At leading order, the same results can be obtained by summing Feynman diagrams, but field integral techniques are much simpler and can be extended to arbitrary orders in $1/N$. Applications include the study of the $(\phi^2)^2$ theory (and the calculation of critical exponents), the Gross–Neveu model, ...

The basic idea is to introduce into the ϕ-field integral the identity

$$1 = \int [\mathrm{d}\lambda \mathrm{d}\rho] \exp\left\{ i \int \mathrm{d}^d x\, \lambda(x) \left[\rho(x) - \phi^2(x)\right] \right\}.$$

For a recent review see Ref. 3.

7.2. *Instantons, vacuum instability and large-order behaviour in quantum field theory*

In simple quantum mechanics, barrier penetration effects can be evaluated in the semi-classical limit by WKB methods. Alternatively, they can be determined in the path-integral framework by looking for finite action solutions of Euclidean (imaginary time) equations of motion (*instantons*). However, the latter method generalizes simply to quantum field theory, unlike methods based on the Schrödinger equation. Important physics phenomena, like the periodic structure of QCD vacuum and the *strong CP problem*, the *solution of the $U(1)$ problem are related to instantons.*

Also, *instantons lead to a determination of the behaviour of the perturbative expansion at large orders.* An important application is the summation of the perturbative expansion, which has provided precise and reliable determination of critical exponents[4] as well as of the Ising scaling equation of state.

7.3. *Instantons and the problem of non-Borel summability*

In the case of *potentials with degenerate classical minima*, instanton calculus applied to the large order behaviour indicates that the perturbative expansion is *non-Borel summable*, that is, does not determine unique functions. In simple quantum mechanics with analytic potentials, the problem can be studied systematically and it can be shown that *all multi-instanton configurations must be taken into account and a generalized summation procedure introduced.*[5,6]

References

1. For references, see for example,
J. Zinn-Justin, *Quantum Field Theory and Critical Phenomena* (Clarendon Press, Oxford, 1989, 4th ed. 2002), and
2. J. Zinn-Justin, *Path Integrals in Quantum Mechanics* (Oxford Univ. Press, Oxford, 2004), Russian translation Fismatlit (Moscow, 2006).
3. M. Moshe and J. Zinn-Justin, *Quantum field theory in the large N limit: a review, Phys. Rept.* **385**, 69 (2003) [arXiv:hep-th/0306133].
4. R. Guida and J. Zinn-Justin, *J. Phys. A* **31**, 8103 (1998) [arXiv:cond-mat/9803240], an improvement over the results published in
J. C. Le Guillou and J. Zinn-Justin, *Phys. Rev. Lett.* **39**, 95 (1977); *Phys. Rev. B* **21**, 3976 (1980).
5. J. Zinn-Justin, *Nucl. Phys. B* **192**, 125–140 (1981), *ibid. B* **218**, 333–348 (1983).
6. U. D. Jentschura and J. Zinn-Justin, *J. Phys. A* **34**, L1–L6 (2001); *Ann. Phys. NY* **313**, 197 (2004).

NONUNIVERSAL FINITE-SIZE EFFECTS NEAR CRITICAL POINTS

V. DOHM

Institute for Theoretical Physics, RWTH Aachen University, D-52056 Aachen, Germany
E-mail: vdohm@physik.rwth-aachen.de

We study the finite-size critical behavior of the anisotropic φ^4 lattice model with periodic boundary conditions in a d-dimensional hypercubic geometry above, at, and below T_c. Our perturbation approach at fixed $d = 3$ yields excellent agreement with the Monte Carlo (MC) data for the finite-size amplitude of the free energy of the three-dimensional Ising model at T_c by Mon [Phys. Rev. Lett. **54**, 2671 (1985)]. Below T_c a minimum of the scaling function of the excess free energy is found. We predict a measurable dependence of this minimum on the anisotropy parameters. Our theory agrees quantitatively with the non-monotonic dependence of the Binder cumulant on the ferromagnetic next-nearest neighbor (NNN) coupling of the two-dimensional Ising model found by MC simulations of Selke and Shchur [J. Phys. A **38**, L739 (2005)]. Our theory also predicts a non-monotonic dependence for small values of the *anti-ferromagnetic* NNN coupling and the existence of a Lifshitz point at a larger value of this coupling. The tails of the large-L behavior at $T \neq T_c$ violate both finite-size scaling and universality even for isotropic systems as they depend on the bare four-point coupling of the φ^4 theory, on the cutoff procedure, and on subleading long-range interactions.

Keywords: Anisotropy; Excess free energy; Binder cumulant; Finite-size scaling; Universality; Monte Carlo simulation; Ising model.

1. Introduction

A major achievement of the renormalization-group (RG) theory is the proof that the *bulk* critical behavior of thermodynamic quantities has the property of scaling and two-scale factor universality.[1] This is summarized by the asymptotic (small reduced temperature $t = (T - T_c)/T_c$, small ordering field h) scaling form of the singular part of the bulk free energy density

$$f_{s,b}(t,h) = A_1 |t|^{d\nu} \, W_{\pm}(A_2 h |t|^{-\beta\delta}) \tag{1}$$

with universal critical exponents ν, β, δ and the universal scaling function $W_{\pm}(z)$ above (+) and below (−) T_c. Equation (1) is valid for both isotropic and anisotropic systems below $d = 4$ dimensions. For *confined* systems with a characteristic length L, two-scale factor universality is no longer valid if the system is spatially anisotropic as described by a $d \times d$ anisotropy matrix $\mathbf{A}$.[2,3] This is summarized by the asymptotic (large L, small t, small h) finite-size scaling form[4] of the singular part of the free energy density

$$f_s(t, h, L) \ = \ L^{-d} \, \mathcal{F}(C_1' t L'^{1/\nu}, C_2' h' L'^{\beta\delta/\nu}; \bar{\mathbf{A}}) \ , \tag{2}$$

with the finite-size scaling function $\mathcal{F}$ where $L' = L(\det \mathbf{A})^{-1/(2d)}$, $h' = h(\det \mathbf{A})^{1/4}$, and with the reduced anisotropy matrix $\bar{\mathbf{A}} = \mathbf{A}/(\det \mathbf{A})^{1/d}$, $\det \mathbf{A} > 0$. In the isotropic case, $\mathbf{A} = c_0 \mathbf{1}$ and $\bar{\mathbf{A}} = \mathbf{1}$. In addition to the two nonuniversal amplitudes C_1', C_2', the matrix $\bar{\mathbf{A}}$ contains up to $d(d+1)/2 - 1$ nonuniversal anisotropy parameters. As a consequence of (2), the critical amplitude $\mathcal{F}(0, 0; \bar{\mathbf{A}})$, the critical Binder cumulant[1,5]

$$U(\bar{\mathbf{A}}) \ = \frac{1}{3} \left[\frac{\partial^4 \mathcal{F}(0, y; \bar{\mathbf{A}})/\partial y^4}{(\partial^2 \mathcal{F}(0, y; \bar{\mathbf{A}})/\partial y^2)^2} \right]_{y=0} , \tag{3}$$

and the critical amplitude of the thermodynamic Casimir force are nonuniversal as well. In this contribution we report very recent results[4] on $\mathcal{F}(\tilde{x}, 0; \bar{\mathbf{A}})$ and $U(\bar{\mathbf{A}})$ based on RG finite-size perturbation theory at fixed $d = 3$. We shall also mention briefly nonuniversal effects on the *bulk* order-parameter correlation function and nonuniversal finite-size effects due to subleading long-range (van der Waals type) interactions in isotropic systems. This diversity of asymptotic critical behavior[4] suggests to distinguish subclasses of interactions within a universality class (see Fig. 1).

2. Anisotropic φ^4 Lattice Model

We consider the $O(n)$ symmetric φ^4 lattice Hamiltonian

$$H = \tilde{a}^d \left[\sum_{i=1}^{N} \left(\frac{r_0}{2} \varphi_i^2 + u_0 (\varphi_i^2)^2 - h \varphi_i \right) + \sum_{i,j=1}^{N} \frac{K_{i,j}}{2} (\varphi_i - \varphi_j)^2 \right] \tag{4}$$

on a simple-cubic lattice with lattice constant $\tilde{a}$ in a hypercubic geometry with volume $V = L^d$ and with periodic boundary conditions. For simplicity we assume $n = 1$. A variety of anisotropies may arise through the couplings $K_{i,j}$. They manifest themselves on macroscopic length scales via the $d \times d$ anisotropy matrix $\mathbf{A} = (A_{\alpha\beta})$ as given by the second moments[3]

$$A_{\alpha\beta} = A_{\beta\alpha} = N^{-1} \sum_{i,j=1}^{N} (x_{i\alpha} - x_{j\alpha})(x_{i\beta} - x_{j\beta}) K_{i,j} \, , \tag{5}$$

where $x_{i\alpha}$ are the components of the lattice points $\mathbf{x}_i$.

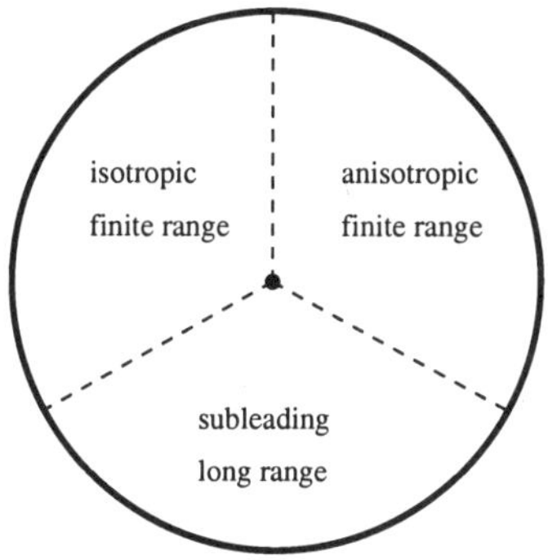

Fig. 1. Schematic representation of subclasses of systems with different types of interactions within a universality class. The subclasses have the same critical exponents and the same thermodynamic bulk scaling functions but, for given geometry and boundary conditions, different finite-size scaling functions, different bulk correlation functions, different bulk amplitude relations,[3] and a different number of nonuniversal parameters entering the asymptotic critical behavior.

As an example we consider isotropic NN couplings $K > 0$ and an anisotropic NNN coupling $J \neq 0$ in the $x - y$ planes, and an additional NN coupling $K_0 > 0$ in the z direction (see Fig. 2 (b)). The corresponding anisotropy matrix is

$$\mathbf{A} = 2\tilde{a}^2 \begin{pmatrix} K+J & J & 0 \\ J & K+J & 0 \\ 0 & 0 & K_0 \end{pmatrix} . \tag{6}$$

The matrix $\mathbf{A}$ enters the long-wavelength form of

$$\delta\widehat{K}(\mathbf{k}) = 2[\widehat{K}(\mathbf{k}) - \widehat{K}(\mathbf{0})] = \sum_{\alpha,\beta=1}^{d} A_{\alpha\beta}\, k_\alpha k_\beta + \; O(k^4) \tag{7}$$

where $\widehat{K}(\mathbf{k})$ is the Fourier transform of the interaction $K_{i,j}$

$$\widehat{K}(\mathbf{k}) \;=\; N^{-1} \sum_{i,j} e^{-i\mathbf{k}\cdot(\mathbf{x}_i - \mathbf{x}_j)} K(\mathbf{x}_i - \mathbf{x}_j)\,. \tag{8}$$

In perturbation theory, $r_0 + \delta\widehat{K}(\mathbf{k})$ plays the role of an inverse propagator.

A characteristic feature of spatial anisotropy with non-cubic symmetry is the fact that there exists no unique bulk (second-moment) correlation length $\xi_\pm$ above and below T_c but rather d different correlation lengths $\xi_\pm^{(\alpha)}$ in the directions $\alpha = 1, ..., d$ of the d principal axes. Such systems still have a single correlation-length exponent ν provided that $\det \mathbf{A} > 0$.

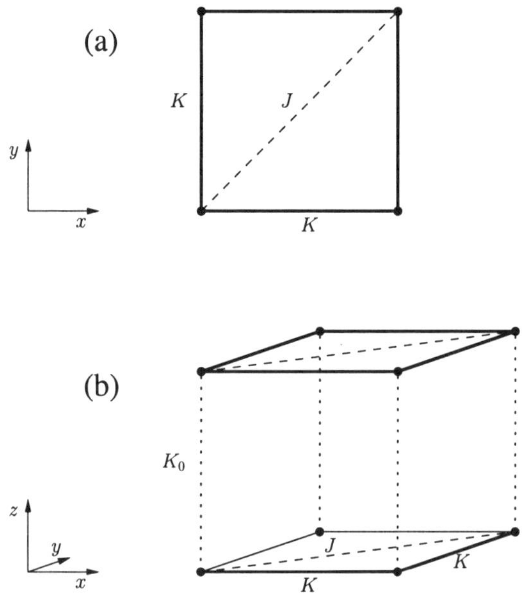

Fig. 2. Lattice points of (a) a square lattice, (b) a simple-cubic lattice, with isotropic NN couplings $K_x = K_y = K$ (solid lines), with an anisotropic NNN coupling J in the $x - y$ planes (dashed lines), and a NN coupling K_0 in the z-direction (dotted lines). The corresponding anisotropic matrix $\mathbf{A}$ is given by Eq. (6).

3. Perturbation Approach

There exist three different types of finite-size critical behavior of f_s: (a) an exponential L dependence for large $L/\xi_+^{(\alpha)} \gg 1$ at fixed temperature $T > T_c$, (b) the power-law behavior $\sim L^{-d}$ for large L at fixed $L/\xi_\pm^{(\alpha)}$,$0 \leq L/\xi_\pm^{(\alpha)} \lesssim O(1)$, above, at and below T_c, (c) an exponential L dependence for large $L/\xi_-^{(\alpha)} \gg 1$ at fixed temperature $T < T_c$. For the cases (a) and (c), ordinary perturbation theory with respect to u_0 is sufficient. For the case (b), a separation of the lowest mode and a perturbation treatment of the higher modes is necessary.[7–11] The case (b) corresponds to the central finite-size region above the dashed lines in Fig. 3. The cases (a) and (c) correspond to the regions below the dashed line. The latter regions are further divided[4] into a scaling region and a non-scaling region. The existence of the non-scaling region (shaded region in Fig. 3) has the following consequence. Unlike the bulk scaling function $W_\pm(z)$, (1), that is valid in the entire range $-\infty \leq z \leq \infty$ of the scaling argument z, the finite-size scaling functions such as $\mathcal{F}(x, y; \bar{\mathbf{A}})$ are valid only in a limited

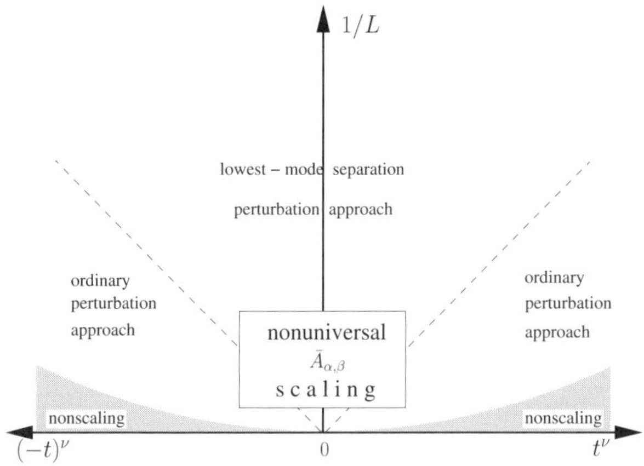

Fig. 3. Asymptotic part of the $L^{-1} - |t|^\nu$ plane at $h = 0$ for the anisotropic φ^4 theory in a cubic geometry with periodic boundary conditions. In the central finite-size region (above the dashed lines), the lowest mode must be separated whereas outside this region ordinary perturbation theory is applicable. Above the shaded region, finite-size scaling is valid but with scaling functions that depend on the anisotropy parameters $\bar{A}_{\alpha\beta}$. In the large - L regime at $t \neq 0$ (shaded region) finite-size scaling and universality are violated for both short-range and subleading long-range interactions and for both isotropic and anisotropic systems. A similar plot is valid for the $L^{-1} - h$ plane at $T = T_c$.

range of x and y, above the shaded region in Fig. 3. In the shaded region, nonuniversal nonscaling effects become nonnegligible and even dominant for sufficiently large $|x|$ and $|y|$ for both short-range and subleading long-range interactions. In this region not only the correlation lengths are relevant but also nonuniversal length scales such as the lattice spacing $\tilde{a}$,[14] the inverse cutoff Λ^{-1} of φ^4 field theory,[15] the length scale $u_0^{-1/\varepsilon}$ set by the four-point coupling,[4] and the van-der-Waals interaction-length $b^{1/(\sigma-2)}$ (see Section 4.3 below).[16,17] Furthermore the anisotropy parameters $\bar{A}_{\alpha\beta}$ are relevant in all regions. This diversity can be traced back[4,14] to a similar diversity of the large-distance ($r \gg \tilde{a}$) behavior of *bulk* correlation functions in the $r^{-1} - |t|^\nu$ plane corresponding to the $L^{-1} - |t|^\nu$ plane of Fig. 3.

It is appropriate to first transform H to a Hamiltonian H' such that the $O(k_\alpha k_\beta)$ terms of $\delta\widehat{K}(\mathbf{k})$ attain an isotropic form. This transformation consists of a rotation and rescaling of lengths in the direction of the principal axes.[3] This rescaling is equivalent to a shear transformation which distorts the geometry, the lattice structure, and the boundary conditions in a nonuniversal way. The advantage of the transformed system is that its bulk renormalizations are well known from the standard isotropic φ^4 field theory. Thus, in order to derive the scaling function $\mathcal{F}$, it is most appro-

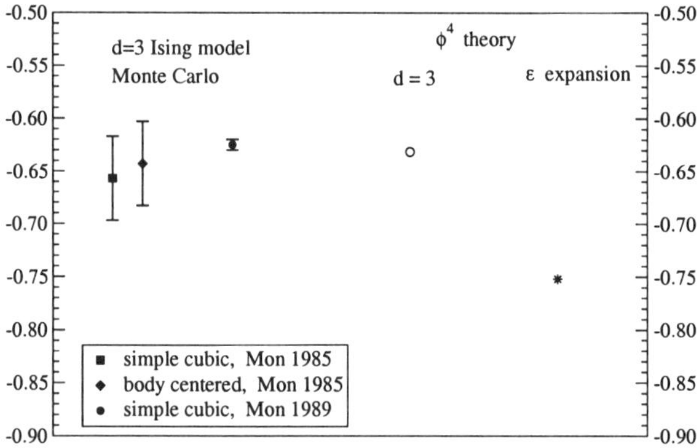

Fig. 4. Finite-size amplitude $\mathcal{F}(0,0;\mathbf{1})$ of the free energy density of isotropic systems in a cubic geometry at T_c in three dimensions. Theoretical prediction[4] at $d=3$ (open circle), and at $\varepsilon=1$ (star) of the ε expansion. MC data for the $d=3$ Ising model on sc and bcc lattices.[12,13]

priate to develop perturbation theory within the transformed system with the Hamiltonian H' for which a unique second-moment correlation length $\xi'_\pm$ is well defined.

4. Results

4.1. *Isotropic case*

For the purpose of calculating the finite-size free energy within the minimal renormalization scheme in three dimensions[6] we have further improved[4] the earlier finite-size perturbation approach[7–11] for the case of a one-component order parameter.

In order to test the reliability of our finite-size theory we first consider the isotropic case $K_0 = K, J = 0$, $\bar{\mathbf{A}} = \mathbf{1}$ where accurate MC data by Mon[12,13] for the $d=3$ Ising model are available. Our theoretical result for the finite-size amplitude at $h=0$ and $T=T_c$

$$\mathcal{F}(0,0;\mathbf{1})_{d=3} \;=\; -\,0.6315 \tag{9}$$

in excellent agreement with the MC results (see Fig. 4). The $\varepsilon = 4-d$ expansion result -0.7520 is in less good agreement.

We have also calculated the scaling function $\mathcal{F}^{ex}(\tilde{x},0;\mathbf{1})$ of the excess free energy density $f_s^{ex}(t,0,L) = f_s(t,0,L) - f_s(t,0,\infty)$ for the isotropic case. The scaling argument is $\tilde{x} = t(L/\xi_{0+})^{1/\nu}$ where ξ_{0+} is the asymptotic amplitude of the second-moment bulk correlation length above T_c. The

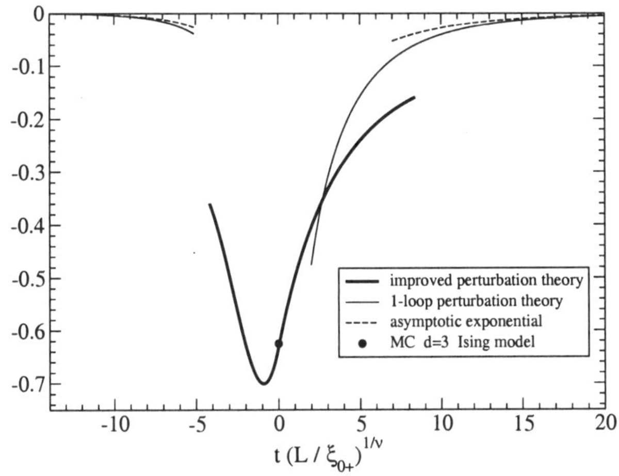

Fig. 5. Theoretical prediction of the scaling function $\mathcal{F}^{ex}(\tilde{x}, 0; \mathbf{1})$ of the excess free energy density of isotropic systems for $d = 3$ (thick solid line). MC result (full circle) for the Ising model on a sc lattice at $t = 0$.[13] No scaling function exists in the large - $|\tilde{x}|$ regions above and below T_c which are sensitive to all nonuniversal details of the model.

result is shown in Fig. 5 (thick solid line). The thin lines are the result of ordinary one-loop perturbation theory that breaks down at T_c.

4.2. *Anisotropic case*

Highly precise numerical information on the nonuniversal anisotropy effect on the critical Binder cumulant U of the anisotropic two-dimensional Ising model, see Fig.2 (a), has been provided recently by MC simulations of Selke and Shchur.[18] In order to mimic the two-dimensional anisotropy within our three-dimensional φ^4 lattice model we choose $K_0 = K + J$. To exhibit the *deviations* from isotropy and for the purpose of a comparison with the MC data[18] for the anisotropic two-dimensional Ising model we have plotted in Fig. 6 our theoretical result for the *difference* $U(\bar{\mathbf{A}}) - U(\mathbf{1})$ as a function of J/K together with the corresponding difference of the MC data.[18,19] We see that for positive J/K there is remarkable agreement.

The non-monotonicity for small *negative* values of J/K and the maximum at $J/K = -0.316$ predicted by our theory was not detected in the preliminary MC simulations by Selke and Shchur[18] who found a *monotonic decrease* of U when taking a weak antiferromagnetic coupling J.[20] We also predict the existence of a Lifshitz point near $J/K \lesssim -1/2$ with a wave vector instability in the $(1, 1, 0)$ direction. Very recently MC simulations have

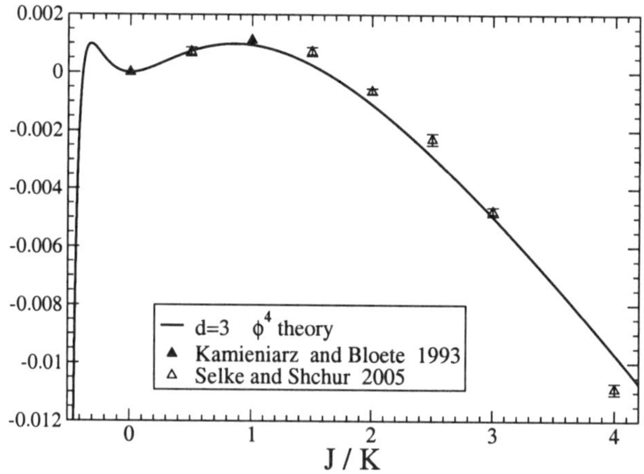

Fig. 6. Difference $U(\bar{\mathbf{A}}) - U(\mathbf{1})$ of the Binder cumulant plotted as a function of J/K together with the corresponding difference of the MC data[18,19] for the two-dimensional Ising model of Fig. 2(a).

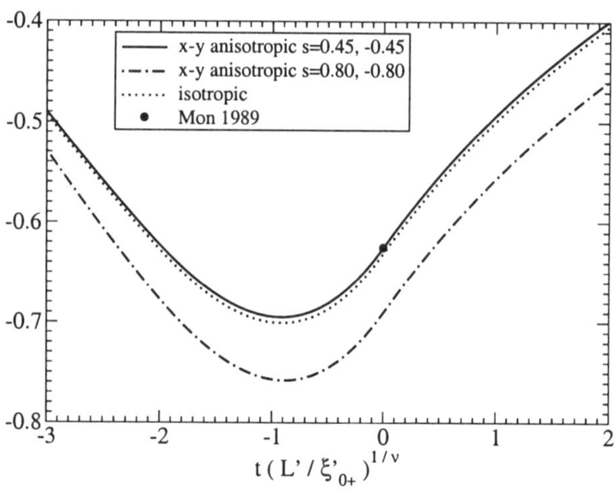

Fig. 7. Scaling function $\mathcal{F}^{ex}(\tilde{x}, 0; \bar{\mathbf{A}})$ of the excess free energy density of the anisotropic model with the anisotropy matrix (6) with $K_0 = K + J$ in a cubic geometry in three dimensions as a function of the scaling variable $\tilde{x} = t(L'/\xi'_{0+})^{1/\nu}$ for several values of the anisotropy parameter $s = (1 + K/J)^{-1}$ with $s = 0.45, -0.45$ (solid line), $s = 0.80, -0.80$ (dot-dashed line), $s = 0$ (dotted line, isotropic case). MC result (full circle) for the three-dimensional Ising model on a sc lattice.[13]

been started by Selke[21] in order to test these predictions in the regime of $J/K < 0$.

We have also calculated the nonuniversal anisotropy effect on the finite-size scaling function of $\mathcal{F}^{ex}$ for the anisotropy matrix (6) with $K_0 = K + J$ near the minimum below T_c as shown in Fig. 7 for several values of

$$s = (1 + K/J)^{-1}. \tag{10}$$

Our theory predicts the finite-size effects to depend on s^2 rather than s. It would be interesting to test this symmetry property by MC simulations. As shown in Fig. 7, the anisotropy effect for $s = \pm 0.80$ corresponding to $J/K = 4$ and $J/K = -4/9$ is far outside the error bars of the MC data by Mon[13] for the isotropic case and may be be detectable in future MC simulations.

4.3. *Subleading long-range interactions*

Finally we discuss the case of an isotropic subleading long-range interaction of the van der Waals type as defined by the long-wavelength form[16,17,22,23]

$$\delta\widehat{K}(\mathbf{k}) = \mathbf{k}^2 \ - b\,|\mathbf{k}|^\sigma \ + \ O(k^4) \tag{11}$$

with $2 < \sigma < 4$, $b > 0$. It was pointed out by Dantchev and Rudnick[23] that it affects the finite-size susceptibility in the regime $L/\xi_+ \gg 1$, similar to the effect caused by a sharp cutoff.[15] The effect of the interaction (11) on the excess free energy f_s^{ex} and on the critical Casimir force in the case of film geometry was first studied in Ref. 16,17. The asymptotic structure for $L/\xi_+ \gg 1$ in one-loop order above T_c at $h = 0$ is[16,17]

$$f_s^{ex}(t, 0, L) \ = L^{-d}\Big[\mathcal{F}^{ex}(L/\xi_+) + bL^{2-\sigma}\Psi(L/\xi_+)\Big]. \tag{12}$$

We have verified that, for $n = 1$, the same structure is valid also for cubic geometry with periodic boundary conditions above and below T_c where the function Ψ_{cube} has an algebraic large-L behavior $\sim (L/\xi_\pm)^{-2}$. The latter is dominant compared to the exponentially decaying scaling part $\mathcal{F}^{ex,\pm}$ in the shaded region of Fig. 3. This implies that, in this region, *two* nonuniversal length scales $b^{1/(\sigma-2)}$ and $\xi_\pm$ at $h = 0$ govern the *leading* singular part of the excess free energy density

$$f_s^{ex,\pm}(t, 0, L) \ \sim \ L^{-d} \ \left[\frac{b^{1/(\sigma-2)}}{L}\right]^{\sigma-2} \ \left[\frac{\xi_\pm}{L}\right]^2, \tag{13}$$

even arbitrarily close to criticality. In addition, there is a nonuniversal u_0 dependent exponential tail[4] of $\mathcal{F}^{ex,\pm}$. In (13), both the amplitude $\sim b$ and the power $-d - \sigma$ of the L dependence are nonuniversal. Thus, for isotropic systems with subleading long-range interactions, there exists no universal finite-size scaling form with only *one* reference length scale in the region $L/\xi_\pm \gg 1$ of the $L^{-1} - |t|^\nu$ plane although such systems are members of the same universality class as, e.g., Ising models with isotropic short-range interactions. The structure of (12) and (13) was confirmed and further studied by several authors.[24–26]

Acknowledgments

I thank B. Kastening and W. Selke for useful discussions. Support by DLR under Grant No. 50WM0443 is gratefully acknowledged.

References

1. V. Privman, A. Aharony, and P. C. Hohenberg, in *Phase Transitions and Critical Phenomena*, edited by C. Domb and J. L. Lebowitz (Academic, New York, 1991), Vol. 14, p. 1.
2. X. S. Chen and V. Dohm, *Phys. Rev. E* **70**, 056136 (2004).
3. V. Dohm, *J. Phys. A* **39**, L 259 (2006).
4. V. Dohm, submitted to *Phys. Rev. E* [arXiv:0801.4096].
5. K. Binder, *Z. Phys. B* **43**, 119 (1981).
6. V. Dohm, *Z. Phys. B* **60**, 61 (1985); R. Schloms and V. Dohm, *Nucl. Phys. B* **328**, 639 (1989).
7. J. Rudnick, H. Guo, and D. Jasnow, *J. Stat. Phys.* **41**, 353 (1985).
8. E. Brézin and J. Zinn-Justin, *Nucl. Phys. B* **25 [FS 14]**, 867 (1985).
9. A. Esser, V. Dohm, M. Hermes, and J. S. Wang, *Z. Phys. B* **97**, 205 (1995).
10. A. Esser, V. Dohm, and X. S. Chen, *Physica A* **222**, 355 (1995).
11. X. S. Chen, V. Dohm, and N. Schultka, *Phys. Rev. Lett.* **77**, 3641 (1996).
12. K. K. Mon, *Phys. Rev. Lett.* **54**, 2671 (1985).
13. K. K. Mon, *Phys. Rev. B* **39**, 467 (1989).
14. X. S. Chen and V. Dohm, *Eur. Phys. J. B* **15**, 283 (2000).
15. X. S. Chen and V. Dohm, *Eur. Phys. J. B* **10**, 687 (2000).
16. X. S. Chen and V. Dohm, *Phys. Rev. E* **66**, 016102 (2002).
17. X. S. Chen and V. Dohm, *Physica B* **329-333**, 202 (2003).
18. W. Selke and L. N. Shchur, *J. Phys. A* **38**, L 739 (2005).
19. G. Kamieniarz and H. W. J. Blöte, *J. Phys. A* **26**, 233 (1988).
20. We have been informed by W. Selke that the preliminary simulations[18] for $J < 0$ were performed at $J/K = -0.5$ and -0.75.
21. W. Selke, private communication.
22. B. Widom, *J. Chem. Phys.* **41**, 74 (1964); R. F. Kayser and H. J. Raveché, *Phys. Rev. A* **29**, 1013 (1984).
23. D. Dantchev and J. Rudnick, *Eur. Phys. J. B* **21**, 251 (2001).
24. D. Dantchev, M. Krech, and S. Dietrich, *Phys. Rev. E* **67**, 066120 (2003).
25. D. Dantchev, H. W. Diehl, and D. Grüneberg, *Phys. Rev. E* **73**, 016131 (2006).
26. D. Dantchev, F. Schlesener, and S. Dietrich, *Phys. Rev. E* **76**, 011121 (2007).

TIME SCALE RATIOS AND CRITICAL DYNAMICS

R. FOLK

Institute for Theoretical Physics, University of Linz,
A-4040 Linz, Austria
E-mail: reinhard.folk@jku.at
http://www.tphys.jku.at/group/folk/folk.html

An overview is given of recent results concerning systems described by a set of at least two slow dynamical variables. The simplest model contains a relaxing order parameter coupled to the energy density (model C). The effects induced by randomness in such a model are discussed. At the superconducting transition the gauge dependence of the critical dynamics is considered for a model of two coupled relaxation equations.

Keywords: Critical dynamics; Randomness; Superconducting transition.

1. Introduction

Near a phase transition one observes the phenomenon of critical slowing down for the order parameter (OP) dynamics. The time scale for reaching the equilibrium state increases when the critical point is approached. Thus the dynamics separates into slow and fast dynamic variables. A correct description of the critical dynamics has to take into account all slow variables besides the OP. These are the densities of conserved quantities (CD). The dynamic universality classes therefore depend on the structure of the system of these variables, namely on the number of CDs and the type of the coupling to the OP. These dynamic unversality classes have been reviewed by Hohenberg and Halperin.[1]

In principle each of the dynamics variables, which have to be taken into account has its own time scale but near the critical point in many cases it was observed that the time scales of all variables behave in the same way and this was the basis for the dynamic scaling hypothesis which characterized the critical dynamics by *one* dynamical critical exponent z defined by the dispersion $\omega_c(k) = Ak^z$ of the OP characteristic frequency at the critical temperature T_c, where k is the wave vector modulus and A a non universal amplitude setting a time scale. Later on it was recognized that there are cases where the critical dynamics cannot be described by only one critical time scale but the time scales of the OP and the CDs might be different. We shall review here some examples where this is the case.

2. Time Scale Ratio in a Simple Dynamic Model

The simplest example where time scale ratios (TSRs) can be studied is a system whose critical dynamics is defined by a nonconserved OP $\vec{\phi}_0$, described by a relaxation equation, coupled to one CD m_0, described by a diffusive equation.[2] The coupling of the OP and the CD is accomplished by a term in the static functional H, which enters the restoring force. This model was named[1] model C and reads explicitly

$$\frac{\partial \vec{\phi}_0}{\partial t} = -\overset{o}{\Gamma} \frac{\delta H}{\delta \vec{\phi}_0} + \vec{\theta}_\phi \qquad \frac{\partial m_0}{\partial t} = \overset{o}{\lambda} \nabla^2 \frac{\delta H}{\delta m_0} + \theta_m \,. \tag{1}$$

The stochastic forces fulfill Einstein relations which assure an approach of an equilibrium described by the static functional

$$H = \int d^d x \Bigg\{ \frac{1}{2} \overset{o}{\tau} (\vec{\phi}_0 \cdot \vec{\phi}_0) + \frac{1}{2} \sum_{i=1}^{n} \vec{\nabla} \phi_{i0} \cdot \vec{\nabla} \phi_{i0} + \frac{\overset{o}{\tilde{u}}}{4!} (\vec{\phi}_0 \cdot \vec{\phi}_0)^2 + \frac{1}{2} a_m m_0^2 + \frac{1}{2} \overset{o}{\gamma} m_0 (\vec{\phi}_0 \cdot \vec{\phi}_0) - \overset{o}{h}_m m_0 \Bigg\} \,. \tag{2}$$

The usual ϕ^4 theory has been extended by a Gaussian part for the CD and the asymmetric coupling $\overset{o}{\gamma}$ between the CD and the OP squared. The important dynamical parameter is the TSR $\overset{o}{w} = \overset{o}{\Gamma} / \overset{o}{\lambda}$

We applied the field-theoretic formalism[3] to this model and calculated the fixed point (FP) value of the TSR as function of space dimension d ($\epsilon = 4 - d$) and number of components n of the OP (the so called 'phase diagram') in two loop order.[5,6] It turns out that the FP value of the TSR might be (i) nonzero and finite, (ii) zero or (iii) infinite. Case (i) is the so called *strong scaling* FP, case (ii) the *weak scaling* FP and (iii) a FP with unclear scaling properties.[2,4] In the region of the (ϵ, n)-space, where the specific heat does not diverge, the FP value of the asymmetric coupling $\overset{o}{\gamma}$ is zero and the two equations decouple. Then the system belongs to the universality class of a simple relaxational model (model A). The CD then may be characterized by a dynamic exponent $z_{CD} = 2$. On the basis of the correct two loop field theoretic functions[5,6] one concludes that the FP

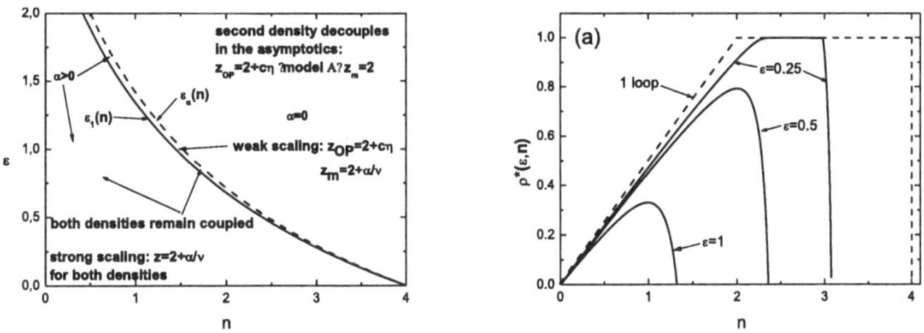

Fig. 1. (a) 'Phase diagram of model C'. (b) Time scale ratio of model C ($\rho^\star = w^\star/(1+w^\star)$).

of case (iii) does not exist (see Fig. 1 (a)). An infinite FP value of the TSR w (quantities without a super or subscript zero are renormalized) is suppressed by a logarithmic term $\ln w$ in the ζ-function for the relaxation rate Γ, however such a term leads to a nonanalytic dependence of the TSR in the limit where ϵ goes to zero ($d \to 4$).

3. Randomness and Time Scale Ratio

One may ask how defects and randomness influence the dynamic critical behavior of model C. Randomness can be induced by several effects, e.g. (i) bond disorder, (ii) site disorder, or (iii) anisoropic axis disorder. This is shown in the following spin-Hamiltonian

$$\mathcal{H} = -\frac{1}{2}\sum_{\mathbf{R},\mathbf{R}'} J(|\mathbf{R}-\mathbf{R}'|)c_{\mathbf{R}}c_{\mathbf{R}'}\vec{S}_{\mathbf{R}}\vec{S}_{\mathbf{R}'} - D_0\sum_{\mathbf{R}}(\hat{x}_{\mathbf{R}}\vec{S}_{\mathbf{R}})^2,$$

where the disorder is defined in case (i) by a distribution of the spin couplings J like $p(J) = \exp\left(-J^2/\Delta\right)$, in case (ii) by probability $p(c)$ of occupation $c = 1$ or vacancy $c = 0$, and in case (iii) by a non-isotropic distribution of the directions of the anisotropy axis $\hat{x}$ $p(\hat{x}) = \frac{1}{2m}\sum_{i=1}^{m}\left[\delta^{(m)}(\hat{x}-\hat{k}_i)+\delta^{(m)}(\hat{x}+\hat{k}_i)\right]$.

From static considerations formulated in the so called Harris criterion[7] one may argue in the following: If the pure system's *specific heat is diverging* then the critical exponents may be changed by disorder (this is the case for the examples given). The disordered system is then characterized by a non-diverging specific heat. Otherwise disorder remains in the universality class of the pure system. If the *specific heat is not diverging* the coupling γ of a

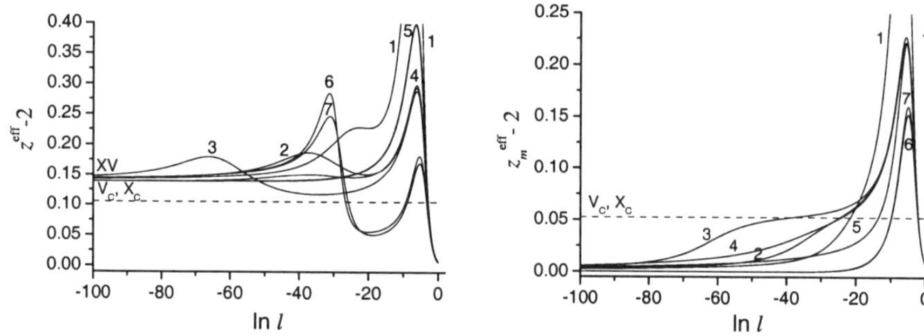

Fig. 2. Effective dynamical critical exponents in model C for a Heisenberg magnet with random anisotropy (from Ref. 11) (a) for the OP (b) for the CD.

CD to the OP goes to zero and model A applies (see Fig. 1(a) left to the dashed line). Therefore the coupling of a conserved density is in any case irrelevant. However this is only an argument which holds in the asymptotics. From statics one knows[8,9] that the static critical behavior observed might be an *effective* one, therefore one may consider also effective dynamical critical behavior.[10,11] Therefore *effective* dynamical critical exponents are defined by the field-theoretic function ζ_Γ of the kinetic coefficient Γ of the OP and ζ_m of the CD m

$$z^{\mathrm{eff}} = 2 + \zeta_\Gamma(\{u_i(\ell)\}, \gamma(\ell), w(\ell)), \qquad z_m^{\mathrm{eff}} = 2 + \zeta_m(\{u_i(\ell)\}, \gamma^2(\ell))\,. \quad (3)$$

They depend on the solution of the flow equations of the static model parameters $u_i(\ell)$, $\gamma(\ell)$, and the TSR $w(\ell)$. It turns out that including a CD leads to a new *small* dynamic transient exponent.[12] Thus nonasymptotic effects might be observable. In such a case the effective scaling of the OP and CD are in general different as shown in Fig. 2. Quite recently the asymptotic critical dynamics of model A for the Ising model has been studied by computer simulations[13] and it has been demonstrated that, as one expected, the dynamics of case (i) and (ii) belong to the same universality class.

4. Gauge Dependence of Time Scale Ratio

The static critical behavior of a superconductor is described by a complex OP $\vec{\psi}$ (generalized to $n/2$ components) and the gauge field $\vec{A}$ coupled to the OP by the minimal coupling. The corresponding static functional reads

$$\mathcal{H} = \int d^dx \Big\{ \frac{1}{2}\mathring{r}|\vec{\psi}_0|^2 + \frac{1}{2}\sum_{i=1}^{n/2} |(\boldsymbol{\nabla} - i\mathring{e}\boldsymbol{A}_0)\psi_{0,i}|^2$$

$$+ \frac{\mathring{u}}{4!}(|\vec{\psi}_0|^2)^2 + \frac{1}{2}(\nabla \times \boldsymbol{A}_0)^2 + \frac{1}{2\mathring{\varsigma}}(\nabla \cdot \boldsymbol{A}_0)\Big\} .$$

This coupling is due to the charge of the electrons building the condensating Cooper pairs. Renormalization group theory calculated a 'charged' FP describing the critical behavior of superconductors of the second kind. Experimental verification has been found by measuring the behavior of the penetration depth.[14] Moreover, RG theory predicts that nonmeasurable quantities may show a dependence on the gauge chosen in the calculation, whereas measurable quantities have to be gauge independent. Thus some of the static critical exponents like γ (OP susceptibility) or η (decay at T_c of the OP correlations) are gauge dependent whereas exponents like ν or α are gauge independent. The usual scaling laws are completely consistent with this behavior.[15]

Recently a dynamical model has been suggested for the critical dynamics of superconductors of the second kind[16] consisting of two coupled relaxational equations for the OP and the vector potential as follows

$$\frac{\partial \psi_{0,i}}{\partial t} = -2\mathring{\Gamma}_\psi \frac{\delta \mathcal{H}}{\delta \psi^+_{0,i}} + \theta_i\,, \qquad \frac{\partial A_{0,\alpha}}{\partial t} = -\mathring{\Gamma}_A \frac{\delta \mathcal{H}}{\delta A_{0,\alpha}} + \theta_\alpha .$$

The TSR is now defined as $w = \Gamma/\Gamma_A$. At the weak scaling FP (here $w^\star = \infty$) different time scales for the characteristic frequencies are obtained

$$\omega_\psi \sim k^{z_\psi} g_\psi(k\xi) \quad \omega_A \sim k^{z_A} g_A(k\xi) . \tag{4}$$

Here z_A is found to be gauge independent as it must be since it can be measured via the frequency dependent conductivity[17] $\sigma(k = 0, \omega) = \xi^{z_A - 2 + \eta_A} \mathcal{G}(\omega \xi^{z_A}) \sim \xi^{z_A + 2 - d}$ in the limit $\xi \to \infty$ with the exact result for the static exponent $\eta_A = 4 - d$ and a finite value of the scaling function $\mathcal{G}(\infty)$. But this FP is dynamically unstable. At the stable strong scaling FP ($w^\star$ finite and gauge dependent) $z = z_\psi = z_A = 2 + \frac{18}{n}\varepsilon - \varsigma\frac{6}{n}\frac{\varepsilon}{1+w^\star}$ thus z_A turns out to be gauge dependent.[18] In consequence the strong scaling FP of this model cannot describe the critical dynamics. Either the model does not apply or the stability of the FPs is changed in higher loop order.

5. Outlook

Two other longstanding problems (i) the dynamical critical behavior at the tricritical point in ^{3}He-^{4}He mixtures[21] and (ii) the dynamical critical scattering above the Neel temperature T_N of the three-dimensional Heisenberg antiferromagnet[22] are related to the FP value of the TSR w. In both cases the critical dynamics is described by a model more complicated than model

C, containing mode coupling terms in the dynamic equations. In (i) the FP value of one of the TSRs turned out to be infinite in a one loop calculation.[19] In two loop order this FP is absent and the mass diffusion is diverging contrary to measurement.[20,21] In the second case the FP value of the TSR between the kinetic coefficients of the staggered magnetization and the magnetization changes in two loop order[22] from roughly $w^\star = 3$ to $w^\star = 1$, which changes the dynamic shape function. This has to be taken into account in the comparison with experiment[23] since it changes the size of the additional elastic component observed in the critical scattering.[24]

Acknowledgment

This work was supported by the Fonds zur Förderung der wissenschaftlichen Forschung under project no P19583. I would like to thank M. Dudka, J. Holovatch, G. Moser and M. Weiretmayr for collaboration, in which the results I reported, have been obtained.

References

1. P. C. Hohenberg and B. I. Halperin, *Rev. Mod. Phys.* **49**, 435 (1977).
2. B. I. Halperin, P. C. Hohenberg, and S.-K. Ma, *Phys. Rev. B* **10**, 139 (1974).
3. R. Bausch, H. K. Janssen, and H. Wagner, *Z. Phys. B* **24**, 113 (1976).
4. E. Brezin and C. De Dominicis, *Phys. Rev. B* **12**, 4945 (1975).
5. R. Folk and G. Moser, *Phys. Rev. Lett.* **91**, 030601 (2003).
6. R. Folk and G. Moser, *Phys. Rev. E* **69**, 036101 (2004).
7. A. B. Harris, *J. Phys. C: Solid State Phys.* **7**, 1671 (1974).
8. A. Perumal *et al.*, *Phys. Rev. Lett.* **91**, 137202 (2003).
9. M. Dudka *et al.*, *J. Magn. Magn. Mater.* **256**, 243 (2003).
10. M. Dudka *et al.*, *J. Phys. A: Math. Gen.* **39**, 7943 (2006).
11. M. Dudka *et al.*, *J. Phys. A: Math. Gen.* **40**, 8247 (2007).
12. M. Dudka *et al.*, *Phys. Rev. B* **72**, 064417 (2005).
13. M. Hasenbusch, A. Pelissetto, and E. Vicari, arXiv: cond-mat/07094179.
14. T. Schneider, R. Khasanov, and H. Keller, *Phys. Rev. Lett.* **94**, 077002 (2005).
15. H. Kleinert and A. M. J. Schakel, *Phys. Rev. Lett.* **90**, 097001 (2003).
16. C. Lannert *et al.*, *Phys. Rev. Lett.* **92**, 097004 (2004).
17. D. S. Fisher, M. P. Fisher, and D. A. Huse, *Phys. Rev. B* **43**, 130 (1991).
18. M. Dudka *et al.*, *Cond. Mat. Phys. (Ukraine)* **10**, 189 (2007).
19. E. D. Siggia and D. R. Nelson, *Phys. Rev. B* **15**, 1427 (1977); L. Peliti, in *Lecture Notes in Physics*, edited by Ch. P. Enz (Springer, Berlin, 1979), Vol. **104**, p. 189.
20. R. P. Behringer and H. Meyer, *J. Low Temp. Phys.* **46**, 407 (1982).
21. R. Folk and G. Moser, *J. Low Temp. Phys.*, to be published.
22. R. Folk and G. Moser, *J. Phys. A: Math. Gen.* **39**, R207 (2006).
23. R. Coldea *et al.*, *Phys. Rev.* **57**, 5281 (1998).
24. M. Weiretmayr, R. Folk, and G. Moser, unpublished.

ON LOCAL SCALE INVARIANCE IN THE PHASE-ORDERING OF THE 2D DISORDERED ISING MODEL

M. HENKEL

Laboratoire de Physique des Matériaux CNRS UMR 7556,
Université Henri Poincaré Nancy I, B.P. 239,
F-54506 Vandœuvre lès Nancy Cedex, France
E-mail: henkel@lpm.u-nancy.fr

The phase-ordering kinetics of the ferromagnetic two-dimensional Ising model with uniform disorder is characterised by a dynamical exponent $z = 2 + \varepsilon/T$ which depends continuously on the disorder and on temperature. This allows for a detailed test of local scale invariance for several distinct values of z.

Keywords: Ageing phenomena; Local scaling; Disordered Ising model.

A ferromagnetic system quenched from an initially disordered state into its coexistence phase with at least two equivalent equilibrium states undergoes phase-ordering kinetics, driven by the surface tension between the ordered domains whose linear size grows as $L = L(t) \sim t^{1/z}$ where z is the dynamical exponent. For a non-conserved order-parameter it is well-known that $z = 2$.[1] Phase-ordering is one of the instances where *physical ageing* occurs, which may be defined by the properties: (i) slow (i.e. non-exponential) dynamics, (ii) breaking of time-translation invariance, and (iii) dynamical scaling. In the ageing regime $t, s \gg t_{\rm micro}$ and $t - s \gg t_{\rm micro}$ one expects the scaling behaviour of the two-time correlation and response functions

$$C(t,s;\boldsymbol{r}) = \langle\phi(t,\boldsymbol{r})\phi(s,\boldsymbol{0})\rangle = s^{-b} f_C\left(\frac{t}{s}, \frac{\boldsymbol{r}}{(t-s)^{1/z}}\right) \;,$$
$$R(t,s;\boldsymbol{r}) = \left.\frac{\delta\langle\phi(t,\boldsymbol{r})\rangle}{\delta h(s,\boldsymbol{0})}\right|_{h=0} = s^{-1-a} f_R\left(\frac{t}{s}, \frac{\boldsymbol{r}}{(t-s)^{1/z}}\right) \;, \tag{1}$$

where $\phi(t,\boldsymbol{r})$ is the space-time-dependent order-parameter, and $h(s,\boldsymbol{r})$ is the conjugate magnetic field.[1–6] Asymptotically, for $y \to \infty$, one expects $f_{C,R}(y;\boldsymbol{0}) \sim y^{-\lambda_{C,R}/z}$.

Recently, the question has been raised[7,8] whether standard dynamical scaling could be extended to a *local* form of dynamical scaling, in some sense analogous to conformal invariance in equilibrium critical phenomena. Such an extension appears to be possible for any given dynamical exponent z and leads, among others, to the following testable predictions[8,9] for the integrated response function $M_{\rm TRM}(t,s;\boldsymbol{r}) := h\int_0^s \mathcal{D}\tau\, R(t,\tau;\boldsymbol{r})$,

$$M_{\rm TRM}(t,s;\boldsymbol{r}) = r_0 M_{\rm age}\left(\frac{t}{s},\frac{\boldsymbol{r}}{s^{1/z}}\right) + r_1 t^{-\lambda_R/z}\mathcal{F}^{(\alpha,\beta)}\left(\frac{\boldsymbol{r}}{t^{1/z}}\right)\,, \qquad (2)$$

where $\mathcal{F}^{(\alpha,\beta)}(\boldsymbol{u}) = \int_{\mathbb{R}^d}\frac{\mathcal{D}\boldsymbol{k}}{(2\pi)^d}|\boldsymbol{k}|^\beta \exp\left(\mathrm{i}\boldsymbol{u}\cdot\boldsymbol{k} - \alpha|\boldsymbol{k}|^z\right)$, $\beta = \lambda_C - \lambda_R$ and r_0, r_1, α are free parameters. The scaling function reads

$$M_{\rm age}\left(\frac{t}{s},\frac{\boldsymbol{r}}{s^{1/z}}\right) = s^{-a} f_M\left(\frac{t}{s},\frac{\boldsymbol{r}}{s^{1/z}}\right)\,, \qquad (3)$$

$$f_M(y,\boldsymbol{w}) = \int_0^1 \mathcal{D}v\,(1-v)^{-1-a}\left(\frac{y}{1-v}\right)^{1+a'-\lambda_R/z} \times\left(\frac{y}{1-v}-1\right)^{-1-a'}\mathcal{F}^{(\alpha,\beta)}\left(\boldsymbol{w}\,(y-1+v)^{-1/z}\right). \qquad (4)$$

On the other hand, the autocorrelator is predicted to read,[8,9] for $T < T_c$

$$f_C(y,\boldsymbol{0}) = c_2 y^{(2\beta+d-\lambda_C)/z}\int_{\mathbb{R}^d}\frac{\mathcal{D}\boldsymbol{k}}{(2\pi)^d}|\boldsymbol{k}|^{2\beta}\exp\left(-\alpha|\boldsymbol{k}|^z(y-1)-\frac{\boldsymbol{k}^2}{4\nu}\right). \qquad (5)$$

Here one takes into account that the correct 'initial' correlation function is *not* delta-correlated, as one might naïvely suppose since simulations start from a fully disordered state, but rather should be identified with the correlator at those late stages[10] where ageing sets in, where the most simple expectation is a Gaussian form[11] $C(s+\tau,s;\boldsymbol{r}) \sim \exp(-\nu\boldsymbol{r}^2/s^{2/z})$ which will be sufficient if $z > 2$. The parameters c_2, ν must be found empirically.

The meaning of a test of these specific scaling functions can be understood as follows. Local scale invariance (LSI)[7,8] is built on the assumption that there is a *single* relevant time-dependent length scale $L(t) \sim t^{1/z}$ which then implies the scaling forms (1). Next, one assumes that the response functions transform co-variantly under a coordinate change $t \mapsto t/(\gamma t+\delta)$, $\boldsymbol{r} \mapsto \boldsymbol{r}(t)$ which can be shown to fix the autoresponse $M_{\rm TRM}(t,s;\boldsymbol{0})$. Furthermore, one requires that the *deterministic* part of the associated Langevin equation is invariant under a 'Galilei-transformation' generalised to $z \neq 2$. From this, extensions of the Bargman superselection rules follow from which one can prove, using the Janssen-de Dominicis functional, that the response functions actually found in concrete models are identical to the response functions calculated from the symmetries of the pure deterministic part. In addition, one has the concrete predictions (2)-(4). Finally, for $z \neq 2$, iterated Galilei-transformations lead to the conservation of higher powers of the momenta of the in- and outgoing particles in a $2n$-point response function. It is known from factorisable relativistic scattering that this implies a factorisable S-matrix.[12] A consequence of this factorisation is the explicit form (5) of the autocorrelator.

We present a test of the predictions (2)-(5) in a model for which the associated Langevin equation cannot be reduced to a linear equation. Consider a two-dimensional (2D) ferromagnetic Ising model with quenched disorder. The nearest-neighbour hamiltonian is given by[13]

$$\mathcal{H} = -\sum_{(i,j)} J_{ij}\sigma_i\sigma_j, \quad \sigma_i = \pm 1 . \tag{6}$$

The random variables J_{ij} are uniformly distributed over $[1-\varepsilon/2, 1+\varepsilon/2]$ where $0 \leq \varepsilon \leq 2$. There is a second-order phase transition at a critical temperature $T_c(\varepsilon) > 0$ between a paramagnetic and a ferromagnetic state. Using heat-bath dynamics with a non-conserved order-parameter and starting from a fully disordered initial state, phase-ordering occurs where the dynamical exponent is given by

$$z = z(T, \epsilon) = 2 + \epsilon/T . \tag{7}$$

This follows from phenomenological scaling arguments which assume that the disorder is creating defects with logarithmically distributed barrier heights parametrised by the constant ϵ.[13] The T-dependence is confirmed in field-theoretical studies in the Cardy-Ostlund model.[14] Simulations of the linear domain size[13] $L(t) \sim t^{1/z}$ and of the scaling of the autoresponse function[15] $R(t, s; \mathbf{0})$ in the disordered Ising model also confirm Eq. (7) and furthermore suggest the empirical identification $\epsilon = \varepsilon$.

Therefore, since z depends continuously on control parameters, the disordered 2D Ising model offers a nice possibility to test universality and especially to test local scale invariance for several values of z. Previously, tests of all three aspects of local scale invariance as mentioned above have been performed for the (non-linear) pure systems ($z = 2$) in the 2D/3D Ising model[16,17] and the 2D q-states Potts model[18] with $q = 2, 3, 8$.

Numerical data were obtained[9] on a 300^2 lattice using the standard heat-bath algorithm. From an uncorrelated initial state, the system's temperature was lowered at time $t = 0$ to the final temperature T, in the presence of a spatially random binary field $h = \pm 0.05$. For the autocorrelation function a lattice of size 600^2 was used. In the interpretation of

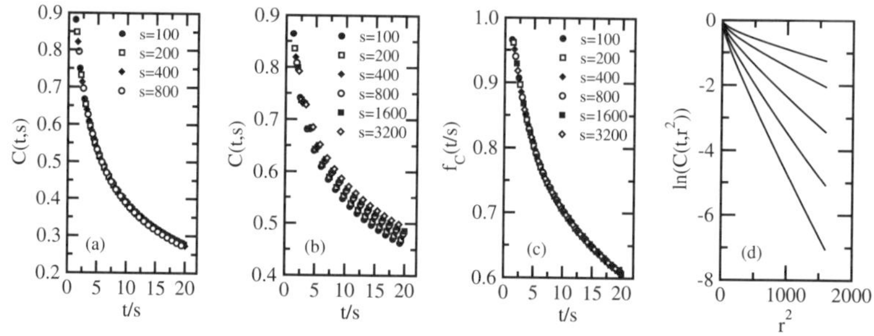

Fig. 1. The autocorrelation[9] as a function of t/s for (a) $\varepsilon = 0.5$ and $T = 1$, yielding $z = 2.5$, and (b) $\varepsilon = 2$ and $T = 1$, yielding $z = 4$. Whereas in case (a) we find the scaling behaviour of simple ageing, strong corrections to scaling behaviour are seen in case (b). Subtracting off the correction term, the scaling behaviour of simple ageing is recovered, as shown in panel (c). In (d) we show the equal-time space-dependent correlator $C(t, t; \boldsymbol{r})$ for $\varepsilon = 0.5$ and $T = 1.0$ for the times $t = [200, 300, 500, 1000, 2000]$ from bottom to top.

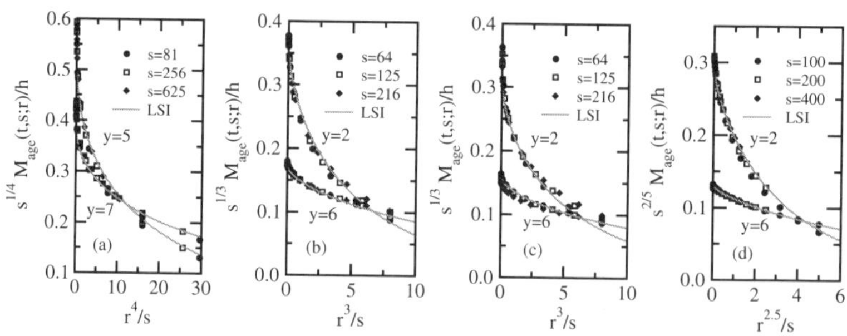

Fig. 2. Scaling behaviour of the ageing part of the integrated spatio-temporal response[9] $M_{\rm TRM}(t, s; \boldsymbol{r})$, for several waiting times s and the values (a) $\varepsilon = 2.0$, $T = 1.0$, (b) $\varepsilon = 1.0$, $T = 1.0$, (c) $\varepsilon = 0.5$, $T = 0.5$ and (d) $\varepsilon = 0.5$, $T = 1.0$, as a function of r^z/s, where z is given by (7) and for two fixed values of $y = t/s$. The full lines are the predictions of LSI, with parameters given in Table 1.

the raw data, it was already established for pure systems that finite-time corrections in $M_{\rm TRM}(t, s; \boldsymbol{r})$ must be subtracted carefully, see (2). Since in disordered magnets z may become considerably larger than in pure systems, see (7), the domain size $L(t) \sim t^{1/z}$ should grow slower and the observables should become more sensitive to finite-time corrections to scaling. Figure 1a shows that a good dynamical scaling is found for $\varepsilon = 0.5$ and $T = 1$, hence $z = 2.5$, while for $\varepsilon = 2$ and $T = 1$, yielding the large value $z = 4$, the expected simple scaling (1) does not seem to hold, see Fig. 1b. Taking the leading corrections to scaling of the form $C(t, s) = f_C(t/s) - s^{-b'} g_C(t/s)$ with $b' = 0.075$ into account, a good scaling is found, see Fig. 1c.

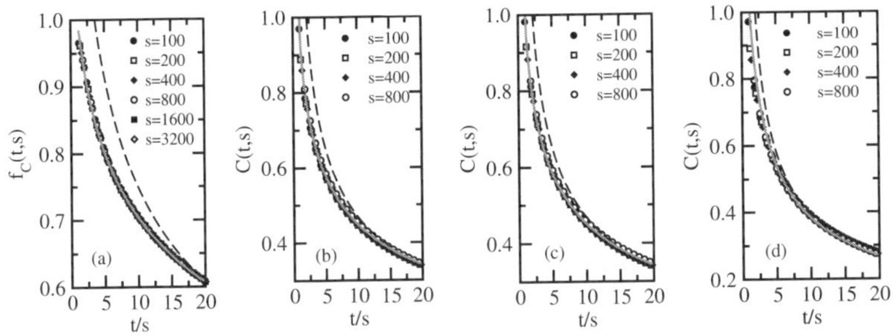

Fig. 3. Scaling behaviour of the autocorrelator $C(t, s)$,[9] for the values (a) $\varepsilon = 2.0$, $T = 1.0$, (b) $\varepsilon = 1.0$, $T = 1.0$, (c) $\varepsilon = 0.5$, $T = 0.5$ and (d) $\varepsilon = 0.5$, $T = 1.0$, as a function of $y = t/s$ and for several waiting times s. The full lines give the LSI prediction (5). The parameter values used are given in Table 2. In panels (b,c,d), $C(t, s) = f_C(t/s)$.

Table 1. Critical exponents and parameters of LSI for the different values of ε and T for the integrated space-time response function.[9]

ε	T	$a = a'$	λ_R/z	α	$\beta = \lambda_C - \lambda_R$	r_0	r_1
0.5	1.0	0.40(3)	0.61(1)	0.24(2)	−0.10(5)	0.0064(1)	0.0025(2)
0.5	0.5	0.33(5)	0.51(1)	0.20(2)	−0.06(7)	0.0049(1)	0.005(2)
1.0	1.0	0.33(5)	0.51(1)	0.20(1)	−0.06(7)	0.00575(2)	−0.02(1)
2.0	1.0	0.25(2)	0.33(1)	0.15(2)	−0.04(10)	0.0365(1)	−0.035(2)

In Figs. 2 and 3 we show the scaling functions for the ageing part of the spatial thermoremanent magnetisation $M_{\rm age}(t, s; \boldsymbol{r})$ and the autocorrelation function $C(t, s)$ for several combinations of T and ε. These extend and confirm our earlier results[6,15] of the scaling of the integrated autoresponse function $M_{\rm TRM}(t, s; \boldsymbol{0})$, notably on the value of z, see Eq. (7) and further strengthen the conclusions of a simple power-law scaling in the disordered Ising model reached earlier.[6,13,15] The suggestion of a 'superageing' behaviour,[19] motivated by the deviation from simple scaling as seen in Fig. 1b and by similar data in the site-disordered Ising model,[19] does not appear to be required (recall that at least one type of 'superageing' has been shown to be incompatible with basic constraints from probability[20]). The numerical data actually suggest a stronger universality in that the entire scaling functions appear to depend only on the *ratio* ε/T.[9]

Finally, we see that the data for both $M_{\rm TRM}(t, s; \boldsymbol{r})$ as well as for $C(t, s)$ are fully compatible with the predictions of local scale invariance Eqs. (2) and (5), respectively. The parameter values are collected in Tables 1 and 2. In the latter case, we point out that the time-evolved 'initial' correlator[11] has to be used, see also Fig. 1d which illustrates the Gaussian behaviour of

Table 2. Critical exponents and parameters of LSI for the different values of ε and T for the autocorrelation function.[9]

ε	T	z	λ_C/z	ν	c_2
0.5	1.0	2.5	0.570(5)	0.31(2)	1.72(1)
0.5	0.5	3.0	0.490(5)	0.35(2)	1.30(1)
1.0	1.0	3.0	0.490(5)	0.36(2)	1.30(1)
2.0	1.0	4.0	0.320(5)	0.48(2)	1.12(1)

$C(t, t; \boldsymbol{r})$. Remarkably, a single symmetry principle appears to reproduce the shapes of the scaling functions for values of the dynamical exponents which vary considerably. This is strong evidence that local scale invariance should capture the essence of the dynamical scaling behaviour of the integrated linear auto- and space-time responses, as well as for the autocorrelator.

Acknowledgments

It is a pleasure to thank M. Pleimling and F. Baumann for the fruitful collaboration whose results are described here and the organisers for their kind invitation to present them at PI07.

References

1. A. J. Bray, *Adv. Phys.* **43**, 357 (1994).
2. C. Godrèche and J.-M. Luck, *J. Phys. Cond. Matt.* **14**, 1589 (2002).
3. L. F. Cugliandolo, arXiv:cond-mat/0210312.
4. C. Chamon *et al.*, *J. Stat. Mech.* **P01006** (2006).
5. C. Chamon and L. F. Cugliandolo, *J. Stat. Mech.* **P07022** (2007).
6. M. Henkel and M. Pleimling, in *Rugged Free-Energy Landscapes*, ed. W. Janke (Springer, Heidelberg, 2008).
7. M. Henkel, *Nucl. Phys. B* **641**, 605 (2002).
8. F. Baumann and M. Henkel, *Nucl. Phys. B*, submitted.
9. F. Baumann, M. Henkel, and M. Pleimling, arXiv:0709.3228.
10. W. Zippold, R. Kühn, and H. Horner, *Eur. Phys. J. B* **13**, 531 (2000).
11. T. Ohta, D. Jasnow, and K. Kawasaki, *Phys. Rev. Lett.* **49**, 1223 (1982).
12. A. B. Zamolodchikov, *Adv. Stud. Pure Math.* **19**, 641 (1989).
13. R. Paul, S. Puri, and H. Rieger, *Europhys. Lett.* **68**, 881 (2004).
14. G. Schehr and P. Le Doussal, *Europhys. Lett.* **71**, 290 (2005); G. Schehr and H. Rieger, *Phys. Rev. B* **71**, 184202 (2005).
15. M. Henkel and M. Pleimling, *Europhys. Lett.* **76**, 561 (2006).
16. M. Henkel and M. Pleimling, *Phys. Rev. E* **68**, 065101(R) (2003).
17. M. Henkel, A. Picone, and M. Pleimling, *Europhys. Lett.* **68**, 191 (2004).
18. E. Lorenz and W. Janke, *Europhys. Lett.* **77**, 10003 (2007).
19. R. Paul, G. Schehr, and H. Rieger, *Phys. Rev. E* **75**, 030104 (2007).
20. J. Kurchan, *Phys. Rev. E* **66**, 017101 (2002).

CRITICAL CASIMIR FORCE SCALING FUNCTIONS OF THE MEAN SPHERICAL MODEL IN $2<d\leq 3$ DIMENSIONS FOR NONPERIODIC BOUNDARY CONDITIONS

B. KASTENING* and V. DOHM†

Institute for Theoretical Physics, Technische Hochschule Aachen, Physikzentrum, D-52056 Aachen, Germany
**E-mail: kastening@physik.rwth-aachen.de*
†E-mail: vdohm@physik.rwth-aachen.de

Finite-size effects are investigated in the mean spherical model in film geometry with nonperiodic boundary conditions above and below bulk T_c. We have obtained exact results for the excess free energy and the Casimir force for antiperiodic, Neumann, Dirichlet, and Neumann-Dirichlet mixed boundary conditions in $2 < d \leq 3$ dimensions. Analytic results are presented in $2 < d < 3$ dimensions for Dirichlet boundary conditions and for $d = 3$ for Neumann-Dirichlet boundary conditions. We find an unexpected leading size dependence $\propto C_{\pm}t/L^2$ of the Casimir force, with different amplitudes C_+ and C_- above and below T_c for large L at fixed $t \equiv (T - T_c)/T_c \neq 0$ for other than periodic boundary conditions.

Keywords: Mean spherical model; Exact solution; Free energy; Critical Casimir force; Finite-size scaling; Scaling function.

1. Introduction

Little is known about finite-size effects of critical systems below the bulk transition temperature T_c for realistic boundary conditions, such as Dirichlet or Neumann boundary conditions. Even for the exactly solvable mean spherical model[1,2] no finite-size investigation has been performed so far below T_c for nonperiodic boundary conditions. Previous studies of this model for the case of Dirichlet (or free) boundary conditions for $T \geq T_c$ have shown that finite-size scaling is violated in $d = 3$ dimensions[1,2] whereas finite-size scaling holds in $2 < d < 3$ dimensions.[2] Here we present exact results of the free energy and the Casimir force of this model above and below bulk T_c in film geometry with various nonperiodic boundary conditions in $2 < d \leq 3$ dimensions. The validity of finite-size scaling below three dimensions is confirmed and the Casimir force finite-size scaling functions for nonperiodic boundary conditions in $d = 2.5$ dimensions are graphically displayed. As unexpected results we find the validity of finite-size scaling in $d = 3$ dimensions for mixed Neumann-Dirichlet boundary conditions and a leading size dependence $\propto C_\pm t/L^2$ of the Casimir force, with different amplitudes C_+ and C_- above and below T_c, for large L at fixed $t \equiv (T - T_c)/T_c \neq 0$ for other than periodic boundary conditions.

2. Model

Consider a d-dimensional simple cubic lattice with lattice spacing a, $\mathcal{N} \equiv \tilde{N}^{d-1} \times N$ sites and volume $V = \tilde{L}^{d-1}L$, where $\tilde{L} \equiv \tilde{N}a$ and $L \equiv Na$. The Hamiltonian of the mean spherical model on this lattice is

$$\mathcal{H} = a^d \left[\frac{J}{2a^2} \sum_{\langle \boldsymbol{x}, \boldsymbol{x}' \rangle} (S_{\boldsymbol{x}} - S_{\boldsymbol{x}'})^2 + \frac{\mu}{2} \sum_{\boldsymbol{x}} S_{\boldsymbol{x}}^2 \right], \tag{1}$$

with $J > 0$ and where $\sum_{\langle \boldsymbol{x}, \boldsymbol{x}' \rangle}$ denotes a double sum over both primed an unprimed coordinates, where only nearest neighbors ($|\boldsymbol{x} - \boldsymbol{x}'| = a$) contribute. The fluctuations of the scalar spin variables $S_{\boldsymbol{x}}$ are subject to the constraint

$$a^{d-2} \sum_{\boldsymbol{x}} \langle S_{\boldsymbol{x}}^2 \rangle = \mathcal{N}, \tag{2}$$

implying that the "spherical field" μ is not an independent quantity but is a function of $\beta = 1/(k_B T)$ and of the geometry of the system.

As we are only interested in the film limit $\tilde{N} \to \infty$, we only need to specify the boundary conditions in the dth direction. After adding two fictitious sites x_0 and x_{N+1} in the negative and positive dth direction for each value of the remaining $d - 1$ coordinates (which we omit from the notation now), the various boundary conditions considered here are defined by

$$\begin{aligned}
&\text{p}: \text{ periodic}, && S_{x_{N+1}} = S_{x_1}, && && \text{(3a)}\\
&\text{a}: \text{ antiperiodic}, && S_{x_{N+1}} = -S_{x_1}, && && \text{(3b)}\\
&\text{NN}: \text{ Neumann-Neumann}, && S_{x_0} = S_{x_1}, && S_{x_{N+1}} = S_{x_N}, && \text{(3c)}\\
&\text{ND}: \text{ Neumann-Dirichlet}, && S_{x_0} = S_{x_1}, && S_{x_{N+1}} = 0, && \text{(3d)}\\
&\text{DD}: \text{ Dirichlet-Dirichlet}, && S_{x_0} = 0, && S_{x_{N+1}} = 0, && \text{(3e)}
\end{aligned}$$

the terminology being in analogy to the corresponding continuum model.

The dimensionless partition function Z and thermodynamic potential Φ are defined by

$$Z(T, \mu, L, \tilde{L}) = \exp[-\beta\Phi(T, \mu, L, \tilde{L})] = \prod_{\boldsymbol{x}} \int_{-\infty}^{+\infty} \frac{dS_{\boldsymbol{x}}}{a^{(2-d)/2}} \exp(-\beta\mathcal{H}). \quad (4)$$

The appropriate (Legendre transformed) reduced free-energy density is

$$f(t, L) = \lim_{\tilde{L}\to\infty} \frac{\beta}{\tilde{L}^{d-1}L} \left[\Phi - \mu \left(\frac{\phi\Phi}{\phi\mu}\right)_{T,L,\tilde{L}}\right], \quad (5)$$

with $\mu(t, L)$ determined by the constraint (2) for $\tilde{N} \to \infty$. For $2 < d \leq 3$ dimensions there is no phase transition at finite temperature for finite L, but there is a transition at a finite T_c in the bulk limit $L \to \infty$.

The excess free-energy density is $f^{\text{ex}}(t, L) = f(t, L) - f(t, \infty)$. We are interested in the thermodynamic Casimir force per unit area[3]

$$F(t, L) = -\frac{\partial[L f^{\text{ex}}(t, L)]}{\partial L}. \quad (6)$$

It is expected[3,4] that $f^{\text{ex}}(t, L)$ and $F(t, L)$ can be decomposed into singular and regular parts, $f^{\text{ex}} = f^{\text{ex}}_{\text{sing}} + f^{\text{ex}}_{\text{reg}}$ and $F = F_{\text{sing}} + F_{\text{reg}}$. If finite-size scaling holds, the singular parts have the asymptotic (large L, small $|t|$) scaling structure[3,4]

$$f^{\text{ex}}_{\text{sing}}(t, L) = L^{-d}\mathcal{F}(s), \qquad F_{\text{sing}}(t, L) = L^{-d}X(s), \quad (7)$$

with the scaling variable $s = t(L/\xi_0)^{1/\nu}$, $\nu = 1/(d-2)$. The nonuniversal reference length ξ_0 can be chosen as the asymptotic amplitude of the second-moment bulk correlation length $\xi = \xi_0 t^{-\nu}$ above T_c. In the presence of surface contributions it is appropriate to further decompose, for $|t| \neq 0$,

$$f^{\text{ex}}_{\text{sing}}(t, L) = f^{a}_{\text{surf,sing}}(t)L^{-1} + f^{b}_{\text{surf,sing}}(t)L^{-1} + L^{-d}\mathcal{G}(s), \quad (8)$$

where a and b denote the two surfaces of the film. The scaling function $X(s)$ is determined by $\mathcal{F}(s)$ and $\mathcal{G}(s)$ according to

$$\begin{aligned} X(s) &= (d-1)\mathcal{F}(s) - (d-2)s\mathcal{F}'(s) \\ &= (d-1)\mathcal{G}(s) - (d-2)s\mathcal{G}'(s). \end{aligned} \quad (9)$$

3. Results

As a cross check, we have successfully compared our expressions for the free energy for periodic boundary conditions with Ref. 5 for $d = 3$ and with Ref. 6 for $2 < d < 3$. As an example with nonperiodic boundary conditions, we provide here the exact result for $X(s)$ for Dirichlet-Dirichlet boundary conditions for $2 < d < 3$,

$$X(s) = -\frac{\Gamma(\frac{2-d}{2})}{2(4\pi)^{d/2}}\left[y_L^2 - \pi^2 - y_\infty^2\right]s - \frac{(d-1)\Gamma(\frac{-d}{2})}{2(4\pi)^{d/2}}\left[y_L^d - y_\infty^d\right]$$
$$- \frac{\Gamma(\frac{3-d}{2})}{2(4\pi)^{(d-1)/2}}y_L^{d-1} - \frac{\pi^2(d-1)\Gamma(\frac{2-d}{2})}{2(4\pi)^{d/2}}y_L^{d-2}$$
$$- \frac{d-1}{2^{d+1}\pi}\int_0^\infty dz\left(\frac{\pi}{z}\right)^{(d+1)/2} e^{-zy_L^2/\pi^2}\left\{e^z[K(z)-1]-\sqrt{\frac{\pi}{z}}+1-\sqrt{\pi z}\right\}, \tag{10}$$

with $y_\infty(s)$ defined by

$$y_\infty \equiv \begin{cases} s^{1/(d-2)} & s > 0, \\ 0 & s \le 0, \end{cases} \tag{11}$$

and with $y_L(s)$ determined by the constraint, which now reads

$$\Gamma(\tfrac{2-d}{2})(y_L^{d-2} - s) = \sqrt{\pi}\Gamma(\tfrac{3-d}{2})y_L^{d-3} - 2(4\pi)^{d/2}\mathcal{E}_d(y_L), \tag{12}$$

where

$$\mathcal{E}_d(y_L) \equiv \frac{1}{2^{d+1}\pi^2}\int_0^\infty dz\left(\frac{\pi}{z}\right)^{(d-1)/2} e^{-zy_L^2/\pi^2}\left\{e^z[K(z)-1]-\sqrt{\frac{\pi}{z}}+1\right\}, \tag{13}$$

and $K(z) \equiv \sum_{n=-\infty}^{\infty} e^{-n^2 z}$. Similar expressions result for the other boundary conditions of Eq. (3). The $d = 2.5$ results are displayed in Figs. 1(a)-(e) for all these boundary conditions.

For $d = 3$, both the reduced free-energy density and the Casimir force exhibit scaling violations for Neumann-Neumann and Dirichlet-Dirichlet boundary conditions, for example in the form of a logarithmic dependence on L/a at bulk T_c, as will be detailed elsewhere. We find, however, that scaling holds not only for periodic and antiperiodic, but also for mixed Neumann-Dirichlet boundary conditions in three dimensions. The reason for this unexpected behavior will be discussed elsewhere. The Casimir force scaling function for Neumann-Dirichlet boundary conditions reads

$$X_3(s) = -\frac{1}{4}E_3\bar{y}_L^2 + \frac{1}{8\pi}\left(\bar{y}_L^2 - y_\infty^2\right)s - \frac{1}{6\pi}\left(\bar{y}_L^3 - y_\infty^3\right)$$
$$- \frac{1}{8\pi}\left[\mathrm{Li}_3(-e^{-2\bar{y}_L}) + 2\bar{y}_L\mathrm{Li}_2(-e^{-2\bar{y}_L})\right], \tag{14}$$

where Li_2 and Li_3 are polylogarithms, with

$$E_3 \equiv \int_0^\infty dy B(y)^2\left[e^{-2y} - B(y)\right] \approx -0.237167, \tag{15}$$

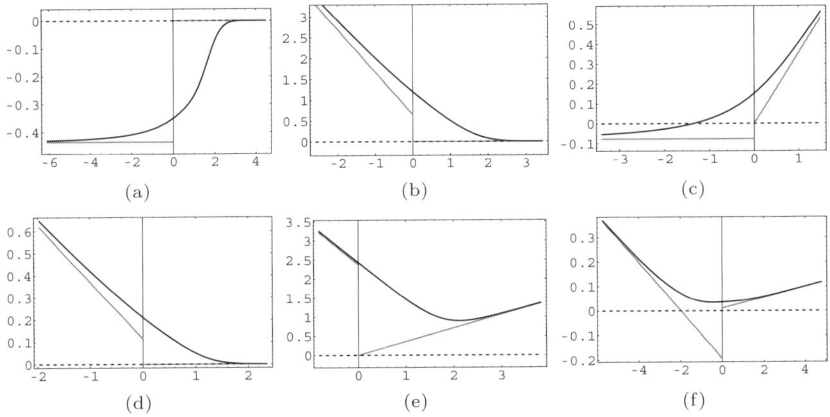

Fig. 1. (a)-(e): Scaling function $X(s)$ of the Casimir force for periodic (a), antiperiodic (b), Neumann-Neumann, (c), Neumann-Dirichlet (d), and Dirichlet-Dirichlet (e) boundary conditions for $d = 2.5$. The grey lines specify the asymptotic ($s \to \pm\infty$) behavior. (f): $X_3(s)$ for Neumann-Dirichlet boundary conditions for $d = 3$.

$y_\infty(s)$ from (11) and $\bar{y}_L(s)$ given by the constraint

$$\bar{y}_L \equiv \sqrt{y_L^2 - \tfrac{\pi^2}{4}} = \text{arcosh}\left(\tfrac{1}{2}e^{s-\pi E_3}\right) \tag{16}$$

for $y_L \geq \pi/2$ and appropriate analytic continuations to $0 < y_L < \pi/2$. $X_3(s)$ is shown in Fig. 1(f). $X_3(s)$ exhibits a linear asymptotic behavior not only for large negative s [as for $d = 2.5$, see Fig. 1(d)], but also for large positive s.

For other than periodic boundary conditions we find at fixed $|t| \neq 0$ for sufficiently large L that the Casimir force has the leading behavior $F \sim C_\pm t/L^2$ with different amplitudes C_+ and C_- above and below T_c, respectively. In Eqs. (10) and (14) we have included these terms in the singular part F_{sing} which then implies the linear asymptotic behavior of the scaling functions shown in Fig. 1(b)-(f) for large positive and/or negative s for the various nonperiodic boundary conditions. We note here, however, that this unexpected behavior cannot uniquely be attributed to the scaling functions because of an ambiguity in defining the regular part of the Casimir force, as will be discussed elsewhere.

Finally, a reservation must be made with regard to the exponential tails of the L-dependence of the Casimir force. As pointed out elsewhere,[6,7] these tails violate finite-size scaling and universality. They are not contained in Eqs. (10) and (14).

Acknowledgments

Support by DLR (German Aerospace Center) is gratefully acknowledged.

References

1. M. N. Barber and M. E. Fisher, *Annals of Physics (N.Y.)* **77**, 1 (1973).
2. X. S. Chen and V. Dohm, *Phys. Rev. E* **67**, 056127 (2003).
3. M. Krech, *The Casimir Effect in Critical Systems* (World Scientific, Singapore, 1994).
4. V. Privman, P. C. Hohenberg, and A. Aharony, in *Phase Transitions and Critical Phenomena*, Vol. 14, eds. C. Domb and J. L. Lebowitz (Academic, NY, 1991), p. 1.
5. D. Dantchev, *Phys. Rev. E* **53**, 2104 (1996).
6. X. S. Chen and V. Dohm, *Phys. Rev. E* **66**, 016102 (2002).
7. X. S. Chen and V. Dohm, *Eur. Phys. J. B* **10**, 687 (1999); V. Dohm, submitted to *Phys. Rev. E.*

PHASE DIAGRAM OF VORTICES IN HIGH-T_c SUPERCONDUCTORS

J. DIETEL*,‡ and H. KLEINERT*,†

*Institut für Theoretische Physik, Freie Universität Berlin, Arnimallee 14, D-14195 Berlin, Germany

†ICRANeT, Piazzale della Repubblica 1, 10-65122, Pescara, Italy

‡Email: dietel@physik.fu-berlin.de

The theory presented is based on a simple Hamiltonian for a vortex lattice in a weak impurity background which includes linear elasticity and plasticity, the latter in the form of integer valued fields accounting for defects. By using the variational approach of Mézard and Parisi established for random manifolds, we obtain the phase diagram including glass transition lines for superconductors with a melting line near H_{c2} like YBCO and also for superconductors with a melting line in the deep H_{c2} region like BSCCO.

Keywords: Phase diagram; Vortex lattice; Superconductors.

1. Introduction

We consider the phase diagram of high-T_c superconductors in the magnetic field vs temperature $H-T$-plane where H is the external magnetic field and T is the temperature. It is dominated by the interplay of thermal fluctuations and disorder.[1] At low magnetic fields near T_c the vortex solid melts into a vortex liquid (VL) via a first-order melting transition. Prominent examples of high-T_c superconductors exhibiting a solid-liquid melting are the anisotropic compound $YBa_2Cu_3O_{7-\delta}$ (YBCO), and the strongly layered compound $Bi_2Sr_2CaCu_2O_8$ (BSCCO). The main difference between these two superconductors comes from the fact that for YBCO the vortex cores are almost non-overlapping but for BSCCO we have large overlapping cores making the elasticity constants of both lattices different.[2]

When including weak pinning, the solid phase becomes a quasi-long-range ordered Bragg glass (BG).[1] At higher magnetic fields, the quasi-long-range order is destroyed and there exists also a vortex glass (VG) phase. The first-order melting line separates the BG phase from the VL phase at high temperatures and from the VG phase at lower temperatures. In the following, we will discuss the phase diagram for YBCO (square vortex lattice) explicitly where a more lengthy discussion can be found in Ref. 3. We will summarize our results for BSCCO (triangular lattice) only briefly at the end.

2. Model

The partition function used here for the vortex lattice without disorder was proposed in Ref. 4. It is motivated by similar melting models for two-dimensional square[5] crystals. The partition function of the disordered flux line lattice can be written in the canonical form as a functional integral

$$Z_{\mathrm{fl}} = \int \mathcal{D}[u_i, \sigma_{im}, n_i] e^{-(H_0[u_i,\sigma_{im},n_i]+H_{\mathrm{dis}}[u_i])/k_B T}, \tag{1}$$

where

$$\begin{aligned}\frac{H_0[u_i,\sigma_{im},n_i]}{k_B T} &= \sum_{\mathbf{x}} \frac{1}{2\beta}\Bigg[\sum_{i<j}\sigma_{ij}^2 + \frac{1}{2}\sum_i \sigma_{ii}^2 \\ &- \Big(\sum_i \frac{\overline{\nabla}_i}{\nabla_i}\sigma_{ii}\Big)\frac{c_{11}-2c_{66}}{4(c_{11}-c_{66})}\Big(\sum_i \frac{\overline{\nabla}_i}{\nabla_i}\sigma_{ii}\Big) + \sum_i \sigma_{i3}\frac{c_{66}}{c_{44}}\sigma_{i3}\Bigg] \\ &- 2\pi i \sum_{\mathbf{x}}\left(\sum_{i,m}\sigma_{im}\nabla_m u_i + \sum_{i\le j}\sigma_{ij}N_{ij}\right)\end{aligned} \tag{2}$$

is the canonical representation of elastic and plastic energies summed over the lattice sites $\mathbf{x}$ of a three-dimensional lattice, and σ_{ij} with $\sigma_{21} \equiv \sigma_{12}$ are stress fields which are canonically conjugate to the distortion fields.[5] The subscripts i, j have the values $1, 2$, and l, m, n run from 1 to 3. The parameter β is proportional to the inverse temperature, $\beta \equiv a^2 a_3 c_{66}/k_B T(2\pi)^2$, where a is the transverse distance of neighboring vortex lines, and a_3 is the persistence length along the dislocation lines introduced in Ref. 4. Note that a_3 is assumed to be independent on the disorder potential on the average. Its value is given by[4] $a_3 \approx 4a\sqrt{2}\,\lambda_{ab}/\lambda_c\sqrt{\pi}(1 - B/H_{c2})^{1/2}$. The volume of the fundamental cell v is equal to $a^2 a_3$ for the square lattice. The matrix $N_{ij}(\mathbf{x})$ in Eq. (2) is a discrete-valued local defect matrix composed of integer-valued defect gauge fields n_1, n_2. It depends on the lattice symmetry. For a square vortex lattice it is given by

$$N_{ij} = \begin{pmatrix} n_1 & n_2 \\ n_2 & -n_1 \end{pmatrix}. \tag{3}$$

The lattice derivatives ∇_m and their conjugate counterparts $\overline{\nabla}_m$ are the lattice differences for a cubic three-dimensional crystal. We have suppressed the spatial arguments of the elasticity parameters, which are functional matrices $c_{ij}(\mathbf{x}, \mathbf{x}') \equiv c_{ij}(\mathbf{x}-\mathbf{x}')$. Their precise forms were first calculated by Brandt[2] and generalized in Ref. 4 by taking into account thermal softening relevant for BSCCO. The second term in the exponent of (1)

$$H_{\text{dis}}[u_i] = \sum_{\mathbf{x}} V(\mathbf{x} + \mathbf{u}), \tag{4}$$

accounts for disorder. The measure of the functional integral is

$$\begin{aligned} \int \mathcal{D}[u_i, \sigma_{im}, n_i] = &\det \left[\frac{c_{66}}{4(c_{11}-c_{66})}\right]^{1/2} \det \left[\frac{1}{2\pi\beta}\right]^{5/2} \\ &\times \left\{ \prod_{\mathbf{x}} \left[\prod_{i \le m} \int_{-\infty}^{\infty} d\sigma_{im} \right] \left[\prod_j \sum_{n_j(\mathbf{x})=-\infty}^{\infty} \right] \left[\int_{-\infty}^{\infty} \frac{d\mathbf{u}}{a} \right] \right\}. \end{aligned} \tag{5}$$

The disorder potential $V(\mathbf{x})$ due to pinning is assumed to possess the Gaussian short-scale correlation function

$$\overline{V(\mathbf{x})V(\mathbf{x}')} = d(T)\, a_3\, \frac{\phi_0^4\, \xi_{ab}^3}{\tilde{\lambda}_{ab}^4} K(x_i - x_i')\, \delta_{x_3, x_3'}\,, \tag{6}$$

where $K(x_i - x_i') \approx 1/(\xi')^2$ for $|\mathbf{x} - \mathbf{x}'| < \xi'$, and is zero elsewhere. ϕ_0 is the magnetic flux quantum $\phi_0 = hc/2e$. The parameter ξ' is the correlation length of the impurity potential which is similar to the coherence length ξ_{ab} in the xy-plane. Furthermore, $\tilde{\lambda}_{ab} = \lambda_{ab}/(1 - B/H_{c2})$ is the screened penetration depth in the xy-plane.

The temperature dependence of the parameter $d(T)$ is mainly due to the temperature dependence of the correlation length and the pinning mechanism where we discuss here in the following only the δ_l-pinning mechanism[1] having its origin in fluctuations in the mean free path coming from fluctuations in the impurity density given by $d(T) = d_0(1-T/T_c)^{3/2}$. Furthermore, we carry out the explicit calculations with an effective disorder correlation function with the Fourier transform

$$\hat{K}(q) = 2\pi \exp(-\xi'^2 q_i^2/2) \tag{7}$$

leading also to an exponentially vanishing of the disorder correlation function in real space. The parameter ξ' in (7) is an effective correlation length which can also include for example screening effects of the impurities in the δ_l-pinning case.

3. Methods

The model has two mutual representations. One can be evaluated efficiently in the low-temperature phase, the other in the high-temperature phase. The lowest approximation to the former contains only elastic fluctuations of the vortex lattice without defects. The dual representation sums over all integer-valued stress configurations, which to lowest approximation are completely frozen out. The transverse part of the vortex fluctuations in the high-temperature approximation corresponds to non-interacting three-dimensional elastic strings where the length in z-direction is discretized with the persistence length as the lattice spacing.[4]

We will calculate the free energy in both temperature regimes non-perturbatively by using once the replica-trick[6] and furthermore the variational approach set up by Mézard and Parisi[7] for random manifolds and spin-glasses. It is based on replacing the non-quadratic part of the replicated Hamiltonian by a quadratic one, with possible mixing of replica fields. A transition line from a liquid to a glass consists within the Mézard-Parisi approach on a boundary in thermodynamical space from a replica symmetric quadratic Hamiltonian to a Hamiltonian which breaks the symmetry in the replica fields. This is the situation we found for the high-temperature liquid phase where the replica symmetric quadratic Hamiltonian corresponds to the VL phase and the replica symmetry broken Hamiltonian to the VG phase. The best quadratic Hamiltonian in the low-temperature solid phase is full replica symmetry broken corresponding to the BG phase. We will calculate the first-order melting line separating the BG from the VL and VG phases by intersecting the free energies of the low-temperature Hamiltonian and the high-temperature Hamiltonian.

4. Results

By using the methods outlined in the last section, we obtain the following disorder terms[3] for the free energies of the BG and the VG,VL phases

$$\Delta f_{\rm var}^{T\to 0} \approx \frac{k_B T}{2}\mathcal{D}(0)\left[1-\frac{3}{20}\mathcal{D}^4(0)(\mathcal{D}(0)A)^{-3}\right] \qquad \text{BG phase}, \tag{8}$$

$$\Delta f_{\rm var}^{T\to \infty} \approx \frac{k_B T}{2}\mathcal{D}(0)\left[1-(\mathcal{D}(0)A)^{-1/3}\right]^3 \Theta[\mathcal{D}(0)A-1] \ \text{VG}-\text{VLphase}$$

with the disorder constant $\mathcal{D}$ defined by

$$\mathcal{D}(2\langle u^2\rangle) = d(T)\frac{a_3}{(k_B T)^2}\frac{\phi_0^4\,\xi_{ab}^3}{\tilde{\lambda}_{ab}^4}\int \frac{d^2 q}{(2\pi)^2}\,\hat{K}(q)\,e^{-\frac{q^2}{2}\langle u^2\rangle}, \tag{9}$$

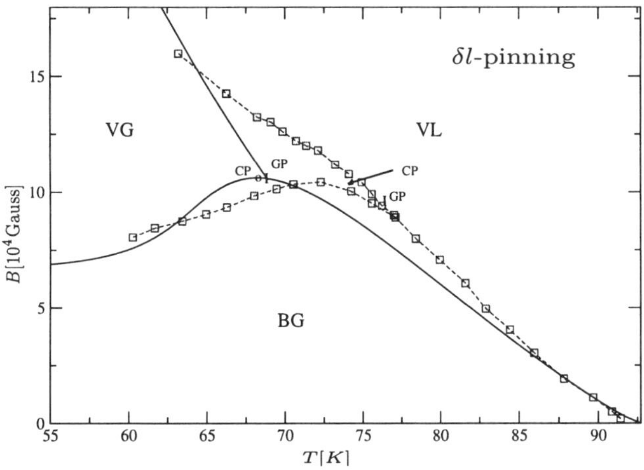

Fig. 1. Phase diagram for YBCO. Solid lines represent the theoretically determined phase transition lines between the various phases calculated for δl-pinning with $2\pi d_0\, \xi_{ab}^2/\xi'^2 = 1.32 \cdot 10^{-6}$ and $\xi_{ab}/\xi' = 1.49$. The solid curve separating the BG phase from the VG and VL phases represents the first-order melting line given by (12). The solid glass transition line between the VG and the VL phase was calculated from (11). Square points represent the experimentally determined phase diagram of Bouquet *et al.*[8]

and the parameter

$$A = \frac{4}{k_B T} \frac{c_{44} a^2\, \xi'^2}{a_3} . \tag{10}$$

We obtain from (8) a third-order glass transition line at

$$\mathcal{D}(0) A = 1 \tag{11}$$

separating the VL and the VG phase. Intersecting the free energies of the low and high-temperature phase we obtain the first-order transition line B_m separating the BG with the VL and VG phases (melting line)

$$B_m \approx \frac{\phi_0^5\, (1 - B_m/H_{c2})^3}{(k_B T)^2 \lambda_{ab}^2 \lambda_c^2} \frac{3.9 \cdot 10^{-5}}{\pi^4}\, e^{-(2/k_B T)(f_{\rm var}^{T\to 0} - f_{\rm var}^{T\to\infty})} . \tag{12}$$

We show in Fig. 1 the calculated phase diagram for YBCO with typical parameters specified in Ref. 3. We compare in the figure our theoretical results with experimental ones (square dots) of Bouquet *et al.*[8]

We now briefly summarize our results for BSCCO. Instead of the third-order line separating the VG and the VL phases, we find a second-order glass transition line separating both phases. This curve does not stop at the point

GP shown in Fig. 1 on the first-order melting line. Instead it goes further dividing the BG phase in two regions. This behaviour is in accordance to the experimentally determined glass transition lines of Beidenkopf *et. al.*[9] In the liquid phase VL, we obtain additionally an almost vertical third-order glass transition line starting near the critical temperature. The position of this line is in accordance to a magnetic anomaly found by Fuchs *et al.*[10] by measuring the vortex penetration through surface barriers.

References

1. G. Blatter, M. V. Feigel'man, V. Geshkenbein, A. Larkin, and V. M. Vinokur, *Rev. Mod. Phys.* **66**, 1125 (1994).
2. E. H. Brandt, *Rep. Prog. Phys.* **58**, 1465 (1995).
3. J. Dietel and H. Kleinert, *Phys. Rev. B* **75**, 144513 (2007).
4. J. Dietel and H. Kleinert, *Phys. Rev. B* **74**, 024515 (2006).
5. H. Kleinert, *Gauge Fields in Condensed Matter*, Vol. II *Stresses and Defects* (World Scientific, Singapore, 1989) (readable online at `www.physik.fu-berlin.de/~kleinert/re.html#b2`).
6. S. F. Edwards and P. W. Anderson, *J. Phys. France* **5**, 965 (1975).
7. M. Mézard and G. Parisi, *J. Phys. I* **1**, 809 (1991).
8. F. Bouquet, C. Marcenat, E. Steep, R. Calemczuk, W. K. Kwok, U. Welp, G. W. Crabtree, R. A. Fisher, N. E. Phillips, and A. Schilling, *Nature* **411**, 448 (2001).
9. H. Beidenkopf, N. Avraham, Y. Myasoedov, H. Shtrikman, E. Zeldov, B. Rosenstein, E. H. Brandt, and T. Tamegai, *Phys. Rev. Lett.* **95**, 257004 (2005).
10. D. T. Fuchs, E. Zeldov, T. Tamegai, S. Ooi, M. Rappaport, and H. Shtrikman, *Phys. Rev. Lett.* **80**, 4971 (1998).

FUNCTIONAL RENORMALIZATION GROUP IN THE BROKEN SYMMETRY PHASE

A. SINNER*, N. HASSELMANN†, and P. KOPIETZ♯

Institut für Theoretische Physik, Universität Frankfurt,
Max-von-Laue Straße 1, D-60438 Frankfurt, Germany
**E-mail: sinner@itp.uni-frankfurt.de*
†E-mail: hasselma@itp.uni-frankfurt.de
♯E-mail: pk@itp.uni-frankfurt.de
URL: http://www.itp.uni-frankfurt.de

We study the critical behavior of the classical φ^4-model approaching criticality from the broken symmetry phase using the functional renormalization group (RG). We derive and solve RG flow equations for the flowing order parameter and the coupling constant. We also calculate the scaling function for the momentum dependent self-energy at the critical temperature T_c.

Keywords: Field theory; Renormalization group; Critical phenomena.

The functional renormalization group (FRG) is a powerful tool for studying critical phenomena.[1,2] In this work we apply the general FRG formalism for systems in the broken symmetry phase developed in Ref. 3 to the Ising universality class at criticality.[4] We start with the action of the classical φ^4-model in D dimensions,

$$S[\varphi] = \int d^D r \left[\frac{1}{2}(\nabla\varphi)^2 + \frac{r_{\Lambda_0}}{2}\varphi^2 + \frac{u_{\Lambda_0}}{4!}\varphi^4 \right], \tag{1}$$

where Λ_0 is an UV cutoff. In the broken symmetry phase, the Fourier transformed fields $\varphi_{\mathbf{k}}$ have non-vanishing vacuum expectation values

$$\varphi_{\mathbf{k}} = \delta\varphi_{\mathbf{k}} + \varphi^0_{\mathbf{k},\Lambda}, \quad \langle \delta\varphi_{\mathbf{k}} \rangle = 0, \quad \varphi^0_{\mathbf{k},\Lambda} = (2\pi)^D \delta(\mathbf{k}) M_\Lambda . \tag{2}$$

Following the strategy explained in Ref. 3, we derive flow equations for the running magnetisation M_Λ and for the irreducible self-energy $\Sigma_\Lambda(\mathbf{k})$,

$$(\partial_\Lambda M_\Lambda)\, \Sigma_\Lambda(0) = -\frac{1}{2} \int_{\mathbf{k}} \dot{G}_\Lambda(\mathbf{k}) \Gamma^{(3)}_\Lambda(\mathbf{k}, -\mathbf{k}, 0), \tag{3}$$

$$\partial_\Lambda \Sigma_\Lambda(\mathbf{k}) = (\partial_\Lambda M_\Lambda)\Gamma_\Lambda^{(3)}(\mathbf{k},-\mathbf{k},0) - \frac{1}{2}\int_{\mathbf{k}'} \dot{G}_\Lambda(\mathbf{k}')\Gamma_\Lambda^{(4)}(\mathbf{k}',-\mathbf{k}';\mathbf{k},-\mathbf{k})$$
$$+ \int_{\mathbf{k}'} \dot{G}_\Lambda(\mathbf{k}')\, G_\Lambda(\mathbf{k}'+\mathbf{k})\,\Gamma_\Lambda^{(3)}(\mathbf{k},-\mathbf{k}-\mathbf{k}',\mathbf{k}')\Gamma_\Lambda^{(3)}(-\mathbf{k}',\mathbf{k}+\mathbf{k}',-\mathbf{k}), \quad (4)$$

where the cutoff dependent full propagator is defined via

$$G_\Lambda^{-1}(\mathbf{k}) = \mathbf{k}^2 + \Sigma_\Lambda(\mathbf{k}) + R_\Lambda(\mathbf{k}), \quad (5)$$

and the single-scale propagator is $\dot{G}_\Lambda(\mathbf{k}) = -\partial_\Lambda R_\Lambda(\mathbf{k})G_\Lambda^2(\mathbf{k})$. We use here an additive IR regulator[5]

$$R_\Lambda(\mathbf{k}) = (1-\delta_{\mathbf{k},0})Z_l^{-1}\left(\Lambda^2-\mathbf{k}^2\right)\Theta\left(\Lambda^2-\mathbf{k}^2\right), \quad (6)$$

where $Z_l^{-1} = 1 + \left.\frac{\partial \Sigma_\Lambda(\mathbf{k})}{\partial \mathbf{k}^2}\right|_{\mathbf{k}^2=0}$ is the wave function renormalization factor and $l = -\ln(\Lambda/\Lambda_0)$ is the logarithmic scale factor. Our aim is to calculate the true momentum dependent self-energy $\Sigma(\mathbf{k}) = \lim_{\Lambda\to 0}\Sigma_\Lambda(\mathbf{k})$.[6,7]

Our truncation procedure consists of replacing the three- and four-point vertex functions on the right-hand sides of Eqs. (3) and (4) by momentum independent constants, which we relate to the momentum independent part of the two-point function as in the local potential approximation (LPA)[1]

$$\Gamma_\Lambda[\varphi] \approx \frac{u_\Lambda}{4!}\int d^D r \left(\varphi^2(\mathbf{r}) - M_\Lambda^2\right)^2. \quad (7)$$

However, in contrast to the LPA, we retain the momentum dependence of the self energy in the full propagator of Eq. (5). Expressed in terms of dimensionless couplings u_l and M_l^2, Eqs. (3) and (4) yield,

$$\partial_l M_l^2 = (D-2+\eta_l)\,M_l^2 - s_l G_l^2(0), \quad (8)$$
$$\partial_l u_l = (4-D-2\eta_l)\,u_l - s_l u_l^2 G_l^3(0), \quad (9)$$
$$\partial_l \gamma_l(\mathbf{q}) = (2-\eta_l - \mathbf{q}\cdot\nabla_{\mathbf{q}})\,\gamma_l(\mathbf{q}) + \dot{\gamma}_l(\mathbf{q}), \quad (10)$$

where $\gamma_l(\mathbf{q})$ is a momentum dependent part of the rescaled self energy, $\dot{\gamma}_l(\mathbf{q})$ is the rescaled right-hand side of Eq. (4), $G_l^{-1}(0) = 1 + u_l M_l^2/3$, $s_l = 6\,(D+2-\eta_l)\,/\,(D\,(2+D))$, and $\eta_l = -\partial_l \ln Z_l$ is the anomalous dimension. In our approximation $\eta_l = u_l^2 M_l^2 G_l^4(0)/D$.

There are two fixed points: the Gaussian fixed point corresponding to the non-interacting system, and the Wilson-Fisher fixed point at finite interaction. In our approximation the Wilson-Fisher fixed point in $D=3$ is at $u_* = 0.942$ and $M_*^2 = 1.022$. To find the flow along the critical surface, we need to fine tune the initial values u_0 and M_0^2. A typical critical flow of the coupling constants M_l^2 and u_l is shown in the Fig. 1. The scale l_c where the perturbative behavior crosses over into the critical one can

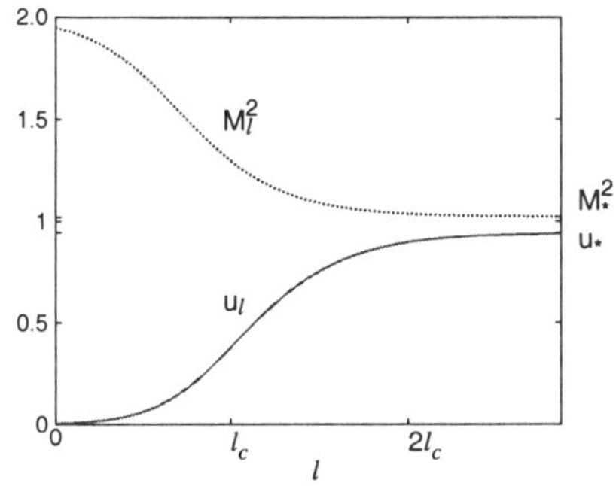

Fig. 1. The critical flow of the coupling constants u_l and M_l^2.

be determined from Eq. (9) assuming $G_l(0) \approx 0$ and $\eta_l \approx 0$; we obtain $l_c \approx (4-D)^{-1} \ln(u_*/u_0)$. The corresponding momentum scale $k_c = \Lambda_0 e^{-l_c}$ measures the size of the Ginzburg critical region. From the linearised flow around the Wilson-Fisher fixed point we obtain in $D = 3$ the critical exponents $\nu \approx 0.553$ and $\eta \approx 0.099$. The comparison of our results to the established values[1,7] shows the limitation of our simple approximation. However, the accuracy of critical exponents can be improved easily by taking higher orders of LPA into account.[8]

While in general the self-energy exhibits dependence on several characteristic length scales, at criticality its momentum dependent part can be parametrised in terms of a one-parameter scaling function[6]

$$\Delta\Sigma(\mathbf{k}) = \Sigma(\mathbf{k}) - \Sigma(0) = k_c^2 \Delta\sigma(k/k_c), \tag{11}$$

where

$$\Delta\sigma(x) = x^2 \int_{xe^{-l_c}}^{\infty} dq\, q^{-3} Z^{-1}_{l_c+\ln(q/x)} \dot{\gamma}_{l_c+\ln(q/x)}(q). \tag{12}$$

For $k \ll k_c$ we can replace the lower limit in Eq. (12) by zero and all flowing quantities by their fixed point values. It is then straightforward to show that $\Delta\sigma_*(x) \sim x^{2-\eta}$. In the opposite limit $k \gg k_c$ our approach reproduces correctly leading order perturbative theory. In Fig. 2 we show the momentum dependent effective anomalous dimension

$$\eta_{\text{eff}}(k) = 2 - \frac{\partial \ln \Delta\sigma(k/k_c)}{\partial \ln k}, \tag{13}$$

which reveals very clearly the crossover region between two regimes.

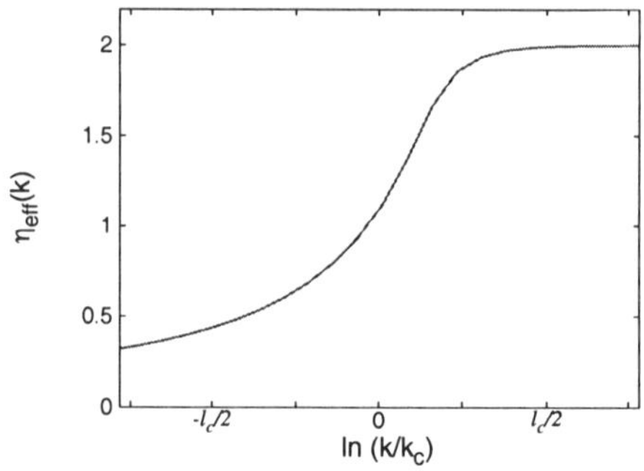

Fig. 2. The effective anomalous dimension.

It is interesting to calculate the quantity

$$c_1 = -\frac{\alpha}{u_{\Lambda_0}} \int \frac{d^3k}{(2\pi)^3} \frac{\Delta\Sigma(\mathbf{k})}{\mathbf{k}^2 (\mathbf{k}^2 + \Delta\Sigma(\mathbf{k}))}. \tag{14}$$

For two-component fields a similar expression is related to the interaction induced shift in the critical temperature of a weakly interacting Bose gas.[9] Here $\alpha = -256\pi^3\zeta(3/2)^{-4/3} \approx -2206.19$. Within our simple truncation we obtain $c_1 \approx 0.705$, which is comparable with the field-theoretical result $c_1 \approx 1.07$ (see Ref. 10) and FRG calculation with more sophisticated truncation yielding $c_1 \approx 1.11$ (see Ref. 7).

References

1. J. Berges, N. Tetradis, and C. Wetterich, *Phys. Rep.* **363**, 223 (2002).
2. C. Wetterich, *Phys. Lett. B* **301**, 90 (1993); T. R. Morris, *Int. J. Mod. Phys. A* **9**, 2411 (1994).
3. F. Schütz, L. Bartosch, and P. Kopietz, *Phys. Rev. B* **72**, 035107 (2005); F. Schütz and P. Kopietz, *J. Phys. A: Math. Gen.* **39**, 8205 (2006).
4. N. Hasselmann, A. Sinner, and P. Kopietz, *Phys. Rev. E* **76**, 040101(R) (2007); A. Sinner, N. Hasselmann, and P. Kopietz, arXiv:0707.4110 (cond-mat.stat-mech).
5. D. F. Litim, *Phys. Rev. D* **64**, 105007 (2001).
6. S. Ledowski, N. Hasselmann, and P. Kopietz, *Phys. Rev. A* **69**, 061601(R) (2004); N. Hasselmann, S. Ledowski, and P. Kopietz, *ibid.* **70**, 063621 (2004).
7. J.-P. Blaizot, R. Méndez-Galain, and N. Wschebor, *Phys. Rev. E* **74**, 051116 and 051117 (2006).
8. L. Canet, B. Delamotte, D. Mouhanna, and J. Vidal, *Phys. Rev. D* **67**, 065004 (2003).
9. G. Baym, J.-P. Blaizot, M. Holtzmann, F. Laloe, and D. Vautherin, *Eur. Phys. J. B* **24**, 107 (2001).
10. B. Kastening, *Phys. Rev. A* **68**, 061601(R) (2003); *ibid.* **69**, 043613 (2004).

PERTURBATIVE RESULTS WITHOUT DIAGRAMS

R. ROSENFELDER

Particle Theory Group, Laboratory for Particle Physics, Paul Scherrer Institut, CH-5232 Villigen PSI, Switzerland
E-mail: roland.rosenfelder@psi.ch

Higher-order perturbative calculations in Quantum (Field) Theory suffer from the factorial increase of the number of individual diagrams. Here I describe an approach which evaluates the total contribution numerically for finite temperature from the cumulant expansion of the corresponding observable followed by an extrapolation to zero temperature. This method (originally proposed by Bogolyubov and Plechko) is applied to the calculation of higher-order terms for the ground-state energy of the polaron. Using state-of-the-art multidimensional integration routines two new coefficients are obtained corresponding to a 4- and 5-loop calculation.

Keywords: High-order perturbative calculations; Cumulant expansion; Monte Carlo integration.

1. Introduction

Highly accurate measurements require precise theoretical calculations which perturbation theory can yield if the coupling constant is small. However, in Quantum Field Theory (QFT) the number of diagrams grows factorially with the order of perturbation theory and they become more and more complicated. The prime example is the anomalous magnetic moment of the electron where new experiments[1] need high-order quantum-electrodynamical calculations but the number of diagrams for them "explodes" as shown by the generating function[2]

$$\Gamma(\alpha) = 1 + \alpha + 7\,\alpha^2 + 72\,\alpha^3 + 891\,\alpha^4 + 12672\,\alpha^5 + 202770\,\alpha^6 + \ldots . \quad (1)$$

There are ongoing efforts[3] to calculate all 12672 diagrams in $\mathcal{O}(\alpha^5)$ – a huge, heroic effort considering the complexity of individual diagrams and the large cancellations among them.

Obviously new and more efficient methods would be most welcome for a cross-check or further progress.

2. A New Method (Applied to the Polaron Ground-State Energy)

Here I present a "new" method which – as I learned during the conference – was already proposed 20 years by Bogolyubov (Jr.) and Plechko (BP) (Ref. 4). However, to my knowledge it has been never applied numerically which turned out to be quite a challenging task.

The BP method is formulated for the polaron problem, a non-relativistic (but non-trivial) field theory describing an electron slowly moving through a polarizable crystal. Due to medium effects its energy is changed and it acquires an effective mass: $E_{\mathbf{p}} = E_0 + \mathbf{p}^2/(2m^\star) + \ldots$. The aim is to calculate the power-series expansion for the ground-state (g.s.) energy

$$E_0(\alpha) =: \sum_{n=1} e_n \, \alpha^n \tag{2}$$

as function of the dimensionless electron-phonon coupling constant α. The lowest-order coefficients are well-known[5] ($e_1 = -1\,,\ e_2 = -0.01591962$), but since Smondyrev's calculation[6] in 1986

$$e_3 = -0.00080607 \tag{3}$$

there has been no progress towards higher-order terms.

This will be remedied by the first numerical application of the BP method. For this purpose the path-integral formulation of the polaron problem will be used where the phonons have been integrated out exactly.[7] For large Euclidean times β this gives the following effective action

$$S_{\text{eff}}[\mathbf{x}] = \int_0^\beta dt \frac{1}{2}\dot{\mathbf{x}}^2 - \frac{\alpha}{\sqrt{2}} \int_0^\beta dt \int_0^t dt' \, e^{-(t-t')} \int \frac{d^3k}{2\pi^2} \frac{\exp\left[i\mathbf{k}\cdot(\mathbf{x}(t)-\mathbf{x}(t'))\right]}{\mathbf{k}^2}\,, \tag{4}$$

which will be split into a free part S_0 and an interaction term S_1. The g.s. energy may be obtained from the partition function

$$Z(\beta) = \oint \mathcal{D}^3x \, e^{-S_{\text{eff}}[\mathbf{x}]} \stackrel{\beta\to\infty}{\longrightarrow} e^{-\beta E_0} \tag{5}$$

at asymptotic values of β, i.e. zero temperature. The central idea is to use the *cumulant expansion* of the partition function

$$Z(\beta) = Z_0 \exp\left[\sum_{n=1} \frac{(-)^n}{n!} \lambda_n(\beta)\right] \tag{6}$$

where the $\lambda_n(\beta)$'s are the cumulants w.r.t. S_1. These are obtained from the *moments*

$$m_n := \mathcal{N} \oint \mathcal{D}^3x \, \left(S_1[x] \right)^n \, e^{-S_0[x]} \;, \quad m_0 = 1 \tag{7}$$

by the recursion relation (see, e.g. Eq. (51) in Ref. 8)

$$\lambda_{n+1} = m_{n+1} - \sum_{k=0}^{n-1} \binom{n}{k} \lambda_{k+1} m_{n-k} \,. \tag{8}$$

Explicitly the first cumulants read

$$\begin{aligned}
&\lambda_1 = m_1 \;,\; \lambda_2 = m_2 - m_1^2 \;,\; \lambda_3 = m_3 - 3\, m_2\, m_1 + 2\, m_1^3, \\
&\lambda_4 = m_4 - 4\, m_3\, m_1 - 3\, m_2^2 + 12\, m_2\, m_1^2 - 6\, m_1^4, \\
&\lambda_5 = m_5 - 5\, m_4\, m_1 - 10\, m_3\, m_2 + 20\, m_3\, m_1^2 + 30\, m_2^2\, m_1 - 60\, m_2\, m_1^3 + 24\, m_1^5.
\end{aligned} \tag{9}$$

By construction $m_n \propto \alpha^n$ and Eq. (8) shows that the cumulants share this property. Thus we immediately obtain

$$e_n = \lim_{\beta \to \infty} \frac{1}{\beta} \frac{(-)^{n+1}}{\alpha^n \, n!} \lambda_n(\beta) \,. \tag{10}$$

The functional integral for the moments can be done since it is Gaussian. The integrals over the phonon momenta $\mathbf{k}_m$, $m = 1 \ldots n$, can also be performed if the mth propagator is written as

$$\frac{1}{\mathbf{k}_m^2} = \frac{1}{2} \int_0^\infty du_m \, \exp\left[-\frac{1}{2} \mathbf{k}_m^2 \, u_m \right] . \tag{11}$$

Then one obtains

$$m_n = \frac{(-)^n \alpha^n}{(4\pi)^{n/2}} \prod_{m=1}^{n} \left(\int_0^\beta dt_m \int_0^{t_m} dt'_m \int_0^\infty du_m \right) \exp\left[- \sum_{m=1}^{n} (t_m - t'_m) \right]$$
$$\cdot \left[\det A\, (t_1 \ldots t_n, t'_1 \ldots t'_n; u_1 \ldots u_n) \right]^{-3/2} \,. \tag{12}$$

Here the $(n \times n)$-matrix A,

$$A_{ij} = \frac{1}{2} \Big[-|t_i - t_j| + |t_i - t'_j| + |t'_i - t_j| - |t'_i - t'_j| \Big] + u_i \, \delta_{ij} \,, \tag{13}$$

is non-analytic in the times t_i, t'_i, but analytic in the auxiliary variables u_i.

3. Numerical Procedures and Results

The task is now to perform the $(3n)$-dimensional integral over t_i, t'_i, u_i for large enough β in the expression for the cumulants/moments. It is clear that any reduction in the dimensionality of the integral will greatly help in obtaining reliable numerical results in affordable CPU time. A closer inspection of the structure of the integrand reveals that two integrations over the auxiliary variables (say u_n, u_{n-1}) can always be done analytically.

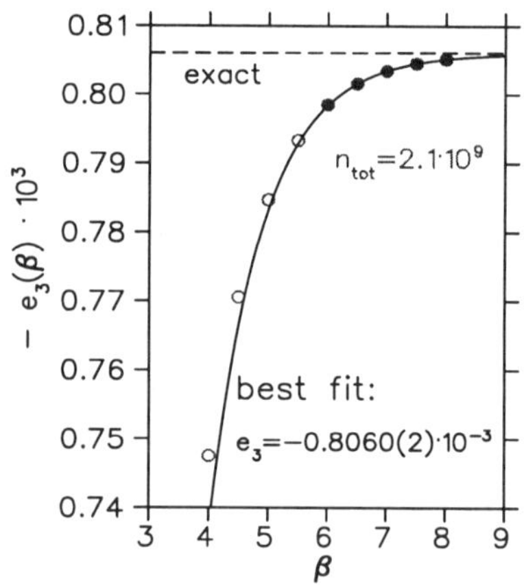

Fig. 1. Monte Carlo results for the derivative of the 3rd cumulant as function of the Euclidean time β. The total number of function calls is denoted by n_{tot} and the full (open) circles are the points used (not used) in the fit.

Furthermore, we do not use Eq. (10) to extract the energy coefficient e_n but

$$e_n = \frac{(-)^{n+1}}{\alpha^n n!} \lim_{\beta\to\infty} \frac{\partial \lambda_n(\beta)}{\partial \beta} =: \lim_{\beta\to\infty} e_n(\beta)\,. \tag{14}$$

This "kills two birds with one stone": first the derivative w.r.t. β takes away one further integration over a time (see Eq. (12) where β appears as upper limit) requiring that only a $(3n-3)$-dimensional integral has to be done numerically. Second, it vastly improves the convergence to $e_n \equiv e_n(\beta=\infty)$ because now

$$e_n(\beta) \overset{\beta\to\infty}{\longrightarrow} \frac{\partial}{\partial\beta}\left[\beta\cdot e_n + \text{const} - \frac{a_n}{\sqrt{\beta}}\, e^{-\beta} + \ldots\right] = e_n + \frac{a_n}{\sqrt{\beta}}\, e^{-\beta} + \ldots\,. \tag{15}$$

In other words: we obtain an *exponential* convergence to the value e_n whereas previously the approach would be very slow, like const$/\beta$. This exponential convergence of the derivative version has been demonstrated analytically for $n = 1, 2$ and numerically for $n = 3$ (see below). In the following we will assume that it holds for *all* n. After mapping to the hypercube $[0, 1]$ the remaining $(3n-3)$-dimensional integral can be evaluated by Monte Carlo techniques utilizing the classic VEGAS program[9] or the more modern programs from the CUBA library.[10]

We first have tested this approach by comparing with the analytical result given in Eq. (3). Figure 1 shows $e_3(\beta)$ and the best fit to the data assuming the β-dependence (15). Since the asymptotic behaviour is not

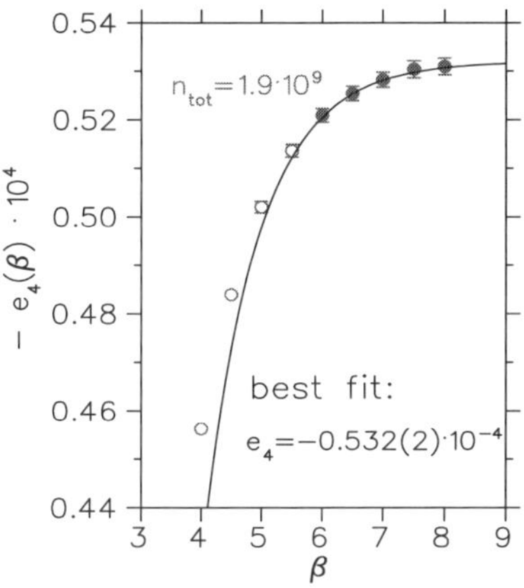

(a)

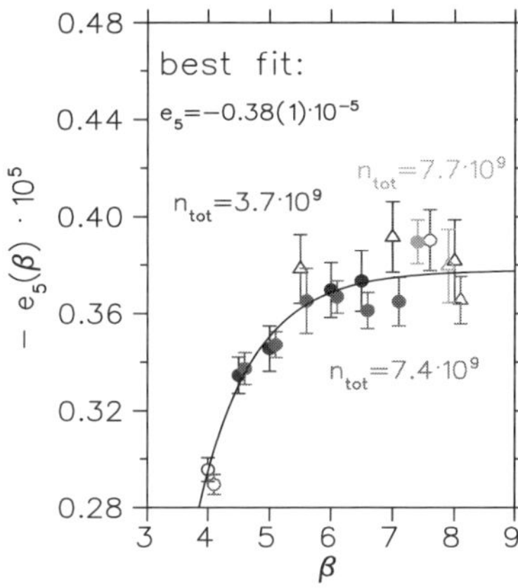

(b)

Fig. 2. (a) Same as Fig. 1 but for the 4th cumulant. (b) Data for the derivative of the 5th cumulant. Open triangles denote results (not used in the fit) which have a $\chi^2 > 1.5$ indicating that successive Monte Carlo iterations are not consistent with each other.

valid for low values of β we have eliminated small-β points successively until the resulting χ^2/dof of the fit reaches a minimum. Excellent agreement with Smondyrev's result (3) is found. If one allows for a different power of β in the prefactor of Eq. (15) then the fit gives an exponent $-0.55(3)$ instead of -0.5 assumed before.

However, when extending these calculations to the case $n = 4$, a very slow convergence of the numerical result with the number of function calls n_{tot} is observed at fixed β. Fortunately, a solution was found by performing the remaining $(n-2)\,u_i$-integrations not by stochastic (Monte Carlo) methods but by deterministic quadrature rules. This is possible since the u_i-dependence of the integrand is analytic (see Eq. (13)). We have used the very efficient "tanhsinh-integration" method[11] but Gaussian quadrature is nearly as good. A dramatic improvement in stability results together with a reduction of n_{tot} needed for the much smaller values of $|e_n|\,, n > 3$. This allows a reliable evaluation of e_4 (see Fig. 2(a)) and also makes the determination of e_5 feasible as shown in Fig. 2(b).

The best fit values for e_4 and e_5 displayed in Figs. 2(a), (b) are still preliminary as a more detailed error analysis has to be made. Also for the $n = 5$ case the Monte Carlo statistics should be improved. Note that each high-statistic point in Fig. 2(b) took about 30 days runtime on a Xeon 3.0 GHz machine.

4. Summary and Outlook

- Two additional perturbative coefficients e_4, e_5 for the polaron g.s. energy have been determined by the method of Bogolyubov and Plechkov (rediscovered independently). This amounts to performing a 4-loop and 5-loop calculation in Quantum Field Theory.
- The method is based on a combination of Monte Carlo integration techniques and deterministic quadrature rules for finite β (temperature) and on a judicious extrapolation to $\beta \to \infty$ (zero temperature). As a check the value of e_3 calculated analytically by Smondyrev has been reproduced with high accuracy.
- The cancellation in $n^{\rm th}$ order is *not* among many individual diagrams but among the much fewer terms in the integrand of the $(3n-3)$-dimensional integral (see Eq. (9)).
- The method can be simply extended to the calculation of higher-order terms in the small-coupling expansion of the effective mass $m^\star(\alpha)$ for a moving polaron.
- Generalizing this approach to relativistic QFT in the worldline representation[12] and calculation of higher-order terms for the anomalous magnetic moment of the electron is under investigation. New challenges arise from the divergences which now occur and the need for renormalization.

References

1. B. C. Odom *et al.*, *Phys. Rev. Lett.* **97**, 030801 (2006); G. Gabrielse *et al.*, *Phys. Rev. Lett.* **97**, 030802 (2006) [Erratum *ibid.* **99**, 039902 (2007)].
2. C. Itzykson and J.-B. Zuber, *Quantum Field Theory* (McGraw-Hill, New York, 1980), Eq. (9-195).
3. T. Kinoshita and M. Nio, *Phys. Rev. D* **73**, 053007 (2006).
4. N. N. Bogolyubov (Jr.) and V. N. Plechko, *Theor. Math. Phys.* **65**, 1255 (1985).
5. G. Höhler and A. Müllensiefen, *Z. Phys.* **157**, 159 (1959); J. Röseler, *Phys. Status Solidi (b)* **25**, 311 (1968).
6. M. A. Smondyrev, *Theor. Math. Phys.* **68**, 653 (1986); O. V. Selyugin and M. A. Smondyrev, *Phys. Status Solidi (b)* **155**, 155 (1989).
7. R. P. Feynman, *Phys. Rev.* **97**, 660 (1955).
8. R. Rosenfelder, *Ann. Phys.* **128**, 188 (1980).
9. G. P. Lepage, *J. Comp. Phys.* **27**, 192 (1978).
10. T. Hahn, *Comp. Phys. Comm.* **168**, 78 (2005).
11. J. Borwein, D. Bailey, and R. Girgensohn, *Experimentation in Mathemathics: Computational Paths to Discovery* (AK Peters, Natick, Massachusetts, 2004), ch. 7.4.3.
12. C. Alexandrou *et al.*, *Phys. Rev. A* **59**, 1762 (1999).

PART VI

Monte Carlo Techniques

PATH INTEGRALS AND SUPERSOLIDS

D. M. CEPERLEY
Department of Physics, University of Illinois Urbana-Champaign,
Urbana, IL 61801, USA
E-mail: ceperley@uiuc.edu

Recent experiments by Kim and Chan on solid ^{4}He have been interpreted as discovery of a supersolid phase of matter. Arguments based on wavefunctions have shown that such a phase exists, but do not necessarily apply to solid ^{4}He. Imaginary time path integrals, implemented using Monte Carlo methods, provide a definitive answer; a clean system of solid ^{4}He should be a normal quantum solid, not one with superfluid properties. The Kim-Chan phenomena must be due to defects introduced when the solid is formed.

Keywords: Monte Carlo simulation; Superfluids; Supersolids.

1. Introduction

Around the year 1970, there were several papers suggesting that a quantum solid, such as ^{4}He could show superfluidity and Bose condensation. First, Andreev and Lifshitz[1] introduced the vacancy model. Assuming that a quantum solid contains a non-zero fraction of mobile vacancies, it is inevitable that at sufficiently low temperature they will Bose condense. The resulting system will have superfluid properties. Chester[2] arrived at the same conclusion using a Jastrow (or pair product) wavefunction. Solids described by such a wavefunction must have vacancies (assuming the Jastrow wavefunction is short ranged), and a non-zero condensate fraction at zero temperature. We can then ask what type of interaction will have a Jastrow wavefunction as its ground state: the interaction will have a very strong three-body component; stronger than the pair interaction and very different from the interaction between helium atoms. The unanswered questions are then *a)* are there enough vacancies in ^{4}He at low temperature, *b)* is a Jastrow wavefunction adequate to describe a quantum solid, and *c)* are there any other mechanisms that can cause a quantum solid to be superfluid?

In 2004 Kim and Chan[3] reported superfluid-like responses in their torsional oscillator (TO) experiments. In these experiments, the moment of rotational inertia of a small cylinder containing the solid helium sample is measured by monitoring its period. They found that below approximately 0.15K, a small fraction (up to 2%) of the helium no longer oscillates with the cylinder, but is presumably at rest. This would be understood within the vacancy model as vacancies being at rest.

For a recent review of the experimental situation, please consult Balibar and Caupin[4] and for the theory, Prokofev.[5] This article does not present new results, nor is it an extensive review. In the context of this path integral conference proceedings, we sketch how path integrals lead to a deeper and unique understanding of the problem of supersolid ^{4}He, an analysis not available with other methods.

2. Path-Integral Methods

Imaginary time path integrals (ITPI) were introduced by Feynman[6] to understand superfluidity of ^{4}He, leading to the ring exchange model. ITPIs are an exact mapping of equilibrium statistical mechanics of quantum bosons onto the classical statistical mechanics of a peculiar "ring polymer" model. Let $R(t) \equiv \{\mathbf{r}_1(t), \mathbf{r}_2(t), \ldots \mathbf{r}_N(t)\}$ be the $3N$ coordinates of N helium atoms at imaginary time t. Then a path consists of a trajectory $R(t)$ where $0 \leq t \leq \beta$ and $\beta = 1/k_B T$ with T the temperature, and, for computational reasons, we work with discrete time; imaginary time is in steps of τ so that $t = k\tau$ and $M = \beta/\tau$. The probability of a path is given by $\exp(-\sum_{k=1}^{M} S(R((k-1)\tau), R(k\tau)))$ where $S(R, R')$ is the link-action. In the so-called primitive approximation, this is simply the kinetic and potential action, while in actual implementation more accurate forms are used. See the review article[9] for a detailed discussion of numerical methods.

What remains is to set the initial and final boundary conditions for the path. For calculations of scalar properties of bosons, one uses periodic boundary conditions in time, but allowing for exchange so that $R(0) = \hat{P}R(\beta)$ where P is a specific permutation of particle labels, all $N!$ permutations being allowed. Then, as shown by Feynman,[6] the lambda transition in liquid ^{4}He occurs because the contributing permutations change from small exchange cycles above T_λ, to macroscopic exchanges below T_λ. There are two other important connections between quantum collective behavior and the ring polymer model, namely, the superfluid density and Bose condensation.

The superfluid density is defined in terms of the free energy needed to place a sample in rotation, for an infinitesimally small angular velocity. (Note that this is steady state rotation, not oscillation as in the TO experiment.) It is possible to show that the superfluid density fraction can be calculated within ITPI as[7]

$$\frac{\rho_s}{\rho} = \frac{\langle (2mA_z)^2 \rangle}{\beta \hbar^2 I} = \frac{m \langle W_z^2 \rangle}{\beta \hbar^2 N} \tag{1}$$

where A_z is the path area projected on the xy plane, I is the classical moment of inertia and m the mass. For periodic boundary conditions, we use the second formula which contains the "winding number" of the path W_z in the z direction. In either case, in the thermodynamic limit, non-zero superfluid fraction can only come from path involving a macroscopic number of atoms forming a long exchange cycle.

The second property is the condensate fraction. Again, it can be shown[8] that the fraction of atoms with precisely zero momentum, n_0, is given by the large distance limit of the single particle density matrix, $n(r)$, and can by calculated in ITPI by the probability that the ends of a cut "polymer" (*i.e.* one not constrained to be periodic in IT) are separated from each other by a large distance. Both of these properties have been demonstrated to work[9] for liquid helium using the same potential as solid helium, but at a lower density.

Now turning to quantum crystals, the path picture of a quantum crystal is that of a "polymer" crystal; a set of N such polymers in a periodic box that spontaneously order spatially. Experimentally, and in computer simulations, we define the solid phase by the presence of a Bragg peak in the structure factor, $S(k)$, that scales with the number of atoms. If the particles are bosons, then they are allowed to cross-link (form ring exchanges), however, they may choose not to. At low enough temperature, one will always find a few isolated local ring exchanges, *e. g.* a neighboring pair of atoms exchanging. Note that this exchange does not contribute to either the superfluid density or to the condensate fraction. It is not a winding exchange because it is local. As in superfluid helium, we need a winding exchange to signal supersolidity, which necessarily involves a macroscopic number of atoms.

Concerning the question of whether the ground state has vacancies, one is faced with a problem of how to define a vacancy in a quantum crystal. Because of quantum fluctuations, many lattice sites may be temporarily vacant, but if the there is a nearby interstitial atom, these are not really vacancies but rather vacancy-interstitial pairs (VIPs). The number of true vacancies equals the number of lattice sites minus the number of atoms;

both can be experimentally measured. For calculations in the canonical ensemble, one adds or subtracts an atom from a perfect crystal, while changing the box size to keep the density the same,[10] while in the grand canonical ensemble one looks at the slope of the single particle Green's function.[11] Both approaches lead to a consistent value of the excitation energy of a vacancy of about 14K and of an interstitial of 23K, for the lowest, most favorable density of solid ^{4}He. Since the energy for these point defects, is non-zero and large, we do not expect them to be present in solid ^{4}He at low temperatures, thus making the vacancy mechanism for supersolidity invalid. Questions have been raised[13] whether there could be finite size effects that invalidates this conclusion.

ITPI uniquely put the questions of supersolidity in a clear light. Methods based on wavefunctions, such as the Chester proof of BEC,[2] are only as valid as long as the wavefunction is correct. It is difficult to make the connection between the wavefunction and the Hamiltonian. Typically a wavefunction is justified on the basis of how well other properties, such as the ground state energy, compare to experiment. However, the important properties we have just mentioned, the condensate fraction and the superfluid density, are far off-diagonal properties of the density matrix, and very different from a local property such as the energy, so that such methods of establishing the accuracy of a wavefunction are not reliable. Let us discuss two qualitative arguments based on ITPI.

Suppose a quantum crystal is superfluid; then it must have a macroscopic winding exchange. One possibility is that all the atoms simply slide along a row in the crystal, however, this process has a very low probability because all the atoms would have to move simultaneously. A simple way for such a process to proceed is for a VIP to form, for the vacancy and interstitial to separate and then diffuse independently; if one or other of the defects winds around the boundary before recombining (as required by the periodicity in time), then the system is a supersolid. However, during this diffusion, it is clear that there must be isolated vacancies and interstitials, and if this process is frequent, then there is no reason that either vacancies or interstitials would not be stable by themselves. So Prokof'ev and Svistunov[14] conclude that stable point defects are both necessary and sufficient for a supersolid. The weak point of the argument is the possible existence of more complicated mechanisms leading to winding exchanges. We searched[22] for the most probable mechanism for winding exchanges, and found that they occur in a "plasma" state of VIPs, however, such a state is not stable.

A second ITPI argument[18] shows that there does exist a normal phase for quantum crystals. On the contrary, it has been suggested that any Bose system at low enough temperature will Bose condense. Take a system heavier than helium that forms a good crystal, *e.g.* solid ^{132}Xe. Because of its heavy mass, the atoms sit at the absolute minimum of the potential energy. If we imagine a winding path, needed to have a non-zero superfluid density, the winding path must necessarily have a higher potential energy. So in the limit of low temperature, this will have an action above the ground state action by an amount at least proportional to a macroscopic length, $N^{1/3}$. Thus in the $\beta \to \infty$ limit the extra action will scale to ∞ and such winding paths will not contribute to the partition function. Hence the ground state of ^{132}Xe has $\rho_s = 0$. Solid helium is unusual in that at zero temperature, the lattice constant is larger than required to sit at the minimum of the potential, so this argument does not apply; it could lower its energy by exchanging.

3. Path-Integral Results

The above discussion was meant to be general. The real utility of ITPI is to apply them on the actual Hamiltonian of many-body helium, since as we have seen there exist Hamiltonians that have a supersolid ground state, and those that have a normal solid ground state. Does solid ^{4}He fall into either of those categories? Now helium is unique in that a helium atom is well-represented as an elementary particle interacting in a pairwise fashion (in a restricted range of temperatures and densities) because the electronic gap is more than $2 \times 10^5 K$. Semi-empirical potentials such as Aziz *et al.*[19] are correct to about 0.1%; the errors concern the three- and higher body terms. Numerous comparisons with experiment bear out this point of view. For example, the Debye-Waller factors[20] of solid helium (both isotopes) are in very good agreement (in some cases better than 1%) with experiment for a wide range of density and pressure. Wc already discussed the results of the vacancy energy. PIMC is in agreement with some of the vacancy data, but the experimental results vary widely.

Two independent ITPIMC calculations[11,15] for the single particle density matrix have been made. Both show the function decays exponentially fast (but with oscillations caused by the lattice), reaching a value of roughly 10^{-6} at the edge of the box 15 Å distant. If it stabilized at this value it would be an incredibly small condensate fraction. We note that this result is different than computed using the shadow wave function technique,[21] which gives stable condensate fraction of 5×10^{-6}. It seems clear that the

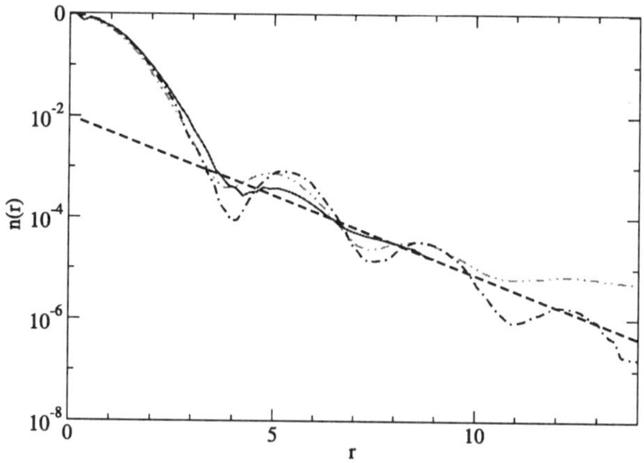

Fig. 1. The single particle density matrix[15] as a function of distance (in Å), estimated with PIMC, in hcp ^{4}He at a density 286 nm^{-3} and a temperature of 0.5K. The solid curve is the spherically averaged $n(r)$ while the dot-dashed curve is $n(x)$. The upper double-dot dashed curve is a variational MC calculation using the shadow wave function Ref. 13. The dashed straight line has a slope determined from the exchange calculations of Ref. 16.

finite value comes because the SWF has inherited the properties of the Jastrow wave function, and recent calculations on 2D solid helium confirm this view.[21]

The PIMC results also do not show any winding exchanges and hence no superfluid. In order to understand the situation, we did a detailed analysis of the probability of a given ring exchange; it is these ring exchange frequencies[17] that give rise to the magnetic ordering in solid ^{3}He. At low temperatures, there are infrequent local exchanges, but the probability of long exchanges drops exponentially fast, thus solid ^{4}He is predicted to be a normal (insulating) quantum solid, not a supersolid. In conclusion, well-controlled numerical calculations predict that solid ^{4}He should not be a supersolid.

4. Conclusion and Prospectives

Because of the excitement over the Kim-Chan experiments, there have been numerous new experiments in the last few years. Several groups have reproduced the TO experiments. There have been several experiments that directly address the issue of phase transition. Lin *et al.*[23] have performed a high precision heat capacity measurement, and did not see any direct evidence of the lambda transition expected for the vacancy mechanism. At low temperature they did see a linear term, possibly due to the heat capacity of lower dimensional structures such as grain boundaries or dislocations. In scattering experiments,[24] it was found that the Bragg peak persisted to low temperature, evidence that the one-body density was not changed significantly. In other experiments there was no evidence of Bose condensation in the momentum distribution.[25]

Day and Beamish[26] found that there was no static superflow as a result of pressure implying that the phase, if it exists, is different from a normal superfluid; *e. g.* a dynamical phenomena. It is also found that there is a strong dependence of the phenomena[27] on ^{3}He concentration even at the concentration of parts per billion. Parts per million of ^{3}He increase the superfluid response, but have smaller effect on the transition temperature. This dependence is understandable if the ^{3}He goes to lower dimensional structures such as dislocations or grain boundaries.

Rittner and Reppy[28] have studied how the superfluid density depends on the sample geometry and sample preparation. By making thin samples, they report NCRI up to 20% much larger than could conceivably generated in even a disordered solid. We believe this shows that the fundamental interpretation of the TO experiments is incorrect. A TO experiment measures the angular momentum response, which in a torus is given by mass flow times velocity. Various estimates imply that the fraction of the mass that could flow is quite small. However, it seems that the flow velocity could be very much greater than imposed on the sample. Critical velocities of superfluid helium in channels[4] are of order of 1m/sec. Hence, a simple model is of a coordinated melting-freezing oscillation, with mass being transported by superfluid through a grain boundary. This oscillation could be tuned to be in resonance with the TO frequency by the motion of ^{3}He impurities and is controlled by the strain fields in the crystal produced by the oscillations. PIMC has been used to look at crystal grain boundaries,[29] but it is not obvious if such calculations which assume equilibrium, are relevant to the experimental situation.

Acknowledgments

The work done here was performed in collaboration with B. Bernu, B. Clark, and S. Khairallah with financial support from the NASA program in

Fundamental Physics, NSF DMR-03 25939 and the CNRS-UIUC exchange program. Computer time was provided by NCSA and the MCC.

References

1. A. F. Andreev and I. M. Lifshitz, *Sov. Phys. JETP* **29**, 1107 (1969).
2. G. V. Chester, *Phys. Rev. A* **2**, 256 (1970).
3. E. Kim and M. H. W. Chan, *Science* **305**, 1941 (2004); *Nature* **427**, 225 (2004).
4. S. Balibar and F. Caupin, *J. Physics CM*, in press (2008).
5. N. Prokof'ev, *Adv. Phys.* **56**, 381 (2007).
6. R. P. Feynman, *Phys. Rev.* **90**, 1116 (1953).
7. E. L. Pollock and D. M. Ceperley, *Phys. Rev. B* **36**, 8343 (1987).
8. D. M. Ceperley and E. L. Pollock, *Phys. Rev. Lett.* **56**, 351 (1986).
9. D. M. Ceperley, *Rev. Mod. Phys.* **67**, 279 (1995).
10. B. K. Clark and D. M. Ceperley, *Comp. Phys. Comm.* **179**, 82 (2008).
11. M. Boninsegni, A. B. Kuklov, L. Pollet, N. V. Prokof'ev, B. V. Svistunov, and M. Troyer, *Phys. Rev. Lett.* **97**, 080401 (2006).
12. D. E. Galli, M. Rossi, and L. Reatto, *Phys. Rev. B* **71**, 140506 (2005).
13. D. E. Galli and L. Reatto, *Phys. Rev. Lett.* **96**, 165301 (2006).
14. N. Prokof'ev and B. Svistunov, *Phys. Rev. Lett.* **94**, 155302 (2005).
15. B. Clark and D. M. Ceperley, *Phys. Rev. Lett.* **96**, 105301 (2006) [arXiv:cond-mat/0512547].
16. D. M. Ceperley and B. Bernu, *Phys. Rev. Lett.* **93**, 155303 (2004) [arXiv:cond-mat/040933].
17. D. M. Ceperley and G. Jacucci, *Phys. Rev. Lett.* **58**, 1648 (1987).
18. B. Svistunov, talk at Stillwater conference (2007).
19. R. A. Aziz, A. R. Janzen, and M. R. Moldover, *Phys. Rev. Lett.* **74**, 1586 (1995).
20. E. W. Draeger and D. M. Ceperley, *Phys. Rev. B* **61**, 12094 (2000).
21. E. Vitali, M. Rossi, F. Tramonto, D. E. Galli, and L. Reatto, arXiv:0801.1753.
22. B. Bernu and D. M. Ceperley, *J. Phys. Chem. Solids* **66**, 1462 (2005) [arXiv:cond-mat/0502486].
23. X. Lin, A. C. Clark, and M. H. W. Chan, *Nature* **449**, 1025 (2007).
24. E. Blackburn *et al.*, *Phys. Rev. B* **76**, 024523 (2007); M.A. Adams *et al.*, *Phys. Rev. Lett.* **98**, 085301 (2007).
25. S. O. Diallo *et al.*, *Phys Rev. Lett.* **98**, 205301 (2007).
26. J. Day and J. Beamish, *Phys. Rev. Lett.* **96**, 105304 (2006)
27. A. C. Clark, J. T. West, and M. H. W. Chan, *Phys. Rev. Lett.* **99**, 135302 (2007).
28. A. S. Rittner and J. D. Reppy, *Phys. Rev. Lett.* **97**, 165301 (2006); *Phys. Rev. Lett.* **98**, 175302 (2007).
29. E. Burovski *et al.*, *Phys. Rev. Lett.* **94**, 165301 (2005); L. Pollet *et al.*, *Phys. Rev. Lett.* **98**, 135301 (2007).

DIFFUSION MONTE CARLO CALCULATIONS FOR THE GROUND STATES OF ATOMS AND IONS IN NEUTRON STAR MAGNETIC FIELDS

S. BÜCHELER, D. ENGEL, J. MAIN, and G. WUNNER*

Institut für Theoretische Physik 1, Universität Stuttgart,
D-70550 Stuttgart, Germany
**E-mail: wunner@itp1.uni-stuttgart.de*

The diffusion quantum Monte Carlo method is extended to solve the old theoretical physics problem of many-electron atoms and ions in intense magnetic fields. The feature of our approach is the use of adiabatic approximation wave functions augmented by a Jastrow factor as guiding functions to initialize the quantum Monte Carlo prodecure. We calculate the ground state energies of atoms and ions with nuclear charges from $Z = 2, 3, 4, \ldots, 26$ for magnetic field strengths relevant for neutron stars.

Keywords: Quantum Monte Carlo simulation; Atoms; Neutron stars; Magnetic fields.

1. Introduction

The discovery of features in the X-ray spectra of the thermal emission spectra of the isolated neutron star 1E 1207 (Refs. 1,2) and three other isolated neutron stars has revived the interest in studies of medium-Z elements in strong magnetic fields. The reason is that the observed features could be due to atomic transitions in elements that are fusion products of the progenitor star. However, to calculate synthetic spectra for model atmospheres, and thus to be in a position to draw reliable conclusions from observed spectra to the elemental composition of the atmosphere and the distribution of elements on different ionization stages, accurate atomic data for these elements at very strong magnetic fields ($\sim 10^7$ to 10^9 T) are indispensable.

While the atomic properties of hydrogen and, partly, helium at such field strengths have been clarified in the literature over the last 25 years (for a detailed list of references see, e. g., Ref. 3), for elements with nuclear charges $Z > 2$ only fragmentary atomic data exist with an accuracy necessary for the calculations of synthetic spectra.

We have tackled[3] the problem by adapting the diffusion Monte Carlo method (DQMC)[4–6] to the case of neutron star magnetic fields. This method has the advantage that ground-state energies can be determined practically free from approximations.

2. DQMC for Neutron Star Magnetic Fields

The basic idea of DQMC is to identify the ground state wave function $\Phi_0(\boldsymbol{R},t)$ $(\boldsymbol{R}=(\boldsymbol{r}_1,\ldots,\boldsymbol{r}_N))$ of an N-body Hamiltonian $\hat{H}$ with a *particle density* whose correct distribution is found by following the random walk of many test particles ("walkers") in imaginary time in $3N$-dimensional configuration space. To reduce fluctuations one works with a density distribution $f(\boldsymbol{R},\tau)\equiv\Psi(\boldsymbol{R},\tau)\Psi_{\mathrm{G}}(\boldsymbol{R})$, where Ψ_{G} is a given guiding function used for importance sampling. The density distribution f obeys a drift-diffusion equation in imaginary time. Because the importance-sampled Green's function is an exponential operator, one can expand it in terms of a Euclidean path integral. For sufficiently small time steps one can write down accurate approximations to the Green's function, and sample it with diffusion Monte Carlo.[3–6]

2.1. *Choice of the guiding functions*

The choice of the guiding function is crucial for the success of the DQMC procedure. We take the guiding function $\Psi_{\mathrm{G}}^{\mathrm{ad}}$ as a Slater determinant of single-particle orbitals each of which is a product of a Landau state in the lowest level with a given magnetic quantum number and an unknown longitudinal wave function ("adiabatic approximation"[7]). The different longitudinal wave functions are obtained selfconsistently by an iterative solution of the Hartree-Fock equations using B-splines on finite elements.

2.2. *Jastrow factor*

To incorporate correlation effects it is common to multiply the guiding function by a Jastrow factor, $\Psi_{\mathrm{G}}=\Psi^{\mathrm{JF}}\Psi_{\mathrm{G}}^{\mathrm{ad}}=\mathrm{e}^{-U(\boldsymbol{R})}\Psi_{\mathrm{G}}^{\mathrm{ad}}$. We adopt the form

$$U=-1/4\sum_{i<j}^{N}r_{ij}/(1+\sqrt{\beta}\,r_{ij})+Z\sum_{i=1}^{N}r_i/(1+\sqrt{\beta}\,r_i)\,, \tag{1}$$

where β is the magnetic field strength in atomic units ($\beta=B/B_0$, $B_0=4.701\times10^5$ T). This leads to modifications of the adiabatic approximation guiding functions only at small distances, of the order of the Larmor radius.

3. Results and Discussion

As a representative example, Fig. 1 shows for the ground state of neutral iron ($Z = 26$) at $B = 5\times10^8$ T the typical flow of a diffusion quantum Monte Carlo simulation. Ions can be treated without additional complication in the same way.[3] The figure depicts the energy offset E_{T}, the block energy E_{B} and the averaged block energy $\langle E_{\mathrm{B}}\rangle$ as a function of the number of blocks performed.

The complete simulation goes through three stages. During the first 100 blocks, a variational quantum Monte Carlo calculation (VQMC) is performed. Since the adiabatic approximation guiding wave function is augmented by the Jastrow factor, the VQMC calculation already lowers the energy in comparison with the initial adiabatic approximation result. This stage is followed, in the next 300 blocks, by a fixed-phase diffusion quantum Monte Carlo (FPDQMC) simulation. It is seen that the onset of the simulation leads to a considerable drop in the energy. Finally, in the last 300 blocks a released-phase diffusion quantum Monte Carlo (RPDQMC) simulation is carried out, which still slightly lowers the averaged block energy, by roughly 0.1 per cent. The dashed vertical lines in Fig. 1 indicate the blocks where dynamical equilibrium of the walkers is reached. The relatively small difference between the fixed-phase and the released-phase results indicates that the phase of the adiabatic approximation wave function already well reproduces the phase of the ground-state wave function. The small fluctuations of the individual block energies E_{B} evident in Fig. 1 are characteristic of diffusion quantum Monte Carlo simulations. It is also seen, however, that the averaged block energies $\langle E_{\mathrm{B}}\rangle$ quickly converge to constant values in all three stages of the simulation. Our final RPDQMC result for the energy is $E_0 = -109.079$ keV and lies well below the density functional (DF) value. The standard deviation of the block energies at the end of the simulation in this case is $\sigma = \pm 0.186$ keV.

Table 1 lists the results for all elements from helium to iron at the magnetic field strength $B = 10^8$ T. The table contains in the first three columns the results of the three stages of the simulation and in the fourth column the energy values in adiabatic approximation calculated with our own Hartree-Fock finite-element method HFFEM. Literature values obtained by Ivanov and Schmelcher[11] (2DHF), by Mori and Hailey[8] (MCPH^3, multi-configurational perturbative hybrid Hartree-Hartree-Fock) and the results of density functional calculations[9,10] (DF) are given in the remaining columns. The numbers in brackets attached to the HFFEM, 2DHF, MCPH^3 and DF results designate the number of electrons occupying an

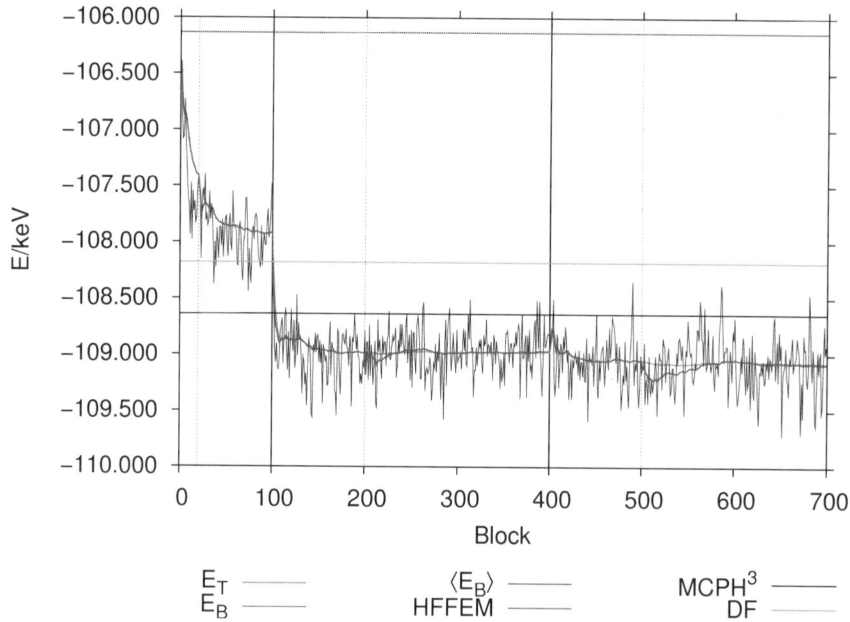

Fig. 1. Behavior of the block energy E_B (ragged curve) and the averaged block energy $\langle E_B \rangle$ (smooth curve) in the DQMC simulation for the ground-state energy of neutral iron ($Z = 26$) at $B = 5 \times 10^8$ T as a function of the number of blocks. In each block, 200 time steps $\Delta\tau = 5 \times 10^{-6}$ a.u. were performed. (HFFEM (top horizontal line): energy value in adiabatic approximation; DF (second horizontal line from top): density functional result of Ref. 9; $MCPH^3$ (third horizontal line from top): result of Ref. 8).

excited hydrogen-like single-particle longitudinal state. It can be seen that already the fixed-phase results lie slightly below the values that were obtained using the 2DHF method. The comparison with the results of the $MCPH^3$ method shows that our RPDQMC energy values generally lie below those results, but there are also exceptions where our results lie above the $MCPH^3$ energies. This may be due to the fact that the hybrid method is not self-consistent, since it evaluates the exchange energy in first-order perturbation theory in a basis of Hartree states and it does not include the back-reaction of the excited Landau states whose admixtures are taken into account perturbatively on the effective interaction potentials. Therefore the method need not necessarily produce an upper bound on the energy.

The comparison with the results of the DF calculations shows that these yield lower ground state energies at small nuclear charge numbers than our RPDQMC results, while for large Z the reverse is the case. The DF

Table 1. Energy values in keV for the ground states from helium to iron at $B = 10^8$ T. Parameters of the QMC simulations: 500 walkers, time steps $\Delta\tau(Z = 2, \ldots, 10) = 10^{-4}$ a.u., $\Delta\tau(Z = 11, \ldots, 19) = 5 \times 10^{-5}$ a.u., $\Delta\tau(Z = 20, \ldots, 26) = 2 \times 10^{-5}$ a.u. (discussion see text).

Z	RPDQMC	FPDQMC	VQMC	HFFEM	2DHF	MCPH[3]	DF
2	−0.5827	−0.5827	−0.5791	−0.5754	−0.57999	−0.5766	−0.6035[b]
3	−1.230	−1.229	−1.220	−1.211	−1.22443	−1.214	
4	−2.081	−2.080	−2.065	−2.044	−2.07309	−2.056	
5	−3.122	−3.119	−3.095	−3.057	−3.10924	−3.085	
6	−4.338	−4.331	−4.294	−4.236	−4.31991	−4.288	−4.341[b]
7	−5.716	−5.712	−5.660	−5.568	−5.69465	−5.657	
8	−7.252	−7.246	−7.173	−7.045	−7.22492	−7.176	
9	−8.938	−8.930	−8.834	−8.658	−8.90360	−8.845	
10	−10.766	−10.753	−10.630	−10.400	−10.72452	−10.664	−10.70[a]
11	−12.725	−12.716	−12.569	−12.266		−12.625	
12	−14.827	−14.817	−14.618	−14.249		−14.745	
13	−17.061	−17.043	−16.813	−16.352[1]		−16.973[1]	
14	−19.480	−19.461	−19.185	−18.619[1]		−19.408[1]	−19.09[a]
15	−22.022	−22.009	−21.665	−21.002[1]		−21.987[1]	
16	−24.700	−24.668	−24.275	−23.482[2]		−24.718[2]	
17	−27.541	−27.523	−27.044	−26.130[2]		−27.618[2]	
18	−30.529	−30.509	−29.950	−28.890[2]		−30.766[2]	
19	−33.650	−33.605	−32.999	−31.756[2]		−34.036[2]	
20	−36.891	−36.881	−36.145	−34.750[3]		−37.500[3]	−35.48[a]
21	−40.296	−40.274	−39.458	−37.865[3]			
22	−43.867	−43.821	−42.900	−41.083[3]			
23	−47.526	−47.490	−46.458	−44.426[4]			
24	−51.360	−51.271	−50.102	−47.877[4]			
25	−55.279	−55.224	−53.915	−51.430[5]			
26	−59.366	−59.311	−57.913	−55.108[5]			−56.01[a]

Note: [a]Ref. 9. [b]Ref. 10.

results listed in Table 1 differ in the choice of the exchange functional. Given this restriction, it cannot be ensured that the DF calculations in all cases produce an upper bound on the ground state energy in magnetic fields as do the ab-initio methods used in this work or in Ref. 11.

4. Conclusions

We have extended the released-phase diffusion Monte Carlo method to the calculation of the ground state energies of atoms and ions from helium to iron neutron star magnetic field strengths by using adiabatic approximation wave functions as guiding wave functions.[3] However, for matching observed thermal spectra from isolated neutron stars, wavelength information, and thus energies of excited states, are requisite. Jones *et al.*[6] have shown a way how to calculate excited states of small atoms in strong magnetic fields using the correlation function Monte Carlo method.[5] The challenge remains to transfer their method to the DQMC simulations presented in this paper, and to calculate excited states of large atoms in intense fields.

Acknowledgments

This work was supported by Deutsche Forschungsgemeinschaft within the SFB 382 "Methods and Algorithms for Simulating Physical Processes on High-Performance Computers" at the Universities of Tübingen and Stuttgart.

References

1. D. Sanwal, G. G. Pavlov, V. E. Zavlin, and A. A. Teter, *Astrophys. J.* **574**, L61 (2002).
2. S. Mereghetti, A. de Luca, P. A. Caraveo, W. Becker, R. Mignami, and G. F. Bignami, *Astrophys. J.* **581**, 1280 (2002).
3. S. Bücheler, D. Engel, J. Main, and G. Wunner, *Phys. Rev. A* **76**, 032501 (2007).
4. P. J. Reynolds, D. M. Ceperley, B. J. Alder, and W. A. Lester Jr., *J. Chem. Phys.* **77**, 5593 (1982).
5. D. M. Ceperley and B. Bernu, *J. Chem. Phys.* **89**, 6316 (1988).
6. M. D. Jones, G. Ortiz, and D. M. Ceperley, *Phys. Rev. E* **55**, 6202 (1997).
7. L. I. Schiff and H. Snyder, *Phys. Rev.* **55**, 59 (1937).
8. K. Mori and C. J. Hailey, *Astrophys. J.* **564**, 914 (2002).
9. P. B. Jones, *Mon. Not. R. Astro. Soc.* **216**, 503 (1985).
10. Z. Medin and D. Lai, *Phys. Rev. A* **74**, 062507 (2006).
11. M. V. Ivanov and P. Schmelcher, *Phys. Rev. A* **61**, 022505 (2000).

PHASE TRANSITIONS AND QUANTUM EFFECTS IN MODEL COLLOIDS AND NANOSTRUCTURES

P. NIELABA* and W. STREPP

Physics Department, University of Konstanz, D-78457 Konstanz, Germany
**E-mail: peter.nielaba@uni-konstanz.de*
www.uni-konstanz.de/nielaba

Hard and soft disks in external periodic (light-) fields show rich phase diagrams including freezing and melting transitions when the density of the system is varied. Monte Carlo simulations for detailed finite-size scaling analysis of various thermodynamic quantities like the order parameter, its cumulants, etc. have been used in order to map the phase diagram of the system for various values of the density and the amplitude of the external potential.

Interpreting hard disks as the simplest model of an atomic fluid, quantum effects on the phase diagram are investigated by path integral Monte Carlo simulations.

Keywords: Monte Carlo simulation; Colloids; Phase transition.

1. Introduction

Colloidal dispersions can roughly be classified as solutions of mesoscopic solid particles with a stable (non-fluctuating), often spherical shape, embedded in a molecular fluid solvent. Examples are aqueous suspensions of polystyrene, latex spheres, or rods. These particles can be prepared and characterized in a controlled way, the interactions are tunable, and many systems can be well approximated as hard sphere- (in three dimensions) and hard disk- (in two dimensions) potentials. Due to their simplicity, these systems can be considered as the simplest complex fluids and as prototypes of soft matter systems. They are nearly ideal model systems in statistical physics and are, thus, sometimes "bridges" to the "world" on nanometer length scales.

Systems such as hard disks are not only models for suspensions of colloidal particles (with diameters on the μm scale) which behave according to classical statistical mechanics under all circumstances of practical interest (for examples see Refs. 1–3), but also are the simplest possible model for fluids formed by atoms or small molecules (with diameters on the Angstrom scale). Using the latter interpretation, it makes sense to explore the extent to which quantum statistical mechanics rather than classical statistical mechanics is needed to understand their behavior. Quantum effects on the phase diagram are analyzed by path integral Monte Carlo simulations (PIMC).

2. Melting of Hard Disks in Two Dimensions

One of the first continuous systems to be studied by computer simulations[4] is the system of hard disks of diameter σ interacting with the two-body potential

$$\phi(r) = \begin{cases} \infty & r \leq \sigma \\ 0 & r > \sigma \,, \end{cases} \tag{1}$$

where, σ, taken to be 1 in the rest of the paper, denotes the hard disk diameter, which sets the length scale for the system, and the energy scale is set by $k_B T = 1$. Despite its simplicity, this system was shown to undergo a phase transition from solid to liquid as the density ρ was decreased. The reason for this phase transition with increasing density roughly is the higher entropy of the solid structure at high densities, where the particles can fluctuate around their average lattice positions, compared to the entropy of a liquid, in which the particles have positional disorder, but are essentially locked and cannot explore much phase space anymore.

The solid-liquid transition of hard disk systems has been discussed controversially in the literature of the last 40 years. By application of a new finite-size scaling procedure[5] for the computation of elastic constants, this transition has been analyzed again.[6,7]

According to our results the behavior of the system is consistent with the predictions of the KTHNY theory.[8,9] The core energy amounts to $E_c > 2.7$ at the transition, so KTHNY perturbation theory is valid, though numerical values of non universal quantities may depend on the order of the perturbation analysis. The solution of the recursion relations shows that a KTHNY transition at $p_c = 9.39$ *preempts* the first-order transition at $p_1 = 9.2$. Since these transitions, as well as the hexatic-liquid KTHNY transition lie so close to each other, the effect of, as yet unknown, higher order corrections to the recursion relations may need to be examined in the future. Due to this caveat, our conclusion that a hexatic phase exists over some region of density exceeding $\rho = 0.899$ still must be taken as preliminary.[7]

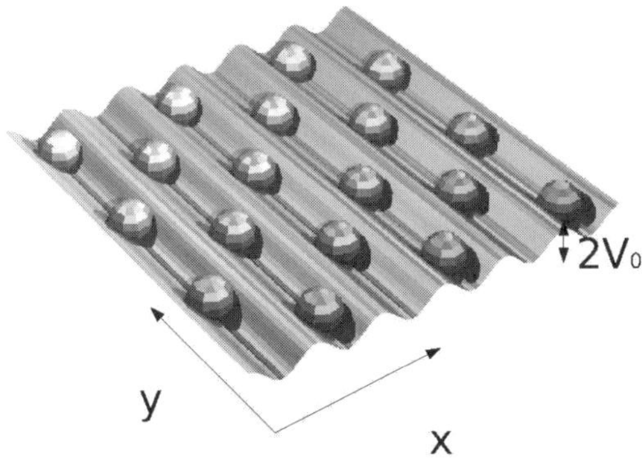

Fig. 1. Schematic picture of the (two-dimensional) system in an external periodic field, periodic in x-direction and with amplitude V_0.

3. Phase Transitions in External Periodic Potentials

3.1. *Model and classical systems*

The liquid-solid transition in two dimensional colloidal systems under the influence of external modulating potentials has recently attracted a fair amount of attention from experiments,[10–16] density functional theory,[17,18] dislocation unbinding calculations[8,9] and computer simulations.[19–23] The field acts on the particles like a commensurate, one dimensional, modulating potential, see Fig. 1.

One of the more surprising results of these studies is the fact that there exist regions in the phase diagram over which one observes reentrant[13–15] freezing/melting behavior. As a function of the laser field intensity the system first freezes from a modulated liquid to a two dimensional triangular solid. This effect is described in the literature as *Laser Induced Freezing* (LIF). A further increase of the intensity confines the particles strongly within the troughs of the external potential, suppressing fluctuations perpendicular to the troughs, which leads to an uncoupling of neighboring troughs and to re-melting. This effect is described in the literature as *Laser Induced Melting* (LIM).

We have analyzed the phase diagrams of two dimensional systems of hard[24–27] and various soft disks[27–29] in an external sinusoidal potential,

$$V(x,y) = V_0 \sin\left(\frac{2\pi}{\lambda}x\right)\,, \qquad (2)$$

see Fig. 1, which is periodic in the x-direction and constant in the y-direction, by Monte Carlo simulations. The system is characterized by the reduced density $\varrho^* = \varrho\sigma^2$ and the reduced potential strength $V_0^* = V_0/(k_B T)$, where k_B is the Boltzmann constant and T the temperature. We find an increase of the freezing region with the range of particle interaction, and a decrease of the reentrance region for the most long ranged potential $1/r^6$. The relative extent of the reentrance region is closest to the experimental data for the DLVO or the $1/r^{12}$- potentials.[31]

3.2. *Quantum effects*

Besides these classical studies we explored the validity of our results on atomic length scales. In this context we were able to investigate the properties of quantum hard disks with a finite particle mass m and interaction diameter σ in an external periodic potential by path integral Monte Carlo simulations (PIMC).[30,31] PIMC simulations[32–37] have been successfully applied to the analysis of quantum effects in solids,[38–40] pore condensates,[41] fluids,[42,43] and adsorbed molecular layers.[44,45]

Canonical averages $< A >$ of an observable A in a N-particle system with Hamiltonian $\mathcal{H} = E_{\rm kin} + V_{\rm pot}$ in the volume V are given by:

$$< A >= Z^{-1} \quad {\rm Sp}\,[A\, e^{-\beta\mathcal{H}}] \quad . \tag{3}$$

Here $Z = {\rm Sp}\, e^{-\beta\mathcal{H}}$ is the partition function and $\beta = 1/k_B T$ the inverse temperature. After the application of the Trotter product formula,

$$e^{-\beta\mathcal{H}} = \lim_{P\to\infty} \left(e^{-\beta E_{\rm kin}/P}\, e^{-\beta V_{\rm pot}/P)} \right)^P \quad , \tag{4}$$

we obtain the path integral formulation of the partition function:

$$Z = \frac{1}{N!} \lim_{P\to\infty} \left(\frac{mP}{2\pi\beta\hbar^2}\right)^{3NP/2} \prod_{S=1}^{P} \int d\{\mathbf{r}^{(S)}\}$$
$$\times\exp\left[-\sum_{J=1}^{N} \frac{mP}{2\hbar^2\beta} (\mathbf{r}_J^{(S)} - \mathbf{r}_J^{(S+1)})^2 - \frac{\beta}{P} V_{\rm pot}(\{\mathbf{r}^{(S)}\}) \right] . \tag{5}$$

Here m is the particle mass, P the Trotter number, and $\mathbf{r}_J^{(S)}$ the coordinate of particle J at Trotter index S, and periodical boundary conditions apply, i.e. $P + 1 = 1$. This form of the partition function makes it possible to perform Monte Carlo simulations with increasing values of P in order to approximate the quantum limit $P \to \infty$. The effects of quantum statistics on the phase diagram are neglected here, since they will play only a significant role at much lower temperatures as studied here.

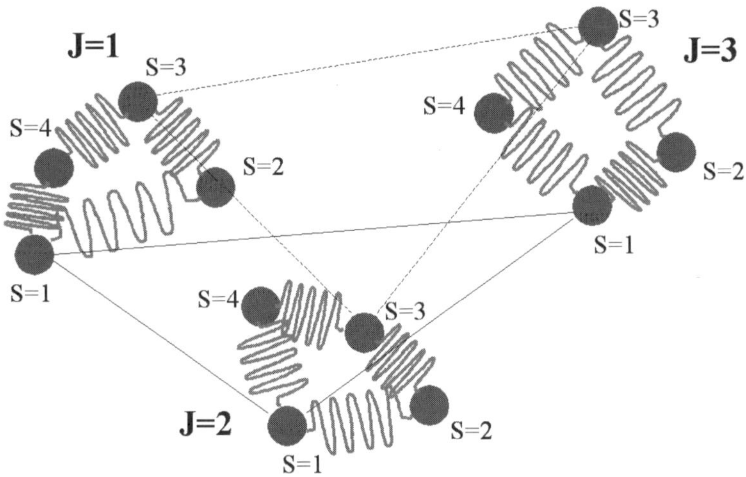

Fig. 2. Schematic picture of a system with $N = 3$ particles and Trotter number $P = 4$. All harmonic chains for each quantum particle are shown, and for the Trotter index $S = 1$ and $S = 3$ the interacting "particles".

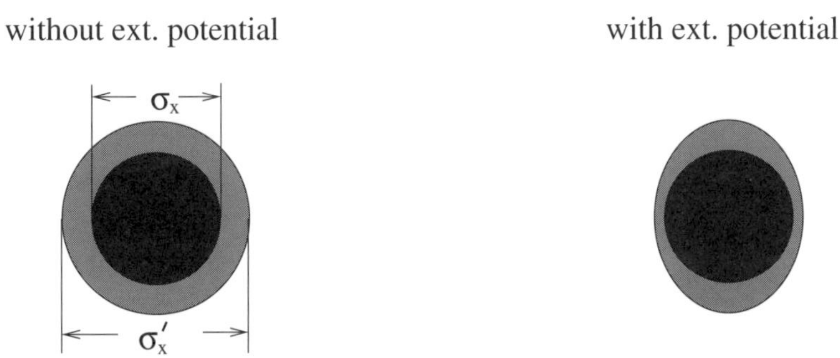

Fig. 3. Schematic picture of the effect of an external periodic potential of the form $V(x, y) = V_0 \sin(x/a)$ on the "effective" diameter of quantum hard disks.

In Eq. (5) we see that in the path integral formalism each quantum particle J (for finite P-values) can be represented by a closed quantum chain of length P in coordinate space, where the classical coordinate of the point $\mathbf{r}_J^{(S)}$ on this chain at Trotter index S has a harmonic interaction to the nearest neighbors on the chain at $\mathbf{r}_J^{(S+1)}$ and $\mathbf{r}_J^{(S-1)}$. An interaction between different quantum particles only happens between particles with coordinates $\{\mathbf{r}^{(S)}\}$ at the same Trotter index S. A schematic picture of a system with $N = 3$ particles and $P = 4$ is shown in Figure 2.

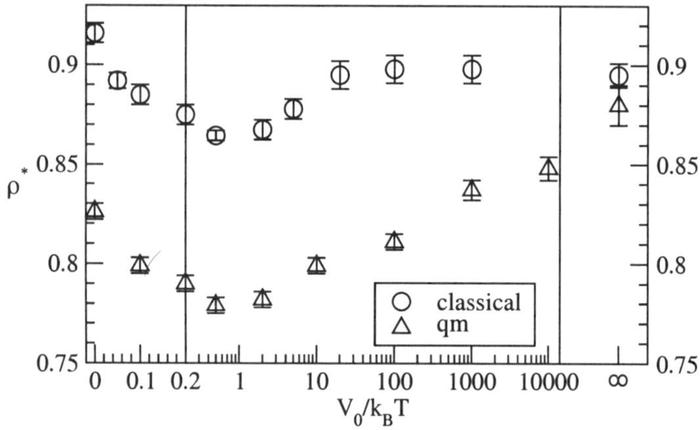

Fig. 4. Phase diagram in the density ($\rho^* = \rho\sigma^2$)- potential amplitude (V_0/k_BT)- plane for a system with $N = 400$ particles, $m^* = mT\sigma^2 = 10.000$ ("qm") and $m^* = \infty$ (classical) and Trotter order $P = 64$.[30,31]

Due to the quantum delocalization effect a larger effective particle diameter results, and in the external potential this delocalization is asymmetrical: in the direction perpendicular to the potential valleys we obtain a stronger particle localization than parallel to the valleys, see Fig. 3. As a result the reentrance region in the phase diagram is significantly modified in comparison to the classical case, see Fig. 4.

Due to the larger quantum "diameter" the transition densities at small potential amplitudes are reduced in comparison to the classical values. At large amplitudes the classical and quantum transition densities merge. This effect is due to the approach of the effective quantum disk size to the classical value in the direction perpendicular to the potential valleys and leads to the surprising prediction that the quantum crystal in a certain density region has a direct transition to the phase of the modulated liquid by an increase of the potential amplitude. This scenario is classically not known in the classical case.[30,31]

Acknowledgments

We gratefully acknowledge useful discussions with K. Binder, F. Bürzle, D. Chaudhuri, K. Franzrahe, P. Henseler, M. Lohrer, M. Rao, A. Ricci, C. Schieback, S. Sengupta, support by SFB 513, SFB 767, and SFB-TR6, as well as granting of computer time from the NIC, the HLRS, and the SSC.

References

1. K. Zahn, A. Wille, G. Maret, S. Sengupta, and P. Nielaba, *Phys. Rev. Lett.* **90**, 155506 (2003).
2. A. Ricci, P. Nielaba, S. Sengupta, and K. Binder, *Phys. Rev. E* **75**, 011405 (2007).
3. M. Köppl, P. Henseler, A. Erbe, P. Nielaba, and P. Leiderer, *Phys. Rev. Lett.* **97**, 208302 (2006).
4. B. J. Alder and T. E. Wainwright, *Phys. Rev.* **127**, 359 (1962).
5. S. Sengupta, P. Nielaba, M. Rao, and K. Binder, *Phys. Rev. E* **61**, 1072 (2000).
6. S. Sengupta, P. Nielaba, and K. Binder, *Phys. Rev. E* **61**, 6294 (2000).
7. K. Binder, S. Sengupta, and P. Nielaba, *J. Phys.: Cond. Mat.* **14**, 2323 (2002).
8. J. M. Kosterlitz and D. J. Thouless, *J. Phys. C* **6**, 1181 (1973); B. I. Halperin and D. R. Nelson, *Phys. Rev. Lett.* **41**, 121(1978); D. R. Nelson and B. I. Halperin, *Phys. Rev. B* **19**, 2457 (1979); A. P. Young, *Phys. Rev. B* **19**, 1855 (1979).
9. E. Frey, D. R. Nelson, and L. Radzihovsky, *Phys. Rev. Lett.* **83**, 2977 (1999); L. Radzihovsky, E. Frey, and D. R. Nelson, *Phys. Rev. E* **63**, 031503 (2001).
10. N. A. Clark, B. J. Ackerson, and A. J. Hurd, *Phys. Rev. Lett.* **50**, 1459 (1983).
11. A. Chowdhury, B. J. Ackerson, and N. A. Clark, *Phys. Rev. Lett.* **55**, 833 (1985).
12. K. Loudiyi and B. J. Ackerson, *Physica A* **184**, 1 and 26 (1992).
13. Q.-H. Wei, C. Bechinger, D. Rudhardt, and P. Leiderer, *Phys. Rev. Lett.* **81**, 2606 (1998).
14. C. Bechinger, Q.-H. Wei, and P. Leiderer, *J. Phys.: Cond. Mat.* **12**, A425 (2000).
15. C. Bechinger, M. Brunner, and P. Leiderer, *Phys. Rev. Lett.* **86**, 930 (2001)
16. J. Baumgartl, M. Brunner, and C. Bechinger, *Phys. Rev. Lett.* **93**, 168301 (2004).
17. J. Chakrabarti, H. R. Krishnamurthy, and A. K. Sood, *Phys. Rev. Lett.* **73**, 2923 (1994).
18. L. L. Rasmussen and D. W. Oxtoby, *J. Phys.: Cond. Mat.* **14**, 12021 (2002).
19. J. Chakrabarti, H. R. Krishnamurthy, A. K. Sood, and S. Sengupta, *Phys. Rev. Lett.* **75**, 2232 (1995).
20. C. Das and H. R. Krishnamurthy, *Phys. Rev. B* **58**, R5889 (1998).
21. C. Das, A. K. Sood, and H. R. Krishnamurthy, *Physica A* **270**, 237 (1999).
22. C. Das, P. Chaudhuri, A. K. Sood, and H. R. Krishnamurthy, *Current Science* **80**, 959 (2001).
23. D. Chaudhuri and S. Sengupta, *Europhys. Lett.* **67**, 814 (2004); *Phys. Rev. E* **73**, 011507 (2006).
24. W. Strepp, S. Sengupta, and P. Nielaba, *Phys. Rev. E* **63**, 046106 (2001).
25. K. Franzrahe and P. Nielaba, *Phys. Rev. E* **76**, 061503 (2007).
26. F. Bürzle and P. Nielaba, *Phys. Rev. E* **76**, 051112 (2007).
27. W. Strepp, S. Sengupta, M. Lohrer, and P. Nielaba, *Comp. Phys. Comm.* **147**, 370 (2002).
28. W. Strepp, S. Sengupta, and P. Nielaba, *Phys. Rev. E* **66**, 056109 (2002).

29. W. Strepp, S. Sengupta, M. Lohrer, and P. Nielaba, *Math. Comp. Simul.* **62**, 519 (2003).
30. W. Strepp, PhD thesis, University of Konstanz (2003).
31. P. Nielaba, S. Sengupta, and W. Strepp, in *Computer Simulations in Condensed Matter Systems: From Materials to Chemical Biology*, edited by M. Ferrario, G. Ciccotti, and K. Binder, Springer, Berlin (2006), 163.
32. K. S. Schweizer, R. M. Stratt, D. Chandler, and P. G. Wolynes, *J. Chem. Phys.* **75**, 1347 (1981).
33. D. M. Ceperley, *Rev. Mod. Phys.* **67**, 279 (1995).
34. K. Nho and D. P. Landau, *Phys. Rev. A* **70**, 053614 (2004); *Phys. Rev. A* **73**, 033606 (2006).
35. C. Zhang, K. Nho, and D. P. Landau, *Phys. Rev. A* **77**, 025601 (2007).
36. S. A. Khairallah, M. B. Sevryuk, D. M. Ceperley, and J. P. Toennies, *Phys. Rev. Lett.* **98**, 183401 (2007).
37. M. Boninsegni, *J. Low. Temp. Phys.* **141**, 27 (2005).
38. M. H. Müser, P. Nielaba, and K. Binder, *Phys. Rev. B* **51**, 2723 (1995).
39. M. Presber, D. Löding, R. Martoňák, and P. Nielaba, *Phys. Rev. B* **58**, 11937 (1998).
40. Chr. Rickwardt, P. Nielaba, M. H. Müser, and K. Binder, *Phys. Rev. B* **63**, 045204 (2001).
41. J. Hoffmann and P. Nielaba, *Phys. Rev E* **67**, 036115 (2003).
42. P. de Smedt, P. Nielaba, J. L. Lebowitz, J. Talbot, and L. Dooms, *Phys. Rev. A* **38**, 1381 (1988).
43. F. Schneider and P. Nielaba, *Phys. Rev. E* **54**, 5826 (1996).
44. M. Kreer and P. Nielaba, in *Monte Carlo and Molecular Dynamics of Condensed Matter Systems*, edited by K. Binder and G. Ciccotti (SIF, Bologna, 1996), p. 501.
45. J. Turnbull and M. Boninsegni, *Phys. Rev. B* **76**, 104524 (2007).

THERMODYNAMICS OF QUANTUM 2D HEISENBERG MAGNETS WITH INTERMEDIATE SPIN

A. CUCCOLI and G. GORI
Dipartimento di Fisica, Università di Firenze,
Via G. Sansone 1, 50019 Sesto Fiorentino (FI), Italy
E-mail: cuccoli@fi.infn.it

R. VAIA
Istituto dei Sistemi Complessi, Consiglio Nazionale delle Ricerche,
Via Madonna del Piano 10, 50019 Sesto Fiorentino (FI), Italy

P. VERRUCCHI
Centro di Ricerca e Sviluppo SMC dell'Istituto Nazionale di Fisica della Materia – CNR, Sezione di Firenze, Via G. Sansone 1, 50019 Sesto Fiorentino (FI), Italy

By Hamiltonian path-integration a purely-quantum, self-consistent, spin-wave approximation can be developed for spin models on a lattice, that finally allows to map the original quantum problem to a classical one ruled by an effective classical spin Hamiltonian. Such approach has revealed especially valuable to investigate systems with $S > 1/2$ which cannot be easily addressed by other methods. This has made possible to quantitatively interpret experimental data for intermediate-spin compounds and to study how different observables reach the classical limit by increasing S. Here, we focus on the spin-flop phase of a quantum 2D antiferromagnet frustrated by an applied magnetic field that acts as an effective easy-plane anisotropy and determines Berezinskii-Kosterlitz-Thouless (BKT) behavior. By acting on the field one can tune the BKT transition temperature, giving a unique opportunity to observe the otherwise elusive BKT critical behavior in real magnetic systems. The calculated data are shown to well concur with the experimental findings for the $S = 5/2$ compound manganese-formate-dihydrate.

Keywords: Hamiltonian path integral; Spin system; Magnet; BKT transition.

1. Introduction

In recent years a huge amount of data has been accumulated on low-dimensional magnetic systems like spin-chains, magnetic multilayers or quasi-2D magnetic compounds. When a quantitative interpretation of experimental data at temperatures of the order of the interaction energy is required, one of the difficulties to be faced is that in many real compounds the magnetic ions carry intermediate spin values: in this case many of the common theoretical approaches become ineffective. For instance, quantum Monte Carlo simulations become very demanding in terms of computing resources and simulation time when values of spin higher than $1/2$ are considered; at the same time, $1/S$ perturbation expansions may require too many terms to reach convergence or even become useless for describing strongly nonlinear effects like those due to soliton or vortex excitations. Indeed, the investigation of quasi-2D compounds is made even more intriguing by the presence of the topological phase transition named after Berezinskii, Kosterlitz and Thouless (BKT), which is the only phase transition expected in a purely 2D easy-plane magnet. The 'pure-quantum self-consistent harmonic' approximation[1] (PQSCHA) overcomes these difficulties by giving an effective classical Hamiltonian which incorporates the effects of quanticity in its renormalized interaction parameters: for quasi-2D isotropic Heisenberg systems such approximation has revealed its power by allowing a unified interpretation of experimental and numerical simulation data in a broad range of temperature and spin values;[2] moreover, its ability to retain the full classical nonlinearity makes it especially suitable to study strongly nonlinear phenomena as the BKT transition.

Despite magnetic systems being the prototypical models usually employed to illustrate this interesting topological critical behavior, an unambiguous observation of the BKT transition in real compounds is hindered, when approaching the critical temperature from above, by the onset of 3D order due to the residual interplane interactions. However, if one considers that a magnetic field H acts as an effective easy-plane anisotropy on an antiferromagnet and gives rise to a peculiar BKT phase diagram, $T_{\rm c}(H)$, it appears that one can prove BKT behavior for a real compound by checking that the critical curve closely follows $T_{\rm c}(H)$.

2. 2D Antiferromagnet in a Field and BKT Transition

This system is described by the Hamiltonian

$$\hat{\mathcal{H}} = \frac{J}{2}\sum_{\langle ij\rangle} \hat{\boldsymbol{S}}_i\cdot\hat{\boldsymbol{S}}_j - g\mu_{\rm B} H \sum_i \hat{S}_i^x \ , \tag{1}$$

where the spins $\{\hat{\boldsymbol{S}}_i\}$ lie on a square lattice and the first sum is over all nearest-neighbor pairs $\boldsymbol{ij}$; for field H larger than the exact saturation value $H_{\rm s} = 8JS/g\mu_{\rm B}$ its ground state is fully aligned.

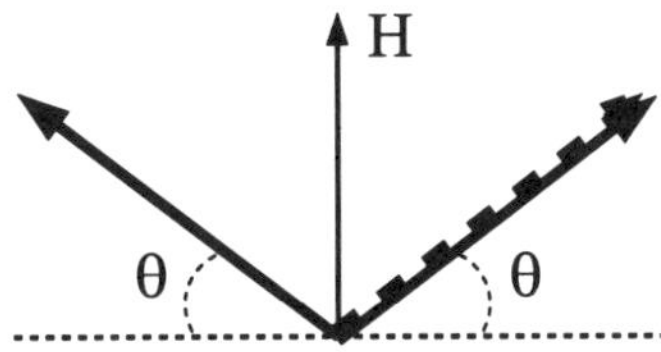

Fig. 1. The minimum energy configuration of the 2D AFM in a field, and the representation of the sublattice reflection, Eq. (2).

In the corresponding classical model the minimum-energy configuration for field $H=0$ corresponds to antiparallel spins in *whatever* direction, while a finite $H<H_{\rm s}$ favors a *canted* configuration, as shown in Fig. 1, with $\sin\theta=H/H_{\rm s}$; the S^x components are frozen and the system symmetry is reduced to $O(2)$ (rotations in the yz-plane), *i.e.*, the field acts as an effective easy-plane anisotropy: therefore, a BKT transition occurs, whose critical temperature has a distinctive dependence on H. Quantum Monte Carlo simulations[3] for $S=1/2$ prove that the quantum antiferromagnet also shows field-tunable BKT behavior: by the effective classical Hamiltonian one can obtain the quantum phase diagram $T_{\rm c}(S,H)$ for any value of S just knowing[4] the classical curve $T_{\rm c}^{(\rm cl)}(H)$.

3. Effective Hamiltonian for Canted Configurations

The approach for deriving the effective Hamiltonian corresponding to Eq. (1) is a little more complicated than that described, *e.g.*, in Ref. 5, since one must apply the Holstein-Primakoff transformation (HPT) after performing two canonical transformations in spin space, namely, a reflection of the in-plane spin components in one sublattice, such that in the minimum-energy configuration all spins are aligned (see Fig. 1),

$$\big(\hat S_i^x,\hat S_i^y,\hat S_i^z\big)\longrightarrow\big(\hat S_i^x,(-)^i\hat S_i^y,(-)^i\hat S_i^z\big) \tag{2}$$

and a rotation such that the spin z-axis points in the canted direction:

$$\hat S_i^x\longrightarrow\cos\theta\,\hat S_i^x+\sin\theta\,\hat S_i^z\,,\qquad\qquad \hat S_i^z\longrightarrow-\sin\theta\,\hat S_i^x+\cos\theta\,\hat S_i^z\,. \tag{3}$$

After these transformations, one can then apply the HPT with 'quantization axis' z, and eventually proceed with the PQSCHA recipe as in Ref. 5, taking into account that the canting angle θ is to be self-consistently determined by requiring the vanishing of the linear terms in the bosonic effective Hamiltonian. The rest of the calculations involves the application of the inverse

HPT to classical spin variables in order to recast the effective Hamiltonian[6] in the surprisingly simple form[a]

$$\mathcal{H}_{\rm eff} = \frac{J}{2}\sum_{\langle ij\rangle} \boldsymbol{S}_i\cdot\boldsymbol{S}_j - g\mu_{\rm B}H\sum_i S_i^x \;, \tag{4}$$

in terms of the classical spins $\boldsymbol{S}_i = (S_i^x, S_i^y, S_i^z)$. Remarkably, all quantum effects are included in the reduction of their classical-limit length $\widetilde{S} = S+1/2$ by a factor $\kappa<1$, i.e. $|\boldsymbol{S}_i| = \widetilde{S}\;\kappa(S,t,h)$. Using the energy scale $J\widetilde{S}^2$ to define the dimensionless temperature and field, $t=T/J\widetilde{S}^2$ and $h=g\mu_{\rm B}H/4J\widetilde{S}$, respectively, one has for κ the following expression

$$\kappa(S,t,h) = 1 - \frac{1}{2\widetilde{S}\,N}\sum_{\boldsymbol{k}} \frac{a_{\boldsymbol{k}}}{b_{\boldsymbol{k}}}\,(1+\gamma_{\boldsymbol{k}})\left(\coth f_{\boldsymbol{k}} - \frac{1}{f_{\boldsymbol{k}}}\right) , \tag{5}$$

where $\gamma_{\boldsymbol{k}} = (\cos k_x + \cos k_y)/2$, $f_{\boldsymbol{k}} = a_{\boldsymbol{k}}\,b_{\boldsymbol{k}}/2\widetilde{S}\,t$, $a_{\boldsymbol{k}}^2 = 4\,\kappa\,(1-\gamma_{\boldsymbol{k}})$, and $b_{\boldsymbol{k}}^2 = 4\,\kappa\,\big[1+(1-h^2/2\kappa^2)\gamma_{\boldsymbol{k}}\big]$. The above system of equations is self-consistent and can be easily solved numerically. Given its classical nature, when dealing with $\mathcal{H}_{\rm eff}$ it is more convenient to express it in terms of unit vectors $\boldsymbol{s}_i$ as

$$\frac{\mathcal{H}_{\rm eff}}{J\widetilde{S}^2} = \frac{j_{\rm eff}}{2}\sum_{\langle ij\rangle} \left(\boldsymbol{s}_i\cdot\boldsymbol{s}_j - 2\,h_{\rm eff}\,s_i^x\right) \;, \tag{6}$$

with the effective exchange $j_{\rm eff}(S,t,h)=\kappa^2$ and field $h_{\rm eff}(S,t,h)=h/\kappa$. The saturation field in reduced units is $h_{\rm s}=2\,S/\widetilde{S}$. It is apparent from Eq. (6) that the quantum critical line $t_{\rm c}(S,h)$ is connected to its known classical counterpart $t_{\rm c}^{\rm (cl)}(h)$ by

$$t_{\rm c}(h) = j_{\rm eff}(t_{\rm c},h)\;\times\; t_{\rm c}^{\rm (cl)}\big(h_{\rm eff}(t_{\rm c},h)\big) \;. \tag{7}$$

This equation can be solved numerically for any spin value using a fitted function for the classical data, and gives results that agree even with those from quantum Monte Carlo[3] in the extreme case $S=1/2$, as can be seen in Fig. 2.

4. Theory Versus Experiment

We also performed Monte Carlo simulations with $\mathcal{H}_{\rm eff}$ and calculated the magnetic specific heath $c_{\rm m}(t,h)$ and the uniform susceptibility $\chi_{\rm u}^x(t,h)$

[a]We omitted an additive free-energy term, irrelevant in calculating averages with $H_{\rm eff}$.

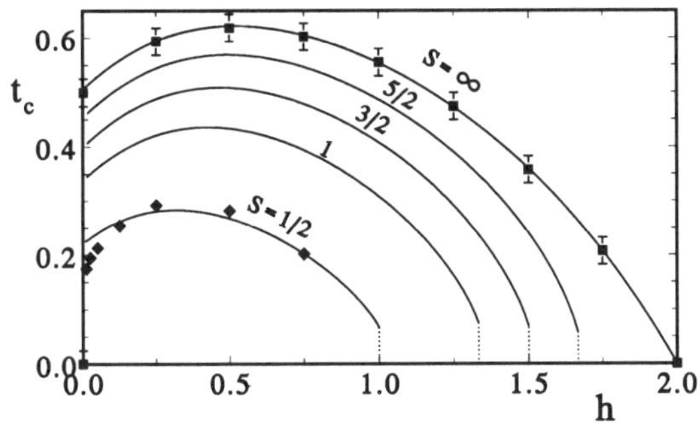

Fig. 2. Theoretical phase diagram for different spin values; classical data from Ref. 4, quantum MC data for $S=1/2$ from Ref. 3.

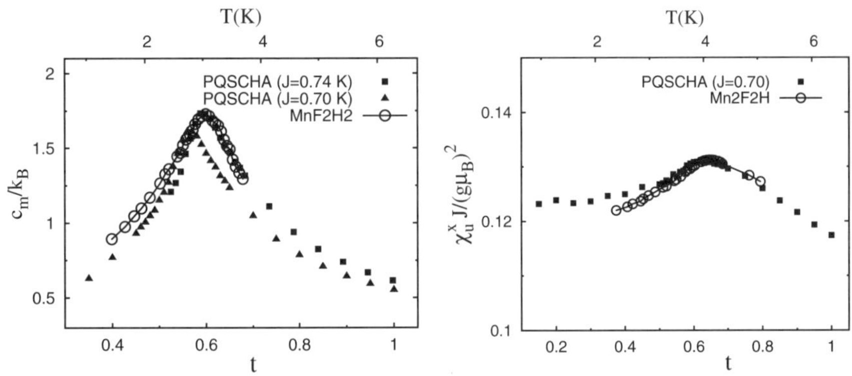

Fig. 3. Theoretical vs experimental data for temperature behavior of the magnetic specific heath $c_{\rm m}$ at $H = 50\,{\rm kG}$ and of the uniform susceptibility $\chi_{\rm u}^x$ at $H = 40\,{\rm kG}$.

for $S=5/2$, in order to compare with experimental data[7] for the compound $Mn(HCOO)_2 2H_2O$ (manganese formate dihydrate, or briefly Mn-f-2h), whose spin structure is composed of alternating antiferromagnetic and paramagnetic Mn layers with a tiny interlayer interaction. The former ones can be approximated as square-lattice QHAF, with a small anisotropy that can be neglected when the magnetic field is applied. The measured saturation field[7] $H_{\rm S} = 105 {\pm} 5$ kG can be used to estimate the exchange constant $J = g\mu_{\rm B} H_{\rm S}/8S = 0.70 \pm 0.03$ K.

In Fig. 3 the measurements at intermediate field values are reported and compared with the corresponding theoretical curves, showing that the agreement is very good if one accounts for the uncertainty in the estimated

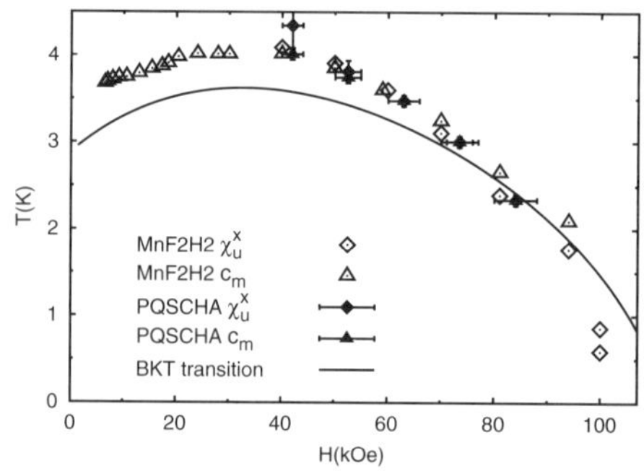

Fig. 4. Location of the maxima of the specific heat $c_m(h)$ and of the susceptibility $\chi_u^x(h)$ measured for Mn-f-2h,[7] compared with the theoretical data for the 2D AFM for $S = 5/2$, assuming $J = 0.70 \pm 0.03$. Generally, in BKT systems these maxima usually occur at a temperature 10%–20% above the critical one (line).

exchange constant. In particular, as proposed in Ref. 7, one can draw a kind of 'phase diagram' by reporting the maxima of $c_m(t,h)$ and $\chi_u^x(t,h)$ in the $h-t$ plane. We remind that in BKT systems such quantities are not critical and display maxima slightly *above* (by 10%-20%) the critical temperature $t_c(h)$. This is indeed what can be observed in Fig. 4, confirming that the overall behavior of the measured data follows the calculated BKT curve and hence that the experiments indeed observed a BKT system.

Acknowledgments

This work was accomplished within the framework of MUR-PRIN2005 research project 2005029421_004.

References

1. A. Cuccoli, V. Tognetti, P. Verrucchi, and R. Vaia, *Phys. Rev. B* **46**, 11601 (1992); A. Cuccoli, R. Giachetti, V. Tognetti, P. Verrucchi, and R. Vaia, *J. Phys.: Condens. Matter* **7**, 7891 (1995).
2. B. B. Beard, A. Cuccoli, P. Verrucchi, and R. Vaia, *Phys. Rev. B* **68**, 104406 (2003).
3. A. Cuccoli, T. Roscilde, R. Vaia, and P. Verrucchi, *Phys. Rev. B* **68**, 060402 (2003).
4. D. P. Landau and K. Binder, *Phys. Rev. B* **24**, 1391 (1981).
5. A. Cuccoli, V. Tognetti, P. Verrucchi, and R. Vaia, *Phys. Rev. B* **56**, 14456 (1997); *Phys. Rev. B* **58**, 14151 (1998).
6. G. Gori, Degree Thesis (Università di Firenze, 2007).
7. K. Takeda and K. Koyama, *J. Phys. Soc. Jap.* **52**, 648 and 656 (1983).

MICROCANONICAL METHOD FOR THE STUDY OF FIRST-ORDER TRANSITIONS

V. MARTIN-MAYOR

Departamento de Física Teórica I, Universidad Complutense de Madrid, Madrid, 28040, Spain
E-mail: victor@lattice.fis.ucm.es

A simple microcanonical strategy for the simulation of first-order phase transitions is presented. The method does not require iterative parameters optimization, nor long waits for tunneling between the ordered and the disordered phases. It is illustrated in the Q-states Potts model in two dimensions for which several exact results are known, and where a cluster method nicely works.

Keywords: Numerical simulations; Order-disorder transformations.

1. Introduction

Due to phase coexistence, systems undergoing (Ehrenfest) first-order phase transitions[1] are a challenge for Monte Carlo simulations,[2] specially for large linear system size, L (or space dimension, D). Correct sampling requires that the system tunnels between coexisting pure phases. It does so by building an interface of size L, with a free-energy cost of ΣL^{D-1} (Σ: surface tension). Therefore, the natural time scale for the simulation grows with L as $\exp[\Sigma L^{D-1}]$, and we suffer *exponential* critical slowing down (ECSD).

The invention of multicanonical Monte Carlo,[3] and the variants known as flat-histogram methods,[4] was a major step forward. The multicanonical probability for the energy density is constant in the energy gap $e^{\mathrm{o}} < e < e^{\mathrm{d}}$ (e^{o} and e^{d}: energy densities of the coexisting low-temperature ordered phase and high-temperature disordered phase). The system performs a random walk in the energy gap. Yet, one-dimensional random walk expectations are misleading: the state of L^D spins cannot be encoded into just one number (the energy). Topologically diverse configurations may have quite the same energy, see Fig. 1. So, flat-histogram methods suffer ECSD.[5]

Recently, a microcanonical method to simulate first-order transitions without iterative parameter optimization nor energy random walk was pro-

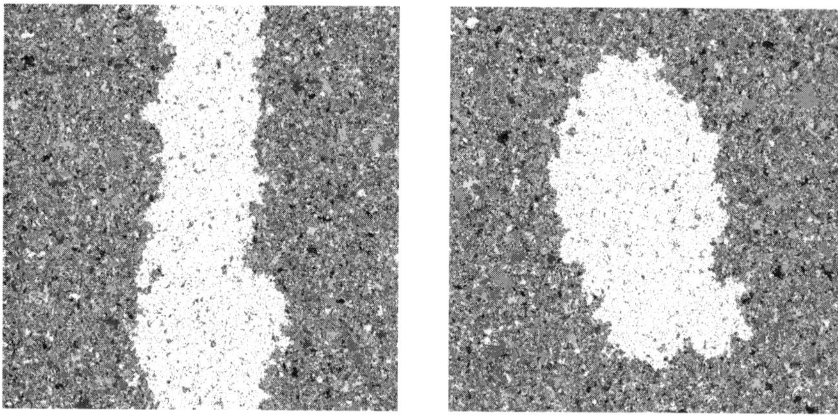

Fig. 1. $L = 1024$ equilibrium configurations for the $Q=10$, $D=2$ Potts model, at the two sides of the droplet-strip transition.[6] Periodic boundary conditions reduce the surface energy of the strip-like ordered domain (left), as compared with the droplet (right).

posed[7] (and tested in the Potts model[8]). There are two practical advantages. First, microcanonical simulations[9] are feasible at any value of e within the gap. Second, Fluctuation-Dissipation Eqs. (5) and (6) define the (inverse) temperature $\hat{\beta}$, a function of e and the spins, playing a role dual to that of e in the canonical ensemble. Since the microcanonical mean value $\langle\hat{\beta}\rangle_e$ is almost constant over the energy gap, it is easily interpolated from a grid. Please mind that, at variance with $\hat{\beta}$, $\beta = 1/T$ is a parameter (rather than a function) in the canonical ensemble.

The microcanonical method can be regarded as an artifact to perform *canonical* computations in large lattices (see also Janke[10]). Yet, a microcanonical point of view suggests a new interpretation of quenched averages in disordered systems,[11] and avoids the rare-events intrinsic to the canonical ensemble.[12] Here, a simple presentation is made of this new approach.

2. Our Microcanonical Setting

As in Hybrid Monte Carlo,[13] we add N real momenta, p_i, to our N variables, σ_i (named spins here). We work in the *microcanonical* ensemble for the $\{\sigma_i, p_i\}$ system. Let U be the spin Hamiltonian. Our total energy is

$$\mathcal{E} = \sum_{i=1}^{N} \frac{p_i^2}{2} + U\,, \quad (e \equiv \mathcal{E}/N\,,\ u \equiv U/N)\,. \tag{1}$$

In the canonical ensemble, the $\{p_i\}$ are a trivial Gaussian bath decoupled from the spins (at inverse temperature β, one has $\langle e\rangle_\beta = \langle u\rangle_\beta + 1/(2\beta)$).

In a microcanonical setting, the crucial role is played by the entropy density, $s(e,N)$, given by ($\sum_{\{\sigma_i\}}$: summation over spin configurations)

$$\exp[Ns(e,N)] = \int_{-\infty}^{\infty} \prod_{i=1}^{N} \mathrm{d}p_i \sum_{\{\sigma_i\}} \delta(Ne - \mathcal{E})\,. \tag{2}$$

Integrating out the $\{p_i\}$ using Dirac's delta function in (2) we get

$$\exp[Ns(e,N)] = \frac{(2\pi N)^{\frac{N}{2}}}{N\Gamma(N/2)} \sum_{\{\sigma_i\}} \omega(e,u,N)\,, \tag{3}$$

($\omega(e,u,N) \equiv (e-u)^{\frac{N-2}{2}}\theta(e-u)$). The step function, $\theta(e-u)$, enforces $e > u$. Equation (3) suggests to define the microcanonical average at fixed e of any function of e and the spins, $O(e,\{\sigma_i\})$, as[9]

$$\langle O \rangle_e \equiv \sum_{\{\sigma_i\}} O(e,\{\sigma_i\})\,\omega(e,u,N) \Big/ \sum_{\{\sigma_i\}} \omega(e,u,N)\,. \tag{4}$$

We use Eq. (3) to compute $\mathrm{d}s/\mathrm{d}e$:[7]

$$\frac{\mathrm{d}s(e,N)}{\mathrm{d}e} = \langle \hat{\beta}(e;\{\sigma_i\}) \rangle_e\,, \quad \hat{\beta}(e;\{\sigma_i\}) \equiv \frac{N-2}{2N(e-u)}\,. \tag{5}$$

Fluctuation-Dissipation follows by taking the derivative of Eq. (4):

$$\frac{\mathrm{d}\langle O \rangle_e}{\mathrm{d}e} = \left\langle \frac{\partial O}{\partial e} \right\rangle_e + N\left[\langle O\hat{\beta} \rangle_e - \langle O \rangle_e \langle \hat{\beta} \rangle_e \right]. \tag{6}$$

A histogram method allows to obtain an integral version of (6), to extrapolate $\langle O \rangle_{e'}$ from simulations at a neighboring e.[7]

3. Microcanonical Data Analysis

The connection between the canonical and the microcanonical ensemble follows from the saddle-point approximation for the partition function, $\langle \hat{\beta} \rangle_{\langle e \rangle_\beta} = \beta$.[a] Yet, at a first-order phase transition $\langle \hat{\beta} \rangle_e$ is not a monotonically decreasing function of e, Fig. 2, and the saddle-point equation $\langle \hat{\beta} \rangle_e = \beta$ has several roots.

We start from the exact relation between the *canonical* probability density function for e, $P_\beta^{(L)}(e) \propto \exp[N(s(e,N) - \beta e)]$, and $\langle \hat{\beta} \rangle_e$, Eq. (5):

$$\log P_\beta^{(L)}(e_2) - \log P_\beta^{(L)}(e_1) = N \int_{e_1}^{e_2} \mathrm{d}e \left(\langle \hat{\beta} \rangle_e - \beta \right). \tag{7}$$

Equation (7) allows us to interpret several saddle points:

[a] A relation best known as the second law of thermodynamics: $T\mathrm{d}s = \mathrm{d}e$.

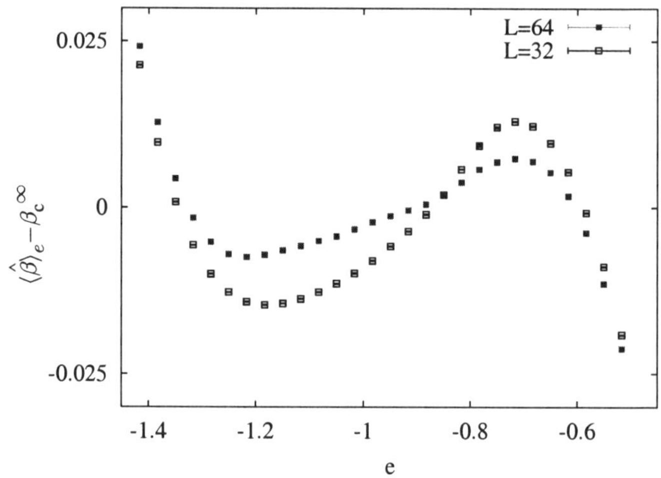

Fig. 2. Excess of $\langle\hat{\beta}\rangle_e$ over $\beta_{\mathrm{c}}^{L=\infty}$ vs. e, for the $Q = 10$, $D = 2$ Potts model and two system sizes. The local extrema of $\langle\hat{\beta}\rangle_e$ are the L-dependent spinodal points.[1]

- The rightmost root, $e^{\mathrm{d}}_L(\beta)$, is a local maximum of $P^{(L)}_\beta$ corresponding to the disordered phase.
- The leftmost root, $e^{\mathrm{o}}_L(\beta)$, is a local maximum of $P^{(L)}_\beta$ corresponding to the ordered phase.
- The *second rightmost* root, $e^*_L(\beta)$, is a local minimum of $P^{(L)}_\beta$.

Maxwell's construction yields the finite-system critical point, β_{c}^L:

$$0 = \int_{e^{\mathrm{o}}_L(\beta^L_{\mathrm{c}})}^{e^{\mathrm{d}}_L(\beta^L_{\mathrm{c}})} \mathrm{d}e \left(\langle\hat{\beta}\rangle_e - \beta^L_{\mathrm{c}} \right) , \tag{8}$$

which coincides with the standard *canonical estimator*,[14] $P^{(L)}_{\beta^L_{\mathrm{c}}}(e^{\mathrm{d}}_L(\beta^L_{\mathrm{c}})) = P^{(L)}_{\beta^L_{\mathrm{c}}}(e^{\mathrm{o}}_L(\beta^L_{\mathrm{c}}))$. In a cubic box the surface tension is estimated as[b]

$$\Sigma^L = \frac{N}{2L^{D-1}} \int_{e^*_L(\beta^L_{\mathrm{c}})}^{e^{\mathrm{d}}_L(\beta^L_{\mathrm{c}})} \mathrm{d}e \left(\langle\hat{\beta}\rangle_e - \beta^L_{\mathrm{c}} \right) . \tag{9}$$

Also the specific heat for each of the coexisting phases can be easily obtained.[7] Of course, the L-dependent estimates for Σ^L, β^L_{c}, etc. need to be extrapolated to $L \to \infty$.[15,16] Our analysis was presented elsewhere.[7]

[b] In the strip phase (Fig. 3) *two* interfaces form, hence[15] $P^{(L)}_{\beta^L_{\mathrm{c}}}(e^{\mathrm{d}}_L(\beta^L_{\mathrm{c}}))/P^{(L)}_{\beta^L_{\mathrm{c}}}(e^*_L(\beta^L_{\mathrm{c}})) = \exp(2\Sigma_L L^{D-1})$.

4. Simulations

We now specialize to the Potts model.[8] The spins $\sigma_i = 0, 1, \ldots, Q-1$, live on the $N = L^D$ nodes of a (hyper)cubic lattice of side L with periodic boundary conditions, and interaction ($\langle ij \rangle$: lattice nearest-neighbors)

$$U = -\sum_{\langle ij \rangle} \delta_{\sigma_i,\sigma_j} \,. \tag{10}$$

The Metropolis simulation of Eq. (4), as applied to (10), is straightforward. Yet, since one needs to reach topologically non-trivial configurations such as those in Fig. 1, it is preferable to have a cluster algorithm.[17–20] This can be achieved also for the microcanonical ensemble.[7] Although not rejection-free, acceptance rates are well above 50% both for a single-cluster variant,[7] or for the Swendsen-Wang-like update.[11]

The cluster method is derived from a canonical probability at inverse temperature κ, $\exp(-\kappa N u)$, that tries to falsify the microcanonical weight. Perhaps unsurprisingly, $\kappa = \langle \hat{\beta} \rangle_e$ maximizes the acceptance.[c] The Swendsen-Wang like algorithm consists of three steps:[d]

(1) Trace clusters canonically: adjacent spins belong to the same cluster only if equal and, in that case, with probability $1-\exp(-\kappa)$.
(2) For each cluster, choose with uniform probability the common value for all spins that belong to it.
(3) Let $w(e,u,\kappa) = (e-u)^{\frac{N}{2}-1} \exp(\kappa N u) \theta(e-u)$. Accept the new spin configuration, generated in step (2), with the Metropolis probability: $p(e,\kappa) = \min\{1, w(e, u^{\text{final}}, \kappa)/w(e, u^{\text{initial}}, \kappa)\}$.

The single-cluster variant,[7] that is also analogous to the canonical algorithm,[18] was used to generate the data in Figs. 2 and 3. For disordered systems,[11] the Swendsen-Wang-like alternative is preferable.

In practice, what one has is a grid of values $\{e_i, \langle \hat{\beta} \rangle_{e_i}\}$, from which the function $\langle \hat{\beta} \rangle_e$ has to be interpolated. A cubic spline,[e] combined with a Jack-knife method to estimate errors,[7] works nicely for small lattices, such as those of Fig. 2. However, in large enough systems clear droplet-strip

[c] A short Metropolis run provides a first working κ estimate.

[d] It is easy to adapt the recursive C routines in Amit and Martin-Mayor[20] to this problem.

[e] You are advised *not* to use the so called natural spline, that enforces $0 = \mathrm{d}^2 \langle \hat{\beta} \rangle_e / \mathrm{d}e^2$, both at the largest and smallest e value in the grid. Instead, we fixed the first derivative at the borders, from a 3-point parabolic fit.

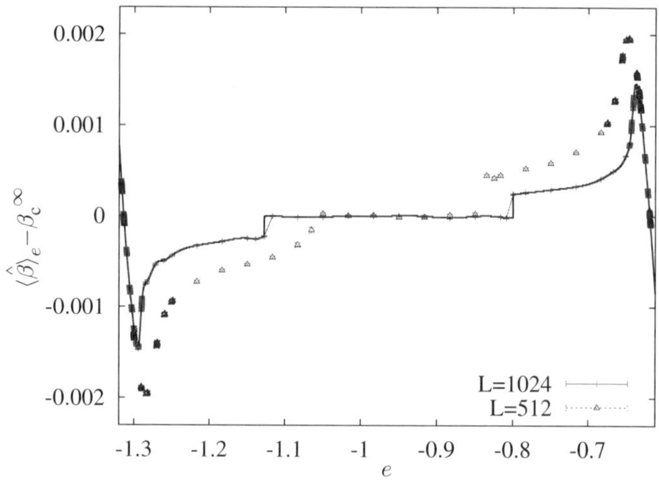

Fig. 3. As in Fig. 2 for $L = 512$ and $L = 1024$. The flat central region is the strip phase (the strip width, see Fig. 1, varies at fixed surface free energy). Lines (shown for $L = 1024$) are the two interpolations used for $L \geq 512$. We connect 3 independent cubic splines, in the strip phase and in its sides, either by a linear function or by a step-like $1/100$ power. Differences among the two interpolations are used to estimate the error induced by the uncertainty in the location of the droplet-strip transitions.

transitions are visible, see Figs. 1 and 3, causing a strong Gibbs oscillation phenomenon close to the discontinuities in $\langle\hat{\beta}\rangle_e$. Fortunately, this is fairly easy to cure by first locating the jumps, and later on performing independent interpolations on the e-intervals where the $\langle\hat{\beta}\rangle_e$ function is smooth (for technical details see caption to Fig. 3).

Following the steps outlined here, one can obtain an estimate for the (inverse) critical temperature $\beta_c^\infty = 1.426066(9)$,[7] to be compared with the exact result $1.4260624389\ldots$.

5. Conclusions

We have briefly presented a general, powerful method to simulate first-order transitions.[7] The method has been already very successful[11] on the study of the highly non-trivial problem of first-order transitions in the presence of quenched disorder.[12] Hopefully, by now the reader will be convinced of the intrinsic simplicity of this approach.

Acknowledgments

We thank D. Yllanes for carefully reading the manuscript. We were partly supported by BSCH—UCM, and by MEC through contract FIS2006-08533.

References

1. J. D. Gunton, M. S. Miguel, and P. S. Sahni, in *Phase Transitions and Critical Phenomena,*, Vol. **8** , eds. C. Domb and J. L. Lebowitz (Academic Press, New York, 1983); K. Binder, *Rep. Prog. Phys.* **50**, 783 (1987).
2. A. D. Sokal, in *Functional Integration: Basics and Applications* (1996 Cargèse school), eds. C. DeWitt-Morette, P. Cartier, and A. Folacci (Plenum, New York, 1997).
3. B. A. Berg and T. Neuhaus, *Phys. Rev. Lett.* **68**, 9 (1992).
4. F. Wang and D. P. Landau, *Phys. Rev. Lett.* **86**, 2050 (2001); Q. Yan and J. J. de Pablo, *Phys. Rev. Lett.* **90**, 035701 (2003); J. Lee, *Phys. Rev. Lett.* **71**, 211 (1993); W. Janke and S. Kappler, *Phys. Rev. Lett.* **74**, 212 (1995); Y. Wu *et al.*, *Phys. Rev. E* **72**, 046704 (2005); S. Trebst, D. A. Huse, and M. Troyer, *Phys. Rev. E* **70**, 046701 (2004); S. Reynal and H. T. Diep, *Phys. Rev. E* **72**, 056710 (2005).
5. T. Neuhaus and J. S. Hager, *J. Stat. Phys.* **113**, 47 (2003).
6. K. T. Leung and R. K. P. Zia, *J. Phys. A* **23**, 4593 (1990).
7. V. Martin-Mayor, *Phys. Rev. Lett.* **98**, 137207 (2007).
8. F. Y. Wu, *Rev. Mod. Phys.* **54**, 235 (1982).
9. R. Lustig, *J. Chem. Phys.* **109**, 8816 (1998).
10. W. Janke, *Nucl. Phys. B (Proc. Suppl.)* **63**, 631 (1998).
11. L. A. Fernandez, A. Gordillo-Guerrero, V. Martin-Mayor, and J. J. Ruiz-Lorenzo, *Phys. Rev. Lett.* **100**, 057201 (2008).
12. C. Chatelain, B. Berche, W. Janke, and P.-E. Berche, *Nucl. Phys. B* **719**, 275 (2005).
13. S. Duane, A. D. Kennedy, B. J. Pendleton, and D. Roweth, *Phys. Lett. B* **195,** 216 (1987).
14. M. S. S. Challa, D. P. Landau, and K. Binder, *Phys. Rev. B* **34**, 1841 (1986); J. Lee and J. M. Kosterlitz, *Phys. Rev. Lett.* **65**, 137 (1990).
15. K. Binder, *Phys. Rev. A* **25**, 1699 (1982).
16. C. Borgs and R. Kotecký, *Phys. Rev. Lett.* **68**, 1734 (1992).
17. R. H. Swendsen and J.-S. Wang, *Phys. Rev. Lett.* **58**, 86 (1987).
18. U. Wolff, *Phys. Rev. Lett.* **62**, 361 (1989).
19. R. G. Edwards and A. D. Sokal, *Phys. Rev. D* **38**, 2009 (1988).
20. See, e.g., D. Amit and V. Martin-Mayor, *Field Theory, the Renormalization Group and Critical Phenomena*, 3rd edition (World Scientific, Singapore, 2005).

MONTE CARLO METHODS FOR GENERATION OF RANDOM GRAPHS

B. WACLAW*,¶, L. BOGACZ†, Z. BURDA‡, and W. JANKE*,§

*Institut für Theoretische Physik, Universität Leipzig, Postfach 100 920, D-04009 Leipzig, Germany

†Department of Information Technologies, Faculty of Physics, Astronomy and Applied Informatics, Jagellonian University, Reymonta 4, 30-059 Kraków, Poland

‡Marian Smoluchowski Institute of Physics and Mark Kac Complex Systems Research Centre, Jagellonian University, Poland

§Centre for Theoretical Sciences (NTZ), Universität Leipzig, Germany

¶E-mail: bartlomiej.waclaw@itp.uni-leipzig.de

Random graphs are widely used for modeling complex networks. Instead of considering many different models, to study dynamical phenomena on networks, it is desirable to design a general algorithm which produces random graphs with a variety of properties. Here we present a Monte Carlo method based on a random walk in the space of graphs. By ascribing to each graph a statistical weight we can generate networks of different types by tuning the weight function. The algorithm allows in particular to perform multicanonical simulations known, e.g., from spin models.

Keywords: Complex networks; Random graphs; Monte Carlo methods; Multicanonical simulations.

Complex networks[1] such as the Internet, citation networks, neural or transportation networks are usually modeled by various classes of random graphs. The word "random" means that the graph is built in some stochastic process and is not fully deterministic. When one repeats the procedure of generation, one obtains different graphs, so it is natural to speak about an ensemble of random graphs. Therefore, in the language of statistical physics we can say that each graph has a certain statistical weight which determines the probability of its occurrence.

Many models were thought to mimic processes leading to the emergence of networks observed in nature, but statistical weights resulting from "physical" processes are usually not easy to guess. Sometimes, however, one only needs to generate a network with known features, in order to study some process taking place on it, e.g., the Ising model. The problem can then be formulated as follows: given a set of desired structural properties of the network, how can one generate it? A good general-purpose algorithm should be flexible to allow for the generation of trees, simple- or degenerated graphs, to adjust specific features like the degree distribution or distribution of component sizes, and be easy to implement. In this short article we present a Monte Carlo (MC) method performing this task, which has been described in details in Ref. 2. Its computer implementation has been presented in Ref. 3. The idea is similar to that previously used for example in simulations of simplicial quantum gravity, where different topologies were obtained in a sort of Markovian process by performing local updates.

To introduce a statistical ensemble of graphs we need two ingredients: a set of graphs and a set of statistical weights. Usually, one is interested in simple graphs, that is graphs without multiple- and self-connections, with given number of nodes N and links L. Thus, the set of all possible simple graphs with labeled nodes is our starting point. This set can be further restricted by demanding, e.g., that only trees, that is graphs without loops, are allowed. Or, one can impose a causal labeling of nodes which leads to so-called causal (growing) networks.[4] On the other hand, one can relax these constraints and allow to have multiple links, fluctuating N, L etc. Some of the possible choices are discussed in Refs. 2 and 5.

The question of statistical weights is in fact a question about what features are desired as "typical" in the ensemble of graphs. If one assumes that all labeled graphs are equiprobable, one obtains the ensemble of maximally random graphs (Erdös-Rényi graphs). Another, non-trivial choice is to ascribe to each graph α a product weight:

$$W(\alpha) = \prod_{i=1}^{N} p(k_i), \tag{1}$$

where k_i is the degree of the ith node and $p(k)$ is some arbitrary function. This leads to graphs with tunable degree distribution $\pi(k) = p(k)k!$, if the mean degree $\bar{k} = 2L/N$ is properly adjusted. By taking $p(k) \cong k^{-\gamma}/k!$ one can produce graphs with heavy tails[2] in $\pi(k)$. In a similar manner one can alter the typical number of triangles T in the graph, by setting $W(\alpha) = \exp(-\mu T)$, or to introduce correlations between node degrees.

As mentioned above, in a computer algorithm graphs can be recursively generated through a Markov chain by consecutive local modifications. In the simplest version, a new graph α_{t+1} is obtained from a previous one α_t by rewiring a single, randomly chosen link, to some randomly chosen node.

The move is accepted with a Metropolis probability

$$P(\alpha_t \to \alpha_{t+1}) = \min\left\{1, \frac{W(\alpha_{t+1})}{W(\alpha_t)}\right\}, \tag{2}$$

which ensures that graphs are generated with correct probabilities $W(\alpha)$. To restrict the ensemble to simple graphs only, we reject moves leading to multiple– or self–connections. To produce trees, only leaves (nodes with degree one) should be rewired. To simulate graphs with fluctuating number of links, or with fixed sequence of degrees, other moves are needed.[2] Because moves are local, consecutive graphs are correlated and one has to wait some number of steps to get two (almost) uncorrelated configurations. The initial graph can be chosen arbitrarily, for example as Erdös-Rényi graph with given N, L.

The approach presented here offers the possibility of using more sophisticated methods known, e.g., from MC simulations of spin systems. One of them is the multicanonical simulation algorithm we consider here to study systems with the degree distribution $\pi(k)$ going over many orders of magnitude. Suppose that in our ensemble (1) we have $p(k) \sim k^{-\gamma}k!$ which in the thermodynamic limit $N \to \infty$ leads to a power law: $\pi(k) \sim k^{-\gamma}$. When N is finite, however, finite-size corrections appear and $\pi(k)$ has a cut-off. How does this cut-off scales with N for various graphs? To answer this question, one needs to sample large degrees k which appear rarely and therefore the tail of $\pi(k)$ has poor statistics. To overcome this difficulty, we introduce an additional weight $r(\alpha)$, so that each graph has now the weight $W(\alpha)r(\alpha)$, and the function $r(\alpha)$ is chosen to increase the probability in the tail of $\pi(k)$. During the simulation, node degrees are collected and their probabilities of occurrence are reweighed by $1/r(\alpha)$ in order to get $\pi(k)$ for the original ensemble. The function $r(\alpha)$ can depend only on the maximal degree $k_{\max}$, because it gives the most significant contribution to the tail. Alternatively, in case of graphs where all nodes are statistically equivalent, it can depend on the degree k of some node chosen at the beginning of the simulation. The form of $r(k_{\max})$ or $r(k)$ can be iteratively obtained "on the fly".[6] This allows one to study $\pi(k)$ that are smaller than 10^{-40}, and networks of size $N \geq 40000$, magnitudes that cannot normally be reached. In Fig. 1 we show a comparison between $\pi(k)$ obtained in a simple MC and in a multicanonical simulation.

To summarize, we have proposed a method for generating graphs, similar in spirit to that used, e.g., in simulations of quantum gravity. A new graph is obtained by a local modification of the previous one, and the modification is accepted with the probability (2), so that graphs appear with

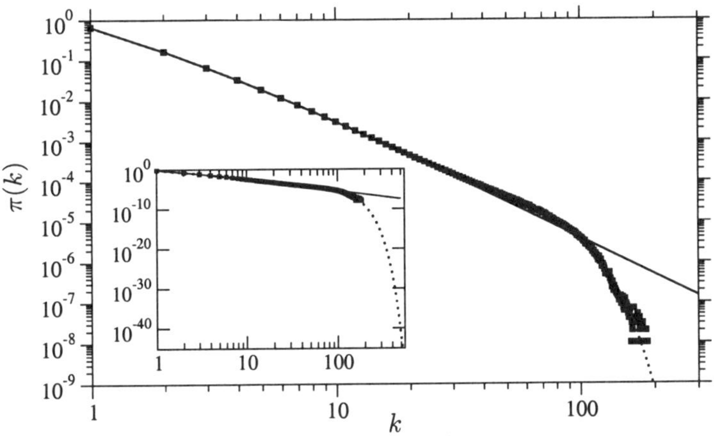

Fig. 1. Example of $\pi(k)$ for simple MC (squares) and multicanonical simulations (dotted line) for $N = 10^3$. The network is defined to be a simple, maximally random graph with Barabási-Albert degree distribution[1] $\pi(k) = 4/(k(k+1)(k+2))$ for $k > 0$, and $\pi(0) = 0$, in the thermodynamic limit (solid line). Inset: the same for a broader range of k.

the prescribed frequencies. By tuning these weights one can modify "typical" properties of a graph. The method can also be easily extended to multicanonical simulations.

Acknowledgments

This work was partially supported by EC-RTN Network "ENRAGE" under grant No. MRTN-CT-2004-005616, an Institute Partnership grant of the Alexander von Humboldt Foundation, and the German Academic Exchange Service (DAAD).

References

1. R. Albert and A.-L. Barabási, *Rev. Mod. Phys.* **74**, 47 (2002).
2. L. Bogacz, Z. Burda, and B. Waclaw, *Physica A* **366**, 587 (2006).
3. L. Bogacz, Z. Burda, W. Janke, and B. Waclaw, *Comp. Phys. Comm.* **173**, 162 (2005).
4. P. Bialas, Z. Burda, J. Jurkiewicz, and A. Krzywicki, *Phys. Rev. E* **67**, 066106 (2003).
5. S. N. Dorogovtsev, J. F. F. Mendes, and A. N. Samukhin, *Nucl. Phys. B* **666**, 396 (2003).
6. W. Janke, in: *Computer Simulations of Surfaces and Interfaces*, NATO Science Series, II. Mathematics, Physics and Chemistry – Vol. **114**, edited by B. Dünweg, D. P. Landau, and A. I. Milchev (Kluwer, Dordrecht, 2003), p. 137.

MONTE CARLO SIMULATIONS OF STOCHASTIC DIFFERENTIAL EQUATIONS AT THE EXAMPLE OF THE FORCED BURGERS' EQUATION

D. HOMEIER

Institut für Theoretische Physik, Westfälische Wilhelms-Universität Münster, Wilhelm-Klemm-Str. 9, D-48149 Münster, Germany
E-mail: homeierd@uni-muenster.de
http://pauli.uni-muenster.de

K. JANSEN*, D. MESTERHAZY†, and C. URBACH†

**DESY Zeuthen, Germany*
†Humboldt-Universität Berlin, Germany
**E-mail: karl.jansen@desy.de*

We investigate the behaviour of stochastic differential equations, especially Burgers' equation, by means of Monte Carlo techniques. By analysis of the produced configurations, we show that direct and often intuitive insight into the fundamentals of the solutions to the underlying equation, like shock wave formation, intermittency and chaotic dynamics, can be obtained. We also demonstrate that very natural constraints for the lattice parameters are sufficient to ensure stable calculations for unlimited numbers of Monte Carlo steps.

Keywords: Monte Carlo methods; Burgers' equation; Shock waves; Intermittency; Turbulence.

1. Introduction and Motivation

Hydrodynamic turbulence remains a basically unsolved problem of modern physics. This is especially noticeable as the fundamentals seem to be fairly easy – the Navier-Stokes equations for the velocity and pressure fields v and p,

$$\partial_t v_\alpha + v_\beta \partial_\beta v_\alpha - \nu \nabla^2 v_\alpha + \frac{1}{\rho} \partial_\alpha p = 0, \tag{1}$$

with the additional constraint

$$\partial_\alpha v_\alpha = 0, \tag{2}$$

express the conservation of momentum in a classical, Newtonian, incompressible fluid of viscosity ν and density ρ. Laminar flows are reproduced very accurately; in the turbulent regime, it still is an open question how the universal characteristics of a flow, the scaling exponents ξ_p of the structure functions S_p of order p, defined by

$$S_p(x) := \langle [|v(r+x) - v(r))|^p \rangle_r \propto |x|^{\xi_p}, \tag{3}$$

can be extracted from the basic equations. Intermittency is reflected by exponents ξ_p that differ from those expected by dimensional analysis. Monte Carlo simulations enable us to analyze turbulent flow patterns in detail, to gain direct insight into the formation of localized structures and their behaviour, and to measure observables like the scaling exponents.

Instead of working with the full Navier-Stokes equations, we decided to elaborate the methods using Burgers' equation

$$\partial_t v_\alpha + v_\beta \partial_\beta v_\alpha - \nu \nabla^2 v_\alpha = 0, \tag{4}$$

which can be interpreted as the flow equation for a fully compressible fluid. A finite viscosity ν and energy dissipation ϵ provide a dissipation length scale λ corresponding to the Kolmogorov-scale in Navier-Stokes turbulence:

$$\lambda = \left(\frac{\nu^3}{\epsilon} \right)^{\frac{1}{4}}. \tag{5}$$

Besides being of interest of its own (e.g. in cosmology), working with Eq. (4) has a number of technical advantages:[a]

- Burgers' equation is local, while the incompressibility condition acts as a nonlocal interaction in Navier-Stokes turbulence.
- The fundamental solutions to Burgers' equation are well-known; in the limit of vanishing viscosity (Hopf equation), these form singular shocks as seen in Fig. (1). The dissipation-scale provides an UV-regularization of the shock structures.
- A huge variety of analytical methods have been applied to Burgers equation, giving results that can directly be compared to numerical measurements; and the origin of intermittency is well understood.

2. Path Integral

For the moment, we restrict our work to 1-dimensional Burgulence which already shows intermittent statistics. A random force f has to be introduced

[a]For a more complete overview, see Ref. 2.

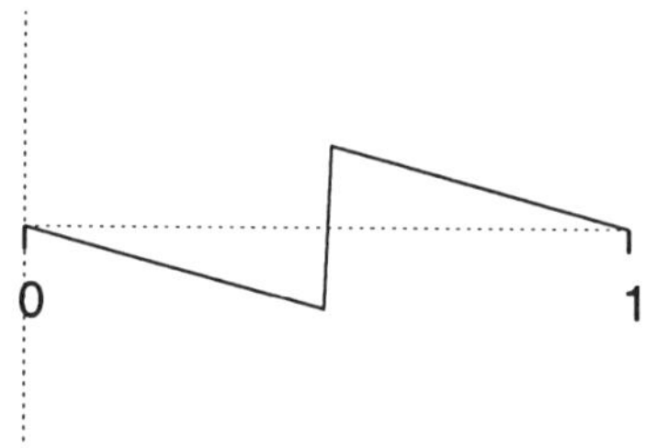

Fig. 1. Typical solution of Burgers' equation in the limit of vanishing viscosity. The graph shows $v(x,t)$ as a function of x in periodic boundary conditions at constant t.

into Eq. (4) so that solutions are statistically homogenous in time:

$$\mathbf{B}[v] \doteq \partial_t v + v\frac{\partial}{\partial x}v - \nu\frac{\partial^2}{\partial x^2}v = f, \tag{6}$$

where we model the stochastic force to be Gaussian distributed with zero mean, energy injection rate ϵ and correlation length Λ:

$$\langle f(x,t)\rangle = 0, \tag{7}$$

$$\chi^{-1}(x,x';t,t') \doteq \langle f(x,t)f(x',t')\rangle = \epsilon\delta(t-t')\exp\left(-\frac{|x-x'|}{\Lambda}\right). \tag{8}$$

We then expect intermittent statistics to be found within the inertial subrange $\lambda \ll x \ll \Lambda$. We can further identify a characteristic velocity at the injection scale, $u_0 = (\epsilon\Lambda)^{1/3}$, and a Reynolds number

$$Re = (\epsilon\Lambda^4/\nu^3)^{1/3}. \tag{9}$$

The path integral is introduced via the standard Martin-Siggia-Rose formalism, giving for the generating functional[b]

$$Z[J] = \int \mathcal{D}v\,\mathcal{D}f\,\delta(\mathbf{B}[v]-f)\exp\left(-\frac{1}{2}\int f\chi f + \int Jv\right) \tag{10}$$

$$= \int \mathcal{D}v\, e^{-S[v;J]}, \tag{11}$$

with the action

$$S[v;J] = \int dxdtdx'dt'\ (\mathbf{B}[v(x,t)]\chi(x,x';t,t')\mathbf{B}[v(x',t')] + J(x,t)v(x,t)). \tag{12}$$

[b]The functional determinant can be shown not to contribute for local theories, see e.g. Ref. 3.

Beginning from an equivalent path integral, it has been shown that intermittent statistics of Burgers' equation can be understood in terms of instanton solutions.[1]

3. Monte Carlo Simulations

We discretized the above action onto a rectangular lattice with L sites in space-, and T sites in time-direction. Derivatives have been written in a symmetric (Stratanovich-) prescription. Lattice spacings will be denoted Δx and Δt, respectively. We mainly used a heat bath algorithm on single nodes.

3.1. *Lattice parameters*

To identify the lattice parameters with the constants of the continuum theory, we first notice that the viscosity has to be defined as[c]

$$\nu = \alpha \frac{(\Delta x)^2}{\Delta t}, \tag{13}$$

in which we define the arbitrary constant $\alpha = 1$. The so-defined ν gives us Re according to Eq. (9). We further find that the dissipation length λ is related to the correlation length Λ simply by

$$\lambda = \frac{\Lambda}{Re}. \tag{14}$$

In any practical application, it is sufficient to define the η and Re, to choose L and Λ, and to calculate from that T and ϵ. Stability considerations lead to further constraints for the lattice size, as will be explained in the following subsection.

3.2. *Stability*

As would be expected, the stability of the simulations over a large number of Monte Carlo (MC) steps is a big issue, due to the shock-like solutions of Eq. (4). Indeed, if certain restrictions to the lattice parameters are not taken care of rigorously, the simulation terminates sooner or later due to divergences.

It is interesting to notice that the occurrence of instabilities in our MC simulations is related to the (non-trivial) existence of the dissipation scale

[c]For example, this can be shown via the continuum limit of the symmetric random walk, leading to the diffusion equation.

λ. We found that to obtain stable simulations, λ has to be resolved on the lattice: $\lambda > \Delta x$. Unstable simulations occur otherwise — we observed that the overall energy of the configurations accumulate in the smallest scale Δx, causing the configuration to separate into two sub-configurations, of which one looks as the expected solution to the Hopf equation, while the other grows beyond any limit, eventually breaking the simulation.

As long as the dissipation length is resolved, the simulations are stable. Having performed several millions of MC steps, no further instabilities occurred. If length scales are measured in units of the system size, this translates into a constraint involving the Reynolds number:

$$Re < \Lambda L, \tag{15}$$

which enforces huge lattices for high Re as $\Lambda \leq 1$.

3.3. *Configurations*

We simulated systems of different sizes, from $(L = 4) \times (T = 16)$ to larger lattices of the same viscosity, as 8×64 and 16×256, and also different viscosities, like 16×16 or 64×32. The Reynolds number ranged from $Re = 0.01$ to $Re = 100$, not respecting the above conditions for stability. This may seem surprising, but we could see that interesting information could also be extracted from the physical sub-configurations. Stable runs have been obtained for $Re = 1$. Long runs of several millions of MC steps, on a single node, are realistic for small lattices like 4×16 or 8×64 only. For larger lattices, a parallelized version of the code will be needed.

We obtained the following results:

- Thermalization and autocorrelation times are very long, up to the order of 10^5 MC steps.
- In the stable runs, after thermalization, the typical shock solutions of the Burgers' equation form and can be observed moving and interacting with each other, see Fig. (2).
- In the unstable runs before occurrence of the instability, configurations resemble the kink-solutions of the Hopf equation.
- The distinction between stable and unstable simulations can directly be related to the existence of a dissipation length scale which is either bigger (stable) or smaller (unstable simulations) than the lattice spacing.

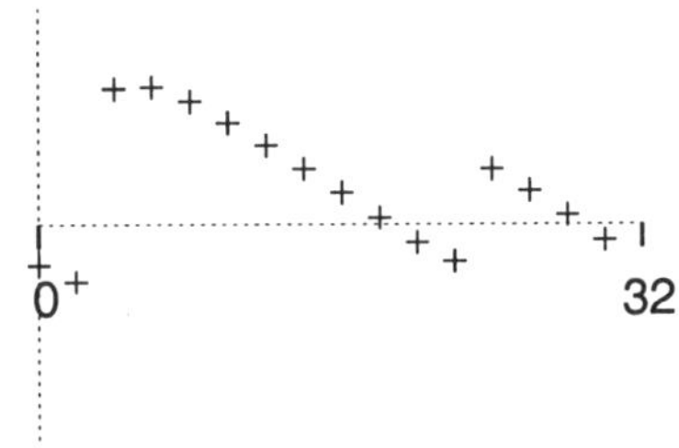

Fig. 2. Calculated configuration $v(x,t)$ of a $(L = 32) \times (T = 8)$-lattice; the plot shows a slice of constant t, periodic boundary conditions in x. Two shock-like structures are clearly visible.

4. Summary and Outlook

We have shown how to perform stable MC simulations of stochastic partial differential equations, like Burgers' or the Hopf equation. The lattice versions of the theories can directly be identified with their continuum counterparts, and, as long as certain constraints on the lattice size are respected, unlimited numbers of configurations produced. Direct insight into the structures leading to intermittency and, thus, multiscaling, can be obtained. Especially, we want to point out that the existence of a dissipation length scale can be observed.

As next steps, complete statistics will be made; especially the scaling exponents of the structure functions have to be compared with the analytic results. Later, we will proceed by analyzing the incompressible Navier-Stokes equations.

Acknowledgments

We especially want to thank Gernot Münster, Westfälische Wilhelms-Universität Münster, for his support in numerous discussions.

References

1. E. Balkovsky, G. Falkovich, I. Kolokolov, and V. Lebedev, *Phys. Rev. Lett.* **78**, 1452 (1997).
2. J. Bec and K. Khanin, arXiv:0704.1611 (nlin.CD).
3. D. Hochberg, C. Molina-París, J. Pérez-Mercader, and M. Visser, *Phys. Rev. E* **60**, 6343 (1999).

PATH INTEGRALS IN LATTICE QUANTUM CHROMODYNAMICS

F. X. LEE
Physics Department, George Washington University, Washington, DC 20052, USA
E-mail: fxlee@gwu.edu

I discuss the use of path integrals to study strong-interaction physics from first principles. The underlying theory is cast into path integrals which are evaluated numerically using Monte Carlo methods on a space-time lattice. Examples are given in progress made in nuclear physics.

Keywords: Path integrals; Lattice QCD; Strong interaction; Monte Carlo.

1. Introduction

Quantum chromodynamics (or QCD) is the underlying theory of the strong interaction (one of the four fundamental interactions in nature besides gravity, electromagnetism and the weak interaction). It is the force that binds quarks and gluons as the nucleus in the heart of the atom. Since all matter in the universe is built that way, unraveling the structure of matter at its deepest level as governed by QCD is key to our understanding of the physical world, and presents one of the most challenging tasks in contemporary nuclear and particle physics. While it is easy to write down the basic equation of QCD, it is very difficult to obtain quantitative solutions in view of the complex quark-gluon dynamics. At present, the only known way to solve QCD directly is by numerically evaluating the path integrals in the theory on a discrete space-time lattice using supercomputers. For textbooks on the subject, see Refs. 1–3.

2. Formalism

All physics can be computed by path integrals in QCD. Take the calculation of the proton mass for example. It requires the fully-interacting quark propagator defined by the Euclidean-space path integrals over gluon field

A_μ and quarks fields ψ and $\bar{\psi}$,

$$\langle M^{-1}\rangle \equiv \langle 0|\psi(x)\bar{\psi}(0)|0\rangle \equiv \frac{\int D\psi D\bar{\psi} DA_\mu e^{-S_{QCD}} M^{-1}}{\int D\psi D\bar{\psi} DA_\mu e^{-S_{QCD}}}. \tag{1}$$

The action is the sum of a gluon part and a quark part

$$S_{QCD} = S_G + S_q = \frac{1}{2}\int dx^4 \mathrm{Tr} F_{\mu\nu} F^{\mu\nu} + \int dx^4 \bar{\psi} M \psi, \tag{2}$$

where $F_{\mu\nu} = \partial_\mu A_\nu - \partial_\nu A_\mu + g[A_\mu, A_\nu]$ is the gluon field strength tensor and $M = \gamma^\mu D_\mu + m_q$ is the Dirac operator for quarks and γ^μ the 4x4 gamma matrices. The interaction between the two is provided by the covariant derivative $D_\mu = \partial_\mu + gA_\mu$. The basic parameters of the theory are the coupling constant g and quark mass m_q. The quark part of the integration can be done analytically using Grassmann variable integration, leading to a path integral over only gluon fields

$$\langle M^{-1}\rangle \equiv \frac{\int DA_\mu \det(M) e^{-S_G} M^{-1}}{\int DA_\mu \det(M) e^{-S_G}} \tag{3}$$

which is evaluated numerically by Monte Carlo methods. In the quenched approximation, the quark determinant $\det(M)$ is ignored (equivalent to suppressing vacuum polarization), making the numerical calculation significantly faster.

To compute the proton mass, one considers the time-ordered, two-point correlation function in the QCD vacuum, projected to zero momentum:

$$G(t) = \sum_{\boldsymbol{x}} \langle 0\,|\, T\{\, \eta(x)\, \bar{\eta}(0)\, \}\,|\, 0\rangle. \tag{4}$$

Here η is called the interpolating field which is built from quark fields with the quantum numbers of the proton (spin-1/2, isospin-1/2, quark content uud, charge +e)

$$\eta(x) = \epsilon^{abc} \left[u^{aT}(x) C\gamma_5 d^b(x) \right] u^c(x) \tag{5}$$

where C is the charge conjugation operator and the superscript T means transpose. Sum over color indices a,b,c is implied and the ϵ^{abc} ensures the proton is color-singlet.

The calculation of $G(t)$ at the quark level proceeds by contracting out all the quark pairs, resulting in

$$G(t) = \sum_{\boldsymbol{x}} \epsilon^{abc} \epsilon^{a'b'c'} \left\{ S_u^{aa'} \gamma_5 C S_d^{cc'^T} C\gamma_5 S_u^{bb'} + S_u^{aa'} \mathrm{Tr}(C S^{cc'^T}_{\ d} C\gamma_5 S_u^{bb'} \gamma_5) \right\}, \tag{6}$$

where S_q^{ij} denotes the fully-interacting quark propagator $\langle M^{-1}\rangle$ in Eq. (3).

On the hadronic level, the correlation function is saturated by the complete spectrum of intermediate states with the proton as the ground state:

$$G(t) = \sum_i \lambda_i^2 \, e^{-m_i t}, \tag{7}$$

where m_i are the masses and λ^2 are the 'amplitudes' which are a measure of the ability of the interpolating field to excite or annihilate the states from the QCD vacuum. At large time, the ground state proton, $\lambda_1^2 \, e^{-m_1 t}$, dominates $G(t)$, with the excited states exponentially suppressed. Other physics quantities are computed in a similar way.

3. Some Examples

It is impossible to give a full account of the achievements made in lattice QCD in the space given here. A good place to gauge the progress in the field (which is not limited to QCD) is the annual gathering[4] by active practitioners. Here I select a few examples relevant to nuclear physics.

Example 1. Fig. 1 shows that the computed light hadron mass spectrum which comes within 10% of the observed spectrum. The remaining discrepancy is attributed to the quenched approximation. Intense efforts are under way to extend the success to the excited sectors of the mass spectrum.[6]

Example 2. Fig. 2 shows the high-precision lattice QCD calculations for selected physics quantities which are within 3% of the observed values. The effects of the quenched approximation are clearly demonstrated.

Example 3. Fig. 3 shows a calculation of the magnetic moment of the proton and neutron which find good agreement between lattice QCD and experiment. Results from a different method can be found in Ref. 9, as well as precise determinations of the strangeness magnetic (G_M^s) (Ref. 10) and electric form factors (G_E^s) (Ref. 11) of the proton.

Example 4. Fig. 4 shows the good agreement between lattice QCD and experiment for pion-pion scattering length,[12] using Lüscher's method.[13] Other hadron-hadron scattering channels have been studied by the same group.

Example 5. Fig. 5 shows a direct lattice QCD calculation of the nucleon-nucleon potential[14] whose features are consistent with the known phenomenological features of the nuclear force.

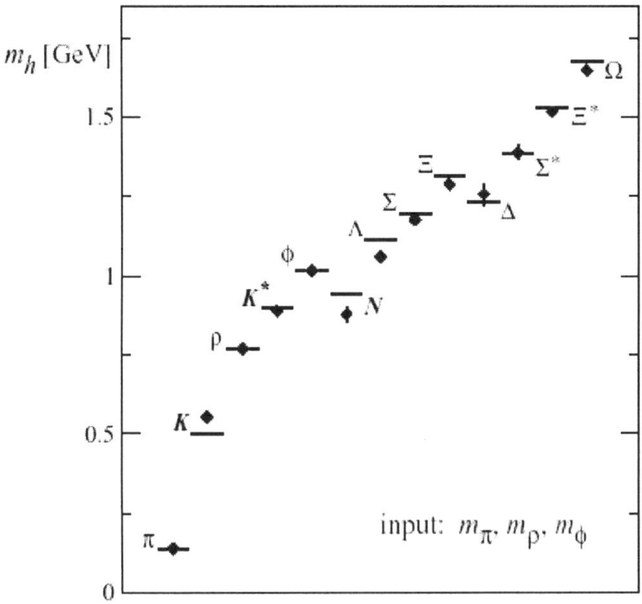

Fig. 1. The computed light hadron mass spectrum in the quenched approximation. See Ref. 5 for details.

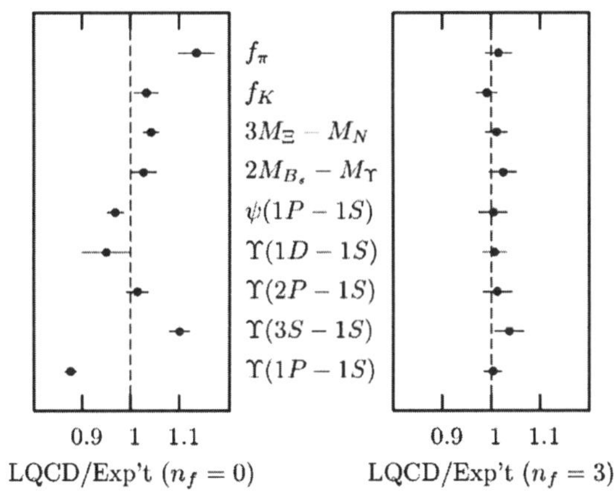

Fig. 2. Lattice QCD results compared to experiment as ratios for selected physics quantities (the dashed line is perfect agreement), in the quenched approximation of QCD (left panel) and full QCD (right panel). See Ref. 7 for details.

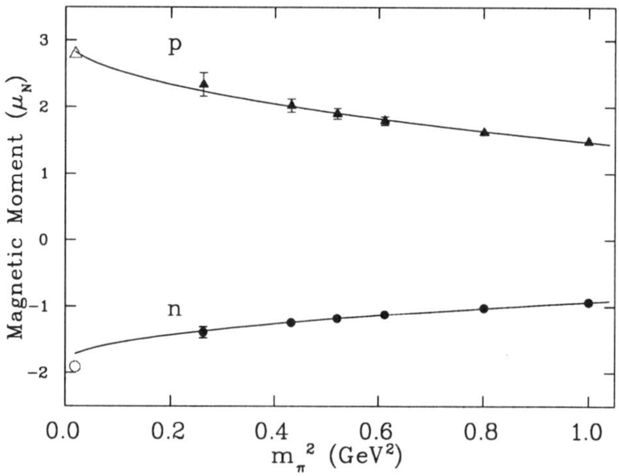

Fig. 3. Magnetic moment of the proton and neutron as a function of the pion mass squared. The physical points correspond to pion mass of 138 MeV. The line is a chiral extrapolation. See Ref. 8 for details.

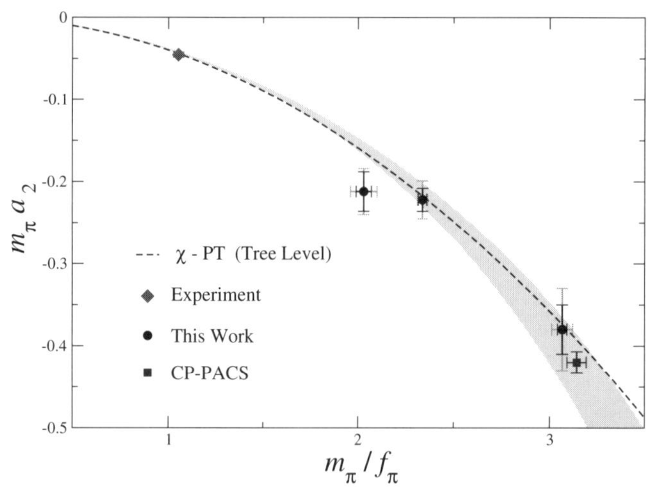

Fig. 4. Lattice QCD calculation of the $\pi - \pi$ scattering length in the isospin 2 channel. The dashed line is a chiral fit. See Ref. 12 for details.

4. Conclusion

The path-integral formulation of QCD, coupled with large-scale numerical simulations, has played a crucial role in the progress of nuclear and particle

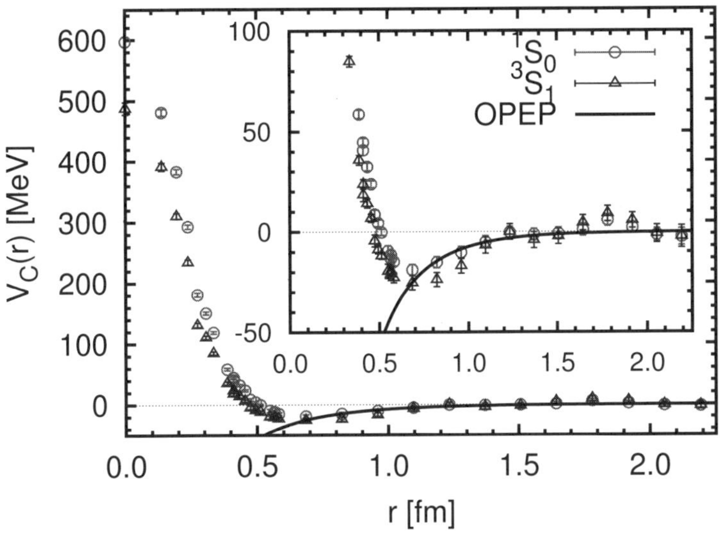

Fig. 5. The lattice QCD result of the central (effective central) part of the NN potential in the $^1S_0$ ($^3S_1$) channels. The inset shows its enlargement. The solid lines correspond to the one-pion exchange potential (Yukawa potential). See Ref. 14 for details.

physics. Its continued prominence as the only way to solve QCD with controlled errors is expected to play out with better results in the foreseeable future.

Acknowledgment

This work is supported in part by U.S. Department of Energy under grant DE-FG02-95ER40907.

References

1. H. J. Rothe, *Lattice Gauge Theories: An Introduction*, 3rd edition (World Scientific, Singapore, 2005).
2. I. Montvay and G. Münster, *Quantum Fields on the Lattice* (Cambridge University Press, Cambridge, 1997).
3. T. DeGrand and C. DeTar, *Lattice Methods for Quantum Chromodynamics* (World Scientific, Singapore, 2006).
4. The proceedings of the annual International Symposium on Lattice Field Theory are freely available at `http://pos.sissa.it`. Another free resource is the dedicated e-print archive `http://arxiv.org/list/hep-lat`, updated daily.
5. S. Aoki *et al.* [CP-PACS Collaboration], *Phys. Rev. Lett.* **84**, 238 (2000).

6. S. Basak *et al.* [LHPC Collaboration], arXiv:0709.0008 (hep-lat).
7. C. T. H. Davies *et al.* [HPQCD, UKQCD, MILC, and Fermilab Lattice Collaboration], *Phys. Rev. Lett.* **92**, 022001 (2004).
8. F. X. Lee, R. Kelly, L. Zhou, and W. Wilcox, *Phys. Lett. B* **627**, 71 (2005).
9. S. Boinepalli *et al.*, *Phys. Rev. D* **74**, 093005 (2006).
10. D. B. Leinweber *et al.*, *Phys. Rev. Lett.* **94**, 212001 (2005).
11. D. B. Leinweber *et al.*, *Phys. Rev. Lett.* **97**, 022001 (2006).
12. S. Beane *et al.* [NPLQCD Collaboration], *Phys. Rev. D* **73**, 054503 (2006).
13. M. Lüscher, *Nucl. Phys. B* **354**, 531 (1991).
14. N. Ishii, S. Aoki, and T. Hatsuda, *Phys. Rev. Lett.* **99**, 022001 (2007).

PART VII

Bose-Einstein Condensation

TACKLING FLUCTUATION CORRECTIONS IN THE BEC/BCS CROSSOVER AT NONZERO TEMPERATURES

J. TEMPERE*,†, S. N. KLIMIN*, and J. T. DEVREESE*

*Theoretische Fysica van de Vaste Stoffen, Departement Natuurkunde, Universiteit Antwerpen, Groenenborgerlaan 171, B-2020 Antwerpen, Belgium

†Lyman Laboratory of Physics, Harvard University, Cambridge MA02138, USA

E-mail: Jacques.Tempere@ua.ac.be

The investigation of superfluid atomic Fermi gases in the regime of strong interactions is conveniently investigated with the path-integral method at temperature zero, or at the critical temperature where the gap vanishes, by taking particle-pair or hole-pair fluctuations into account. Here, we also take the particle-hole excitations into account, which is important to investigate intermediate temperatures. The additional terms in the fluctuation propagator are identified, and a contour integral representation is used to calculate the contribution of these terms to the free energy and to the density of noncondensed fermions.

Keywords: BEC/BCS transition; Fermi superfluid; Fluctuations.

1. Introduction

Ultracold atomic Fermi gases provide a 'quantum laboratory' for studying the crossover physics between the Bardeen-Cooper-Schrieffer (BCS) state and a Bose-Einstein condensate (BEC) of molecules. In the experimental realization of the crossover superfluid,[1] many parameters are experimentally tunable: the interaction strength, the temperature, the geometry, and the individual amount of fermions in each hyperfine spin state contributing to pairing. These experimental advances have renewed the theoretical interest in this particular problem. In this contribution, we use the functional integral description of the system and consider the problem of incorporating quantum fluctuations at finite temperatures in the description of the thermodynamics of the superfluid. A recent review of the state-of-the-art in the calculation of Gaussian fluctuations is given in Ref. 2. In these calculations, the terms in the fluctuation propagator that vanish at either zero temperature or zero order parameter are omitted. The former allows to derive the temperature-zero results, and the latter allows to calculate the critical temperature. However, to describe the superfluid at the intermediate temperatures, we need to retain these terms. This is the subject of the present calculation.

2. Fluctuation Propagators

The partition sum of an interacting Fermi gas with two hyperfine states is characterized by an imaginary time functional integral over Grassmann fields $\bar{\psi}$ and ψ,

$$Z = \int \mathcal{D}\bar{\psi}_{x\sigma}\mathcal{D}\psi_{x\sigma} \exp\left\{-\mathcal{S}[\bar{\psi}_{x\sigma}, \psi_{x\sigma}]\right\} \tag{1}$$

where σ is the hyperfine spin coordinate and $x = \{\mathbf{r}, \tau\}$ with $\mathbf{r}$ the spatial coordinate and τ the imaginary time, and the action functional is given by

$$\mathcal{S} = \int dx \left\{\sum_\sigma \bar{\psi}_{x\sigma}\left[\partial_\tau - \frac{1}{2m}\nabla^2_{\mathbf{x}} - \mu\right]\psi_{x\sigma} + g\bar{\psi}_{x,\uparrow}\bar{\psi}_{x,\downarrow}\psi_{x,\downarrow}\psi_{x,\uparrow}\right\} \tag{2}$$

Here, μ is the chemical potential fixing the density $n = k_F^3/(3\pi^2)$ of fermionic atoms and

$$\int dx = \int d^3\mathbf{x} \int_0^\beta d\tau \tag{3}$$

with $\beta = 1/(k_B T)$ the inverse temperature. The interactions between the fermions occur only between atoms in different hyperfine states (indicated as $\sigma =\uparrow$ and $\sigma =\downarrow$, respectively), and are well characterized as contact interactions with a bare coupling strength g. This bare coupling strength g is related to the s-wave scattering length a_s through a renormalization relation

$$\frac{m}{4\pi a_s} = \frac{1}{g} + \sum_{|k|<\Lambda} \frac{m}{k^2} \tag{4}$$

where Λ is a cut-off wavevector that can safely be taken to infinity at the end of the calculation. Performing the Hubbard-Stratonovic transformation introduces the bosonic pair fields $\Delta_x, \bar{\Delta}_x$ which couple to $\bar{\psi}_{x,\uparrow}\bar{\psi}_{x,\downarrow}$ and $\psi_{x,\downarrow}\psi_{x,\uparrow}$, respectively. Integrating over the Grassmann variables then yields the well-known result[3,4]

$$Z = \int \mathcal{D}\bar{\Delta}_x\mathcal{D}\Delta_x \exp\left\{-\mathcal{S}[\Delta_x, \bar{\Delta}_x]\right\} \tag{5}$$

where the original action functional is transformed into a new action functional for the bosonic pair field

$$\mathcal{S}[\Delta_x, \bar{\Delta}_x] = \int dx \left\{ \mathrm{tr}\left[\ln\left(-\mathbb{G}^{-1}\right)\right] - \frac{\bar{\Delta}_x \Delta_x}{g} \right\} \tag{6}$$

where $\mathbb{G}^{-1}$ is the inverse Nambu Green's function

$$\mathbb{G}^{-1} = \begin{pmatrix} -\partial_\tau + \frac{1}{2m}\nabla_{\mathbf{x}}^2 + \mu & \Delta_x \\ \bar{\Delta}_x & -\partial_\tau - \frac{1}{2m}\nabla_{\mathbf{x}}^2 - \mu \end{pmatrix} \tag{7}$$

and the trace is taken over the spin variables. Expressions (5)–(7) provide an exact result for the partition sum, and hence the thermodynamics of the Fermi gas. However, the remaining functional integral over the bosonic fields cannot be performed analytically. It is possible to use a saddle-point result, approximating the functional integral by the value where its integrand becomes maximal. This gives good results at low temperature, and in the BCS regime ($a_s < 0$), but fails to incorporate the quantum fluctuations necessary for nonzero temperatures and in the strong coupling regime. The obvious way out, introducing a second collective field to improve the description of the noncondensed bosonic pairs, fails due to double-counting,[5] which can be remedied using the variational perturbation method.[6] Here, we use a different approach,[7,8] based on the Bogoliubov shift:

$$\Delta_x = \left(\bar{\Delta}_x\right)^* = \Delta + \eta_x \tag{8}$$

Here Δ is a saddle-point value, and η_x is a bosonic field representing the fluctuations around the saddle point. The saddle point value is independent of x, since it is interpreted as the order parameter of the condensate of pairs, which condense in the momentum zero, i.e. uniform density state. The saddle point serves to factorize out the largest contribution. The fluctuations are assumed to be small so that an expansion of the action, Eq. (6), up to second order in η_x, captures the essence of the thermodynamics of the noncondensed pairs. For the quadratic expansion, the remaining functional integral over η_x can be performed analytically. The result can be written as

$$Z = \exp\left\{-\mathcal{S}_{sp} - \mathcal{S}_{fl}\right\}. \tag{9}$$

The action is the sum of a saddle point action $\mathcal{S}_{sp} = \beta F_{sp}$, where F_{sp} is the saddle-point free energy (for the condensed pairs), and a fluctuation action $\mathcal{S}_{fl} = \beta F_{fl}$, where F_{fl} is the contribution to the free energy coming from the noncondensed pairs. The saddle-point free energy is

$$F_{fl} = -\frac{m\left|\Delta\right|^2}{4\pi a_s} - \sum_{\mathbf{k}} \left(\frac{2}{\beta} \ln\left[2\cosh(\beta E_{\mathbf{k}}/2)\right] - \xi_{\mathbf{k}} \right) \tag{10}$$

with

$$\xi_{\mathbf{k}} = k^2/(2m) - \mu,$$
$$E_{\mathbf{k}} = \sqrt{\xi_{\mathbf{k}}^2 + |\Delta|^2}.$$

The fluctuation action is given by

$$\mathcal{S}_{fl} = \frac{1}{2} \sum_{\mathbf{q}} \sum_{i\Omega_m} \ln \left\{ \det \left[\mathbb{M}\left(\mathbf{q}, i\Omega_m \right) \right] \right\} \tag{11}$$

with $\mathbb{M}$ the inverse Nambu fluctuation propagator, a 2×2 matrix like $\mathbb{G}^{-1}$, as a function of the wave vector $\mathbf{q}$ and the bosonic Matsubara frequency $i\Omega_m = i2\pi m/\beta$. The diagonal elements of the propagator are related by $\mathbb{M}_{22}(\mathbf{q}, z) = \mathbb{M}_{11}(\mathbf{q}, -z)$ with

$$\begin{aligned} \mathbb{M}_{11}(\mathbf{q}, z) = -\frac{1}{g} + \int \frac{d\mathbf{k}}{(2\pi)^3} \Bigg[& \frac{E_{\mathbf{k}} + \xi_{\mathbf{k}}}{2E_{\mathbf{k}}} \frac{E_{\mathbf{k+q}} + \xi_{\mathbf{k+q}}}{2E_{\mathbf{k+q}}} \frac{1}{z - E_{\mathbf{k}} - E_{\mathbf{k+q}}} \\ & - \frac{E_{\mathbf{k}} - \xi_{\mathbf{k}}}{2E_{\mathbf{k}}} \frac{E_{\mathbf{k+q}} + \xi_{\mathbf{k+q}}}{2E_{\mathbf{k+q}}} \frac{1}{z + E_{\mathbf{k}} - E_{\mathbf{k+q}}} \\ & + \frac{E_{\mathbf{k}} + \xi_{\mathbf{k}}}{2E_{\mathbf{k}}} \frac{E_{\mathbf{k+q}} - \xi_{\mathbf{k+q}}}{2E_{\mathbf{k+q}}} \frac{1}{z - E_{\mathbf{k}} + E_{\mathbf{k+q}}} \\ & - \frac{E_{\mathbf{k}} - \xi_{\mathbf{k}}}{2E_{\mathbf{k}}} \frac{E_{\mathbf{k+q}} - \xi_{\mathbf{k+q}}}{2E_{\mathbf{k+q}}} \frac{1}{z + E_{\mathbf{k}} + E_{\mathbf{k+q}}} \Bigg] \tanh(\beta E_{\mathbf{k}}/2) \end{aligned} \tag{12}$$

Here we have used the analytical continuation of $\mathbb{M}_{11}\left(\mathbf{q}, i\Omega_m\right)$ in the complex plane. The off-diagonal elements $\mathbb{M}_{21}(\mathbf{q}, z) = \mathbb{M}_{12}(\mathbf{q}, -z)$ are given by

$$\begin{aligned} \mathbb{M}_{12}(\mathbf{q}, z) = |\Delta|^2 \int \frac{d\mathbf{k}}{(2\pi)^3} \frac{\tanh(\beta E_{\mathbf{k}}/2)}{4E_{\mathbf{k}}E_{\mathbf{k+q}}} \Bigg[& \frac{1}{z - E_{\mathbf{k}} - E_{\mathbf{k+q}}} - \frac{1}{z - E_{\mathbf{k}} + E_{\mathbf{k+q}}} \\ & + \frac{1}{z + E_{\mathbf{k}} - E_{\mathbf{k+q}}} - \frac{1}{z + E_{\mathbf{k}} + E_{\mathbf{k+q}}} \Bigg] = \mathbb{M}_{12}(\mathbf{q}, -z) \end{aligned} \tag{13}$$

The current treatments of the BEC/BCS crossover[2,9–12] are valid for temperature zero or order parameter zero, and can be rewritten in the current formalism. However, they neglect in the above formula the terms that contain a denominator $z + E_{\mathbf{k}} - E_{\mathbf{k+q}}$ or $z - E_{\mathbf{k}} + E_{\mathbf{k+q}}$, and only keep the terms with denominators $z - E_{\mathbf{k}} - E_{\mathbf{k+q}}$ and $z + E_{\mathbf{k}} + E_{\mathbf{k+q}}$ (compare the above equations for example with Eqs. (21), (22) of Ref. 2). The neglected terms, however, are important to evaluate the nonzero temperature results. However, care should be taken in evaluating these integrals. We outline our procedure in the next section and focus on the density of the noncondensed pairs as an example.

3. Fluctuation Density

The free energy contribution from the fluctuations can be written as

$$F_{fl} = \frac{1}{2\beta}\sum_{i\Omega_m}\int\frac{d\mathbf{q}}{(2\pi)^3}\ln\left[\Gamma(\mathbf{q},z)\right]_{z=i\Omega_m} \tag{14}$$

with

$$\Gamma(\mathbf{q},z) = \mathbb{M}_{11}(\mathbf{q},z)\mathbb{M}_{11}(\mathbf{q},-z) - \left[\mathbb{M}_{12}(\mathbf{q},z)\right]^2 \tag{15}$$

The density of fermions contributing to the condensed pairs is $n_{sp} = -\partial F_{fl}/\partial\mu$ and the density of fluctuations is then given by

$$n_{fl} = -\frac{\partial F_{fl}}{\partial\mu} = \frac{1}{2\beta}\sum_{i\Omega_m}\int\frac{d\mathbf{q}}{(2\pi)^3}J(\mathbf{q},z) \tag{16}$$

with

$$J(\mathbf{q},z) = \frac{1}{\Gamma(\mathbf{q},z)}\left[\mathbb{M}_{11}(\mathbf{q},-z)\frac{\partial\mathbb{M}_{11}(\mathbf{q},z)}{\partial\mu} - \mathbb{M}_{12}(\mathbf{q},-z)\frac{\partial\mathbb{M}_{12}(\mathbf{q},z)}{\partial\mu}\right] \tag{17}$$

The function $J(\mathbf{q},z)$, needed for the calculation of n_{fl}, is analytical in the complex z-plane except for poles on the real axis. In the calculation of F_{fl}, the function $\ln[\Gamma(\mathbf{q},z)]$ also has poles on the real axis, and moreover a branch line on the real axis. So, when transforming the Matsubara summation into its complex integral representation, one should avoid the real axis. We found that the optimal contour is a large circle in the upper complex plane, bounded by $z = i\pi/\beta$, complemented by a large circle in the lower complex plane, bounded by the line $z = -i\pi/\beta$. This choice of contour (cf. Ref. 2) leads in our case to

$$n_{fl} = -\int\frac{d\mathbf{q}}{(2\pi)^3}\left\{\frac{1}{\pi}\int_{-\infty}^{\infty}\mathrm{Im}\left[\frac{J(\mathbf{q},\omega+i\pi/\beta)}{e^{\beta(\omega+i\pi/\beta)}-1}\right]d\omega + \frac{1}{\beta}J(\mathbf{q},0)\right\} \tag{18}$$

which lends itself to numerical evaluation. We then find the gap equation from $\partial F_{sp}/\partial|\Delta| = 0$, and the number equation from

$$n_{sp} + n_{fl} = k_F^3/(3\pi^2) \tag{19}$$

These two equations allow us to evaluate the gap $|\Delta|$ and the chemical potential μ, when the interaction strength a_s and the density (through k_F) are given.

4. Results and Conclusion

The saddle-point results for the gap and the chemical potential are shown in the top panel of Figure 1; the bottom panel shows the results when fluctuations are included. In the present treatment, not only the particle-particle and the hole-hole contributions are taken into account, but also particle-hole contributions are kept in Eqs. (12) and (13). The result is that the smooth temperature dependence of $|\Delta|$ and μ that follows from the saddle-point result, is changed into more abrupt changes near the critical temperature. Note that even when $|\Delta| = 0$, fluctuation effects are present; in the normal state they reflect the effect of the interactions on the normal Fermi gas. So, whereas the normal state in the saddle-point treatment is the ideal gas, it is an interacting Fermi gas when Gaussian fluctuations beyond mean field are taken into account.

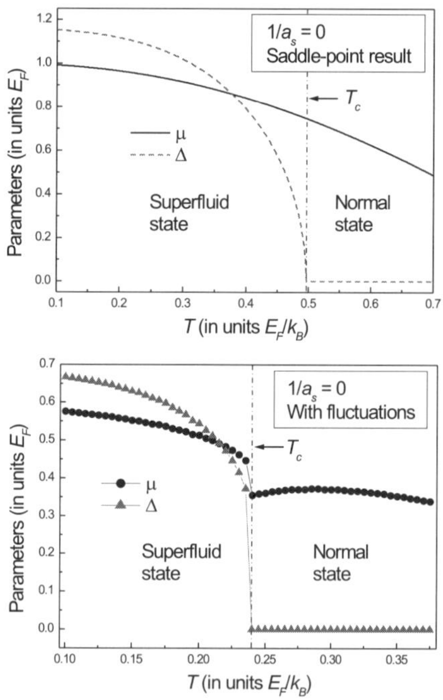

Fig. 1. The superfluid order parameter $|\Delta|$ and the chemical potential μ can be determined as a function of temperature from the gap equation $\partial F_{sp}/\partial |\Delta| = 0$ and the number equation $n_{sp} = k_F^3/(3\pi^2)$, yielding the saddle point results (top panel). When fluctuations are included, the number equation is modified to $n_{sp} + n_{fl} = k_F^3/(3\pi^2)$, and we find the results of the bottom panel.

Acknowledgments

Discussions with M. Wouters and D. Lemmens are gratefully acknowledged. This work has been supported by the FWO-Vl. projects Nos. G.0356.06, G.0115.06, G.0435.03, G.0306.00 and G.0449.04, and the WOG WO.035.04N (Belgium). J. T. gratefully acknowledges support of the Special Research Fund of the University of Antwerp, BOF NOI UA 2004.

References

1. C. A. Regal, M. Greiner, and D. Jin, *Phys. Rev. Lett.* **92**, 040403 (2004); M. W. Zwierlein *et al.*, *Phys. Rev. Lett.* **92**, 120403 (2004); T. Bourdel *et al.*, *Phys. Rev. Lett.* **93**, 050401 (2004); G. B. Partridge *et al.*, *Phys. Rev. Lett.* **95**, 020404 (2005); M. Greiner, C. A. Regal, and D. Jin, *Nature* **426**, 537 (2003); S. Jochim *et al.*, *Science* **302** (2003); M. W. Zwierlein *et al.*, *Phys. Rev. Lett.* **91**, 120403 (2003).
2. R. B. Diener, R. Sensarma, and M. Randeria, arXiv:0709.2653.
3. H. Kleinert, *Path Integrals in Quantum Mechanics, Statistics, Polymer Physics, and Financial Markets*, Fourth Edition (World Scientific, Singapore, 2006).
4. N. Nagaosa, *Quantum Field Theory in Condensed Matter Physics* (Springer, Berlin, 1999).
5. H. Kleinert, *Fortschr. Physik* **26**, 565 (1978).
6. H. Kleinert, *Annals of Physics* **266**, 135 (1998).
7. P. Nozières and S. Schmitt-Rink, *J. Low Temp. Phys.* **59**, 195 (1985).
8. C. A. R. Sá de Melo, M. Randeria, and J. R. Engelbrecht, *Phys. Rev. Lett.* **71**, 3202 (1993).
9. A. Perali, P. Pieri, L. Pisani, and G. C. Strinati, *Phys. Rev. Lett.* **92**, 220404 (2004).
10. Q. Chen, Y. He, C. C. Chien, and K. Levin, *Phys. Rev. B* **75**, 014521 (2007).
11. Y. Ohashi and A. Griffin, *Phys. Rev. A* **67**, 063612 (2003).
12. G. M. Falco and H. T. C. Stoof, *Phys. Rev. Lett.* **92**, 130401 (2004).

EFFECTIVE FIELD THEORY FOR THE BEC/BCS TRANSITION

R. J. RIVERS

Blackett Laboratory, Imperial College London,
London SW7 2AZ, UK
E-mail: r.rivers@imperial.ac.uk

D.-S. LEE* and C.-Y. LIN†

Department of Physics, National Dong Hwa University,
Hua-Lien, Taiwan 974, Republic of China
** E-mail: dslee@mail.ndhu.edu.tw*
† E-mail: lcyong@mail.ndhu.edu.tw

We construct an effective field theory for a condensate of cold Fermi atoms whose scattering is controlled by a Feshbach resonance, with particular emphasis on the speed of sound and its hydrodynamic description.

Keywords: Condensate; Feshbach resonance; Speed of sound; Hydrodynamics.

1. Condensates of Ultracold Fermions

Cold alkali atoms whose scattering is controlled by a Feshbach resonance can form diatomic molecules with tunable binding energy on applying an external magnetic field. For weak fermionic pairing we have a BCS theory of Cooper pairs, whereas strong fermionic pairing gives a BEC theory of diatomic molecules. The transition is characterised by the divergence of the s-wave scattering length. However, most properties of the condensate (e.g. the speed of sound) vary continuously across the transition.

A considerable theoretical and experimental effort has been expended on understanding such macroscopic quantum systems. Although this is sufficient motivation for our work we are also interested in possible analogies[1] between condensed matter physics and the early universe. For example, it has been suggested that the 'naturalness' of Lorentz violation (in LV models) has a counterpart in the dispersion relations of condensate modes. In a different direction, changing the speed of sound changes the causal horizons (metric), with implications for particle production.

Our goals are to represent the ($T = 0$) condensate by a single effective bosonic field theory across the BEC/BCS regimes, from which we can i) derive the dispersion relations (most simply, the speed of sound) ii) provide a hydrodynamic description in the semiclassical limit. In the light of our earlier comments this permits a realisation in terms of an analogue gravitational metric, if we so wish, although we shall not do so.

The literature is huge – our approach is idiosyncratic. Many references are given in our paper[2] upon which much of this article is based.

2. Idealised Theory of Cold Fermi Atoms ($T = 0$)

Consider the action for cold Fermi fields ψ_σ, with spin label $\sigma = (\uparrow, \downarrow)$,

$$
\begin{aligned}
S[\psi_\sigma, \psi^*_\sigma, \phi, \phi^*] = \int dt\, d\mathbf{x} \Big\{ & \sum_{\uparrow,\downarrow} \psi^*_\sigma(x) \left[i\,\partial_t + \frac{\nabla^2}{2m} + \mu_\sigma \right] \psi_\sigma(x) \\
& + U\, \psi^*_\uparrow(x)\, \psi^*_\downarrow(x)\, \psi_\downarrow(x)\, \psi_\uparrow(x) \\
& + \phi^*(x) \left[i\,\partial_t + \frac{\nabla^2}{2M} + (\mu_\uparrow + \mu_\downarrow) - \nu \right] \phi(x) \\
& - g \left[\phi^*(x)\, \psi_\downarrow(x)\, \psi_\uparrow(x) + \phi(x)\, \psi^*_\downarrow(x)\, \psi^*_\uparrow(x) \right] \Big\}.
\end{aligned}
$$

We have displayed explicit coupling to a (narrow) resonant diatomic field ϕ with $M = 2m$ and binding energy ν, which can be varied by applying an external magnetic field. External fields lead to a spin imbalance[3] in that the chemical potentials μ_σ also carry the $\sigma = (\uparrow, \downarrow)$ label, with $\mu_\uparrow \neq \mu_\downarrow$, which we write as $\mu_\uparrow = \mu + \xi, \quad \mu_\downarrow = \mu - \xi$. At zero external energy-momentum the effective coupling strength varies with ν as $U_{\text{eff}} = U + g^2/(\nu - 2\mu)$.

To construct the condensate action we consider

$$
Z = \int D\psi\, D\psi^*\, D\phi\, D\phi^* \exp iS[\psi_\sigma, \psi^*_\sigma, \phi, \phi^*].
$$

Introducing auxiliary fields $\Delta(x) = U\psi_\downarrow(x)\psi_\uparrow(x)$, $\Delta^*(x) = U\psi^*_\downarrow(x)\psi^*_\uparrow(x)$ leads to the equivalent action

$$
\begin{aligned}
\bar{S}[\Psi_\sigma, \Psi^*_\sigma, \phi, \phi^*, \Delta, \Delta^*] = \int dt\, d\mathbf{x} \Big\{ & \Psi^\dagger(x) G^{-1}_\xi \Psi(x) \\
& + \phi^*(x) \left[i\,\partial_t + \frac{\nabla^2}{2M} + 2\mu - \nu \right] \phi(x) - \frac{1}{U}|\Delta|^2 \Big\},
\end{aligned}
$$

in terms of the total condensate $\tilde{\Delta}(x) = \Delta(x) - g\,\phi(x)$. $\Psi^\dagger(x) = (\psi^\dagger_\uparrow, \psi_\downarrow)$ is the Nambu spinor, and G^{-1}_ξ is the inverse Nambu Green function

$$G^{-1}_\xi = \begin{pmatrix} i\partial_t - \varepsilon_\uparrow & \tilde{\Delta}(x) \\ \tilde{\Delta}^*(x) & i\partial_t + \varepsilon_\downarrow \end{pmatrix} \tag{1}$$

with $\varepsilon_\sigma = -\frac{\nabla^2}{2m} - \mu_\sigma$. On integrating out the Fermi fields, we have

$$Z = \int D\phi\, D\phi^*\, D\Delta\, D\Delta^* \exp iS_{NL}[\phi, \phi^*, \Delta, \Delta^*]$$

where the non-local effective condensate action is

$$S_{NL} = -i\,Tr \ln G^{-1}_\xi - \frac{1}{U}|\Delta|^2.$$

If we write $\Delta(x) = |\Delta(x)|\, e^{i\theta_\Delta(x)}$, $\phi(x) = -|\phi(x)|\, e^{i\theta_\phi(x)}$ then $\tilde{\Delta}(x) = |\tilde{\Delta}(x)|\, e^{i\theta_{\tilde{\Delta}}(x)}$ follows by definition.

We conclude these preliminary comments with the observation that it is not necessary[3] to include the bound-state field ϕ explicitly in (1). However, the 'double' linearisation that we have effected here in the construction of $\tilde{\Delta}$ is, to our way of thinking, more intuitive in the BEC regime and more amenable to analytic approximation.

3. The Effective Action S_{NL}

Invariance under the $U(1)$ transformations $\theta_\Delta(x) \to \theta_\Delta(x) + \alpha(x)$, $\theta_\phi(x) \to \theta_\phi(x) + \alpha(x)$ is spontaneously broken: $\delta\, S_{NL} = 0$ implies spacetime constant $|\Delta(x)| = |\Delta_0| \neq 0$ and $|\phi(x)| = |\phi_0| \neq 0$ whereby $|\tilde{\Delta}(x)| = |\tilde{\Delta}_0| \neq 0$, the constant-field *gap* equations. The mean-field, or semiclassical, approximation is the general solution to $\delta S_{NL} = 0$: In this we ignore the subtle issue of multivaluedness[4] (and that of the possibility of a non-trivial Jacobian). To expand S_{NL} about the gap solutions, we use the gauge transformation

$$e^{-i\sigma_3\theta_{\tilde{\Delta}}(x)/2} G^{-1}_\xi e^{i\sigma_3\theta_{\tilde{\Delta}}(x)/2} = \bar{G}^{-1}_\xi - \Sigma = \bar{G}^{-1}_0 - (\Sigma - \xi I),$$

where

$$\bar{G}^{-1}_\xi = \begin{pmatrix} i\partial_t - \varepsilon_\uparrow & |\tilde{\Delta}_0| \\ |\tilde{\Delta}_0| & i\partial_t + \varepsilon_\downarrow \end{pmatrix} = \bar{G}^{-1}_0 + \xi I \tag{2}$$

and

$$\begin{aligned} \Sigma = {} & (-i\nabla^2\theta_{\tilde{\Delta}}/4m + (\nabla\theta_{\tilde{\Delta}})(-i\nabla)/2m)\mathrm{I} \\ & + (\dot{\theta}_{\tilde{\Delta}}/2 + (\nabla\theta_{\tilde{\Delta}})^2/8m)\sigma_3 - \delta|\tilde{\Delta}|\,\sigma_1, \end{aligned} \tag{3}$$

with $|\tilde{\Delta}| = |\tilde{\Delta_0}| + \delta|\tilde{\Delta}|$. We perform a simultaneous expansion in derivatives and ξ as

$$S_{NL} = -i\mathrm{TrLn}(\bar{G}_0^{-1} + \xi I) + i\mathrm{Tr}\sum_{n=1}^{\infty}\frac{[\bar{G}_0(\Sigma - \xi I)]^n}{n} - \int d^4x \frac{|\Delta|^2}{U}$$
$$+ \int d^4x \phi^*(x)\left(i\partial_t + \frac{\nabla^2}{2M} + 2\mu - \nu\right)\phi(x).$$

By construction this is Galilean invariant term-by-term.

Our approximation is to keep $n = 1, 2$ terms only, to give a *local* effective action $S^{(2)} = \int dt\, d\mathbf{x} L$. It is convenient to express $L(\theta_{\tilde{\Delta}}, \theta_\phi, \epsilon, \delta\phi)$ in terms of the phase angles $\theta_{\tilde{\Delta}}$, θ_ϕ, $\delta\phi = \phi - |\phi_0|$ and $\epsilon \propto \delta|\tilde{\Delta}|$. The outcome is

$$\begin{aligned}
&L(\theta_{\tilde{\Delta}}, \theta_\phi, \epsilon, \delta\phi) = \\
&-\frac{1}{2}\rho_B^0\left[\dot{\theta}_\phi + \frac{(\nabla\theta_\phi)^2}{4m}\right] - 2|\phi_0|\delta\phi\left[\dot{\theta}_\phi + \frac{(\nabla\theta_\phi)^2}{4m}\right] - \frac{1}{2}\Omega^2(\theta_{\tilde{\Delta}} - \theta_\phi)^2 \\
&-\frac{1}{2}\rho_F^0\left[\dot{\theta}_{\tilde{\Delta}} + \frac{(\nabla\theta_{\tilde{\Delta}})^2}{4m} + \frac{(\nabla\epsilon)^2}{4m}\right] + \frac{N_0}{4}\left[\dot{\theta}_{\tilde{\Delta}} + \frac{(\nabla\theta_{\tilde{\Delta}})^2}{4m} + \frac{(\nabla\epsilon)^2}{4m}\right]^2 \\
&-\alpha\,\epsilon\left[\dot{\theta}_{\tilde{\Delta}} + \frac{(\nabla\theta_{\tilde{\Delta}})^2}{4m}\right] + \frac{1}{4}\gamma\left(\dot{\epsilon} + \frac{\nabla\theta_{\tilde{\Delta}}.\nabla\epsilon}{2m}\right)^2 - \frac{1}{4}\mathcal{M}^2\epsilon^2 \\
&+\delta\phi\left(\frac{\nabla^2}{4m} + (2\mu - \nu)\frac{U_{\mathrm{eff}}}{U}\right)\delta\phi + \zeta\epsilon\delta\phi,
\end{aligned} \tag{4}$$

where ρ_F^0 and $\rho_B^0 = 2|\phi_0|^2$ are the densities of fermions in the Cooper pair and molecular parts of the condensate for the gap solutions and N_0 measures the density of states. Since $Tr(\Sigma\bar{G}_0\Sigma) = 0$ the form of the action is *ξ-independent* but the coefficients depend on ξ. Renormalisation is required which, because of IR divergences that follow from the gapless mode, is non-trivial.[5]

Superficially, there are four dynamical variables $\theta_{\tilde{\Delta}}, \theta_\phi, \epsilon, \delta\phi$. However, both θ_ϕ and $\delta\phi$ are not 'fully' dynamical variables, because they do not possess relevant time derivatives in the action. Hence, we have one gapless mode and one 'massive' (gap) mode.

4. Speed of Sound – the Linearised EL Equations

For the speed of sound we only need the quadratic Lagrangian density

$$L_{qu}(\theta_{\tilde{\Delta}}, \theta_\phi, \epsilon, \delta\phi) = -\frac{1}{2}\rho_B^0\frac{(\nabla\theta_\phi)^2}{4m} - 2|\phi_0|\delta\phi\dot{\theta}_\phi - \frac{1}{2}\Omega^2(\theta_{\tilde{\Delta}} - \theta_\phi)^2 \tag{5}$$

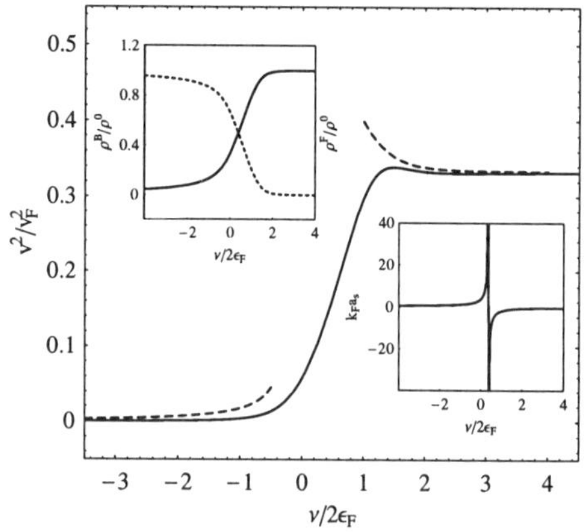

Fig. 1. The behavior of the sound velocity v throughout BEC and BCS regimes for $\xi = 0$ as a function of the threshold energy for $U = 7.54\,\epsilon_F/k_F^3$ and $g = 4.62\,\epsilon_F/k_F^{3/2}$. The lower inset shows the scattering length a_s, while the upper inset shows the fermion density ρ_0^F (solid line) and the molecule density ρ_0^B (dotted line).

$$-\frac{1}{2}\rho_F^0\left[\frac{(\nabla\theta_{\tilde{\Delta}})^2}{4m}+\frac{(\nabla\epsilon)^2}{4m}\right]-\alpha\,\epsilon\,\dot{\theta}_{\tilde{\Delta}}+\frac{1}{4}\gamma\dot{\epsilon}^2-\frac{1}{4}\mathcal{M}^2\epsilon^2$$
$$+\delta\phi\left(\frac{\nabla^2}{4m}+(2\mu-\nu)\frac{U_{\rm eff}}{U}\right)\delta\phi+\zeta\epsilon\delta\phi.$$

In general, once we have imposed conservation of fermion number, v^2 is just a ratio of combinations of simple integrals.[2] For example, in Fig. 1 we show the speed of sound over the whole BCS/BEC range, in agreement with Ref. 9, for example.

In the deep BCS regime ($|\phi_0| \approx 0$), for which $\alpha \approx 0; \gamma \approx N_0$, we recover $v^2 = 2\rho_F^0/N_0 (= v_F^2/3$ for $\xi = 0)$ and in the deep BEC regime, where $|\Delta_0| \approx 0$, v^2 vanishes as $v^2 \simeq |\tilde{\Delta_0}|^2/8m\mu$ (the dashed lines in Fig. 1).

Note that we can essentially reconstruct the full action from the quadratic action by invoking Galilean invariance,[7,8] replacing $\dot{\theta} \to \dot{\theta} + (\nabla\theta)^2/4m$ and $(\nabla\theta)^2 \to (\nabla\theta)^2 + 4m\dot{\theta}$.

5. Hydrodynamics

A discussion of the hydrodynamics of the condensate requires a restoration of the non-linearity of the action. As a first step we take $g = 0$, $\tilde{\Delta} = \Delta$, to give the BCS Lagrangian[6,8,10]

$$L(\theta_\Delta,\epsilon)=-\frac{1}{2}\rho_F^0\left[\dot{\theta}_\Delta+\frac{(\nabla\theta_\Delta)^2}{4m}+\frac{(\nabla\epsilon)^2}{4m}\right]+\frac{N_0}{4}\left[\dot{\theta}_\Delta+\frac{(\nabla\theta_\Delta)^2}{4m}+\frac{(\nabla\epsilon)^2}{4m}\right]^2$$

$$+\frac{1}{4}\gamma\left(\dot{\epsilon}+\frac{\nabla\theta_\Delta.\nabla\epsilon}{2m}\right)^2-\frac{1}{4}\mathcal{M}^2\epsilon^2. \tag{6}$$

To make this more familiar set $\epsilon = 0$, whence

$$L(\theta_\Delta)=-\frac{1}{2}\rho_F^0\left[\dot{\theta}_\Delta+\frac{(\nabla\theta_\Delta)^2}{4m}\right]+\frac{N_0}{4}\left[\dot{\theta}_\Delta+\frac{(\nabla\theta_\Delta)^2}{4m}\right]^2. \tag{7}$$

To show that L of (7) describes a classical fluid we observe that the Euler-Lagrange equation is the continuity equation

$$\partial_t\rho_\Delta+\nabla\cdot(\rho_\Delta\,\mathbf{v}_\Delta)=0,$$

where $\mathbf{v}_\Delta=\nabla\theta_\Delta/2m$ and ρ_Δ is defined by $\rho_\Delta=\rho_0-N_0G$. Rewritten as

$$\dot{\theta}_\Delta+\left[\frac{1}{N_0}(\rho_\Delta-\rho_0)+\frac{(\nabla\theta_\Delta)^2}{4m}\right]=0$$

this leads to

$$m\frac{\nabla\dot{\theta}_\Delta}{2m}+\nabla\left[\frac{1}{2N_0}\delta\rho+\frac{1}{2}m\left(\frac{\nabla\theta_\Delta}{2m}\right)^2\right]=0,$$

i.e., Euler's equation

$$m\dot{\mathbf{v}}_\Delta+\nabla\left[\delta\mu+\frac{1}{2}m\,v_\Delta^2\right]=0,\qquad \delta p=\rho_0\delta\mu.$$

Equivalently, if $\mathbf{j}=m\rho\mathbf{v}_\Delta$, then combining the continuity and Euler's equations gives

$$\frac{\partial j_i}{\partial t}+\frac{\partial}{\partial x_j}t_{ij}=0,$$

in which $t_{ij}=m\,\rho\,v_i\,v_j+p\,\delta_{ij}$ is the stress tensor.

Restoring density fluctuations ($\epsilon\neq 0$) into the classical picture leads to a characteristic doubling of fields: Introducing $\theta_\pm=\theta_\Delta\pm\epsilon$ enables $L(\theta_\Delta,\epsilon)$ to be written as

$$L(\theta_+,\theta_-)=\frac{1}{2}L(\theta_+)+\frac{1}{2}L(\theta_-)+\frac{1}{2}L_I(\theta_+,\theta_-)$$

where $L(\theta_\pm)$ is the classical fluid action (7) and $L_I(\theta_+,\theta_-)=-(1/4)\mathcal{M}^2(\theta_+-\theta_-)^2$. The gap $\mathcal{M}$ determines the strength of the 'spring' that couples the phases. If we now define $\mathbf{v}_\pm=\nabla\theta_\pm/2m$ and $\rho_\pm=\rho_\Delta\pm\rho_\epsilon$, where

$$\rho_\Delta=\rho_0-N(0)\left(\dot{\theta}+\frac{(\nabla\theta)^2}{4m}+\frac{(\nabla\epsilon)^2}{4m}\right),\quad \rho_\epsilon=-N(0)\left(\dot{\epsilon}+\frac{\nabla\epsilon\cdot\nabla\theta}{2m}\right), \tag{8}$$

then translation invariance gives ($p_I = L_I$)

$$\frac{\partial}{\partial t}(m\rho_+\mathbf{v}_+ + m\rho_-\mathbf{v}_-)_i + \nabla_j[(m\rho_+ v_{+,j}v_{+,i} + p_+\delta_{ji}) + (m\rho_- v_{-,j}v_{-,i} + p_-\delta_{ji}) + p_I\delta_{ji}] = 0,$$

a coupled two-fluid system.

The density fluctuations act as sources and sinks in the continuity equations but as long as ϵ is small (e.g. we are away from vortices) and the gap is not too small, one can argue for a single BCS fluid description.

Let us now consider[2,11] the complete system described by (4). The EL 'continuity' equations are

$$\frac{\partial}{\partial t}\rho_{\tilde{\Delta}} + \nabla\cdot(\rho_{\tilde{\Delta}}\mathbf{v}_{\tilde{\Delta}} + \rho_\epsilon\mathbf{v}_\epsilon) - 2\Omega^2(\theta_{\tilde{\Delta}} - \theta_\phi) = 0$$

and

$$\frac{\partial}{\partial t}\rho_\phi + \nabla\cdot(\rho_\phi\mathbf{v}_\phi) + 2\Omega^2(\theta_{\tilde{\Delta}} - \theta_\phi) = 0,$$

where $\mathbf{v}_{\tilde{\Delta}} = \nabla\theta_{\tilde{\Delta}}/2m$ and $\mathbf{v}_\phi = \nabla\theta_\phi/2m$, with

$$\rho_{\tilde{\Delta}} = \rho_F^0 - N_0(\dot{\theta}_{\tilde{\Delta}} + m\mathbf{v}_{\tilde{\Delta}}^2 + m\mathbf{v}_\epsilon^2) + 2\alpha\,\epsilon, \tag{9}$$

and $\rho_\phi = 2|\phi|^2 \approx \rho_B^0 + 4\delta\phi|\phi_0|$. In a small fluctuation approximation in which $\rho_\epsilon\mathbf{v}_\epsilon$ can be ignored the continuity equations can be combined as

$$\frac{\partial}{\partial t}(\rho_\phi + \rho_{\tilde{\Delta}}) + \nabla\cdot(\rho_\phi\mathbf{v}_\phi + \rho_{\tilde{\Delta}}\mathbf{v}_{\tilde{\Delta}}) = 0,$$

the continuity equation for two coupled fluids.

Since $\Omega^2\,(\theta_{\tilde{\Delta}} - \theta_\phi) \propto |\Delta_0||\phi_0|\,(\theta_\Delta - \theta_\phi)$ vanishes in the deep BCS and BEC regimes, we recover an approximate single fluid description in each of these cases (of which the BCS single fluid (7) described above is one).

The definition of density (9) is again equivalent to Euler's equation

$$m\dot{\mathbf{v}}_{\tilde{\Delta}} + \nabla\left[\delta\mu + \frac{1}{2}m\,v_{\tilde{\Delta}}^2\right] = 0,$$

over the *whole* regime. If $\delta\mu$ is the *total* specific enthalpy, then the EOS is

$$\frac{dp}{d\rho_{\tilde{\Delta}}} = \rho_{\tilde{\Delta}}\,\frac{d\mu}{d\rho_{\tilde{\Delta}}}.$$

Given $p \propto \rho_{\tilde{\Delta}}^{\gamma+1}$, then γ is completely determined, varying from $\gamma = 2/3$ in the deep BCS regime to $\gamma = 1$ in the deep BEC regime (e.g., see Ref. 12). We have ignored the fluctuations in $\delta\phi$, which give rise to a fluctuation pressure $\delta p = O(\delta\rho_\phi)$.

6. Conclusions

We have shown that, beginning from a microscopic theory of interacting Fermi atoms, with an explicit Feshbach resonance, we can straightforwardly derive the speed of sound across the BCS/BEC transition. However, we stress that, whereas phase fluctuations and density fluctuations decouple in the deep BCS regime, in general we are only able to tune the speed of sound because of the coupling between them. Nonetheless, a classical hydrodynamic description exists when density fluctuations are small. It is a coupled two-fluid system, collapsing to a single fluid in both deep BCS and BEC regimes.

Acknowledgments

We thank Adriaan Schakel for helpful discussions on the programme outlined here.

References

1. C. Barcelo, S. Liberati, and M. Visser, *Analogue gravity*, arXiv:gr-qc/0505065 (rolling arXiv review article, unpublished).
2. R. J. Rivers, D.-S. Lee, and C.-Y. Lin, *Phys. Rev. Lett.* **98**, 020603 (2007) and references therein.
3. J. Tempere, M. Wouters, and J. P. Devreese, *Phys. Rev. B* **75**, 184526 (2007); and J. P. Devreese, these proceedings.
4. A. M. J. Schakel, *Int. J. Mod. Phys. B* **10**, 999 (1996). See also H. Kleinert, these proceedings.
5. A. M. J. Schakel, *Int. J. Mod. Phys. B* **8**, 202 (1994).
6. M. P. Kemoklidze and L. P. Pitaevskii, *Zh. Eksp. Teor.* **50**, 160 (1966).
7. M. Greiter, F. Wilczek, and E. Witten, *Mod. Phys. Lett. B* **3**, 903 (1989).
8. A. M. J. Schakel, *Mod. Phys. Lett.* **4**, 927 (1990).
9. H. T. C. Stoof and M. Romans, *Phys. Rev. Lett.* **95**, 260407 (2005).
10. I. J. R. Aitchison, P. Ao, D. J. Thouless, and X.-M. Zhu, *Phys. Rev. B* **51**, 6531 (1995).
11. R. J. Rivers, D.-S. Lee, and C.-Y. Lin, in preparation.
12. H. Heiselberg, *Phys. Rev. Lett.* **93**, 040402 (2004).

FUNCTIONAL-INTEGRAL APPROACH TO DISORDERED BOSONS

R. GRAHAM and A. PELSTER

Universität Duisburg-Essen, Campus Duisburg, Fachbereich Physik, Lotharstraße 1, D-47048 Duisburg, Germany
Email: robert.graham@uni-due.de, axel.pelster@uni-due.de

The appearance of coherence and order is studied for ultracold bosonic atoms in the presence of additional disorder potentials. These arise either naturally like in current carrying wire traps, or artificially and controllably like in laser speckle fields. The description of such disordered bosons within a suitably generalized Bogoliubov theory, first given by Huang and Meng, is rederived here within a functional integral approach for replicated bosonic fields. The superfluidity in homogeneous Bose systems with condensates depleted by weak interactions and disorder can thereby be discussed.

Keywords: Bogoliubov theory; Disordered bosons; Functional integral.

1. Introduction

Usually, one studies ultracold quantum gases moving in a one-particle potential $V(\mathbf{x})$ which is fixed by an external magneto-optical trap. Here, however, we consider a different physical situation where the one-particle potential $V(\mathbf{x})$ is fluctuating at each space point $\mathbf{x}$. Such a frozen disorder potential, which consists of a random distribution of hills and valleys, was considered some time ago for modelling superfluid helium in porous media.[1] By doing so, one assumes that the pores can be modelled by statistically distributed local scatterers so that the ensemble average of the disorder potential vanishes

$$\overline{V(\mathbf{x}_1)} = 0 \tag{1}$$

and their correlation function

$$\overline{V(\mathbf{x}_1)V(\mathbf{x}_2)} = R^{(2)}(\mathbf{x}_1, \mathbf{x}_2)\,. \tag{2}$$

decays with a characteristic correlation length ξ which models the average pore size. Nowadays such random potentials can be created artificially and controllably by laser speckles,[2–5] incommensurate lattices,[6] or different localized atomic species.[7] However, they can also arise naturally via the spatial fluctuations of the electric currents generating the magnetic wire traps.[8–10]

Thus, one can raise the fundamental question how the phenomenon of Bose-Einstein condensation is affected by an additional weak disorder. To this end Huang and Meng in 1992 generalized the Bogoliubov theory of ultracold Bose gases[11] for frozen random potentials.[12–16] It turned out that the formation of local condensates in the minima of the random potential reduces the superfluid component of the fluid even at zero temperature, where, in the absence of disorder, the whole fluid would be superfluid.[17] Although this finding agrees qualitatively with the earlier experiments in porous media, the predictions of this Bogoliubov theory for disordered bosons have not yet been tested experimentally in a more quantitative manner. Recently, we have worked out a proposal to test this theory by measuring the disorder-induced shifts in the frequencies of the collective excitations in trapped Bose-Einstein condensed gases,[18] as these can be measured with an accuracy of a few fractions of a percent.[19]

Here we study the Bogoliubov theory of disordered bosons within a functional integral approach. In order to perform ensemble averages with respect to the random potential, we apply the replica method which turned out to be quite useful for studying spin glasses.[20,21] In particular, we find that a replica-symmetric ansatz for the background of the replicated bose fields is sufficient to completely rederive the Huang-Meng theory. There is no indication that replica-symmetry breaking occurs for disordered bosons in contrast to the replica theory of spin glasses. This result coincides with the general physical reasoning that the tunneling of bosons between the local minima of the frozen disorder potential leads to a quantum state with a well-defined global phase where any frustration is absent. Contrary to that replica-symmetry breaking occurs within the classical theory of spin glasses as the spins trapped in the respective local minima of the energy landscape are frustrated.

2. Theoretical Description

We start with the thermodynamical properties of a disordered homogeneous Bose gas. The functional integral for the grand-canonical partition function reads

$$\mathcal{Z} = \oint \mathcal{D}\psi^* \oint \mathcal{D}\psi \, e^{-\mathcal{A}[\psi^*,\psi]/\hbar}\,, \tag{3}$$

where the integration is performed over all Bose fields $\psi^*(\mathbf{x},\tau), \psi(\mathbf{x},\tau)$ which are periodic in imaginary time τ. The euclidean action is given by

$$\mathcal{A}[\psi^*,\psi] = \int_0^{\hbar\beta} d\tau \int d^D x \left\{ \psi^*(\mathbf{x},\tau) \left[\hbar\frac{\partial}{\partial\tau} - \frac{\hbar^2}{2M}\boldsymbol{\Delta} + V(\mathbf{x}) - \mu \right] \psi(\mathbf{x},\tau) + \frac{g}{2}\,\psi^{*2}(\mathbf{x},\tau)\psi^2(\mathbf{x},\tau) \right\}, \tag{4}$$

where M denotes the particle mass, μ the grand-canonical potential, and g the strength of the contact interaction. As the grand-canonical partition function represents a functional of the disorder potential $V(\mathbf{x})$, the corresponding thermodynamic potential follows from the expectation value

$$\Omega = -\frac{1}{\beta}\overline{\ln \mathcal{Z}}\,. \tag{5}$$

In general it is not possible to explicitly evaluate expression (5), as the averaging with respect to the disorder potential $V(\mathbf{x})$ and the nonlinear function of the logarithm do no commute:

$$\overline{\ln \mathcal{Z}} \neq \ln \overline{\mathcal{Z}}\,. \tag{6}$$

An important method to perform the averaging procedure prescribed by (5) is provided by investigating the Nth power of the grand-canonical partition function $\mathcal{Z}$ in the limit $N \to 0$. Indeed, from

$$\mathcal{Z}^N = e^{N \ln \mathcal{Z}} = 1 + N \ln \mathcal{Z} + \ldots \tag{7}$$

we deduce for the thermodynamic potential (5):

$$\Omega = -\frac{1}{\beta} \lim_{N\to 0} \frac{\overline{\mathcal{Z}^N} - 1}{N}\,. \tag{8}$$

The N-fold replication of the disordered Bose gas (3), (4) and a subsequent averaging with respect to the disorder potential $V(\mathbf{x})$ corresponds to the characteristic functional

$$I[j] = \overline{\exp\left\{ i \int d^D x\, j(\mathbf{x}) V(\mathbf{x}) \right\}} \tag{9}$$

with the auxiliary current field

$$j(\mathbf{x}) = \frac{i}{\hbar} \int_0^{\hbar\beta} d\tau \sum_{\alpha=1}^{N} \psi_\alpha^*(\mathbf{x},\tau)\psi_\alpha(\mathbf{x},\tau)\,. \tag{10}$$

Due to the above assumptions (1) and (2), the characteristic functional is of the general form

$$I[j] = \exp\left\{ \sum_{n=2}^{\infty} \frac{i^n}{n!} \int d^D x_1 \cdots \int d^D x_n R^{(n)}(\mathbf{x}_1,\ldots,\mathbf{x}_n) j(\mathbf{x}_1)\cdots j(\mathbf{x}_n) \right\} \tag{11}$$

By definition the cumulant functions $R^{(n)}(\mathbf{x}_1, \ldots, \mathbf{x}_n)$ are symmetric with respect to their arguments $\mathbf{x}_1, \ldots, \mathbf{x}_n$. Therefore, the disordered Bose gas is described by the disorder averaged, replicated grand-canonical partition function

$$\overline{\mathcal{Z}^N} = \left\{ \prod_{\alpha=1}^{N} \oint \mathcal{D}\psi_\alpha^* \oint \mathcal{D}\psi_\alpha \right\} e^{-\mathcal{A}^{(N)}[\psi^*, \psi]/\hbar} \tag{12}$$

with the replica action

$$\mathcal{A}^{(N)}[\psi^*, \psi] = \int_0^{\hbar\beta} d\tau \int d^D x \sum_{\alpha=1}^{N} \left\{ \psi_\alpha^*(\mathbf{x}, \tau) \left[\hbar \frac{\partial}{\partial \tau} - \frac{\hbar^2}{2M} \mathbf{\Delta} - \mu \right] \psi_\alpha(\mathbf{x}, \tau) \right.$$
$$\left. + \frac{g}{2} \left| \psi_\alpha(\mathbf{x}, \tau) \right|^4 \right\} + \sum_{n=2}^{\infty} \frac{1}{n!} \left(\frac{-1}{\hbar} \right)^{n-1} \int_0^{\hbar\beta} d\tau_1 \cdots \int_0^{\hbar\beta} d\tau_n \int d^D x_1 \cdots \int d^D x_n$$
$$\int d^D x_n \sum_{\alpha_1=1}^{N} \cdots \sum_{\alpha_n=1}^{N} R^{(n)}(\mathbf{x}_1, \ldots, \mathbf{x}_n) \left| \psi_{\alpha_1}(\mathbf{x}_1, \tau_1) \right|^2 \cdots \left| \psi_{\alpha_n}(\mathbf{x}_n, \tau_n) \right|^2 . \tag{13}$$

Thus, in leading order $n = 2$ the random potential leads to a residual attractive interaction between the replica fields $\psi_\alpha^*(\mathbf{x}, \tau)$, $\psi_\alpha(\mathbf{x}, \tau)$ which is, in general, bilocal in both space and imaginary time.

3. Bogoliubov Theory

Now we apply the field-theoretic background method[23–25] in order to derive the effective potential for the replicated action (13) within a Bogoliubov theory.[11] As we restrict ourselves to a homogeneous Bose gas, the background fields are assumed to be independent of space and imaginary time. Furthermore, as the replica action (13) has a global $U(1) \times O(N)$-symmetry, we assume that the background fields are replica symmetric. Thus, we arrive at the decomposition

$$\psi_\alpha(\mathbf{x}, \tau) = \sqrt{n_0} + \delta\psi_\alpha(\mathbf{x}, \tau) , \tag{14}$$

where n_0 denotes the condensate density. Inserting this decomposition in the replica action (13), we have only to take into account the fluctuation fields $\delta\psi_\alpha^*(\mathbf{x}, \tau)$, $\delta\psi_\alpha(\mathbf{x}, \tau)$ in zeroth and second order due to the background method. Performing the replica limit $N \to 0$ according to (8), then yields the effective potential defined by

$$V_{\text{eff}}(n_0) = -\frac{1}{V\beta} \lim_{N \to 0} \frac{\ln \overline{\mathcal{Z}^N}}{N} . \tag{15}$$

A lengthy but straight-forward calculation shows that, within the Bogoliubov theory, all higher contributions $n > 2$ of the cumulant functions $R^{(n)}(\mathbf{x}_1, \ldots, \mathbf{x}_n)$ do not contribute to the effective potential due to the replica limit $N \to 0$. Thus, we end up with

$$V_{\text{eff}} = -\mu n_0 + \frac{g}{2} n_0^2 + \int \frac{d^D k}{(2\pi)^D} \left\{ \frac{E(\mathbf{k})}{2} + \frac{1}{\beta} \ln\left[1 - e^{-\beta E(\mathbf{k})}\right] - \frac{\epsilon(\mathbf{k}) - \mu + g n_0}{E(\mathbf{k})^2} R(\mathbf{k}) n_0 \right\} \tag{16}$$

with the Fourier transformed

$$R(\mathbf{k}) = \int d^D x \, e^{-i\mathbf{k}\mathbf{x}} R^{(2)}(\mathbf{0}, \mathbf{x}) \tag{17}$$

as well as the dispersions $\epsilon(\mathbf{k}) = \hbar^2 \mathbf{k}^2 / 2M$ and

$$E(\mathbf{k}) = \sqrt{\left[\epsilon(\mathbf{k}) - \mu + g n_0\right]^2 + 2\left[\epsilon(\mathbf{k}) - \mu + g n_0\right] g n_0} \, . \tag{18}$$

4. Condensate Density

Extremizing the effective potential (16) with respect to the condensate density n_0 then yields the particle density via $n = -\partial V_{\text{eff}} / \partial \mu$:

$$n = n_0 + \int \frac{d^D k}{(2\pi)^D} \left\{ \frac{\epsilon(\mathbf{k}) + gn}{E(\mathbf{k})} \left[\frac{1}{2} + \frac{1}{e^{\beta E(\mathbf{k})} - 1} \right] + \frac{n R(\mathbf{k})}{\left[\epsilon(\mathbf{k}) + 2gn\right]^2} \right\} . \tag{19}$$

Here $E(\mathbf{k})$ denotes the Bogoliubov dispersion

$$E(\mathbf{k}) = \sqrt{\epsilon(\mathbf{k})^2 + 2\epsilon(\mathbf{k}) gn} \, , \tag{20}$$

which does not depend on the disorder. This finding of the original Huang-Meng theory[12] here stems from the fact that the disorder average in (13) is taken before analyzing the quantum fluctuations. From (19) we read off at $T = 0$ that the depletion of the condensate density consists of two terms. The interaction-induced depletion is given by the UV-divergent expression

$$\Delta n_0^{(\text{int})} = -\int \frac{d^D k}{(2\pi)^D} \frac{\epsilon(\mathbf{k}) + gn}{2\sqrt{\epsilon(\mathbf{k})^2 + 2\epsilon(\mathbf{k}) gn}} \, , \tag{21}$$

which is calculated by applying dimensional regularization.[26] In $D = 3$ dimension Eq. (21) reduces with $g = 4\pi\hbar^2 a / M$ and the s-wave scattering length a to the well-known result of Bogoliubov:[11]

$$\Delta n_0^{(\text{int})} = -\frac{8}{3\sqrt{\pi}} (an)^{3/2} \, . \tag{22}$$

Correspondingly, the disorder-induced depletion reads

$$\Delta n_{\mathbf{0}}^{(\mathrm{dis})} = -\int \frac{d^D k}{(2\pi)^D} \frac{R(\mathbf{k})n}{\left[\epsilon(\mathbf{k}) + 2gn\right]^2}, \tag{23}$$

which can be further evaluated for a specific Fourier-transformed disorder correlation $R(\mathbf{k})$.[14,18]

5. Superfluid Density

In order to determine the superfluid density of a disordered Bose gas, we transform our system to an inertial frame with moves uniformly with the velocity $\mathbf{u}$ with respect to the laboratory.[27] A corresponding Galilei boost leads to the additional action

$$\delta\mathcal{A} = \int_0^{\hbar\beta} d\tau \int d^D x \psi^*(\mathbf{x},\tau)\, \mathbf{u}\frac{\hbar}{i}\boldsymbol{\nabla}\psi(\mathbf{x},\tau)\,. \tag{24}$$

Going again through the replica and the Bogoliubov formalism yields the effective potential

$$V_{\mathrm{eff}} = -\mu n_{\mathbf{0}} + \frac{g}{2} n_{\mathbf{0}}^2 + \int \frac{d^D k}{(2\pi)^D} \left\{ \frac{E(\mathbf{k})}{2} + \frac{1}{\beta} \ln\left[1 - e^{-\beta E(\mathbf{k})}\right] - R(\mathbf{k}) n_{\mathbf{0}} \right. \tag{25}$$

$$\left. \times \frac{\epsilon(\mathbf{k}) - \mu + g n_{\mathbf{0}}}{E(\mathbf{k})^2} - \left[\frac{\beta\hbar^2\mathbf{k}^2\, e^{\beta E(\mathbf{k})}}{2D\left[e^{\beta E(\mathbf{k})} - 1\right]^2} + \frac{R(\mathbf{k}) n_{\mathbf{0}} \hbar^2 \mathbf{k}^2 \left[\epsilon(\mathbf{k}) - \mu + g n_{\mathbf{0}}\right]}{D E(\mathbf{k})^4} \right] \mathbf{u}^2 \right\}$$

Its extremum with respect to the condensate density coincides with the grand-canonical potential Ω whose explicit dependence on the boost velocity $\mathbf{u}$ defines the momentum $\mathbf{p} = -\partial\Omega/\partial\mathbf{u}$ of the system. This yields

$$\mathbf{p} = MVn\mathbf{v} + V \int \frac{d^D k}{(2\pi)^D} \left\{ \frac{\hbar^2 \beta \mathbf{k}^2 e^{\beta E(\mathbf{k})}}{D\left[e^{\beta E(\mathbf{k})} - 1\right]^2} + \frac{2\hbar^2 R(\mathbf{k}) n \mathbf{k}^2 \epsilon(\mathbf{k})}{D E(\mathbf{k})^4} \right\} \mathbf{u}\,. \tag{26}$$

Thus, the system momentum (26) is of the form $\mathbf{p} = MVn_n\mathbf{u} + \ldots$, which defines the normal density n_n. With this we obtain

$$n_n = \int \frac{d^D k}{(2\pi)^D} \left\{ \frac{\hbar^2 \beta \mathbf{k}^2 e^{\beta E(\mathbf{k})}}{DM\left[e^{\beta E(\mathbf{k})} - 1\right]^2} + \frac{4 R_0 n}{D\left[\epsilon(\mathbf{k}) + 2gn\right]^2} \right\}, \tag{27}$$

whose complement defines the superfluid density $n_s = n - n_n$. At $T = 0$ we read off from (27) that the superfluid density has no interaction-induced depletion[17] but a depletion due to disorder

$$\Delta n_s^{(\mathrm{dis})} = -\frac{4n}{D} \int \frac{d^D k}{(2\pi)^D} \frac{R(\mathbf{k})}{\left[\epsilon(\mathbf{k}) + 2gn\right]^2}\,. \tag{28}$$

A comparison of (23) with (28) shows that the disorder-induced depletions of the condensate and the superfluid density are related via

$$\Delta n_s^{(\mathrm{dis})} = \frac{4}{D} \Delta n_{\mathbf{0}}^{(\mathrm{dis})} \tag{29}$$

irrespective of the Fourier-transformed disorder correlation $R(\mathbf{k})$. This means in $D = 3$ dimensions that the depletion of the superfluid density exceeds the corresponding one of the condensate density by a factor 4/3. On general physical grounds it is understandable that this factor is larger than 1. The localized condensates, which form in the randomly distributed minima of the random potential, do not contribute to and thereby "hamper" the superfluid motion.

6. Conclusion and Outlook

Here we have rederived the Huang and Meng theory[12] for a homogeneous three-dimensional hard-sphere Bose gas in a random external potential within a functional integral approach. In qualitative agreement with the experiments in porous media,[1] the formation of local condensates in the minima of the random potential reduces the superfluid component of the fluid even at zero temperature, where, in the absence of disorder, the whole fluid would be superfluid. The recent experimental advances in trapping Bose-Einstein condensates in a disordered medium[3,4] make it interesting to test in a more quantitative manner the predictions of the model considered by Huang and Meng. In Ref. 18 we, therefore, extended the latter approach and the approach of the present work to include a harmonic trapping potential in addition to the weak external random potential. We considered there a condensate in the limit of a large number of particles N and in the presence of disorder with a correlation length shorter than the healing length of the superfluid. These conditions allowed for a simple hydrodynamical formulation of the problem similar to the theory of wave propagation in random elastic media.[28]

Acknowledgment

This work was supported by the German Research Foundation (DFG) within the program SFB/TR 12.

References

1. B. C. Crooker, B. Hebral, E. N. Smith, Y. Takano, and J. D. Reppy, *Phys. Rev. Lett.* **51**, 666 (1983).

2. J. C. Dainty (Ed.), *Laser Speckle and Related Phenomena* (Springer, Berlin, 1975).
3. J. E. Lye, L. Fallani, M. Modugno, D. S. Wiersma, C. Fort, and M. Inguscio, *Phys. Rev. Lett.* **95**, 070401 (2005).
4. D. Clément, A. F. Varón, M. Hugbart, J. A. Retter, P. Bouyer, L. Sanchez-Palencia, D. M. Gangardt, G. V. Shlyapnikov, and A. Aspect, *Phys. Rev. Lett.* **95**, 170409 (2005).
5. T. Schulte, S. Drenkelforth, J. Kruse, W. Ertmer, J. Arlt, K. Sacha, J. Zakrzewski, and M. Lewenstein, *Phys. Rev. Lett.* **95**, 170411 (2005).
6. B. Damski, J. Zakrzewski, L. Santos, P. Zoller, and M. Lewenstein, *Phys. Rev. Lett.* **91**, 080403 (2003).
7. U. Gavish and Y. Castin, *Phys. Rev. Lett.* **95**, 020401 (2005).
8. D.-W. Wang, M. D. Lukin, and E. Demler, *Phys. Rev. Lett.* **92**, 076802 (2004).
9. T. Schumm, J. Esteve, C. Figl, J.-B. Trebbia, C. Aussibal, H. Nguyen, D. Mailly, I. Bouchoule, C. I. Westbrook, and A. Aspect, *Eur. Phys. J. D* **32**, 171 (2005).
10. R. Folman, P. Krüger, J. Schmiedmayer, J. Denschlag, and C. Henkel, *Adv. At. Mol. Opt. Phys.* **48**, 263 (2002).
11. N. Bogoliubov, *J. Phys. USSR* **11**, 23 (1947).
12. K. Huang and H. F. Meng, *Phys. Rev. Lett.* **69**, 644 (1992).
13. S. Giorgini, L. Pitaevskii, and S. Stringari, *Phys. Rev. B* **49**, 12938 (1994).
14. M. Koboyashi and M. Tsubota, *Phys. Rev. B* **66**, 174516 (2002).
15. A. V. Lopatin and V. M. Vinokur, *Phys. Rev. Lett.* **88**, 235503 (2002).
16. G. M. Falco, A. Pelster, and R. Graham, *Phys. Rev. A* **75**, 063619 (2007).
17. I. M. Kalatnikov, *An Introduction to the Theory of Superfluidity* (W. A. Benjamin, Inc., New York, 1965).
18. G. M. Falco, A. Pelster, and R. Graham, *Phys. Rev. A* **76**, 013624 (2007).
19. D. M. Stamper-Kurn, H.-J. Miesner, S. Inouye, M. R. Andrews, and W. Ketterle, *Phys. Rev. Lett.* **81**, 500 (1998).
20. M. Mezard, G. Parisi, and M. A. Virasoro, *Spin Glass Theory and Beyond* (World Scientific, Singapore, 1987).
21. K. H. Fischer and J.A. Hertz, *Spin Glasses* (Cambridge University Press, Cambridge, 1991).
22. H. Risken, *The Fokker-Planck Equation*, 2nd Ed. (Springer, Berlin, 1989).
23. C. Morette, *Phys. Rev.* **81**, 848 (1951).
24. B. S. DeWitt, *Theory of Dynamical Groups and Fields* (Gordon and Breach, New York, 1965).
25. H. Kleinert, *Path Integrals in Quantum Mechanics, Statistics, Polymer Physics, and Financial Markets*, Fourth Edition (World Scientific, Singapore, 2006).
26. H. Kleinert and V. Schulte-Frohlinde, *Critical Properties of Φ^4-Theories* (World Scientific, Singapore, 2001).
27. A. Schakel, arXiv:cond-mat/9805152.
28. A. Ishimaru, *Wave Propagation and Scattering in a Random Media*, Vols. 1 and 2 (Academic, New York, 1978).

FUNCTIONAL-INTEGRAL REPRESENTATION OF ATOMIC MIXTURES

O. FIALKO* and K. ZIEGLER

Institut für Physik, Universität Augsburg, Germany
**E-mail: fialkool@physik.uni-augsburg.de*
www.physik.uni-augsburg.de/~fialkool

A mixture of spin-1/2 fermionic atoms and molecules of paired fermionic atoms is studied in an optical lattice. The molecules are formed by an attractive nearest-neighbor interaction. A functional integral is constructed for this many-body system and analyzed in terms of a mean-field approximation and Gaussian fluctuations. This provides a phase diagram with the two merging Mott insulators and an intermediate superfluid. The Gaussian fluctuations give rise to an induced repulsive dimer-dimer interaction mediated by the unpaired fermions. The effect of an unbalanced distribution of spin-up and spin-down fermions is also discussed.

Keywords: Functional integral; Atomic mixture; Phase transition.

1. Introduction

A wide new field for investigating complex many-body systems has been opened by the idea that clouds of atoms can be cooled to very low temperatures by sophisticated cooling techniques.[1] Such an atomic cloud can be brought into a periodic potential which is created by counter-propagating laser fields.[2] This potential mimics the lattice of core atoms of a solid-state system and is called optical lattice due to its origin. It allows the simulation of conventional condensed-matter systems as well as the creation of new many-body systems. New quantum states can emerge due to the interplay of tunneling and interaction between the atoms.

In the following, a cloud of spin-1/2 fermionic atoms in an optical lattice is considered, where an attractive interaction between atoms in nearest-neighbor lattice wells is assumed. Our aim is to study different quantum phases that can appear due to the formation and dissociation of molecules, and condensation of the (bosonic) molecules.

2. Model

Our atomic cloud in an optical lattice is described by a grand-canonical system of spin-1/2 fermions at temperature $1/\beta$ and chemical potentials μ_1 and μ_2, referring to the two possible projections of the spin. Its partition function is defined as a functional integral with respect to Grassmann fields as:[3,4]

$$Z = \int e^{-S(\psi,\bar{\psi})} D[\psi,\bar{\psi}] \;, \tag{1}$$

where S is the action

$$\begin{aligned} S(\psi,\bar{\psi}) = \int_0^\beta \Bigg[& \sum_r (\psi_r^1 \partial_\tau \bar{\psi}_r^1 + \psi_r^2 \partial_\tau \bar{\psi}_r^2) - \sum_r (\mu_1 \psi_r^1 \bar{\psi}_r^1 + \mu_2 \psi_r^2 \bar{\psi}_r^2) \\ & - \frac{t}{2d} \sum_{\langle r,r' \rangle} (\psi_r^1 \bar{\psi}_{r'}^1 + \psi_r^2 \bar{\psi}_{r'}^2) - \frac{J}{2d} \sum_{\langle r,r' \rangle} \psi_r^1 \bar{\psi}_{r'}^1 \psi_r^2 \bar{\psi}_{r'}^2 \Bigg] d\tau . \end{aligned} \tag{2}$$

Here t is the tunneling rate of single fermions, whereas J is the rate for tunneling of a pair of fermions, located at nearest-neighbor sites in the optical lattice. The J-term represents an attractive interaction. In contrast to a local interaction, it provides a dynamics for the molecules. Depending on the ratio t/J, there is a competition between the individual fermion dynamics, which is dominating for $t/J \gg 1$, and the dynamics of molecules, which is dominating for $t/J \ll 1$.

3. Mean-Field Approximation and Gaussian Fluctuations

We decouple the J-term in the functional integral by two complex fields ϕ and χ.[3,4] Here ϕ is related to the order parameter for the formation of molecules, and χ is required for stabilizing the complex integral. A subsequent integration over the Grassmann fields leads to

$$S_{\text{eff}} = \int_0^\beta \left\{ \sum_{r,r'} \bar{\phi}_r \hat{v}_{r,r'}^{-1} \phi_{r'} + \frac{1}{2J} \sum_r \bar{\chi}_r \chi_r - \ln \det \hat{\mathbf{G}}^{-1} \right\} d\tau \;, \tag{3}$$

where

$$\hat{\mathbf{G}}^{-1} = \begin{pmatrix} -i\phi - \chi & \partial_\tau + \mu_1 + t\hat{w} \\ \partial_\tau - \mu_2 - t\hat{w} & i\bar{\phi} + \bar{\chi} \end{pmatrix},$$

and a nearest-neighbor matrix $\hat{w}$ whose elements are $1/2d$ on the d-dimensional lattice. Moreover, we have $\hat{v} = J(\hat{w} + 2\hat{1})$. The saddle-point

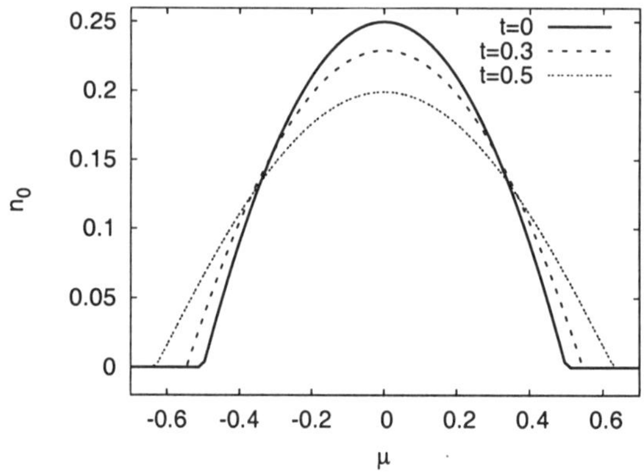

Fig. 1. Density of a molecular condensate for different single-fermion tunneling rates t. Here μ and t are given in units of J.

(SP) approximation $\delta S_{\rm eff} = 0$ for uniform fields and equal chemical potentials $\mu_1 = \mu_2$ leads to the BCS-type mean-field result:

$$\chi = -\frac{2i\phi}{3}, \quad \frac{1}{J} = \frac{1}{\beta}\sum_{\omega_n}\int_{-1}^{1}\frac{\rho(x)\,dx}{|\phi|^2/9 - (i\omega_n + \mu_1 + tx)(i\omega_n - \mu_2 - tx)}, \tag{4}$$

where ρ denotes the density of states of free particles in the optical lattice. The mean-field calculations gives three phases: an empty phase, a Mott insulator, and a Bose-Einstein condensate (BEC) of the molecules whose condensate density is $n_0 = |\phi|^2/9J^2$. The latter is plotted in Fig. 1, and the phase diagram is shown in Fig. 2.

Excitations out of the molecular BEC can be described by Gaussian fluctuations around the SP solution by complex fields ϕ and χ. Using $\Delta = i\phi + \chi$ and $\bar{\Delta} = i\bar{\phi} + \bar{\chi}$, the corresponding action is

$$S_{\rm eff} = S^0_{\rm eff} + \delta S_{\rm eff} \tag{5}$$

with

$$\delta S_{\rm eff} = \int_0^\beta \left\{\sum_{r,r'}\delta\bar{\phi}_r \hat{v}^{-1}_{r,r'}\delta\phi_{r'} + \frac{1}{2J}\sum_r \delta\bar{\chi}_r\delta\chi_r\right\}d\tau - \frac{1}{2}\mathrm{tr}\left[\hat{\mathbf{G}}_0\begin{pmatrix}-\delta\Delta & 0\\ 0 & \delta\bar{\Delta}\end{pmatrix}\right]^2. \tag{6}$$

The above result reads in terms of Fourier coordinates

$$\delta S_{\rm eff} = \sum_{q,\omega}\langle\delta\bar{\phi}_{q,\omega}, \hat{\mathbf{G}}^{-1}_{\rm eff}(q, i\omega)\delta\phi_{q,\omega}\rangle, \tag{7}$$

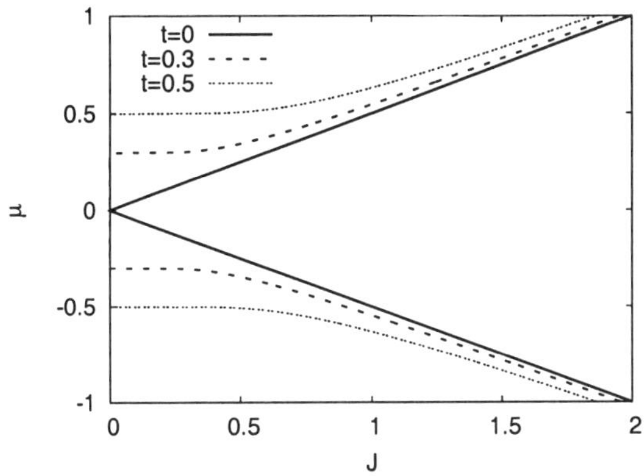

Fig. 2. Phase diagram for different values of t.

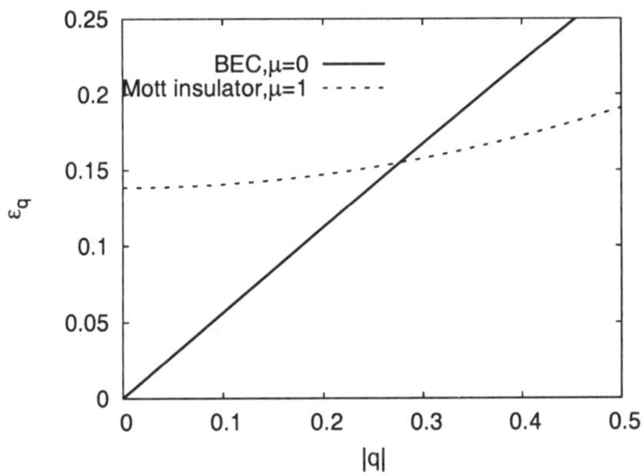

Fig. 3. Quasiparticle excitations for $t = 0.5$ and $J = 1$.

where $\hat{\mathbf{G}}_{\text{eff}}^{-1}$ is a 4 by 4 propagator and $\delta\phi$ is a four-component spinor. More details of the calculation can be found in Ref. 4. The excitation spectrum $\epsilon_q \equiv i\omega(q)$ is the solution of

$$\det \hat{\mathbf{G}}_{\text{eff}}^{-1}(q, i\omega) = 0. \tag{8}$$

This gives the Bogoliubov spectrum in the molecular BEC and a gapped spectrum outside the BEC (see Fig. 3).

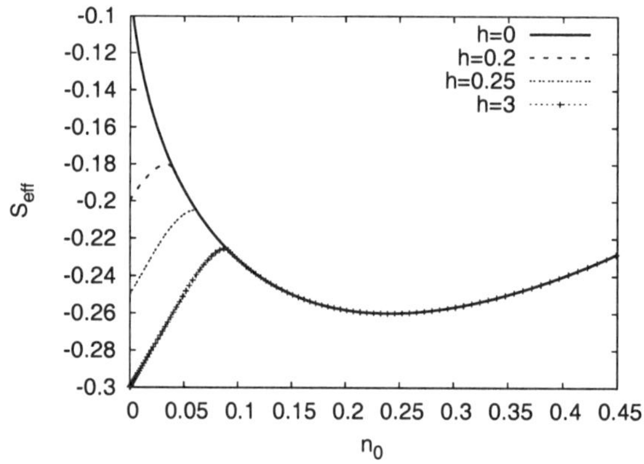

Fig. 4. Mean-field action of an unbalanced system, indicating a first-order transition from the molecular condensate with increasing $h = (\mu_1 - \mu_2)/2$. For $t = 0.2, \mu = 0$ and zero temperature the transition takes place around $h = 0.25$.

4. Discussion and Conclusions

Our mean-field approach also allows us to consider an unbalanced molecular condensate[5,6] by imposing different chemical potentials for the two spin projections of the fermions, $\mu_1 = \mu + h$, $\mu_2 = \mu - h$. Although we cannot address questions about nonlocal properties, like phase separation,[7–9] directly within our mean-field approach, the effect of two chemical potentials provides interesting effects even in a uniform system. In particular, the existence of a first-order phase transition, usually leading to phase separation, can be studied with the mean-field action of the unbalanced system

$$S_{\text{eff}} \sim \frac{|\phi|^2}{9J} - \frac{1}{\beta}\int_{-1}^{1} \rho(x) \ln \left[\cosh\left(\frac{E_+(x)\beta}{2}\right) \cosh\left(\frac{E_-(x)\beta}{2}\right)\right] dx, \quad (9)$$

where

$$E_\pm(x) = -h \pm \sqrt{|\phi|^2/9 + (\mu + tx)^2}\ . \quad (10)$$

This gives a first order-phase transition due to two separated minima in the mean-field action for small μ or for larger μ, depending on h and t for fixed J (cf. Fig. 4).

For small single-fermion tunneling rate t, there is a spin-polarized phase simply because one spin projection has a negative chemical potential. This can happen for small μ (cf. Fig. 5). If t is larger, there is a shift of the chemical potential in Eq. (10) by the single-fermion tunneling rate. This prevents the appearance of a spin-polarized state for small μ but it leads to

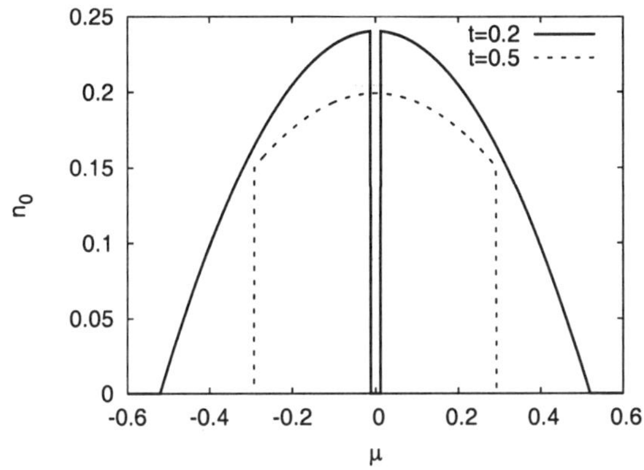

Fig. 5. Density of a molecular condensate as a function of $(\mu_1 + \mu_2)/2$ for $h = 0.26$, $J = 1$, and $T = 0$.

a sudden disappearance of the molecular condensate at large $|\mu|$, as shown in Fig. 5. This is accompanied by a first-order phase transition. At the point $h = \mu$ there might be a coexistence of molecules and spin polarized fermions.[10,11]

In conclusion, our mean-field approach to the model of Eq. (2) reduces to a BSC-type theory of molecules for spin-1/2 fermions with attractive interaction. It also provides a Mott insulator and a spin-polarized phase. The Gaussian fluctuations describe Bogoliubov-type excitations of a molecular condensate and the gapped excitation spectrum of a Mott insulator.

References

1. W. D. Phillips, *Rev. Mod. Phys.* **70**, 721 (1998).
2. M. Lewenstein *et al.*, *Advances in Physics*, **56**, 243 (2007).
3. K. Ziegler, *Laser Physics* **15**, 650 (2005).
4. O. Fialko, Ch. Moseley, and K. Ziegler, *Phys. Rev. A* **75**, 053616 (2007).
5. Y. Shin, *et al.*, *Phys. Rev. Lett.* **97**, 030401 (2006).
6. L. Wenhui *et al.*, *Nucl. Phys. A* **790**, 88C (2007).
7. G. Sarma, *J. Phys. Chem. Solids* **24**, 1029 (1963).
8. P. Bedaque, H. Caldas, and G. Rupak, *Phys. Rev. Lett.* **91**, 247002 (2003).
9. M. Haque and H. Stoof, *Phys. Rev. A* **74**, 011602(R) (2006).
10. P. Pieri and G. Strinati, *Phys. Rev. Lett.* **96**, 150404 (2006).
11. E. Taylor and A. Griffin, *Phys. Rev. A* **76**, 023614 (2007).

ANDERSON LOCALIZATION IN ATOMIC MIXTURES

K. ZIEGLER* and O. FIALKO

Institut für Physik, Universität Augsburg, Germany
** E-mail: ziegler@physik.uni-augsburg.de*
www.physik.uni-augsburg.de/~klausz

A mixture of two types of fermions with different mass in an optical lattice is considered. The light fermions are subject to thermal and quantum fluctuations, the heavy fermions only to thermal fluctuations. We derive the distribution of heavy fermions and study numerically the localization length of the light fermions in a two-dimensional lattice. Depending on the temperature of the system, a transition from extended states at low temperatures to localized states at high temperatures is found.

Keywords: Functional integral; Atomic mixture; Phase transition.

1. Introduction

Recently developed cooling and trapping techniques for ultracold gases has allowed to study complex quantum systems.[1,2] Of particular interest are mixtures of different types of atoms and their interaction. Here we will discuss a mixture of fermions, where one type of fermion is light, the other one is heavy. This mixture is brought into an optical lattice and its behavior is studied. Since the heavy fermions are localized to their wells in the optical lattice, the light particles are scattered elastically by them. This leads to a distribution of heavy fermions which is driven by thermal fluctuations and the interplay with the light fermions. On the other hand, the light fermions experience complex interference effects during their scattering events, leading to localization. Such a system can be described by a functional-integral approach, where the heavy fermions represent classical degrees of freedom and the light fermions quantum degrees of freedom.

2. Functional-Integral Representation

A grand-canonical ensemble of a mixture of light and heavy fermions at the inverse temperature $\beta = 1/k_B T$ can be defined by the partition function in terms of a functional integral on a Grassmann algebra.[3,4] For the latter the integration over a Grassmann field $\Psi_\sigma(r,t)$ and its conjugate $\bar{\Psi}_\sigma(r,t)$ is performed, where we use a pseudospin formalism with $\sigma=\uparrow$ for heavy and $\sigma=\downarrow$ for light fermions. A discrete time is used with $t=\Delta, 2\Delta, ..., \beta$, implying that the limit $\Delta\to 0$ has to be taken in the end. $\bar{\Psi}_\sigma(r,t)$ and $\Psi_\sigma(r,t)$ ($\sigma=\uparrow,\downarrow$) are independent Grassmann fields which satisfy antiperiodic boundary conditions in time, i.e. $\Psi_\sigma(r,\beta+\Delta) = -\Psi_\sigma(r,\Delta)$ and $\bar{\Psi}_\sigma(r,\beta+\Delta) = -\bar{\Psi}_\sigma(r,\Delta)$. For the subsequent calculations it is convenient to rename $\Psi_\sigma(r,t)\to\Psi_\sigma(r,t+\Delta)$. Then the partition function Z of the grand-canonical ensemble of light and heavy fermions reads

$$Z=\int e^{-S}\mathcal{D}[\Psi_\uparrow,\Psi_\downarrow]\,,\qquad \left(\mathcal{D}[\Psi_\uparrow,\Psi_\downarrow]=\prod_{r,t,\sigma} d\Psi_\sigma(r,t)d\bar{\Psi}_\sigma(r,t)\right) \tag{1}$$

and the Green's function that describes the dynamics of a light fermion, starting at (r',t') and terminating at (r,t), is

$$G(r,t;r',t')=\int \Psi_\downarrow(r,t)\bar{\Psi}_\downarrow(r',t')e^{-S}\mathcal{D}[\Psi_\uparrow,\Psi_\downarrow]/Z\equiv\langle\Psi_\downarrow(r,t)\bar{\Psi}_\downarrow(r',t')\rangle\,. \tag{2}$$

The dynamics of fermions is described by the action $S=S_\downarrow+S_\uparrow+S_I$, where light fermions with chemical potential μ and nearest-neighbor tunneling rate J have the action

$$\begin{aligned} S_\downarrow=\sum_t\Big\{&\sum_r\left[\bar{\Psi}_\downarrow(r,t)\Psi_\downarrow(r,t+\Delta)-\bar{\mu}\bar{\Psi}_\downarrow(r,t)\Psi_\downarrow(r,t)\right]\\ &-\Delta J\sum_{\langle r,r'\rangle}\bar{\Psi}_\downarrow(r,t)\Psi_\downarrow(r',t)\Big\}\end{aligned}$$

with $\bar{\mu}=1+\Delta\mu$. Heavy fermions with the same chemical potential μ but no tunneling term have the action

$$S_\uparrow=\sum_t\sum_r\left[\bar{\Psi}_\uparrow(r,t)\Psi_\uparrow(r,t+\Delta)-\bar{\mu}\bar{\Psi}_\uparrow(r,t)\Psi_\uparrow(r,t)\right]\,.$$

The repulsive interaction of strength U between the two types of fermions is given by

$$S_I=\Delta U\sum_{r,t}\bar{\Psi}_\uparrow(r,t)\Psi_\uparrow(r,t)\bar{\Psi}_\downarrow(r,t)\Psi_\downarrow(r,t)\,.$$

The Grassmann field $\Psi_\uparrow$ appears in a quadratic form in S and, therefore, the $\Psi_\uparrow$ integration can be performed in Z and in Eq. (2). It leads to a space-diagonal determinant

$$\int e^{-S_\uparrow-S_I}\prod_{r,t}d\Psi_\uparrow(r,t)d\bar{\Psi}_\uparrow(r,t)=\det(-\partial_t+\bar{\mu}-\Delta U\bar{\Psi}_\downarrow\Psi_\downarrow), \tag{3}$$

where ∂_t is the time-shift operator

$$\partial_t \Psi(r,t) = \begin{cases} \Psi(r,t+\Delta) & \text{for } \Delta \le t < \beta \\ -\Psi(r,\Delta) & \text{for } t = \beta\,. \end{cases} \tag{4}$$

The determinant can also be written as

$$\det(-\partial_t + \bar\mu - \Delta U \bar\Psi_\downarrow \Psi_\downarrow)\mathcal{D}[\Psi_\downarrow] = \prod_r \left[1 + \prod_t (\bar\mu - \Delta U \bar\Psi_\downarrow(r,t)\Psi_\downarrow(r,t))\right]$$

because the matrix is diagonal in space and has a triangular structure in time. This allows us to expand the product on the right-hand side as

$$\begin{aligned}
&\prod_r \left[1 + \prod_t (\bar\mu - \Delta U \bar\Psi_\downarrow(r,t)\Psi_\downarrow(r,t))\right] \\
&\quad = \prod_r \sum_{n(r)=0,1} \prod_t \times [\bar\mu - \Delta U \bar\Psi_\downarrow(r,t)\Psi_\downarrow(r,t)]^{n(r)} \\
&\quad = \prod_r \sum_{n(r)=0,1} \bar\mu^{\beta n(r)/\Delta} e^{-(\Delta U/\bar\mu) n(r) \sum_t \bar\Psi_\downarrow(r,t)\Psi_\downarrow(r,t)}\,,
\end{aligned}$$

where the last equation is due to the fact that Grassmann variables are nilpotent. This result allows us to combine the remaining $\Psi_\downarrow$–depending terms to the new action

$$S'_\downarrow = -\frac{\log(\bar\mu)\beta}{\Delta} \sum_r n(r) + \bar\Psi_\downarrow \cdot (\partial_t - \bar\mu - \hat t + \frac{\Delta U}{\bar\mu} n)\Psi_\downarrow\,, \tag{5}$$

where $\hat t$ is the nearest-neighbor tunneling matrix. After performing the $\Psi_\downarrow$–integration in Z, we obtain for a fixed realization of $\{n(r)\}$

$$Z(\{n(r)\}) = \bar\mu^{(\beta/\Delta)\sum_r n(r)} \det(-\partial_t + \bar\mu + \hat t - (\Delta U/\bar\mu) n).$$

Following the same procedure for the Green's function, we obtain for G in Eq. (2) the averaged resolvent matrix

$$G = \sum_{\{n(r)\}} \left(-\partial_t + \bar\mu + \hat t - (\Delta U/\bar\mu) n\right)^{-1} P(\{n(r)\}) \tag{6}$$

with the probability $P(\{n(r)\})$ for finding the realization $\{n(r)\}$:

$$P(\{n(r)\}) = Z(\{n(r)\}) / \sum_{\{n(r)\}} Z(\{n(r)\})\,. \tag{7}$$

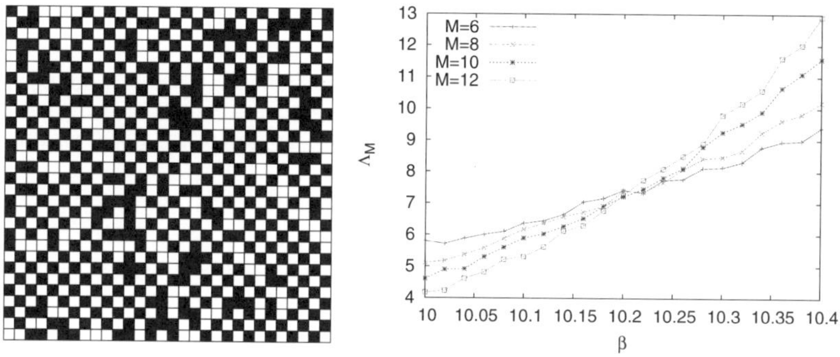

Fig. 1. Properties of the fermionic mixture at half filling on a square lattice with $U = 9$ and $\mu = U/2$ (in units of the tunneling rate J). (a) Typical distribution of heavy fermions at $\beta \approx 10.2$. (b) Scaling of the localization length with respect to the linear system size M for different temperatures $T = 1/k_B\beta$.

3. Discussion and Conclusions

The configuration $\{n(r)\}$ is the distribution of heavy fermions in the optical lattice, given by the distribution function of Eq. (7). The latter is related to a distribution of Ising spins $S(r) = [1 - 2n(r)]/2$, which are subject to an effective symmetry-breaking field, except for the symmetry point $\mu = U/2$, and an antiferromagnetic spin-spin interaction.[4] The distribution is controlled by thermal fluctuations. This is a well-known result for the model, which is also known as the Falicov-Kimball model,[5] and has been studied in detail by Monte-Carlo simulations.[6] Fig. 1a shows a typical distribution of heavy fermions. The light fermions move in the bath of heavy fermions due to tunneling and scatter by the heavy fermions with rate U. This is described by the Green's function in Eq. (6). The dynamics of the light particles is strongly affected by the scattering which can even lead to their localization. The latter is studied numerically with a transfer-matrix method[7] in a two-dimensional lattice by evaluating the localization ξ from

$$\xi^{-1} = -\lim_{|r|\to\infty} \frac{1}{|r|} \log\left(\sum_{t\geq 0} |G(r,t;0,0)|^2\right) .$$

Fig. 1b shows the change of the normalized localization length $\Lambda_M = \xi/M$ with respect to a change of linear system size M at different temperatures. Delocalized (localized) states are characterized by an increasing (decreasing) Λ_M with respect to an increased system size M. In the half-filled system there is a critical temperature with $\beta \approx 10.2$, with delocalized (localized) states in the low- (high-) temperature regime.

References

1. W. D. Phillips, *Rev. Mod. Phys.* **70**, 721 (1998).
2. M. Lewenstein, A. Sanpera, V. Ahufinger, B. Damski, A. Sen, and U. Sen, *Advances in Physics* **56**, 243 (2007).
3. J. W. Negele and H. Orland, *Quantum Many-Particle Systems* (Addison-Wesley, New York, 1988).
4. C. Ates and K. Ziegler, *Phys. Rev. A* **71**, 063610 (2005); K. Ziegler, *Nucl. Phys. A* **790**, 718c (2007).
5. R. Ramirez, L. M. Falicov, and J. C. Kimball, *Phys. Rev. B* **2**, 3383 (1970).
6. M. M. Maska and K. Czajka, *Phys. Rev. B* **74**, 035109 (2006).
7. A. MacKinnon, *J. Phys.: Condens. Matter* **6**, 2511 (1994).

FUNCTIONAL-INTEGRAL APPROACH TO THE LARGE-N LIMIT OF DILUTE BOSE GASES

F. S. NOGUEIRA
Institut für Theoretische Physik, Freie Universität Berlin,
Arnimallee 14, D-14195 Berlin, Germany
E-mail: nogueira@physik.fu-berlin.de

The dilute Bose gas is studied in the large-N limit using functional integration.

Keywords: Functional integral; Bose gas; Large-N limit.

1. Introduction

The random phase approximation (RPA) to interacting Bose gases[1,2] appeared in the past one of the most basic approximations underlying the so called dielectric approach to the superfluid transition.[3] It was also noticed already long time ago that the RPA formalism emerges naturally from a theory containing N bosons species interacting through an $U(N)$-invariant potential in the large-N limit.[4,5]

More recently the $1/N$-expansion was applied to the dilute Bose gas in a different context, namely, the computation of the critical temperature shift due to fluctuation effects.[6] However, such a calculation does not require too much knowledge of dynamical effects, since the T_c-shift is dominated by the zero Matsubara mode.[7] In the framework of the $1/N$-expansion this critical temperature shift was also calculated beyond the leading order in $1/N$.[8] Interestingly, although the $1/N$-expansion is in many respects physically unrealistic, the result for the T_c-shift extrapolated to $N = 1$ was found to agree reasonably well with more accurate approaches.[9]

Another example is the RPA approach discussed recently for the two-dimensional case.[10] In this case the spectrum has the very interesting property of exhibiting a minimum for a nonzero value of the momentum. This effect occurs only at finite temperature and can be thought as a kind of roton-like behavior.

In this paper we revisit the $1/N$-expansion using functional integral techniques. This will allow us to recover and provide further insight on some known results.

2. The $1/N$-Expansion

2.1. *The saddle-point approximation*

Let us consider the following action for an N-component interacting Bose gas:

$$S = \int_0^\beta d\tau \int d^d r \left[\sum_{\alpha=1}^N b_\alpha^* \left(\partial_\tau - \mu - \frac{\nabla^2}{2m} \right) b_\alpha + \frac{g}{2} \left(\sum_{\alpha=1}^N |b_\alpha|^2 \right)^2 \right], \tag{1}$$

where b_α and b_α^* are complex commuting fields. The partition function is then given by

$$Z = \int \left[\prod_\alpha \mathcal{D} b_\alpha^* \mathcal{D} b_\alpha \right] e^{-S}. \tag{2}$$

In order to perform the $1/N$-expansion we introduce an auxiliary field $\lambda(\tau, \mathbf{r})$ via a Hubbard-Stratonovich transformation:

$$S' = \int_0^\beta d\tau \int d^d r \left[\sum_{\alpha=1}^N b_\alpha^* \left(\partial_\tau - \mu - \frac{\nabla^2}{2m} + i\lambda \right) b_\alpha + \frac{1}{2g} \lambda^2 \right]. \tag{3}$$

Now we integrate out $N-1$ Bose fields to obtain the effective action

$$\begin{aligned} S_{\text{eff}} = {} & (N-1) \operatorname{Tr} \ln \left(\partial_\tau - \mu - \frac{\nabla^2}{2m} + i\lambda \right) \\ & + \int_0^\beta d\tau \int d^d r \left[b^* \left(\partial_\tau - \mu - \frac{\nabla^2}{2m} + i\lambda \right) b + \frac{1}{2g} \lambda^2 \right], \end{aligned} \tag{4}$$

where we have called b the unintegrated Bose field.

Next we extremize the action according to the saddle-point approximation (SPA), which is exact for $N \to \infty$. This is done by making the replacement $i\lambda \to \lambda_0$ and $b \to b_0$, with λ_0 and b_0 being constant fields, followed by extremization with respect to these constant background fields. From this SPA we obtain the equations

$$(\lambda_0 - \mu) b_0 = 0, \tag{5}$$

$$\lambda_0 = g|b_0|^2 - \frac{Ng}{\beta} \sum_{n=-\infty}^{\infty} \int \frac{d^d p}{(2\pi)^d} \frac{1}{i\omega_n + \mu - \lambda_0 - \frac{\mathbf{p}^2}{2m}}, \tag{6}$$

where $\omega_n = 2\pi n/\beta$ is the standard bosonic Matsubara frequency. The large-N limit is taken with Ng fixed. Below the critical temperature T_c we have $b_0 \neq 0$, and thus from Eq. (5) $\lambda_0 = \mu$. Therefore, Eq. (6) becomes

$$|b_0|^2 = \frac{\mu}{g} - N \left(\frac{m}{2\pi\beta} \right)^{d/2} \zeta(d/2), \tag{7}$$

provided $d > 2$. The particle density is obtained as usual $n = -\partial f/\partial\mu$, where $f = -\ln Z/(NV\beta)$ is the free energy density. This gives us

$$n = \frac{|b_0|^2}{N} + \int \frac{d^d p}{(2\pi)^d} \frac{1}{\exp\left[\beta\left(\frac{\mathbf{p}^2}{2m} + \lambda_0 - \mu\right)\right] - 1}. \tag{8}$$

By setting $\lambda_0 = \mu$ in Eq. (8) and using Eq. (7), we obtain

$$n = \frac{\mu}{Ng}, \tag{9}$$

and therefore we obtain the condensate density

$$n_0 \equiv \frac{|b_0|^2}{N} = n \left[1 - \left(\frac{T}{T_c} \right)^{d/2} \right], \tag{10}$$

where

$$T_c = \frac{2\pi}{m} \left[\frac{n}{\zeta(d/2)} \right]^{2/d}. \tag{11}$$

We see that the SPA does not change the value of T_c with respect to the non-interacting Bose gas. Indeed, the SPA corresponds to the Hartree approximation and it is well known that it gives a zero T_c shift.

2.2. *Gaussian fluctuations around the saddle-point approximation: beyond Bogoliubov theory*

In order to integrate out λ approximately, we consider the $1/N$-corrections to the SPA by computing the fluctuations around the constant background fields b_0 and λ_0. By setting

$$b = b_0 + \tilde{b}, \qquad i\lambda = \lambda_0 + i\tilde{\lambda}, \tag{12}$$

and expanding the effective action (4) up to quadratic order in the $\tilde{\lambda}$ field, we obtain

$$\begin{aligned} S_{\rm eff} &= S_{\rm eff}^{\rm SPA} \\ &+ \int_\tau \int_r \left[\tilde{b}^* \left(\partial_\tau - \mu + \lambda_0 - \frac{\nabla^2}{2m}\right)\tilde{b} + i\tilde{\lambda}(b_0^*\tilde{b} + b_0\tilde{b}^* + |\tilde{b}|^2) + \frac{1}{2g}\tilde{\lambda}^2\right] \\ &- \frac{N}{2}\int_\tau \int_{\tau'} \int_r \int_{r'} \tilde{\lambda}(\tau, \mathbf{r}) G_0(\tau - \tau', \mathbf{r} - \mathbf{r}') G_0(\tau' - \tau, \mathbf{r}' - \mathbf{r}) \tilde{\lambda}(\tau', \mathbf{r}'), \end{aligned} \tag{13}$$

where $\int_\tau \equiv \int_0^\beta d\tau$ and $\int_r \equiv \int d^d r$, $S_{\rm eff}^{\rm SPA}$ is the effective action (4) in the SPA and

$$G_0(\tau, \mathbf{r}) = \frac{1}{\beta} \sum_{n=-\infty}^{\infty} \int \frac{d^d p}{(2\pi)^d} e^{i(\mathbf{p}\cdot\mathbf{r} - \omega_n \tau)} \hat{G}_0(i\omega_n, \mathbf{p}), \tag{14}$$

with

$$\hat{G}_0(i\omega_n, \mathbf{p}) = \frac{1}{i\omega_n + \mu - \lambda_0 - \frac{\mathbf{p}^2}{2m}}. \tag{15}$$

After integrating out $\tilde{\lambda}$ the effective action for the quadratic fluctuations in the Bose fields is given by

$$\begin{aligned} S_{\rm eff} &= S_{\rm eff}^{\rm SPA} \\ &+ \frac{1}{2}{\rm Tr}\ln\left[\delta(\tau - \tau')\delta^d(\mathbf{r} - \mathbf{r}') - Ng\, G_0(\tau - \tau', \mathbf{r} - \mathbf{r}')G_0(\tau' - \tau, \mathbf{r}' - \mathbf{r})\right] \\ &+ \frac{1}{2}\int_\tau \int_r \int_{\tau'} \int_{r'} \Psi^\dagger(\tau, \mathbf{r})\mathbf{M}(\tau - \tau', \mathbf{r} - \mathbf{r}')\Psi(\tau, \mathbf{r}'), \end{aligned} \tag{16}$$

where we have introduced the two-component fields

$$\Psi^\dagger(\tau, \mathbf{r}) = \left[\tilde{b}^*(\tau, \mathbf{r})\ \tilde{b}(\tau, \mathbf{r})\right], \qquad \Psi(\tau, \mathbf{r}) = \begin{bmatrix} \tilde{b}(\tau, \mathbf{r}) \\ \tilde{b}^*(\tau, \mathbf{r}) \end{bmatrix}, \tag{17}$$

which satisfy $\Psi^\dagger\Psi = 2|\tilde{b}|^2$. The matrix $\mathbf{M}(\tau - \tau', \mathbf{r} - \mathbf{r}')$ has a Fourier transform given by

$$\hat{\mathbf{M}}(i\omega_n,\mathbf{p}) = \begin{bmatrix} -i\omega_n-\mu+\lambda_0+\frac{\mathbf{p}^2}{2m}+|b_0|^2\hat{\Gamma}(i\omega_n,\mathbf{p}) & b_0^2\,\hat{\Gamma}(i\omega_n,\mathbf{p}) \\ (b_0^*)^2\hat{\Gamma}(i\omega_n,\mathbf{p}) & i\omega_n-\mu+\lambda_0+\frac{\mathbf{p}^2}{2m}+|b_0|^2\hat{\Gamma}(i\omega_n,\mathbf{p}) \end{bmatrix}, \tag{18}$$

where

$$\hat{\Gamma}(i\omega_n,\mathbf{p}) = \frac{g}{1-Ng\hat{\Pi}(i\omega_n,\mathbf{p})} \tag{19}$$

is the Fourier transform of the effective interaction $\Gamma(\tau-\tau',\mathbf{r}-\mathbf{r}')$, and

$$\hat{\Pi}(i\omega_n,\mathbf{p}) = \frac{1}{\beta}\sum_{m=-\infty}^{\infty}\int\frac{d^dq}{(2\pi)^d}\hat{G}_0(i\omega_n+i\omega_m,\mathbf{p}+\mathbf{q})\hat{G}_0(i\omega_m,\mathbf{q}) \tag{20}$$

is the polarization bubble. The effective interaction can be represented in terms of Feynman diagrams as in Fig. 1. Physically $\tilde{\lambda}$ corresponds to the fluctuation of the particle density and thus the effective interaction (19) gives in fact the density-density correlation function. An effective interaction like the one in Eq. (19) was already obtained some time ago by a number of authors.[2–4] Thus, the $1/N$-expansion is actually equivalent to a random phase approximation (RPA) considered previously in the literature.[4] Explicit evaluation of the Matsubara sum in Eq. (20) yields

$$\begin{aligned}\hat{\Pi}(i\omega_n,\mathbf{p}) = \int\frac{d^dq}{(2\pi)^d}\frac{1}{i\omega_n-\frac{1}{2m}(\mathbf{p}^2+2\mathbf{p}\cdot\mathbf{q})}\left\{n_{\mathrm{B}}\left(\frac{\mathbf{q}^2}{2m}+\lambda_0-\mu\right)\right.\\ \left. -\,n_{\mathrm{B}}\left[\frac{(\mathbf{p}+\mathbf{q})^2}{2m}+\lambda_0-\mu\right]\right\},\end{aligned} \tag{21}$$

where $n_{\mathrm{B}}(x) = 1/(e^{\beta x}-1)$ is the Bose distribution function.

By inverting the matrix (18) we obtain the propagator

$$\hat{\mathbf{G}}(i\omega_n,\mathbf{p}) = \begin{bmatrix} \hat{\mathbf{G}}_{\tilde{b}^*\tilde{b}}(i\omega_n,\mathbf{p}) & \hat{\mathbf{F}}_{\tilde{b}^*\tilde{b}^*}(i\omega_n,\mathbf{p}) \\ \hat{\mathbf{F}}_{\tilde{b}\tilde{b}}(i\omega_n,\mathbf{p}) & \hat{\mathbf{G}}_{\tilde{b}^*\tilde{b}}(-i\omega_n,\mathbf{p}) \end{bmatrix}, \tag{22}$$

where

$$\hat{\mathbf{G}}_{\tilde{b}^*\tilde{b}}(i\omega_n,\mathbf{p}) = \frac{i\omega_n+\lambda_0-\mu+\frac{\mathbf{p}^2}{2m}+|b_0|^2\hat{\Gamma}(i\omega_n,\mathbf{p})}{\omega_n^2+\left[\frac{\mathbf{p}^2}{2m}+\lambda_0-\mu+|b_0|^2\hat{\Gamma}(i\omega_n,\mathbf{p})\right]^2-|b_0|^4\hat{\Gamma}^2(i\omega_n,\mathbf{p})}, \tag{23}$$

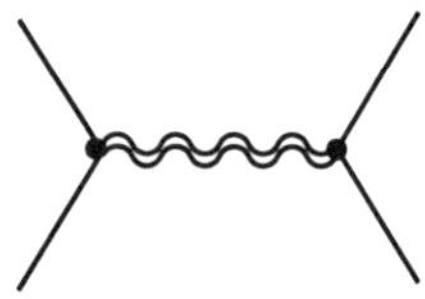

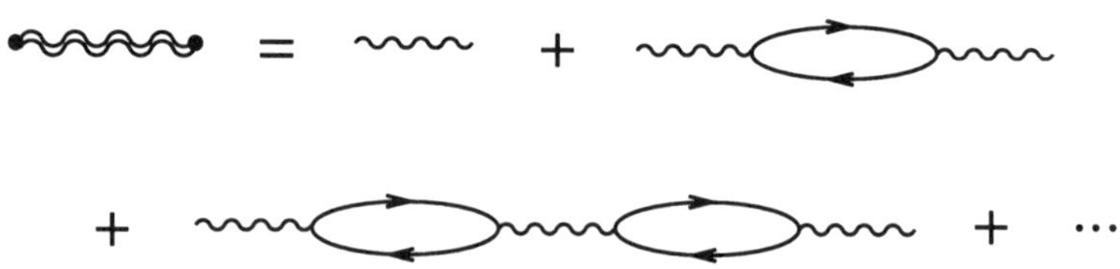

Fig. 1. Feynman diagram representation of the effective interaction Eq. (19). The line wiggle represents the bare λ-field propagator while the double line wiggle represents the dressed $1/N$-corrected λ-field propagator. Continuous lines represent $\tilde{b}$-fields and each loop is the polarization bubble Eq. (20) formed by two $\tilde{b}$-field propagators in convolution. The effective interaction is obtained as a geometric series of polarization bubbles.

and

$$\hat{\mathbf{F}}_{\tilde{b}^*\tilde{b}^*}(i\omega_n,\mathbf{p}) = -\frac{(b_0^*)^2\hat{\Gamma}(i\omega_n,\mathbf{p})}{\omega_n^2 + \left[\frac{\mathbf{p}^2}{2m} + \lambda_0 - \mu + |b_0|^2\hat{\Gamma}(i\omega_n,\mathbf{p})\right]^2 - |b_0|^4\hat{\Gamma}^2(i\omega_n,\mathbf{p})} \tag{24}$$

is the anomalous propagator.

2.3. *The excitation spectrum below T_c*

From the pole of the matrix propagator (22) we obtain that the energy spectrum $E(\mathbf{p})$ satisfies the equation

$$\begin{aligned} E^2(\mathbf{p}) &= \left\{\frac{\mathbf{p}^2}{2m} + \lambda_0 - \mu + |b_0|^2\hat{\Gamma}[E(\mathbf{p}) + i\delta,\mathbf{p}]\right\}^2 - |b_0|^4\hat{\Gamma}^2[E(\mathbf{p}) + i\delta,\mathbf{p}] \\ &= \left(\frac{\mathbf{p}^2}{2m} + \lambda_0 - \mu\right)^2 + 2\left(\frac{\mathbf{p}^2}{2m} + \lambda_0 - \mu\right)|b_0|^2\hat{\Gamma}[E(\mathbf{p}) + i\delta,\mathbf{p}], \end{aligned} \tag{25}$$

where $\delta \to 0^+$. Note that Eq. (25) can be written as the product of two elementary excitations, $E^2(\mathbf{p}) = E_l(\mathbf{p})E_t(\mathbf{p})$, where

$$E_l(\mathbf{p}) = \frac{\mathbf{p}^2}{2m} + \lambda_0 - \mu + 2|b_0|^2\hat{\Gamma}[E(\mathbf{p}) + i\delta, \mathbf{p}], \tag{26}$$

$$E_t(\mathbf{p}) = \frac{\mathbf{p}^2}{2m} + \lambda_0 - \mu \tag{27}$$

are the spectrum of the longitudinal and transverse modes, respectively. When $\lambda_0 = \mu$ we obtain that the transverse mode is gapless, consistent with Goldstone's theorem.

By inserting the saddle-point value $\lambda_0 = \mu$, we obtain the following self-consistent equation for the excitation spectrum

$$\widetilde{E}(\mathbf{p}) = \sqrt{\frac{\mathbf{p}^4}{4m^2} + \frac{|b_0|^2}{m}\mathbf{p}^2\, \hat{\Gamma}[\widetilde{E}(\mathbf{p}) + i\delta, \mathbf{p}]}, \tag{28}$$

where the notation $\widetilde{E}(\mathbf{p}) \equiv E(\mathbf{p})|_{\lambda_0=\mu}$ is used. Note that the above spectrum corresponds to a generalization of the well-known Bogoliubov spectrum.[11] The difference lies in the fact that in the $1/N$-expansion the coupling constant g is replaced by the effective interaction $\hat{\Gamma}[E(\mathbf{p}) + i\delta, \mathbf{p}]$.[2] At zero temperature $\Pi(i\omega, \mathbf{p})$ vanishes and the excitation spectrum corresponds to the usual Bogoliubov spectrum. Such a modification of the spectrum by the effective interaction accounts for thermal fluctuation effects at higher temperatures and leads to a consistent treatment of critical fluctuations near T_c.[12]

3. Conclusions

We have seen how the functional integration formalism recovers in a natural and physically intuitive way some of the earlier results of the RPA (or dielectric) approach to dilute Bose systems. The field theoretic approach discussed here can be used to compute many physical quantities, like for example, the condensate and superfluid densities. These results are new and will be published elsewhere.[12]

References

1. Yu. A. Tserkovnikov, *Dokl. Akad. Nauk SSSR* **159**, 1023 (1964) [Engl. transl.: *Soviet Phys. Dokl.* **9**, 1095 (1965)].
2. I. Kondor and P. Szépfalusy, *Phys. Lett. A* **33**, 311 (1970); P. Szépfalusy and I. Kondor, *Ann. Phys. (N.Y.)* **82**, 1 (1974).

3. T. H. Cheung and A. Griffin, *Canad. J. Phys.* **48**, 2135 (1970); *Phys. Lett. A* **35**, 141 (1971).
4. I. Kondor and P. Szépfalusy, *Phys. Lett. A* **47**, 393 (1974); P. Szépfalusy and I. Kondor, in: *Proceedings of the International School of Physics "Enrico Fermi": Local Properties of Phase Transitions*, ed. K. A. Müller (North-Holland, Amsterdam, 1976), p. 806.
5. R. Abe, *Prog. Theor. Phys.* **52**, 1135 (1974); R. Abe and Hikami, *Prog. Theor. Phys.* **52**, 1463 (1974).
6. G. Baym, J.-P. Blaizot, and J. Zinn-Justin, *Europhys. Lett.* **49**, 150 (2000).
7. G. Baym, J.-P. Blaizot, M. Holzmann, F. Laloë, and D. Vautherin, *Phys. Rev. Lett.* **83**, 1703 (1999).
8. P. Arnold and B. Tomášik, *Phys. Rev. A* **62**, 063604 (2000).
9. P. Arnold and B. Tomášik, *Phys. Rev. A* **64**, 053609 (2001); J.-L. Kneur, M. B. Pinto, and R. O. Ramos, *Phys. Rev. Lett.* **89**, 210403 (2002); H. Kleinert, *Mod. Phys. Lett. B* **17**, 1011 (2003); B. Kastening, *Phys. Rev. A* **68**, 061601(R) (2003); *Phys. Rev. A* **69**, 043613 (2004).
10. F. S. Nogueira and H. Kleinert, *Phys. Rev. B* **73**, 104515 (2006).
11. N. N. Bogoliubov, *Izv. Akad. Nauk SSSR, Ser. Fiz.* **11**, 77 (1947) [Engl. transl.: *J. Phys. (U.S.S.R.)* **11**, 23 (1947)].
12. F. S. Nogueira and A. Pelster, to be published.

THERMODYNAMICAL PROPERTIES FOR WEAKLY INTERACTING DIPOLAR GASES WITHIN CANONICAL ENSEMBLES

K. GLAUM* and H. KLEINERT†

Fachbereich Physik, Freie Universität Berlin,
Arnimallee 14, D-14195 Berlin, Germany
**E-mail: glaum@physik.fu-berlin.de, †E-mail: kleinert@physik.fu-berlin.de*

A. PELSTER

Fachbereich Physik, Universität Duisburg-Essen, Campus Duisburg,
Lotharstraße 1, D-47048 Duisburg, Germany
E-mail: axel.pelster@uni-duisburg-essen.de

We set up a recursion relation for the partition function of a fixed number of harmonically confined bosons. For an ideal Bose gas this leads to the well-known results for the temperature dependence of the specific heat and the ground-state occupancy. Due to the diluteness of the gas, we include both the isotropic contact interaction and the anisotropic dipole-dipole interaction by an infinite-bubble sum of the lowest-order perturbative results. Due to the anisotropy of the dipole-dipole interaction, the thermodynamic quantities of interest crucially depend on the trap configuration.

Keywords: Dipolar gas; Bose-Einstein condensation; Canonical ensemble.

1. Many-Body Path-Integral Formalism

The aim of this work is to give a path-integral description for a system of a fixed particle number N. To this end, we start with the imaginary-time evolution amplitude which is given by the following N-fold path integral

$$(\boldsymbol{x}_1,...,\boldsymbol{x}_N;\tau_b\,|\,\boldsymbol{x}'_1,...,\boldsymbol{x}'_N;\tau_a)^B \equiv \frac{1}{N!}\sum_P\prod_{n=1}^N\left[\int_{\boldsymbol{x}_n(\tau_a)=\boldsymbol{x}'_n}^{\boldsymbol{x}_n(\tau_b)=\boldsymbol{x}_{P(n)}}\mathcal{D}^3x_n(\tau)\right] \times \exp\left\{-\frac{1}{\hbar}\left(\mathcal{A}^{(0)}[\boldsymbol{x}_1,...,\boldsymbol{x}_N]+\mathcal{A}^{(\text{int})}[\boldsymbol{x}_1,...,\boldsymbol{x}_N]\right)\right\}. \quad (1)$$

Therein we have to sum over all possible N-particle orbits which begin at $(\boldsymbol{x}'_1,...,\boldsymbol{x}'_N)$ and go to $(\boldsymbol{x}_1,...,\boldsymbol{x}_N)$ during the imaginary-time interval

$[\tau_a, \tau_b]$. All these orbits are weighted with the help of the euclidean action which contains the interaction-free part

$$\mathcal{A}^{(0)}[\boldsymbol{x}_1, ..., \boldsymbol{x}_N] \equiv \sum_{n=1}^{N} \int_{\tau_a}^{\tau_b} d\tau \left[\frac{M}{2}\, \dot{\boldsymbol{x}}_n^2(\tau) + V(\boldsymbol{x}_n(\tau))\right] \tag{2}$$

and the interaction

$$\mathcal{A}^{(\text{int})}[\boldsymbol{x}_1, ..., \boldsymbol{x}_N] \equiv \frac{1}{2} \sum_{n,m=1}^{N}{}' \int_{\tau_a}^{\tau_b} d\tau \; V^{(\text{int})}(\boldsymbol{x}_n(\tau) - \boldsymbol{x}_m(\tau)) \, . \tag{3}$$

Here, M is the particle mass, V denotes the harmonic background potential

$$V(\boldsymbol{x}) \;=\; \frac{M}{2}\left(\omega_x^2 x^2 + \omega_y^2 y^2 + \omega_z^2 z^2\right), \tag{4}$$

$V^{(\text{int})}$ stands for the interaction potential which is specified later on, and the prime over the sum in (3) indicates that any self interaction is avoided. Since we deal with indistinguishable bosons, the evolution amplitude (1) contains a sum over all $N!$ possible permutations P. Furthermore, for describing the thermodynamics of the N-particle ensemble, we have to calculate its partition function

$$Z_N^B(\beta) \equiv \int d^3x_1 ... d^3x_N \; (\boldsymbol{x}_1,...,\boldsymbol{x}_N; \hbar\beta \mid \boldsymbol{x}_1,...,\boldsymbol{x}_N; 0)^B \, , \tag{5}$$

where only the $\hbar\beta$-periodical path configurations in (1) contribute. Thereby, $\beta \equiv 1/k_B T$ represents the reciprocal temperature. The requirement of indistinguishability of bosons complicates further calculations considerably even for the ideal Bose gas, as is explicitly shown in the next section.

2. Canonical Ensemble for Ideal Particles

Here, we discuss the situation for a non-interacting N-boson system and omit therefore the interaction (3). For this case, the action is given by (2) as the sum of N single-particle actions, so that our N-particle evolution amplitude factorizes into N single-particle ones. This yields for the partition function (5)

$$Z_N^{(0)B}(\beta) = \frac{1}{N!} \sum_P \int d^3x_1 ... d^3x_N (\boldsymbol{x}_{P(1)}; \hbar\beta | \boldsymbol{x}_1; 0)^{(0)} ... (\boldsymbol{x}_{P(N)}; \hbar\beta | \boldsymbol{x}_N; 0)^{(0)} . \tag{6}$$

Because of the occurrence of non-trivial permutations single-particle amplitudes are not necessarily periodical and one is led to multiple cycles as represented in Fig. 1. In general, such an n-cycle is defined by

$$h_n(\beta) \equiv \int d^3x_1 ... d^3x_n (\boldsymbol{x}_1; \hbar\beta | \boldsymbol{x}_n; 0)^{(0)} ... (\boldsymbol{x}_2; \hbar\beta | \boldsymbol{x}_1; 0)^{(0)} = Z_1(n\beta) \, . \tag{7}$$

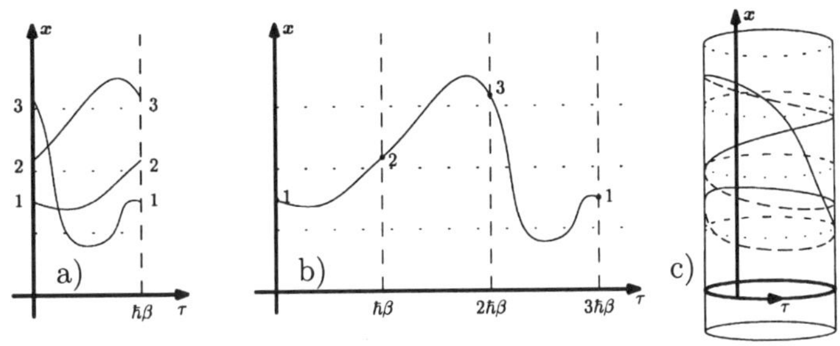

Fig. 1. Example of paths contributing to (6) for a cycle of length $n = 3$: a) final points coincide with initial points of the other particles, b) and c) show the same situation in an extended zone scheme and wrapped upon a cylinder.

The last equation follows from the imaginary-time translationary invariance of a single-particle amplitude and its group properties. This shows that the contribution of an n-cycle is simply the partition function of a single particle with a temperature lowered by a factor n. The full bosonical N-particle partition function (6) can be decomposed in such multiple cycles. The only problem is that the cycle structure strongly depends on the given permutation. One needs to know explicitly all possible cycle numbers with respect to the constraint of the fixed particle number N. In praxis, this becomes problematical for larger N's. But, according to Refs. 1,2, the partition function also follows from the recursion relation

$$Z_N^{(0)B}(\beta) = \frac{1}{N}\sum_{n=1}^{N} Z_1(n\beta)\, Z_{N-n}^{(0)B}(\beta) \quad \text{with} \quad Z_1(n\beta) = \sum_{\boldsymbol{k}} e^{-\beta E_{\boldsymbol{k}}} \,. \tag{8}$$

The partition function in vacuum $Z_0^{(0)B}(\beta) = 1$ serves here for the starting point. We apply, furthermore, the results (8) for the heat capacity

$$C_N^{(0)B} = k_B T \, \frac{\partial^2}{\partial T^2}\left\{T \ln Z_N^{(0)B}(1/k_B T)\right\} \tag{9}$$

and the ground-state occupancy, which is the probability for a particle being in the ground state $E_{\mathbf{0}}$,[2]

$$w_N^{(0)B} = \frac{1}{N}\sum_{n=1}^{N} e^{-n\beta E_0}\, \frac{Z_{N-n}^{(0)B}(\beta)}{Z_N^{(0)B}(\beta)} \,. \tag{10}$$

The results in an isotropic harmonic trap (4) are plotted for different particle numbers N in Fig. 2.

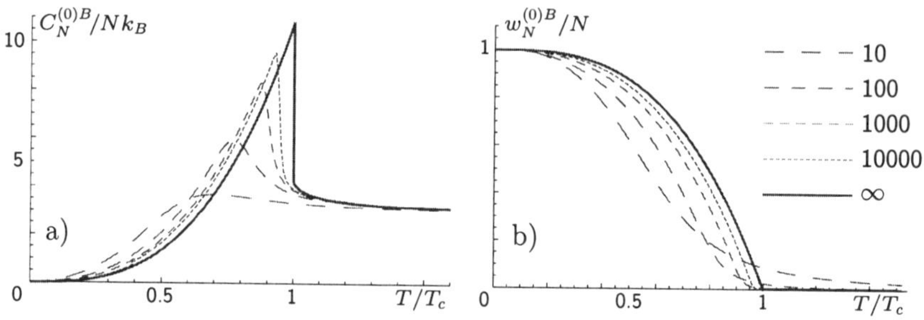

Fig. 2. a) Specific heat capacity and b) ground-state occupancy for an ideal Bose gas in an isotropic harmonic trap for different particle numbers N versus reduced temperature. Here $T_c \equiv \hbar\omega\,[N/\zeta(3)]^{1/3}/k_B$ is the critical temperature in the thermodynamic limit.

3. Dipolar Interacting System

Now we describe the effect of an interaction upon thermodynamical properties of a Bose gas within the canonical ensemble theory. The system to be described is typically a gas of ^{52}Cr atoms which has recently been condensed in Stuttgart in T. Pfau's group.[3] The interaction between the atoms can be modelled by a contact s-wave scattering plus a dipolar interaction

$$V^{(\mathrm{int})}(\boldsymbol{x}) = \frac{4\pi\hbar^2 a_s}{M}\,\delta(\boldsymbol{x}) + \frac{\mu_0 m^2}{4\pi}\left(\frac{1}{|\boldsymbol{x}|^3} - \frac{3z^2}{|\boldsymbol{x}|^5}\right), \tag{11}$$

the latter being caused by the large magnetic dipole moment of these atoms, $m = 6\mu_{\mathrm{Bohr}}$. In the experiment, the gas is trapped in a harmonic potential of a general form (4) with two almost equal frequencies giving rise to a cylindrically symmetric configuration. For aligned dipole moments two different configurations are possible. In the first configuration (I), the dipoles sit on top of each other and experience an attraction. In the second configuration (II), they are placed side by side and repel each other. The dipole forces are expected to distort slightly the condensate whose main interparticle forces come from an s-wave repulsion which is independent of the trap orientation. In Refs. 4,5, we have calculated the shift of the Bose-Einstein condensation temperatures caused by the dipolar forces. Here we discuss the thermodynamic properties of the anisotropic system at the entire low-temperature regime.

We begin with the full N-particle evolution amplitude (1) and Taylor expand the interaction factor $\exp\left\{-\mathcal{A}^{(\mathrm{int})}/\hbar\right\}$ around the zeroth order result of Section 2. The first-order contribution to the partition function (5) has the following cycle decomposition:

$$Z_N^{(1)B}(\beta) = -\frac{1}{2\hbar}\sum_{k=2}^{N}\sum_{l=1}^{k-1}\left[I_{l,k-l}^{(D)}(\beta) + I_{l,k-l}^{(E)}(\beta)\right] Z_{N-n}^{(0)B}(\beta)\,. \tag{12}$$

Thereby many non-interacting cycles (7) occur which are combined to the non-interacting part $Z_{N-n}^{(0)B}$. Furthermore two different kinds of interacting cycles are always present. These are the direct and the exchange contributions, in contrast to the interaction-free one from Fig. 1c) represented by

$I_{1,2}^{(D)}(\beta) \equiv$ 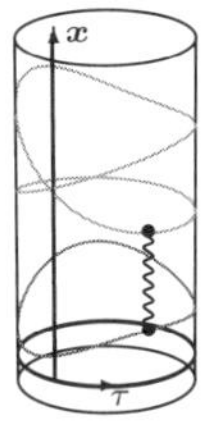 , $I_{1,2}^{(E)}(\beta) \equiv$ 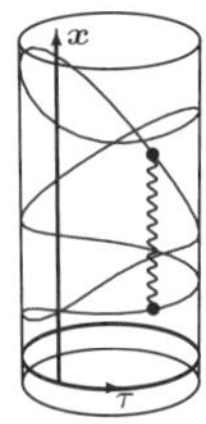 ,

where the wiggly line stands for the interaction. Their simplified pictures represent the Hartree- and the Fock-like Feynman diagrams

$I_{1,2}^{(D)} \equiv$ 

, $I_{1,2}^{(E)} \equiv$ 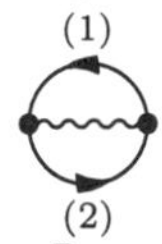

.

Combining both partition functions (8) and (12) to $Z_N^B = Z_N^{(0)B} + Z_N^{(0)B} + \ldots$ and performing the cumulant resummation one obtains the following new recursion relation for the full partition function

$$Z_N^B(\beta) = \frac{1}{N}\sum_{n=1}^{N}\left\{Z_1(n\beta) - \frac{n}{\hbar}\sum_{l=1}^{n-1}\left[I_{l,n-l}^{(D)}(\beta) + I_{l,n-l}^{(E)}(\beta)\right] + \ldots\right\} Z_{N-n}^B(\beta). \tag{13}$$

Calculating both interacting contributions yields the proportionality $I^{(D),(E)} \propto \beta$. Hence, for low temperatures, the second term in the curly brackets of (13) is much larger than the first interaction-free summand leading to negative full partition function. The perturbative result must therefore be resummed. This can be done self-consistently by using the renormalized cycle contributions $\tilde{Z}_1(n\beta) = e^{-\beta n \tilde{E}_{\boldsymbol{k}}^{(n)}}$ instead of the interaction-free terms $Z_1(n\beta)$, where

$$\tilde{E}_{\boldsymbol{k}}^{(n)} = E_{\boldsymbol{k}} - \left[\Sigma_n^{(D)}(\boldsymbol{k}) + \Sigma_n^{(E)}(\boldsymbol{k})\right]/\hbar\beta \tag{14}$$

are new energy levels shifted by the self-energies. They are represented by Feynman diagrams related to (13) by cutting one line:

$\Sigma_n^{(D)}(\boldsymbol{k}) =$ 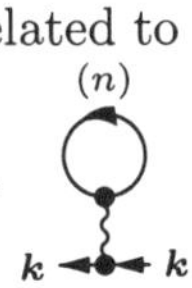

, $\Sigma_n^{(E)}(\boldsymbol{k}) =$ 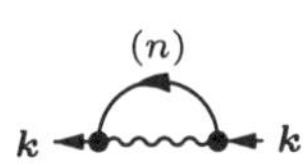

. (15)

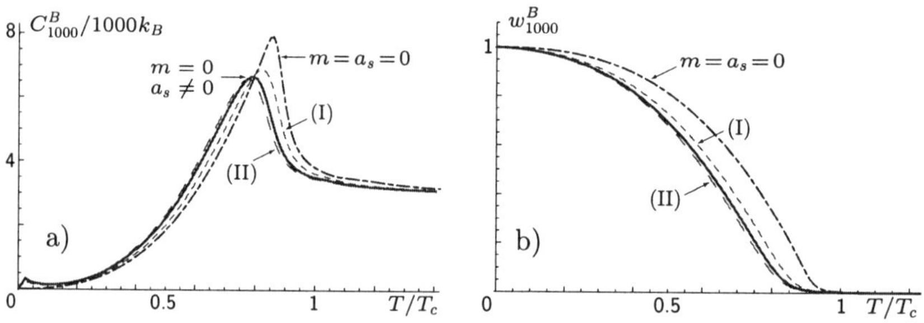

Fig. 3. a) Specific heat capacity and b) ground-state occupancy for $N = 1000$ ideal (dashed-dotted curves), contact (solid curves) and dipolar interacting (dashed curves) bosons within a cylindrically symmetric harmonic trap. The configurations (I) or (II) correspond to dipoles which sit on each other or are placed side by side, respectively.

All interaction terms are given by functions of the form

$$\Sigma_n^{(D,E)}(\boldsymbol{k}) \, , \, I_{l,n-l}^{(D,E)} \sim \hbar\beta \left\{ \frac{4\pi\hbar^2 a_s}{M} - \frac{\mu_0 m^2}{3} f\left(\kappa_{l,n}^{(D,E)}[\omega_\perp, \omega_\|]\right)\right\} \tag{16}$$

with certain anisotropy parameters $\kappa_{l,n}^{(D,E)}[\omega_\perp, \omega_\|]$, which are smaller than 1 for prolate configurations and larger than 1 for oblate ones. The first term in the curly brackets of (16) represents the contribution of the contact interaction irrespective of the trap anisotropy. The second term corresponds to the shift of the dipolar interaction and crucially depends on the anisotropy. The configurations (I) and (II) are described by the anisotropy functions

$$f^{(\mathrm{I})}(\kappa) = -2f^{(\mathrm{II})}(\kappa) = \frac{2\kappa+1}{1-\kappa} - \frac{3\kappa}{(1-\kappa)^{3/2}} \,\mathrm{Artanh}\sqrt{1-\kappa}\,. \tag{17}$$

We apply these results to calculate the canonical N-particle full partition function. From this, the specific heat capacity and the ground-state occupancy for N interacting particles have been obtained by analogy with the non-interacting case. The results for $N = 1000$ are presented in Fig. 3.

References

1. F. Brosens, J. T. Devreese, and L. F. Lemmens, *Phys. Rev. E* **55**, 227 (1997).
2. K. Glaum, H. Kleinert, and A. Pelster, *Phys. Rev. A* **76**, 063604 (2007).
3. A. Griesmaier, J. Werner, S. Hensler, J. Stuhler, and T. Pfau, *Phys. Rev. Lett.* **94**, 160401 (2005).
4. K. Glaum, A. Pelster, H. Kleinert, and T. Pfau, *Phys. Rev. Lett.* **98**, 080407 (2007).
5. K. Glaum and A. Pelster, *Phys. Rev. A* **76**, 023604 (2007).

DENSITY EXCITATIONS OF WEAKLY INTERACTING BOSE GAS

J. BOSSE* and T. SCHLIETER

Institute of Theoretical Physics, Freie Universität Berlin,
D-14195 Berlin, Germany
**E-mail: juergen.bosse@physik.fu-berlin.de*

The dynamic structure factor of a homogeneous, weakly interacting BOSE gas has been calculated for finite temperatures within RPA and compared to corresponding results for a Fermi gas. Approaching T_c from high temperatures, the boson $S(q,\omega)$ exhibits a noteworthy BEC precursor. Temperature–dependent renormalization of the Bogoliubov frequency and the Landau damping of density excitations in the BOSE gas are discussed in detail. The isothermal compressibility of a Bose gas in the condensed phase turns out to vary only slightly with temperature.

Keywords: Quantum gas; Excitations; Bogoliubov dispersion; Landau damping Isothermal compressibility.

1. Dynamical Structure Factor

Within the random–phase approximation (RPA), the van Hove function determining the double–differential inelastic–scattering cross section of a homogeneous quantum gas takes the form ($\omega \neq 0$)[1]

$$S(q,\omega) = \frac{\chi_0''(q,\omega)/(1-\mathrm{e}^{-\omega/T})}{\left|1 + v_q\,\chi_0'(q,\omega) + \mathrm{i}\,v_q\,\chi_0''(q,\omega)\right|^2}\,, \tag{1}$$

where $\chi'(q,\omega) = \pi^{-1} ⨍_{-\infty}^{\infty} \mathrm{d}\bar{\omega}\;\chi''(q,\bar{\omega})/(\bar{\omega}-\omega)$ is Kramers–Kronig related to the known density excitation spectrum of the *non–interacting* gas obeying Bose–Einstein ($\eta = -1$) or Fermi–Dirac ($\eta = 1$) statistics:[1,2]

$$\chi_0''(q,\omega) = 2C_0\,\omega\,\delta\big(\omega^2 - q^4\big) + \frac{3T}{q}\ln\left(\frac{1+\eta\exp\left[\frac{\mu_0}{T} - \frac{(\omega-q^2)^2}{4Tq^2}\right]}{1+\eta\exp\left[\frac{\mu_0}{T} - \frac{(\omega+q^2)^2}{4Tq^2}\right]}\right)^{\frac{1}{\eta}} \tag{2}$$

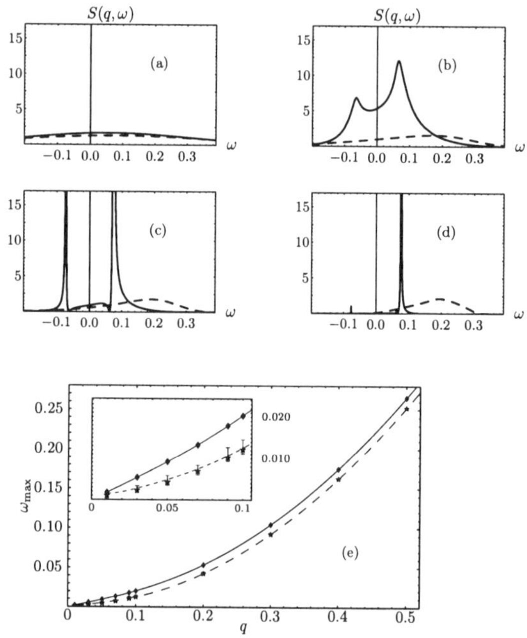

Fig. 1. (a)–(d): $S(q,\omega)$ within RPA for $q=0.25$, $v_q=0.05$ for bosons (full line) and fermions (dashed line) at $T/T_c=5.0\,(a)$, $1.05\,(b)$, $0.5\,(c)$, $0.1\,(d)$. (e): Boson dispersion for $v_q=0.05$; roots of $1+v_q\chi_0'(q,\omega)=0$ for $T/T_c=0.1$ (full line), 0.9 (dashed line), and $\omega_q^{\rm B}=\sqrt{q^4+c^2q^2}$ (full line). Dots: peak positions $\omega_{\rm max}(q)$ taken from $S(q,\omega)$–plots not shown. Inset: same magnified for interval $0<q\le 0.1$; vertical bars: full width at half maximum of boson peak taken from $S(q,\omega)$–plots not shown.

with the chemical potential μ_0 determined from the implicit equation

$$1 = C_0 + 6\sqrt{\pi}T^{\frac{3}{2}}\frac{\zeta_{\frac{3}{2}}\left(-\eta\,{\rm e}^{\mu_0/T}\right)}{(-\eta)}\,,\quad C_0=\left[1-\left(\frac{T}{T_c}\right)^{\frac{3}{2}}\right]\Theta(T_c-T)\delta_{\eta,-1} \quad (3)$$

where $\zeta_\nu(x) := \sum_{j=1}^{\infty} x^j/j^\nu$. The condensed fraction C_0 is non–zero for bosons below the BEC critical temperature $T_c=\left[6\sqrt{\pi}\zeta(3/2)\right]^{-2/3}$, only, and $v_q := \pi n\int {\rm d}^3r\ {\rm e}^{-{\rm i}\mathbf{q}\cdot\mathbf{r}}v(|\mathbf{r}|)$, for contact interactions considered here, will reduce to a constant proportional to the scattering length a: $v_q=8\pi^2na$.

Dimensionless quantities are used throughout with $k_{\rm u}^{-1}$ and $\hbar/\epsilon_{\rm u}$ as units of length and time, resp., where $k_{\rm u} := 2\left(6\pi^2n/(2s+1)\right)^{1/3}$ and $\epsilon_{\rm u} := \hbar^2k_{\rm u}^2/(2m)$. Thus, in the above equations and below, quantities are to be interpreted as $S \to S\ \epsilon_{\rm u}/\hbar$, $\chi_0'' \to \chi_0''\ \epsilon_{\rm u}/\hbar$, (Note the definition $\chi''(q,\omega) := \int_{-\infty}^{\infty}{\rm d}t\ {\rm e}^{{\rm i}t\omega} < [N_{\mathbf{q}}^\dagger(t), N_{\mathbf{q}}(0)] > /(2\pi N)$ used here.) $q \to q/k_{\rm u}$,

$z \to \hbar z/\epsilon_{\rm u}$, $T \to k_{\rm B}T/\epsilon_{\rm u}$, $v_q \to v_q/\epsilon_{\rm u}$, $a \to ak_{\rm u}$, $n \to nk_{\rm u}^{-3}$, $\kappa_T \to \kappa_T\epsilon_{\rm u}k_{\rm u}^3$, etc. according to their physical dimensions. Thus for a given gas, the (dimensionless) excitation spectrum χ_0'' is a function of 3 (dimensionless) variables (q, ω, T), only. Mass m and spin s of particles as well as the gas density n are absorbed in the units.

Figures 1 (a)–(d) display $S(q,\omega)$ calculated for $v_q = 0.05$ and $q = 0.25$ from Eqs. (1)–(3) for a series of temperatures ranging from a high ($T = 5T_c$) to a very low temperature ($T = 0.1T_c$). As expected, the results for both bosons and fermions at high T nearly agree with the van Hove function of a Maxwell–Boltzmann gas. As T is lowered, however, there is not only an overall decrease of intensity for $\omega < 0$ in accordance with the detailed–balance relation $S(q,-\omega) = \exp(-\omega/T)S(q,\omega)$, but boson– and fermion–excitation spectra develop qualitative differences due to the statistics. For $T \to T_c^+$, the boson $S(q,\omega)$ increases strongly displaying a precursor of the resonant peak which appears at $T < T_c$ reflecting long–lived density excitations associated with the Bose–Einstein condensed fraction. In contrast, the fermion $S(q,\omega)$ remains broad and feature–less gradually acquiring the well-known zero–temperature form made up of piecewise straight lines and parabolas.[3]

2. Dispersion, Damping, and Isothermal Compressibility

As displayed in Fig. 1 (b)–(d), the boson $S(q,\omega)$ develops a pronounced peak at $\omega \approx \omega_q$ for $T \to T_c^+$. While the width of this peak narrows with decreasing T, its position is almost fixed for $0 \le T \le T_c$. The gross features of the long–lived boson density excitations are well understood. In the *non–interacting* gas, owing to the δ–contribution in Eq. (2), an excitation of infinite lifetime and of spectral weight C_0 will appear at $\omega = \pm q^2$. This resonance peak is shifted and broadened as a result of particle interactions. The dispersion of the peak position and its width at half–maximum are determined within RPA by the complex root $\tilde\omega = \omega_q + {\rm i}\gamma_q$, for which the denominator on the right–hand side of Eq. (1) will vanish. Introducing the Bogoliubov excitation frequency $\omega_q^{\rm B} := \sqrt{q^4 + c^2q^2}$ with the low–T isothermal sound velocity $c := 1/\sqrt{n\kappa_{T=0}/2} = \sqrt{2v_0/\pi}$, this root may approximately be written as

$$\omega_q = \omega_q^{\rm B}\,[1 - 2v_0\Delta(q,T)]^{\frac{1}{2}} \stackrel{q\to 0}{\longrightarrow} cq\,, \tag{4}$$

$$\gamma_q = \omega_q^{\rm B}\,\frac{c^2/2}{c^2+q^2}\,\chi_{\rm ex}''(q,\omega_q^{\rm B})\,v_0\,[1 - 2v_0\Delta(q,T)]^{-\frac{1}{2}} \stackrel{q\to 0}{\longrightarrow} q^2\,\Gamma_q\,, \tag{5}$$

$$\Delta(q,T) = \frac{1}{\pi}\frac{(T/T_c)^{\frac{3}{2}}}{c^2+q^2} + \frac{c^2/2}{c^2+q^2}\,\int_{-\infty}^{\infty}\frac{{\rm d}\epsilon}{\pi}\,\frac{\chi_{\rm ex}''(q,\epsilon)}{\epsilon - \omega_q^{\rm B}}\,, \tag{6}$$

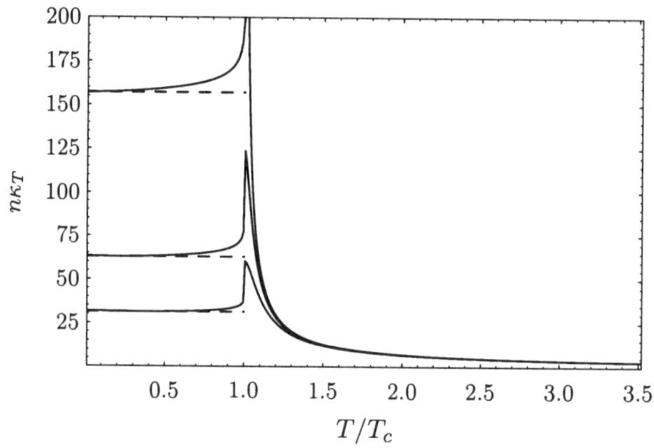

Fig. 2. Reduced isothermal compressibility $n\kappa_T = S(q\!=\!0)/T$ of interacting Bose gas within RPA for $v_q\!=\!0.02$, 0.05, and 0.10 (top to bottom, full lines). Dashed lines: low–T asymptote $n\kappa_{T\to 0} = 2/c^2 = \pi/v_0$.

where $\chi''_{\rm ex}$ denotes the excited–states contribution, i. e. the *ln*–term in Eq. (2) for $\eta\!=\!-1$. The approximate ω_q in Eq. (4) is indistinguishable from the numerically determined roots (full and dashed lines in Fig. 1(e)). It is remarkable that (i) $\omega_q \approx \omega_q^{\rm B}$, since $|\Delta(q,T) \lesssim 0.1|$ for $T \lesssim T_c$ (see Figs. 1 (e) and 2), also reflected by nearly fixed peak positions in Fig. 1 (b)–(d). (ii) the sound–damping "constant" extracted from the Landau damping in Eq. (5),

$$\Gamma_q \overset{q\to 0}{=} q^{-1}\, \frac{3v_0c^2}{2}\, \left[e^{(c^2/4-\mu_0)/T} - 1 \right]^{-1} \overset{T\to 0}{\propto} e^{-c^2/(4T)} \,, \qquad (7)$$

is anomalous: diverging as q^{-1} for fixed T; however, exponentially vanishing with decreasing T for fixed small q. (iii) the isothermal compressibility κ_T will be T–independent for $0 \le T \lesssim T_c$, if v_q is sufficiently large, see Fig.(2). Here the static structure factor $S(q) := \int d\omega\, S(q,\omega)$ was determined from numerical integration of $S(q,\omega)$ in Eq. (1) and $\kappa_T = S(0)/(nT)$ by subsequent extrapolation $q \to 0$.

References

1. J. Bosse and T. Schlieter, publication in preparation.
2. K. Baerwinkel, *Phys. Kondens. Materie* **12**, 287 (1971).
3. A. L. Fetter and J. D. Walecka, *Quantum Theory of Many-Particle Systems* (McGraw-Hill, Inc., New York, 1971).

STOCHASTIC FIELD EQUATION FOR A GRAND CANONICAL BOSE GAS

S. HELLER* and W. T. STRUNZ

Theoretische Quantendynamik, Physikalisches Institut, University of Freiburg, Hermann-Herder-Str. 3, D-79104 Freiburg, Germany
** E-mail: sigmund.heller@physik.uni-freiburg.de*

We introduce a stochastic field equation based on the P-representation of the grand canonical density operator of a Bose gas which is free from ultraviolet problems. Numerical simulations for a harmonic trap potential are presented. Although strictly valid for an ideal gas only, we argue that the behavior of weakly interacting Bose gases at finite temperatures may also be described.

Keywords: Bose gas; Quantum field equation; Grand-canonical ensemble; P-function.

1. Introduction

The fascinating experimental possibilities to manipulate relevant parameters of ultracold quantum gases in traps allow us to investigate fundamental concepts of quantum many particle theory[1] in great detail. Our goal here is to simulate the thermal state of an atomic Bose gas at low temperatures, possibly including the atomic interaction. Well below the critical temperature, and neglecting thermal and quantum fluctuations, such a gas is well described by the Gross-Pitaevskii (GP) equation[1] for the mean field wave function (we set $\hbar = 1$ throughout)

$$\partial_t \psi = -i \left(\frac{p^2}{2m} + V(x) - \mu + g|\psi(x)|^2 \right) \psi, \tag{1}$$

with μ the chemical potential, m the atomic mass, $V(x)$ an external potential, and g an interaction constant reflecting two-body collisions. Viewed as a classical field equation (as it should), the GP equation (1) can be seen as Hamilton's equation of motion for a classical matter field with field energy[1]

$$\mathcal{H} = \int dx\, \psi^*(x) \left(\frac{p^2}{2m} + V(x) + \frac{g}{2}|\psi(x)|^2 - \mu \right) \psi(x). \tag{2}$$

We need to include damping and thermal fluctuations $\xi(x,t)$ in order to describe the thermal (grand canonical) state of the field. In this purely classical framework stochastic field equations are well known, and exist for any given Hamilton functional.[2] In our case, it takes the form

$$\partial_t \psi(x) = -(\Lambda + i)H\psi(x) + \sqrt{2\Lambda k_B T}\,\xi(x,t), \tag{3}$$

with the mean field "Hamiltonian" $H = \frac{p^2}{2m} + V(x) + g|\psi(x)|^2 - \mu$, a damping constant Λ, the Boltzmann constant k_B and temperature T. The complex white noise field $\xi(x,t)$ satisfies the usual condition $\langle \xi(x_1,t_1)\xi^*(x_2,t_2)\rangle = \delta(x_1 - x_2)\delta(t_1 - t_2)$. Formally – using the corresponding Fokker-Planck equation – it is easy to see that in the asymptotic limit $t \to \infty$, the distribution of the stochastic field $\psi(x,t)$ is indeed the classical (grand canonical) thermal field state $\rho_{gc} = \frac{1}{Z_{gc}} e^{-\mathcal{H}/k_B T}$ with the Hamilton functional (2), and the grand canonical partition function Z_{gc}.

In this classical framework we face the old problem of the ultraviolet catastrophe, due to the infinite number of degrees of freedom of a field – simply invoking the equipartition theorem. In order to give some meaning to a stochastic field equation like (3), one has to introduce a cutoff. In the context of ultracold bosonic gases, see Ref. 3 for recent investigations on these matters. Closely related to the problem of the ultraviolet catastrophe is the simple observation that such a "classical" description of a quantum field can only make sense as long as the field modes are occupied by a macroscopic number of quanta. In the case of a Bose gas at very low temperatures, the energetically lowest states are indeed highly populated. However, for modes with high enough energy (the "ultraviolet" ones) this no longer holds true. We establish a "quantum" version of such a stochastic field equation that will not suffer from all these problems.

2. Quantum Field Equation

We begin with an ideal Bose gas, where the energy is simply the sum over all independent one-particle field modes with energy E_n, weighted with the occupation number of that mode. The grand canonical *quantum* density operator may be written as a mixture of coherent states using the (Glauber-Sudarshan) P-function,[4] $\hat{\rho}_{gc} = \int \frac{d^2\psi}{\pi} P_{gc}(\psi^*, \psi)|\psi\rangle\langle\psi|$ with coherent state products $|\psi\rangle = |\psi_0\rangle|\psi_1\rangle \cdots |\psi_n\rangle \cdots$ of all modes. In this representation, quantum expectation values of normally-ordered operators may simply be obtained as phase-space averages over P_{gc},[4] which thus replaces the classical phase-space density ρ_{gc}.

It is now straightforward to find a stochastic field equation that recovers the proper P-function in the long-time limit just as the classical field equation (3) gave us the classical distribution ρ_{gc}. The P-function for the thermal state of a harmonic oscillator is well known;[4] we chose to write it in the form $P_{gc}(\psi^*, \psi) = \exp\left(-\sum_n \frac{(E_n-\mu)}{k_B T_n}|\psi_n|^2\right)/\bar{n}$, with a "local" temperature $k_B T_n = \frac{E_n-\mu}{e^{(E_n-\mu)/k_B T}-1}$ and the product $\bar{n}^{-1} = \prod_n \left(e^{(E_n-\mu)/k_B T} - 1\right)$ over all mean occupation numbers. Compared to the corresponding classical density ρ_{gc}, we see that the only difference between ρ_{gc} and P_{gc} is the occurrence of the "global" temperature $k_B T$ in the classical density, while there is a "local" temperature $k_B T_n$ for each mode in the P-function. The corresponding "quantum" stochastic field equation – now written in a representation independent form – is

$$\partial_t|\psi\rangle = -(\Lambda + i)\hat{H}|\psi\rangle + \sqrt{2\Lambda \widehat{k_B T}}|\xi\rangle, \tag{4}$$

with $\hat{H} = \frac{\hat{p}^2}{2m} + V(\hat{x}) - \mu$ and, crucially, an *operator* $\widehat{k_B T} = \frac{\hat{H}}{e^{\hat{H}/k_B T}-1}$.

We see that if the energy may be considered bounded such that $k_B T \gg \hat{H}$, then indeed, $\widehat{k_B T} \approx k_B T$ and the "quantum" field equation (4) reduces to the "classical" field equation (3). However, the "quantum" field equation (4) does not suffer from the ultraviolet problem of the "classical" equation (3): for energetically high states, such that $\hat{H} \gg k_B T$, the operator $\widehat{k_B T}$ tends to zero, reflecting the built-in cutoff.

It appears more than tempting to use Eq. (4) even in the case of a weakly interacting gas. We simply include a mean-field interaction energy and replace $V(x) \to V(x) + g|\psi(x)|^2$. As we have discussed, the resulting equation is free from ultraviolet problems and coincides with the classical equation (3) (including interaction) in the infinite-temperature limit.

3. First Simulation with our Stochastic Field Equation

We study a 2D ideal Bose gas of $\langle N\rangle = 500$ atoms in an isotropic harmonic oscillator trap. In Fig. 1 the occupation number of the ground state is determined with our stochastic field equation (crosses) and compared with both, an exact direct calculation (full line) and the well-known result of the thermodynamic limit[1] $\frac{\langle N_0\rangle}{N} = 1 - \left(\frac{T}{T_c}\right)^2$ with $T_c = \sqrt{\frac{2N}{\Gamma(2)\zeta(2)\hbar^2\omega^2}}$ (dashed line). We see good agreement (see also Ref. 5).

It is obvious from Fig. 1 that the stochastic results suffer from some remaining statistical fluctuations. This is due to the large fluctuations of the particle number in the grand canonical ensemble for an ideal gas.[6] It

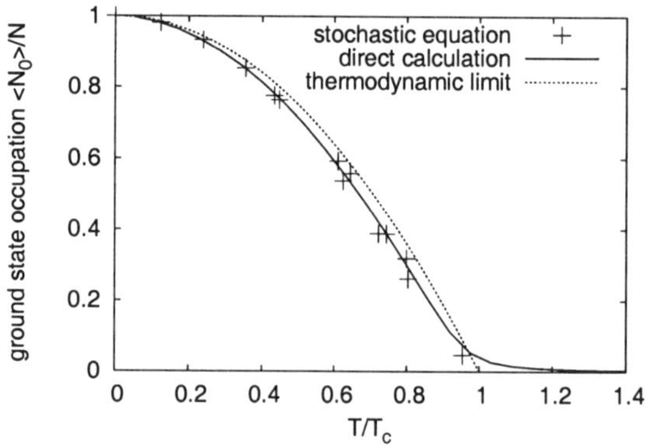

Fig. 1. Ground state occupation from a simulation (crosses) of an ideal 2D Bose gas of 500 particles on average, compared with the direct calculation (full line) and the thermodynamic limit (dashed line).

is therefore desirable to seek a stochastic field equation that describes a *canonical* ensemble.[7]

To summarize, based on a P-representation of the thermal (quantum) density operator, we present a stochastic field equation for an ideal Bose gas. The equation naturally overcomes problems connected with the ultraviolet catastrophe. We are very optimistic that the equation may also be applied to the weakly interacting case – in an approximate mean field sense. It should be emphasized that the stochastic field formulation of a Bose gas not only allows us to determine simple thermodynamic quantities, but it opens the door to (spatial) correlation functions of arbitrary order. Such investigations will be pursued in the near future.[7]

References

1. C. J. Pethik and H. Smith, *Bose-Einstein Condensation in Dilute Gases* (Cambridge University Press, Cambridge, 2002); L. Pitaevskii and S. Stringari, *Bose-Einstein Condensation* (Oxford University Press, Oxford, 2003).
2. P. C. Hohenberg and B. I. Halperin, *Rev. Mod. Phys.* **49**, 435 (1977).
3. M. J. Davis *et al.*, *Phys. Rev. Lett.* **87**, 0160402 (2001); C. W. Gardiner *et al.*, *J. Phys. B: At. Mol. Opt. Phys.* **35**, 1555 (2002).
4. C. C. Gerry and P. L. Knight, *Introductory Quantum Optics* (Cambridge University Press, Cambridge, 2005).
5. S. Grossmann and M. Holthaus, *Phys. Lett. A* **208**, 188 (1995).
6. R. M. Ziff, G. E. Uhlenbeck, and M. Kac, *Phys. Rep.* **32**, 169 (1977).
7. S. Heller and W. T. Strunz, to be published.

CRITICAL TEMPERATURE OF A BOSE-EINSTEIN CONDENSATE WITH $1/r$ INTERACTIONS

M. SCHÜTTE
Fachbereich Physik, Freie Universität Berlin,
Arnimallee 14, D-14195 Berlin, Germany
Email: schuett@physik.fu-berlin.de

A. PELSTER
Fachbereich Physik, Universität Duisburg-Essen, Campus Duisburg,
Lotharstraße 1, D-47048 Duisburg, Germany
Email: axel.pelster@uni-duisburg-essen.de

We investigate a trapped Bose-Einstein condensate underlying besides a short-range contact interaction especially a long-range $1/r$ interaction. The latter one can artificially be created by the radiation field of a certain laser configuration via induced electric dipoles. For this system we calculate the leading shift of the critical temperature with respect to the ideal gas.

Keywords: Bose-Einstein condensation; Long-range interaction; Critical temperature.

After its first experimental realization in 1995[1] the field of Bose-Einstein condensation (BEC) has experienced an enormous increase of interest both experimentally and theoretically. Although BEC is actually an atomic physics phase transition, it represents such a versatile and controllable many-particle system, that the research in BEC could be extended to several other fields of physics. In all early BEC experiments the atoms were alkali metals. So it was sufficient to model the van-der-Waals interaction in the gas by using the short-range pseudopotential

$$V_\delta^{(\mathrm{int})}(\mathbf{r}-\mathbf{r}') = \frac{4\pi\hbar^2 a}{M}\,\delta(\mathbf{r}-\mathbf{r}')\,. \tag{1}$$

Its strength is characterized by the s-wave scattering length a. In 2005, the first dipolar BEC was realized in a gas of ^{52}Cr, where besides (1) also a magnetic dipole-dipole interaction[2] is present. An even more dramatic long-range interaction would be supplied by a $1/r$ potential

$$V_{1/r}^{(\mathrm{int})}(\mathbf{r}-\mathbf{r}') = -\frac{4\pi^2\hbar^2}{Ma_{\mathrm{G}}}\frac{1}{|\mathbf{r}-\mathbf{r}'|}\,. \tag{2}$$

Here, we basically follow an idea of Ref. 3 to artificially create an attractive $1/r$ interaction due to a certain laser configuration which is similar to gravitation but up to 17 orders of magnitude stronger. The characteristic length scale a_{G} contains all information about the laser setup. As the interaction is attractive, a self-binding situation can occur where the attractive $1/r$ interaction balances both the kinetic energy and the contact interaction. Such tremendous attractive gravitational forces are only possible in nature on stellar scales. Thus, such ultracold quantum gases can be used to simulate cosmology in the laboratory and, in particular, to investigate the possibility of a Bose star which, so far, has only been discussed theoretically.[4]

An effective $1/r$ interaction results from the interaction of neutral atoms with a radiation field via induced electric dipoles. A quantum-electrodynamic fourth-order perturbative calculation leads to the interatomic contribution[5]

$$U(\mathbf{r}) = \frac{I}{\varepsilon_0 c}\,\hat{e}_i^{(\lambda)*}(\mathbf{k})\,\hat{e}_j^{(\lambda)}(\mathbf{k})\,\alpha^2(k)\,V_{ij}(\mathbf{r},k)\cos(\mathbf{k}\cdot\mathbf{r})\,. \tag{3}$$

Here, $\alpha(k)$ denotes the polarizability, I the intensity of the radiation field, $\hat{e}_i^{(\lambda)*}(\mathbf{k})$, $\hat{e}_j^{(\lambda)}(\mathbf{k})$ are the polarization vectors, and $\lambda = -, +$ stands for right or left circular polarization. Furthermore, the retarded dipole-dipole interaction tensor $V_{ij}(\mathbf{r},k)$ turns out to be

$$\frac{1}{4\pi\varepsilon_0 r^3}\left[(\delta_{ij} - 3\,\hat{r}_i\hat{r}_j)(\cos kr + kr\sin kr) - (\delta_{ij} - \hat{r}_i\hat{r}_j)\,k^2r^2\cos kr\right]\,. \tag{4}$$

In the near zone $kr \ll 1$ and a rotational average, Eq. (3) reduces to an attractive $1/r$ interaction.[5]

The first experimental proposal, how such an orientation average could be realized for ultracold gases, has been made in Ref. 3. The simplest model that suppresses the $1/r^3$ part consists of three orthogonal, circularly polarized laser beams: $\mathbf{k}_1 = k\hat{\mathbf{e}}_{\mathrm{x}}$, $\mathbf{k}_2 = k\hat{\mathbf{e}}_{\mathrm{y}}$, $\mathbf{k}_3 = k\hat{\mathbf{e}}_{\mathrm{z}}$. In the near zone $kr \ll 1$ we simply superpose the three terms to get the resulting potential

$$U_3(\mathbf{r}) = -\frac{3Ik^2\alpha^2}{16\pi c\varepsilon_0^2}\frac{1}{r}\left[\frac{7}{3} + (\sin\vartheta\,\cos\varphi)^4 + (\sin\vartheta\,\sin\varphi)^4 + (\cos\vartheta)^4\right]\,. \tag{5}$$

The angles φ and ϑ describe the orientation of the atoms with respect to the incident beam. Although the perturbative derivation of the interaction is only valid for static perturbations, we suggest here as a second experimental proposal a setup of rotating lasers within a quasi-static frame, under the constraint $\omega_{\text{excitation}} \ll \omega_{\text{rot}} \ll \omega_{\text{laser}}$. Instead of a static wave vector $\mathbf{k}$ we now use the time-dependent one $\mathbf{k}(t) = k\,(\sin\gamma\,\cos\omega t, \sin\gamma\,\sin\omega t, \cos\gamma)$ with a free parameter γ for which the best choice turns out to be the magic angle $\gamma = \arccos\sqrt{1/3}$.[6,8] Thus, the averaged potential reduces to

$$U_{1\times\text{rot}}(\mathbf{r}) = -\frac{I\alpha^2 k^2}{96\pi c\varepsilon_0^2}\,\frac{1}{r}\,(17 + 6\cos^2\vartheta - 7\cos^4\vartheta). \tag{6}$$

The rotating setup does not depend on φ and is more spherical than (5). A totally spherical-symmetric $1/r$ potential is obtained either from 18 static lasers as suggested in Ref. 3 or from three rotating lasers as proposed here.[7] Using the functional-integral formalism of many-body physics, we apply Feynman's diagrammatic perturbation expansion to calculate the leading shift of the critical temperature for a Bose gas in an isotropic harmonic trap. As there occurs an infrared divergence in the exchange contribution, it is not possible to apply the usual semiclassical equal-time correlation function. Instead we have to use the full quantum-mechanical one $G^{(0)}(\mathbf{x}, 0; \mathbf{x}', 0)$:

$$\sum_{n=1}^{\infty} e^{n\beta(\mu-3\hbar\omega/2)}\left[\frac{M\omega}{\pi\hbar(1-e^{-2\hbar\beta\omega n})}\right]^{3/2} e^{-\frac{M\omega}{2\hbar\sinh\hbar\beta\omega n}\left[(\mathbf{x}^2+\mathbf{x}'^2)\cosh\hbar\beta\omega n - 2\mathbf{x}\cdot\mathbf{x}'\right]}. \tag{7}$$

Herewith we solve the Feynman diagrams of the free energy

$\mathcal{F} = \mathcal{F}^{(0)} - \frac{1}{\beta}\Big\{ \frac{1}{2}$  $+ \frac{1}{2}$ 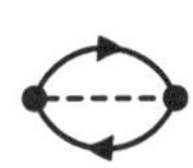 $+ \ldots \Big\}$ (8)

in order to determine the particle number $N = -\partial\mathcal{F}/\partial\mu$ and the self-energy

$\Sigma(\mathbf{x}, \tau; \mathbf{x}', \tau') =$ $\mathbf{x},\tau$  $\mathbf{x}',\tau'$ $+$ $\mathbf{x},\tau$  $\mathbf{x}',\tau'$ $+ \cdots$, (9)

which defines the critical chemical potential.[8] With this we obtain for the leading shift of the critical temperature caused by both interactions[7]

$$\frac{\Delta T_{\text{c}}}{T_{\text{c}}^{(0)}} = -c_\delta\,(N)\,\frac{a}{\lambda_{T_{\text{c}}^{(0)}}} + \frac{\lambda_{T_{\text{c}}^{(0)}}}{a_{\text{G}}}\left[c_{\text{E}}\,(N) + c_{\text{D}}\,(N)\,\frac{1}{\left(\hbar\beta_{\text{c}}^{(0)}\omega\right)^2}\right] \tag{10}$$

with the thermodynamic de Broglie wavelength $\lambda_{T_{\text{c}}^{(0)}} = \sqrt{2\pi\hbar^2\beta_{\text{c}}^{(0)}/M}$. The coefficients c_δ, c_{D}, c_{E} are shown as functions of the particle number in Fig. 1. In the semiclassical approximation we execute the limit $\hbar\beta_{\text{c}}^{(0)}\omega \ll 1$. For the contact interaction we obtain $c_\delta = 3.42603$ which is known from literature and has also been confirmed experimentally.[9] The direct term becomes

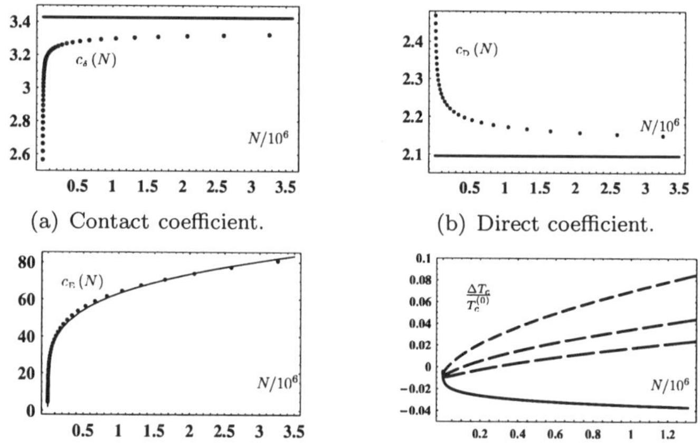

(a) Contact coefficient. (b) Direct coefficient.

(c) Exchange coefficient. (d) Total shift.

Fig. 1. (a)–(c): Numerical evaluation of the coefficients in (10). (a), (b): The straight lines show the corresponding semiclassical results. Due to the divergence there is no one in the exchange case; here we added a fitting line $2.929\,N^{2/9}$ that reproduces the values quite well and shows the behavior of the divergence. (d): The straight line comes from the result of a contact interaction only, $a_G \to \infty$, while the dashed lines give the complete shift (10) for different values of a_G; short to long dashes: $a_G = 0.05,\,0.075,\,0.1\,\mathrm{m}$. Chosen values for all plots: mass of ^{87}Rb, $a = 3\,\mathrm{nm}$, and $\omega = 2\pi \cdot 100\,\mathrm{Hz}$.

$c_D = 2.0951$ but for the exchange contribution c_E we find the divergence $\sum_{m=1}^{\infty} 1/\sqrt{m}$ which was avoided by our full quantum-mechanical calculation and renders $c_E(N)$ finite. For a realistic experiment the total shift is of the order of about $5\% - 10\%$, see Fig. 1(d), and should be measurable.

References

1. M. H. Anderson *et al.*, *Science* **269**, 198 (1995); K. B. Davis *et al.*, Phys. Rev. Lett. **75**, 3969 (1995).
2. A. Griesmaier *et al.*, *Phys. Rev. Lett.* **94**, 160401 (2005).
3. D. O'Dell *et al.*, *Phys. Rev. Lett.* **84**, 5687 (2000).
4. R. Ruffini and S. Bonazzola, *Phys. Rev.* **187**, 1767 (1969); G. Ingrosso and R. Ruffini, *Il Nuovo Cimento* **101 B**, 369 (1988).
5. D. P. Craig and T. Thirunamachandran, *Molecular Quantum Electrodynamics* (Academic Press, London, 1984).
6. H. Günther, *NMR-Spektroskopie*, 3rd Edition (Georg Thieme Verlag, Stuttgart, 1992); S. Giovanazzi *et al.*, Phys. Rev. Lett. **89**, 130401 (2002).
7. M. Schütte, *Diploma thesis*, Freie Universität Berlin (2007).
8. K. Glaum and A. Pelster, *Phys. Rev. A* **76**, 023604 (2007).
9. S. Giorgini *et al.*, *Phys. Rev. A* **54**, R4633 (1996); F. Gerbier *et al.*, *Phys. Rev. Lett.* **92**, 030405 (2004).

CRITICAL TEMPERATURE OF DIRTY BOSONS

B. KLÜNDER*, A. PELSTER**, and R. GRAHAM†

Fachbereich Physik, Universität Duisburg-Essen, Campus Duisburg,
Lotharstraße 1, D-47048 Duisburg, Germany
** E-mail: kluender@theo.phys.uni-duisburg-essen.de*
*** E-mail: axel.pelster@uni-duisburg-essen.de*
† E-mail: robert.graham@uni-duisburg-essen.de

We consider a dilute Bose gas moving in a harmonic trap with a superimposed frozen random potential which arises in experiments either naturally in wire traps or artificially and controllably with laser speckles. The critical temperature, which characterises the onset of Bose-Einstein condensation, depends on the disorder realization within the ensemble. Therefore, we introduce an effective grand-canonical potential from which we determine perturbatively the disorder averages of both the first and the second moment of the critical temperature in leading order. We discuss our results for a finite number of particles by assuming a Gaussian spatial correlation for the quenched disorder potential.

Keywords: Disorder; Bose-Einstein condensation; Critical temperature.

1. Introduction

This work is motivated by recent experiments which study disorder effects on Bose-Einstein condensates (BECs). Frozen random potentials for BECs arise naturally in wire traps due to surface roughness.[1,2] For a better understanding of these effects one needs to create disorder in a more controllable fashion. A prominent example is provided by laser speckles where both strength and correlation length of disorder realizations can be tuned.[3,4]

Here we investigate the effect of disorder on the critical temperature of a harmonically trapped dilute ideal Bose gas. We assume that the disorder potential is spatially Gaussian correlated and that it is weak enough to treat it in a perturbative manner. The aim of this work is to determine the disorder-induced shift of the critical temperature and the mean deviation around its average. To this end we calculate the critical temperature as a functional of the disorder potential and evaluate then the first and second disorder averaged moments.

2. Effective Action

We start with the usual functional integral representation of the grand-canonical partition function

$$Z = \oint \mathcal{D}\psi^* \oint \mathcal{D}\psi \, e^{-\mathcal{A}/\hbar}, \tag{1}$$

where one integrates over all possible bosonic Schrödinger fields which are periodic in imaginary time. The corresponding action reads

$$\mathcal{A}=\int_0^{\hbar\beta} d\tau \int d^3x \, \psi^*(\mathbf{x},\tau)\left[\hbar\frac{\partial}{\partial\tau}-\frac{\hbar^2}{2M}\Delta+\frac{M}{2}\omega^2\mathbf{x}^2+U(\mathbf{x})-\mu\right]\psi(\mathbf{x},\tau), \tag{2}$$

where $U(\mathbf{x})$ and ω denote the disorder potential and the frequency of the isotropic harmonic trap.

To find the average influence of the disorder we consider a large ensemble of macroscopically identical realizations and define an ensemble average as

$$\overline{\quad\bullet\quad} = \prod_{\mathbf{x}} \int_{\infty}^{\infty} dU(\mathbf{x}) \bullet P[U], \qquad \overline{1} = 1. \tag{3}$$

For our purposes it is not necessary to know the concrete form of the probability distribution $P[U]$ apart from the condition that it has to be bounded from below. Furthermore, we assume that the disorder potential vanishes on the average and that the disorder is spatially Gaussian correlated:[7]

$$\overline{U(\mathbf{x}_1)} = 0, \qquad \overline{U(\mathbf{x}_1)U(\mathbf{x}_2)} = \frac{R}{(2\pi\epsilon^2)^{\frac{3}{2}}} e^{-\frac{(\mathbf{x}_1-\mathbf{x}_2)^2}{2\epsilon^2}}. \tag{4}$$

Here R is the disorder strength and ϵ the disorder correlation length.

In the next step we calculate the effective action and make use of the so-called background method.[5,6] Thus, we use the decomposition $\psi(\mathbf{x},\tau)=\Psi(\mathbf{x},\tau)+\delta\psi(\mathbf{x},\tau)$ for the Schrödinger field, where $\Psi(\mathbf{x},\tau)$ and $\delta\psi(\mathbf{x},\tau)$ denote field expectation value and fluctuations, respectively. The remaining functional integral in the partition function is expanded for a small disorder potential $U(\mathbf{x})$ up to second order. After applying the logarithm to the grand-canonical partition function we get for the effective action

$$\Gamma[\Psi^*,\Psi] = -\sum_{k=1}^{\infty} \frac{e^{\beta(\mu-E_0)k}}{\beta k}\left[\frac{1}{(1-e^{-\hbar\omega\beta k})^3}-1\right]$$
$$+\frac{1}{\hbar\beta}\int_0^{\hbar\beta} d\tau \int d^3x \, \Psi^*(\mathbf{x},\tau)\left[\hbar\frac{\partial}{\partial\tau}-\frac{\hbar^2}{2M}\Delta+\frac{M}{2}\omega^2\mathbf{x}^2+U(\mathbf{x})-\mu\right]\Psi(\mathbf{x},\tau)$$

$$-\frac{1}{2\hbar^2\beta}\int_0^{\hbar\beta} d\tau_{1,2}\int d^3x_{1,2}\; U(\mathbf{x}_1)U(\mathbf{x}_2)G_0(\mathbf{x}_1,\tau_1;\mathbf{x}_2,\tau_2)G_0(\mathbf{x}_2,\tau_2;\mathbf{x}_1,\tau_1)$$
$$+\int d^3x\; U(\mathbf{x})G_0(\mathbf{x},0;\mathbf{x},0)+\dots\,. \tag{5}$$

Here $G_0(\mathbf{x}_1,\tau_1;\mathbf{x}_2,\tau_2) = \langle\psi(\mathbf{x}_1,\tau_1)\psi^*(\mathbf{x}_2,\tau_2)\rangle$ stands for the unperturbed correlation function of the Schrödinger fields. The grand-canonical potential Ω can be found by extremizing the effective action (5) with respect to the field expectation values:

$$\frac{\delta\Gamma[\Psi_e^*,\Psi_e]}{\delta\Psi_e(\mathbf{x},\tau)} = \frac{\delta\Gamma[\Psi_e^*,\Psi_e]}{\delta\Psi_e^*(\mathbf{x},\tau)} = 0 \quad\Rightarrow\quad \Omega(T,V,\mu) = \Gamma[\Psi_e^*,\Psi_e]\,. \tag{6}$$

In the symmetry-broken phase we expand the critical chemical potential μ_c and the extremal field expectation value $\Psi_e(\mathbf{x},\tau)$ with respect to $U(\mathbf{x})$ up to second order to determine the respective disorder-induced corrections:

$$\mu_c = \frac{3}{2}\hbar\omega+\mu_c^{(1)}+\mu_c^{(2)}+\dots\,,\quad \Psi_e(\mathbf{x},\tau)=\sqrt{N_0}\psi_0(\mathbf{x})+\Psi_e^{(1)}(\mathbf{x})+\Psi_e^{(2)}(\mathbf{x})+\dots\,. \tag{7}$$

Here $\psi_0(\mathbf{x})$ denotes the ground-state wave function and $3\hbar\omega/2$ the ground-state energy of the three-dimensional harmonic oscillator. Both corrections allow us in the following to calculate the critical temperature as a functional of the disorder potential.

3. Critical Temperature

We determine the critical temperature from the particle number equation N which follows immediately from the grand-canonical potential Ω. To this end we set the ground-state occupation number N_0 equal to zero, the chemical potential equal to its critical value μ_c, and expand the critical temperature: $T_c[U] = T_c^{(0)} + T_c^{(1)}[U] + T_c^{(2)}[U] + \dots$ with $T_c^{(0)} = \hbar\omega N^{1/3}/\zeta(3)^{1/3}k_B$. Using the disorder averages (4) one finds the following equations for the average shift and the variance:

$$\frac{\overline{\Delta T_c[U]}}{T_c^{(0)}} = \frac{\overline{T_c^{(2)}[U]}}{T_c^{(0)}} + \dots\,,\qquad \frac{\overline{\Delta T_c[U]^2}}{T_c^{(0)2}} = \frac{\overline{T_c^{(1)}[U]^2}}{T_c^{(0)2}} + \dots\,. \tag{8}$$

The respective results can be seen in Fig. 1. The average shift of the critical temperature in Fig. 1a) is positive and has a maximum when the correlation length ϵ is about the same size than the thermal wave length $\lambda_c^{(0)}$ of the bosons at the critical temperature. Both the critical temperature and the mean deviation are proportional to $N^{-1/3}$ when $\tilde{\epsilon}$ is kept constant. In the related Ref. 8 it is assumed that the critical temperature is self-averaging.

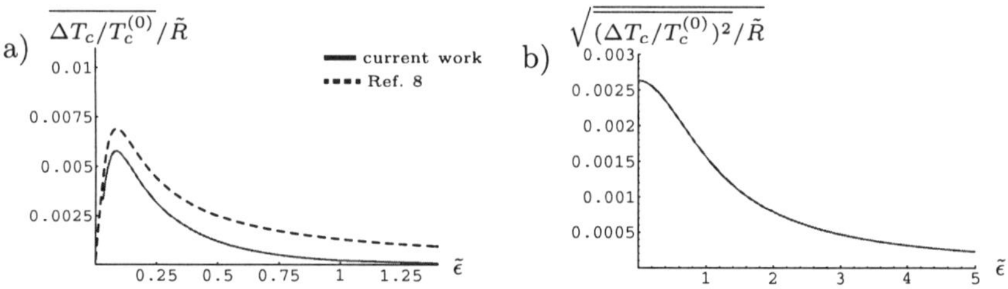

Fig. 1. Average shift of critical temperature a) and mean deviation b) versus $\tilde{\epsilon}=\epsilon\sqrt{M\omega/\hbar}$ for $N=10^5$ and $\tilde{R}=R\ M^{3/2}\omega^{-1/2}\hbar^{-7/2}$.

This approximation leads qualitatively to the same result but differs quantitatively as is shown in Fig. 1a) for an experimentally realistic particle number. One also finds in Fig. 1b) that the relative variance does not vanish for disorder correlation lengths much smaller than the oscillator length. This implies that the critical temperature is not a self-averaging quantity with respect to the correlation length.

Acknowledgments

This work was supported by the SFB/TR 12 of the German Research Foundation (DFG) and by the Wilhelm and Else Heraeus Foundation.

References

1. P. Krüger *et al.*, *Phys. Rev. A* **76**, 063621 (2007) [arXiv:cond-mat/0504686].
2. J. Fortagh, H. Ott, S. Kraft, and C. Zimmermann, *Phys. Rev. A* **66**, R041604 (2002).
3. J. E. Lye *et al.*, *Phys. Rev. Lett.* **95**, 070401 (2005).
4. D. Clément *et al.*, *Phys. Rev. Lett.* **95**, 170409 (2005).
5. C. Morette, *Phys. Rev.* **81**, 848 (1951).
6. B. S. DeWitt, *Theory of Dynamical Groups and Fields* (Gordon and Breach, New York, 1965).
7. M. Kobayashi and M. Tsubota, *Phys. Rev. B* **66**, 174516 (2002).
8. M. Timmer, A. Pelster, and R. Graham, *Europhys. Lett.* **76**, 760 (2006).

DENSITY AND STABILITY IN ULTRACOLD DILUTE BOSON-FERMION MIXTURES

S. RÖTHEL

Institut für Theoretische Physik, Westfälische Wilhelms-Universität Münster, Wilhelm-Klemm-Straße 9, D-48149 Münster, Germany
E-mail: steffen.roethel@uni-muenster.de

A. PELSTER

Fachbereich Physik, Universität Duisburg-Essen, Lotharstraße 1, D-47048 Duisburg, Germany
E-mail: axel.pelster@uni-duisburg-essen.de

We analyze in detail recent experiments on ultracold dilute ^{87}Rb–^{40}K mixtures in Hamburg and in Florence within a mean-field theory. To this end we determine how the stationary bosonic and fermionic density profiles in this mixture depend in the Thomas-Fermi limit on the respective particle numbers. Furthermore, we investigate how the observed stability of the Bose-Fermi mixture with respect to collapse is crucially related to the value of the interspecies s-wave scattering length.

Keywords: Bose-Fermi-mixture; Density; Stability.

1. Introduction

Six years after the first experimental achievement of Bose-Einstein condensation (BEC) of trapped atomic gases in 1995 fermionic atomic gases were brought together with bosonic atoms to quantum degeneracy in a ^{7}Li–^{6}Li mixture,[1,2] ^{23}Na–^{6}Li mixture,[3] and ^{87}Rb–^{40}K mixture.[4] In such mixtures one investigates, in particular, how the two-particle interaction influences the system properties. Depending on the nature of the interspecies interaction, a repulsion between bosons and fermions tends to a demixing in order to minimize the overlapping region,[5] whereas in the case of an attraction the mixture can collapse as long as the particle numbers are sufficiently large.[6,7] Ultracold trapped boson-fermion mixtures were investigated with respect to a demixing of the components[8–10] and to a collapse due to the interspecies attraction.[9–13] Furthermore, the time-dependent dynamics of the collapse[14] and finite-temperature effects on the stability in a boson-fermion mixture were also studied.[15]

Our theoretical investigation is based on the experiments with a ^{87}Rb–^{40}K boson-fermion mixture in a harmonic trap, which were performed in Hamburg[6] and in Florence.[4,7] The parameters of both experiments are summarized in Ref. 16. There our attention is mainly focused on the value of the interspecies s-wave scattering length a_{BF} since this parameter is of great importance for the system, especially for the stability of the mixture against collapsing. It turns out that different determination methods for a_{BF} lead, surprisingly, to incompatible values in the literature.[16] Thus, further investigations in this field are needed.

2. Density Profiles

A boson-fermion mixture at zero temperature, where all bosons are condensed in the single-particle ground state and the fermions occupy every state below the Fermi energy, is described by the stationary Gross-Pitaevskii equation[16]

$$\left[-\frac{\hbar^2}{2m_B}\Delta + V_B(\mathbf{x}) - \mu_B + g_{BB}|\Psi_B(\mathbf{x})|^2 + g_{BF}\, n_F(\mathbf{x})\right] \Psi_B(\mathbf{x}) = 0 \quad (1)$$

with the condensate wave function $\Psi_B(\mathbf{x})$ and the fermionic particle density

$$n_F(\mathbf{x}) = \kappa\, \Theta\left(\mu_F - V_F(\mathbf{x}) - g_{BF}\, n_B(\mathbf{x})\right) \left[\mu_F - V_F(\mathbf{x}) - g_{BF}\, n_B(\mathbf{x})\right]^{3/2}, \quad (2)$$

which modulates the condensate density $n_B(\mathbf{x}) = |\Psi_B(\mathbf{x})|^2$ and vice versa. Here $V_i(\mathbf{x})$ and μ_i $(i = B, F)$ denote the external trap potential and the chemical potential for bosons and fermions, respectively, g_{BB} describes the strength of the contact interaction between two bosons and g_{BF} stands for the corresponding one between a boson and a fermion. Assuming that the potential and interaction energy are larger than the kinetic energy, one can use the Thomas-Fermi approximation, where the kinetic term in the Gross-Pitaevskii equation (1) is neglected, so that the latter reduces to an algebraic equation with respect to the bosonic particle density $n_B(\mathbf{x})$:[17]

$$V_B(\mathbf{x}) - \mu_B + g_{BB}\, n_B(\mathbf{x}) + g_{BF}\, n_F(\mathbf{x}) = 0. \quad (3)$$

The bosonic and the fermionic density distribution solving this equation is plotted in Fig. 1 for the parameters of the Hamburg experiment.[6]

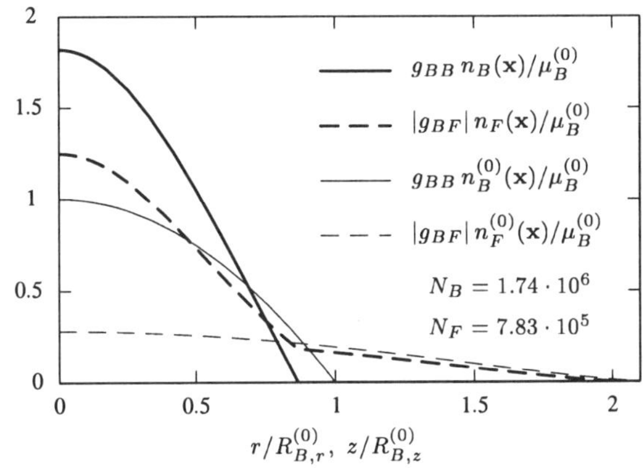

Fig. 1. Comparison of the dimensionless particle densities for bosons and for fermions between mixtures with interspecies interaction (thick lines) and without interspecies interaction (thin lines). The densities are plotted versus the coordinates r at the plane $z = 0$ and versus z at the line $r = 0$, respectively, in units of the Thomas-Fermi radii of the noninteracting BEC for an example of typical particle numbers N_B, N_F of the Hamburg experiment.[6]

3. Stability Against Collapse

In order to determine the border between stability and instability we extremized the grand-canonical free energy[16]

$$\mathcal{F} = \int d^3x \left\{ \Psi_B^*(\mathbf{x}) \left[-\frac{\hbar^2}{2m_B}\Delta + V_B(\mathbf{x}) - \mu_B + \frac{g_{BB}}{2}\,|\Psi_B(\mathbf{x})|^2 \right] \Psi_B(\mathbf{x}) \right.$$
$$\left. - \frac{2}{5}\,\kappa\,\Theta(\tilde{\mu}_F(\mathbf{x}))\,\tilde{\mu}_F^{5/2}(\mathbf{x}) \right\} \quad (4)$$

of the mixture by varying the widths $\alpha\, L_{B,k}$ in the test function

$$\Psi_B(\mathbf{x}) = \sqrt{\frac{N_B \lambda^{1/2}}{\pi^{3/2}\alpha^3 L_{B,r}^3}}\, \exp\left\{ -\frac{r^2 + \lambda z^2}{2\alpha^2 L_{B,r}^2} \right\}, \qquad \lambda = (\omega_z/\omega_r)^2, \quad (5)$$

which has the form of the ground-state wave function of a three-dimensional anisotropic harmonic oscillator. Here the variational parameter α has to fulfill the conditions

$$N_B = N_{B\mathrm{crit}} \qquad \Leftrightarrow \qquad \left.\frac{d\mathcal{F}(\alpha)}{d\alpha}\right|_{\alpha=\alpha_{\mathrm{crit}}} = \left.\frac{d^2\mathcal{F}(\alpha)}{d\alpha^2}\right|_{\alpha=\alpha_{\mathrm{crit}}} = 0, \quad (6)$$

which yield the critical number of bosons N_B and the corresponding one for fermions N_F by integrating out the fermionic particle density (2). Comparing them with the data of the Hamburg experiment allows us to fit the s-wave scattering length to $a_{BF} = (-16.82 \pm 030)$ as shown in Fig. 2.

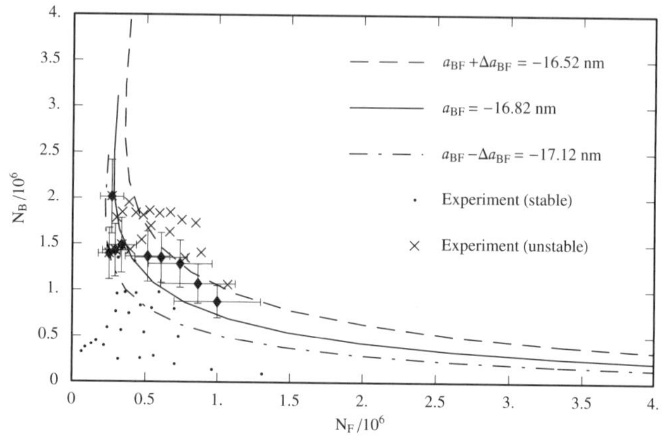

Fig. 2. Stability border for a new value $a_{BF} = (-16.82 \pm 0.30)$ nm of the ^{87}Rb–^{40}K mixture in the Hamburg experiment.[6] All three lines are obtained by numerically integrating the fermionic density (2). Those stable (unstable) points, which are located above (below) the solid line, are equipped with error bars indicating a relative uncertainty of 20% and 30% for N_B and N_F, respectively, according to Fig. 3 of Ref. 6.

References

1. A. G. Truscott, K. E. Strecker, W. I. McAlexander, G. B. Partridge, and R. G. Hulet, *Science* **291**, 2570 (2001).
2. F. Schreck, L. Khaykovich, K. L. Corwin, G. Ferrari, T. Bourdel, J. Cubizolles, and C. Salomon, *Phys. Rev. Lett.* **87**, 080403 (2001).
3. Z. Hadzibabic, C. A. Stan, K. Dieckmann, S. Gupta, M. W. Zwierlein, A. Görlitz, and W. Ketterle, *Phys. Rev. Lett.* **88**, 160401 (2002).
4. G. Roati, F. Riboli, G. Modugno, and M. Inguscio, *Phys. Rev. Lett.* **89**, 150403 (2002).
5. S. Ospelkaus, C. Ospelkaus, L. Humbert, K. Sengstock, and K. Bongs, *Phys. Rev. Lett.* **97**, 120403 (2006).
6. C. Ospelkaus, S. Ospelkaus, K. Sengstock, and K. Bongs, *Phys. Rev. Lett.* **96**, 020401 (2006).
7. G. Modugno, G. Roati, F. Riboli, F. Ferlaino, R. J. Brecha, and M. Inguscio, *Science* **297**, 2240 (2002).
8. N. Nygaard and K. Molmer, *Phys. Rev. A* **59**, 2974 (1999).
9. R. Roth and H. Feldmeier, *Phys. Rev. A* **65**, 021603 (2002).
10. R. Roth, *Phys. Rev. A* **66**, 013614 (2002).
11. T. Miyakawa, T. Suzuki, and H. Yabu, *Phys. Rev. A* **64**, 033611 (2001).
12. S. T. Chui and V. N. Ryzhov, *Phys. Rev. A* **69**, 043607 (2004).
13. S. T. Chui, V. N. Ryzhov, and E. E. Tareyeva, *JETP Lett.* **80**, 274 (2004).
14. S. K. Adhikari, *Phys. Rev. A* **70**, 043617 (2004).
15. X. Liu, M. Modugno, and H. Hu, *Phys. Rev. A* **68**, 053605 (2003).
16. S. Röthel and A. Pelster, *Eur. Phys. J. B* **59**, 343 (2007).
17. K. Molmer, *Phys. Rev. Lett.* **80**, 1804 (1998).

SPINOR FERMI GASES

A. R. P. LIMA
Fachbereich Physik, Freie Universität Berlin,
Arnimallee 14, D-14195 Berlin, Germany
E-mail: lima@physik.fu-berlin.de

A. PELSTER
Fachbereich Physik, Universität Duisburg-Essen, Campus Duisburg,
Lotharstraße 1, D-47048 Duisburg, Germany
E-mail: axel.pelster@uni-duisburg-essen.de

At first, we consider an ideal gas of harmonically trapped fermions with total angular momentum $F = 3/2$ and calculate, e.g., the temperature dependence of the heat capacity for a fixed magnetization. Afterwards, the isotropic short-range contact-interaction is treated perturbatively and its influence on the ground-state energy is worked out. Such spinor Fermi gases are important, for example, in the context of a ^{52}Cr-^{53}Cr boson-fermion mixture.[1]

Keywords: Spinor Fermi gas; Bose-Einstein condensation.

1. Introduction

Since the first observations of Bose-Einstein condensation (BEC) in ultracold dilute gases of alkali metals,[2,3] the interest in fermionic gases has also increased. Superfluidity in fermionic species has been shown to be deeply related to the BEC of Cooper pairs, as originally introduced in the context of the Bardeen-Cooper-Schrieffer (BCS) theory of superconductivity. The recent observation of the BCS-BEC-crossover[4] in ^{6}Li, as theoretically predicted,[5] has emphasised how important the underlying fermionic degrees of freedom are for the physics of ultracold degenerate quantum gases.

Here, we are mainly interested in the spinorial character of Fermi gases with total angular momentum, denoted by F, larger than 1/2. From now on, we assume that fermionic atoms with large F are cooled and trapped in the harmonic potential of an optical trap, where the spinorial character becomes particularly enhanced. One could choose fermionic isotopes of ytterbium, potassium, chromium, etc. Let us consider ^{53}Cr for definiteness. We remark that its bosonic counterpartner ^{52}Cr has been Bose-condensed[6] and that its large magnetic moment of $6\mu_B$, where μ_B denotes the Bohr magneton, makes it an interesting system in itself. Consider the ground-state of ^{53}Cr. The nuclear spin $I = 3/2$ added to the electronic spin $S = 3$ renders it four-fold degenerate with F ranging from $F = |I - S| = 3/2$ to $F = I + S = 9/2$. This suggests to investigate, at first, ideal Fermi gases with large F.

2. Ideal Fermi Gases

The Euclidean action $\mathcal{A}^{(0)}$ of a non-interacting system is given by

$$\mathcal{A}^{(0)} = \int_0^{\hbar\beta} \mathrm{d}\tau \int \mathrm{d}^D x \, \Psi^\dagger \left[\hbar \frac{\partial}{\partial \tau} - \frac{\hbar^2 \nabla^2}{2M} + U(\mathbf{x}) - \mu - \eta F_z \right] \Psi. \tag{1}$$

Here, $\Psi(\mathbf{x}, \tau)$ is a Grassmannian spinor field with the ad-joint $\Psi^\dagger(\mathbf{x}, \tau) = (\psi^*_{+F}(\mathbf{x}, \tau), \cdots, \psi^*_{-F}(\mathbf{x}, \tau))$ and F_z is the z-component of the spin matrix. Both the chemical potential μ and the magneto-chemical one η are functions of the temperature and represent Lagrange parameters for fixing the particle number and the magnetization in z-direction, respectively.

The grand-canonical partition function $\mathcal{Z}^{(0)}$ is defined by

$$\mathcal{Z}^{(0)} = \oint \mathcal{D}\Psi^\dagger \mathcal{D}\Psi \; e^{-\mathcal{A}^{(0)}/\hbar}, \tag{2}$$

where the functional integration runs over all spinor fields, which are anti-periodic in imaginary time. Treating the harmonic potential $U(\mathbf{x}) = M \sum_{i=1}^{3} \omega_i^2 x_i^2 / 2$ semi-classically, the grand-canonical free energy $\mathcal{F}^{(0)} = -(\log \mathcal{Z}^{(0)})/\beta$ is given by

$$\mathcal{F}^{(0)} = \frac{-1}{\beta(\beta\bar{\omega}\hbar)^3} H_0^4(z, \alpha). \tag{3}$$

Here, the function $H_n^k(z, \alpha) = \sum_{f=-F}^{+F} \frac{f^k}{\Gamma(n)} \int_0^\infty dy \frac{y^{n-1}}{1+\exp(y)(z\alpha^f)^{-1}}$ is introduced with z (α) being the fugacity (magnetic fugacity). The geometric mean of the trap frequencies is denoted by $\bar{\omega} = (\omega_x \omega_y \omega_z)^{1/3}$.

From Eq. (3) follow all thermodynamic properties of the system. For example, the total particle number $N = -\partial \mathcal{F}/\partial \mu$ and the magnetization $\mathcal{M} = -\partial \mathcal{F}/\partial \eta$ are given by

$$N = \frac{1}{(\beta\bar{\omega}\hbar)^3} H_3^0(z, \alpha), \qquad \mathcal{M} = \frac{1}{(\beta\bar{\omega}\hbar)^3} H_3^1(z, \alpha). \tag{4}$$

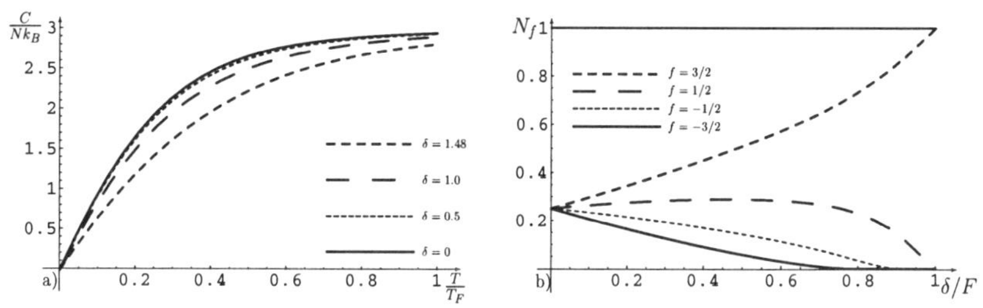

Fig. 1. Spinor Fermi gas with $F = 3/2$: *Left*: Heat capacity versus temperature for some values of δ. *Right*: Occupation numbers against δ at $T = 0$.

A similar expression can be derived for the heat capacity

$$\begin{aligned}\frac{C}{Nk_B} &= 12\frac{H_4^0(z,\alpha)}{H_3^0(z,\alpha)} - 9\frac{H_3^1(z,\alpha)H_2^1(z,\alpha) - H_3^0(z,\alpha)H_2^2(z,\alpha)}{H_2^1(z,\alpha)^2 - H_2^0(z,\alpha)H_2^2(z,\alpha)} \\ &\quad +9\delta\frac{H_3^1(z,\alpha)H_2^0(z,\alpha) - H_3^0(z,\alpha)H_2^1(z,\alpha)}{H_2^1(z,\alpha)^2 - H_2^0(z,\alpha)H_2^2(z,\alpha)}, \end{aligned} \qquad (5)$$

where $\delta = \mathcal{M}/N$ characterises the magnetization per particle.

Evaluating thermodynamic quantities for given F, δ necessitates to solve Eq. (4) for z, α. Some examples are presented in Fig. 1 for $F = 3/2$. The left part shows the heat capacity (5) against temperature in units of the Fermi temperature $T_F \equiv E_F/k_B$, where $E_F = \hbar\overline{\omega}(3N/2)^{1/3}$ is the Fermi energy. The right part depicts the occupation numbers for $T = 0$ against δ.

3. Weakly Interacting Fermi Gases at $T = 0$

The energy differences between the different F substates in the ground-state of ^{53}Cr is much larger[1] than the expected trapping frequencies. This, combined with the diluteness of the system, allows to describe the low-energy dynamics of the system through a pairwise contact interaction which is rotationally invariant in the hyperfine spin space and preserves the hyperfine spin of the individual atoms. Such an interaction is described through[7]

$$V_{ij i'j'}^{(\mathrm{contact})}(\mathbf{x},\mathbf{x}') = \delta(\mathbf{x}-\mathbf{x}')\left[c_0\,(\mathbf{F}_{ij}\cdot\mathbf{F}_{i'j'})^0 + c_1\,(\mathbf{F}_{ij}\cdot\mathbf{F}_{i'j'})^1\right], \qquad (6)$$

where the coefficients c_0 and c_1 depend only on a_0 and a_2, the scattering lengths in the $|\mathbf{F}+\mathbf{F}'| = 0, 2$-channels. The ground-state energy, up to first order in (6) with $c_0' = 5a_2 - a_0$ and $c_1' = (a_2 - a_0)$, reads (see Fig. 2)

$$\frac{E}{NE_F} = \frac{3}{4} + 7\frac{N^{1/6}}{\sqrt{2\pi}}\left(\frac{3}{2}\right)^{7/6}\left\{\sum_{f,f'\neq f} A(y_<,y_>)\left[\frac{c_0'}{4\overline{a}_{\mathrm{osc}}} + f(f'+1)\frac{c_1'}{3\overline{a}_{\mathrm{osc}}}\right]\right.$$

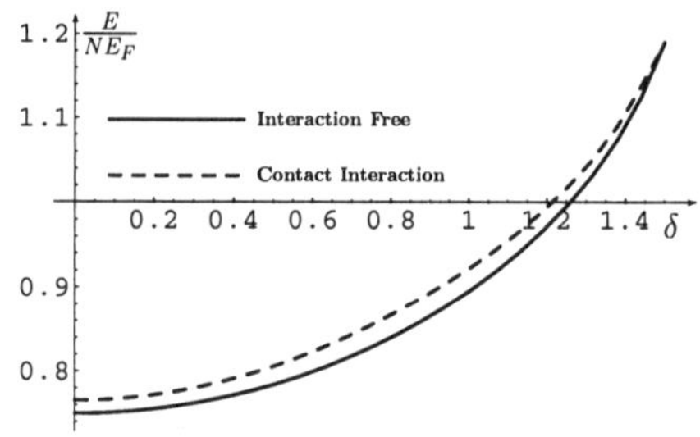

Fig. 2. Spinor Fermi gas with $F = 3/2$: Ground-state energy for $\overline{\omega} \approx 2\pi \times 100$ Hz, $a_0 \approx a_2 = 10^{-9}$m, and $N \approx 10^5$.

$$+ \frac{c_1'}{6\overline{a}_{\rm osc}} \sum_{f=-3/2}^{3/2} \left[\left(\frac{3}{2} - f\right) \left(\frac{5}{2} + f\right) A(y_f, y_{f+1}) + \left(\frac{3}{2} + f\right) \left(\frac{5}{2} - f\right) A(y_{f-1}, y_f) \right] \Bigg\}, \tag{7}$$

where $A(y_<, y_>) = {}_2F_1\left(-\frac{3}{2}, \frac{3}{2}; 4; \frac{y_<}{y_>}\right) \frac{y_<^3}{\Gamma(4)} \frac{y_>^{3/2}}{\Gamma(5/2)} \Theta(y_<)$ and $y_<$ $(y_>)$ is the smallest (largest) of $y_f = \mu + f\eta$ and $y_{f'} = \mu + f'\eta$.

4. Outlook

We have studied some properties of the weakly interacting spinor Fermi gas. Future studies are necessary to understand, e.g., multi-channel superfluidity,[8] the role of the dipole-dipole interaction,[9] and their combination.

Acknowledgment

A. Lima would like to acknowledge the support from the German Academic Exchange Service (DAAD).

References

1. R. Chicireanu *et al.*, *Phys. Rev. A* **73**, 053406 (2006).
2. M. H. Anderson *et al.*, *Science* **269**, 198 (1995).
3. K. B. Davis *et al.*, *Phys. Rev. Lett.* **75**, 3969 (1995).
4. M. Bartenstein *et al.*, *Phys. Rev. Lett.* **92**, 120401 (2004).
5. A. J. Leggett, *J. Phys. (Paris)* **41**, 19 (1980).
6. A. Griesmaier *et al.*, *Phys. Rev. Lett.* **94**, 160401 (2005).
7. T. L. Ho, *Phys. Rev. Lett.* **81**, 742 (1998).
8. T. L. Ho and S. Yip, *Phys. Rev. Lett.* **82**, 247(1999).
9. S. Yi, L. You, and H. Pu, *Phys. Rev. Lett.* **93**, 040403 (2004).

PART VIII
Condensed Matter

COUNTING ELECTRICAL CHARGES: A PROBLEM OF THERMAL ESCAPE AND QUANTUM TUNNELING IN PRESENCE OF NON-GAUSSIAN NOISE

J. ANKERHOLD

Institut für Theoretische Physik, Universität Ulm,
Albert-Einstein-Allee 11, D-89069 Ulm, Germany
E-mail: joachim.ankerhold@uni-ulm.de

Electrical noise produced when charges flow through a mesoscopic conductor is non-Gaussian, where higher order cumulants carry valuable information about the microscopic transport process. In these notes we briefly discuss the experimental situation and show that theoretical descriptions for the detection of electrical noise with Josephson junctions lead to generalizations of classical and quantum theories, respectively, for decay rates out of metastable states.

Keywords: Non-Gaussian noise; Mesoscopic systems; Thermal activation; Quantum tunneling.

1. Introduction

Typically noise is an annoying phenomenon. This is particularly true for mesoscopic systems since they are inevitably in contact with external leads, substrates, gates etc., which only allow for control and manipulation of the system under investigation. However, this is not the whole story. In fact, noise may even carry information that cannot be obtained by standard measurements detecting mean values. The relevance of the shot noise power of the electrical current was already discovered by Schottky, who showed that one gains direct access to the effective charge carrying the current by observing the shot noise and the mean current. Based on this seed, within the last decade electrical noise has moved into the focus of research activities on electronic transport in nanostructures[1,2] since it provides information on microscopic mechanisms of the transport not available from the voltage dependence of the average current. Lately, attention has turned from the noise auto-correlation function (shot noise) to higher order cumulants of the current fluctuations characterizing non-Gaussian statistics.[2,3] While theoretical attempts to predict these cumulants for a variety of devices are quite numerous,[2] experimental observation is hard because of small signals, large bandwidth detection, and strict filtering demands. A first pioneering measurement by Reulet *et al.*[4] of the third cumulant of the current noise from a tunnel junction has intensified efforts and several new proposals for experimental set-ups have been put forward very recently, some of which are based on Josephson junctions (JJ) as on-chip noise detectors.[5]

In all experimental set-ups to measure higher order cumulants realized and proposed so far, heating is one of the major experimental obstacles.[4,6] Thus, experiments have primarily attempted to establish just the unspecified non-Gaussian nature of the noise or to measure the third cumulant (skewness). The latter one is particularly accessible since it can be discriminated from purely Gaussian noise due to its asymmetry, e.g. when inverting the current through the conductor. This way, very recently the skewness of current noise generated by a tunnel junction was measured in the adiabatic limit (low frequency noise) of the quantum tunneling regime of a JJ.[7] In the even more interesting regime of finite frequencies,[8] first data have been obtained in Ref. 9, where the JJ stays in the classical domain of thermally activated switching. This latter situation requires a new theoretical framework generalizing the well-established Kramers' theory[10] to escape in presence of non-Gaussian noise.[11] We will present the main results in the first part of these notes.

The problem for a corresponding quantum theory is that electrical noise is produced by an environment in a steady state but far from equilibrium. A consistent general theory for quantum tunneling in such a situation is still elusive. What one could think about is to place the mesoscopic conductor in parallel to a current biased JJ in the zero voltage state.[12] Then, no net current flows through the sample prior to the read out and the generated electrical noise is equilibrium noise (with non-Gaussian cumulants though). Accordingly, on the one hand only even higher order cumulants exist so that in particular the fourth order cumulant (sharpness) becomes accessible that due to heating effects may always be hidden behind the second and third order ones. On the other hand, a generalization of the standard ImF approach,[13] which is based on a proper evaluation of the partition function of the total system, is possible as we will briefly discuss in the second part.

2. Preliminaries

The complete statistics of current noise generated by a mesoscopic conductor can be gained from the generating functional

$$G[\chi] = \mathrm{e}^{-S_G[\chi]} = \left\langle \mathcal{T} \exp\left[\frac{i}{e}\int_C dt I(t)\chi(t)\right]\right\rangle , \tag{1}$$

where $I(t) = e\int_0^t ds N(s)$ is the current operator, N the number of transferred charges, χ the counting field, and $\mathcal{T}$ the time ordering operator along the Kadanoff-Baym contour $\mathcal{C}$. Time correlation functions of arbitrary order of the current are determined from functional derivatives of $G[\chi]$, namely,

$$C_n(t_1,\ldots,t_n) = -(-ie)^n\, \partial^n S_G[\chi]/\partial\chi(t_1)\cdots\partial\chi(t_n)|_{\chi=0} , \tag{2}$$

where cumulants $C_n, n \geq 3$ display non-Gaussian properties of the noise. We remark that the functional $S_G[\chi]$ carries the full frequency dependence of all current cumulants and not just their time averaged zero frequency values usually studied in the field of full counting statistics.[3]

3. Classical Thermal Activation

Let us consider the experimental situation summarized in Fig. 1, where a tunnel junction represents the noise generating sample and a JJ the detector. We assume the underlying stochastic processes to be classical, both for the dynamics of the detector and for the electrical noise originating from the mesoscopic conductor. Hence, the standard RSJ model applies, where now the total noise consists of the Johnson-Nyquist noise $\delta I_b = I_b - \langle I_b\rangle$, in the simplest case produced by a resistor R in parallel to the JJ biased by I_b, and of the weak stationary mesoscopic noise δI_m from the fluctuating part of the mesoscopic current $I_m = \langle I_m\rangle + \delta I_m$. If the phase φ of the JJ is initially trapped in one of the wells of the tilted washboard potential $U(\varphi) = -E_J\, \cos(\varphi) - (\langle I_b\rangle(\hbar/2e)\varphi$ (zero-voltage state), it may for sufficiently large $\langle I_b\rangle < I_0 = (2e/\hbar)E_J$ escape so that the JJ switches to a finite voltage state. Further, since the third cumulant vanishes in equilibrium due to time-reversal symmetry, experimentally, the mesoscopic conductor is at low temperatures driven far from equilibrium, where no fluctuation-dissipation theorem applies. Hence, the switching of the JJ can be visualized as the diffusive dynamics of a fictitious particle in a metastable well with non-Gaussian continuous fluctuations acting as *external* random driving force. In a refined version of this theory also back-action effects can be taken into account.

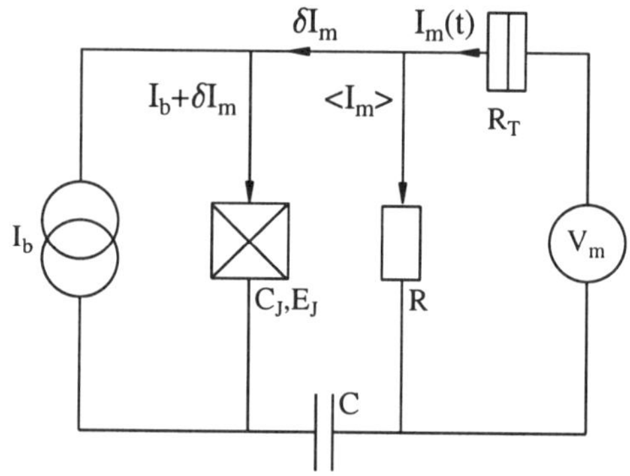

Fig. 1. Simplified scheme of the experiment reported in Ref. 9. The high frequency ($f > 1/RC$) fluctuations δI_m of the current through a voltage-biased tunnel junction (tunnel resistance R_T) pass through a Josephson junction (capacitance C_J, Josephson energy E_J). The switching of the Josephson junction to the finite-voltage state depends on the sum of δI_m and of the pulsed current bias I_b. The switching probability of the junction during one pulse can be related to the current fluctuations.

According to the Kirchhoff rules and the Josephson relations we have for the currents entering the detector (cf. Fig. 1)

$$\langle I_b \rangle + \delta I_b + \delta I_m(t) = \frac{\hbar}{2e}\frac{\dot{\varphi}}{R} + I_0 \sin\varphi + C_J \frac{\hbar}{2e}\ddot{\varphi}\,, \tag{3}$$

where φ denotes the phase difference of the JJ. The two noisy forces are assumed to be Markovian, which is an accurate approximation since the experiment is operated in a regime, where the typical correlations times are much smaller than the typical time scales of the junction, e.g. the inverse of the plasma frequency Ω. This situation also ensures that we can work with continuous noise processes. The thermal noise δI_b has a vanishing mean and a standard δ-correlated second cumulant. The statistical properties of the stationary non-Gaussian noise δI_m are determined by the generating functional (1) with the current operator $I(t)$ replaced by the classical noise $\delta I_m(t)$. In particular, one has $\langle \delta I_m(t)\delta I_m(0)\rangle = F_2 e^2 \langle I_m\rangle \delta(t)$ and and $\langle \delta I_m(t_2)\delta I_m(t_1)\delta I_m(0)\rangle = F_3 e^3\langle I_m\rangle \delta(t_2)\delta(t_1)$. The factors F_2 and F_3 denote the Fano-factors, which for a simple tunnel junction turn out to be $F_2 = F_3 = 1$. It is important to realize that the average mesoscopic current can be large so that the second moment can be comparable to or lager than that of the thermal noise, while the third moment is still small due to $e\Omega/I_0 \ll 1$. In fact, the Gaussian component of the mesoscopic noise effectively adds to the thermal noise to determine the effective temperature of the junction (heating). Equation (3) is a Langevin equation for a fictitious particle in the tilted Josephson potential $U(\varphi)$ in presence of thermal Gaussian and electrical non-Gaussian noise.

As usual in rate theory, for analytical treatments it is much more convenient to work with phase space probabilities rather than individual stochastic trajectories.[10] The general problem for non-Gaussian noise is then that the corresponding Fokker-Planck equation (FPE) based on a Kramers-Moyal expansion contains diffusion coefficients up to infinite order. The basic idea for weak non-Gaussian noise with a leading third cumulant is thus, to derive an effective, finite order Fokker-Planck equation. Based on a cumulant expansion of the noise generating functional (1) with the counting field proportional to the momentum derivative such a generalized FPE has been derived in Ref.[11] It leads to a FPE with a momentum dependent diffusion term, where the momentum dependence is weighted by the third cumulant.

The thermal rate expression derived from the steady state solution of the standard FPE looks as $\Gamma = A\exp(-\beta U_b)$ and is dominated by the exponential (activation factor) being identical to the probability to reach the barrier top from the well bottom (energy difference U_b) by a thermal fluctuation.[10] Within the theoretical framework of the extended FPE an analytical expression for the exponent including leading corrections due to a third cumulant have been obtained.[11] The rate takes the form $\Gamma_\pm \propto \exp[-\beta U_b(1 \mp |g_3|)]$ with the correction g_3 due to the third cumulant such that Γ_+ corresponds to $\langle I_m \rangle > 0$ and Γ_- to $\langle I_m \rangle < 0$. The rate asymmetry $R_\Gamma = \Gamma_+/\Gamma_- - 1$ is found as being proportional to F_3 and strongly depends on the effective temperature, damping, and bias current. Hence, a measurement of R_Γ gives direct information about the third cumulant of the electrical noise. This has been discussed in detail in Ref. 9. There also results of numerical simulations have been presented which are in agreement with the analytical expressions.

To complete this discussion, we remark that an analytical expression for the asymmetry of the escape rate in the limits of low and high friction has been also derived in Ref. 14 leading up to minor deviations to identical results.

4. Quantum Tunneling

As mentioned above, a general theory for quantum tunneling in presence of steady state non-Gaussian noise has not been formulated yet. The idea is thus to place the mesoscopic conductor in parallel to a current biased JJ

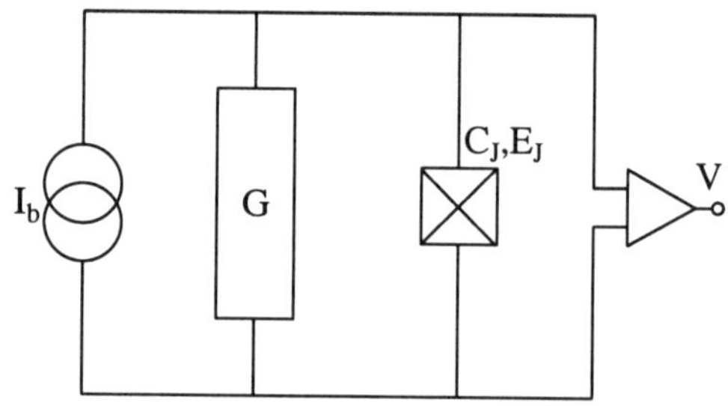

Fig. 2. Electrical circuit containing a mesoscopic conductor G in parallel to a JJ with capacitance C_J and coupling energy E_J biased by an external current I_b. The switching out of the zero voltage state of the JJ by MQT is detected as a voltage pulse V.

as depicted in the circuit diagram of Fig. 2. For a bias current I_b below the critical current I_0, the JJ is in its zero voltage state and the bias current flows as a supercurrent entirely through the JJ branch of the circuit. Consequently, no heating occurs in the conductor and the total system can easily be kept at low temperatures, where the decay of the zero voltage state occurs through Macroscopic Quantum Tunneling (MQT).[12] The rate of this process depends with exponential sensitivity on the current fluctuations of the conductor so that the JJ acts as a noise detector.

The MQT rate Γ can be calculated in the standard way[13,15] from the imaginary part of the free energy F, i.e., $\Gamma = (2/\hbar)\,\mathrm{Im}\{F\}$, where $F = -(1/\beta)\ln(Z)$ is related to the partition function $Z = \mathrm{Tr}\{\mathrm{e}^{-\beta H}\}$. In the path integral representation one has

$$Z = \int \mathcal{D}[\varphi]\mathrm{e}^{-S[\varphi]}\,,$$

which is a sum over all imaginary time paths with period $\hbar\beta$ where each paths is weighted by the dimensionless effective action of the JJ. In the present case, we have $S[\varphi] = S_{JJ}[\varphi] + S_G[\varphi/2]$, where the first term is the bare action of the JJ, i.e. the phase dynamics in the tilted Josephson potential, while the second one describes the influence of the environment. The factor of 2 in the argument of S_G arises from the fact that the voltage across the conductor equals the voltage $(\hbar/e)(\dot{\varphi}/2)$ across the JJ. In the standard theory only thermal Gaussian noise is present, which is described by an ohmic resistor in parallel to the JJ. The corresponding action $S_G[\varphi] \equiv S_R[\varphi]$ can be calculated exactly and gives the well-known Feynman-Vernon influence functional.[15] Now, if we consider a tunnel junction as environment generating stationary electrical non-Gaussian noise an exact expression for the action S_T is found as well,[16] where the granularity of the charge appears as a periodic dependence on φ. The current generating functional thus determines a non-Gaussian influence functional for the system with the detector degree of freedom being the counting field.

In the MQT regime the partition function of the isolated JJ is dominated by the so-called bounce trajectory, an extremal $\delta S[\varphi] = 0$ periodic path in the inverted barrier potential. In the limit of vanishing temperature one finds an analytical solution of the bounce φ_B so that for zero temperature the bare rate coincides with the usual WKB result, namely, $\Gamma_0 \propto \exp[-(36/5)U_b/\hbar\Omega]$ with U_b being the barrier height and Ω the frequency for oscillations near the well bottom (plasma frequency). Following the theory of the effect of an electromagnetic environment on MQT,[13] the partition function can now be calculated for arbitrary coupling between detector and conductor based on a numerical scheme developed in Ref. 15. Analytical progress is made when the noise generating element has a dimensionless conductance $g_T \equiv \hbar/(2e^2R_T) \ll I_0/2e\Omega$ so that the influence of the noise can be calculated by expanding about the unperturbed bounce which gives

$$\Gamma = \Gamma_0 \, \mathrm{e}^{-S_G[\varphi_B/2]} \, . \tag{4}$$

For a tunnel junction with tunnel resistance R_T, the correction $S_G[\varphi_B/2]$ can be represented as a series in even order cumulants, which is usually dominated by the second cumulant C_2 and the fourth cumulant C_4, see Ref. 12. Note that this treatment still contains the full dynamics of detector and noise source.

For the on-chip detection circuit proposed here, the impact of the fourth order cumulant needs to be clearly discriminated from effects of purely Gaussian noise. This is achieved by considering the function $B(x) = -\ln[\Gamma(x)/\Gamma_0(x)]$ with the variable $x = (1-s^2)/s^2$ ($s = \langle I_b\rangle/I_0$) which allows to discriminate between weak Gaussian and non-Gaussian noise due to a qualitatively different scaling behavior with varying x. For purely Gaussian noise $B(x)$ results essentially in a straight line, while non-Gaussian noise displays a nonlinear behavior. Even more pronounced are the differences in the slopes $dB(x)/dx$, which saturate for larger x-values when only Gaussian noise is present, but strongly decrease with increasing x in presence of a nonlinear conductor. This scaling property is robust against additional Gaussian noise present in the wiring incorporated by an additional resistor with resistance $R \ll R_T$: it merely shifts $dB(x)/dx$ and thus does *not* spoil the scaling behavior originating from C_4 (cf. Ref. 12).

Acknowledgments

Fruitful discussions with D. Esteve, H. Grabert, B. Huard, J. Pekola, H. Pothier, and X. Waintal are gratefully acknowledged.

References

1. Y. M. Blanter and M. Büttiker, *Phys. Rep.* **336**, 1 (2000).
2. *Quantum Noise in Mesoscopic Physics*, edited by Y. V. Nazarov, NATO Science Series in Mathematics, Physics and Chemistry (Kluwer, Dordrecht, 2003).
3. L. S. Levitov, H. B. Lee, and G. B. Lesovik, *J. Math. Phys.* **37**, 4845 (1996).
4. B. Reulet, J. Senzier, and D. E. Prober, *Phys. Rev. Lett.* **91**, 196601 (2003).
5. J. Tobiska and Y. V. Nazarov, *Phys. Rev. Lett.* **93**, 106801 (2004); T. T. Heikkila *et al.*, *Phys. Rev. Lett.* **93**, 247005 (2004); R. K. Lindell *et al.*, *Phys. Rev. Lett.* **93**, 197002 (2004); E. B. Sonin, *Phys. Rev. B* **70**, 140506(R) (2004); Yu. Bomze *et al.*, *Phys. Rev. Lett.* **95**, 176601 (2005); J. P. Pekola, *Phys. Rev. Lett.* **93**, 206601 (2004); S. Gustavsson *et al.*, *Phys. Rev. Lett.* **96**, 076605 (2006).
6. J. P. Pekola *et al.*, *Phys. Rev. Lett.* **95**, 197004 (2005).
7. A. V. Timofeev *et al.*, *Phys. Rev. Lett.* **98**, 207001 (2007).
8. K. E. Nagaev, S. Pilgram, M. Büttiker, *Phys. Rev. Lett.* **92**, 176804 (2004).
9. B. Huard *et al.*, *Ann. Phys.* **16**, 736 (2007).
10. For a review see: P. Hänggi, P. Talkner, and M. Borkovec, *Rev. Mod. Phys.* **62**, 251 (1990).
11. J. Ankerhold, *Phys. Rev. Lett.* **98**, 036601 (2007).
12. J. Ankerhold and H. Grabert, *Phys. Rev. Lett.* **95**, 186601 (2005).
13. A. O. Caldeira and A. J. Leggett, *Ann. Phys. (New York)* **149**, 374 (1983).
14. E. V. Sukhorukov and A. N. Jordan, *Phys. Rev. Lett.* **98**, 136803 (2007).
15. H. Grabert, P. Olschowski, and U. Weiss, *Phys. Rev. B* **36**, 1931 (1987).
16. G. Schön and A.D. Zaikin, *Phys. Rep.* **198**, 237 (1990).

ANSWERS AND QUESTIONS ON PATH INTEGRALS FOR SUPERCONDUCTIVITY IN A WEDGE

F. BROSENS*, V. M. FOMIN*,†, J. T. DEVREESE*,†, and V. V. MOSHCHALKOV‡

*Theoretische Fysica van de Vaste Stoffen, Departement Natuurkunde, Universiteit Antwerpen, Groenenborgerlaan 171, B-2020 Antwerpen, Belgium
†Technische Universiteit Eindhoven, P.O.Box 513, 5600 MB Eindhoven, The Netherlands
‡INPAC-Institute for Nanoscale Physics and Chemistry, Katholieke Universiteit Leuven, Celestijnenlaan 200 D, B-3001 Leuven, Belgium

E-mail: Fons.Brosens@ua.ac.be, Vladimir.Fomin@ua.ac.be, Jozef.Devreese@ua.ac.be, Victor.Moshchalkov@fys.kuleuven.ac.be

Some time ago, DeWitt-Morette[1] *et al.* discussed the problem of the propagation of radiation or particles in the presence of a wedge. Their treatment includes path integral solutions of the wedge problem with Dirichlet or Neumann boundary conditions for various situations. Recently, the superconducting phase in a wedge has gained increasing interest. The linearized Ginzburg-Landau equation for the order parameter is formally similar to the Schrödinger equation. But the conditions of no normal current at the boundary pose very specific problems, which are discussed in the present paper.

Keywords: Superconductivity; Wedge; Ginzburg-Landau equation.

1. Introduction

We discuss some problems related to the application of path integrals to treat the first linearized Ginzburg-Landau equation[2,3] for describing the superconducting phase in a wedge. This work was motivated by the increasing possibilities of engineering phase boundaries of mesoscopic superconductors. A comprehensive review of some recent advances can, e.g., be found in Ref. 4.

For superconducting samples with sizes of the order of the coherence length, the Ginzburg-Landau equations provide good agreement between the experimental[5,6] and the theoretical[7,8] phase boundary in the temperature versus magnetic field plane. We here concentrate on the superconducting wedge, which has already been the subject of intense studies, in particular in the limit of a small wedge angle $\alpha \ll \pi$. E.g., some earlier variational efforts[9–12] concentrated on a lower bound to the nucleation field H_{c3} for $\alpha \to 0$. Some of them found the correct limit $H_{c3} = \sqrt{3}H_{c2}/\alpha$ (with H_{c2} the bulk upper critical field) which was analytically obtained.[13,14] Also the asymptotic behavior of the lowest eigenvalue and corresponding eigenfunction for $\alpha \to 0$ has been fully investigated.[15–17] Furthermore, for the semi-infinite superconductor ($\alpha = \pi$) the nucleation field is accurately known.[18,19]

No analytical solutions are known for general α, but an analytical variational approach[11,20,21] provides a lower bound for the nucleation field as a function of α.

Inspired by the interesting path integral study of the Schrödinger equation in a wedge by DeWitt-Morette[1] *et al.*, we here investigate whether the path integral treatment can elucidate the superconducting properties of the wedge. But serious difficulties arise from the boundary conditions on the superconducting current.

2. Ginzburg-Landau Equation and Boundary Conditions

Consider a wedge shaped superconductor with angle α, infinitely extended along its edge which is considered to be the z direction. A uniform magnetic field with vector potential $\mathbf{A}$ is applied parallel to the edge. We thus can eliminate the free motion along the z direction, and restrict the treatment to the (x, y) dependence of the order parameter Ψ. In the Ginzburg-Landau equation $\frac{1}{2m}\left(\mathbf{p} - \frac{e}{c}\mathbf{A}\right)^2 \Psi + a\Psi + b\Psi\left|\Psi\right|^2 = 0$, two phenomenological parameters a and b occur, and e and m denote the charge and the mass of a Cooper pair. In the *vicinity of the phase boundary*, when the superconductivity is just beginning to nucleate, $\left|\Psi\right|^2$ is small, and the Ginzburg-Landau equation can be used in its linearized form

$$\frac{1}{2m}\left(\mathbf{p} - \frac{e}{c}\mathbf{A}\right)^2 \Psi = \mathcal{E}\Psi \tag{1}$$

with a given coefficient $\mathcal{E} = -a$ related to the coherence length $\xi = \sqrt{\hbar^2/(2m\mathcal{E})}$. The radius of the wedge is assumed large enough for radial current effects to be negligible. The absence of a normal superconducting current through the boundaries of the wedge can then be limited to the angular boundaries:

$$J_\perp = (\mathbf{r} \times \mathbf{p})_z - \frac{e}{c}(\mathbf{r} \times \mathbf{A})_z\,; \qquad J_\perp \Psi|_{\varphi=0,\alpha} = 0. \tag{2}$$

The lowest eigenvalue *in bulk* would be $\frac{1}{2}\frac{eH_{c2}}{mc} = \frac{1}{2}\hbar\omega$, where H_{c2} is the bulk critical magnetic field, and ω the effective cyclotron frequency. The nucleation (or 3rd critical) magnetic field H_{c3}, which is compatible with the boundary problem in the wedge, is determined by the relation[12]

$$\min(\mathcal{E}) = \frac{1}{2}\frac{eH_{c3}}{mc}, \tag{3}$$

The minimum of the eigenvalues $\mathcal{E}$ of the Schrödinger equation (1) thus immediately determines the nucleation field H_{c3}.

2.1. *The case* $\alpha = \pi$

The semi-infinite superconductor deserves special attention as a reference system for further treatments or approximations. In the gauge $\mathbf{A} = H\begin{pmatrix} 0 \\ x \end{pmatrix}$, the Schrödinger equation (1) and the current condition (2) can be cast in the form

$$\left[\frac{p_x^2}{2m} + \frac{m\omega^2}{2}(x-X)^2\right]\Psi = \mathcal{E}\Psi, \qquad \text{with } \left.\frac{\partial \Psi}{\partial x}\right|_{x=0} = 0, \tag{4}$$

where $X \equiv p_y/(m\omega)$ is a constant of the motion. The order parameter is thus even in x, which is equivalent to treating a reflected potential $V(x) = \frac{m\omega^2}{2}(|x| - X)^2$. For $X > 0$, e.g., this leads to a variant of a "Mexican hat" potential. The eigenfunctions can be expressed in terms of Weber functions, and minimizing the lowest eigenvalue with respect to X gives[18]

$$\min(\mathcal{E}_{\alpha=\pi}) = 0.59010\,\frac{\hbar\omega}{2}. \tag{5}$$

Although we know of no path integral solution for this problem, a direct application of the Jensen-Feynman inequality[22] is rather satisfactory. The action S becomes (in the Euclidean time variable $\tau = it/\hbar$; $\dot{x} = \frac{dx}{d\tau}$)

$$S = -\frac{m}{2}\int_0^\beta \left(\dot{x}^2/\hbar^2 + \omega^2 (|x(\tau)| - X)^2\right) d\tau. \tag{6}$$

Using $Z = \oint \mathcal{D}x e^S$, the Jensen-Feynman inequality reads

$$Z = Z_0 \left\langle e^{S-S_0}\right\rangle_0 \geq Z_0 e^{\langle S-S_0\rangle_0} \text{ with } \langle\bullet\rangle_0 = \frac{1}{Z_0}\oint \mathcal{D}x\,(\bullet)\, e^{S_0}, \tag{7}$$

where S_0 is a real trial action with corresponding partition function Z_0. Consider an harmonic trial action S_0 with variational parameter Ω

$$S_0 = -\frac{m}{2}\int_0^\beta \left(\dot{x}^2/\hbar^2 + \Omega^2 x^2(\tau)\right) d\tau \tag{8}$$

with a symmetrized propagator K_0, based on the propagator K_Ω of the harmonic oscillator

$$K_0(x,\beta|x_0) = \frac{1}{2}\left(K_\Omega(x,\beta|x_0) + K_\Omega(x,\beta|-x_0)\right) \tag{9}$$

$$K_\Omega(x,\beta|x_0) = \sqrt{\frac{m\Omega}{2\pi\hbar\sinh\Omega\beta\hbar}}\exp\left(-\frac{m\Omega}{2\hbar}\frac{\left(x^2+x_0^2\right)\cosh\Omega\beta\hbar - 2xx_0}{\sinh\Omega\beta\hbar}\right). \tag{10}$$

Elementary algebra reveals

$$\begin{aligned}
Z_0 &= \tfrac{e^{-\frac{1}{2}\Omega\beta\hbar}}{1-e^{-2\Omega\beta\hbar}} && \underset{\beta\to\infty}{\to} e^{-\frac{1}{2}\Omega\beta\hbar} \\
\left\langle x^2\right\rangle_0 &= \tfrac{1}{Z_0}\tfrac{\hbar\sinh\Omega\beta\hbar}{4\sqrt{2}m\Omega}\left(\tfrac{1}{(\cosh\Omega\beta\hbar-1)^{\frac{3}{2}}} + \tfrac{1}{(\cosh\Omega\beta\hbar+1)^{\frac{3}{2}}}\right) && \underset{\beta\to\infty}{\to} \tfrac{\hbar}{2m\Omega} \\
\left\langle |x|\right\rangle_0 &= \tfrac{1}{Z_0}\sqrt{\tfrac{\hbar\sinh\Omega\beta\hbar}{8m\Omega\pi}}\left(\tfrac{1}{\cosh\Omega\beta\hbar-1} + \tfrac{1}{\cosh\Omega\beta\hbar+1}\right) && \underset{\beta\to\infty}{\to} \sqrt{\tfrac{\hbar}{m\Omega\pi}}
\end{aligned}$$

$$\to \min\left(\mathcal{E}_{\alpha=\pi}\right) \leq \frac{\Omega\hbar}{2} - \frac{m}{2}\left(\frac{\hbar\left(\Omega^2-\omega^2\right)}{2m\Omega} + 2\omega^2 X\sqrt{\frac{\hbar}{m\Omega\pi}} - \omega^2X^2\right) \tag{11}$$

which is readily minimized with respect to X and Ω

$$\min\left(\mathcal{E}_{\alpha=\pi}\right) \leq \sqrt{\frac{\pi-2}{\pi}}\frac{\hbar\omega}{2} = 0.602\,81\,\frac{\hbar\omega}{2}, \tag{12}$$

giving the optimal estimate at $\Omega = \omega\sqrt{1-2/\pi}$ and $X = \sqrt{\hbar/(\pi m\Omega)}$. This estimate differs by only 2% from the exact value in (5), and can be easily improved, e.g., from Chapter 5 of Ref. 23. This seems encouraging for a path integral approach to a superconducting wedge.

2.2. *Variational approach*

Inspired by the wave function for $\alpha = \pi$, we developed a variational approach[20,21] with variants of trial functions of the form

$$\Psi_{var}(r,\varphi) = \exp\left(i\frac{m\omega}{2\hbar}\frac{\sin\frac{2\pi n\varphi}{\alpha}}{2\pi n/\alpha}r^2 + iqr\cos\frac{\pi\varphi}{\alpha} - r^2\lambda\left(1-\gamma\cos^2\frac{\pi\varphi}{\alpha}\right)\right) \tag{13}$$

where q, λ γ, and n (integer) are variational parameters. Ψ_{var} satisfies the boundary conditions in the symmetric gauge for $\varphi \in [0,\alpha]$. For $\alpha \ll \pi$ the variational order parameter turns out to be concentrated in the origin, with the variational energy proportional to α. With increasing α, Ψ_{var} concentrates near the surface, and the energy becomes independent of α.

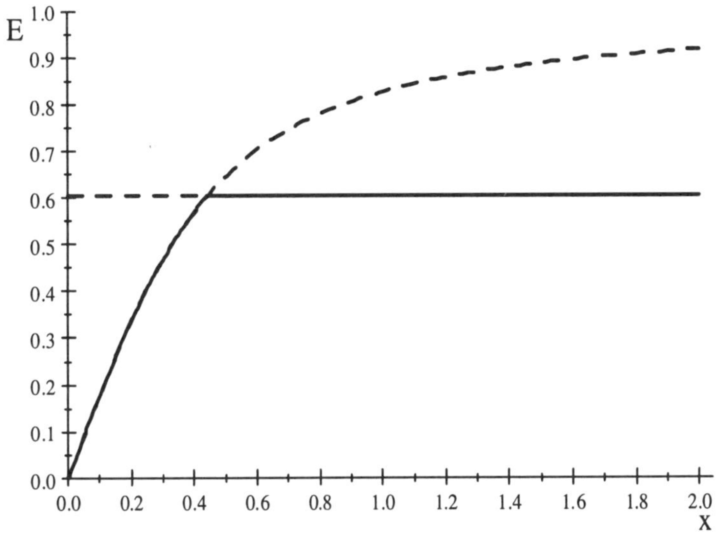

Fig. 1. $E = \mathcal{E}_{var}/\,(\hbar\omega/2)$ as a function of $x = \alpha/\pi$. The dashed line shows the extension of the two constituting expressions beyond the critical angle $\alpha = 0.4411\pi$. The case $\alpha \leq 0.4411\pi$ is found in Ref. 11, whereas the upper bound for $\alpha \geq 0.4411\pi$ results from our approach.

Recently we became aware of an alternative variational approach[11] (also in the symmetric gauge):

$$\Psi_{\text{Simonov}}(r,\varphi) = \exp\left(i\frac{m\omega}{2\hbar}\left(\varphi - \frac{\sinh(2\varphi-\alpha)}{2\cosh\alpha}\right)r^2\right) \qquad (14)$$

which for sufficiently small α gives a lower trial energy than ours. Combining both results one finds

$$\min(\mathcal{E}) \leq \mathcal{E}_{var} = \begin{cases} \frac{1}{2}\hbar\omega\sqrt{1-\frac{\sinh\alpha}{\alpha\cosh\alpha}} & \text{for } \alpha/\pi \leq 0.4411 \\ \frac{1}{2}\hbar\omega\sqrt{1-2/\pi} & \text{for } \alpha/\pi \geq 0.4411 \end{cases}, \qquad (15)$$

as represented in Fig. 1.

3. Path-Integral Considerations

The path integral for the Schrödinger equation in a wedge was treated in Ref. 1. Dirichlet or Neumann boundary conditions were translated in absorption or reflection, respectively. Apart from algebraic complications, this treatment is a natural extension of the one-dimensional infinite square well problem, like, e.g., formulated in Chapter 7 of Ref. 24. But for the superconducting wedge under consideration, the boundary condition (2) is more complicated.

3.1. *Symmetric gauge*

The symmetric gauge

$$\mathbf{A}_S = \frac{H}{2}\begin{pmatrix} -y \\ x \end{pmatrix} \tag{16}$$

seems a useful start, because the propagator $G_S(\mathbf{x}, T|\mathbf{x}_0)$ for an electron in a magnetic field nicely reflects the classical circular motion

$$G_S(\mathbf{x}, T|\mathbf{x}_0) = \frac{m\omega \exp\left[\frac{im\omega}{2\hbar \sin\frac{\omega T}{2}}\begin{pmatrix} \frac{1}{2}\left(r^2 + r_0^2\right)\cos\frac{\omega T}{2} \\ -rr_0 \cos\left(\varphi - \varphi_0 + \frac{\omega T}{2}\right)\end{pmatrix}\right]}{4i\pi\hbar \sin\frac{\omega T}{2}}. \tag{17}$$

The eigenfunctions and energy values follow from the spectral representation

$$\begin{aligned} G_S(\mathbf{x}, T|\mathbf{x}_0) &= \textstyle\sum_{n=0}^{\infty}\sum_{j=-\infty}^{\infty} \Psi_{n,j}^*(\mathbf{x})\,\Psi_{n,j}(\mathbf{x}_0)\,e^{-iE_{n,j}T/\hbar}; \\ \to \Psi_{n,j}(\mathbf{x}) &\sim e^{ij\varphi} \times r^{|j|} L_n^{(|j|)}\left(\tfrac{\hbar}{m\omega} r^2\right) \times e^{-\frac{\hbar r^2}{2m\omega}}; \\ E_{n,j} &= \left(1 + 2n + |j| - j\right)\tfrac{\hbar\omega}{2}. \end{aligned} \tag{18}$$

A detailed derivation can, e.g., be found in Chapter 9.5 of Ref. 23 (but be careful with the sign of the indices of the Bessel functions in the derivation.) The operator $J_{\perp,S}$ for the normal current in this gauge is

$$J_{\perp,S} = \frac{1}{mr}\left[\frac{\hbar}{i}\frac{\partial}{\partial\varphi} - \frac{m\omega}{2}r^2\right], \tag{19}$$

resulting in neither Dirichlet nor Neumann boundary conditions. But the wave functions $\{\Psi_{n,j}(\mathbf{x})\}$ do not form a complete set for $\varphi \in [0, \alpha]$, because linear combinations of the states $\Psi_{n,j}(\mathbf{x})$ give energy expectation values $\mathcal{E} \geq \hbar\omega/2$.

3.2. *Gauge transformation(s)*

The operator $J_\perp$ (2) for the normal current in a point $\mathbf{r}$ contains a term from the vector potential, which in the symmetric gauge is $(\mathbf{x} \times \mathbf{A}_S)_z = \frac{1}{2}Hr^2$. A gauge transformation introduces a phase factor in the propagator

$$\mathbf{A} = \mathbf{A}_S + \nabla\Lambda \to G(\mathbf{x}, T|\mathbf{x}_0) = \exp\left(i\frac{e}{\hbar c}\left(\Lambda(\mathbf{x}) - \Lambda(\mathbf{x}_0)\right)\right) G_S(\mathbf{x}, T|\mathbf{x}_0). \tag{20}$$

Imposing $(\mathbf{x} \times \mathbf{A})_z|_{\varphi=0,\alpha} = 0$ leads to a partial differential equation $\frac{1}{2}Hr^2 + \left.\frac{d\Lambda(\mathbf{x})}{d\varphi}\right|_{\varphi=0,\alpha} = 0$. Among the many solutions, we mention

$$\Lambda_n(\mathbf{x}) = -\frac{Hr^2}{2}\frac{\sin\frac{2n\pi\varphi}{\alpha}}{2n\pi/\alpha} \text{ with } n \text{ integer.} \tag{21}$$

In these gauges we obtain

$$\left.\frac{\partial}{\partial\varphi}G(\mathbf{x}, t|\mathbf{x}_0, t_0)\right|_{\varphi=0,\alpha} = 0. \tag{22}$$

But the formalism for Neumann boundary conditions, developed in Ref. 1, would give a minimal energy $\hbar\omega/2$, which is much too large. The problem is that $\frac{\hbar}{i}\frac{\partial}{\partial\varphi}$ does not commute with the transformed Hamiltonian. So far we did not succeed in finding a solution for such a situation, or to construct an appropriate trial action*.

4. Concluding Remarks

The solution of the linearized Ginzburg-Landau equation in a wedge is seriously hindered by the boundary conditions which forbid a current component normal to the boundary. For a semi-infinite superconductor the minimal energy is known to be substantially lower than the bulk energy $\hbar\omega/2$. And for general α a variational approach assures that the minimal energy does never exceed 0.602 81 $\hbar\omega/2$, and decreases with decreasing α.

A variational path integral treatment with the Jensen-Feynman inequality is promising for $\alpha = \pi$. But for general α, the boundary conditions prohibit the path integral treatment as developed in Ref. 1, even after a gauge transformation which imposes Neumann boundary conditions. Fortunately, these reflections seem numerically tractable with diffusion Monte Carlo. But the numerical treatment is not yet sufficiently developed and analyzed to allow for definite conclusions.

Acknowledgments

This work has been supported by the FWO-Vl. projects No. G.0306.00 and G.0449.04, and the WOG WO.035.04N (Belgium).

*For instance, it is possible to calculate the action for classical trajectories with reflection of the normal current, given the initial velocity and position. But we did not succeed in finding it for given initial and final position.

References

1. C. DeWitt-Morette, S. G. Low, L. S. Schulman, and A. Y. Shiekh, *Foundations of Physics* **16**, 311 (1986).
2. V. L. Ginzburg and L. D. Landau, *Zh. Eksp. i Teor. Fiz.* **20**, 1064 (1950).
3. L. D. Landau and E. M. Lifshitz, *Course of Theoretical Physics,* Vol. 9 (*Statistical Physics,* Vol. 2) (Pergamon, Oxford, 1989).
4. V. V. Moshchalkov, *J. Supercond. Novel Magn.* **19,** 409 (2006).
5. V. V. Moshchalkov, L. Gielen, C. Strunk, R. Jonckheere, X. Qiu, C. Van Haesendonck, and Y. Bruynseraede, *Nature* **373**, 319 (1995).
6. C. Strunk, V. Bruyndoncx, C. Van Haesendonck, V.V. Moshchalkov, and Y. Bruynseraede, *J. Physique* **6**, Suppl. No. 4, C-253 (1996).
7. V. M. Fomin, V. R. Misko, J. T. Devreese, and V. V. Moshchalkov, *Solid State Commun.* **101**, 303 (1997).
8. V. M. Fomin, V. R. Misko, J. T. Devreese, and V. V. Moshchalkov, *Phys. Rev. B* **58**, 11703 (1998).
9. A. Houghton and F. B. McLean, *Phys. Lett.* **19**, 172 (1965).
10. A. P. van Gelder, *Phys. Rev. Lett.* **20**, 1435 (1968).
11. A. Yu. Simonov, A. S. Melnikov, and S. V. Sharov, *Fizika Nizkikh Temperatur* **15**, 1206 (1989).
12. V. M. Fomin, J. T. Devreese, and V. V. Moshchalkov, *Europhys. Lett.* **42**, 553 (1998); Erratum **46**, 118 (1999).
13. S. N. Klimin, V. M. Fomin, J. T. Devreese, and V. V. Moshchalkov, *Solid State Commun.* **111**, 589 (1999).
14. V. A. Schweigert and F. M. Peeters, *Phys. Rev. B* **60**, 3084 (1999).
15. V. Bonnaillie-Noël, *Comptes Rendus Math.* **336**, 135 (2003).
16. V. Bonnaillie-Noël, *Asymptotic Analysis* **41**, 215 (2005).
17. V. Bonnaillie-Noël and M. Dauge, *Annales Henri Poincaré* **7**, 899 (2006).
18. D. Saint-James and P.-G. de Gennes, *Phys. Lett.* **7**, 306 (1963).
19. P. G. de Gennes, *Superconductivity of Metals and Alloys* (Addison-Wesley Publishing Company, Reading, 1989); equation (6-74).
20. F. Brosens, V. M. Fomin, J. T. Devreese, and V. V. Moshchalkov, *Solid State Commun.* **111**, 565 (1999).
21. F. Brosens, V. M. Fomin, J. T. Devreese, and V. V. Moshchalkov, *Solid State Commun.* **144**, 494 (2007).
22. R. P. Feynman, *Phys. Rev.* **97**, 660 (1955).
23. H. Kleinert, *Path Integrals in Quantum Mechanics, Statistics, Polymer Physics, and Financial Markets*, Fourth Edition (World Scientific, Singapore, 2006).
24. L. S. Schulman, *Techniques and Applications of Path Integration* (J. Wiley & Sons, New York, 1981).

LOW-ENERGY EFFECTIVE REPRESENTATION OF THE PROJECTED BCS HAMILTONIAN CLOSE TO HALF FILLING

E. A. KOCHETOV

Bogolyubov Theoretical Laboratory, Joint Institute for Nuclear Research, 141980 Dubna, Russia
E-mail: kochetov@unb.br

It has recently been shown (K. Park, *Phys. Rev. Lett.* **95**, 027001 (2005)) that the ground state of the t–J model at half filling is entirely equivalent to the ground state of the Gutzwiller-projected Bardeen-Cooper-Schrieffer (BCS) Hamiltonian with strong pairing. Here we extend this result to finite doping. We show that in the immediate vicinity of half filling the projected 2D BCS Hamiltonian with strong pairing still develops the antiferromagnetically (AF) ordered ground state.

Keywords: Strongly correlated electrons; BCS Hamiltonian.

1. Introduction

In a recent paper, Park discussed a close connection between the t–J model and the Gutzwiller-projected BCS Hamiltonian.[1] It was shown both numerically and analytically that the ground state of the t–J model at half filling (i.e. the 2D anti-ferromagnetic Heisenberg model) is equivalent to the ground state of the strong-pairing BCS Hamiltonian. Moreover, at sufficiently small doping, there is numerical evidence for a strong overlap between the two ground state wave functions, which provides support for the existence of superconductivity in the t–J model. It would be interesting to analytically address this issue at finite hole concentration. As is known, slightly away from half filling the long-range AF order is still observed in the cuprate superconductors. Within the t–J model representation this phase can be accounted for by the AF ordered state described by the Heisenberg magnetic J-term slightly disturbed by the kinetic t-term. If the projected BCS Hamiltonian is indeed believed to contain the same low-energy physics close to half filling of the t–J Hamiltonian, its ground state must also exhibit the AF order in the immediate vicinity of half filling. This manifests itself as a quite nontrivial necessary condition for the low-energy physics described by the Gutzwiller-projected BCS Hamiltonian to be considered identical to that of the t–J Hamiltonian at sufficiently low doping.

The purpose of the present report is to analytically investigate the Gutzwiller-projected BCS Hamiltonian close to half filling. We derive the low-energy long-wavelength effective action for the 2D projected BCS Hamiltonian on a bipartite lattice in this region. The action obtained is shown to be identical to that of the 2D quantum antiferromagnetic Heisenberg model explicitly represented by the 3D nonlinear σ-model. Since the conventional BCS Hamiltonian does not exhibit any magnetic ordering and always displays superconductivity, the previous results follow as an immediate consequence of the Gutzwiller projection that eliminates the doubly occupied electron states.

2. Gutzwiller-Projected BCS Hamiltonian

We start with the Gutzwiller-projected BCS Hamiltonian on a bipartite lattice,

$$H^G_{BCS} = -t\sum_{ij\sigma}(X_i^{\sigma 0}X_j^{0\sigma} + h.c.) + \mu\sum_i X_i^{00}$$
$$+ \sum_{ij}\Delta_{ij}\left(X_i^{\uparrow 0}X_j^{\downarrow 0} - X_i^{\downarrow 0}X_j^{\uparrow 0} + h.c.\right), \qquad (1)$$

where we have introduced the chemical potential term to control the total number of doped holes. The local NDO constraint is rigorously taken into account at the expense of the introduction of the Hubbard operators with more complicated commutation relations than those of the standard fermion algebra. In fact, fermionic operators $X_i^{\sigma 0}$ together with the bosonic ones, $X_i^{\sigma\sigma'}$ form, on every lattice site, a basis of the fundamental representation of the graded (supersymmetrical) Lie algebra su(2|1) given by the (anti)commutation relations

$$\{X_i^{ab}, X_j^{cd}\}_\pm = (X_i^{ad}\delta^{bc} \pm X_j^{bc}\delta^{ad})\delta_{ij}, \qquad (2)$$

where the (+) sign should be used only when both operators are fermionic.

In the strong-pairing limit ($|\Delta| >> t$) the projected BCS Hamiltonian (1) reduces to

$$H^G_\Delta = \sum_{ij}\Delta_{ij}\left(X_i^{\uparrow 0}X_j^{\downarrow 0} - X_i^{\downarrow 0}X_j^{\uparrow 0} + h.c.\right) + \mu\sum_i X_i^{00}. \qquad (3)$$

3. su(2|1) Coherent-State Path-Integral Representation of the Partition Function

In the su(2|1) coherent-state basis the partition function

$$Z_\Delta = tr\ \exp(-\beta H_\Delta^G)$$

takes the form of the $su(2|1)$ coherent-space phase-space path integral

$$Z_\Delta = \int D\mu(z,\xi)\ e^{S_\Delta}, \tag{4}$$

where

$$D\mu(z,\xi) = \prod_{i,t} \frac{d\bar{z}_i(t)dz_i(t)}{2\pi i(1+|z_i|^2)^2}\, d\bar{\xi}_i(t)d\xi_i(t).$$

Here z_i is a complex number that keeps track of the spin degrees of freedom, while ξ_i is a complex Grassmann parameter that describes the charge degrees of freedom.

The effective action reads

$$S_\Delta = i\sum_i \int_0^\beta a_i(t)dt - \sum_i \int_0^\beta \bar{\xi}_i\left(\partial_t + ia_i\right)\xi_i dt - \int_0^\beta H_\Delta^{G,cl}dt, \tag{5}$$

where

$$ia = -\langle z|\partial_t|z\rangle = \frac{1}{2}\frac{\dot{\bar{z}}z - \bar{z}\dot{z}}{1+|z|^2},$$

with $|z\rangle$ being the su(2) coherent state. This term is frequently referred to as the Berry connection. The dynamical part of the action takes the form

$$H_\Delta^{G,cl} = \sum_{ij}\left(\Delta_{ij}\xi_i\xi_j \frac{\bar{z}_j - \bar{z}_i}{\sqrt{(1+|z_i|^2)(1+|z_j|^2)}} + h.c\right) + \mu\sum_i \bar{\xi}_i\xi_i. \tag{6}$$

Here $z_i(t)$ and $\xi_i(t)$ are the dynamical fields. This representation rigorously incorporates the constraint of no double occupancy.

4. Effective Action

The fermionic degrees of freedom in Eq. (4) can formally be integrated out to yield

$$\int D\bar{\xi}D\xi \exp\left(\sum_{ij}\int_0^\beta \bar{\xi}_i(t)G_{ij}^{-1}(t,s)\xi_j(s)dtds\right) = \exp Tr\log G^{-1}$$

$$= \exp\left(Tr \log G_{(0)}^{-1} + Tr \log(1 - G_{(0)} ia + G_{(0)}\Sigma)\right), \tag{7}$$

where

$$G_{ij}^{-1}(t,s) = G_{(0)ij}^{-1}(t,s) - ia_i(t)\delta_{ij}\delta(t-s) + \Sigma_{ij}(t)\delta(t-s),$$

with $\Sigma_{ij} = \Delta_{ij}\langle z_j|z_i\rangle$ and

$$G_{(0)ij}^{-1}(t,s) = \delta_{ij}(-\partial_t - \mu)\delta(t-s),\ i \in A,$$

$$G_{(0)ij}^{-1}(t,s) = \delta_{ij}(-\partial_t + \mu)\delta(t-s),\ i \in B.$$

Here the trace has to be carried out over both space and time indices.

Up to this point no approximation has been made in the derivation of the effective action. In fact, we are interested in a derivation of an effective action to describe a low-energy dynamics of the spin degrees of freedom of the projected strong-pairing Hamiltonian close to half-filling. For that purpose we deduce an effective action in the spin degrees of freedom by performing a perturbative expansion of the expression $Tr \log(1 - G_{(0)} ia + G_{(0)}\Sigma)$ in powers of $|\Delta|/\mu \ll 1$. Physically, this corresponds to a highly underdoped region of the phase diagram. The second step consists in expanding thus obtained representation up to first order in ∂_t and second order in Δ_{ij} implying that eventually we will set $i \to j$. This amounts to the so-called gradient expansion that corresponds to the low-energy and long-wavelength limit of the action. In this way we obtain

$$Z_\Delta^{\text{eff}}/Z_0 = \int D\mu(z,\bar{z})\ e^{S_\Delta^{\text{eff}}}, \tag{8}$$

with the SU(2) invariant measure factor

$$D\mu(z,\bar{z}) = \prod_{i,t} \frac{d\bar{z}_i(t) dz_i(t)}{2\pi i(1+|z_i|^2)^2}$$

and

$$S_\Delta^{\text{eff}} = i\sum_i \int_0^\beta a_i(t)dt - \sum_{ij} J_{ij} \int_0^\beta \left(|\langle z_i|z_j\rangle|^2 - 1\right) dt, \tag{9}$$

where the long-wavelength limit $(j \to i)$ is implied. This action describes the anti-ferromagnetic Heisenberg model with the effective coupling $J_{ij} = |\Delta_{ij}|^2/2\mu > 0$.

5. Conclusion

We conclude by discussing the physical implications of the close connection between the Gutzwiller-projected BCS Hamiltonian and the t–J model of the high-T$_c$ superconductors. Our result shows that the ground state of the Gutzwiller-projected BCS Hamiltonian can in principle be considered the reference state of a slightly doped Mott insulator. Note, however, that this state does not coincide with the Gutzwiller-projected BCS ground state which is just the short-range RVB state proposed by Anderson. The RVB state is known to show no long-range order even at half filling. In contrast, right at half filling as well as in the immediate vicinity of half filling the ground state of the Gutzwiller-projected Hamiltonian exhibits long-ranged AF order observed in the cuprates superconductors. Note also that the low-energy action that corresponds to the strong-pairing BCS Hamiltonian cannot in itself account for the weakening as well as the eventual disappearance of the magnetic ordering as the hole concentration increases. This effect is produced by the growing influence of the kinetic t-term that gradually destroys the long-range ordered state. Therefore, one needs to include the kinetic t-term into consideration to regain the full Gutzwiller-projected BCS Hamiltonian, H^G_{BCS}, to describe the actual behavior of the high-T$_c$ phase diagram close to half filling.

At a moderate, non-zero doping the RVB wave function yields good agreement with experiments[2] as well as with numerical studies[3] and is conjectured to be a good ansatz wave function for the t–J model.[4] In doped regimes sufficiently away from half filling, the RVB state turns out indeed to be qualitatively similar to the ground state of H^G_{BCS}.[1] One can therefore conclude that the Gutzwiller-projected BCS Hamiltonian qualitatively captures the essential low-energy physics of the high-T$_c$ superconductors close to half filling, and beyond.

References

1. K. Park, *Phys. Rev. Lett.* **95**, 027001 (2005); *Phys. Rev. B* **72**, 245116 (2005).
2. A. Paramekanti, M. Randeiria, and N. Trivedi, *Phys. Rev. Lett.* **87**, 217002 (2001).
3. S. Sorella *et al.*, *Phys. Rev. Lett.* **88**, 17002 (2002).
4. F. C. Zhang, C. Gros, T. M. Rice, and H. Shiba, *Supercond. Sci. and Tech.* **1**, 36 (1988).

BATH-INDEPENDENT TRANSITION PROBABILITIES IN THE DISSIPATIVE LANDAU-ZENER PROBLEM

S. KOHLER* and P. HÄNGGI

Institut für Physik, Universität Augsburg,
Universitätsstr. 1, D-86135 Augsburg, Germany
**E-mail: sigmund.kohler@physik.uni-augsburg.de*

M. WUBS

Niels Bohr Institute, Copenhagen University,
DK 2100, Denmark
E-mail: wubs@nbi.dk

We study Landau-Zener transitions of a two-level system that is coupled to a quantum heat bath at zero temperature. In particular, we reveal that for a whole class of models, the probability for a nonadiabatic transition is bath-independent.

Keywords: Nonadiabatic transitions; Quantum dissipation.

1. Introduction

Nonadiabatic transitions at avoided level crossings play an essential role in numerous dynamical phenomena in physics and chemistry. They have been studied both theoretically and experimentally in various contexts like spin-flip processes in nano-scale magnets,[1,2] molecular collisions,[3] optical systems,[4] quantum-dot arrays,[5] Bose-Einstein condensates,[6] and recently also in quantum information processing.[7–10]

The "standard" Landau-Zener problem describes the ideal situation in which the dynamics is restricted to two levels that are coupled by a constant tunnel matrix element and cross at a constant velocity. The quantity of primary interest is the probability that finally the system ends up in the one or the other of the two states. This classic problem was solved independently by several authors in 1932.[11–14]

In an experiment, the two-level system will be influenced by its environment, which may affect the quantum phase of the superposition, alter the effective interaction between the levels, or may cause spontaneous decay. The environment of a quantum system can often be described as a bath of harmonic oscillators.[15–19] In some situations, it is known that the dominant environmental effects can best be modelled as a spin bath instead,[20–22] for example for molecular magnets[1] and for Josephson phase qubits.[7]

In the presence of a heat bath, the Landau-Zener dynamics will sensitively depend on the qubit operator to which the bath couples.[23,24] Ao and Rammer[25] studied Landau-Zener transitions for the special case in which an ohmic heat bath couples to the same operator as the driving and derived the transition probabilities in the limit of high and of low temperatures. In the limits of very fast and very slow sweeps at zero temperature, they found that the transition probability is the same as in the absence of the heat bath, as was confirmed by numerical studies.[26,27]

This zero-temperature result was recently proven to hold exactly for *arbitrary* Landau-Zener sweep speeds, as a special case of an exact expression for arbitrary qubit-bath couplings and spectral densities.[23] An exact solution is also possible if the decoherence stems from the coupling of the system to a spin bath.[24]

2. The Dissipative Landau-Zener Problem

The dissipative Landau-Zener problem is specified by the system-bath Hamiltonian

$$H(t) = H_{\mathrm{LZ}}(t) + H_{\text{q-env}} + H_{\mathrm{env}}, \tag{1}$$

where H_{env} and $H_{\text{q-env}}$ describe the environment and its coupling to the two-level system, henceforth termed qubit. The time-dependent qubit Hamiltonian reads

$$H_{\mathrm{LZ}}(t) = \frac{vt}{2}\sigma_z + \frac{\Delta}{2}\sigma_x, \tag{2}$$

which defines the "standard" Landau-Zener problem. The adiabatic energies, i.e. the eigenstates of $H_{\mathrm{LZ}}(t)$ form at time $t = 0$ an avoided crossing between the diabatic states $|\uparrow\rangle$ and $|\downarrow\rangle$. The latter are the eigenstates of $H_{\mathrm{LZ}}(t)$ at large times.

If the qubit starts at time $t = -\infty$ in state $|\uparrow\rangle$, one finds that finally at time $t = \infty$, the qubit will be in state $|\uparrow\rangle$ with a probability given by the classic expression[11–13]

$$P_{\uparrow\to\uparrow} = \exp\Big(-\frac{\pi\Delta^2}{2\hbar v}\Big). \tag{3}$$

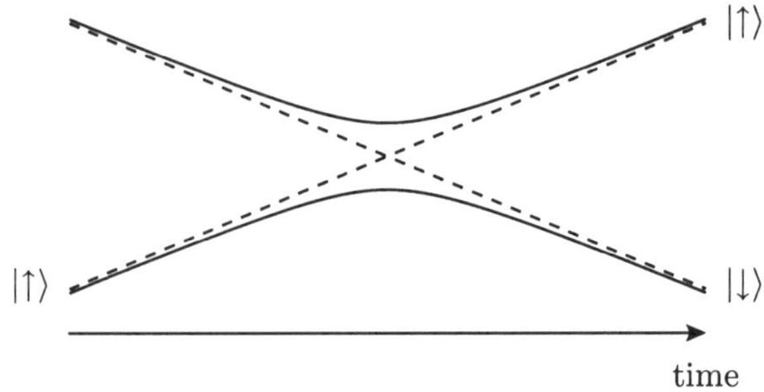

Fig. 1. Adiabatic (solid) and diabatic (dashed) energy levels of "standard" Landau-Zener Hamiltonian (2).

In general, this transition probability is modified by a coupling to the environment[23,24] for which we assume the form $H_{\text{q-env}} = \boldsymbol{\sigma} \cdot \boldsymbol{n}\xi$, where $\boldsymbol{n}$ determines the "direction" of the coupling and ξ is a collective coordinate of the bath. In the following, we explore for which types of coupling the opposite holds, namely that $P_{\uparrow\to\uparrow}$ still is given by Eq. (3), despite the coupling to the bath.

3. Multi-Level Landau-Zener Dynamics

The two-level Landau-Zener problem defined by the Hamiltonian (1) can be mapped to the multi-level Landau-Zener problem sketched in Fig. 2 which has been solved in Ref. 24. It is defined by the Hamiltonian

$$\begin{aligned} H(t) = & \sum_a (\varepsilon_a + \frac{vt}{2})|a\rangle\langle a| + \sum_b (\varepsilon_b - \frac{vt}{2})|b\rangle\langle b| \\ & + \sum_{a,b} (X_{ab}|a\rangle\langle b| + X_{ab}^*|b\rangle\langle a|), \end{aligned} \tag{4}$$

which describes a group of levels $|a\rangle$ whose energy increases linearly in time, while the energy of the levels $|b\rangle$ deceases. In the limit $t \to \pm\infty$, the states $|a\rangle$, $|b\rangle$ become eigenstates of the Hamiltonian (4), which means that they represent the diabatic eigenstates. The off-diagonal part of the Hamiltonian is such that it only couples states of different groups while states within one group are uncoupled.

If now the system starts in any non-degenerate state $|a\rangle$, one can derive for the transition to a state $|a'\rangle$ the following two statements:[24] First, the

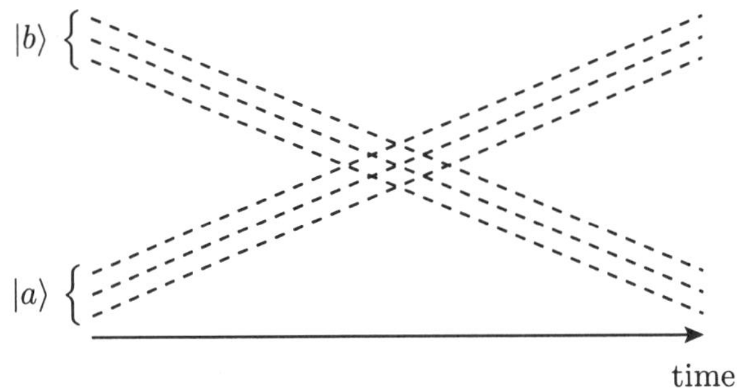

Fig. 2. Diabatic levels of the multi-level Hamiltonian to which we map the dissipative Landau-Zener problem.

probability to end up in the initial state is given by

$$P_{a\to a} = \exp\Big(- \frac{2\pi \langle a|X^2|a\rangle}{\hbar v} \Big), \tag{5}$$

while, second, all a-states with higher energy are finally not populated, i.e.

$$P_{a\to a'} = 0 \quad \text{for } \varepsilon_{a'} > \varepsilon_a. \tag{6}$$

If in particular, the initial state is the a-state with the lowest energy, relation (6) implies that $1 - P_{a\to a}$ denotes the probability to end up in any state of group b.

4. Bath-Independent Landau-Zener Probability

In order to make use of the results of the last section, we have to identify the diabatic states of the dissipative Landau-Zener Hamiltonian (1). Since at large times, the time-dependent part of the Hamiltonian dominates, the diabatic qubit states are the eigenstates of σ_z, $|\uparrow\rangle$ and $|\downarrow\rangle$. The bath states corresponding to $|\uparrow\rangle$ are determined by the Hamiltonian

$$H_{\text{env}\uparrow} = \langle\uparrow|\boldsymbol{n}\cdot\boldsymbol{\sigma}|\uparrow\rangle\xi + H_{\text{env}} = n_z\xi + H_{\text{env}} \tag{7}$$

and will be denoted by $|\nu_+\rangle$ with $|0_+\rangle$ being the ground state. Thus the states $|\uparrow,\nu_+\rangle$ correspond to group a while the accordingly defined states $|\downarrow,\nu_-\rangle$ form group b. The coupling operator X then becomes

$$X = \frac{\Delta}{2}\sigma_x + (n_x\sigma_x + n_y\sigma_y)\xi. \tag{8}$$

Note that $n_z\sigma_z\xi$ does not couple states from different groups and, thus, is not contained in X.

At zero temperature, the natural initial state of the qubit coupled to the bath is the diabatic state $|\uparrow, 0_+\rangle$, which is the ground state in the limit $t \to -\infty$. Since Eq. (6) implies that all states $|\uparrow, \nu_+\rangle$ with $|\nu_+\rangle \neq |0_+\rangle$ will finally be unoccupied, we find

$$P_{\uparrow\to\uparrow} = \sum_{\nu_+} P_{\uparrow,0_+\to\uparrow,\nu_+} = \exp\Big(- \frac{2\pi\langle\uparrow,0_+|X^2|\uparrow,0_+\rangle}{\hbar v}\Big), \tag{9}$$

where the coupling operator X is given by Eq. (8).

A particular case is now $\boldsymbol{n} = \boldsymbol{e}_z$ for which $X = \frac{\Delta}{2}\sigma_x$, such that the bath couples only via the Pauli matrix that determines the diabatic states. Then $\langle\uparrow, \nu_+|(\frac{\Delta}{2}\sigma_x)^2|\uparrow, \nu_+\rangle = \Delta^2/4$ and, consequently, the transition probability for a diabatic transition in the presence of a heat bath at zero temperature, Eq. (3), becomes identical to the "standard" Landau-Zener result (9). This result holds true at zero temperature whenever the bath couples to the qubit via σ_z, irrespective of the nature and the spectral density of the bath.

An important experimentally relevant case for which this prediction applies is the measurement of tiny tunnel splittings Δ in nanomagnets[1] for which laboratory experience tells us that at temperatures well below 1K, Landau-Zener tunneling is robust against dephasing. Recent theories[2] for multiple Landau-Zener transitions in such systems presumed that during the individual Landau-Zener transitions, dephasing does not play a role, which is in accordance with our results. Our results show that these theories should be more widely applicable than guessed previously.

5. Conclusions

We have investigated the dissipative Landau-Zener problem for a qubit with a qubit-bath coupling that commutes with the time-dependent part of the qubit Hamiltonian. For large time, this bath coupling causes pure dephasing, while it can induce spin flips at the center of the avoided crossing of the adiabatic levels. As a central result, we have shown that at zero temperature, the Landau-Zener transition probability is dissipation independent. This result holds true for all quantum heat baths with a non-degenerate ground state.

References

1. W. Wernsdorfer and R. Sessoli, *Science* **284**, 133 (1999).
2. M. N. Leuenberger and D. Loss, *Phys. Rev. B* **61**, 12200 (2000).
3. M. S. Child, *Molecular Collision Theory* (Academic Press, London, 1974).

4. R. J. C. Spreeuw, N. J. van Druten, M. W. Beijersbergen, E. R. Eliel, and J. P. Woerdman, *Phys. Rev. Lett.* **65**, 2642 (1990).
5. K. Saito and Y. Kayanuma, *Phys. Rev. B* **70**, 201304(R) (2004).
6. D. Witthaut, E. M. Graefe, and H. J. Korsch, *Phys. Rev. A* **73**, 063609 (2006).
7. K. B. Cooper, M. Steffen, R. McDermott, R. W. Simmonds, S. Oh, D. Hite, D. P. Pappas, and J. M. Martinis, *Phys. Rev. Lett.* **93**, 180401 (2004).
8. G. Ithier, E. Collin, P. Joyez, D. Vion, D. Esteve, J. Ankerhold, and H. Grabert, *Phys. Rev. Lett.* **94**, 057004 (2005).
9. W. D. Oliver, Y. Yu, J. C. Lee, K. K. Berggren, L. S. Levitov, and T. P. Orlando, *Science* **310**, 1653 (2005).
10. M. Sillanpää, T. Lehtinen, A. Paila, Y. Makhlin, and P. Hakonen, *Phys. Rev. Lett.* **96**, 187002 (2006).
11. L. D. Landau, *Phys. Z. Sowjetunion* **2**, 46 (1932).
12. C. Zener, *Proc. R. Soc. London, Ser. A* **137**, 696 (1932).
13. E. C. G. Stueckelberg, *Helv. Phys. Acta* **5**, 369 (1932).
14. E. Majorana, *Nuovo Cimento* **9**, 43 (1932).
15. V. B. Magalinskii, *Zh. Eksp. Teor. Fiz.* **36**, 1942 (1959) [Sov. Phys. JETP **9**, 1381 (1959)].
16. A. J. Leggett, S. Chakravarty, A. T. Dorsey, M. P. A. Fisher, A. Garg, and W. Zwerger, *Rev. Mod. Phys.* **59**, 1 (1987).
17. P. Hänggi, P. Talkner, and M. Borkovec, *Rev. Mod. Phys.* **62**, 251 (1990).
18. M. Grifoni and P. Hänggi, *Phys. Rep.* **304**, 229 (1998).
19. U. Weiss, *Quantum Dissipative Systems*, 2nd edition (World Scientific, Singapore, 1998).
20. J. Shao and P. Hänggi, *Phys. Rev. Lett.* **81**, 5710 (1998).
21. N. V. Prokof'ev and P. C. E. Stamp, *Rep. Prog. Phys.* **63**, 669 (2000).
22. A. Hutton and S. Bose, *Phys. Rev. A* **69**, 042312 (2004).
23. M. Wubs, K. Saito, S. Kohler, P. Hänggi, and Y. Kayanuma, *Phys. Rev. Lett.* **97**, 200404 (2006).
24. K. Saito, M. Wubs, S. Kohler, Y. Kayanuma, and P. Hänggi, *Phys. Rev. B* **75**, 214308 (2007).
25. P. Ao and J. Rammer, *Phys. Rev. Lett.* **62**, 3004 (1989).
26. Y. Kayanuma and H. Nakayama, *Phys. Rev. B* **57**, 13099 (1998).
27. H. Kobayashi, N. Hatano, and S. Miyashita, *Physica A* **265**, 565 (1999).

CORRELATED NONEQUILIBRIUM CHARGE TRANSPORT THROUGH IMPURITIES

A. HERZOG* and U. WEISS†

II. Institut für Theoretische Physik, Universität Stuttgart,
D-70550 Stuttgart, Germany
**E-mail: herzog@theo2.physik.uni-stuttgart.de*
†*E-mail: weiss@theo2.physik.uni-stuttgart.de*
www.theo2.physik.uni-stuttgart.de

We review recent advances in the full counting statistics (FCS) of transport of charge through an impurity in a 1D quantum wire. The model also applies to a coherent conductor in a resistive environment and to a resistively shunted Josephson device. Starting out from the path integral Coulomb gas representation both the weak and strong tunneling series representations for the cumulant generating function can be found in analytic form in particular regions of the parameter space. The zero temperature case is discussed in some detail.

Keywords: Nonequilibrium quantum transport; Correlated electron transport; Coulomb gas representation; Self-duality; Full counting statistics.

1. Introduction

The physics of 1D metals is drastically different from the physics of conductors in two or three dimensions since the Fermi liquid behavior is abrogated by the electron interaction. The generic features of 1D interacting systems are well described in terms of the Tomonaga-Luttinger liquid (TLL) of which a field-theoretical formulation has been given by Haldane.[1] In the TLL universality class, all effects of the spinless electron-electron interaction are captured by a single dimensionless parameter g.

A sensitive experimental probe of a Luttinger liquid state is the tunneling conductance through a point contact in a 1D quantum wire.[2] The differential conductance is a power law of the applied voltage or temperature, depending on what is larger. This is due to the formation of a highly correlated collective state. The exponent of the power law is related to the interaction parameter g. Of interest are also the dc nonequilibrium current noise and cumulants of higher order. The generic model is a point-like impurity embedded in a Luttinger liquid environment (QI-TLL model). Depending on the parameter regime, the impurity may operate as a weak or strong barrier. Tunneling of edge currents in the fractional quantum Hall (FQH) regime provides another realization of a Luttinger phase. The nature of the entity transferred may depend on the parameter region. In FQH samples, the tunneling entities in the strong-backscattering regime are physical electrons. In the weak-backscattering limit, on the other hand, Laughlin quasiparticles of fractional charge are tunneling independently.

Analytic solutions for the QI-TLL model are available since the underlying field theory is integrable and the regions of weak and strong tunneling are related by self-duality.[3] The existence of analytic solutions makes 1D quantum wires very interesting for the growing field of full counting statistics. Here we review the charge transfer statistics at $T = 0$.

We introduce in Sec. 2 the quantum impurity model. In Sec. 3, we give exact formal expressions for the current statistics and moment generating function. The full counting statistics at $T = 0$ is discussed in Sec. 4.

2. Quantum Impurity in a Tomonaga-Luttinger Liquid

The low-energy modes of the 1D interacting electron liquid are conveniently treated in the framework of bosonization.[4] The creation operator for spinless fermions may be written in terms of bosonic field operators $\phi(x,t)$ and $\theta(x,t)$ obeying $[\,\phi(x,t),\, \theta(x',t)\,] = -(\mathrm{i}/2)\,\mathrm{sgn}(x-x')$ as $\psi^\dagger(x) \propto \sum_{n\,\mathrm{odd}} \exp\{\mathrm{i}\, n[k_\mathrm{F} x + \sqrt{\pi}\,\theta(x)]\} \exp[\mathrm{i}\,\sqrt{\pi}\phi(x)]$. The ground state of 1D interacting spinless electrons is a Tomonaga-Luttinger liquid (TLL). The generic interaction of the TLL universality class is effectively given by a δ-potential yielding the interaction term $H_\mathrm{I} = U \int \mathrm{d}x\, \mathrm{d}y\, \rho(x)\, \delta(x-y)\, \rho(y)$. In the TLL, the electron interaction is characterized by a single dimensionless parameter $g = 1/\sqrt{1 + U/\pi v_\mathrm{F}}$. We have $g < 1$ for repulsive interaction. Excitations of the liquid are described by the generic harmonic Hamiltonian

$$H_\mathrm{L}(g) \;=\; \frac{v_\mathrm{F}}{2} \int \mathrm{d}x \; \left[\, (\partial_x \theta)^2/g \,+\, g\,(\partial_x \phi)^2 \,\right]. \tag{1}$$

A hopping term of the form $H_\mathrm{I}' = -\Delta[\,\psi^\dagger(x=0^+)\,\psi(x=0^-) + \mathrm{h.c.}]$ represents a single strong point-like impurity. This term induces a jump of the ϕ-field at the impurity, $\bar\phi = \frac{1}{2}[\,\phi(x=0^+) - \phi(x=0^-)\,]$. In the ϕ-representation, the tunneling term reads

$$H_\mathrm{I}'(\bar\phi) \;=\; -\,\Delta \cos[\,2\sqrt{\pi}\bar\phi \,+\, eV_\mathrm{a} t/\hbar\,]\,. \tag{2}$$

Here we have included a voltage term. Adding the terms describing the harmonic liquid in the right/left ($\pm$) lead, we get the weak-link Hamiltonian[2]

$$H_\phi = H_{\rm L,+}(g) + H_{\rm L,-}(g) + H_{\rm I}'(\bar\phi) \,. \qquad (3)$$

A single weak impurity is modelled by the Hamiltonian $H_{\rm sc} = \int {\rm d}x\, V(x)\psi^\dagger(x)\psi(x)$, where $V(x)$ is the scattering potential. For a point-like scatterer at $x = 0$, the contribution describing $2k_{\rm F}$-backscattering takes the form $H_{\rm sc}(\bar\theta) = -V_0 \cos[\,2\sqrt{\pi}\bar\theta\,]$, where $\bar\theta = \theta(0)$. With the lead and voltage term added, the weak-impurity or strong-tunneling Hamiltonian is

$$H_\theta = -V_0 \cos[\,2\sqrt{\pi}\bar\theta\,] - eV_{\rm a}\bar\theta/\sqrt{\pi} + H_{\rm L}(g) \,. \qquad (4)$$

The harmonic modes in the leads of the ϕ- and θ-model away from $x = 0$ act as reservoir modes and may be integrated out. Then, the correlators of the tunneling degree of freedom $\bar\phi$, $J_{\bar\phi}(t; g) = 4\pi\langle[\,\bar\phi(0) - \bar\phi(t)]\bar\phi(0)\rangle_\beta$, and of the scattering degree of freedom $\bar\theta$, $J_{\bar\theta}(t; g) = 4\pi\langle[\,\bar\theta(0) - \bar\theta(t)]\bar\theta(0)\rangle_\beta$, are found in thermal equilibrium as

$$J_{\bar\theta}(t; g) = J_{\bar\phi}(t; 1/g) = 2g \ln\left[\frac{\beta\omega_c}{\pi}\sinh\left(\frac{\pi|t|}{\beta}\right)\right] + {\rm i}\,\pi g\,{\rm sgn}(t) \,. \qquad (5)$$

The Hamiltonians (3) and (4) are analogous to the tight-binding (TB) and weak-binding (WB) Hamiltonians of a quantum Brownian particle in a tilted periodic potential. From the form (5) we see that the Luttinger modes in the leads act as an Ohmic thermal reservoir with $g = 1/K$, where K is the usual dimensionless Ohmic coupling parameter.[5] The TB and WB representations of the Brownian particle model are related by an exact self-duality, as shown first by A. Schmid.[6] According to the analogy, also the weak-tunneling model (3) and the strong-tunneling model (4) are related by an exact duality symmetry in which the interaction parameter g maps on the inverse of it, as can be seen from the dual equilibrium correlation functions given in Eq. (5).

In the theory of FCS, the most natural quantity to consider is the generating function $\mathcal{Z}(\lambda, \tau)$, which is the Fourier transform of the probability distribution $P_n(\tau)$ of the amount of charge ne crossing the impurity during time τ, $\mathcal{Z}(\lambda, \tau) = \sum_n {\rm e}^{{\rm i}\lambda e n}\, P_n(\tau)$. The function $\mathcal{Z}(\lambda, \tau)$ captures all moments of the charge $Q_\tau = \int_0^\tau {\rm d}t\, I(t)$ transferred during time τ,

$$\mathcal{Z}(\lambda, \tau) = \sum_k \frac{({\rm i}\lambda)^k}{k!}\,\langle Q_\tau^k\rangle = \exp\left\{\tau \sum_k \frac{({\rm i}\lambda)^k}{k!}\,\langle \delta^k Q\rangle\right\} . \qquad (6)$$

Here, $I(t)$ is the current through the barrier, $\langle Q_\tau^k\rangle = \sum_n (en)^k P_n(\tau)$ is the kth moment, and $\tau\langle\delta^k Q\rangle$ is the kth cumulant of the distribution. The dynamics of the charge transfer is governed by the set of master equations

$$\dot{P}_n(t) = \sum_{\ell=1}^{\infty} \left[k_\ell^+ P_{n-\ell}(t) + k_\ell^- P_{n+\ell}(t) - (k_\ell^+ + k_\ell^-) P_n(t) \right] . \tag{7}$$

The quantity $k_n^\pm$ is the weight per unit time for joint forward/backward $(+/-)$ transfer of charge en through the barrier.

The moment generating function (MGF) $\mathcal{Z}(\lambda, \tau)$ is found from Eq. (7) as

$$\mathcal{Z}(\lambda, \tau) = \exp\left\{ \tau \sum_{n=1}^{\infty} \left[(\mathrm{e}^{\mathrm{i}\lambda en} - 1) k_n^+ + (\mathrm{e}^{-\mathrm{i}\lambda en} - 1) k_n^- \right] \right\} . \tag{8}$$

Now observe that the curly bracket is just the cumulant generating function. Thus, the Nth cumulant is given in terms of the transition weights $k_n^\pm$ as

$$\langle \delta^N Q \rangle = \sum_{n=1}^{\infty} (en)^N \left[k_n^+ + (-1)^N k_n^- \right] . \tag{9}$$

The first cumulant yields the average particle current, $\langle I \rangle = \langle \delta Q \rangle$. The second one gives the nonequilibrium diffusion coefficient or dc current noise. The third cumulant (skewness) provides information about the leading asymmetric deviation from the Gaussian distribution, while the fourth cumulant is a measure for the sharpness of the distribution compared to the standard distribution.

3. Exact Formal Expressions for the Current Statistics

The most powerful tool to calculate the reduced density matrix (RDM) and the MGF is the nonequilibrium Keldysh or Feynman-Vernon method. Let us now apply this approach to the discrete ϕ-model (3). The double path for the population $P_n(\tau)$ is conveniently parametrized in terms of charges $\{\eta_j = \pm 1\}$ describing forward/backward moves parallel to the diagonal of the RDM and in terms of charges $\{\xi_j = \pm 1\}$ describing moves off and towards the diagonal of the RDM. Then the effects of the Luttinger modes for a path with $2m$ moves parametrized with 2m ξ-charges and 2m η-charges are represented by the influence factor

$$\mathcal{F}_m = \mathcal{G}_m \exp\left\{ -\mathrm{i}\frac{\pi}{g} \sum_{j=1}^{2m-1} p_j \eta_j \right\}, \tag{10}$$

$$\mathcal{G}_m = \exp\left\{ \frac{2}{g} \sum_{j>k=1}^{2m} \xi_j \xi_k \ln\left[\frac{\beta\omega_{\mathrm{c}}}{\pi} \sinh\left(\frac{\pi(t_j - t_k)}{\beta} \right) \right] \right\} . \tag{11}$$

The cumulative charge $p_j = \sum_{k=1}^{j} \xi_k = -\sum_{k=j+1}^{2m} \xi_k$ measures how far the system is off-diagonal after j transitions. For the MGF the η-sums can be done explicitly in each order of Δ. The expansion in powers of the transition amplitude Δ then leads us to the so-called Coulomb gas representation of the MGF[5,7] (we put $eV_{\rm a} = \epsilon$),

$$\mathcal{Z}(\lambda,\tau) = 1 + \sin\left(\frac{\lambda e}{2}\right) \sum_{m=1}^{\infty} (-1)^m \Delta^{2m} \int_0^{\tau} \mathcal{D}_{2m}\{t_j\} \qquad (12)$$

$$\times \sum_{\{\xi_j\}'_m} \exp\left\{ \mathrm{i}\epsilon \sum_{j=1}^{2m-1} p_j (t_{j+1} - t_j) \right\} \mathcal{G}_m \prod_{j=1}^{2m-1} \sin\left(\frac{\lambda e}{2} - \frac{\pi}{g} p_j\right) ,$$

where the integration symbol denotes time-ordered integration,

$$\int_0^t \mathcal{D}_{2m}\{t_j\} \times \ldots := \int_0^t \mathrm{d}t_{2m} \int_0^{t_{2m}} \mathrm{d}t_{2m-1} \cdots \int_0^{t_2} \mathrm{d}t_1 \times \ldots . \qquad (13)$$

The sum $\{\xi_j\}'_m$ is over all sequences of $2m$ charges $\xi_j = \pm 1$ with over-all charge neutrality, $\sum_{j=1}^{2m} \xi_j = 0$, which is indicated by the prime. Charge neutrality is, because each path contributing to the MGF starts out from a diagonal state and returns to a diagonal state of the reduced density matrix (RDM). Every individual charge sequence together with its charge-conjugate counterpart gives a real contribution to $\mathcal{Z}(\lambda, \tau)$.

Explicit forms for moments follow from (12) upon differentiation,

$$\langle Q_\tau^N \rangle = \frac{e^N}{2} \sum_{m=1}^{\infty} (-1)^{m-1} \Delta^{2m} \int_0^t \mathcal{D}_{2m}\{t_j\} \sum_{\{\xi_j\}'} \mathcal{G}_m \, b_m^{(N)} . \qquad (14)$$

For the most interesting cases $N = 1$ and $N = 2$ we get

$$b_m^{(1)} = i \prod_{j=1}^{2m-1} \sin(\pi p_j/g) , \qquad b_m^{(2)} = \sum_{k=1}^{2m-1} \cos(\pi p_k/g) \prod_{j=1,\, j\neq k}^{2m-1} \sin(\pi p_j/g) .$$

The Laplace transform $\hat{\mathcal{Z}}(\lambda, z)$ is the analogue of an isobaric ensemble of a classical gas of charges $\xi_j = \pm 1$ with charge interactions (11). The expression (8) follows from a cluster decomposition of $\hat{\mathcal{Z}}(\lambda, z)$ resulting from the series (12). The clusters are the z-independent (irreducible) path sections which start and end in diagonal states of the RDM. They have total charge zero and are noninteracting with each other. Path segments with intermediate visits of diagonal states are reducible, i.e, factorize into clusters of lower order. One finds that after subtraction of the reducible part an irreducible part is left. The transition weights $k_n^{\pm}$, can be identified as the sum of all clusters which interpolate between the (arbitrary) diagonal state

m and the diagonal state $m \pm n$. As a consequence of the thermal equilibrium property $J(t - \mathrm{i}\beta) = J^*(t)$ satisfied by the correlation function (5), the transition weights are found to obey detailed balance, $k_n^- = \mathrm{e}^{-n\,\epsilon/T}\,k_n^+$.[8]

Evaluation of the Coulomb gas representation (12) in analytic form is possible in special regions of the parameter space: (*i*) in the classical limit $g \to 0$ for general T and V_a, (*ii*) at the so-called Toulouse point $g = \frac{1}{2}$ (and by self-duality also at $g = 2$) for general T and V_a, and (*iii*) at $T = 0$ for general g and V_a.[5]

4. Full Counting Statistics at Zero Temperature

At $T = 0$, it is possible to determine the expansions of the CGF in analytic form for strong and weak tunneling. The calculation has been carried out first in the integrable field theory approach with the thermodynamic Bethe ansatz. Subsequently, the results have been confirmed with the rigorous Keldysh method.[8,9]

There are clear physical pictures in the limits of weak and strong tunneling.[9] When the barrier is very high, the true ground state is that of two completely disconnected leads. Evidently, then only physical electrons with integer charge may tunnel between these subsystems. On the other hand, for a weak barrier a strongly coupled system with a collective state between the edges is formed, which in the case of a FQH bar has fractionally charged Laughlin quasiparticles for elementary excitations.

In the weak-tunneling limit, the clusters formally appear as the series

$$k_n^+ \;=\; \frac{\epsilon}{2\pi}\,x^{2n}\sum_{\ell=0}^{\infty} x^{2\ell} f_n^{(\ell)}\,, \qquad x \;=\; \frac{\Delta}{\epsilon}\left(\frac{\epsilon}{\omega_\mathrm{c}}\right), \tag{15}$$

in which the coefficients are given in terms of a sum of $2(n+\ell) - 1$-fold integrals, each of them representing a particular sequence of ξ-charges. A detailed analysis has revealed that there are extensive cancellations among the various charge sequences, so that only paths with the minimal number $2n$ of hops contribute to the transition weight k_n^+. Thus we have

$$k_n^+ \;=\; \frac{\epsilon}{2\pi}\,x^{2n}\,f_n^{(0)}\,, \qquad k_n^- \;=\; 0\,. \tag{16}$$

It is convenient to introduce an energy scale ϵ_0 analogous to the Kondo energy in the Kondo model and to the scale T_B' in QIPs.[9] The relations of ϵ_0 to the bare parameters of the ϕ- and θ-model are found to read

$$\epsilon_0^{2-2/g} \;=\; \frac{2^{2-2/g}\,\pi^2}{\Gamma^2(1/g)}\,\frac{\Delta^2}{\omega_\mathrm{c}^{2/g}}\,, \qquad \epsilon_0^{2-2g} \;=\; g^{2g-2}\,\frac{2^{2-2g}\,\pi^2}{\Gamma^2(g)}\,\frac{V_0^2}{\omega_\mathrm{c}^{2g}}\,. \tag{17}$$

For strong backscattering or weak tunneling, one readily obtains

$$\ln \mathcal{Z}(\lambda, \tau) = \tau \sum_{m=1}^{\infty} (e^{\mathrm{i}\, e\lambda m} - 1) k_m^+ \,. \tag{18}$$

The transition weight k_m^+ is

$$k_m^+ \;=\; \frac{1}{m}\, a_m(1/g)\, G_0 V_{\rm a}\, (\epsilon/\epsilon_0)^{2m(1/g-1)} \tag{19}$$

with $G_0 = e^2/2\pi$, and the g-dependent prefactor is[3,10]

$$a_m(g) \;=\; (-1)^{m+1} \frac{\Gamma(3/2)\Gamma(mg)}{\Gamma(m)\,\Gamma[3/2 + m(g-1)]} \,. \tag{20}$$

The physical meaning of the expression (18) is quite illuminating. Suppose that k_m is the probability per unit time for transfer of a particle of charge me (or joint transfer of m particles of charge e) through the impurity barrier. Then the charge transferred in the time interval τ is the result of a Poisson process for particles of charge e crossing the barrier, contributing a current $I_1 = ek_1$, plus a Poisson process for particles of charge two contributing a current $I_2 = 2ek_2$, etc. The logarithm of the Fourier transform of the probability distribution of all these Poisson processes would then be the expression (18). The only subtle point now is that the signs of the rates k_m are not quite right since a classical process would require that all rates are positive. Certainly, the $m = 1$–term is indeed a Poisson process for the tunneling of single electrons, but the joint tunneling of pairs of electrons (and of multiples thereof) $m = 2,\, 4,\, \ldots$ comes with the different sign, which is an effect of quantum interference.

In the opposite limit of weak backscattering, the CGF is found as

$$\ln \mathcal{Z}(\lambda, \tau) \;=\; \tau \left[\mathrm{i}\,\lambda\, g\, G_0 V_{\rm a} + \sum_{m=1}^{\infty} (e^{-\mathrm{i}\, g e\lambda m} - 1) \widetilde{k}_m \right] , \tag{21}$$

where

$$\widetilde{k}_m \;=\; \frac{1}{m}\, a_m(g)\, g\, G_0 V_{\rm a}\, (\epsilon/\epsilon_0)^{2m(g-1)} \,. \tag{22}$$

This form is quite similar to (18), but there are subtle differences. The first term in the expression (21) represents the current of fractionally charged quasiparticles in the absence of the barrier. The exponential factor $e^{-\mathrm{i} g e\lambda\, m}$ indicates that now we have tunneling of quasiparticles of charge ge and of multiples thereof, and the minus sign in the exponent means that the tunneling diminishes the current instead of building it up as in the strong-backscattering regime. The sign of the tunneling rate $\widetilde{k}_m$ is now that of

$\cos(m\pi g)$. Therefore, the perception of clusters of quasiparticles with fractional charge ge tunneling independently with a classical Poisson process is quite appropriate when g is small. The classical limit is reached as g goes to zero. Then all the rates $\widetilde{k}_m$ are positive, but the quantum fluctuations $\langle \delta^n Q \rangle$ with $n > 1$ have faded away,

$$\begin{aligned} \ln \mathcal{Z}(\lambda, \tau) &= \mathrm{i}\lambda\tau\, g\, G_0 \epsilon \left[1 - \sum_{m=1}^{\infty} \frac{\Gamma(m-\frac{1}{2})}{2\sqrt{\pi}m!} \left(\frac{\epsilon_0}{\epsilon}\right)^{2m} \right] \\ &= \mathrm{i}\lambda\tau\, g\, G_0 V_{\mathrm{a}} \sqrt{1 - \left(\frac{\epsilon_0}{\epsilon}\right)^2}\,. \end{aligned} \tag{23}$$

Finally, we show that every single cumulant can be calculated once the current is known. First, we see from the relation (9) that cumulants correspond to moments of the current. Secondly, the property (16) yields $(\Delta/2)\,\partial/\partial\Delta\, k_n^+ = n\, k_n^+$. Thus we get in the weak-tunneling regime

$$\langle \delta^N Q \rangle \;=\; \left(e\frac{\Delta}{2}\frac{\partial}{\partial \Delta}\right)^{N-1} \langle I \rangle\,. \tag{24}$$

The corresponding relation for weak backscattering or strong tunneling is

$$\langle \delta^N Q \rangle \;=\; \left(-\, ge\frac{V_0}{2}\frac{\partial}{\partial V_0}\right)^{N-1} \langle I \rangle\,. \tag{25}$$

We conclude with the remark that strict self-duality implies that the entire expansions around weak and strong tunneling are related to each other. In the FQH system, the crossover from strong tunneling to weak tunneling comes along with a crossover from Laughlin quasiparticle tunneling to electron tunneling.

References

1. F. D. M. Haldane, *J. Phys. C: Solid State Phys.* **14**, 2585 (1981).
2. C. L. Kane and M. P. A. Fisher, *Phys. Rev. Lett.* **68**, 1220 (1992); *Phys. Rev. B* **46**, 15 233 (1992).
3. P. Fendley, A. W. W. Ludwig, and H. Saleur, *Phys. Rev. B* **52**, 8934 (1995).
4. A. O. Gogolin, A. A. Nersesyan, and A. M. Tsvelik, *Bosonization and Strongly Correlated Systems* (Cambridge University Press, Cambridge, 1998).
5. U. Weiss *Quantum Dissipative Systems*, Series in Condensed Matter Physics, Vol. 10, 2nd ed. (World Scientific, Singapore, 1999).
6. A. Schmid, *Phys. Rev. Lett.* **51**, 1506 (1983).
7. U. Weiss and M. Wollensak, *Phys. Rev. B* **37**, 2729 (1988); U. Weiss, M. Sassetti, Th. Negele, and M. Wollensak, *Z. Phys. B* **84**, 471 (1991).
8. H. Baur, A. Fubini, and U. Weiss, *Phys. Rev. B* **70**, 024302 (2004).
9. H. Saleur and U. Weiss, *Phys. Rev. B* **63**, 201302(R) (2001).
10. U. Weiss, *Solid State Comm.* **100**, 281 (1996).

HOW TO MEASURE THE EFFECTIVE ACTION FOR DISORDERED SYSTEMS

K. J. WIESE* and P. LE DOUSSAL

Laboratoire de Physique Théorique de l'Ecole Normale Superieure,
24 rue Lhomond, 75005 Paris, France
** E-mail: wiese@lpt.ens.fr*

In contrast to standard critical phenomena, disordered systems need to be treated via the Functional Renormalization Group. The latter leads to a coarse grained disorder landscape, which after a finite renormalization becomes non-analytic, thus overcoming the predictions of the seemingly exact dimensional reduction. We review recent progress on how the non-analytic effective action can be measured both in simulations and experiments, and confront theory with numerical work.

Keywords: Functional renormalisation; Disordered systems; Effective action.

1. Introduction

When talking to his experimental colleagues about the marvels of field theory and his recent achievements in computing the effective action of his favorite model, the conversation is likely to resemble this:

Theorist: I have a wonderful field theory, I can even calculate the effective action!
Experimentalist: Can I see it in an experiment? Can I measure it?
Theorist: ... well, that's difficult, but it tells you all you want to know ...
Experimentalist: okay, I understand, another of these unverifiable predictions ...

Here we will see how to measure it, considering the explicit and far-from-trivial example of elastic manifolds in a disordered environment. Due to a lack of space, we will not be able to give all arguments in the necessary details. We recommend that the reader consults the recent "Basic Recipes and Gourmet Dishes",[1] to which we also refer for a more complete list of references.

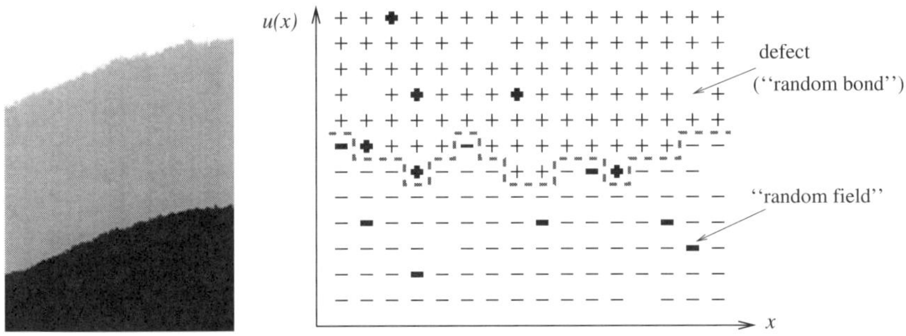

Fig. 1. An Ising magnet at low temperatures forms a domain wall described by a function $u(x)$ (right). An experiment on a thin Cobalt film (left);[2] with kind permission of the authors.

2. The Disordered Systems Treated Here – Our Model

Let us first give some physical realizations. The simplest one is an Ising magnet. Imposing boundary conditions with all spins up at the upper and all spins down at the lower boundary (see Fig. 1), at low temperature T, a domain wall will form in between. In a pure system at $T = 0$, this domain wall is completely flat; it will be roughened by disorder. Two types of disorder are common: random bond (which on a coarse-grained level represents missing spins) and random field (coupling of the spins to an external random magnetic field). Figure 1 shows, how the domain wall is described by a displacement field $u(x)$. Another example is the contact line of water (or liquid hydrogen), wetting a rough substrate. A realization with a two-parameter field $\boldsymbol{u}(\boldsymbol{x})$ is the deformation of a vortex lattice: the position of each vortex is deformed from $\boldsymbol{x}$ to $\boldsymbol{x} + \boldsymbol{u}(\boldsymbol{x})$. A three-dimensional example are charge density waves.

All these models are described by a displacement field

$$x \in \mathbb{R}^d \longrightarrow \boldsymbol{u}(x) \in \mathbb{R}^N . \tag{1}$$

For simplicity, we now set $N = 1$. After some initial coarse-graining, the energy $\mathcal{H} = \mathcal{H}_{\mathrm{el}} + \mathcal{H}_{\mathrm{DO}}$ consists of two parts: the elastic energy, and the disorder:

$$\mathcal{H}_{\mathrm{el}}[u] = \int \mathrm{d}^d x \, \frac{1}{2} \left(\nabla u(x)\right)^2 , \qquad \mathcal{H}_{\mathrm{DO}}[u] = \int \mathrm{d}^d x \, V(x, u(x)) . \tag{2}$$

In order to proceed, we need to specify the correlations of disorder:[3]

$$\overline{V(x,u)V(x',u')} := \delta^d(x - x') R(u - u') . \tag{3}$$

Fluctuations u in the transversal direction will scale as

$$\overline{[u(x)-u(y)]^2} \sim |x-y|^{2\zeta} . \tag{4}$$

There are several useful observables. We already introduced the roughness-exponent ζ. The second is the renormalized (effective) disorder function $R(u)$, and it is this object we want to measure here. Introducing replicas and averaging over disorder, we can write down the bare action or *replica-Hamiltonian*

$$\mathcal{H}[u] = \frac{1}{T}\sum_{a=1}^{n}\int \mathrm{d}^d x\,\frac{1}{2}\left(\nabla u_a(x)\right)^2 - \frac{1}{2T^2}\sum_{a,b=1}^{n}\int \mathrm{d}^d x\, R(u_a(x)-u_b(x)) . \tag{5}$$

Let us stress that one could alternatively pursue a dynamic or a supersymmetric formulation. Since our treatment is perturbative in $R(u)$, the result is unchanged.

3. Dimensional Reduction

There is a beautiful and rather mind-boggling theorem relating disordered systems to pure systems (i.e. without disorder), which applies to a large class of systems, e.g. random-field systems and elastic manifolds in disorder. It is called dimensional reduction and reads as follows:[4]

Theorem: *A d-dimensional disordered system at zero temperature is equivalent to all orders in perturbation theory to a pure system in d−2 dimensions at finite temperature.*

Experimentally, one finds that this result is wrong, the question being why? Let us stress that there are no missing diagrams or any such thing, but that the problem is more fundamental: As we will see later, the proof makes assumptions, which are not satisfied. Before we try to understand why this is so and how to overcome it, let us give one more example. We know that the width u of a d-dimensional manifold at finite temperature in the absence of disorder scales as $u \sim x^{(2-d)/2}$. Making the dimensional shift implied by dimensional reduction leads to

$$\overline{[u(x)-u(0)]^2} \sim x^{4-d} \equiv x^{2\zeta} \quad \text{i.e.} \quad \zeta = \frac{4-d}{2} . \tag{6}$$

4. The Larkin Length

To understand the failure of dimensional reduction, let us turn to an interesting argument given by Larkin.[5] He considers a piece of an elastic manifold of size L. If the disorder has correlation length r, and characteristic potential energy $\bar{f}$, this piece will typically see a potential energy of strength $E_{\mathrm{DO}} = \bar{f}\left(\frac{L}{r}\right)^{\frac{d}{2}}$. On the other hand, there is an elastic energy, which scales like $E_{\mathrm{el}} = c\,L^{d-2}$. These energies are balanced at the *Larkin length* $L = L_c$ with $L_c = \left(\frac{c^2}{\bar{f}^2}r^d\right)^{\frac{1}{4-d}}$. More important than this value is the observation that in all physically interesting dimensions $d < 4$, and at scales $L > L_c$, the membrane is pinned by disorder; whereas on small scales the elastic energy dominates. Since the disorder has a lot of minima which are far apart in configurational space but close in energy (metastability), the manifold can be in either of these minimas, and the ground state is no longer unique. However exactly this is assumed in the proof of dimensional reduction.

5. The Functional Renormalization Group (FRG)

Let us now discuss a way out of the dilemma: Larkin's argument suggests that $d = 4$ is the upper critical dimension. So we would like to make an $\epsilon = 4 - d$ expansion. On the other hand, dimensional reduction tells us that the roughness is $\zeta = \frac{4-d}{2}$ (see Eq. (6)). Even though this is systematically wrong below four dimensions, it tells us correctly that at the critical dimension $d = 4$, where disorder is marginally relevant, the field u is dimensionless. This means that having identified any relevant or marginal perturbation (as the disorder), we can find another such perturbation by adding more powers of the field. We can thus not restrict ourselves to keeping solely the first moments of the disorder, but have to keep the whole disorder-distribution function $R(u)$. Thus we need a *functional renormalization group* treatment (FRG). Functional renormalization is an old idea, and can e.g. be found in Ref. 6. For disordered systems, it was first proposed in 1986 by D. Fisher.[7] Performing an infinitesimal renormalization, i.e. integrating over a momentum shell à la Wilson, leads to the flow $\partial_\ell R(u)$, with $(\epsilon = 4 - d)$

$$\partial_\ell R(u) = (\epsilon - 4\zeta)\,R(u) + \zeta u R'(u) + \frac{1}{2}R''(u)^2 - R''(u)R''(0)\,. \qquad (7)$$

The first two terms come from the rescaling of R and u, respectively. The last two terms are the result of the 1-loop calculations, see e.g. Refs. 1,7.

More important than the form of this equation is it actual solution, sketched in Fig. 2. After some finite renormalization, the second derivative of the disorder $R''(u)$ acquires a cusp at $u = 0$; the length at which this happens is the Larkin length. How does this overcome dimensional reduction? To understand this, it is interesting to study the flow of the second

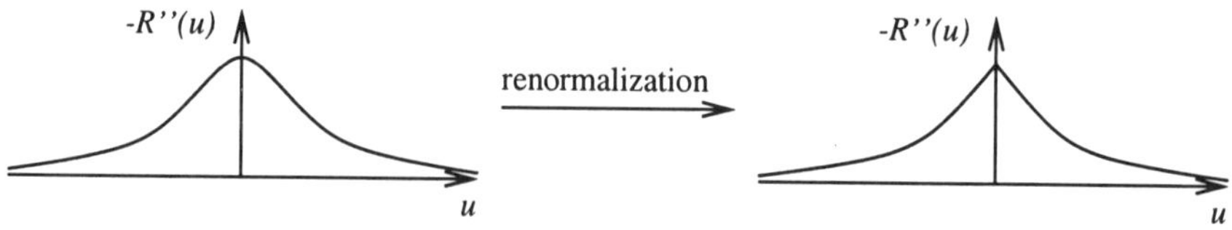

Fig. 2. Change of $-R''(u)$ under renormalization and formation of the cusp.

and forth moment. Taking derivatives of (7) w.r.t. u and setting u to 0, we obtain

$$\partial_\ell R''(0) = (\epsilon - 2\zeta)\, R''(0) + R'''(0)^2 \;\longrightarrow\; (\epsilon - 2\zeta)\, R''(0) \tag{8}$$
$$\partial_\ell R''''(0) = \epsilon R''''(0) + 3R''''(0)^2 + 4R'''(0)R'''''(0) \longrightarrow \epsilon R''''(0) + 3R''''(0)^2 .$$

Since $R(u)$ is an even function, and moreover the microscopic disorder is smooth (after some initial averaging, if necessary), $R'''(0)$ and $R'''''(0)$ are 0, which we have already indicated. The above equations for $R''(0)$ and $R''''(0)$ are in fact closed. The former tells us first that the flow of $R''(0)$ is trivial and that $\zeta = \epsilon/2 \equiv \frac{4-d}{2}$. This is exactly the result predicted by dimensional reduction. The appearance of the cusp can be inferred from the second one. Its solution is $R''''(0)\big|_\ell = \frac{c\,e^{\epsilon\ell}}{1-3\,c(e^{\epsilon\ell}-1)/\epsilon}$, with $c = R''''(0)\big|_{\ell=0}$. Thus after a finite renormalization $R''''(0)$ becomes infinite: The cusp appears. By analyzing the solution of the flow equation (7), one also finds that beyond the Larkin length $R''(0)$ is no longer given by (8) with $R'''(0)^2 = 0$. The correct interpretation of (8), which remains valid after the cusp-formation, is $\partial_\ell R''(0) = (\epsilon - 2\zeta)\, R''(0) + R'''(0^+)^2$. Renormalization of the whole function thus overcomes dimensional reduction. The appearance of the cusp also explains why dimensional reduction breaks down: The simplest way to see this is by redoing the proof for elastic manifolds in disorder, which in the absence of disorder is a simple Gaussian theory. Terms contributing to the 2-point function involve $R''(0)$, $TR''''(0)$ and higher derivatives of $R(u)$ at $u = 0$, which all come with higher powers of T. To obtain the limit of $T \to 0$, one sets $T = 0$, and only $R''(0)$ remains. This is the dimensional-reduction result. However we just saw that $R''''(0)$ becomes infinite. Not surprisingly $R''''(0)T$ may also contribute; indeed one can show that it does, hence the proof fails.

6. The Cusp and Shocks

Let us give a simple argument of why a cusp is a physical necessity, and not an artifact. The argument is quite old and appeared probably first in

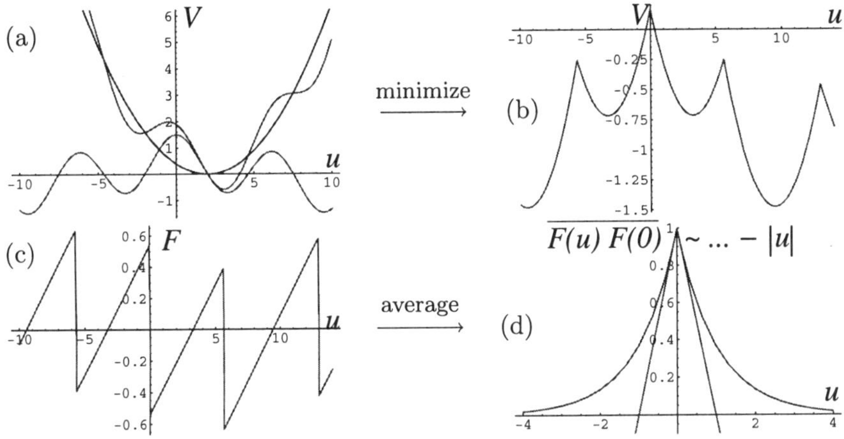

Fig. 3. Generation of the cusp, as explained in the main text.

the treatment of correlation functions by shocks in Burgers turbulence. It was nicely illustrated in Ref. 8. Suppose, we want to integrate out a single degree of freedom coupled with a spring. This harmonic potential and the disorder term are represented by the parabola and the lowest curve in Fig. 3(a), respectively; their sum is the remaining curve. For a given disorder realization, the minimum of the potential as a function of u is reported in Fig. 3(b). Note that it has non-analytic points, which mark the transition from one minimum to another. Taking the derivative of the potential leads to the force in Fig. 3(c). It is characterized by almost linear pieces, and shocks (i.e. jumps). Calculating the force-force correlator, the dominant contribution for small distances is due to shocks. Their contribution is proportional to their probability, i.e., to the distance between the two observable points. This leads to $\overline{F(u)F(0)} = \overline{F(0)^2} - c|u|$, with some numerical coefficient c.

7. The Field-Theoretic Version

The above toy model can be generalized to the field theory.[9] Consider an interface in a random potential, and add a quadratic potential well, centered around w:

$$\mathcal{H}^{w}_{\text{tot}}[u] = \int \mathrm{d}^d x \, \frac{m^2}{2}(u(x) - w)^2 + \mathcal{H}_{\text{el}}[u] + \mathcal{H}_{\text{DO}}[u] \, . \tag{9}$$

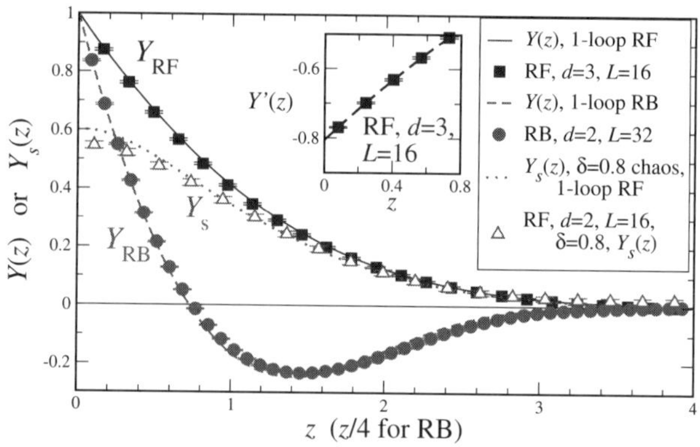

Fig. 4. Filled symbols show numerical results for $Y(z)$, a normalized form of the interface displacement correlator $-R''(u)$ [Eq. (12)], for $D = 2+1$ random field (RF) and $D = 3+1$ random bond (RB) disorders. These suggest a linear cusp. The inset plots the numerical derivative $Y'(z)$, with intercept $Y'(0) \approx -0.807$ from a quadratic fit (dashed line). Open symbols plot the cross-correlator ratio $Y_s(z) = \Delta_{12}(z)/\Delta_{11}(0)$ between two related copies of RF disorder. It does not exhibit a cusp. The points are for confining wells with width given by $M^2 = 0.02$. Comparisons to 1-loop FRG predictions (curves) are made with no adjustable parameters. Reprinted from Ref. 11.

In each sample (i.e. disorder configuration), and given w, one finds the minimum energy configuration. This ground state energy is

$$\hat{V}(w) := \min_{u(x)} \mathcal{H}_{\text{tot}}^{w}[u] \ . \tag{10}$$

It varies with w as well as from sample to sample. Its second cumulant

$$\overline{\hat{V}(w)\hat{V}(w')}^{c} = L^d R(w-w') \tag{11}$$

defines a function $R(w)$ which is proven[9] to be the same function computed in the field theory, defined from the zero-momentum effective action.[10] Physically, the role of the well is to forbid the interface to wander off to infinity. The limit of small m is taken to reach the universal limit. The factor of volume L^d is necessary, since the width $\overline{u^2}$ of the interface in the well cannot grow much more than $m^{-\zeta}$. This means that the interface is made of roughly L/L_m pieces of internal size $L_m \approx m$ pinned independently: (11) expresses the central-limit theorem and $R(w)$ measures the second cumulant of the disorder seen by each piece.

The nice thing about (11) is that it can be measured. One varies w and computes (numerically) the new ground-state energy; finally averaging over many realizations. This has been performed recently in Ref. 11 using a powerful exact-minimization algorithm, which finds the ground state in a time

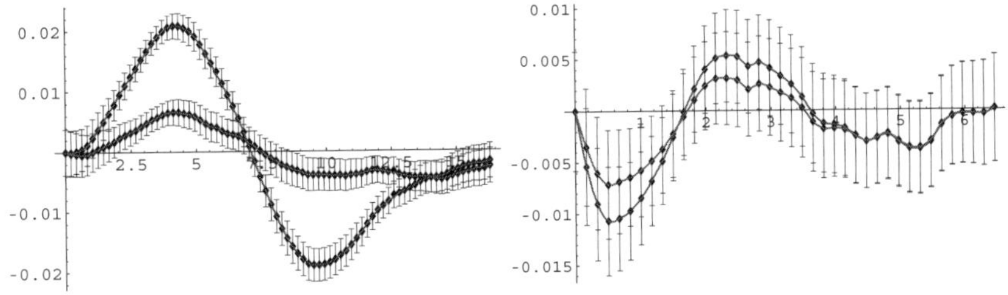

Fig. 5. The measured $Y(u)$ with the 1- and 2-loop corrections subtracted. Left: RB-disorder, right: RF-disorder. One sees that the 2-loop corrections improve the precision.

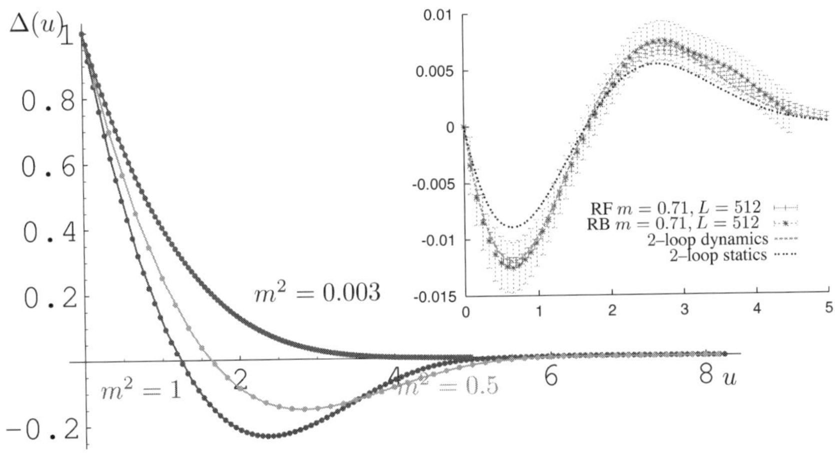

Fig. 6. Running the RG in a numerical simulation: Crossover from RB disorder to RF for a driven particle (left).[11] Residual errors for $Y(u)$ for a driven string[12] which show that statics and depinning are controlled by different fixed points.

polynomial in the system size. In fact, what was measured there are the fluctuations of the center of mass of the interface $u(w) = L^{-d} \int \mathrm{d}^d x\, u_0(x; w)$:

$$\overline{[w - u(w)][w' - u(w')]}^{c} = m^{-4} L^{-d} \Delta(w - w') \tag{12}$$

which measures directly the correlator of the pinning force $\Delta(u) = -R''(u)$. To see why it is the total force, write the equilibrium condition for the center of mass $m^2[w - u(w)] + L^{-d} \int \mathrm{d}^d x\, F(x, u) = 0$ (the elastic term vanishes if we use periodic b.c.). The result is represented in Fig. 4. It is most convenient to plot the function $Y = \Delta(u)/\Delta(0)$ and normalize the u-axis to eliminate all non-universal scales. The plot in Fig. 4 is free of any parameter. It has several remarkable features. First, it clearly shows that a linear cusp exists in any dimension. Next it is very close to the 1-loop prediction. Even more remarkably the statistics is good enough[11] to reliably compare the deviations to the 2-loop predictions of Ref. 13.

When we vary the position w of the center of the well, it is not a real motion. It means to find the new ground state for each w. Literally "moving" w is another interesting possibility: It measures the universal properties of the so-called "depinning transition".[12,14] This was recently implemented numerically (see Fig. 6).

Acknowledgments

It is a pleasure to thank Wolfhard Janke and Axel Pelster, the organizers of PI-2007, for the opportunity to give this lecture. We thank Alan Middleton and Alberto Rosso for fruitful collaborations which allowed to measure the FRG effective action in numerics and Andrei Fedorenko and Werner Krauth for stimulating discussions. This work has been supported by ANR (05-BLAN-0099-01).

References

1. K. Wiese and P. Le Doussal, arXiv:cond-mat/0611346.
2. S. Lemerle, J. Ferré, C. Chappert, V. Mathet, T. Giamarchi, and P. Le Doussal, *Phys. Rev. Lett.* **80**, 849 (1998).
3. This is explained in detail in Ref. 1.
4. K. Efetov and A. Larkin, *Sov. Phys. JETP* **45**, 1236 (1977).
5. A. Larkin, *Sov. Phys. JETP* **31**, 784 (1970).
6. F. Wegner and A. Houghton, *Phys. Rev. A* **8**, 401 (1973).
7. D. Fisher, *Phys. Rev. Lett.* **56**, 1964 (1986).
8. L. Balents, J. Bouchaud, and M. Mézard, *J. Phys. I (France)* **6**, 1007 (1996).
9. P. Le Doussal, *Europhys. Lett.* **76**, 457 (2006).
10. P. Le Doussal, K. Wiese, and P. Chauve, *Phys. Rev. E* **69**, 026112 (2004).
11. A. Middleton, P. Le Doussal, and K. Wiese, *Phys. Rev. Lett.* **98**, 155701 (2007).
12. A. Rosso, P. Le Doussal, and K. Wiese, *Phys. Rev. B* **75**, 220201 (2007).
13. P. Chauve, P. Le Doussal, and K. Wiese, *Phys. Rev. Lett.* **86**, 1785 (2001).
14. P. Le Doussal and K. Wiese, *Europhys. Lett.* **77**, 66001 (2007).

A FUNCTIONAL RENORMALIZATION GROUP APPROACH TO SYSTEMS WITH LONG-RANGE CORRELATED DISORDER

A. A. FEDORENKO

CNRS-Laboratoire de Physique Théorique de l'Ecole Normale Supérieure,
24 rue Lhomond, 75231 Paris, France
E-mail: andrei.fedorenko@lpt.ens.fr

We studied the statics and dynamics of elastic manifolds in disordered media with long-range correlated disorder using functional renormalization group (FRG). We identified different universality classes and computed the critical exponents and universal amplitudes describing geometric and velocity-force characteristics. In contrast to uncorrelated disorder, the statistical tilt symmetry is broken resulting in a nontrivial response to a transverse tilting force. For instance, the vortex lattice in disordered superconductors shows a new glass phase whose properties interpolate between those of the Bragg and Bose glasses formed by pointlike and columnar disorder, respectively. Whereas there is no response in the Bose glass phase (transverse Meissner effect), the standard linear response expected in the Bragg glass gets modified to a power law response in the presence of disorder correlations. We also studied the long distance properties of the $O(N)$ spin system with random fields and random anisotropies correlated as $1/x^{d-\sigma}$. Using FRG we obtained the phase diagram in (d, σ, N)-parameter space and computed the corresponding critical exponents. We found that below the lower critical dimension $4+\sigma$, there can exist two different types of quasi-long-range-order with zero order-parameter but infinite correlation length.

Keywords: Functional renormalization group; Depinning; Random field.

1. Elastic Manifolds in Disordered Media

Elastic objects in disordered media is a fruitful concept to study diverse physical systems such as domain walls (DW) in ferromagnets, charge density waves (CDW) in solids, vortices in type II superconductors.[1] In all these systems the interplay between elastic forces which tend to keep the system ordered, i.e., flat or periodic, and quenched disorder, which promotes deformations of the local structure, forms a complicated energy landscape with numerous metastable states. Most studies of elastic objects in disordered media restricted to uncorrelated pointlike disorder. Real systems, however, often contain extended defects in the form of linear dislocations, planar grain boundaries, three-dimensional cavities, etc. We studied the statics and dynamics of elastic manifolds in the presence of (power-law) long-range (LR) correlated disorder.[2] The power-law correlation of defects in d-dimensional space with exponent $a = d - \varepsilon_d$ can be ascribed to randomly distributed extended defects of internal dimension ε_d with random orientation.

The configuration of elastic object can be parametrized by displacement field u_x, where x denotes the d dimensional internal coordinate of the elastic object. The Hamiltonian is given by

$$\mathcal{H}[u] = \int d^d x \left[\frac{c}{2} (\nabla u_x)^2 + V(x, u_x) \right] . \tag{1}$$

Here c is the elasticity, $V(x,u)$ a random potential with zero mean and variance

$$\overline{V(x,u)V(x',u')} = R_1(u-u')\delta^d(x-x') + R_2(u-u')g(x-x'), \tag{2}$$

where $g(x) \sim x^{-a}$. For interfaces one has to distinguish two universality classes: random bond (RB) disorder described by a short-range function $R(u)$ and random field (RF) disorder corresponding to a function which behaves as $R(u) \sim |u|$ at large u. Random periodic (RP) universality class corresponding to a periodic function $R(u)$ describes systems such as CDW or vortices in $d = 1+1$ dimensions. Disorder makes the interfaces rough with displacements growing with the distance x as $C(x) \sim x^{2\zeta}$, where ζ is the roughness exponent. Elastic periodic structures loose their strict translational order and may exhibit a slow logarithmic growth of displacements, $C(x) = \mathcal{A}_d \ln |x|$. The driven dynamics at zero temperature is described by the over-damped equation of motion

$$\eta \partial_t u_{xt} = c \nabla^2 u_{xt} + F(x, u_{xt}) + f. \tag{3}$$

Here η is the friction coefficient, f the applied force, and $F = -\partial_u V(x,u)$ the pinning force with zero mean and correlator

$$\overline{F(x,u)F(x',u')} = \Delta_1(u-u')\delta^d(x-x') + \Delta_2(u-u')g(x-x'), \tag{4}$$

where $\Delta_i(u) = -R_i''(u)$. The system undergoes the so-called depinning transition at the critical force f_c, which separates sliding and pinned states. Upon approaching the depinning transition from the sliding state $f \to f_c^+$ the center-of-mass velocity $v = L^{-d} \int_x \partial_t u_{xt}$ vanishes as a power law $v \sim (f - f_c)^\beta$ and the time and length scales are related by $t \sim x^z$.

Let us firstly discuss the case of short range (SR) correlated disorder (Δ_2=0). The problem turns out to be notably difficult due to the so-called dimensional reduction (DR). However, simple Imry-Ma arguments show that DR is wrong. The metastability renders the zero temperature perturbation theory useless breaking it down on scales larger than the so-called Larkin length. To overcome these difficulties one can apply functional renormalization group (FRG). Scaling analysis gives that the large-scale properties of a d-dimensional elastic system are governed by uncorrelated disorder in $d < d_{\rm uc} = 4$. The peculiarity of the problem is that there is an infinite set of relevant operators which can be parametrized by a function. The latter is nothing but the disorder correlator. We generalized the FRG approach to systems with LR correlated disorder. Using a double expansion in $\varepsilon = 4-d$ and $\delta = 4 - a$ we derived the flow equations to one-loop order:

$$\partial_\ell \Delta_1(u) = (\varepsilon - 2\zeta)\Delta_1(u) + \zeta u \Delta_1'(u) - \frac{1}{2}\frac{d^2}{du^2}[\Delta_1(u) + \Delta_2(u)]^2 + A\Delta_1''(u),$$
$$\partial_\ell \Delta_2(u) = (\delta - 2\zeta)\Delta_2(u) + \zeta u \Delta_2'(u) + A\Delta_2''(u), \qquad (5)$$

where $A = [\Delta_1(0) + \Delta_2(0)]$. The scaling behavior of the system is controlled by a stable fixed point $[\Delta_1^*(u), \Delta_2^*(u)]$ of flow equations (5). Note that for uncorrelated disorder the elasticity remains uncorrected to all orders due to the statistical tilt symmetry (STS). LR disorder destroys the STS and allows for the renormalization of elasticity introducing a new exponent ψ. The SR part of disorder correlator $\Delta_1(u)$ becomes a nonanalytic function beyond the Larkin scale, while the LR part $\Delta_2(u)$ remains analytic along the flow. The appearance of non-analyticity in the form of a cusp at the origin is related to metastability, and nicely accounts for the generation of threshold force f_c at the depinning transition. The actual values of critical exponents are fixed by the disorder correlator at the stable FP, for instance, $\psi = \frac{1}{4}(\delta - \varepsilon)\Delta_2^{*\prime\prime}(0)$ and $z = 2 - \Delta_1^{*\prime\prime}(0) - \Delta_2^{*\prime\prime}(0)$. Let us summarize the results obtained for different universality classes.

RB disorder. The crossover from the SR FP ($\Delta_2 = 0$) to the LR FP ($\Delta_2 \neq 0$) takes place for $\delta > 1.041\varepsilon$. At the LR RB FP there is an exact relation between exponents: $\zeta_{\rm LR} = (\delta + 2\psi)/5$, where $\psi = \mathcal{O}(\varepsilon^2, \delta^2, \varepsilon\delta)$ to one-loop order. We also computed the amplitude of height-height correlation function which turns out to be universal up to the strength of disorder.

RF disorder. The large-scale properties of the system is controlled by the LR FP for $\delta > \varepsilon$. The roughness exponent is given by the exact relation $\zeta_{\rm LR} = (\delta + 2\psi)/3$. Surprisingly, the function $\Delta_1^*(u)$ satisfies $\int_0^\infty du\ \Delta_1^*(u) =$

0, characteristic for RB type correlations along the u direction. In other words, the LR RF FP is in fact of mixed type: RB for the SR part and RF for the LR part of the disorder correlator. The RF FP describes the depinning transition of interfaces in the presence of LR correlated disorder. The corresponding dynamic exponents read: $z = 2 - \varepsilon/3 + \delta/9$ and $\beta = 1 - \varepsilon/6 + \delta/18$. It is remarkable that for $\delta > 3\varepsilon$ the exponent β is larger than 1, and z larger than 2.

Random periodic. For $\delta > \pi^2\varepsilon/9$ the periodic system is controlled by the LR RP FP. The system exhibits a slow growth of the displacements: $\overline{(u_x - u_0)^2} = \frac{\delta}{2\pi^2} \ln|x|$. As an example, we consider vortices in the presence of three types of disorder: uncorrelated, LR, and columnar. The correlation of disorder violates the STS resulting in a highly nonlinear response to tilt. In the presence of columnar disorder vortices exhibit a transverse Meissner effect: disorder generates the critical field h_c below which there is no response to tilt and above which the tilt angle behaves as $\vartheta \sim (h - h_c)^{\phi}$ with the universal exponent $\phi < 1$. The RP case describes a weak Bose glass which is expected in type II superconductors with columnar disorder at small temperatures and at high vortex density which exceeds the density of columnar pins. The weak Bose glass is pinned collectively and shares features of the Bragg glass, such as a power-law decay of the translation order, and features of the strong Bose glass such as a transverse Meissner effect. For isotropically LR correlated disorder the linear tilt modulus vanishes at small fields leading to a power-law response $\vartheta \sim h^{\phi}$ with $\phi > 1$. The response of the system with LR correlated disorder interpolates between responses of systems with uncorrelated and columnar disorder. We argued that in the presence of LR correlated disorder vortices can form a strong Bragg glass which exhibits Bragg peaks and vanishing linear tilt modulus without transverse Meissner effect.[3]

2. LR Random Field and Random Anisotropy $O(N)$ Models

The large-scale behavior of the $O(N)$ symmetric spin system at low temperatures can be described by the nonlinear σ model with the Hamiltonian

$$\mathcal{H}[\boldsymbol{s}] = \int d^d x \left[\frac{1}{2}(\nabla \boldsymbol{s})^2 + V(x, \boldsymbol{s})\right], \tag{6}$$

where $\boldsymbol{s}(x)$ is the N-component classical spin with a fixed-length constraint $\boldsymbol{s}^2 = 1$. $V(x, \boldsymbol{s})$ is the random disorder potential, which can be expanded in spin variables: $V(x,\boldsymbol{s}) = -\sum_{\mu=1}^{\infty}\sum_{i_1\ldots i_\mu} h^{(\mu)}_{i_1\ldots i_\mu}(x)s_{i_1}(x)\ldots s_{i_\mu}(x)$. The corresponding coefficients have simple physical interpretation: $h_i^{(1)}$ is a random field, $h_{ij}^{(2)}$ is a random second-rank anisotropy, and $h^{(\mu)}$ are general μth tensor anisotropies. We studied system (6) with LR correlated disorder given by cumulants $\overline{h^{(\mu)}_{i_1\ldots i_\mu}(x)h^{(\nu)}_{i_1\ldots j_\nu}(x')} = \delta^{\mu\nu}\delta_{i_1 j_1}\ldots\delta_{i_\mu j_\nu}[r_1^{(\mu)}\delta(x-x') + r_2^{(\mu)}g(x-x')]$ with $g(x)\sim |x|^{\sigma-d}$. The corresponding replicated Hamiltonian reads[4]

$$\mathcal{H}_n = \int d^dx \Bigg[\frac{1}{2}\sum_a(\nabla \boldsymbol{s}_a)^2 - \frac{1}{2T}\sum_{a,b}R_1\big(\boldsymbol{s}_a(x)\cdot\boldsymbol{s}_b(x)\big) - \frac{1}{2T}\sum_{a,b}\int d^dx'\, g(x-x')\, R_2\big(\boldsymbol{s}_a(x)\cdot\boldsymbol{s}_b(x')\big)\Bigg], \quad (7)$$

where $R_i(z) = \sum_\mu r_i^{(\mu)} z^\mu$ are arbitrary for RF and even for RA. Power counting suggests that $d_{\mathrm{lc}} = 4+\sigma$ is the lower critical dimension for both models. At criticality or in the quasi-long-range-ordered (QLRO) phase the connected and disconnected two-point function scale with different exponents η and $\bar{\eta}$. The FRG equations to first order in $\varepsilon = 4-d$ and σ are given by

$$\partial_\ell R_1(\phi) = -\varepsilon R_1(\phi) + \frac{1}{2}\big[R_1''(\phi)+R_2''(\phi)\big]^2 - (N-2)\Big\{2AR_1(\phi) - AR_1''(\phi) + AR_1'(\phi)\cot\phi - \frac{1}{2\sin^2\phi}\big[R_1'(\phi)+R_2'(\phi)\big]^2\Big\}, \quad (8)$$

$$\partial_\ell R_2(\phi) = -(\varepsilon-\sigma)R_2(\phi) - \Big\{(N-2)\big[2R_2(\phi)+R_2'(\phi)\cot\phi\big] + R_2''(\phi)\Big\}A.$$

Here $A = R_1''(0)+R_2''(0)$ and $z=\cos\phi$. An attractive FP of the flow equations describes a QLRO phase, while a singly unstable FP describes the critical behavior. The critical exponents are determined by the FRG flow in the vicinity of the FP and to one-loop order are given by $\eta = -A^*$ and $\bar{\eta} = -\varepsilon-(N-1)A^*$. A singly unstable FP has only one positive eigenvalue λ_1 which determines the third independent exponent $\nu = 1/\lambda_1$ characterizing the divergence of the correlation length at the transition. For the RF model the critical exponents are given by

$$\eta_{\mathrm{LR}} = \frac{\varepsilon-\sigma}{N-3}, \qquad \bar{\eta}_{\mathrm{LR}} = \frac{2\varepsilon-(N-1)\sigma}{N-3}. \quad (9)$$

The stability regions of different FPs and the phase diagrams for the RF and RA models are shown in Figure 1. For the LR RF model we found that the truncated RG developed in work 5 for the RF model gives the correct

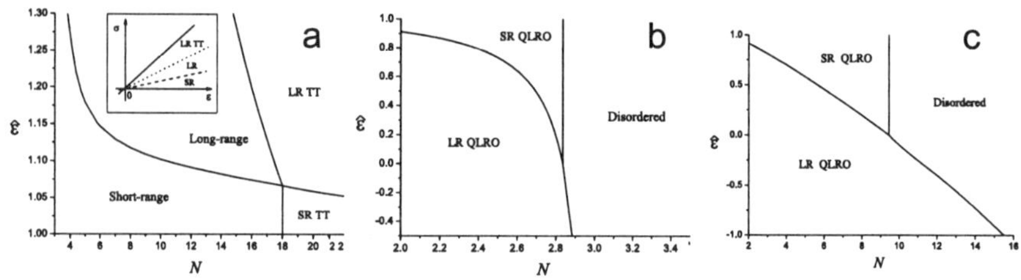

Fig. 1. (a) Stability regions of various FPs for RF $O(N)$ model corresponding to different patterns of the critical behavior above the lower critical dimension, $\varepsilon > \sigma$. Insert: Schematic phase diagram on the (ε, σ) plane for a particular value of $N \in (3, 18)$. (b) Phase diagram of the RF model below the lower critical dimension, $d < 4 + \sigma$. (c) Phase diagram of the RA model below the lower critical dimension, $d < 4 + \sigma$.

one-loop values of exponents η and $\bar{\eta}$ above the $d_{\rm lc}$, but not the phase diagram and the critical exponent ν except for the region controlled by the weakly nonanalytic LR TT FP. Thus, although the truncated RG overcomes the dimensional reduction it fails to reproduce all properties which can be obtained using FRG. We found a new LR QLRO phase existing in the LR RF model below the $d_{\rm lc}$ for $N < 3$ and determined the regions of its stability in the (ε, σ, N) parameter space. We obtained that the weak LR correlated disorder does not change the critical behavior of the RA model above $d_{\rm lc}$ for $N > N_c = 9.4412$, but can create a new LR QLRO phase below $d_{\rm lc}$. The existence of two QLRO phases in the LR RA system may be relevant to two different states of ^{3}He-A in aerogel observed recently in NMR experiments.[6]

Acknowledgments

I would like to thank Kay Wiese, Pierre Le Doussal and Florian Kühnel for fruitful collaboration and useful discussions. Support from the European Commission through a Marie Curie Postdoctoral Fellowship under contract number MIF1-CT-2005-021897 is acknowledged.

References

1. S. Brazovskii and T. Nattermann, *Adv. Phys.* **53**, 177 (2004).
2. A. A. Fedorenko, P. Le Doussal, and K. J. Wiese, *Phys. Rev. E* **74**, 061109 (2006).
3. A. A. Fedorenko, in preparation.
4. A. A. Fedorenko and F. Kühnel, *Phys. Rev. B* **75**, 174206 (2007).
5. M. C. Chang and E. Abrahams, *Phys. Rev. B* **29**, 201 (1984).
6. G. E. Volovik, *JETP Lett.* **84**, 455 (2006).

PART IX

Spin Models

THE CRITICAL BEHAVIOR OF THE RANDOM ISING FERROMAGNETS

B. N. SHALAEV

A. F. Ioffe Physical and Technical Institute, RAS,
Polytechnicheskaya str., 26, St. Petersburg, Russia
E-mail: shalaev@mail.ioffe.ru

The critical behavior of the d-dimensional uniaxial ferromagnetic systems with weak quenched disorder is studied. In the critical region this model is known to be described by both the conventional Ginzburg-Landau Hamiltonian and the two-dimensional fermionic Gross-Neveu model in the $n = 0$ limit. Renormalization group calculations are used to obtain the temperature dependences near the critical point of some thermodynamic quantities, the large distance behavior of the two-spin correlation function and the equation of state at criticality. The ϵ expansion of the critical exponents is also discussed as well as the Kramers-Wannier duality of two-dimensional dilute systems. Most important questions of the theory of critical phenomena are the problem of universality of the critical behavior of random systems and the role of Griffiths singularities.

Keywords: Renormalization group; Ising model; Kramers-Wannier duality; Universality.

1. Introduction

The critical properties of random spin systems have been extensively studied since more than 30 years (see the review textbook and articles in Refs. 1–5). Perfect crystals are the exception rather than the rule, quenched disorder always existing in different degrees. Even weak disorder may drastically effect the critical behavior. Despite many efforts, this problem is a complex and poorly understood phenomenon.

The conventional field-theoretic renormalization group (RG) being a highly-developed approach based on the standard ϕ^4 theory in $(4 - \epsilon)$-dimensions[1] firstly was applied to study properties of disordered systems in the papers of Refs. 6–8.

In the beginning of eighties of the last century, the Dotsenko brothers[2] initiated a considerable progress in the study of the two-dimensional (2D) random bond Ising model (IM) by exploiting the remarkable equivalence between this system and the $n = 0$ Gross-Neveu model. For weak dilution the new temperature dependence of the specific heat was found to be as $C \sim \ln \ln \tau$, with $\tau = \frac{T-T_c}{T_c}$ being a reduced deviation from the critical temperature. However, their results concerning the two-spin correlation function at the Curie point were later questioned in Refs. 9–11.

2. Spin Correlation Function

We begin with the classical Hamiltonian of the 2D IM with random bonds defined on a square lattice with periodic boundary conditions:

$$H = -\sum_{i=1,j=1}^{N} [J_1(i,j) s_{ij} s_{ij+1} + J_2(i,j) s_{ij} s_{i+1j}], \tag{1}$$

where i, j label sites of the square lattice, $s_{ij} = \pm 1$ are spin variables, $J_1(i,j)$ and $J_2(i,j)$ are horizontal and vertical random independent couplings having the same probability distribution, which reads:

$$P(x) = (1-p)\delta(x-J) + p\delta(x-J'), \tag{2}$$

where p is the concentration of impurity bonds and both J and J' are assumed to be positive so that the Hamiltonian favors aligned spins. Both antiferromagnetic couplings (creating frustration) and broken bonds ($J' = 0$) leading to ambiguities in the transfer matrix are excluded in this treatment. The partition function of the IM reads:

$$Z = \sum \exp\left(-\frac{H}{T}\right) \tag{3}$$

where H is defined in Eq. (1) and the sum runs over all 2^{N^2} possible spin configurations. The partition function is known to be represented as the trace of the product of the row-to-row transfer matrices $\hat{T}_i$.[12,13] It is very important that the transfer-matrix of the Ising model with random bonds possesses self-duality if $p = 0.5$. Using the replica-trick one obtains the well-known identity for the averaged free energy:

$$\bar{F} = -T\overline{\ln Z} = -T \lim_{n\to 0} \frac{1}{n}\overline{(Z^n - 1)}. \tag{4}$$

It is shown that near the Curie point the model under consideration can be described by the $O(n)$-symmetric Lagrangian of the Gross-Neveu model:[2]

$$L = \int d^2x [i\bar{\psi}_a \hat{\partial} \psi_a + m_0 \bar{\psi}_a \psi_a + u_0 (\bar{\psi}_a \psi_a)^2], \tag{5}$$

where $\gamma_\mu = \sigma_\mu, \hat{\partial} = \gamma_\mu \partial_\mu, \mu = 1, 2, \bar{\psi} = \psi^T \gamma_0$ and

$$m_0 \sim \tau = \frac{T - T_c}{T_c}, \qquad u_o \sim p. \tag{6}$$

Here m_0, u_0 are the bare mass of the fermions and their quartic coupling constant, respectively.

The one-loop RG equations and initial conditions are given by:

$$\begin{aligned} &\frac{du}{dt} = \beta(u) = -\frac{(n-2)u^2}{\pi}, \qquad \frac{d\ln F}{dt} = -\gamma_{\bar{\psi}\psi}(u) = \frac{(1-n)u}{\pi}, \\ &u(t=0) = u_0, \qquad F(t=0) = 1, \end{aligned} \tag{7}$$

where u is the dimensionless quartic coupling constant, $\beta(u)$ is the Gell-Mann-Low function, $\gamma_{\bar{\psi}\psi}(u)$ is the anomalous dimension of the composite operator $\bar{\psi}_a \psi_a$, being actually the energy density operator $\epsilon(x)$, $t = \ln \frac{\Lambda}{m}$, $\Lambda = \frac{1}{a}$ is an ultraviolet cutoff, a and m are the lattice spacing and renormalized mass, respectively, and F is a Green function with zero external momenta.

The solution of the above equations yields the temperature dependence of the correlation length ξ and specific heat C in the asymptotic region $t \to \infty$, $n = 0$:[2]

$$\begin{aligned} &u = \frac{\pi}{2t}, \qquad F \sim t^{-\frac{1}{2}}, \\ &\xi = m^{-1} \sim \tau^{-1} \left[\ln \frac{1}{\tau}\right]^{\frac{1}{2}}, \\ &C \sim \int dt F(t)^2 \sim \ln \ln \frac{1}{\tau}. \end{aligned} \tag{8}$$

The main conclusion of our consideration is that the critical behavior of the 2D random bond IM is governed by the pure Ising fixed point confirmed by a good deal of numerical calculations.[15]

The Ising model exhibits dual relationship in 2D, and in the special case $p = 0.5$ duality imposes strong restrictions, in particular it gives the exact value of the Curie temperature T_c which is believed to be unique![14] (see also Ref. 3):

$$\tanh\left(\frac{J}{T^*}\right) = \exp\left(-\frac{2J}{T}\right). \tag{9}$$

We conjecture that there are only two phases divided by a single critical point given by the self-duality equation likewise Eq. (9).

The square of the spin-spin correlation function of the pure 2D IM in terms of the Dirac fermions can be written as a path integral over the real bosonic field ϕ of the quantum sine-Gordon model:[3,21]

$$G(x-y)^2 = Z^{-1}\frac{1}{2\pi^2 a^2}\int D\phi \sin(\sqrt{4\pi}\phi(x))\sin(\sqrt{4\pi}\phi(y))\exp(-S),$$
$$S = \frac{1}{2}\int d^2x\left[(\partial_\mu\phi)^2 + \frac{2m_0}{\pi a}\cos(\sqrt{4\pi}\phi)\right],$$
$$Z = \int D\phi\exp(-S). \tag{10}$$

The temperature dependences of the homogeneous susceptibility, spontaneous magnetization and the equation of state at criticality are described by power-law functions of the correlation length ξ (Refs. 16,17):

$$\chi \sim \xi^{2-\eta} \sim \tau^{-\frac{7}{4}}\left[\ln\frac{1}{\tau}\right]^{\frac{7}{8}},$$
$$M \sim \xi^{\frac{\eta}{2}} \sim (-\tau)^{\frac{1}{8}}\left[\ln\frac{1}{(-\tau)}\right]^{\frac{1}{16}}, \tag{11}$$

and

$$H \sim M^{\frac{4+\eta}{\eta}} \sim M^{15}. \tag{12}$$

3. Critical Exponents

The powerful machinery of the RG equations[1] yields an asymptotic result for the ϵ-expansion of critical exponents, these being the five-loop ϵ-expansion results for the critical exponents and a marginal spin dimensionality giving us exceptionally bad asymptotic expansions of the critical exponents.[22]

Critical properties of the three-dimensional Ising system with weak quenched disorder were discussed in the good review paper[5] where experimental, numerical and theoretical results were reviewed. Recent theoretical results and MC simulations showed that the IM fixed point of the three-dimensional system is stable,[5,18–20] giving numerical values of critical exponents which are close to the known estimates. The critical exponents obtained by MC simulations are in agreement with reasonable modern estimates: $\nu = 0.682(3), \eta = 0.036(1)$.[5]

4. Griffiths Singularities

All abovementioned results were obtained by using the perturbations theory without taking into account non-perturbative effects, more precisely, Griffiths singularities.[23,24] That problem has originated from the celebrated theorem of T. D. Lee and C. N. Yang.[25]

The central idea of the Griffiths's theory is that above the ferromagnetic transition point in a disordered system there is always a finite probability of finding an arbitrary large ferromagnetic cluster. These rare-fluctuation clusters give singular contribution to the magnetization, which can be proven to be a non-analytic function of the external magnetic field. The corresponding phase being not purely paramagnetic, is called the Griffiths phase. Experimenters have probably observed large deviations from standard scaling relations.[26]

It has been recently considered a large influence of quenched disorder on diluted systems between the completely disordered paramagnets in the high-temperature regime and the magnetically ordered state.[26–28]

5. Conclusions

It is known that random systems behave like the unhappy families in Tolstoy's novel "All happy families look the same, all unhappy ones are unhappy in their own way".[29] The most interesting question concerning random ferromagnets reads: Is the critical behavior of random systems as universal as in perfect ones (as "all happy families") or not?[1]

It was shown that the critical behavior of the random 2D Ising ferromagnets is governed by the IM fixed point as well as the 3D Ising models. In the 2D case, Kramers-Wannier duality intimately connected with symmetries of the IM with $p = 0.5$ saves universality!

In conclusion, we should like to quote B. M. McCoy:[24] "There are still many open questions in the theory of random Ising model and their study will continue to enlarge our intuition of the effects of impurities on phase transitions."

Acknowledgments

The author is most grateful to the Physical Department of the University of Aveiro for support and warm hospitality. He has also benefitted from numerous discussions with his colleaugues working in LETI, PTI and Aveiro university: A. I. Sokolov, S. A. Ktitorov, F. Mendes, G. Jug, A. V. Goltsev, S. N. Dorogovtsev and A. N. Samukhin.

References

1. J. Zinn-Justin, *Quantum Field Theory and Critical Phenomena*, 3rd edition (Oxford, Clarendon Press, 1999).
2. Vl. S. Dotsenko and Vik. S. Dotsenko, *Adv. in Phys.* **32**, 129 (1983).

3. B. N. Shalaev, *Phys. Rep.* **237**, 129 (1994).
4. Vik. S. Dotsenko, *Uspekhi Fiz. Nauk* **165**, 481 (1995).
5. R. Folk, Yu. Holovatch, and T. Yavorskii, *Uspekhi Fiz. Nauk* **173**, 175 (2003).
6. A. B. Harris, *J. Phys. C* **7**, 1671 (1974).
7. A. B. Harris and T. C. Lubensky, *Phys. Rev. Lett.* **33**, 1540 (1974).
8. D. E. Khmelnitskii, *Soviet Phys. JETP* **68**, 1960 (1975).
9. B. N. Shalaev, *Soviet Physics-Solid State* **26**, 1811 (1984).
10. R. Shankar, *Phys. Rev. Lett.* **58**, 2466 (1987).
11. A. W. W. Ludwig, *Phys. Rev. Lett.* **61**, 2388 (1988).
12. R. J. Baxter, *Exactly Solved Models in Statistical Mechanics* (Academic Press, London 1982).
13. J. B. Kogut, *Rev. Mod. Phys.* **51**, 659 (1979).
14. R. Fisch, *J. Stat. Phys.* **18**, 111 (1978).
15. J.-S. Wang, W. Selke, Vl. S. Dotsenko, and V. B. Andreichenko, *Nucl. Phys. B* **344**, 531 (1990); J.-S. Wang, W. Selke, Vl. S. Dotsenko, and V. B. Andreichenko, *Physica* **A 164**, 221 (1990).
16. R. Kenna, D. A. Johnston, and W. Janke, *Phys. Rev. Lett.* **96**, 115701 (2006).
17. R. Kenna, D. A. Johnston, and W. Janke, *Phys. Rev. Lett.* **97**, 155702 (2006); *Erratum*, *ibid.* **97**, 169901 (E) (2006).
18. M. Hasenbusch, F.P. Toldin, A. Pelissetto, and E. Vicari, *The 3D* $\pm J$ *Ising model at the ferromagnetic transition line*, arXiv:0704.0427 (cond-mat.dis-nn).
19. P.-E. Berche, C. Chatelain, B. Berche, and W. Janke, *Eur. Phys. J. B* **38**, 463 (2004).
20. B. N. Shalaev, *Sov. Phys.-Solid State* **31**, 51 (1989).
21. J.-B. Zuber and C. Itzykson, *Phys. Rev. D* **15**, 2875 (1977).
22. B. N. Shalaev, S. A. Antonenko, and A. I. Sokolov, *Phys. Lett. A* **230 N 1-2** 105 (1997).
23. R. B. Griffiths, *Phys. Rev. Lett.* **23**, 17 (1969).
24. B. M. McCoy, *The connection between statistical mechanics and quantum field theory*, arXiv:hep-th/9403084.
25. T. D. Lee and C. N. Yang, *Phys. Rev.* **87**, 410 (1952).
26. J. Deisenhofer, D. Braak, H.-A. Krug von Nidda, J. Hemberger, R. M. Eremina, V. A. Ivanshin, A. M. Balbashov, G. Jug, A. Loidl, T. Kimura, and Y. Tokura, *Phys. Rev. Lett.* **95**, 257202 (2005).
27. V. S. Dotsenko, *J. Stat. Phys.* **122**, 197 (2006) [arXiv:cond-mat/0505233].
28. P. Y. Chan, N. Goldenfeld, and M. Salamon, *Phys. Rev. Lett.* **97**, 137201 (2006).
29. L. N. Tolstoy, Anna Karenina.

VORTEX-LINE PERCOLATION IN A THREE-DIMENSIONAL COMPLEX $|\psi|^4$ THEORY

E. BITTNER and W. JANKE

Institut für Theoretische Physik, Universität Leipzig,
Postfach 100 920, D-04009 Leipzig, Germany
E-mail: bittner@itp.uni-leipzig.de, janke@itp.uni-leipzig.de
www.physik.uni-leipzig.de/CQT.html

The functional measure of the three-dimensional complex $|\psi|^4$ theory allows for line-like topological excitations which can be related to vortex lines in superfluid helium by universality arguments. Upon approaching the λ point, these lines proliferate and destroy the superfluidity. To study the phase transition from this geometrical point of view, we investigated the statistical properties of the emerging vortex-loop network in the vicinity of the critical point by means of high-precision Monte Carlo simulations. For comparison the standard magnetic properties of the system were considered as well. Using sophisticated embedded cluster update techniques we examined if both of them exhibit the same critical behaviour leading to the same critical exponents and therefore to a consistent description of the phase transition. We find that different definitions for constructing the vortex-loop network lead to slightly (but statistically significantly) different results in the thermodynamic limit, and that the percolation thresholds are close to but do not really coincide with the thermodynamic phase transition point.

Keywords: Phase transition; Universality class; Vortex network.

The partition function of the Villain model in three dimensions, a particular spin model with global $O(2)$ symmetry due to the 2π-periodicity in the Hamiltonian, can be equivalently represented as a partition function of a dual theory in which the "spin" configurations are integer-valued and sourceless.[1,2] These configurations can be interpreted as line-like excitations forming closed networks which can be identified with the vortex loops of the original theory. At the transition point, where the broken $O(2)$ symmetry in the low-temperature phase is restored, loops of infinite length become important which provides the basis for attempting a percolational treatment.[3] The question arises whether the percolational threshold coincides with the thermodynamic critical point, or under which conditions such a coincidence can be established.[4] Therefore we connect the obtained vortex line elements to closed lines, i.e. loops, which are geometrically defined objects. When a branching point, where $n \geq 2$ junctions are encountered, is reached, a decision on how to continue has to be made. This step involves a certain ambiguity. We want to investigate the influence of treating such a branching point as a knot.

We start form the standard complex or two-component Ginzburg-Landau theory defined by the Hamiltonian

$$H[\psi] = \int \mathrm{d}^d r \left[\alpha|\psi|^2 + \frac{b}{2}|\psi|^4 + \frac{\gamma}{2}|\nabla\psi|^2 \right], \quad \gamma > 0 , \tag{1}$$

where $\psi(\boldsymbol{r}) = \psi_x(\boldsymbol{r}) + i\psi_y(\boldsymbol{r}) = |\psi(\boldsymbol{r})|e^{i\phi(\boldsymbol{r})}$ is a complex field, and α, b and γ are temperature independent coefficients derived from a microscopic model. In order to carry out Monte Carlo simulations we put the model (1) on a d-dimensional hypercubic lattice with spacing a. Using the notation of Ref. 5, we introduce scaled variables $\tilde{\psi} = \psi/\sqrt{(|\alpha|/b)}$ and $\boldsymbol{u} = \boldsymbol{r}/\xi$, where $\xi^2 = \gamma/|\alpha|$ is the mean-field correlation length at zero temperature. This leads to the normalized lattice Hamiltonian

$$H[\tilde{\psi}] = k_B \tilde{V}_0 \sum_{n=1}^{N} \left[\frac{\tilde{\sigma}}{2}(|\tilde{\psi}_n|^2 - 1)^2 + \frac{1}{2}\sum_{\mu=1}^{d} |\tilde{\psi}_n - \tilde{\psi}_{n+\mu}|^2 \right], \tag{2}$$

with $\tilde{V}_0 = |\alpha|\gamma a^{d-2}/bk_B$, $\tilde{\sigma} = a^2 \xi^2$, where μ denotes the unit vectors along the d coordinate axes, $N = L^d$ is the total number of sites, and an unimportant constant term has been dropped. The parameter $\tilde{V}_0$ merely sets the temperature scale and can thus be absorbed in the definition of the reduced temperature $\tilde{T} = T/\tilde{V}_0$. After these rescalings and omitting the tilde on ψ, σ, and T for notational simplicity in the rest of the paper, the partition function Z considered in the simulations is given by

$$Z = \int D\psi D\bar{\psi}\, e^{-H/T} , \tag{3}$$

where the functional measure $\int D\psi\, D\bar{\psi} \equiv \int D\,\mathrm{Re}\psi\, D\,\mathrm{Im}\psi$ stands short for integrating over all possible complex field configurations.

The main focus in this work is on the properties of the geometrically defined vortex-loop network. The standard procedure to calculate the vorticity on each plaquette is by considering the quantity

$$m = \frac{1}{2\pi}([\phi_1 - \phi_2]_{2\pi} + [\phi_2 - \phi_3]_{2\pi} + [\phi_3 - \phi_4]_{2\pi} + [\phi_4 - \phi_1]_{2\pi}) , \tag{4}$$

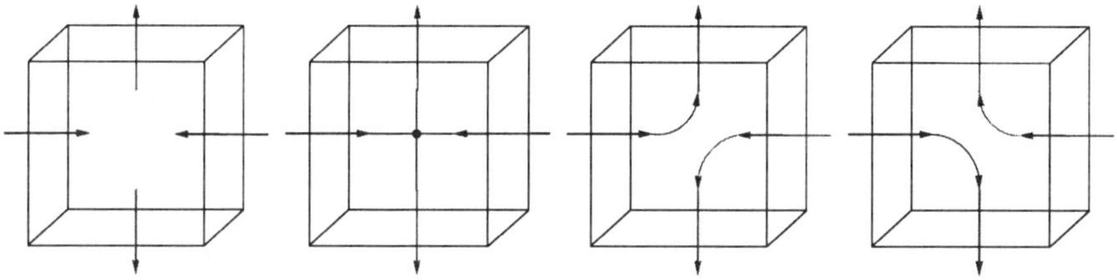

Fig. 1. If two (or three) vortex lines pass through one cell, the vortex tracing algorithm must decide how to connect them, and this leads to an ambiguity in the length distribution. Using the "maximum" rule we connect all line elements (forming a knot). Following the "stochastic" rule the connections are made stochastically.

where $\phi_1, \ldots, \phi_4$ are the phases at the corners of a plaquette labelled, say, according to the right-hand rule, and $[\alpha]_{2\pi}$ stands for α modulo 2π: $[\alpha]_{2\pi} = \alpha + 2\pi n$, with n an integer such that $\alpha + 2\pi n \in (-\pi, \pi]$, hence $m = n_{12} + n_{23} + n_{34} + n_{41}$. If $m \neq 0$, there exists a topological charge which is assigned to the object dual to the given plaquette, i.e., the (oriented) line elements $*l_\mu$ which combine to form closed networks ("vortex loops"). With this definition, the vortex "currents" $*l_\mu$ can take three values: $0, \pm 1$ (the values ± 2 have a negligible probability and higher values are impossible). The quantity $v = \frac{1}{N} \sum_{n,\mu} |*l_{\mu,n}|$ serves as a measure of the vortex density.

In order to study percolation observables we connect the obtained vortex line elements to closed loops. Following a single line, there is evidently no difficulty, but when a branching point, where $n \geq 2$ junctions are encountered, is reached, a decision on how to continue has to be made. This step involves a certain ambiguity. If we connect all in- and outgoing line elements ("maximal" rule), knots will be formed. Another choice is to join only one incoming with one outgoing line element ("stochastic" rule), with the outgoing direction chosen randomly. These two possibilities are shown in Fig. 1.

We can thus extract from each lattice configuration a set of vortex loops, which have been glued together by one of the connectivity definitions above. For each loop in the network, we measure the "extent" of a vortex loop in 3 dimensions, $\mathcal{O}_{3\mathrm{D}}$. This means simply to project the loop onto the three axes and record whether the projection covers the whole axis, or to be more concrete, whether one finds a vortex-line element of the loop in all planes perpendicular to the eyed axis. If there is a loop fulfilling this requirement, then this loop is percolating and we record 1 in the time series of measurements; if not, a value of 0 is stored. This quantity can thus be interpreted as percolation probability[3] which is a convenient quantity for locating the percolation threshold β_p. Furthermore we measure the "susceptibilities", $\chi_i = N(\langle \mathcal{O}_i^2 \rangle - \langle \mathcal{O}_i \rangle^2)$, for the vortex density v and for the extent of a vortex loop $\mathcal{O}_{3D}$ defined above.

In order to perform Monte Carlo simulations, we employed the single-cluster algorithm[6] to update the direction of the field.[7] The modulus of ψ is updated with a Metropolis algorithm.[8] Here some care is necessary to treat the measure in (3) properly (see Ref. 5). One sweep consisted of N spin flips with the Metropolis algorithm and N_{sc} single-cluster updates. We performed simulations for lattices with linear lattice size $L = 6$ up to 40, subject to periodic boundary conditions. After an initial equilibration time of 20 000 sweeps we took about 100 000 measurements, with ten sweeps between the measurements. All error bars are computed with the Jackknife method. For further details concerning the simulation, see Ref. 9.

In order to be able to compare standard, thermodynamically obtained results (working directly with the original field variables) with the percolative treatment of the geometrically defined vortex-loop networks considered here, we used the same value for the parameter $\sigma = 1.5$ as in Ref. 10 for which we determined by means of standard finite-size scaling (FSS) analyses a critical coupling of $\beta_c = 0.780\,08(4)$. Focussing here on the vortex loops, we performed new simulations at this thermodynamically determined critical value, $\beta = 0.780\,08$, as well as additional simulations at $\beta = 0.79$, 0.80, and 0.81. The latter β values were necessary because of the spreading of the pseudo-critical points of the vortex loop related quantities. As previously we recorded the time series of the energy H, the magnetization M, the mean modulus $\overline{|\psi|}$, and the mean-square amplitude $\overline{|\psi|^2}$, and the vorticity v. In the present simulations, however, we saved in addition also the field configurations in each measurement. This enabled us to perform the time-consuming analyses of the vortex-loop networks after finishing the simulations.

The FSS ansatz for the pseudo-critical inverse temperatures $\beta_i(L)$, defined as the points where the various χ_i attain their maxima, is taken as usual as

$$\beta_i(L) = \beta_{i,c} + c_1 L^{-1/\nu} + c_2 L^{-1/\nu - \omega} + \dots \,, \tag{5}$$

where $\beta_{i,c}$ denotes the infinite-volume limit, and ν and ω are the correlation length and confluent correction critical exponents, respectively. Here we have deliberately retained the subscript i on $\beta_{i,c}$.

Let us start with the susceptibility χ_v of the vortex density. Note that this quantity, while also being expressed entirely in terms of vortex ele-

Table 1. The critical exponents of the 3D XY model universality class as reported in Ref. 11 and the correction-to-scaling exponent ω of Ref. 7.

α	β	γ	ν	ω
−0.0146(8)	0.3485(2)	1.3177(5)	0.67155(27)	0.79(2)

ments, plays a special role in that it is locally defined, i.e., does *not* require a decomposition into individual vortex loops (which, in fact, is the time-consuming part of the vortex-network analysis). Assuming the XY model values for ν and ω compiled in Table 1, and fitting only the coefficients $\beta_{i,c}$ and c_i, we arrive at the estimate $\beta_{v,c} = 0.779\,7(14)$ with a goodness-of-fit parameter $Q = 0.20$. This value is perfectly consistent with the previously obtained "thermodynamic" result,[10] derived from FSS of the magnetic susceptibility and various (logarithmic) derivatives of the magnetization. On the basis of this result one would indeed conclude that the phase transition in the three-dimensional complex Ginzburg-Landau field theory can be explained in terms of vortex-line *proliferation*.[12,13]

To develop a purely geometric picture of the mechanism governing this transition, however, one should also consider for example the quantity $\mathcal{O}_{3\mathrm{D}}$ introduced above that focus on the *percolative* properties of the vortex-loop network. We show in Fig. 2(a) the resulting scaling behaviour of the maxima locations $\beta_{3\mathrm{D}}(L)$ for the susceptibility $\chi_{3\mathrm{D}}$ of $\mathcal{O}_{3\mathrm{D}}$ for the "stochastic" and the "maximal" rule. The lines indicate fits according to Eq. (5) with exponents fixed again according to Table 1. We obtain $\beta_{3\mathrm{D},c} = 0.7824(1)$ with $\chi^2/\mathrm{dof} = 1.14$ $(Q = 0.32, L \geq 8)$ for the "stochastic" and $\beta_{3\mathrm{D},c} = 0.8042(4)$ with $\chi^2/\mathrm{dof} = 0.75$ $(Q = 0.58, L \geq 20)$ for "maximal" rule. While for the "stochastic" rule the infinite-volume limit of $\beta_{3\mathrm{D}}(L)$ is at least close to β_c, it is clearly significantly larger than β_c for the fully knotted vortex networks.

With these remarks in mind we nevertheless performed tests whether at least for the "stochastic" rule the critical behaviour of the vortex-loop network may consistently be described by the 3D XY model universality class. As an example for a quantity that is *a priori* expected to behave as a percolation probability we picked again the quantity $\mathcal{O}_{3\mathrm{D}}$. As is demonstrated in Fig. 2(b) for the "stochastic" rule, by plotting the raw data of $\mathcal{O}_{3\mathrm{D}}$ as a function of β for the various lattice sizes, one obtains a clear crossing point so that the interpretation of $\mathcal{O}_{3\mathrm{D}}$ as percolation probability is nicely confirmed. To test the scaling behaviour we rescaled the raw data in the FSS master plot shown in the inset of Fig. 2(b), where the critical exponent ν has the XY model value given in Table 1 and $\beta_c(\mathcal{O}_{3\mathrm{D}}) = 0.7842$ was independently determined by optimizing the data collapse, i.e., virtually this

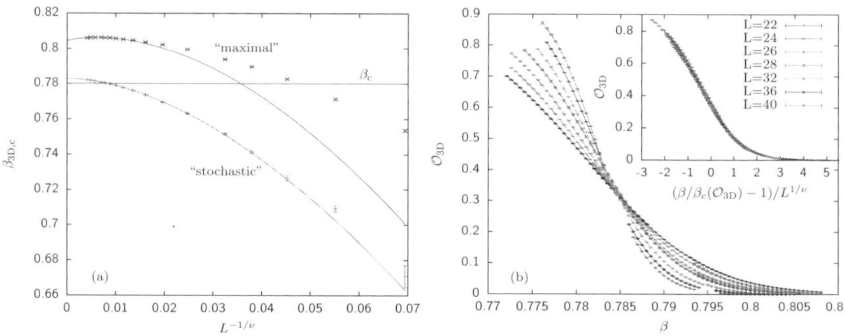

Fig. 2. (a) Location of the percolation thresholds determined from the maximum of susceptibility of $\mathcal{O}_{3D}$ for the "stochastic" and "maximal" rule as a function of $L^{-1/\nu}$, respectively. The lines indicate fits according to Eq. (5) with ν and ω fixed according to Table 1. The horizontal dashed line shows the thermodynamically determined critical coupling $\beta_c = 0.78008(4)$. (b) $\mathcal{O}_{3D}$ as a function of β for the "stochastic" rule. Inset: Rescaled data with ν fixed at the 3D XY model value (cf. Table 1) and choosing $\beta_c(\mathcal{O}_{3D}) = 0.7842$ to be the location of the crossing point in (a) for the best data collapse.

is the location of the crossing point in Fig. 2(b). The collapse turns out to be quite sharp which we explicitly judged by comparison with similar plots for standard bond and site percolation (using there the proper percolation exponent, of course).

In this work[9] we have found for the three-dimensional complex Ginzburg-Landau field theory that the geometrically defined percolation transition of the vortex-loop network is close to the thermodynamic phase transition, but does not coincide with it for any observable we have used. Nevertheless we believe that it should be possible to bring the percolation transition closer to the thermodynamic one by using different vortex-loop network definitions, e.g., using a temperature-dependent or a size-dependent connectivity parameter. To verify this presumption would be an interesting future project, but thereby one should first investigate the XY model which is much less CPU time-consuming.

Acknowledgments

We would like to thank Adriaan Schakel for many useful discussions. Work supported by the Deutsche Forschungsgemeinschaft (DFG) under grants No. JA483/22-1 and No. JA483/23-1 and in part by the EU RTN-Network 'ENRAGE': "Random Geometry and Random Matrices: From Quantum Gravity to Econophysics" under grant No. MRTN-CT-2004-005616. Supercomputer time at NIC Jülich under grant No. hlz10 is also greatfully acknowledged.

References

1. T. Banks, R. Myerson, and J. Kogut, *Nucl. Phys. B* **129**, 493 (1977).
2. H. Kleinert, *Gauge Fields in Condensed Matter*, Vol. I: *Superflow and Vortex Lines* (World Scientific, Singapore, 1989).
3. D. Stauffer and A. Aharony, *Introduction to Percolation Theory*, 2nd ed. (Taylor and Francis, London, 1994).
4. K. Kajantie, M. Laine, T. Neuhaus, A. Rajantie, and K. Rummukainen, *Phys. Lett. B* **482**, 114 (2000).
5. E. Bittner and W. Janke, *Phys. Rev. Lett.* **89**, 130201 (2002).
6. U. Wolff, *Phys. Rev. Lett.* **62**, 361 (1989); *Nucl. Phys. B* **322**, 759 (1989).
7. M. Hasenbusch and T. Török, *J. Phys. A* **32**, 6361 (1999).
8. N. Metropolis, A. W. Rosenbluth, M. N. Rosenbluth, A. H. Teller, and E. Teller, *J. Chem. Phys.* **21**, 1087 (1953).
9. E. Bittner, A. Krinner, and W. Janke, *Phys. Rev. B* **72**, 094511 (2005).
10. E. Bittner and W. Janke, *Phys. Rev. B* **71**, 024512 (2005).
11. M. Campostrini, M. Hasenbusch, A. Pelissetto, P. Rossi, and E. Vicari, *Phys. Rev. B* **63**, 214503 (2001).
12. N. D. Antunes, L. M. A. Bettencourt, and M. Hindmarsh, *Phys. Rev. Lett.* **80**, 908 (1998).
13. N. D. Antunes and L. M. A. Bettencourt, *Phys. Rev. Lett.* **81**, 3083 (1998).

ENVIRONMENTAL EFFECTS ON THE THERMODYNAMICS OF QUANTUM SPIN SYSTEMS

R. VAIA

Istituto dei Sistemi Complessi, Consiglio Nazionale delle Ricerche,
Via Madonna del Piano 10, 50019 Sesto Fiorentino (FI), Italy
E-mail: ruggero.vaia@isc.cnr.it

A. CUCCOLI, A. FUBINI, and V. TOGNETTI

Dipartimento di Fisica, Università di Firenze,
Via G. Sansone 1, 50019 Sesto Fiorentino (FI), Italy

A magnetic system is usually described in terms of the exchange coupling between neighboring spins lying on the sites of a given lattice. Our goal here is to account for the unavoidable quantum effects due to the further coupling with the vibrations of the ions constituting the lattice. A Caldeira-Leggett scheme allows one to treat such effects through the analysis of the associated influence action, obtained after tracing-out the phonons. In a physically sound model, it turns out that one must deal with an environmental coupling which is nonlinear in the system's variables. The corresponding path integral can be dealt with by suitably extending the pure-quantum self-consistent harmonic approximation. In this way one can obtain extended phase diagrams for magnetic phase transitions, accounting for the environmental interaction.

Keywords: Quantum environment; Magnetic systems; Phase diagram.

1. Introduction

It is well known that real physical systems are unavoidably coupled with their environment, although their modeling as isolated systems is a very useful and often good approximation. However, there are cases where environmental coupling displays significant effects, as in the case of Josephson junction arrays, where the inclusion of shunt resistances across the junctions enhances the coherence of the array and stabilizes the globally superconducting phase.[1]

Quantum magnetic systems, by the very reason of having the spins sitting on the sites of a real lattice, are no exception as their quantum environment is well identified with the thermalized vibrations of the lattice.

It is therefore natural to investigate whether and under which conditions environmental effects could be displayed by magnetic materials. This is the original motivation for the present study, that has led us to extend the *system plus reservoir* approach[2–4] in order to include coupling forms more general than those considered so far.[5] As for the spin-phonon coupling, we consider both a spin-orbit-mediated mechanism and the direct influence of the bond length upon the exchange constant, which give rise to a quadratic and to a quartic influence action, respectively.

2. The 2D XXZ Ferromagnet

Although the treatment we are going to propose is valid for whatever spin Hamiltonian, it will be more clearly explained by choosing a reference model: let's then assume the easy-axis ferromagnet[a] on the square lattice, i.e.,

$$\hat{\mathcal{H}} = -J \sum_{\langle ij \rangle} \left[\mu (\hat{S}_i^x \hat{S}_j^x + \hat{S}_i^y \hat{S}_j^y) + \hat{S}_i^z \hat{S}_j^z \right] , \tag{1}$$

where $J > 0$ is the exchange constant and $\mu < 1$ is the easy-axis anisotropy, while $\hat{\boldsymbol{S}}_i$ are spin-S operators with $\hat{\boldsymbol{S}}_i^2 = S(S{+}1)$. This system undergoes an *Ising* phase transition, that was characterized both in the classical $(S \to \infty)$[6,7] and in the quantum case.[8,9] The latter study relied upon the *effective classical Hamiltonian* method,[10] also called *pure-quantum self-consistent harmonic approximation* (PQSCHA),[11] an approach that generalized the variational effective classical potential[12,13] and that can be naturally extended to account for environmental coupling.[14]

Basically, obtaining the effective classical Hamiltonian for a magnet entails (*i*) a spin-boson transformation, such as the Holstein-Primakoff (HP),

$$\hat{S}_i^z = S - \hat{a}_i^\dagger \hat{a}_i , \qquad \hat{S}_i^+ = \hat{S}_i^x + i\hat{S}_i^y = (2S - \hat{a}_i^\dagger \hat{a}_i)^{1/2} \, \hat{a}_i ; \tag{2}$$

(*ii*) applying the PQSCHA recipe for phase space variables $(\hat{p}_i, \hat{q}_i)$ defined by $\hat{a}_i = (\widetilde{S}/2)^{1/2}(\hat{q}_i + i\hat{p}_i)$; the *classical spin value* $\widetilde{S} = S + \frac{1}{2}$ is such that the Weyl symbols (required in the Hamiltonian path integral) for the spin operators read

$$S_i^z = \widetilde{S}\Big(1 - \frac{q_i^2{+}p_i^2}{2}\Big) , \qquad S_i^\pm = \widetilde{S} \sqrt{1 - \frac{p_i^2{+}q_i^2}{4}} \, (q_i \pm i p_i) + O(\widetilde{S}^{-1}) ; \tag{3}$$

[a]The antiferromagnet can be treated with slightly more complicated expressions.

this transformation is eventually used to recast the effective Hamiltonian in terms of classical spins, i.e., unit vectors $\boldsymbol{s}_i$:

$$\mathcal{H}_{\rm eff} = E(T) - J\widetilde{S}^2 j_{\rm eff} \sum_{\langle ij \rangle} \Big[\mu_{\rm eff} \big(s_i^x s_j^x + s_i^y s_j^y \big) + s_i^z s_j^z \Big] \ . \tag{4}$$

Using the reduced temperature $t \equiv T/J\widetilde{S}^2$, the effective exchange $j_{\rm eff}(t,\mu,S) = \theta_1^2$ and anisotropy $\mu_{\rm eff}(t,\mu,S) = \mu\,\theta_2/\theta_1$ depend, through $\theta_1 = 1 - D + \mu D'$ and $\theta_2 = 1 - D + D'/\mu$, on the renormalization parameters

$$D = \frac{1}{2\widetilde{S}\,N} \sum_{\boldsymbol{k}} \mathcal{L}_{\boldsymbol{k}} \ , \qquad D' = \frac{1}{2\widetilde{S}\,N} \sum_{\boldsymbol{k}} \gamma_{\boldsymbol{k}}\, \mathcal{L}_{\boldsymbol{k}} \ , \tag{5}$$

where $\gamma_{\boldsymbol{k}} = (\cos k_x + \cos k_y)/2$; they describe the (on-site and nearest-neighbor) pure-quantum fluctuations of the spins, being

$$\mathcal{L}_{\boldsymbol{k}} = \coth \frac{\omega_{\boldsymbol{k}}}{2t\widetilde{S}} - \frac{2t\widetilde{S}}{\omega_{\boldsymbol{k}}} \ . \tag{6}$$

D and D' are evaluated self-consistently with the (dimensionless) frequencies $\omega_{\boldsymbol{k}} = z(\theta_1 - \mu\,\theta_2 \gamma_{\boldsymbol{k}})$.

3. Linear Environmental Coupling

A model that entails a linear coupling between spins and environment has been used in spin-boson systems[15,16] and for the related Ising chain in transverse field.[17,18] The underlying physical mechanism can be identified in the spin-orbit mediated coupling that for a static lattice can give rise to the magnetic anisotropy. It is reasonable to neglect spin correlations mediated by the environment, i.e., to assume the latter has a fast thermalization dynamics. Each spin is therefore subject to an *independent* environment, and for simplicity let us assume that that this holds for any *component* (a generalization is possible along the lines of Ref. 19). For the spin component $\hat{S}_i^\sigma$ the environmental Hamiltonian reads then

$$\hat{\mathcal{H}}_{\rm E}^{(i\sigma)} = \frac{1}{2} \sum_{\alpha} \Big[a_{\sigma\alpha}^2\, \hat{p}_\alpha^2 + b_{\sigma\alpha}^2\, (\hat{q}_\alpha - \hat{S}_i^\sigma)^2 \Big] \ , \tag{7}$$

where $\{\hat{p}_\alpha, \hat{q}_\alpha\}$ are the environmental variables. The corresponding influence action is obtained by *tracing out* the environment from the path integral[4]

$$\mathcal{S}_{\rm I}^{(i\sigma)} = -\frac{J}{2} \int_0^\beta du\, du'\ \kappa_\sigma(u{-}u')\ S_i^\sigma(u)\, S_i^\sigma(u') \ , \tag{8}$$

where the kernel $\kappa_\sigma(u) = \beta^{-1}\sum_n \kappa_{\sigma n}\, e^{i\nu_n u}$ has Matsubara components that are related[4] to the Laplace transform of the memory function $\gamma(t)$

$$\kappa_{\sigma n} = \frac{1}{J}\sum_\alpha \frac{b_\alpha^2\,\nu_n^2}{\nu_n^2 + a_\alpha^2 b_\alpha^2} = |\nu_n|\;\gamma(z=|\nu_n|)\ . \tag{9}$$

Inserting the HP variables (3) and using the property $\kappa_0 = \int_0^\beta du\,\kappa(u) = 0$, it turns out that the lowest-order contributions to $\mathcal{S}_\mathrm{I}$ are quadratic and arise from S^x and S^y (by symmetry, $\kappa_{xn} = \kappa_{yn} = \kappa_n$). The inclusion of the lowest order terms of Eq. (8) in the action for the isolated magnet eventually gives the effective Hamiltonian of Eqs. (4) and (5), but with Eq. (6) replaced by

$$\mathcal{L}_{\boldsymbol{k}} = 4t\widetilde{S}\sum_{n=1}^{\infty} \frac{\omega_{\boldsymbol{k}} + \kappa_n}{\tilde{\nu}_n^2 + (\omega_{\boldsymbol{k}} + \kappa_n)^2}\ , \tag{10}$$

where $\tilde{\nu}_n \equiv \nu_n/J\widetilde{S}^2 = 2\pi t\widetilde{S}\,n$. Assuming a phenomenological *Drude* memory function, the corresponding kernel is characterized by two parameters: the strength γ and the cutoff frequency ω_D (of the order of the lattice Debye frequency), so that

$$\kappa_n = \frac{1}{J\widetilde{S}^2}\,\frac{\gamma\,\omega_\mathrm{D}\,\nu_n}{\omega_\mathrm{D} + \nu_n} = \gamma\;\frac{\sigma_\mathrm{D}\,\tilde{\nu}_n}{\sigma_\mathrm{D} + \tilde{\nu}_n}\ , \tag{11}$$

where the dimensionless cutoff $\sigma_\mathrm{D} \equiv \omega_\mathrm{D}/J\widetilde{S}^2$ is, essentially, the ratio between the Debye and the Curie temperature. In Fig. 1 the typical environmental effect upon the phase diagram is reported: it appears that the transition temperature is lower as a result of larger pure-quantum fluctuations, and that the effect is very sensitive to the value of σ_D, which can be very large in magnetic materials with small magnetic exchange, like the organic ones.

However, the fact that the anisotropies of real materials are always very small ($\mu \lessapprox 1$) suggests that spin-orbit coupling is usually weak: consequently, γ is expected to be very small. Indeed, a different spin-lattice coupling mechanism we are now going to discuss can be more effective.

4. δJ-Coupling Model

It is well known that exchange coupling is a quantum mechanical effect arising from the fermionic character of electrons, and essentially depends on the overlap of orbitals, hence on the separation of neighboring ions; for this reason the exchange energy J over the bond connecting lattice sites $\boldsymbol{i}$ and $\boldsymbol{j}$ depends on $\hat{r}_{\boldsymbol{ij}} = |\boldsymbol{d} + \delta\hat{\boldsymbol{r}}_{\boldsymbol{ij}}|$, where $\boldsymbol{d} \equiv \boldsymbol{j} - \boldsymbol{i}$,

$$J(\hat{r}_{\boldsymbol{ij}}) \simeq J + \delta J_{\boldsymbol{ij}} = J + J'\,\boldsymbol{d}\cdot\delta\hat{\boldsymbol{r}}_{\boldsymbol{ij}}\ . \tag{12}$$

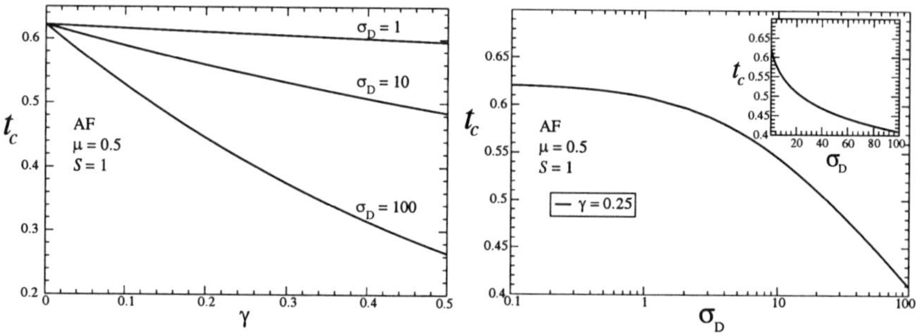

Fig. 1. Transition temperature vs coupling strength (left) and vs cutoff frequency (right) for the S=1 XXZ antiferromagnet with anisotropy $\mu = 0.5$.

If one considers the isotropic part of Hamiltonian

$$\hat{\mathcal{H}}_{\rm is} = -\sum_{\langle \boldsymbol{ij} \rangle} (J + \delta J_{\boldsymbol{ij}})\, \hat{\boldsymbol{S}}_{\boldsymbol{i}} \cdot \hat{\boldsymbol{S}}_{\boldsymbol{j}} \tag{13}$$

it appears that one has to consider a coupling model which is quadratic in the spin operators

$$\hat{\mathcal{H}}_{\rm E}^{(\boldsymbol{ij})} = \frac{1}{2} \sum_{\alpha} \left[a_\alpha^2\, \hat{p}_{\boldsymbol{ij}\alpha}^2 + b_\alpha^2\, (\hat{q}_{\boldsymbol{ij}\alpha} - \hat{\boldsymbol{S}}_{\boldsymbol{i}} \cdot \hat{\boldsymbol{S}}_{\boldsymbol{j}})^2 \right] , \tag{14}$$

where, as in Sec. 3, an independent phonon environment for each bond is assumed. The above environmental coupling model is so effective that it can even lead to a *spin-Peierls* transition,[20] where a distorted lattice configuration permits a gain in magnetic energy; the same model is used to account for environmental effects on quantum phase transitions,[21] on thermal-transport anomalies,[22] and on the magnetic phase diagrams.[23,24] It is immediate to obtain the influence action as

$$\mathcal{S}_{\rm I} = -\frac{J}{2\widetilde{S}^2} \sum_{\langle \boldsymbol{ij} \rangle} \int_0^\beta du\, du'\; K(u{-}u')\; \big[\hat{\boldsymbol{S}}_{\boldsymbol{i}}(u) \cdot \hat{\boldsymbol{S}}_{\boldsymbol{j}}(u)\big]\; \big[\hat{\boldsymbol{S}}_{\boldsymbol{i}}(u') \cdot \hat{\boldsymbol{S}}_{\boldsymbol{j}}(u')\big] \;. \tag{15}$$

The Hamiltonian of lattice vibrations together with Eq. (12) can be used to evaluate the spectral density of spin-phonon coupling $\mathcal{J}(\omega) = \sum_\alpha\, b_\alpha^2\, \omega_\alpha\, \delta(\omega^2 - \omega_\alpha^2)$, and the related kernel $K(u)$. If the vibrating lattice coincides with the magnetic one one gets

$$\mathcal{J}(\omega) = \frac{2J'^2}{m\Omega^2}\;\; \mathcal{W}(\omega) \;, \tag{16}$$

where Ω is the maximum phonon frequency and $\mathcal{W}(\omega)$ is the phonon density of states. As $\mathcal{W}(\omega)$ is peaked at large ω, it is reasonable to use the *Einstein approximation*, i.e., $\mathcal{W}_{\rm E}(\omega) \approxeq \delta(\omega - \Omega)$, so $\mathcal{J}(\omega) \approxeq \frac{4\,J'^2}{m\,\Omega}\;\delta(\omega^2 - \Omega^2)$ and the corresponding kernel is

$$K_n = \gamma\,\frac{\nu_n^2}{\nu_n^2 + \Omega^2}\;, \qquad\qquad \gamma = \frac{4\,J\widetilde{S}^2}{m\,\Omega^2\,R^2}\;, \tag{17}$$

where the magnetic length scale $R \equiv J/J' = \big[\partial_r \ln J(r)\big]^{-1}$ enters the coupling strength γ. At difference with Sec. 3, it appears in this case that the environmental effect is expected to increase with J.

If one is happy with a model that unambiguously gives all interaction parameters, there is however a significant complication: the influence action (15) is quartic in the spins, and going over to HP variables, Eq. (3), one finds at lowest order 36 quartic terms in the integrand of (15), such as

$$q_i(u)\,q_j(u)\,q_i(u')\,q_j(u') \;+\; q_i(u)\,q_j(u)\,p_i(u')\,p_j(u') \;+\; \dots\;. \tag{18}$$

It is therefore necessary to extend the previously used approach in such a way to account for nonlinear environmental coupling.

5. PQSCHA with Nonlinear Environmental Coupling

The framework of the PQSCHA can be extended in a natural way to nonlinearities appearing in the nonlocal action: one introduces a *trial influence action* in terms of the pure-quantum fluctuations of the paths, $\xi_i(u) = q_i(u) - \bar{q}$ and $\eta_i(u) = p_i(u) - \bar{p}$, where $(\bar{p},\bar{q}) = \beta^{-1}\int du\big(p(u),q(u)\big)$,

$$\begin{aligned}\mathcal{S}_{\rm I0} = -J\widetilde{S}^2 \sum_{\langle ij\rangle} \int_0^\beta du\,du'\,\Big\{\kappa(u{-}u')\,\big[\xi_i(u)\,\xi_i(u') + \eta_i(u)\,\eta_i(u')\big] \\ -\kappa'(u{-}u')\,\big[\xi_i(u)\,\xi_j(u') + \eta_i(u)\,\eta_j(u')\big]\Big\}\,.\end{aligned} \tag{19}$$

Its Fourier-Matsubara transform reads

$$\mathcal{S}_{\rm I0} = -\frac{z}{2t}\sum_{\boldsymbol{k}}\sum_{n\neq 0}\big(\kappa_n - \gamma_{\boldsymbol{k}}\,\kappa'_n\big)\,\big(p_{\boldsymbol{k}n}\,p^*_{\boldsymbol{k}n} + q_{\boldsymbol{k}n}\,q^*_{\boldsymbol{k}n}\big)\,. \tag{20}$$

By symmetry there are two possible components (self and nn) of the trial kernel which has to be optimized; they correspond to the only nonvanishing pure-quantum variances

$$\begin{aligned} D(u) &= \langle\!\langle \xi_i(0)\xi_i(u)\rangle\!\rangle = \langle\!\langle \eta_i(0)\eta_i(u)\rangle\!\rangle\;, \\ D'(u) &= \langle\!\langle \xi_i(0)\xi_j(u)\rangle\!\rangle = \langle\!\langle \eta_i(0)\eta_j(u)\rangle\!\rangle\;, \end{aligned} \tag{21}$$

where the 'double-bracket' is the Gaussian average taken with respect to the full (quadratic) trial action $\mathcal{S}_0$. All parameters are then optimized by imposing the generalized (nonlocal) PQSCHA conditions,

$$\langle\!\langle \mathcal{S} - \mathcal{S}_0 \rangle\!\rangle = 0 \ , \qquad \left\langle\!\!\left\langle \frac{\delta^2(\mathcal{S} - \mathcal{S}_0)}{\delta\xi_{\boldsymbol{i}}(u)\,\delta\xi_{\boldsymbol{j}}(u')} \right\rangle\!\!\right\rangle = 0 \ , \tag{22}$$

which are equivalent to *decouple* all nonlinear terms, i.e., extracting the averages (21) of all possible pairs and leaving a quadratic expression. In this way several terms vanish and the effective kernel turns out to be given by

$$\kappa(u) = \kappa'(u) = K(u)\ \mathcal{D}(u) \ , \tag{23}$$

where $\mathcal{D}(u) \equiv 2\big[D(u) - D'(u)\big]$. In spite of the simple appearance, one has to account for the involute self-consistency that is more evident if one expresses the Matsubara components,

$$\kappa_n = \sum_{\ell\neq 0} K_{n-\ell}\ \mathcal{D}_\ell \ , \tag{24}$$

$$\mathcal{D}_n = \frac{1}{\widetilde{S} N} \sum_{\boldsymbol{k}} \,(1-\gamma_{\boldsymbol{k}})\, \frac{4t\widetilde{S}\,[\omega_{\boldsymbol{k}} + z(1-\gamma_{\boldsymbol{k}})\,\kappa_n]}{\tilde{\nu}_n^2 + [\omega_{\boldsymbol{k}} + z(1-\gamma_{\boldsymbol{k}})\,\kappa_n]^2} \ . \tag{25}$$

Indeed, Eq. (24) is a nonlinear system in the numerable components κ_n. It can be solved numerically by iteration, using a finite number of coefficients $(\kappa_1, ..., \kappa_M)$ such that $\kappa_M \ll \tilde{\nu}_M$ and exploiting the fact that one has a rapidly convergent sum in terms of $\delta K_n = K_\infty - K_n \sim n^{-2}$.

6. Conclusions

We have derived the expressions accounting for the effects of linear environmental coupling in the XXZ model, which give a decrease of the transition temperature as the coupling strength and the cutoff frequency are increased.

From the point of view of the general method, facing the problem of δJ-coupling has lead us to a nontrivial extension of the pure-quantum self-consistent harmonic approximation (PQSCHA), which permits to embody both pure-quantum fluctuations and environmental effects in the effective classical Hamiltonian. It will be interesting to use this approach also for simpler problems that were previously afforded with an inappropriate trial action.[25,26]

In the case of δJ-coupling it is possible to identify the environmental coupling parameters, i.e., in the Einstein approximation, the coupling strength γ and the cutoff frequency Ω, in terms of the parameters of the vibrational Hamiltonian. The work in progress concerns the actual evaluation of the related environmental effects and the determination of the phase-diagram, in view of possibly identifying evidences of such effects in experimental data.

References

1. L. Capriotti, A. Cuccoli, A. Fubini, V. Tognetti, and R. Vaia, *Phys. Rev. Lett.* **91**, 247004 (2003).
2. A. O. Caldeira and A. J. Leggett, *Phys. Rev. Lett.* **46**, 211 (1981).
3. P. Ullersma, *Physica* (Amsterdam) **32**, 27, 56, 74, 90 (1966).
4. U. Weiss, *Quantum Dissipative Systems* (World Scientific, Singapore, 1999).
5. A. Cuccoli, A. Fubini, V. Tognetti, and R. Vaia, *Phys. Rev. E* **64**, 066124 (2001).
6. P. A. Serena, N. García, and A. Levanyuk, *Phys. Rev. B* **47**, 5027 (1993).
7. M. E. Gouvêa, G. M. Wysin, S. A. Leonel, A. S. T. Pires, T. Kamppeter, and F. G. Mertens, *Phys. Rev. B* **59**, 6229 (1999).
8. A. Cuccoli, T. Roscilde, V. Tognetti, R. Vaia, and P. Verrucchi, *Phys. Rev. B* **62**, 3771 (2000).
9. A. Cuccoli, T. Roscilde, V. Tognetti, R. Vaia, and P. Verrucchi, *Eur. Phys. J. B* **20**, 55 (2001).
10. A. Cuccoli, V. Tognetti, P. Verrucchi, and R. Vaia, *Phys. Rev. A* **45**, 8418 (1992).
11. A. Cuccoli, R. Giachetti, V. Tognetti, P. Verrucchi, and R. Vaia, *J. Phys.: Condens. Matter* **7**, 7891 (1995).
12. R. Giachetti and V. Tognetti, *Phys. Rev. Lett.* **55**, 912 (1985).
13. R. P. Feynman and H. Kleinert, *Phys. Rev. A* **34**, 5080 (1986).
14. A. Cuccoli, A. Rossi, V. Tognetti, and R. Vaia, *Phys. Rev. E* **55**, 4849 (1997).
15. A. J. Leggett *et al.*, *Rev. Mod. Phys.* **59**, 1 (1987).
16. J. S. Lim, M.-S. Choi, and M. Y. Choi, arXiv:cond-mat/0610449.
17. P. Werner, K. Völker, M. Troyer, and S. Chakravarty, *Phys. Rev. Lett.* **94**, 047201 (2005).
18. G. Schehr and H. Rieger, *Phys. Rev. Lett.* **96**, 227201 (2006).
19. H. Kohler and F. Sols, *New J. Phys.* **8**, 149 (2006).
20. R. J. Bursill, R. H. McKenzie, and C. J. Hamer, *Phys. Rev. Lett.* **83**, 408 (1999).
21. C. Raas, U. Löw, G. Uhrig, and R. W. Kühne, *Phys. Rev. B* **65**, 144438 (2002).
22. A. L. Chernyshev and A. V. Rozhkov, *Phys. Rev. B* **72**, 104423 (2005).
23. A. Kumar, *Indian J. Pure Appl. Phys.* **9**, 141 (1971).
24. N. N. Chan, *Czech. J. Phys.* **40**, 341 (1990).
25. J.-D. Bao, *Phys. Rev. A* **65**, 052120 (2002).
26. J.-D. Bao, *Phys. Rev. A* **69**, 022102 (2004).

A PATH-INTEGRATION APPROACH TO THE CORRELATORS OF XY HEISENBERG MAGNET AND RANDOM WALKS

N. M. BOGOLIUBOV* and C. MALYSHEV†

Steklov Institute of Mathematics, St.-Petersburg Department,
Fontanka 27, St.-Petersburg, 191023, Russia
**E-mail: bogoliub@pdmi.ras.ru*
†E-mail: malyshev@pdmi.ras.ru

The path integral approach is used for the calculation of the correlation functions of the XY Heisenberg chain. The obtained answers for the two-point correlators of the XX magnet are of the determinantal form and are interpreted in terms of the generating functions for the random turns vicious walkers.

Keywords: Path integration; XY Heisenberg magnet; Random walk.

1. Introduction

The problem of enumeration of paths of *vicious walkers* on a one-dimensional lattice was formulated by M. E. Fisher[1] and since then continues to attract much attention (see Refs. 2,3). The walkers are called 'vicious' because they annihilate each other at the same lattice site, and their trajectories are thus non-intersecting. Similar problems appear in the theory of domain walls,[4] directed percolation,[5] self-organized criticality,[6] and polymer theory.[7] It has been proposed in Ref. 2 to use the XX Heisenberg chain to enumerate the paths of the random turns vicious walkers.

Our approach based on path integration was developed in Refs. 8,9 to calculate thermal correlation functions of the XY Heisenberg magnet. Dependence of the integration variables on the imaginary time is defined by special quasi-periodicity conditions. In the present paper, this method is used for the calculation of the two-point correlation functions of the XX model and the interpretation of the obtained answer in terms of the generating functions of the random turns vicious walkers is given.

2. The Problem

The Hamiltonian of the periodic XY Heisenberg chain of "length" M (M is chosen to be even) in transverse magnetic field $h > 0$ is:

$$H = H_0 + \gamma H_1 - h\, S^z\,, \qquad H_0 \equiv -\sum_{n,m=1}^{M} \Delta^{(+)}_{nm} \sigma^+_n \sigma^-_m\,,$$
$$H_1 \equiv -\frac{1}{2}\sum_{n,m=1}^{M} \Delta^{(+)}_{nm} (\sigma^+_n \sigma^+_m + \sigma^-_n \sigma^-_m)\,, \quad S^z \equiv \frac{1}{2}\sum_{n=1}^{M} \sigma^z_n\,. \tag{1}$$

Here S^z is the z-component of the total spin operator, and the entries of the so-called *hopping matrix* $\Delta^{(s)}_{nm}$ ($s = \pm$) are:

$$2\Delta^{(s)}_{nm} \equiv \delta_{|n-m|,1} + s\delta_{|n-m|,M-1}\,, \tag{2}$$

where $\delta_{n,l}$ is the Kronecker symbol. The Pauli matrices $\sigma^\pm_n = (1/2)(\sigma^x_n \pm i\sigma^y_n)$ and σ^z_n, where $n \in \mathcal{M} \equiv \{1, \ldots, M\}$, satisfy the commutation relations: $\left[\sigma^+_k, \sigma^-_l\right] = \delta_{kl}\, \sigma^z_l$ and $\left[\sigma^z_k, \sigma^\pm_l\right] = \pm 2\, \delta_{kl}\, \sigma^\pm_l$. The periodic boundary condition reads: $\sigma^\alpha_{n+M} = \sigma^\alpha_n$, $\forall n$. The Hamiltonian H (1), taken at $\gamma = 0$ (the case of XX magnet), commutes with S^z.

Time-dependent thermal correlation functions are defined as follows:

$$G_{ab}(m,t) \equiv Z^{-1}\mathrm{Tr}\left(\sigma^a_{l+m}(0)\, \sigma^b_l(t)\, e^{-\beta H}\right), \quad Z \equiv \mathrm{Tr}\left(e^{-\beta H}\right), \tag{3}$$

where $\sigma^b_l(t) \equiv e^{itH}\, \sigma^b_l\, e^{-itH}$, $\beta = 1/T$ is inverse temperature, and t is time. This correlator may be rewritten in terms of the canonical lattice Fermi fields c_i, $c^\dagger_j$, where i, $j \in \mathcal{M}$, by means of the Jordan-Wigner map:

$$\sigma^+_n = \left(\prod_{j=1}^{n-1} \sigma^z_j\right) c_n\,, \quad \sigma^-_n = c^\dagger_n \left(\prod_{j=1}^{n-1} \sigma^z_j\right), \qquad n \in \mathcal{M}\,,$$

where $\sigma^z_j = 1 - 2c^\dagger_j c_j$. The periodic conditions for the spin operators result in the boundary conditions for the fermions:

$$c_{M+1} = (-1)^{\mathcal{N}} c_1\,, \qquad c^\dagger_{M+1} = c^\dagger_1\, (-1)^{\mathcal{N}}\,, \tag{4}$$

where $\mathcal{N} = \sum_{n=1}^{M} c^\dagger_n c_n$ is the operator of the total number of particles. In the fermionic representation, H in Eq. (1) will take a form $H = H^+ P^+ + H^- P^-$, where $P^\pm = (1/2)(\mathbb{I} \pm (-1)^{\mathcal{N}})$ are projectors.[8] The operators H^s are of identical form with $s = \pm$ pointing out a correspondence between these operators and an appropriate specification of the conditions (4): $c_{M+1} = -\, s\, c_1$, $c^\dagger_{M+1} = -\, s\, c^\dagger_1$.

Equation (3) for the z-components of spins, for instance, becomes:

$$G_{zz}(m,t) = 1 - 2\, Z^{-1}\mathrm{Tr}\left(c^\dagger_{l+m}\, c_{l+m}\, e^{-\beta H}\right) - 2\, Z^{-1}\mathrm{Tr}\left(c^\dagger_l\, c_l\, e^{-\beta H}\right)$$

$$+4\,Z^{-1}\mathrm{Tr}\left(c^{\dagger}_{l+m}\,c_{l+m}\,e^{itH}\,c^{\dagger}_{l}\,c_{l}\,e^{-(\beta+it)H}\right). \tag{5}$$

To evaluate (5), it is convenient to consider the generating functional:

$$\mathcal{G}\equiv\mathcal{G}(S,T|\,\mu,\nu)\;=\;Z^{-1}\mathrm{Tr}\left(e^{S}\,e^{-\mu H}\,e^{T}\,e^{-\nu H}\right), \tag{6}$$

where μ, ν are the complex parameters, $\mu+\nu=\beta$. Two operators, $S\equiv c^{\dagger}\widehat{S}c$ and $T\equiv c^{\dagger}\widehat{T}c$, are defined through the matrices $\widehat{S}=\mathrm{diag}\,\{S_1,S_2,\dots,S_M\}$, $\widehat{T}=\mathrm{diag}\,\{T_1,T_2,\dots,T_M\}$. For instance, the last term on the r.h.s. of (5) is obtained from (6) in the following way:

$$\frac{\partial}{\partial S_k}\,\frac{\partial}{\partial T_l}\,\mathcal{G}(S,T|\,\mu,\nu)\Bigg|_{\substack{S_n,\,T_n,\,\forall n\;\longrightarrow\;0\\ \mu,\,\nu\;\longrightarrow\;-it,\,\beta+it}}.$$

As a result, we express the trace on the r.h.s. of (6) in the form:[8]

$$\mathrm{Tr}\left(e^{S}\,e^{-\mu H}\,e^{T}\,e^{-\nu H}\right)\;=\;\frac{1}{2}\left(\mathcal{G}^{+}_{\mathrm{F}}\,Z^{+}_{\mathrm{F}}\,+\,\mathcal{G}^{-}_{\mathrm{F}}\,Z^{-}_{\mathrm{F}}\,+\,\mathcal{G}^{+}_{\mathrm{B}}\,Z^{+}_{\mathrm{B}}\,-\,\mathcal{G}^{-}_{\mathrm{B}}\,Z^{-}_{\mathrm{B}}\right), \tag{7}$$

where

$$\begin{aligned}\mathcal{G}^{\pm}_{\mathrm{F}}\,Z^{\pm}_{\mathrm{F}}&=\mathrm{Tr}\left(e^{S}\,e^{-\mu H^{\pm}}\,e^{T}\,e^{-\nu H^{\pm}}\right),\\ \mathcal{G}^{\pm}_{\mathrm{B}}\,Z^{\pm}_{\mathrm{B}}&=\mathrm{Tr}\left(e^{S}\,e^{-\mu H^{\pm}}\,e^{T}\,(-1)^{\mathcal{N}}\,e^{-\nu H^{\pm}}\right),\end{aligned} \tag{8}$$

and $Z^{\pm}_{\mathrm{F}}=\mathrm{Tr}\,(e^{-\beta H^{\pm}})$, $Z^{\pm}_{\mathrm{B}}=\mathrm{Tr}\left((-1)^{\mathcal{N}}e^{-\beta H^{\pm}}\right)$.

3. The Path Integral

We use the coherent states $|z\rangle\equiv\exp(c^{\dagger}z)|0\rangle$ and $\langle z^*|\equiv\langle 0|\exp(z^*c)$ generated from the Fock vacuum $|0\rangle$, $c_k|0\rangle=0$, $\forall k$. We use the short-hand notations for the M-component objects, say, $z^*\equiv(z^*_1,\dots,z^*_M)$ and $z\equiv(z_1,\dots,z_M)$ formed by the independent Grassmann parameters z_k, z^*_k ($k\in\mathcal{M}$). Besides, $\sum_{k=1}^{M}c^{\dagger}_k z_k\equiv c^{\dagger}z$, $\prod_{k=1}^{M}\mathrm{d}z_k\equiv\mathrm{d}z$, etc. Then, we shall represent[9] the trace of the operator in $\mathcal{G}^{\pm}_{\mathrm{F}}\,\mathcal{Z}^{\pm}_{\mathrm{F}}$ in Eq. (8) by means of the Grassmann integration over $\mathrm{d}z$, $\mathrm{d}z^*$:

$$\mathcal{G}^{\pm}_{\mathrm{F}}\,Z^{\pm}_{\mathrm{F}}\;=\;\int\mathrm{d}z\,\mathrm{d}z^*\,e^{z^*z}\,\langle z^*|\,e^{S}\,e^{-\mu H^{\pm}}\,e^{T}\,e^{-\nu H^{\pm}}\,|z\rangle\,. \tag{9}$$

For the sake of simplicity we shall consider the XX model only and take those $H^{\pm}$ that correspond to H in Eq. (1) at $\gamma=0$.

To represent the r.h.s. of (9) as a path integral, we first introduce new coherent states $|x(I)\rangle$, $\langle x^*(I)|$, where $2L\times M$ independent Grassmann parameters are arranged in the form of $2L$ "vectors" $x^*(I)$, $x(I)$ ($I\in\{1,\dots,L\}$).

It allows to insert L times the decompositions of unity

$$\int \mathrm{d}x^*(I)\mathrm{d}x(I)\, \exp\bigl(-x^*(I)x(I)\bigr)\, |x(I)\rangle\langle x^*(I)|$$

into the r.h.s. of (9). We define then the additional variables satisfying the quasi-periodicity conditions:

$$-\widehat{E}\, x(0) = x(L+1) \equiv z\,, \qquad -x^*(L+1) = x^*(0)\, \widehat{E}^{-1} \equiv z^*\,. \tag{10}$$

Here, $\widehat{E} \equiv e^{\widehat{S}}\, e^{-\mu \widehat{H}^{\pm}}\, e^{\widehat{T}}$ with the matrices $\widehat{H}^{\pm}$ expressed[9] through the hopping matrices (2): $\widehat{H}^{\pm} = -\widehat{\Delta}^{(\mp)} + h\widehat{I}$, where $\widehat{I}$ is a unit $M \times M$ matrix. The described procedure allows to pass in the limit $L \to \infty$ from $(L+1)$-fold integration to the *continuous* one over the "infinite" product of the measures $\mathrm{d}x^*(\tau)\mathrm{d}x(\tau)$ on a space of trajectories $x^*(\tau)$, $x(\tau)$, where $\tau \in \mathbb{R}$:

$$\mathcal{G}_{\mathrm{F}}^{\pm}\, Z_{\mathrm{F}}^{\pm} = \int e^{S}\, \mathrm{d}\lambda^*\, \mathrm{d}\lambda \prod_{\tau} \mathrm{d}x^*(\tau)\mathrm{d}x(\tau)\,.$$

The integration over the auxiliary Grassmann variables λ^*, λ guarantees the fulfilment of the continuous version of the constraints (10). The action functional is $S \equiv \int L(\tau)\, d\tau$, where $L(\tau)$ is the Lagrangian:

$$L(\tau) \equiv x^*(\tau)\Bigl(\frac{\mathrm{d}}{\mathrm{d}\tau} - \widehat{H}^{\pm}\Bigr)\, x(\tau) + J^*(\tau)x(\tau) + x^*(\tau)J(\tau)\,,$$

$$J^*(\tau) \equiv \lambda^*\bigl(\delta(\tau)\,\widehat{I} + \delta(\tau-\nu)\,\widehat{E}^{-1}\bigr)\,, \quad J(\tau) \equiv \bigl(\delta(\tau)\,\widehat{I} + \delta(\tau-\nu)\,\widehat{E}\bigr)\,\lambda\,.$$

The δ-functions reduce τ to the segment $[0,\beta]$. The stationary phase requirements $\delta S/\delta x^* = 0$, $\delta S/\delta x = 0$ yield the regularized answer:[9]

$$\mathcal{G}_{\mathrm{F}}^{\pm} = \det\left(\widehat{I} + \frac{e^{(\beta-\nu)\widehat{H}^{\pm}}\, e^{\widehat{S}}\, e^{-\mu\widehat{H}^{\pm}}\, e^{\widehat{T}} - \widehat{I}}{\widehat{I} + e^{\beta\,\widehat{H}^{\pm}}}\right),$$

The remainder correlators $G_{ab}(m,t)$ (3) (with $a, b \in \{+,-\}$) are obtained analogously.

4. Random Walks

The evolution of the states obtained by selective flipping of the spins governed by the XX Hamiltonian H_0 in Eq. (1) is related to a model of random turns vicious walkers.[2,3] Indeed, let us consider the following average over the ferromagnetic state vectors $\langle \Uparrow |$, $| \Uparrow \rangle$:

$$F_{j;\,l}(\lambda) \equiv \langle \Uparrow | \sigma_j^+ e^{-\lambda H_0} \sigma_l^- | \Uparrow \rangle\,, \tag{11}$$

where $| \Uparrow \rangle \equiv \otimes_{n=1}^{M} | \uparrow \rangle_n$, i.e., all spins are up, and λ is an "evolution" parameter. Spin up (or down) corresponds to an empty (or occupied) site.

Differentiating $F_{j;\,l}(\lambda)$ and applying the commutator $[H_0, \sigma_j^+]$, we obtain the differential-difference equation (*master equation*):

$$\frac{\mathrm{d}}{\mathrm{d}\lambda}\, F_{j;\,l}(\lambda) \,=\, \frac{1}{2}\left(F_{j+1;\,l}(\lambda) \,+\, F_{j-1;\,l}(\lambda)\right). \tag{12}$$

The average $F_{j;\,l}$ may be considered as the generating function of paths made by a random walker travelling from l^{th} to j^{th} site. Really, its K-th derivative has the form

$$\frac{\mathrm{d}^K}{\mathrm{d}\lambda^K}\, F_{j;\,l}(\lambda)\bigg|_{\lambda=0} = \langle\Uparrow|\, \sigma_j^+(-\,H_0)^K \sigma_l^-\, |\Uparrow\rangle = \sum_{n_1,\dots,n_{K-1}} \Delta^{(+)}_{j n_{K-1}} \dots \Delta^{(+)}_{n_2 n_1} \Delta^{(+)}_{n_1 l}\,.$$

A single step to one of the nearest sites is prescribed by the hopping matrix (2) with $s = +$. After K steps, each path connecting l^{th} and j^{th} sites contributes into the sum. The N-point correlation function ($N \le M$),

$$F_{j_1,j_2,\dots,j_N;\,l_1,l_2,\dots,l_N}(\lambda) \,=\, \langle\Uparrow|\, \sigma_{j_1}^+\sigma_{j_2}^+ \dots \sigma_{j_N}^+\, e^{-\lambda H_0}\, \sigma_{l_1}^-\sigma_{l_2}^- \dots \sigma_{l_N}^-\, |\Uparrow\rangle\,, \tag{13}$$

enumerates the nests of the lattice paths of N random turns vicious walkers being initially located at the positions $l_1 > l_2 > \dots > l_N$ and, eventually, at $j_1 > j_2 > \dots > j_N$. It is expressed in the form:[2]

$$F_{j_1,\dots,j_N;\,l_1,\dots,l_N}(\lambda) = \det\bigl(F_{j_r;\,l_s}(\lambda)\bigr)_{1\le r,\, s\le N}\,. \tag{14}$$

The ground state and the excited states of the XX chain at $h = 0$ with the total spin equal to $(M/2) - N$ are decomposed over a basis of states $\sigma_{l_1}^-\sigma_{l_2}^- \dots \sigma_{l_N}^-\, |\Uparrow\rangle$ with N spins flipped.[10] Therefore, the trace $\widetilde{F}_{m+1;\,1}(\lambda) \equiv \mathrm{Tr}\,(\sigma_{m+1}^+ e^{-\lambda H_0} \sigma_1^-)$ is a linear combination of the generating functions (13) describing the evolution of $N+1$ random turns walkers. The initial and the final positions of one of them are fixed at $l_1 = 1$ and $j_1 = m+1$, respectively, while for the remaining ones these positions are random. In the thermodynamic limit, the number of the *virtual* walkers tends to infinity. We apply the procedure described in Sec. 3 to calculation of $\widetilde{F}_{m+1;\,1}(\lambda)$ in the limit when M and N are large enough. In this limit, the contributions with the subindex 'B' become, with regard of (8), negligible in (7):

$$\begin{aligned}\widetilde{F}_{m+1;\,1}(\lambda) &= \left[\mathrm{tr}\,\bigl(e^{-\lambda \widehat{H}^0}\widehat{e}_{1,m+1}\bigr) - \frac{\mathrm{d}}{\mathrm{d}\alpha}\right] \det\Bigl(\widehat{I} + \widehat{\mathcal{U}}_m + \frac{\alpha}{M}\widehat{\mathcal{V}}_m\Bigr)\bigg|_{\alpha=0} \\ &= \det\bigl(\widehat{I} + \widehat{\mathcal{U}}_m\bigr) \left[\mathrm{tr}\,\bigl(e^{-\lambda \widehat{H}^0}\widehat{e}_{1,m+1}\bigr) - \frac{1}{M}\,\mathrm{tr}\left(\frac{\widehat{\mathcal{V}}_m}{\widehat{I} + \widehat{\mathcal{U}}_m}\right)\right],\end{aligned}$$

where $\widehat{e}_{1,m+1} \equiv (\delta_{1,n}\delta_{m+1,l})_{1\le n,l\le M}$, and the matrix $\widehat{H}^0$ is used instead of $\widehat{H}^{\pm}$ since s can be taken zero at large enough M. The traces of λ-dependent $M \times M$ matrices $\widehat{\mathcal{U}}_m$ and $\widehat{\mathcal{V}}_m$ are given below. A differential equation analogous to (12) is fulfilled by $\widetilde{F}_{m+1;\,1}(\lambda)$. At large separation m it takes the form:

$$\frac{\mathrm{d}}{\mathrm{d}\lambda}\widetilde{F}_{m+1;\,1}(\lambda) = \frac{1}{2}\big(\widetilde{F}_{m;\,1}(\lambda) + \widetilde{F}_{m+2;\,1}(\lambda)\big) - \mathrm{Tr}\left(H_0\, \sigma^+_{m+1}\, e^{-\lambda H_0}\, \sigma^-_1\right). \quad (15)$$

We expand formally $\widetilde{F}_{m+1;\,1}(\lambda)$ with respect to $\widehat{\mathcal{U}}_m$ and obtain the answer in the two lowest orders as follows:

$$\begin{aligned} \widetilde{F}_{m+1;\,1}(\lambda) &\approx F_{m+1;\,1}(\lambda) \,+\, F_{m+1;\,1}(\lambda) \times \mathrm{tr}\,\widehat{\mathcal{U}}_m \,-\, \frac{1}{M}\,\mathrm{tr}\,\widehat{\mathcal{V}}_m\,, \\ \mathrm{tr}\,\widehat{\mathcal{U}}_m &= (M \,-\, 2m)\, F_{1;\,1}(\lambda)\,, \\ \frac{1}{M}\,\mathrm{tr}\,\widehat{\mathcal{V}}_m &= F_{m+1;\,1}(2\lambda) \,-\, 2\sum_{l=1}^{m} F_{m+1;\,l}(\lambda)\, F_{l;\,1}(\lambda)\,. \end{aligned} \quad (16)$$

Although M and m are chosen to be large in this expansion, the ratio m/M is kept bounded. In each order the master equation (15) is fulfilled by (16). The contribution of the second order can be re-expressed through the two-point functions $F_{m+1,l;\,l,1}(\lambda)$ (see Eqs. (13) and (14)). Thus, summation over intermediate positions (of a virtual walker located at the lth site) arises in the second order. A similar picture is expected in the next orders.

Acknowledgments

One of us (C. M.) is grateful to the Organizers of the 9th International Conference "Path Integrals – New Trends and Perspectives". This work was partially supported by the RFBR, No. 07–01–00358.

References

1. M. E. Fisher, *J. Stat. Phys.* **34**, 667 (1984).
2. N. M. Bogoliubov, *J. Math. Sci.* **138**, 5636 (2006).
3. N. M. Bogoliubov, *J. Math. Sci.* **143**, 2729 (2007).
4. D. Huse and M. Fisher, *Phys. Rev. B* **29**, 239 (1984).
5. E. Domany and W. Kinzel, *Phys. Rev. Lett.* **53**, 311 (1984).
6. P. Bak, C. Tang, and K. Wiesenfeld, *Phys. Rev. A* **38**, 364 (1988).
7. J. W. Essam and A. J. Guttmann, *Phys. Rev. E* **52**, 5849 (1995).
8. C. Malyshev, *Functional integration with "automorphic" boundary conditions and correlators of z-components of spins in the XY and XX Heisenberg chains*, in *New Developments in Mathematical Physics Research*, ed. C. V. Benton (Nova Science Publishers, New York, 2004), pp. 85–116.
9. C. Malyshev, *J. Math. Sci.* **136**, 3607 (2006).
10. F. Colomo, A. G. Izergin, V. E. Korepin, and V. Tognetti, *Theor. Math. Phys.* **94**, 19 (1993).

CRITICAL EXPONENTS OF MIXED QUANTUM SPIN CHAIN

R. BISCHOF* and W. JANKE

Institut für Theoretische Physik, Universität Leipzig, Postfach 100 920, D-04009 Leipzig, Germany
** E-mail: rainer.bischof@itp.uni-leipzig.de*

With continuous time quantum Monte Carlo simulations we investigate a continuous quantum phase transition in the mixed quantum spin chain with spin arrangement $-S^{\mathrm{a}} - S^{\mathrm{a}} - S^{\mathrm{b}} - S^{\mathrm{b}}-$, with $S^{\mathrm{a}} = 1/2$ and $S^{\mathrm{b}} = 1$. By finite-size scaling analysis we calculate estimates of the critical control parameter as well as estimates of critical exponents.

Keywords: Quantum spin chain; Critical exponent; Quantum Monte Carlo simulation.

1. Introduction

Antiferromagnetic quantum spin chains with bond alternation show a rich variety of (quantum) effects.[1] Among many applications, they serve as generic models to study an interesting type of topological quantum phase transitions.[2] Depending on the strength of bond alternation, topologically different phases appear that can be understood qualitatively within the picture of valence bonds.[3]

Our model Hamiltonian reads

$$\boldsymbol{H} = J\sum_i \left(\boldsymbol{S}^{\mathrm{a}}_{4i}\boldsymbol{S}^{\mathrm{a}}_{4i+1} + \alpha\boldsymbol{S}^{\mathrm{a}}_{4i+1}\boldsymbol{S}^{\mathrm{b}}_{4i+2} + \boldsymbol{S}^{\mathrm{b}}_{4i+3}\boldsymbol{S}^{\mathrm{b}}_{4i+4} + \alpha\boldsymbol{S}^{\mathrm{b}}_{4i+4}\boldsymbol{S}^{\mathrm{a}}_{4i+5}\right), \quad (1)$$

with α being the parameter that controls bond alternation. For a coupling constant $J > 0$, the model is antiferromagnetic. For $S = S^{\mathrm{a}} = S^{\mathrm{b}} = 1/2$, the model becomes the antiferromagnetic Heisenberg model (AFHM) which is critical at the point $\alpha = 1$, i.e., there is no excitation gap. The critical point separates two different quantum phases.[4]

In contrast to the AFHM, the quantum spin chain with $S = 1$ is not critical. This fundamental difference between uniform quantum spin chains with $S = 1/2$ and $S = 1$, and even more generally between uniform quantum spin chains of half-odd integer sized spin and integer sized spin, has become famous as *Haldane's conjecture.*[5] Yet, by tuning the control parameter α one can drive the system with $S = 1$ towards criticality at $\alpha \approx 0.578$.[6]

The different phases that can be accessed by tuning the control parameter α, can be conveniently pictured by valence bonds.[3] Spins of magnitude $S \geq 1$ are represented by subspins of size $S = 1/2$. Then subspins at different sites couple to form antiferromagnetic singlets.

The model we study in this paper, with $S^{\mathrm{a}} = 1/2$ and $S^{\mathrm{b}} = 1$, serves as a natural interpolation between the two cases discussed above. The critical value of the control parameter has been calculated previously by means of quantum Monte Carlo simulations,[7] but no attempts to calculate critical exponents have yet been published.

2. Observables

We apply the continuous imaginary time loop algorithm in the path integral representation[8,9] to create, after thermalisation, $N = 100\,000$ configurations. The expectation value of an observable $\mathcal{O}$ reads

$$\langle \mathcal{O} \rangle = \frac{1}{N} \sum_{i=1}^{N} \mathcal{O}_i \,, \tag{2}$$

with $\mathcal{O}_i$ denoting the value calculated from configuration i.

We measure, among other quantities, the dynamic susceptibilities[a] $\chi(\pi, 0)$ and $\chi(\pi, 2\pi/\beta)$ by improved estimators[8]

$$\langle \chi(k,\omega)_{\mathrm{imp}} \rangle = \frac{1}{4\beta L} \left\langle \sum_{\text{all loops } i} |\mathcal{L}_i(k,\omega)|^2 \right\rangle , \tag{3}$$

where β is the inverse temperature, L the chain length, and $\mathcal{L}_i(k,\omega)$ denotes the generalised "size" of loop i with respect to k and ω. The imaginary time correlation length ξ_τ is then estimated by the second-moment method[10]

$$\xi_\tau = \frac{\beta}{2\pi} \sqrt{\frac{\chi(\pi,0)}{\chi(\pi, 2\pi/\beta)} - 1} \,. \tag{4}$$

The inverse of ξ_τ gives the energy gap between ground state and lowest lying excited states and thus diverges at the critical point.

We, furthermore, measure the twist order parameter z, defined by the ground state expectation value of the unitary operator

$$\boldsymbol{U}_{\mathrm{T}} = \exp\left(\mathrm{i} \frac{2\pi}{L} \sum_{j=1}^{L} j \boldsymbol{S}_j^z \right) . \tag{5}$$

[a] Fourier modes of the two-point correlation function $\langle \boldsymbol{S}_{i_0}^z \boldsymbol{S}_{i_0+i}^z \rangle - \langle \boldsymbol{S}_{i_0}^z \rangle \langle \boldsymbol{S}_{i_0+i}^z \rangle$ averaged over all points i_0, with $i = (x, \tau)$.

Table 1. Finite-size scaling analysis. Results are obtained by fits of the pseudo-critical points $\alpha^*(L)$ to Eq. (6). Pseudo-critical points are the maxima of the imaginary time correlation length ξ_τ and the zero points of the twist order parameter z, respectively.

	ξ_τ			z		
L	α_c	θ	χ^2_{pdof}	α_c	θ	χ^2_{pdof}
8...128	0.76230(4)	2.38(6)	0.68	0.76244(8)	1.66(2)	1.63
8...128	0.76230(4)	2.38(6)	0.68	0.76244(8)	1.66(2)	1.63
12...128	0.76228(4)	2.46(8)	0.61	0.76237(9)	1.67(3)	1.60
16...128	0.76228(4)	2.46(13)	0.68	0.76236(10)	1.68(3)	1.69
20...128	0.76226(4)	2.63(17)	0.52	0.76225(8)	1.73(3)	0.82
24...128	0.76226(5)	2.63(40)	0.60	0.76227(9)	1.72(4)	0.93
32...128	0.76227(7)	2.47(48)	0.70	0.76233(15)	1.68(8)	1.04

It has been shown that z signals the transition from one valence bond configuration to another by a change of sign.[6]

To extrapolate finite-size data to the thermodynamic limit we apply a finite-size scaling analysis.[11] Pseudo-critical points, $\alpha^*(L)$, for a given chain length L are defined as the maxima of $\xi_\tau(\alpha; L)$, and the zero points of $z(\alpha; L)$, respectively. All simulations are performed at a constant aspect ratio of $\beta/L = 2$, so in first approximation we can assume that the pseudo-critical points obey

$$\alpha^*(L) = \alpha_c + bL^{-\theta}\,, \tag{6}$$

where α_c is the critical point of the infinite chain and $\theta = 1/\nu$ the shift exponent. The maxima of $\xi_\tau(\alpha; L)$ scale with $L^{-z\nu}$. In our case the dynamic critical exponent takes the value $z = 1$,[2] and we can expect to get an estimate for ν from finite-size scaling analysis of the maxima of $\xi_\tau(\alpha; L)$. This, however, is not necessarily true for the pseudo-critical points of $z(\alpha; L)$.

3. Results and Discussion

Taking all chain lengths into account we get $\alpha_c = 0.76230(4)$ from $\xi_\tau(\alpha; L)$ and $\alpha_c = 0.76244(8)$ from $z(\alpha; L)$. This slight discrepancy vanishes if we neglect small chain lengths in the analysis of the twist order parameter. In Table 1 we list the results of finite-size scaling fits when successively omitting small chain lengths.

The shift of the maxima of $\xi_\tau(\alpha; L)$ is controlled by a shift exponent $\theta \approx 2.40$. This implies $\nu = 1/\theta \approx 0.42$ for the critical exponent of the correlation length. We, however, clearly see that the shift of the zero points of $z(\alpha; L)$ is controlled by a different shift exponent θ (see also Fig. 1).

Studies of uniform quantum spin chains with bond alternation and *exchange anisotropy* for $S = 1/2$,[4] and $S = 1$,[12] imply also for mixed quantum

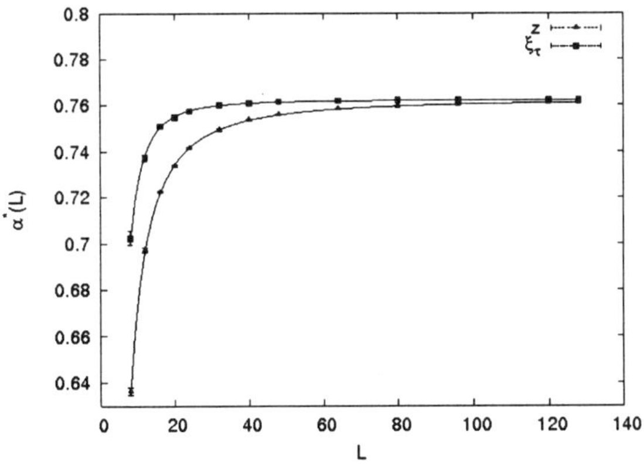

Fig. 1. Pseudo-critical points of the imaginary time correlation length ξ_τ, and the twist order parameter z. Lines show the fits to Eq. (6) performed with pseudo-critical points of all chain lengths L.

spin chains that the critical point we are interested in is the end point – or at least very close to – of a Gaussian type critical line with continuously varying critical exponents. Moreover they show that this line bifurcates exactly at the point with no exchange anisotropy. Thus, in fact, we are at – or close to – a multicritical point, and this might spoil our simple finite-size scaling ansatz of Eq. (6). Future investigation of the whole phase diagram of mixed quantum spin chains in the parameter space of bond alternation and exchange anisotropy will reveal more.

References

1. A. Auerbach, *Interacting Electrons and Quantum Magnetism* (Springer, New York, 1994); W. J. Caspers, *Spin Systems* (World Scientific, Singapore, 1989).
2. S. Sachdev, *Quantum Phase Transitions* (Cambridge University Press, Cambridge, 1999).
3. I. Affleck, T. Kennedy, E. H. Lieb, and H. Tasaki, *Phys. Rev. Lett.* **59**, 799 (1987).
4. M. Yamanaka, Y. Hatsugai, and M. Kohmoto, *Phys. Rev. B* **48**, 9555 (1993).
5. F. D. M. Haldane, *Phys. Rev. Lett.* **50**, 1153 (1983).
6. M. Nakamura and S. Todo, *Phys. Rev. Lett.* **89**, 077 204 (2002).
7. Z. Xu, J. Dai, H. Ying, and B. Zheng, *Phys. Rev. B* **67**, 214 426 (2003).
8. H. G. Evertz, *Adv. Phys.* **52**, 1 (2003).
9. B. B. Beard and U. J. Wiese, *Phys. Rev. Lett.* **77**, 5130 (1996).
10. S. Todo and K. Kato, *Phys. Rev. Lett.* **87**, 047 203 (2001).
11. M. E. Fisher and M. N. Barber, *Phys. Rev. Lett.* **28**, 1516 (1972).
12. Y. Kato and A. Tanaka, *J. Phys. Soc. Japan.* **63**, 1277 (1994).

NEW RESULTS ON THE PHASE DIAGRAM OF THE FFXY MODEL: A TWISTED CFT APPROACH

G. CRISTOFANO* and V. MAROTTA

Dipartimento di Scienze Fisiche and INFN, Universitá di Napoli "Federico II", Via Cintia, Compl. Universitario M. Sant'Angelo, 80126 Napoli, Italy
**E-mail: cristofano@na.infn.it*

P. MINNHAGEN

Department of Theoretical Physics, Umea University, 901 87 Umea, Sweden
E-mail: Petter.Minnhagen@physics.umu.se

A. NADDEO

CNISM, Unità di Ricerca di Salerno, Via Salvador Allende, 84081 Baronissi (SA), Italy
E-mail: naddeo@sa.infn.it

G. NICCOLI

LPTM, Université de Cergy-Pontoise, 2 avenue Adolphe Chauvin, 95302 Cergy-Pontoise, France
E-mail: Giuliano.Niccoli@ens-lyon.fr

The issue of the number, nature and sequence of phase transitions in the fully frustrated XY (FFXY) model is a highly non trivial one due to the complex interplay between its continuous and discrete degrees of freedom. In this contribution we attack such a problem by means of a twisted conformal field theory (CFT) approach[1] and show how it gives rise to the $U(1) \otimes Z_2$ symmetry and to the whole spectrum of excitations of the FFXY model.[2]

Keywords: Fully frustrated XY model; Phase diagram; Conformal field theory.

1. Introduction: The State of Art

The phase diagram of the FFXY model has been the subject of intensive studies in the last thirty years, due to the presence of the mixed symmetry $U(1) \otimes Z_2$. But a full and definitive answer to such a question still lacks; in this paper we address the problem by means of a twisted CFT approach.[1] The XY model on a square lattice in the presence of an external magnetic field transversal to the lattice plane is described by the action:

$$H = -\frac{J}{kT}\sum_{\langle ij\rangle}\cos\left(\varphi_i - \varphi_j - A_{ij}\right), \tag{1}$$

where $\{\varphi\}$ are the phase variables on the sites, the sum is over nearest neighbors, $J > 0$ is the coupling constant and the bond variables $A_{ij} = \frac{2e}{\hbar c}\int_i^j A\cdot dl$ satisfy the full frustration condition $\sum_{plaquette} A_{ij} = \pi$. Choosing the Landau gauge in such a way to get a lattice where each plaquette displays one antiferromagnetic and three ferromagnetic bonds, we obtain two ground states with opposite chiralities. The discrete Z_2 symmetry of the FFXY model is broken at low temperature and will be restored beyond a certain temperature after the formation of domain walls separating islands of opposite chirality. The Ising transition overlaps to a vortex-unbinding transition, which is associated with the continuous $U(1)$ symmetry.[3] The action (1) can be rewritten as a fractionally charged Coulomb gas (CG) defined on the dual lattice, $H_{CG} = -\frac{J}{kT}\sum_{r,r'}\left(m\left(r\right)+\frac{1}{2}\right)G\left(r,r'\right)\left(m\left(r'\right)+\frac{1}{2}\right)$, where $\lim_{|r-r'|\to\infty} G\left(r,r'\right) = \log|r-r'| + \frac{1}{2}\pi$ and the neutrality condition $\sum_r\left(m\left(r\right)+\frac{1}{2}\right) = 0$ holds. Such a model exhibits two possible phase transitions, an Ising and a vortex-unbinding one. The issue whether there are two distinct phase transitions, $T_V > T_{dw}$ or $T_V < T_{dw}$ with T_V and T_{dw} marking respectively the breaking of $U(1)$ and of Z_2 symmetry,[4] or a single transition with the simultaneous breaking of both symmetries[5] has been widely investigated. The model allows for the existence of two topological excitations: vortices and domain walls. Vortices are point-like defects such that the phase rotates by $\pm 2\pi$ in going around them.[3] A domain wall can be viewed as a line on the square lattice, each segment of which separates two cells with the same chirality.[6] If the domain walls form a right angle, such corners must behave as fractional vortices with topological charge $\pm 1/4$.[3] There exists a temperature T_{dw} such that, when $T > T_{dw}$, dissociation of bound pairs of fractional vortices is allowed, which triggers the dissociation of pairs of ordinary vortices. The system undergoes two phase transitions with temperatures such that $T_V < T_{dw}$.[7] At finite temperatures kinks may appear on the domain wall. Simple kinks must behave as fractional vortices with topological charge $\pm 1/2$ while a double kink does not introduce mismatches in the phase distribution. At low temperatures, all simple kinks are bound into neutral pairs. As the temperature increases, a phase transition in the gas of logarithmically interacting kinks leads to pair dissociation and emergence of free simple kinks.[6] That takes place[8] at $T_K < T_V$ and produces two distinct bulk transitions with $T_V < T_{dw}$. A more general conclusion is reached by studying the coupled XY-Ising model,[9,10] which is in the same universality class of the FFXY model. Such a model can be introduced starting with a system of two XY models coupled through a symmetry breaking term:[9]

$$H = A\left[\sum_{i=1,2}\sum_{\langle r,r'\rangle} \cos\left(\varphi^{(i)}(r)-\varphi^{(i)}(r')\right)\right] + h\sum_{r}\cos 2\left(\varphi^{(1)}(r)-\varphi^{(2)}(r)\right). \tag{2}$$

The limit $h \to 0$ corresponds to a full decoupling of the fields $\varphi^{(i)}$, $i = 1,2$, while the $h \to \infty$ limit corresponds to the phase locking $\varphi^{(1)}(r) - \varphi^{(2)}(r) = \pi j$, $j = 1,2$; as a consequence the model gains a symmetry $U(1) \otimes Z_2$ and its Hamiltonian renormalizes towards the XY-Ising model.[9] Its phase diagram[9,10,12] is built up with three branches which meet at a multi-critical point P. Two branches describe separate Kosterlitz-Thouless (KT) and Ising transitions while the third (PT) corresponds to single transitions with simultaneous loss of XY and Ising order. It becomes a first order one at a tricritical point T and seems to be non-universal;[9] in fact the numerical estimate for the central charge, $c \sim 1.60$, is higher than the value $c = 3/2$, pertinent to a superposition of critical Ising and Gaussian models.[11] Indeed the central charge seems to vary continuously from $c \approx 1.5$ near P to $c \approx 2$ at T.[10] The system lacks conformal invariance,[9] so one could consider suitable perturbations of the XY-Ising model as a starting point to study the vicinity of the point P.[12] Instead, recent numerical simulations on huge lattices[13] lead to two very close but separate transitions on the PT line. A possible solution could be[7] the addition of an antiferromagnetic coupling ($\overline{J}$) term to the Coulomb gas Hamiltonian H_{CG}; for $\overline{J} \neq 0$, the two transitions on the PT line separate with the KT one occurring at a lower temperature.[7] On the other side, by adding higher harmonics contributions to the potential,[11,14] the possibility of a merging critical point T will be provided here in the context of a twisted CFT approach,[1] which extends the results of Ref. 11, so recovering the whole phase diagram.[9,10,12]

2. m-Reduction Procedure

In this section we recall those aspects of the twisted model (TM) which are relevant for the FFXY model. We focus in particular on the m-reduction procedure for the special $m = 2$ case,[1] since we are interested in a system with $U(1) \otimes Z_2$ symmetry. Such a theory describes well a system consisting of two parallel layers of 2D electron gas in a strong perpendicular magnetic field, with filling factor $\nu^{(a)} = \frac{1}{2p+2}$ for each of the two $a = 1, 2$ layers.[1] Regarding the integer p, characterizing the flux attached to the particles, we choose the "bosonic" value $p = 0$, since it enables us to describe the highly correlated system of vortices with flux quanta $\frac{hc}{2e}$. Let us start from the "filling" $\nu = \frac{1}{2}$, described by a CFT with $c = 1$ in terms of a scalar chiral field $Q(z) = q - i\, p\, lnz + i \sum_{n\neq 0} \frac{a_n}{n} z^{-n}$, compactified on a circle with radius $R^2 = 1/\nu = 2$; here a_n, q and p satisfy the commutation relations $[a_n, a_{n'}] = n\delta_{n,n'}$ and $[q, p] = i$. From such a CFT (mother theory), using the m-reduction procedure, which consists in considering the subalgebra generated only by the modes in $Q(z)$ which are a multiple of an integer m, we get a $c = m$ orbifold CFT (the TM), which is symmetric under a discrete Z_m group and, for $m = 2$, will be shown to describe the whole phase diagram of the FFXY model. Its primary fields content, for the special $m = 2$ case, can be expressed in terms of two scalar fields given by:

$$X(z) = \frac{1}{2}\left(Q(z) + Q(-z)\right), \; \phi(z) = \frac{1}{2}\left(Q(z) - Q(-z)\right); \tag{3}$$

$X(z)$ is Z_2-invariant and describes the electrically "charged" sector of the new theory, while $\phi(z)$ satisfies the twisted boundary conditions $\phi(e^{i\pi} z) = -\phi(z)$ and describes the "neutral" sector.[1] The TM primary fields are composite vertex operators $V(z) = U_X(z)\,\psi(z)$, where $U_X(z)$ is the vertex describing its "charge" content and $\psi(z)$ describing the "neutral" one. In the neutral sector it is useful to introduce the two chiral operators $\psi(z)\,(\overline{\psi}(z)) = \frac{1}{2\sqrt{z}}\left(: e^{i\alpha\phi(z)} : \pm : e^{i\alpha\phi(-z)} :\right)$, with only the first one obeying the boundary conditions. In a fermionized theory they correspond to two $c = 1/2$ Majorana fermions with Ramond and Neveu-Schwartz boundary conditions[1] and, in the TM, they appear to be not completely equivalent. In fact the whole TM theory decomposes into a tensor product of two CFTs, a Z_2 invariant one with $c = 3/2$ and symmetry $U(1) \otimes Z_2$ and the remaining $c = 1/2$ one realized by a Majorana fermion in the twisted sector. Furthermore the energy-momentum tensor of the Ramond part of the neutral sector develops a cosine term, $T_\psi(z) = -\frac{1}{4}(\partial\phi)^2 - \frac{1}{16z^2}\cos\left(2\sqrt{2}\phi\right)$, a signature of a tunneling phenomenon which selects out the new stable $c = 3/2$ vacuum; we identify such a theory with the one describing the FFXY model conjectured in Ref. 11.

3. FFXY Phase Diagram from TM Model

In this section we will derive the FFXY phase diagram in terms of the RG flow which originates from perturbing our TM model with relevant operators. We observe that the limit $h \to 0$ in the Hamiltonian of Eq. (2) gives rise, in the continuum, to a CFT with two scalar boson fields $\varphi^{(i)}$ and with central charge $c = 2$. Now a good candidate to describe the FFXY model at criticality around the point T of the phase diagram is a CFT, with $c = 2$, which accounts for the full spectrum of excitations of the model: vortices, domain walls, and kinks. The role of the boundary conditions in the description of the excitation spectrum is crucial. In fact, by imposing the coincidence between opposite sides of the square lattice, we obtain a closed geometry, which is the discretized analogue of a torus and gives rise, for the ground state, to two topologically inequivalent configurations, one for even and the other one for odd number of plaquettes. So the ground state on the square lattice maps into the ground state for the even case while it generates two straight domain walls along the two cycles of the torus for the odd case. Such a behaviour has to be taken into account by non trivial boundary conditions on the field $\varphi^{(i)}$ at the borders of the finite lattice. To this aim, let $(-L/2, 0)$, $(L/2, 0)$, $(L/2, L)$, $(-L/2, L)$ be the corners of the square lattice $\mathcal{L}$ and assume that the fields $\varphi^{(i)}$ satisfy the boundary conditions $\varphi^{(1)}(r) = \varphi^{(2)}(r)$ for $r \in \mathcal{L} \cap \mathbf{x}$, $\mathbf{x}$ being the x axis. That allows us to consider the two fields $\varphi^{(1)}$ and $\varphi^{(2)}$ on the square lattice $\mathcal{L}$ as the folding of a single field $\mathcal{Q}$, defined on the lattice $\mathcal{L}_0$ with corners $(-L/2, -L)$, $(L/2, -L)$, $(L/2, L)$, $(-L/2, L)$. We can implement a discrete version of the 2-reduction procedure by defining the fields $\mathcal{X}(r) = \frac{1}{2}\left(\mathcal{Q}(r) + \mathcal{Q}(-r)\right)$, $\Phi(r) = \frac{1}{2}\left(\mathcal{Q}(r) - \mathcal{Q}(-r)\right)$ $(r \in \mathcal{L}_0)$, which are symmetric and antisymmetric under the action of the group Z_2. The Hamiltonian (2) can be rewritten in terms of these fields and, for $h = 0$, it gives rise, in the continuum, to the TM action $\mathcal{A} = \int \left[\frac{1}{2}\left(\partial\mathcal{X}\right)^2 + \frac{1}{2}\left(\partial\Phi\right)^2\right] d^2x$. Let us now show how the phase diagram of the FFXY model can be described by the action:

$$\mathcal{A} = \int \left[\frac{1}{2}\left(\partial\mathcal{X}\right)^2 + \frac{1}{2}\left(\partial\Phi\right)^2 + \mu\cos\left(\beta\Phi\right) + \lambda\cos\left(\frac{\beta}{2}\Phi + \delta\right)\right] d^2x, \quad (4)$$

which embodies the higher harmonic potential term conjectured in Refs. 7, 11,14. We assume the constraints $\beta^2 < 8\pi$, which characterizes both the cosine terms as relevant perturbations, and $|\delta| \leq \pi/2$.[15] Thus the "neutral" sector is a two-frequency sine-Gordon theory that can be viewed as a deformation of a pure sine-Gordon one with the perturbing term $\lambda\cos\left(\beta\Phi/2 + \delta\right)$. The ultraviolet (UV) fixed point $\mu = 0$, $\lambda = 0$ of the action (4) corresponds to the TM model with central charge $c = 2$, describing the fixed point T in the FFXY phase diagram. In order to study the RG flow in the "neutral" sector let us define the dimensionless variable $\eta \equiv \lambda\mu^{-(8\pi-(\beta/2)^2)/(8\pi-\beta^2)}$; when $\eta = 0$, the "neutral" sector reduces to a sine-Gordon model with a particle spectrum built of solitons and antisolitons and, for $\beta^2 < 4\pi$, some breathers. Switching on the perturbation a confinement of solitons into states with zero topological charge takes place and packets formed by two of the original solitons survive as stable excitations for generic values of $|\delta| < \pi/2$. In the limit $\eta \to \infty$ the 2-soliton evolves into the 1-soliton of the pure sine-Gordon model with $\mu = 0$. An unbinding phenomenon takes place in the case $\delta = \pm\pi/2$ for finite η and the 2-soliton decomposes into a sequence of two kinks K_1. So the existence of an intermediate critical value $\eta = \eta_c$ is required at which a phase transition takes place and the RG flow ends into the infrared (IR) fixed point described by a CFT with central charge $c = 1/2$, the Ising model. The central charge of the full model (4) so changes from $c = 2$ of the UV fixed point to $c = 3/2$ of the IR fixed point, i.e. we recover early known Monte Carlo results.[10] Such an IR fixed point coincides with the $U(1) \otimes Z_2$ symmetric component of the TM model, which results then to properly describe[11] the fixed point P in the phase diagram of the FFXY model.

References

1. G. Cristofano, G. Maiella, and V. Marotta, *Mod. Phys. Lett. A* **15**, 1679 (2000); G. Cristofano, G. Maiella, V. Marotta, and G. Niccoli, *Nucl. Phys. B* **641**, 547 (2002).
2. G. Cristofano, V. Marotta, P. Minnhagen, A. Naddeo, and G. Niccoli, *J. Stat. Mech.* **P11009** (2006).
3. T. C. Halsey, *J. Phys. C* **18**, 2437 (1985); S. Teitel and C. Jayaprakash, *Phys. Rev. B* **27**, 598 (1983).
4. J. Lee, *Phys. Rev. B* **49**, 3317 (1994); P. Olsson, *Phys. Rev. B* **55**, 3585 (1997).
5. J. M. Thijssen and H. J. F. Knops, *Phys. Rev. B* **42**, 2438 (1990); E. Granato and M. P. Nightingale, *Phys. Rev. B* **48**, 7438 (1993).
6. S. E. Korshunov, *Phys. Rev. Lett.* **88**, 167007 (2002).
7. J. M. Thijssen and H. J. F. Knops, *Phys. Rev. B* **37**, 7738 (1988).
8. P. Olsson and S. Teitel, *Phys. Rev. B* **71**, 104423 (2005).
9. E. Granato and J. M. Kosterlitz, *Phys. Rev. B* **33**, 4767 (1986); J. Lee, E. Granato, and J. M. Kosterlitz, *Phys. Rev. B* **44**, 4819 (1991).
10. M. P. Nightingale, E. Granato, and J. M. Kosterlitz, *Phys. Rev. B* **52**, 7402 (1995).
11. O. Foda, *Nucl. Phys. B* **300**, 611 (1988).
12. P. Baseilhac, *Nucl. Phys. B* **636**, 465 (2002).

13. M. Hasenbusch, A. Pelissetto, and E. Vicari, *J. Stat. Mech.* **P12002** (2005).
14. K. Sokalski, T. Ruijgrok, and B. Schoenmaker, *Physica A* **144**, 322 (1987).
15. G. Delfino and G. Mussardo, *Nucl. Phys. B* **516**, 675 (1998).

EVAPORATION/CONDENSATION OF ISING DROPLETS

A. NUßBAUMER*, E. BITTNER, and W. JANKE
Institut für Theoretische Physik, Universität Leipzig,
Postfach 100 920, D-04009 Leipzig, Germany
* *E-mail: andreas.nussbaumer@itp.uni-leipzig.de*

We performed Monte Carlo simulations of the two-dimensional Ising model in the low-temperature phase at $T = 1.5$ at constant magnetisation (Kawasaki dynamics) and measured the size of the largest minority droplet, i.e., the largest cluster and all overturned spins within. The measured values are compared to theoretical predictions by Biskup *et al.*, which can explain a jump of the droplet size in the vicinity of the spontaneous magnetisation.

Keywords: Ising model; Phase transition; Finite-size correction.

For our investigation we use the two-dimensional (2D) Ising model on a square $L \times L$ lattice in zero field with Hamiltonian

$$H = -J \sum_{\langle ij \rangle} \sigma_i \sigma_j , \qquad \sigma_i = \pm 1 , \tag{1}$$

where $\langle ij \rangle$ denotes all pairs of nearest-neighbour spins. For temperatures T below the critical temperature T_c the magnetisation $m = M/V = \sum_i \sigma_i / V$ has in the infinite-volume limit ($V = L^2 \to \infty$) a value of $m_0(T) \neq 0$. This value is called the spontaneous magnetisation and is given by the famous Onsager formula[1]. Assuming that the majority of spins has a positive sign, i.e. $\sigma_i = 1$, then, on a microscopic level, at every given time there is a fixed amount of spins $V(1-m_0)/2$ in the system that are overturned and therefore have a negative sign. If we artificially increase the number of overturned spins by a macroscopic amount, then the magnetisation decreases and we can pose the question what happens to the extra -1-spins. One possible answer is that the system forms a droplet of the "wrong phase" that has the same magnitude of the spontaneous magnetisation but the opposite sign, i.e., this phase consists of negative majority spins with some overturned

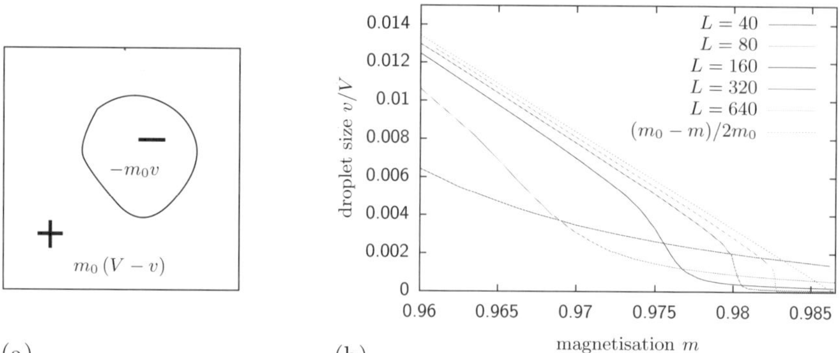

(a) (b)

Fig. 1. (a) Sketch of the situation considered in Eq. (2): a droplet of size v and magnetisation $-m_0$ inside a system of size V and magnetisation m_0. (b) Relative droplet volume v/V for the 2D Ising model at the temperature $T = 1.5$ and for different system sizes from $L = 40$ to $L = 640$. The abscissa ranges from $m_{\mathrm{min}} = 0.96$ to the value of the spontaneous magnetisation $m_{\mathrm{max}} = m_0(1.5) = 0.9865$. The dashed line shows Eq. (3); for the measured values a combination of Hoshen-Kopelman and flood-fill algorithm was used. Similar plots can be found in Refs. 2 and 3.

spins within. If all additional -1-spins are "absorbed" into this phase, and if this phase is compact, then there is only one (large) droplet with negative magnetisation as shown in Fig. 1 (a). The total magnetisation is then given by a contribution $m_0(V-v)$ from the positive phase (background) of volume $V-v$ and a contribution $-m_0 v$ from the negative phase of volume v giving

$$M = m_0(V-v) - m_0 v = m_0 V - 2m_0 v \,. \tag{2}$$

We can measure the volume of this droplet and it must hold

$$\frac{v}{V} = \frac{m_0 - m}{2m_0} \,. \tag{3}$$

To check this result, we performed several simulations of the 2D Ising model at constant magnetisation, equivalent to Kawasaki dynamics.[4] All simulation were performed at the temperature $T = 1.5$. After every simulation sweep the spin field was decomposed into geometric clusters where all spins within a cluster have the same value ($+1$ or -1) using the Hoshen-Kopelman algorithm.[5] As we are interested in the *volume* of the second largest droplet (the largest droplet is the background), which includes all the overturned spins within the cluster, in a second step, a flood-fill routine[6] was employed. Starting at one spin in the interior of the cluster, recursively all spins are marked to belong to the droplet. This process stops only in directions where a spin belongs to the largest cluster but not necessarily if it is overturned.

Figure 1 (b) shows the droplet size for various values of the magnetisation m (every point is a single simulation). Clearly, only for large system sizes and small values of the magnetisation m the theoretical value (3) of the droplet is approached. For $L \geq 160$ and large values of m, a kink in the droplet size becomes visible and the droplet size rapidly approaches zero. The position of this drop-off is moving for larger system sizes towards m_0 while the height of the drop-off is decreasing. Apparently, the assumption that *all* extra spins form a droplet is not correct but only a part of them form the droplet while the rest stays in the background. For values of the magnetisation larger than the drop-off value, there is no droplet at all, i.e., in this case all extra spins contribute to the fluctuations in the background and the maximal droplet volume is of the order one.

In recent work Biskup *et al.*[7,8] were able to proof this behaviour in the case of the 2D Ising model rigorously. They give an analytic expression for $\lambda_L(m)$, the fraction of the additionally overturned spins that help to form the droplet. Then, the actual droplet volume is not $v(m)$ but $\lambda_L(m)v(m)$. Furthermore, for values $m > m_c$ there is no large minority droplet at all and consequently $\lambda_L(m) = 0$. At $m = m_c$ the value of λ_L jumps to 2/3, marking the position where the system makes a transition from a one-phase state (evaporated) to a two-phase state (evaporated/condensed), thereby absorbing 2/3 of the extra -1-spins into the droplet. For lower values of the magnetisation the fraction $\lambda_L(m)$ gradually increases to 1 and the actual droplet size approaches that of Eq. (3). In Fig. 2 we show the data for $L = 160$ and $L = 640$ (inset) from Fig. 1 again but additionally the (red) solid curve shows the finite-size corrected theoretical value of the droplet size $\lambda_L(m)v(m)$.

For large but finite systems, there exists a jump in the size of the largest minority droplet. Our simulations confirm that already for systems of moderate size the theoretical value is approached and there exists a transition from an evaporated state with many small bubbles of the "wrong" orientation to a mixed state with a condensed and an evaporated phase. In the thermodynamic limit the transition point coincides with the spontaneous magnetisation m_0 and the jump in the relative droplet size v/V approaches zero. Therefore the effect cannot be seen in the thermodynamic limit, where Eq. (3) gets (trivially) exact and $\lambda_L = 1$ holds everywhere.

More details can be found in Ref. 9 and the forthcoming detailed work in Ref. 10.

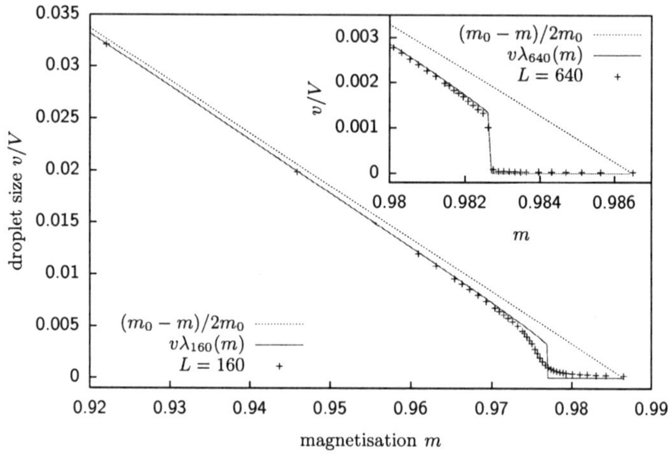

Fig. 2. Relative droplet volume v/V for the 2D Ising model at the temperature $T = 1.5$ and system sizes $L = 160$ and $L = 640$ (inset). The dashed line shows Eq. (3) and the '+' symbols indicate the measured values. The solid line represents the theoretical curve modified by the factor $\lambda_L(m)$.

Acknowledgments

Work supported by the Deutsche Forschungsgemeinschaft (DFG) under grants No. JA483/22-1 and No. JA483/23-1 and in part by the EU RTN-Network 'ENRAGE' under grant No. MRTN-CT-2004-005616. Supercomputer time at NIC Jülich under grant No. hlz10 is also gratefully acknowledged.

References

1. L. Onsager, *Nuovo Cim. (Suppl.)* **6**, 261 (1949).
2. M. Pleimling and W. Selke, *J. Phys. A: Math. Gen.* **33**, L199 (2000).
3. T. Neuhaus and J. S. Hager, *J. Stat. Phys.* **113**, 47 (2003).
4. K. Kawasaki, *Phys. Rev.* **145**, 224 (1966).
5. J. Hoshen and R. Kopelman, *Phys. Rev. B* **14**, 3438 (1976).
6. M. K. Agoston, *Computer Graphics and Geometric Modelling* (Springer, London, 2004).
7. M. Biskup, L. Chayes, and R. Kotecký, *Europhys. Lett.* **60**, 21 (2002).
8. M. Biskup, L. Chayes, and R. Kotecký, *Comm. Math. Phys.* **242**, 137 (2003).
9. A. Nußbaumer, E. Bittner, T. Neuhaus, and W. Janke, *Europhys. Lett.* **75**, 716 (2006).
10. A. Nußbaumer, E. Bittner, and W. Janke, *Phys. Rev. E* **77**, 041109 (2008).

PART X

Biophysics and Stochastics

CONFORMATIONAL TRANSITIONS IN MOLECULAR SYSTEMS*

M. BACHMANN and W. JANKE

Institut für Theoretische Physik, Universität Leipzig,
Postfach 100 920, D-04009 Leipzig, Germany
E-mail: bachmann@itp.uni-leipzig.de, janke@itp.uni-leipzig.de
www.physik.uni-leipzig.de/CQT.html

Proteins are the "work horses" in biological systems. In almost all functions specific proteins are involved. They control molecular transport processes, stabilize the cell structure, enzymatically catalyze chemical reactions; others act as molecular motors in the complex machinery of molecular synthetization processes. Due to their significance, misfolds and malfunctions of proteins typically entail disastrous diseases, such as Alzheimer's disease and bovine spongiform encephalopathy (BSE). Therefore, the understanding of the trinity of amino acid composition, geometric structure, and biological function is one of the most essential challenges for the natural sciences. Here, we glance at conformational transitions accompanying the structure formation in protein folding processes.

Keywords: Conformational transition; Protein folding; Monte Carlo simulation.

1. Conformational Mechanics of Proteins

Structural changes of polymers and, in particular, proteins in collapse and crystallization processes, but also in cluster formation and adsorption to substrates, require typically collective and cooperative rearrangements of chain segments or monomers. Structure formation is essential in biosystems as in many cases the function of a bioprotein is connected with its three-dimensional shape (the so-called "native fold"). Proteins are linear chains of amino acids linked by a peptide bond (see Fig. 1). Twenty different amino acids occur in biologically relevant, i.e., functional proteins. The amino acid residues differ in physical (e.g., electrostatic) and chemical (e.g., hydrophobic) properties. Hence, the sequence of amino acids typically

*Work supported by Deutsche Forschungsgemeinschaft under grant No. JA483/24-1/2.

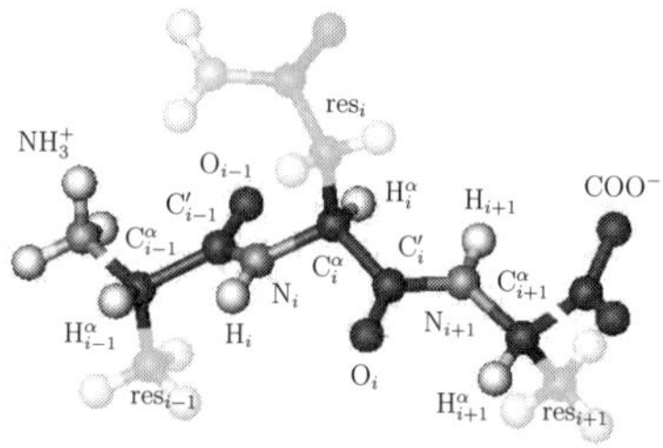

Fig. 1. Atomic composition of the protein backbone. Amino acids are connected by the peptide bond between C'_{i-1} and N_i. Side chains or amino acid residues ("res") are usually connected to the backbone by a bond with the C_i^α atom (except proline which has a second covalent bond to its backbone nitrogen).

entails a unique heterogeneity in geometric structure and, thus, a nonredundant biological function.

Proteins are synthesized by the ribosomes in the cell, where the genetic code in the DNA is translated into a sequence of amino acids. The folding of a synthesized protein into its three-dimensional structure is frequently a spontaneous process. In a complex biological system, the large variety of processes which are necessary to keep an organism alive requires a large number of different functional proteins. In the human body, for example, about 100 000 different proteins fulfil specific functions. However, this number is extremely small, compared to the huge number of possible amino acid sequences ($= 20^N$, where N is the chain length and is typically between 100 and 3000). The reason is that bioproteins have to obey very specific requirements. Most important are stability, uniqueness, and functionality.

Under physiological conditions, flexible protein degrees of freedom are the dihedral angles, i.e., a subset of backbone and side-chain torsional angles (see Fig. 2). Denoting the set of dihedral angles of the nth amino acid in the chain by $\boldsymbol{\xi}_n = \{\phi_n, \psi_n, \omega_n, \chi_n^{(1)}, \chi_n^{(2)}, \ldots\}$, the conformation of an N residue protein is then entirely defined by $\mathbf{X} = \mathbf{X}(\boldsymbol{\xi}_1, \boldsymbol{\xi}_2, \ldots, \boldsymbol{\xi}_N)$. Therefore, the partition function can formally be written as a path integral over all possible conformations:

$$Z = \int \mathcal{D}\mathbf{X} \exp\left[-E(\mathbf{X})/k_BT\right], \quad \int \mathcal{D}\mathbf{X} = \prod_{n=1}^{N} \left[\int d\boldsymbol{\xi}_n\right], \tag{1}$$

where $E(\mathbf{X})$ is the energy of the conformation $\mathbf{X}$ in a typically semiclassical all-atom protein model. A precise modeling is intricate because of the importance of quantum effects in this complex macromolecular system, which are "hidden" in the parameterization of the semiclassical model. Another

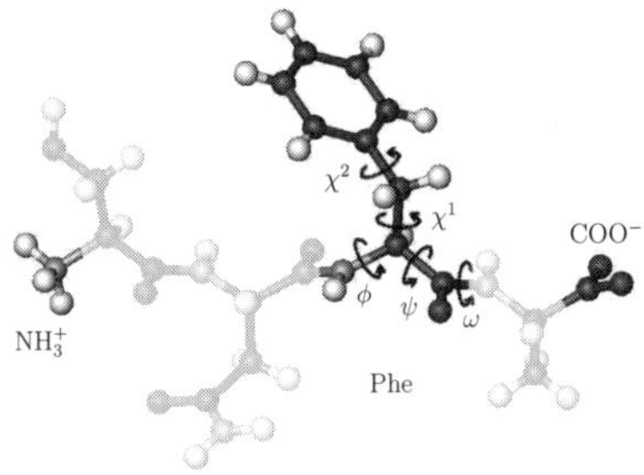

Fig. 2. Definition of the backbone dihedral angles ϕ, ψ, and ω. Exemplified for phenylalanine, also the only two side-chain degrees of freedom χ^1 and χ^2 are denoted. The convention is that the torsional angles can have values between $-180°$ and $+180°$, counted from the N-terminus (NH_3^+) to the C-terminus (COO^-) according to the right-hand rule and in the side chains starting from the C^α atom.

important problem is the modeling of the surrounding, strongly polar solvent. The hydrophobic effect that causes the formation of a compact core of hydrophobic amino acids screened from the polar solvent by a shell of polar residues is expected to be the principal driving force towards the native, functional protein conformation.[1–3] Conformational transitions accompanying molecular structuring processes, however, exhibit similarities to thermodynamic phase transitions and it should thus be possible to characterize these transitions by means of a strongly reduced set of effective degrees of freedom, in close correspondence to order parameters that separate thermodynamic phases. Assuming that a single "order" parameter Q is sufficient to distinguish between two (pseudo)phases, its mean value $\langle Q\rangle = Z^{-1}\int \mathcal{D}\mathbf{X}\, Q(\mathbf{X}) \exp\left[-E(\mathbf{X})/k_BT\right]$ should possess significantly different values in these phases. In typical first-order-like nucleation transitions such as helix formation[4] or tertiary two-state folding,[5] the free-energy landscape $F(Q) \sim -k_BT \ln \langle \delta(Q - Q_0(\mathbf{X}))\rangle$ exhibits a single folding barrier.

2. From Microscopic to Mesoscopic Modeling

If the characterization of conformational macrostates by low-dimensional parameter spaces is possible, it should also be apparent to introduce coarse-grained substructures and thus to reduce the complexity of the model to a minimum. Such minimal models for proteins have indeed been introduced[1,2] and have proven useful in thermodynamic analyzes of folding, adsorption, and aggregation of polymers and proteins.[3,5–8]

In the simplest approaches,[1,2] only two types of amino acids are considered: hydrophobic and polar residues. This is plausible as most of the 20

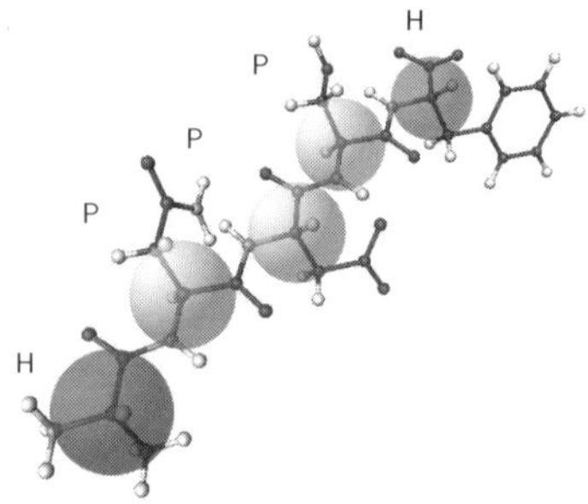

Fig. 3. Coarse-graining proteins in a "united atom" approach. Each amino acid is contracted to a single "C^α" interaction point. The effective distance between adjacent, bonded interaction sites is about 0.38 nm. In the class of so-called hydrophobic-polar models, only hydrophobic (H) and polar (P) amino acid residues are distinguished.

amino acids occurring in natural bioproteins can be classified with respect to their hydrophobicity. Amino acids with charged side chains or with residues containing polar groups (amide or hydroxylic) are soluble in the aqueous environment, because these groups are capable of forming hydrogen bonds with water molecules. Nonpolar amino acids do not form hydrogen bonds and, if exposed to water, would disturb the hydrogen-bond network. This is energetically unfavorable. In fact, hydrophobic amino acids effectively attract each other and typically form a compact hydrophobic core in the interior of the protein.

Figure 3 shows an example how the complexity of a protein segment can be reduced by coarse-graining. On one hand, the residual complexity is limited by only distinguishing hydrophobic (H) and polar (P) amino acids. On the other hand, the steric extension of the side chains is mesoscopically rescaled and the whole side chain is contracted into a single interaction point. Volume exclusion in the interaction of different side chains is then energetically modeled by short-range repulsion. For this reason, lattice proteins are modeled as self-avoiding walks[1] and in off-lattice models Lennard-Jones-like potentials[2] satisfy this constraint.

Systematic enumeration studies of simplified hydrophobic-polar lattice models have indeed qualitatively revealed characteristic features of real proteins, such as the small number of amino acid sequences possessing a unique native fold, but also the comparatively small number of native topologies proteins fold into.[9] It is also remarkable that typical protein folding paths known from nature are also identified by employing coarse-grained models. This regards, in particular, folding landscapes with characteristic barriers – from the simple two-state characteristics with a single kinetic barrier,[5] over folding across several barriers via weakly stable intermediate structures, to folding into degenerate native states.[6] Metastable conformations as in the latter case are important for biological functions, where the local refolding of protein segments is essential, as, e.g., in molecular motors.

3. A Particularly Simple Example: Two-State Folding

A few years ago, experimental evidence was found that classes of proteins show particular simple folding characteristics, single exponential and two-state folding.[10] In the two-state folding process, the peptide is either in an unfolded, denatured state or it possesses a native-like, folded structure. In contrast to the barrier-free single-exponential folding, there exists an unstable transition state to be passed in the two-state folding process. This can nicely be seen in the exemplified chevron plot shown in Fig. 4, obtained from Monte Carlo computer simulations of folding and unfolding events of a mesoscopic protein model.[5] In this plot, the mean first passage (MFP) time $\tau_{\rm MFP}$ (in Monte Carlo steps) is plotted versus temperature. The MFP time is obtained by averaging the times passed in the folding process from a random conformation to the stable fold over many folding trajectories. MFT times for unfolding events can be estimated in a like manner, but one starts from the native conformation and waits until the protein has unfolded. A structure is defined to be folded, if it is structurally close to the native conformation. A frequently used measure is the fraction Q of already established native contacts (i.e., the number of residue pairs that reside within the optimal van der Waals distance), compared to the total number of contacts the native fold possesses. Thus, if $Q > 0.5$, the structure is folded and unfolded if $Q < 0.5$. For $Q = 0.5$, the conformation is in the transition state. Apparently, Q serves as a sort of order parameter.

The two branches in Fig. 4 belong to the folding and unfolding events. With increasing temperature folding times grow, and unfolding is getting slower with decreasing temperature. These two processes are in competition with each other and the intersection point defines the folding transition temperature. The whole process exhibits characteristics of first-order-like phase transitions. At the intersection point, the ensembles of folded and unfolded conformations coexist with equal weight. In the transition region, both branches exhibit exponential behavior. Thus, $\tau_{\rm MFP}$ is directly related to exponential folding and unfolding rates $k_{f,u} \approx 1/\tau_{\rm MFP}^{f,u} \sim \exp(-\varepsilon_{f,u}/k_BT)$, respectively, where the constants $\varepsilon_{f,u}$ determine the kinetic folding (unfolding) propensities. The dashed lines in Fig. 4 are tangents to the logarithmic folding and unfolding curves at the transition state temperature.

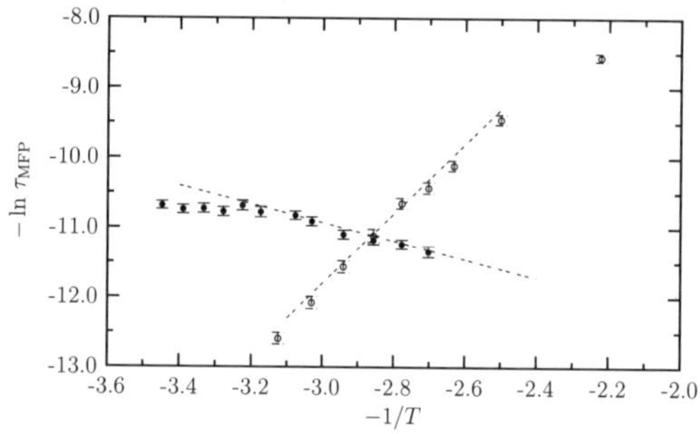

Fig. 4. Chevron plot of the mean-first passage times from folding (•) and unfolding (○) events at different temperatures. The hypothetic intersection point corresponds to the transition state.[5]

4. Conclusion

Conformational transitions of macromolecular systems, in particular, proteins, exhibit clear analogies to phase transitions in thermodynamics. The main difference is that proteins are finite systems and a thermodynamic limit does not exist. Nonetheless, the analysis of structure formation processes in terms of an "order" parameter is also a very useful approach to a better understanding of conformational transitions. In this context it also turns out to be reasonable to introduce coarse-grained models where the reduction to only relevant degrees of freedom allows for a more systematic analysis of characteristic features of protein folding processes than it is typically possible with models containing specific properties of all atoms.

References

1. K. F. Lau and K. A. Dill, *Macromolecules* **22**, 3986 (1989).
2. F. H. Stillinger and T. Head-Gordon, *Phys. Rev. E* **52**, 2872 (1995).
3. M. Bachmann and W. Janke, *Lect. Notes Phys.* **736**, 203 (2008).
4. G. Gökoğlu, M. Bachmann, T. Celik, and W. Janke, *Phys. Rev. E* **74**, 041802 (2006).
5. A. Kallias, M. Bachmann, and W. Janke, *J. Chem. Phys.* **128**, 055102 (2008).
6. S. Schnabel, M. Bachmann, and W. Janke, *Phys. Rev. Lett.* **98**, 048103 (2007); *J. Chem. Phys.* **126**, 105102 (2007).
7. M. Bachmann and W. Janke, *Phys. Rev. Lett.* **95**, 058102 (2005); *Phys. Rev. E* **73**, 041802 (2006); *Phys. Rev. E* **73**, 020901(R) (2006).
8. C. Junghans, M. Bachmann, and W. Janke, *Phys. Rev. Lett.* **97**, 218103 (2006); *J. Chem. Phys.* **128**, 085103 (2008).
9. R. Schiemann, M. Bachmann, and W. Janke, *J. Chem. Phys.* **122**, 114705 (2005); *Comp. Phys. Comm.* **166**, 8 (2005).
10. S. E. Jackson and A. R. Fersht, *Biochemistry* **30**, 10428 (1991).

DYNAMICS OF STICKY POLYMER SOLUTIONS

J. GLASER*, C. HUBERT, and K. KROY
Institut für Theoretische Physik, Universität Leipzig,
Postfach 100 920, D-04009 Leipzig, Germany
**E-mail: jens.glaser@itp.uni-leipzig.de*

We give an overview of the glassy wormlike chain model, which describes the dynamics of a semiflexible polymer interacting with a surrounding sticky solution. The model is then generalized to the Rouse chain and we compare the macroscopic shear modulus.

Keywords: Wormlike chain; Glassy wormlike chain; Glassy rouse chain.

1. Introduction

The dynamics of single polymers in solutions is well understood in terms of coarse grained models such as the Rouse/Zimm chain for flexible polymers[1] and the wormlike chain (WLC) for (locally) stiff polymers.[2] While the former is known to yield very generic universal predictions, the latter has recently been demonstrated to encode a very rich dynamical behavior as a consequence of the mechanical anisotropy and the ensuing competition of transverse and longitudinal friction.[3] Steric and enthalpic interactions with a surrounding polymer solution lead to substantial modifications of the free dynamics. For stiff polymers these could recently be captured impressively[4] by a very simple scheme based on the single polymer spectrum, which we call the glassy wormlike chain (GWLC).[5] The latter is obtained by an exponential stretching of the WLC spectrum argued to represent the effect of free energy barriers due to excluded volume constraints and sticky interactions. Both types of interactions can be complex and are in fact much more common in colloid and protein solutions than generally appreciated by most physicists, who are mainly trained to deal with Lennard-Jones-type interactions. Figure 1 depicts qualitatively what is meant by "sticky" or "friction-type" interactions, which may be thermodynamically almost insignificant while causing pronounced dynamical effects. In the following, we summarize some salient consequences of such "unusual" interactions for the

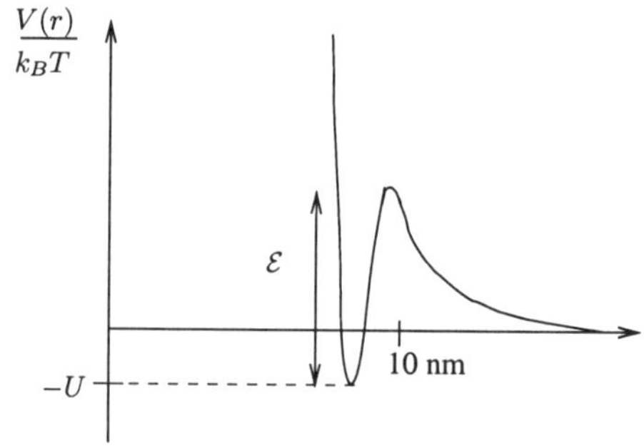

Fig. 1. Schematic potential for sticky interactions between polymers. The depth U of the potential gives $\Lambda/L_e = 1 + \exp(-U)$.

WLC dynamics, before we explore the corresponding predictions for the Rouse chain.

2. Dynamics of Flexible and Semiflexible Polymers

2.1. *The wormlike chain*

In the wormlike chain (WLC) model the polymer contour is represented as a continuous space curve $\mathbf{r}(s)$. The nonlinear inextensibility constraint is taken care of by introducing into the Hamiltonian a Lagrange multiplier function $f(s)$, which has the physical interpretation of a backbone tension. In the weakly-bending rod limit,[3] the polymer contour is parameterized by small transverse deflections around the ground state of a straight line (chosen as the z-axis): $\mathbf{r}(s,t) = [\mathbf{r}_\perp(s,t), s - r_\parallel(s,t)]$. The arc length constraint implies $r'_\parallel = \mathbf{r}'^2_\perp/2$. To lowest order in an expansion in terms of the transverse variables, longitudinal fluctuations can therefore be neglected, i.e. the longitudinal equation of motion reduces to $f =$ const. ($= 0$ for a free polymer). The transverse equation of motion is given by[3]

$$\zeta_\perp \partial_t \mathbf{r}_\perp(s,t) = -\kappa \mathbf{r}''''_\perp(s,t) + [f(s,t)\mathbf{r}'_\perp(s,t)]' + \boldsymbol{\xi}_\perp(s,t), \tag{1}$$

with the bending rigidity κ, the transverse friction constant $\zeta_\perp$ and the transverse Gaussian white noise $\boldsymbol{\xi}_\perp$. We note that the effect of an external prestress σ can be included under the assumption that the tension has equilibrated under a constant value $f = 5\sigma\xi^2$ set by the external force.[5] Introducing normal modes of (half) wavelength λ_p, the correlator of the mode amplitudes a_p is calculated,

$$\langle a_{p,i}(t) a_{q,j}(0)\rangle = \delta_{ij}\delta_{pq}\langle a_p^2\rangle \exp(-t/\tau_p). \tag{2}$$

In the last equation, the equilibrium mode amplitudes follow from the equipartition theorem as $\langle a_p^2 \rangle = k_B T/(\pi^4 \lambda_p^{-4} \kappa + \pi^2 \lambda_p^{-2} f)$, and the relaxation times are defined as $\tau_p = \zeta_\perp/(\pi^4 \lambda_p^{-4} \kappa + \pi^2 \lambda_p^{-2} f)$.

2.2. *The Rouse chain*

The equation of motion of the Rouse chain is[1]

$$\zeta \frac{\partial \mathbf{r}_n(t)}{\partial t} = k \frac{\partial^2 \mathbf{r}_n(t)}{\partial n^2} + \boldsymbol{\xi}_n(t), \tag{3}$$

where ζ, k and $\boldsymbol{\xi}$ represent effective coarse grained friction and spring coefficients and thermal Gaussian noise, respectively, and n is a fictitious monomer index. The equation is solved by introducing normal modes with amplitudes $\mathbf{a}_p$ obeying

$$\langle a_{p,i}(t) a_{q,j}(0) \rangle = \delta_{pq} \delta_{ij} \frac{N^2 k_B T}{\pi^2 k p^2} \exp(-t/\tau_p), \qquad (p > 0) \tag{4}$$

with the mode relaxation time $\tau_p = N^2 \zeta / 2\pi^2 k p^2$.

3. The Glassy Wormlike Chain (GWLC)

The GWLC model is obtained from the WLC by an exponential stretching of the relaxation spectrum. The strategy is reminiscent of the generic trap models[6] underlying soft glassy rheology,[7] but concerns the equilibrium dynamics of a test chain in solution, here. We distinguish two major contributions to the slowdown of the polymer dynamics by the surroundings: steric (excluded volume) interactions and sticky interactions with the neighboring chain. The GWLC is a WLC with the relaxation times for all its eigenmodes of wavelength $\lambda_p \equiv L/p > L_e \equiv L/p_e$ (L_e is the entanglement length) – modified according to

$$\tau_p \to \tilde{\tau}_p = \begin{cases} \tau_p & (p > p_e) \\ \tau_p \exp(N'_p \mathcal{F}) & (p < p_e) \\ \tau_p \exp(N_p \mathcal{E} + N'_p \mathcal{F}) & (p < p_\Lambda \le p_e) \end{cases} . \tag{5}$$

Here

$$N'_p \equiv (p_e/p)^4 - 1 = (\lambda_p/L_e)^4 - 1, \tag{6}$$

$$N_p \equiv p_\Lambda/p - 1 = \lambda_p/\Lambda - 1. \tag{7}$$

One can think of N'_p and N_p as the number of entanglements or sticky interactions per wavelength respectively. The dimensionless parameters $\mathcal{F}$ and $\mathcal{E}$ represent the corresponding free energy barriers (cf. Fig. 1) in units of

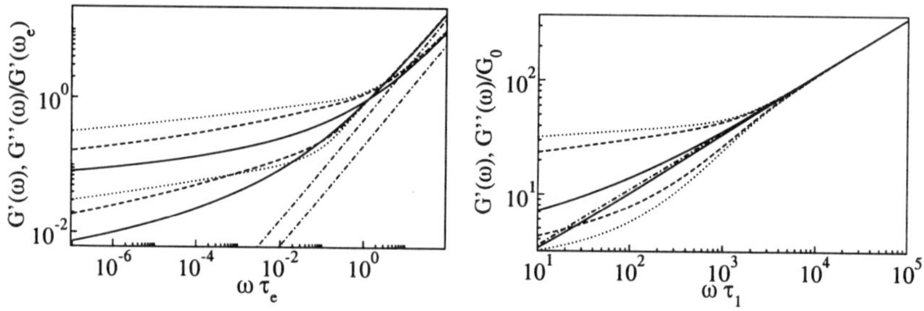

Fig. 2. *Left*: $G', G''(\omega)$ of a GWLC with steric interactions and $L_e/\Lambda = 0.7$. *Right*: GRC with $\xi/R_g = 0.15$. Both figures show the same values of $\mathcal{E}$ and $\mathcal{F}$: $\mathcal{F} = 0$, $\mathcal{E} = 0$ (dash-dotted lines), $\mathcal{F} = 0.1$ and $\mathcal{E} = 0$ (solid lines), 10 (dashed lines) and 25 (dotted lines).

thermal energy k_BT. The relaxation of a mode necessitates the creation of free volume, $V_\lambda \propto \lambda^4$, the number N_p' of effective entanglements is defined as $(V_\lambda/V_{L_e}) - 1$. In general the polymer can be completely sticky, in which case $\Lambda = L_e$, or its contacts with the surroundings may be partially sticky, which corresponds to $\Lambda > L_e$. Under equilibrium conditions, one expects $L_e/\Lambda = 1 + \exp(-U)$.

In the following we compare the shear modulus for the GWLC which derives from the complex susceptibility $\alpha_f^\star(\omega)$ of the test polymer (of length L) under affine deformation on scales longer than Λ.[8] The expression for the shear modulus under a constant external prestress f is $G_f^\star(\omega) = L/5\xi^2\alpha_f^\star(\omega)$, where ξ^2 is the mesh size, L the length of the polymer and

$$\alpha_f^\star(\omega) = \frac{L^4}{k_BT\ell_p^2\pi^4}\sum_{p=1}^{\infty}\frac{p^4}{(p^4+p^2f/f_L)^2}\frac{1}{(1+i\omega\tilde{\tau}_p/2)}\,. \tag{8}$$

We denote by $f_L = \kappa\pi^2/L^2$ the Euler force of a polymer of length L. In the remainder of the discussion we assume $f = 0$.

As a consequence of the strong exponential stretching, the dependence of the complex shear modulus on frequency is essentially logarithmic for small frequencies. However, over small frequency ranges the modulus *locally* looks like a power-law, as familiar from the soft glassy rheology,[7,9] where $G, G''(\omega) \propto \omega^{x-1}$ with x very close to 1. The asymptotic relation between $\mathcal{E}$ and the apparent power-law exponent $x - 1$ for $\mathcal{E} \to \infty$ and $\omega\tau_\Lambda \ll 1$ is $x - 1 \sim 3/\mathcal{E}$, and for the loss angle $\delta \sim 5/\mathcal{E}$ (we remark that the limits $\mathcal{E} \to \infty$ and $\omega \to 0$ do not commute). Figure 2 shows the frequency-dependent macroscopic shear modulus of a GWLC with and without steric interactions.

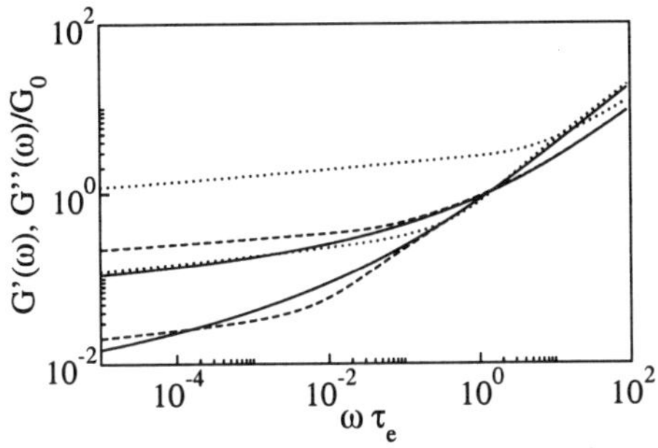

Fig. 3. $G'(\omega)$, $G''(\omega)$ of a GWLC for varying L_e/Λ and for $\mathcal{E} = 30$, $\mathcal{F} = 0.1$. $L_e/\Lambda = 0$ (solid lines), 0.5 (dashed lines), 1.0 (dotted lines).

In Fig. 3 the dependence of the moduli on the ratio L_e/Λ is shown. It can be seen that the finite value of this ratio ($0 < L_e/\Lambda < 1$) determines the width of the frequency interval where the modulus exhibits a strongly curved shape for intermediate values of ω. This interval is followed (for $\omega\tau_e \ll \exp\{-\mathcal{F}[(\Lambda/L_e)^4 - 1]\}$) by a regime, where for $L_e \simeq \Lambda$ the local power-law behavior is clearly visible (due to the dominant contribution of $\mathcal{E} \gg \mathcal{F}$).

4. The Glassy Rouse Chain

In the spirit of the above discussed GWLC model it is straightforward to apply the same prescriptions for the relaxation times to the Rouse model of flexible polymers (the Glassy Rouse chain), to include the interactions of the test chain with the surrounding polymer network. The role of the entanglement length is now played by the mesh size ξ (which is also the wavelength of mode number p_ξ). The well known expression for the shear modulus[1] is easily generalized, by the modification of

$$\tau_p \to \tilde{\tau}_p = \begin{cases} \tau_p & (p > p_\xi) \\ \tau_p \exp(N_p\mathcal{E} + N'_p\mathcal{F}) & (p < p_\xi). \end{cases} \tag{9}$$

We define

$$N_p = \frac{\lambda_p}{\xi} - 1, \tag{10}$$

$$N'_p = \frac{V_p}{\xi^3} - 1, \tag{11}$$

as the number of sticky contacts and the number of entanglements, respectively. The relations $\lambda_p \equiv R_g/\sqrt{p}$ and $V_p \equiv (R_g/\sqrt{p})^3$ hold. p_ξ is related to the radius of gyration via $p_\xi = R_g^2/\xi^2$. Figure 2 shows the predicted modulus of a GRC.

5. Conclusion

Traditionally the viscoelastic rheological response of flexible and semiflexible polymers has been discussed on the basis of the tube idea.[1,10–12] More recently, the GWLC idea has been shown to provide a very successful alternative description of the plateau regime for stiff biopolymers. In the present contribution we have compared these predictions to predictions obtained with the model of a glassy Rouse chain, which we constructed in close analogy to the GWLC. It appears to be a promising task to compare the results of the GRC to experiments in the future.

References

1. M. Doi and S. F. Edwards, *The Theory of Polymer Dynamics* (Oxford University Press, 1988).
2. O. Kratky and G. Porod, *Rec. Trav. Chim.* **68**, 1105 (1949).
3. O. Hallatschek, E. Frey, and K. Kroy, *Phys. Rev. E* **75**, 31905 (2007).
4. C. Semmrich, T. Storz, J. Glaser, R. Merkel, A. R. Bausch, and K. Kroy, *Proc. Natl. Acad. Sci. USA* **104**, 20199 (2007).
5. K. Kroy and J. Glaser, *New J. Phys.* **9**, 416 (2007).
6. C. Monthus and J. Bouchaud, *J. Phys. A: Math. Gen.* **29**, 3847 (1996).
7. P. Sollich, F. Lequeux, P. Hébraud, and M. E. Cates, *Phys. Rev. Lett.* **78**, 2020 (1997).
8. F. Gittes and F. C. MacKintosh, *Phys. Rev. E* **58**, R1241 (1998).
9. B. Fabry, G. Maksym, J. Butler, M. Glogauer, D. Navajas, and J. Fredberg, *Phys. Rev. Lett.* **87**, 148102 (2001).
10. P. G. de Gennes, *Scaling Concepts in Polymer Physics* (Cornell University Press, 1979).
11. D. Morse, *Phys. Rev. E* **63**, 31502 (2001).
12. W. Vandoolaeghe and E. Terentjev, *J. Phys. A: Math. Theor.* **40**, 14725 (2007).

DRIFT OF A POLYMER IN SOLVENT BY FORCE APPLIED AT ONE POLYMER END

S. STEPANOW[*,‡] and N. KIKUCHI[*,†]

[*]*Institut für Physik, Martin-Luther-Universität Halle-Wittenberg, D-06099 Halle, Germany*

[†]*Centre for Condensed Matter Theory, Department of Physics, Indian Institute of Science, Bangalore 560 012, India*

[‡]*E-mail: semjon.stepanow@physik.uni-halle.de*

We investigate the effect of hydrodynamic interactions on the non-equilibrium dynamics of an ideal flexible polymer pulled by a constant force applied at one polymer end using the perturbation theory. For moderate force, if the polymer elongation is small, the hydrodynamic interactions are not screened and the velocity and the longitudinal elongation of the polymer are computed using the renormalization group method. For large chain lengths and a finite force the hydrodynamic interactions are only partially screened, which in three dimensions results in unusual logarithmic corrections to the velocity and the longitudinal elongation.

Keywords: Polymer elasticity and dynamics; Driven diffusive systems; Renormalization group.

1. Introduction

Recent advances in experimental techniques make it possible to explore the motion of individual polymers under hydrodynamic flow, thermal noise, or external fields. For instance, Chu *et al.*[1–3] experimentally studied the behavior of polymers in different flow conditions with the emphasis on biological applications. See also related theoretical studies.[4,5]

In neglecting the hydrodynamic interactions, i.e., in the Rouse theory, the average velocity and the elongation of a chain pulled at one end are given by expressions

$$\mathrm{v}^z \simeq \frac{F}{f_0 N}, \quad \langle r^z(0,t) - r^z(L,t)\rangle = \frac{FLl}{2dk_BT}, \tag{1}$$

where d is the dimensionality, F is the pulling force, and f_0 denotes the monomer friction coefficient. For small forces the polymer is expected to have the shape of coil with the consequence that the hydrodynamic interaction is not screened, and the velocity is described by Stokes formula

$$\mathrm{v}^z \simeq \frac{1}{6\pi\eta_s R_h} F \tag{2}$$

with $R_h \simeq l\sqrt{N}$ being the hydrodynamic radius of the polymer, and η_s is the solvent viscosity. In the present work we calculate the first-order corrections to the velocity and the longitudinal elongation. The renormalization group method allows us to establish the behavior of the quantities under consideration for moderate forces. We also show that in $d = 3$ the first-order corrections to the velocity and the longitudinal size of the polymer in powers of the hydrodynamic interactions logarithmically depend on the parameter $\beta F\sqrt{lL/2d}$, which demonstrates that hydrodynamic interactions are only partially screened for a long polymer pulled at finite force.

2. Computation of the Velocity and the Longitudinal Elongation

Using the Kirkwood diffusion equation (see Ref. 6 for details) we can calculate the velocity and the longitudinal elongation of the polymer chain up to the first order in hydrodynamic interaction. To first order in powers of the hydrodynamic interaction the velocity is given by

$$\mathrm{v}_c^z(t) = \frac{\langle \xi_{k=0}^z(t)\rangle}{\sqrt{L}t} = \frac{F}{f_0 N}\left(1 + \frac{1}{2^2}\frac{f_0}{d\eta_s}\left(\frac{d}{2\pi l}\right)^{d/2} L^{2-d/2} \right.$$
$$\left. \times \int_0^1 dx_2 \int_0^{x_2} dx_1 \frac{A(y)}{(x_2 - x_1)^{d/2-1}} + \cdots \right), \tag{3}$$

where the function $A(y)$ and its argument y are respectively given by

$$A(y) = \frac{(d-1)(d-2)}{y^4} e^{-y^2} - \frac{(d-1)\left(d-2-2y^2\right)}{y^d}$$
$$\times \left(\frac{4-d}{2}\Gamma\left(\frac{d-4}{2}, y^2\right) + \Gamma\left(\frac{d-2}{2}\right)\right), \tag{4}$$

$$y = \beta F\sqrt{\frac{lL}{2d}}(x_2 - x_1)^{1/2}(1 - \frac{x_2 + x_1}{2}). \tag{5}$$

$\Gamma(a, z) = \int_z^\infty dt\, t^{a-1} e^{-t}$ is the incomplete gamma function. The function $A(y)$ behaves for small and large arguments as

$$A(y) \simeq \begin{cases} \frac{8(d-1)}{d(d-2)} - \frac{8(d-1)}{d(d+2)} y^2 + \frac{4(d-1)}{(d+2)(d+4)} y^4 + \cdots, & y \ll 1, \\ \frac{(d-1)(2y^2+2-d)}{y^d}\Gamma\left(\frac{d-2}{2}\right) + \frac{2(d-1)}{y^4} e^{-y^2} + \cdots, & y \gg 1. \end{cases} \tag{6}$$

Integrations over x_1 and x_2 in Eq. (3), which can be carried out in the limit $F \to 0$, yield

$$\int_0^1 dx_2 \int_0^{x_2} dx_1 \frac{A(0)}{(x_2 - x_1)^{d/2-1}} = \frac{4}{(6-d)(4-d)} A(0) \overset{d \sim 4}{\to} \frac{2}{4-d} A(0). \quad (7)$$

Note that the behavior of the first-order correction to the velocity in the vicinity of four dimensions, which is given by the last expression, plays an important role in the renormalization group analysis of the velocity.

The longitudinal size of the polymer is computed as

$$\delta r_z = \frac{FLl}{2dk_BT} \left(1 - \frac{1}{2^2} \frac{f_0}{d\eta_s} \left(\frac{d}{2\pi l} \right)^{d/2} L^{2-d/2} B_r(\tilde{F}) + \cdots \right), \quad (8)$$

where $\delta r_z = \langle r^z(0,t) - r^z(L,t) \rangle$, $\tilde{F} = \beta F \sqrt{\frac{lL}{2d}}$, and

$$B_r(\tilde{F}) = \int_0^1 dx_2 \int_0^{x_2} dx_1 \frac{(2x_2 - 1)A(y)}{(x_2 - x_1)^{d/2-1}}. \quad (9)$$

The computation of integrals in Eq. (9) over x_1 and x_2 for a small force yields $B_r(F \to 0) = \frac{4}{(8-d)(6-d)} A(0)$. The finiteness of the latter at four dimensions means in the context of the renormalization group method that the parameters F, L, l, and T appearing in the prefactor of Eq. (8) do not renormalize. The only quantity which renormalizes is the monomer friction coefficient.

3. Results

3.1. *Small elongation*

The first-order perturbational correction to the velocity is the starting point to perform the renormalization group (RG) analysis, which enables one to take into account the effect of hydrodynamic interaction beyond the first-order. To regularize the theory the $1/(4-d)$ poles in perturbation expansions have to be removed by an appropriate renormalization of the friction coefficient. In the limit of a small pulling force the renormalized friction coefficient is derived from (3) as

$$f = f_0 (1 - \frac{3}{2\varepsilon} \xi_0 (L^{\varepsilon/2} - \lambda^{\varepsilon/2}) + \cdots) \quad (10)$$

where $\xi_0 = (f_0/(\eta_s d))(d/(2\pi l))^{d/2}$ is the expansion parameter of perturbation series in powers of the hydrodynamic interactions, and $\varepsilon = 4 - d$. The cutoff excludes the hydrodynamic interactions between monomers separated along the chain by the contour length less than λ, $|s_1 - s_2| < \lambda$.

The renormalization of the friction coefficient (10) obtained from Eq. (3) coincides with that obtained by studying different problems in polymer dynamics.[7–9] The renormalization of the friction coefficient and the strength of the hydrodynamic interaction $w_0 = \frac{f_0}{\eta_s d}(\frac{d}{2\pi l})^{d/2} L^{\varepsilon/2}$ due to an infinitesimal change of the cutoff are given by differential equations. To the one-loop order one obtains

$$\lambda' \frac{\partial f}{\partial \lambda'} = \frac{3}{4} w, \quad \lambda' \frac{\partial w}{\partial \lambda'} = \frac{\varepsilon}{2} w - \frac{3}{4} w^2 + ... \equiv \beta(w) \tag{11}$$

with the dimensionless effective coupling constant $w = (f/(\eta_s d))(d/(2\pi l))^{d/2} \lambda'^{\varepsilon/2}$. The solutions of the equations in (11) read

$$f = \frac{f_0}{1 + \frac{3}{2\varepsilon}\xi_0(\lambda_m^{\varepsilon/2} - \lambda^{\varepsilon/2})}, \quad w = \frac{f}{\eta_s d}\left(\frac{d}{2\pi l}\right)^{d/2} \lambda_m^{\varepsilon/2}. \tag{12}$$

It follows from (12) that at large λ_m, w approaches the fixed-point value $w^* = 2\varepsilon/3$, which corresponds to the zero point of the Gell-Mann–Low function $\beta(w)$. At the fixed-point the effective friction coefficient depends on λ_m as power law $\lambda_m^{-\varepsilon/2}$. At low forces, λ_m is equal to L, so that the renormalized (effective) friction coefficient scales as $f = w^* \eta_s l^{d/2} L^{-\varepsilon/2}$. The drift velocity behaves consequently as

$$\mathrm{v}_c^z \simeq \frac{F}{fN} \simeq \frac{F}{\eta_s (Ll)^{d/2-1}}, \tag{13}$$

which agrees in $d = 3$ with the Stokes formula (2).

The inspection of the first-order correction to the longitudinal size of the polymer yields that it is finite in four dimensions. This is what is expected, because the friction coefficient, which is the only quantity to renormalize, does not appear in the zero-order correction to the longitudinal size of the polymer. Thus, the RG prediction for the longitudinal size consists in replacing the bare expansion parameter in Eq. (8) by the renormalized one.

For small forces the velocity and the elongation are given in the renormalized theory by expressions

$$\mathrm{v}_c^z = \frac{F}{fN}\left(1 + \frac{1}{4} w B_{\mathrm{v}}(\beta F \sqrt{lL/2d}) + \cdots\right), \tag{14}$$

$$\delta r_z = \frac{FLl}{2dk_B T}\left(1 - \frac{1}{4} w B_{\mathrm{r}}(\beta F \sqrt{lL/2d}) + \cdots\right). \tag{15}$$

The function $B_{\mathrm{v}}(\tilde{F})$ in Eq. (14) is defined by the expression $B_{\mathrm{v}}(\tilde{F}) = \int_0^1 dx_2 \int_0^{x_2} dx_1 \frac{A(y)-A(0)}{(x_2-x_1)^{d/2-1}}$. The effective expansion parameter w is a small number ($\sim O(\epsilon)$), so that the expansions (14) and (15) are reliable.

3.2. *Large elongation*

To estimate the integrals in (3) and (9) for large forces we use the asymptotic expression of $A(y)$ given by Eq. (6). To prevent the divergence in the integral over $t = x_2 - x_1$ from the integration region $t = 0$, we introduce a cutoff $t_0 \simeq 6/Ll\beta^2F^2$. The evaluation of the integrals over t and x_2 can be performed analytically and gives for $d = 3$

$$\mathrm{v}_c^z = \frac{F}{f_0 N}\left(1 + c_{\mathrm{v}}\xi_0 L^{1/2}\frac{\ln(\beta F\,(Ll)^{1/2})\ln\frac{\beta F(Ll)^{1/2}}{3}}{\beta F\,(Ll)^{1/2}} + \cdots\right), \tag{16}$$

where c_{v} is a numerical constant. We would like to stress that the integral in Eq. (3) computed numerically can be approximately fitted with $(\ln\tilde{F})^{1.72}/\tilde{F}$ instead of the estimate given by Eq. (16). The difference is due to the complicated form of the expression for y in Eq. (5).

For large F and finite L we arrive at the Rouse result. However, for large L and finite F the correction increases logarithmically with L, and will become large for large L. Unfortunately, there are no analytical means to study the effect of the whole perturbation expansion on v_c^z in this case. The extrapolation of (16), which consists in disregarding in the bracket of Eq. (16), yields

$$\mathrm{v}_c^z \simeq \frac{k_B T}{\eta_s N l^2}\ln(\beta F\,(Ll)^{1/2})\ln\frac{\beta F\,(Ll)^{1/2}}{3}. \tag{17}$$

Equation (17) shows that in this regime (finite F and large L) the hydrodynamic interactions determine the behavior of the polymer. The friction coefficient drops in the expression of the velocity in this regime. Note that the logarithmic dependencies of v_c^z on the force does not allow to write Eq. (17) in the form of Eq. (1) with some effective friction coefficient. Equation (17) can be formally obtained from Eq. (1) by simultaneous replacements

$$f_0 \to f \simeq l\eta_s/\ln(\beta F\,(Ll)^{1/2}),\;\; F \to \frac{k_B T}{l}\ln\frac{\beta F\,(Ll)^{1/2}}{3}. \tag{18}$$

The "renormalized" friction coefficient in Eq. (18) resembles to some extent the friction coefficient, $L\eta_s/\ln(L/l)$, of a slender body (a cylinder of length L and cross section radius l, $L \gg l$) in a flow.[10,11] However, the elongated polymer chain is different from a rod, and the comparison is only qualitative.

A similar computation of the first-order correction to the chain size yields

$$\delta r_z = \frac{FLl}{2dk_BT}\left(1 - c_r\xi_0 L^{1/2}\frac{(\ln\frac{\beta F(Ll)^{1/2}}{3} - 2)\ln\left(\beta F\,(Ll)^{1/2}\right)}{\beta F\,(Ll)^{1/2}} + \cdots\right).$$

Note that the integral in Eq. (9) computed numerically can be approximately fitted with $(\ln \tilde{F})^{2.48}/\tilde{F}$ instead of the analytical estimate in the above equation. For large L and finite F the first-order correction to the longitudinal size depends logarithmically on L.

4. Summary

To summarize, we have studied the drift of an ideal polymer driven by a constant force applied at one polymer end using perturbation expansions in powers of hydrodynamic interactions. The renormalization group method permits to compute the velocity and the longitudinal elongation of the polymer for moderate forces. For large L but finite force, the regime which we have studied for $d = 3$, the hydrodynamic interactions are partially screened. The first-order corrections to v_c^z and $\langle r^z(0,t) - r^z(L,t)\rangle$ depend logarithmically on L.

Acknowledgments

A financial support from the Deutsche Forschungsgemeinschaft, SFB 418, is gratefully acknowledged.

References

1. R. G. Larson, T. T. Perkins, D. E. Smith, and S. Chu, *Phys. Rev. E* **55**, 1794 (1997).
2. T. T. Perkins, D. E. Smith, and S. Chu, *Science* **276**, 2016 (1997).
3. C. M. Schroeder, H. P. Babcock, E. S. G. Shaqfeh, and S. Chu, *Science* **301**, 1515 (2003).
4. F. Brochard-Wyart, *Europhys. Lett.* **23**, 105 (1993).
5. R. Rzehak, D. Kienle, T. Kawakatsu, and W. Zimmermann, *Europhys. Lett.* **46**, 821 (1999).
6. S. Stepanow and N. Kikuchi, arXiv:0710.1779 (cond-mat.stat-mech).
7. S. Q. Wang and K. F. Freed, *J. Chem. Phys.* **85**, 6210 (1986).
8. Y. Oono, *Adv. Chem. Phys.* **61**, 301 (1985).
9. S. Stepanow and G. Helmis, *Phys. Rev. A* **39**, 6037 (1989).
10. G. K. Batchelor, *J. Fluid Mech.* **44**, 419 (1970).
11. M. Doi and S. F. Edwards, *The Theory of Polymer Dynamics* (Clarendon, Oxford, 1986).

STAR POLYMERS IN CORRELATED DISORDER

V. BLAVATSKA

Institute for Condensed Matter Physics, National Academy of Sciences of Ukraine, Lviv 79011, Ukraine &
Institut für Theoretische Physik, Universität Leipzig, Postfach 100 920, D-04009 Leipzig, Germany
E-mail: viktoria@icmp.lviv.ua

C. VON FERBER

Applied Mathematics Research Centre, Coventry University, Coventry CV1 5FB, UK &
Physikalisches Institut, Universität Freiburg, D-79104 Freiburg, Germany
E-mail: C.vonFerber@coventry.ac.uk

YU. HOLOVATCH

Institute for Condensed Matter Physics, National Academy of Sciences of Ukraine, Lviv 79011, Ukraine &
Institut für Theoretische Physik, Johannes Kepler Universität Linz, A-4040 Linz, Austria
E-mail: hol@icmp.lviv.ua

We analyze the impact of a porous medium (structural disorder) on the scaling of the partition function of a star polymer immersed in a good solvent. We show that corresponding scaling exponents change if the disorder is long-range-correlated and calculate the exponents in the new universality class. A notable finding is that star and chain polymers react in *qualitatively* different manner on the presence of disorder: the corresponding scaling exponents *increase* for chains and *decrease* for stars. We discuss the physical consequences of this difference.

Keywords: Polymer; Polymer networks; Quenched disorder; Scaling laws.

1. Introduction

Polymer theory may serve as an archetype of an approach where the application of the path integral formalism leads both to a quantitative understanding of a whole range of physical, chemical, and biological phenomena

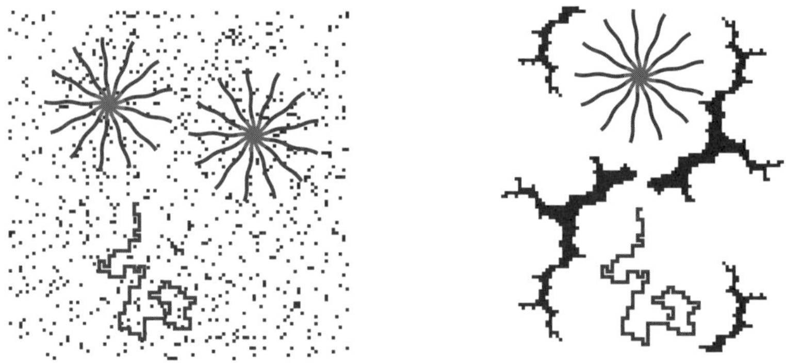

a. b.

Fig. 1. Solution of star and chain polymers in a good solvent immersed in a porous medium. We consider the disorder that characterizes pores (impurities) distribution as uncorrelated (**a**) or correlated (**b**). For a d-dimensional system, disorder is called long range correlated if the impurity-impurity pair correlation function decays for large distances r as $g(r) \sim r^{-a}$ with $a < d$.

as well as to their accurate quantitative description.[1–3] Most directly this is shown by the Edwards model that describes a polymer chain in terms of a path integral and takes into account chain connectivity and self-avoiding interaction.[3] A textbook derivation maps this simple two-parameter model to the $m = 0$ de Gennes limit[4] of the $O(m)$ symmetric field theory. Standard field theoretical tools explain the origin of the scaling laws that govern polymer structural behaviour and allow to calculate the exponents that govern these scaling laws with high accuracy.

One of the generalizations of the above approach extends the theory to describe polymers of complex structure that form networks of interconnected polymer chains.[5] The intrinsic exponents that govern the scaling of a polymer network are uniquely defined by those of its constituents, star-like subunits known as star polymers[6] (see Fig. 1). The exponents that govern the scaling of star polymers are universal in that they depend on space dimension d and star functionality f only (the number of chains attached to a common center).[a] Currently, star polymers are synthesized with high functionalities and form well-defined objects with interesting industrial, technological, and experimental applications.[7]

In this paper we attract attention to another recent development in the analysis of the scaling properties of branched polymers.[8,9] Our analysis con-

[a]Note, that for $f = 1, f = 2$ a chain polymer is recovered.

cerns the impact of structural disorder on the scaling of polymer stars and chains. For lattice models, where polymers are viewed as self-avoiding walks (SAW) this type of disorder may be realized by forbidding the SAW to visit certain lattice sites, which might be interpreted as lattice dilution. For real polymers in solvents, structural disorder may be implemented by filling a porous medium by a solvent with immersed polymers as shown in Fig. 1. As follows from our analysis sketched below, for long-range-correlated disorder the polymer behaviour displays a new universality, different from that of a SAW on a regular (undiluted) lattice. A notable consequence is that star and chain polymers react in *qualitatively* different manner on the presence of disorder: the corresponding scaling exponents *increase* for the chains and *decrease* for the stars. We discuss to which physical consequences such a difference in behaviour may lead.

2. Model and Method

The starting point of our analysis is the Edwards continuous chain model, generalized to describe a branched polymer structure, a star polymer. We describe the conformation of each arm of the star by a path $\mathbf{r}_a(s)$, parameterized by $0 \leq s \leq S_a$, $a = 1, 2, \ldots, f$ (S_a, the Gaussian surface of the a-th arm is related to the contour length of the chain), $\mathbf{r}_a(0)$, corresponds to the central point. The partition function of the system is defined by the path integral:[5]

$$\mathcal{Z}\{S_a\} = \int D[\mathbf{r}_1, \ldots, \mathbf{r}_f] \exp\left[-\frac{H_f}{k_B T}\right] \prod_{a=2}^{f} \delta^d(\mathbf{r}_a(0) - \mathbf{r}_1(0)). \tag{1}$$

Here, the product of δ-functions ensures the star-like configuration of the set of f polymer chains described by the Hamiltonian:

$$\frac{H_f}{k_B T} = \frac{1}{2}\sum_{a=1}^{f} \int_0^{S_a} \mathrm{d}\, s \left(\frac{\mathrm{d}\,\mathbf{r}(s)}{\mathrm{d}s}\right)^2 + \frac{u_0}{4!} \sum_{a,b=1}^{f} \int_0^{S_a} \mathrm{d}s \int_0^{S_b} \mathrm{d}s' \delta^d(\mathbf{r}_a(s) - \mathbf{r}_b(s')). \tag{2}$$

The first term on the r.h.s. of (2) presents chain connectivity whereas the second term describes an excluded volume interaction. Instead of introducing structural disorder directly into Eq. (1), we make use of its field theoretical representation. The corresponding derivations are described in details in Refs. 5,8. The relevant steps read:

(**i**) we map the continuous chain model (1) onto the $m = 0$ limit of $O(m)$ symmetric field theory by a familiar Laplace transform[1] in the Gaussian surface variables S_a to conjugated chemical potentials (mass variables) μ_a.

In this procedure, the product of δ-functions in (1) is represented by a composite operator of a product of f m-component fields

$$\sum_{\{k\}} \sum_{i_1,\dots,i_f=1}^{m} N^{i_1,\dots,i_f} \phi_{k_1}^{i_1} \dots \phi_{k_f}^{i_f}. \tag{3}$$

Here, $N^{i_1,\dots,i_f}$ is a traceless tensor, ϕ^i is the i-th component of the m-vector field ϕ and in the sum over wave vectors $\{k\}$ is restricted by momentum conservation;
(ii) we introduce quenched random-temperature-like disorder shifting $\mu_a \to \mu_a + \delta\mu_a(x)$ by random variables $\delta\mu_a(x)$. These have zero mean and correlations that decay at large distances as a power law:[11]

$$\langle \delta\mu_a(x)\delta\mu_a(y)\rangle \sim |x-y|^{-a}. \tag{4}$$

Here, $\langle \dots \rangle$ stands for the configurational average over spatially homogeneous and isotropic disorder and the exponent a governs the correlation decay. As seen below, this leads to long range correlated disorder effects for $a < d$;
(iii) to perform the configurational average of the free energy, we make use of the replica method resulting in a field theoretical Lagrangean of two couplings u_0, w_0:

$$\mathcal{L} = \frac{1}{2}\sum_{\alpha=1}^{n}\sum_{k}(\mu_0^2 + k^2)(\phi_k^\alpha)^2 + \frac{u_0}{4!}\sum_{\alpha=1}^{n}\sum_{\{k\}}\left(\phi_{k_1}^\alpha \cdot \phi_{k_2}^\alpha\right)\left(\phi_{k_3}^\alpha \cdot \phi_{k_4}^\alpha\right)$$
$$+ \frac{w_0}{4!}\sum_{\alpha,\beta=1}^{n}\sum_{\{k\}}|k_1+k_2|^{a-d}\left(\phi_{k_1}^\alpha \cdot \phi_{k_2}^\alpha\right)\left(\phi_{k_3}^\beta \cdot \phi_{k_4}^\beta\right). \tag{5}$$

Note that the evaluation of the theory (5) involves a simultaneous polymer ($m = 0$) and replica ($n = 0$) limit. It is this anticipated double limit that allows us[8] to write the Lagrangean in terms of two coupling only. A third coupling appears[11] for $m \neq 0$.

We apply the field theoretical renormalization group (RG) approach to extract the universal content of (5). In this approach, the change of the couplings u, w under renormalization defines a flow in parametric space, governed by corresponding β-functions $\beta_u(u,w)$, $\beta_w(u,w)$. The fixed points (FPs) u^*, w^* of this flow are the solutions to the system of equations: $\beta_u(u^*,w^*) = 0$, $\beta_w(u^*,w^*) = 0$. If a stable FP exists and is accessible, it determines the scaling behaviour of the polymer system. In particular, for a single polymer star of f arms of equal length N in a good solvent the

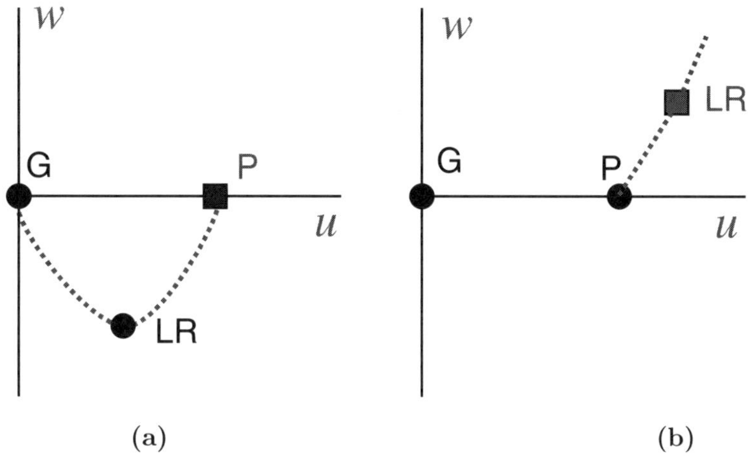

Fig. 2. Fixed points G (Gaussian), P (pure) and LR (long range) of the RG flow in the plane of the two couplings u, w. The stable FP is shown by a square. (**a**) uncorrelated disorder $a > d$: P ($u \neq 0, w = 0$) is stable. (**b**) correlated disorder $a < d$: LR ($u \neq 0, w \neq 0$) is stable. Crossover occurs at $a = d$ (c.f. Fig. 1).

partition sum (number of possible configurations) scales as

$$Z_{N,f} \sim \mathrm{e}^{\mu N f} N^{\gamma_f - 1}, \qquad N \to \infty, \tag{6}$$

with a non-universal fugacity e^{μ} and the universal star exponent γ_f.[6] The latter is uniquely defined by the stable FP value of the anomalous dimension associated with the composite operator (3). We make use of two complementary perturbation theory expansions to calculate coordinates of the FPs and values of the exponents. In a first approximation we apply an expansion in[11] $\varepsilon = 4 - d$ and $\delta = 4 - a$ which allows for a qualitative description of the phenomena. In a further approach we apply perturbation theory in the renormalized couplings u and w evaluated at fixed dimension $d = 3$ for a series of fixed values of the correlation parameter a.[12] In the latter case we proceed within a two-loop approximation refining the analysis by a resummation of the divergent RG series (see Refs. 8–10 for details). The FP picture that arises from our calculations is shown qualitatively in Fig. 2. Both calculation schemes display a range $a_{\text{lower}}(d) < a < a_{\text{upper}}(d)$ of values for a where the long-range-correlated FP (LR, $u \neq 0$, $w \neq 0$) is stable and governs polymer scaling. For $a < a_{\text{lower}}$, no stable FP is found. This has been interpreted[8] as a collapse of a polymer coil for strongly correlated disorder. For $a > a_{\text{upper}}$ the pure FP (P, $u \neq 0$, $w = 0$) is stable and polymer scaling is not perturbed by disorder. As far as power counting

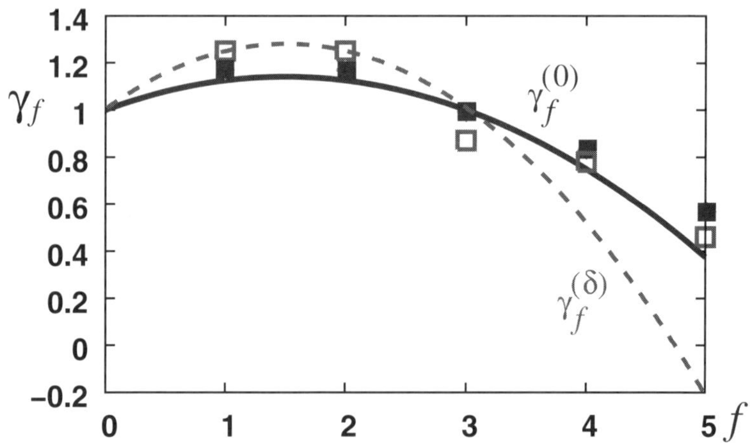

Fig. 3. Exponent γ_f of 3D polymer stars with f arms in a good solvent. Solid lines and filled squares indicate uncorrelated disorder, dashed lines $\delta = 1.1$ and open squares $a = 2.9$ correlated disorder; lines: first-order ε, δ results; open, resp. filled, squares: second- resp. third-order resummed results for fixed d, a.

implies that the w-term in (5) is irrelevant in the RG sense for $a \geq d$ it is natural to identify $a_{\text{upper}} = d$. Nonetheless, in first-order approximation of the ε, δ-expansion one finds that the LR FP is stable in the unphysical region $d < a < 2 + d/2$. Our two-loop calculations at fixed d, a for $d = 3$, however, result in $a_{\text{upper}} = 3 = d$, $a_{\text{lower}} = 2.2$ allowing for direct physical interpretations.

3. Scaling Exponents

Qualitatively, the impact of disorder can be seen already from the first-order ε, δ results. Comparing the γ_f exponent (6) for the cases when the structural disorder is absent (or is short-range-correlated, Fig. 1a),[6] $\gamma_f^{(0)} = 1 - \varepsilon f(f-3)/16$, and when it is long-range-correlated (Fig. 1b),[9] $\gamma_f^{(\delta)} = 1 - \delta f(f-3)/8$ we find:

$$\Delta\gamma_f \equiv \gamma_f^{(\delta)} - \gamma_f^{(0)} = \frac{f(f-3)}{16}(\varepsilon - 2\delta), \qquad \varepsilon/2 < \delta < \varepsilon. \tag{7}$$

As one can see from this estimate, the exponent difference changes sign at $f_\pm = 3$. The two-loop calculations slightly shift this result towards $f_\pm \leq 3$ while otherwise confirming the overall picture with more accurate numerical values for the exponents.[9,10] In Fig. 3 we plot the first-order ε, δ curves for γ_f together with the resummed two-loop estimates for $d = 3$. A complete account of the numerical values of the exponents is given in Ref. 9. A prominent feature that follows from these results is that the effect of long-range-correlated disorder on polymer chains and polymer stars is qualitatively different. Whereas correlated disorder leads to an *increase* of γ_f for chain polymers (i.e. for $f = 1$, $f = 2$), the same type of disorder *decreases* the γ_f exponent for the proper star polymers ($f \geq 3$). Below we discuss some possible consequences of this difference. For $f > 2$ we also observe that γ_f decreases monotonically as function of f for any valid value of a in the same way as it does for polymers in a good solvent: $\gamma_{f_1}^{(a)} > \gamma_{f_2}^{(a)}$ for $2 < f_1 < f_2$. Furthermore, within the accuracy the data confirms a convexity from below of this function for valid values of a as well as for the ε, δ expansion. The latter property ensures that polymer stars remain mutually repulsive in correlated disorder.[13,14]

4. Conclusions and Outlook

It has been recognized by now that star exponents come into play for the qualitative description and quantitative of different phenomena, where statistics of branched self-avoiding and random walks is involved. The examples of such phenomena include short-range interaction of branched polymers in a good solvent,[15] diffusion-controlled reactions in the presence of polymers,[16] and, more generally, they are part of a multifractal description of diffusion limited growth in a Laplacian field.[17] Recently, star exponents have been used to estimate the thermal denaturation transition of DNA.[18] Our analysis opens a way to consider the impact of long-range-correlated disorder on the above phenomena.

A somewhat surprising effect that results from our calculations concerns the static separation in a solution of diluted chains and star polymers of equal molecular weight inside a porous medium. Following the estimates of the star exponents we predict that in a correlated medium star polymers will exert a higher osmotic pressure than chain polymers and in general higher branched star polymers will be expelled more strongly from the correlated porous medium. On the opposite, polymer chains will prefer a stronger correlated medium to a less or uncorrelated medium of the same density.[9]

A generalization of our approach to the case of star polymers built from chains of different species[14] will be presented elsewhere.[19]

Acknowledgments

We acknowledge support by the EU Programme "Marie Curie International Incoming Fellowship" (V.B.) and the Austrian Fonds zur Förderung der wissenschaftlichen Forschung under Project P 19583 - N20 (Yu.H.).

References

1. J. des Cloizeaux and G. Jannink, *Polymers in Solution* (Clarendon Press, Oxford, 1990).
2. H. Kleinert, *Path Integrals in Quantum Mechanics, Statistics and Polymer Physics* (World Scientific, Singapore, 1995).
3. L. Schäfer, *Universal Properties of Polymer Solutions as Explained by the Renormalization Group* (Springer, Berlin, 1999).
4. P.-G. de Gennes, *Scaling Concepts in Polymer Physics* (Cornell University Press, Ithaca and London, 1979).
5. L. Schäfer, C. von Ferber, U. Lehr, and B. Duplantier, *Nucl. Phys. B* **374**, 473 (1992).
6. B. Duplantier, *J. Stat. Phys.* **54**, 581 (1989).
7. G. S. Grest, L. J. Fetters, J. S. Huang, and D. Richter, *Adv. Chem. Phys.* **XCIV**, (1996); C. N. Likos, *Phys. Rep.* **348**, 267 (2001); *Star Polymers*, eds. C. von Ferber and Yu. Holovatch (*Condens. Matter Phys.* **5**, pp. 1–305, 2002).
8. V. Blavatska, C. von Ferber, and Yu. Holovatch, *Phys. Rev. E* **64**, 041102 (2001); *J. Phys.: Condens. Matter* **14**, 9465 (2002).
9. V. Blavatska, C. von Ferber, and Yu. Holovatch, *Phys. Rev. E* **74**, 031801 (2006).
10. V. Blavatska, C. von Ferber, R. Folk, and Yu. Holovatch, *Renormalization group approaches to polymers in disordered media*, in: *Statistics of Linear Polymers in Disordered Media*, ed. B. K. Chakrabarti (Elsevier, Amsterdam, 2005), pp. 103–147.
11. A. Weinrib and B. I. Halperin, *Phys. Rev. B* **27**, 417 (1983).
12. V. V. Prudnikov, P. V. Prudnikov, and A. A. Fedorenko, *J. Phys. A* **32**, L399 (1999); *J. Phys. A* **32**, 8587 (1999); *Phys. Rev. B* **62**, 8777 (2000).
13. B. Duplantier and A. W. Ludwig, *Phys. Rev. Lett.* **66**, 247 (1991).
14. C. von Ferber and Yu. Holovatch, *Europhys. Lett.* **39**, 31 (1997); *Phys. Rev. E* **56**, 6370 (1997).
15. M. Watzlawek, C. N. Likos, and H. Löwen, *Phys. Rev. Lett.* **82**, 5289 (1999); C. von Ferber, Yu. Holovatch, A. Jusufi, C. N. Likos, H. Löwen, and M. Watzlawek, *Journ. Mol. Liq.* **93**, 151 (2001).
16. T. Witten and M. Cates, *Science* **232**, 1607 (1986); C. von Ferber and Yu. Holovatch, *Journ. Mol. Liq.* **93**, 155 (2001).
17. M. E. Cates and T. A. Witten, *Phys. Rev. A* **35**, 1809 (1987); C. von Ferber and Yu. Holovatch, *Phys. Rev E* **59**, 6914 (1999).
18. Y. Kafri, D. Mukamel, and L. Peliti, *Phys. Rev. Lett.* **85**, 4988 (2000); E. Carlon, E. Orlandini, and A. Stella, *Phys. Rev. Lett.* **88**, 198101 (2002).
19. V. Blavatska, C. von Ferber, and Yu. Holovatch, in preparation (2007).

GENERALIZED NONLINEAR SIGMA MODELS AND PATH-INTEGRAL APPROACH TO POLYMER DYNAMICS

F. FERRARI* and J. PATUREJ†

Institute of Physics, University of Szczecin,
Wielkopolska 15, 70451 Szczecin, Poland
** E-mail: ferrari@fermi.fiz.univ.szczecin.pl*
† E-mail: jpaturej@univ.szczecin.pl

T. A. VILGIS

Max Planck Institute for Polymer Research,
10 Ackermannweg, D-55128 Mainz, Germany
E-mail: vilgis@mpip-mainz.mpg.de

In this work the problem of describing the dynamics of a continuous chain with rigid constraints is treated using a path integral approach.

Keywords: Statistical field theory; Brownian motion; Chain dynamics.

1. Introduction

In this work the dynamics of a chain consisting of a set of small particles attached to the ends of segments of fixed length is studied. The chain fluctuates at constant temperature in some viscous medium. For simplicity, all interactions among the particles have been switched off and the number of space dimensions has been limited to two. A chain of this kind may be regarded as a system of particles performing a random walk and subjected to rigid constraints which simulate the presence of the segments. In the limit in which the chain becomes a continuous system, its behavior may be described by a probability function Ψ given in the path integral form. Ψ measures the probability that a fluctuating chain passes from an initial spatial conformation to a final one in a given interval of time. An expression of the probability function is obtained by inserting in the path integral of a system of brownian particles suitable Dirac delta functions which impose the rigid constraints on the trajectories of the particles. It is shown that

Ψ coincides with the partition function of a field theory which is a generalization of the nonlinear sigma model in two dimensions. The probability function is then computed explicitly in the semi-classical approximation for a ring-shaped chain. The behavior of the chain at both long and short scales of time and distances is investigated. The connection between the generalized nonlinear sigma model presented here and the Rouse model is briefly discussed.

2. Behavior of a Chain at Short and Long Scales of Time and Distance

In this section it will be addressed the problem of describing the dynamics of a random chain subjected to thermal fluctuations at constant temperature T. The strategy in order to treat this problem is to regard the chain as a set of N particles of equal mass m and performing a random walk while subjected to constraints of the kind $\mathcal{R}_n = 0$. Here $\mathcal{R}_n$ is a given functional of the coordinates of the particles. For instance, if we imagine that the particles are attached at the ends of segments of constant length a, then

$$\mathcal{R}_n = \frac{|\mathbf{R}_n(t) - \mathbf{R}_{n-1}(t)|^2}{a^2} - 1 \qquad n = 2, \ldots, N\,, \tag{1}$$

where the $\mathbf{R}_n(t)$'s for $n = 1, \ldots, N$ are the radius vectors of the particles. We require that at the instants $t = t_i, t_f$ the $n-$th particle is located respectively at the initial point $\mathbf{R}_{i,n}$ and at the final point $\mathbf{R}_{f,n}$. For simplicity, the interactions among particles are switched off. As a consequence, the probability function Ψ_N of the system under consideration is obtained following an approach which is commonly applied in the statistical mechanics of polymers.[1–3] The approach consists in the insertion in the path integral of N particles performing a free random walk a set of Dirac $\delta-$functions which enforce the constraints $\mathcal{R}_n = 0$:

$$\Psi_N = \int_{\substack{\mathbf{R}_1(t_f)=\mathbf{R}_{f,1}\\ \mathbf{R}_1(t_i)=\mathbf{R}_{i,1}}} d\mathbf{R}_1(t) \cdots \int_{\substack{\mathbf{R}_N(t_f)=\mathbf{R}_{f,N}\\ \mathbf{R}_N(t_i)=\mathbf{R}_{i,N}}} d\mathbf{R}_n(t) \exp\left\{-\sum_{n=1}^{N}\int_{t_i}^{t_f} \frac{\dot{\mathbf{R}}_n^2(t)}{4D}\,dt\right\} \times \prod_n \delta(\mathcal{R}_n)\,. \tag{2}$$

The coefficient D in the above equation is the diffusion constant.

The continuous limit,[4] in which N goes to infinity, while the masses m and the average distance between two neighboring particles on the chain tend to zero, gives as a result the probability function $\Psi = \lim_{N\to\infty} \Psi_N$ of a continuous chain:

$$\Psi = \int_{\substack{\mathbf{R}(t_f,s)=\mathbf{R}_f(s)\\ \mathbf{R}(t_i,s)=\mathbf{R}_i(s)}} \mathcal{D}\mathbf{R}(t,s) \exp\left\{-c\int_{t_i}^{t_f} dt \int_0^L ds\dot{\mathbf{R}}^2\right\} \prod_s \delta(\mathcal{R}) \tag{3}$$

with $c = M/4kT\tau L$. M is the total mass of the chain, L is its length and k denotes the Boltzmann constant. Finally the relaxation time τ characterizes the rate of the decay of the drift velocity of the particles composing the chain. Let us note that in the continuous limit the discrete vector bonds $\mathbf{R}_n(t)$ have been replaced by the continuous fields $\mathbf{R}(t,s)$, where s represents the arc-length of the chain. Physically, Ψ gives the probability that a fluctuating chain passes from an initial spatial conformation $\mathbf{R}_i(s)$ to a final one $\mathbf{R}_f(s)$ in a given interval of time $\Delta t = t_f - t_i$.

In the following, we will choose a continuous version of the constraints of the form:

$$\mathcal{R} = \left|\frac{1}{s_0}\int_{-s_0}^{s_0} ds' A(s')\left(\mathbf{R}(t,s+s') - \mathbf{R}(t,s)\right)\right|^2 - a^2 = 0\,. \tag{4}$$

In the above equation we have introduced two new lengths scales. a is the smallest scale of lengths at our disposal. It can be identified with the length of the segments of the chain. s_0 is an intermediate length scale such that

$$a \ll s_0 \ll L\,. \tag{5}$$

Finally, $A(s')$ is a function of s' normalized in a such a way that:

$$\frac{1}{s_0}\int_{-s_0}^{s_0} A(s')ds' = 1\,. \tag{6}$$

The insertion of the δ-function $\delta(\mathcal{R})$ in the path integral (3) is related to the internal forces among the beads composing the chain which are needed to impose the constraints. The function $A(s')$ determines the range r of these interactions. For $A(s') = 1/2$, for instance, $r = 2s_0$. If we choose instead $A(s') = s_0\delta(s'-a)$, then $r = a$. This corresponds to very short range interactions.

It is possible to show that the Rouse model can be obtained from the path integral (3) by decreasing the resolution with which the spatial structure of the chain is resolved and looking simultaneously at the fluctuations of the chain over long time-scales. The resolution is decreased by putting $A(s') = 1/2$ in Eq. (4). This is tantamount to consider interactions between chain segments which are averaged over the intermediate scale of distance $2s_0$. To restrict oneself to the long time-scale behavior of the chain, one may proceed as in Ref. 5, Section 4.1, p. 93. We skip the derivation of the Rouse model[6] starting from Eq. (3). The details can be found in Ref. 7.

Since the Rouse model has already been extensively studied in the literature,[5] we will concentrate instead to the behavior of the chain at short scales of times and distances. To observe the motion of the chain at the smallest available scale of distances, *i. e.* the segment length a, we choose $r = a$ by putting: $A(s') = s_0\delta(s' - a)$. As a consequence, the constraint (4) becomes $|\mathbf{R}(t, s + a) - \mathbf{R}(t, s)|^2 - a^2 = 0$. Dividing both members of the above equation by a^2 and supposing that a is very small, we get up to higher order terms in a the relation: $\left|\frac{\partial \mathbf{R}}{\partial s}\right|^2 - 1 = 0$. In the limit $a = 0$, this is exactly the continuous version of the condition which has been imposed to the chain in Eq. (1). It is also easy to check that the probability function Ψ becomes in this special case:

$$\Psi = \int_{\substack{\mathbf{R}(t_f,s)=\mathbf{R}_f(s)\\ \mathbf{R}(t_i,s)=\mathbf{R}_i(s)}} \mathcal{D}\mathbf{R}\mathcal{D}\lambda \exp\left\{-c\int_{t_i}^{t_f} dt \int_0^L ds\dot{\mathbf{R}}^2\right\}$$
$$\times \exp\left\{i\int_{t_i}^{t_f} dt \int_0^L ds\lambda\left(\left|\frac{\partial \mathbf{R}}{\partial s}\right|^2 - 1\right)\right\}. \tag{7}$$

In Eq. (7) the Lagrange multiplier λ has been introduced in order to impose the constraints using the Fourier representation of the δ−function $\delta(\left|\frac{\partial \mathbf{R}}{\partial s}\right|^2 - 1)$. The model described by Eq. (7) will be called here the generalized nonlinear sigma model (GNLSM) due to its close resemblance to a two-dimensional nonlinear sigma model.

3. GNLSM in Semi-Classical Approximation

In the following, we study small gaussian fluctuations of the field $\mathbf{R}(t, s)$ on the background of a classical solution. For simplicity, we treat only the case of a two dimensional chain. Due to the compatibility with the constraints, it is possible to show that, in absence of external sources, the classical equations of motion admit just two kinds of solutions:

(1) Static solutions in which $\lambda = 0$ and $\mathbf{R}(t, s) = \int_0^s du(\cos\varphi(u), \sin\varphi(u))$, $\varphi(u)$ being an arbitrary differentiable function of u.
(2) Static solutions in which $\lambda = \text{const}$ and $\mathbf{R}(t, s) = (0, s)$. This corresponds to a static classical configuration confined on the y−axis.

In computing the probability function Ψ of the chain in the semi-classical approximation, boundary terms may appear in the action functional, which complicate the calculations considerably. To avoid such terms, we restrict ourselves to the following two cases: a) Closed chains satisfying the boundary conditions $\mathbf{R}(t,0) = \mathbf{R}(t,L)$ and b) Open chains in which both ends are fixed.

Let us give at this point the explicit formula of the probability function Ψ in case a):

$$\Psi = \int \mathcal{D}\delta\lambda \; e^{-\int_{t_i}^{t_f} dtdt' \int_0^L dsG(t,t')\left[-4\kappa^2\delta\lambda(t,s)\left(\frac{\partial^2}{\partial s^2} - (\varphi'(s))^2\right)\delta\lambda(t',s)\right]}$$
$$\times\, e^{-\frac{1}{\kappa}\int_{t_i}^{t_f} dt \int_0^L ds\left[\frac{M}{2L}\dot{\mathbf{R}}_{cl}^2\right]} . \qquad (8)$$

In the above equation the constant $\kappa = 2kT\tau$ fulfills the same role as the Planck constant in quantum mechanics. Moreover, $\delta\lambda$ is a small fluctuation of the Lagrange multiplier λ on the background of the classical solution $\lambda = 0$ and it satisfies trivial boundary conditions in t and s. Finally $G(t,t')$ is a Green function satisfying the equation $\frac{M}{L}\frac{\partial^2 G(t,t')}{\partial t^2} = -\delta(t-t')$. If we choose, for instance, Dirichlet boundary conditions for $G(t,t')$ at both initial and final time t_i and t_f, Ψ measures the probability that starting from some static conformation $\mathbf{R}_i(s)$ the fluctuating chain ends up at the time t_f in the same conformation. This is useful to test the stability of a given conformation with respect to the thermodynamic fluctuations which attempt to reshape the chain.

4. Conclusions

This work may be considered as the ideal continuation of the seminal paper of Edwards and Goodyear of Ref. 8, in which the problem of a chain subjected to the constraints (1) has been investigated using the Langevin equations. With respect to Ref. 8, our approach based on the Fokker–Planck–Smoluchowski equations provides a path integral and field theoretical formulation of the dynamics of a freely jointed chain in the continuous limit. The GNLSM obtained here makes possible the application of field theoretical techniques to the study of the fluctuations of a freely jointed chain. We recall that in deriving the GNLSM the contribution of the hydrodynamic interactions has been neglected. This limits the validity of this model to the following cases:

a) The viscosity of the fluid is large, so that the motion of the particles composing the chain is slow.
b) The temperature is low, so that once again the motion of the chain is slow.

The semi-classical approximation used in order to derive the probability function Ψ is valid in the case in which the parameter κ defined at the end of the previous section is small. This parameter depends essentially on the temperature T and on the relaxation time τ. Since τ is inversely proportional to the viscosity, it is reasonable to assume that the semiclassical approach can be applied to a cold isolated chain or to an isolated chain in a very viscous solution.

Acknowledgments

This work has been financed in part by the Polish Ministry of Science, scientific project N202 156 31/2933. F. Ferrari gratefully acknowledges also the support of the action COST P12 financed by the European Union. F. Ferrari and J. Paturej are also grateful to W. Janke, A. Pelster and to the Max Planck Institute for the Physics of Complex Systems for allowing them to participate in such a nice and fruitful conference as the International Conference on "Path Integrals – New Trends and Perspectives" has been.

References

1. S. Edwards, *Proc. Phys. Soc.* **91**, 513 (1967); *J. Phys. A* **1**, 15 (1968).
2. F. Ferrari, H. Kleinert, and I. Lazzizzera, *Int. J. Mod. Phys B* **14**, 3881 (2000).
3. H. Kleinert, *Path Integrals in Quantum Mechanics, Statistics, Polymer Physics, and Financial Markets*, Fourth Edition (World Scientific, Singapore, 2006).
4. H. Kleinert, *Gauge Fields in Condensed Matter*, Vol. 1 (World Scientific, Singapore, 1989).
5. M. Doi and S. F. Edwards, *The Theory of Polymer Dynamics* (Clarendon Press, Oxford, 1986).
6. P. E. Rouse, *J. Chem. Phys.* **21**, 1272 (1953).
7. F. Ferrari, J. Paturej, and T. A. Vilgis, arXiv:cond-mat/0705.4182.
8. S. F. Edwards and A. G. Goodyear, *J. Phys. A: Gen. Phys.* **5**, 965 (1972).

DESCRIPTION OF THE DYNAMICS OF A RANDOM CHAIN WITH RIGID CONSTRAINTS IN THE PATH-INTEGRAL FRAMEWORK

F. FERRARI* and J. PATUREJ†

Institute of Physics, University of Szczecin,
Wielkopolska 15, 70451 Szczecin, Poland
**E-mail: ferrari@fermi.fiz.univ.szczecin.pl*
† E-mail: jpaturej@univ.szczecin.pl

T. A. VILGIS

Max Planck Institute for Polymer Research,
Ackermannweg 10, D-55128 Mainz, Germany
E-mail: vilgis@mpip-mainz.mpg.de

In this work we discuss the dynamics of a three-dimensional chain. It turns out that the generalized sigma model presented in Ref. 1 may be easily generalized to three dimensions. The formula of the probability distribution of two topologically entangled chains is provided. The interesting case of a chain which can form only discrete angles with respect to the $z-$axis is also presented.

Keywords: Statistical field theory; Brownian motion; Chain dynamics, Topological entanglement.

1. Introduction

In this brief report the dynamics of a chain fluctuating in some medium at fixed temperature T is discussed. The three-dimensional case is particularly interesting, because it allows to study the topological entanglement of two or more chains. The problem of the topological entanglement of two chains has been investigated for a long time in the statistical mechanics of polymers, see for instance Ref. 2 and references therein. If the topological constraints which limit the fluctuations of the chains are described by using the Gauss linking number, the probability distribution of the system turns out to be equivalent to the partition function of a zero-component Landau–Ginzburg model interacting with a pair of Chern–Simons fields. The analogous problem in polymer dynamics has not yet been solved. Here we show that the probability function of the system of two chains in the presence of topological constraints may be simplified thanks to the introduction of a Chern–Simons field theory also in the case of dynamics. Finally, we provide a formula for the probability function of a chain which can form only discrete angles with respect to the $z-$axis.

2. A Model of Two Topologically Entangled Chains

We would like to treat the dynamics of a chain fluctuating in some medium at constant temperature T. In Ref. 1 (see also Ref. 3) the case of a two-dimensional chain has been discussed. The approach presented in those references can however be extended to any dimension. Let us consider the probability function Ψ which measures the probability that a $D-$dimensional continuous chain starting from a given spatial configuration $\mathbf{R}_i(s)$ arrives after a time $t_f - t_i$ to a final configuration $\mathbf{R}_f(s)$. The chain is regarded as the continuous limit of a discrete chain consisting of particles connected together with segments of fixed length. In the continuous limit, the constraints arising due to the presence of the segments take the form: $\left|\frac{\partial \mathbf{R}(t,s)}{\partial s}\right|^2 = 1$. Then, an expression of Ψ in terms of path integrals may be written as follows:

$$\Psi = \int_{\substack{\mathbf{R}(t_f,s)=\mathbf{R}_f(s)\\ \mathbf{R}(t_i,s)=\mathbf{R}_i(s)}} \mathcal{D}\mathbf{R}\mathcal{D}\lambda \exp\left\{-c\int_{t_i}^{t_f} dt \int_0^L ds \dot{\mathbf{R}}^2\right\}$$
$$\times \exp\left\{i\int_{t_i}^{t_f} dt \int_0^L ds \lambda \left(\left|\frac{\partial \mathbf{R}}{\partial s}\right|^2 - 1\right)\right\} \tag{1}$$

where the fields $\mathbf{R}(t,s)$ represent $D-$dimensional vectors. Moreover $c = M/4k_BT\tau L$, where M is the total mass of the chain, L is its length and k_B denotes the Boltzmann constant. Finally, the relaxation time τ characterizes the rate of the decay of the drift velocity of the particles composing the chain. In Eq. (1) the Lagrange multiplier λ has been introduced in order to impose the constraints using the Fourier representation of the $\delta-$function $\delta(\left|\frac{\partial \mathbf{R}}{\partial s}\right|^2 - 1)$. The model described by Eq. (1) will be called here the generalized nonlinear sigma model (GNLSM) due to its close resemblance to a two-dimensional nonlinear sigma model. Let us note that the holonomic constraint $\mathbf{R}^2 = 1$ of the nonlinear sigma model has been replaced here by a nonholonomic constraint.

In the following, we will restrict ourselves to the physically relevant case $D = 3$. The three-dimensional case is particularly interesting because it allows the introduction of topological relations. To this purpose, let us imagine two closed chains C_1 and C_2 of lengths L_1 and L_2, respectively. The trajectories of the two chains are described by the two vectors $\mathbf{R}_1(t, s_1)$ and $\mathbf{R}_2(t, s_2)$ where $0 \leq s_1 \leq L_1$ and $0 \leq s_2 \leq L_2$. The simplest way to impose topological constraints on two closed trajectories is to use the Gauss linking number χ: $\chi(t, C_1, C_2) = \frac{1}{4\pi} \oint_{C_1} d\mathbf{R}_1 \cdot \oint_{C_2} d\mathbf{R}_2 \times \frac{(\mathbf{R}_1 - \mathbf{R}_2)}{|\mathbf{R}_1 - \mathbf{R}_2|^3}$. If the trajectories of the chains were impenetrable, then χ would not depend on time, since it is not possible to change the topological configuration of a system of knots if their trajectories are not allowed to cross themselves. However, since we are not going to introduce interactions between the two chains, we just require that, during the time $t_f - t_i$, the average value of the Gauss linking number is an arbitrary constant m, i. e. $m = \frac{1}{t_f - t_i} \int_{t_i}^{t_f} \chi(t, C_1, C_2) dt$. It is possible to show that the probability function of two chains whose trajectories satisfy the above topological constraint is given by:

$$\Psi(C_1, C_2) = \int \mathcal{D}(\text{fields}) e^{-(S_1+S_2)} e^{iS_{CS} + \frac{i\mu}{2} \int_{-\infty}^{+\infty} d\xi \int d^3x \mathbf{J}^i \mathbf{A}_i}, \tag{2}$$

where $\mathcal{D}(\text{fields}) = \prod_{i=1}^{2} \mathcal{D}\mathbf{R}_i \mathcal{D}\lambda_i \mathcal{D}\mathbf{A}_i$,

$$S_i = \int_{t_i}^{t_f} dt \int_0^L ds_i \left[c\dot{\mathbf{R}}_i^2 + i\lambda_i \left(\left| \frac{\partial \mathbf{R}_i}{\partial s} \right|^2 - 1 \right) \right], \quad i = 1, 2, \tag{3}$$

$$S_{CS} = \frac{1}{t_f - t_i} \int_{-\infty}^{+\infty} d\xi \int d^3x \mathbf{A}_1(\xi, \mathbf{x}) \cdot (\nabla_{\mathbf{x}} \times \mathbf{A}_2(\xi, \mathbf{x})), \tag{4}$$

$$\mathbf{J}^i(\xi, \mathbf{x}) = \int_{t_i}^{t_f} dt \int_0^{L_i} ds_i \delta(\xi - t) \frac{\partial \mathbf{R}_i(t, s_i)}{\partial s_i} \delta^{(3)}(\mathbf{x} - \mathbf{R}_i(t, s_i)), \quad i = 1, 2. \tag{5}$$

3. Chain with Constant Angles of Bending

To conclude, we would like to mention the interesting case in which the chain is forced to form with the $z-$axis only the two fixed angles α and $\pi - \alpha$.[4] If there are no interactions depending on the z degree of freedom, it turns out that this problem can be reduced to a two-dimensional one. Since in this work interactions are not considered, the probability function of the chain may be written as follows:

$$\Psi_{\alpha,\pi-\alpha}^{3d} = \int \mathcal{D}x \mathcal{D}y \exp\{-S_{\alpha,\pi-\alpha}\} \delta((\partial_s x)^2 + (\partial_s y)^2 - \tan^2 \alpha) \tag{6}$$

where $S_{\alpha,\pi-\alpha} = c \sin^2 \alpha \int_{t_i}^{t_f} dt \int_0^L ds \left[\dot{x}^2 + \dot{y}^2 \right]$.

4. Conclusions

In this work the dynamics of a $D-$dimensional chain has been investigated. The probability function Ψ of this system is equivalent to the partition function of a generalized nonlinear sigma model. Next, the fluctuations of two topologically entangled chains have been discussed. Analogously to what happens in the case of statistical mechanics,[5,6] the complexities connected with the handling of the Gauss linking number may be partly eliminated with the introduction of Chern-Simons fields, which decouple the interactions of topological origin between the chains.[7] Still, one has to perform a path integration over the trajectories of each chain separately. In statistical mechanics, this is equivalent to compute the path integral of a particle immersed in a magnetic field. In dynamics, the particle is replaced by a two-dimensional field $\mathbf{R}(t,s)$. To evaluate such path integral is a complicated task. Finally, the problem of a three-dimensional chain admitting only fixed angles with respect to the $z-$axis is reduced to the problem of a two-dimensional chain, in a way which is similar to the reduction of the statistical mechanics of a directed polymer to the random walk of a two-dimensional particle.

Acknowledgments

This work has been financed in part by the Polish Ministry of Science, scientific project N202 156 31/2933 and by the action COST P12 financed by the European Union. F. Ferrari and J. Paturej are also grateful to W. Janke, A. Pelster and to the Max Planck Institute for the Physics of Complex Systems for the nice hospitality in Dresden.

References

1. F. Ferrari, J. Paturej, and T. A. Vilgis, *A path integral approach to the dynamics of a random chain with rigid constraints*, arXiv:cond-mat/0705.4182.
2. A. L. Kholodenko and T. A. Vilgis, *Phys. Rep.* **298** (5), 251 (1998).
3. F. Ferrari, J. Paturej, and T. A. Vilgis, *Generalized nonlinear sigma models and path integral approach to polymer dynamics*, these Proceedings.
4. F. Ferrari, J. Paturej, and T. A. Vilgis, *A path integral approach to the dynamics of random chains*, arXiv:cond-mat/0607788.
5. F. Ferrari, *Annalen der Physik* (Leipzig) **11**, 255 (2002).
6. H. Kleinert, *Path Integrals in Quantum Mechanics, Statistics, Polymer Physics, and Financial Markets*, 3rd edition (World Scientific, Singapore, 2003).
7. F. Ferrari, *Topological field theories with non-semisimple gauge group of symmetry and engineering of topological invariants*, in *Current Topics in Quantum Field Theory Research*, ed. O. Kovras (Nova Science, Hauppauge, 2006).

ON A STOCHASTIC PATH-INTEGRAL APPROACH TO BIOPOLYMER CONFORMATIONS

C. C. BERNIDO and M. V. CARPIO-BERNIDO

Research Center for Theoretical Physics, Central Visayan Institute Foundation, Jagna, Bohol 6308, Philippines
E-mail: cbernido@mozcom.com

Biopolymer conformations are investigated using the white noise path integral approach. Analytical evaluation of the path integral exhibits various features such as chirality, overwinding of biopolymers when stretched, and the helix-turn-helix structure.

Keywords: Biopolymer; Conformations; Random walk; White noise calculus; Chirality; Winding probability.

1. Introduction

Proteins and DNA's are observed to assume a unique shape or conformation before they can carry out their biological functions. Although it is known that information encoded along the length of a biopolymer is responsible for a particular conformation, the theory and dynamics of how a distinct conformation could arise is still not completely understood. It remains a theoretical challenge to come up with a model that could elucidate how a linear information can give rise to a three-dimensional shape for chain-like macromolecules.

A biopolymer with endpoints at $\mathbf{r}_0$ and $\mathbf{r}_1$ may be viewed as composed of N repeating units or monomers, each of length l (e.g., the amino acids that make up a protein). Each repeating unit is allowed to rotate as a freely hinged rod to allow for bending and twisting, such that the biopolymer of length $L = Nl$ is flexible enough to assume different conformations. A mathematical description of this pictorial representation of a biopolymer has, in fact, been investigated in the early works of S. F. Edwards,[1] as well as by S. Prager and H. L. Frisch,[2] where various paths of a random walk, or Brownian motion, represent possible conformations of a biopolymer. In this paper, we follow this approach for a biopolymer modelled by Brownian paths[3–5] using the conditional probability density function $P(\mathbf{r}_1, t \mid \mathbf{r}_0, 0)$ which solves the Fokker-Planck equation.

2. Path-Integral Representation

Designating the length $\triangle s$ travelled in time, $\tau << 1$ as, $\triangle s = v\tau << 1$, where v is the average speed of the random motion, the solution of the Fokker-Planck equation is written as a path integral,

$$P(\mathbf{r}_1, L \mid \mathbf{r}_0, 0) = \int \exp\left\{ -\frac{3}{2l} \int_0^L \left[\frac{d\mathbf{r}}{ds} - \frac{l}{3D}\mathbf{A} \right]^2 ds \right\} \mathcal{D}[\mathbf{r}], \qquad (1)$$

where we defined, $l = 3D/v$, with a dimension of length, and L the total path length. The probability density function, Eq. (1), is evaluated over all paths $\mathbf{r}(s)$ such that $\mathbf{r}(0) = \mathbf{r}_0$ and $\mathbf{r}(L) = \mathbf{r}_1$, with $0 \leq s \leq L$.

To investigate helical structures, let us employ the circular cylindrical coordinates $\mathbf{r} = (\rho, \vartheta, z)$. In the paper by F. W. Wiegel[6] for a polymer represented by a random walk, he showed that a polymer subjected to a potential $V(\rho)$ winds around the z-axis at a distance $\rho = R$, where R is a minimum of the potential. This holds true for any interaction potential $V(\rho)$ at low temperatures. Employing this result, we simplify the study of the winding behavior of polymers by projecting the random walk on the $\rho - \vartheta$ plane. One then considers a chain which lies on the plane perpendicular to the z-axis, with endpoints at $\rho_0 = (\rho_0, \vartheta_0)$ and $\rho_1 = (\rho_1, \vartheta_1)$. In this scenario, Wiegel noted that a potential $V(\rho)$, with a minimum at $\rho = R$, leads to polymer conformations on the plane confined to a narrow strip in the immediate vicinity of a circle of radius R. For $\rho = (R, \vartheta)$, the probability function corresponding to Eq. (1) becomes,

$$P(\vartheta_1, \vartheta_0) = \int \exp\left[-\frac{1}{l} \int_0^L \left[R\left(\frac{d\vartheta}{ds}\right) - \frac{l}{2D} f(s) \right]^2 ds \right] \mathcal{D}[d\vartheta]. \qquad (2)$$

Here, the drift coefficient is taken to be length-dependent, i.e., $\mathbf{A} = f(s)$ to account for effects arising from interaction of the amino acid with the aqueous environment at each length segment. The variable ϑ parameterized in terms of a Brownian motion B is given by,

$$\vartheta(L) = \vartheta_0 + \left(\sqrt{l}/R\right) \int_0^L \omega(s)\, ds, \qquad (3)$$

where, $\omega(s) = dB/ds$, is a Gaussian white noise variable.[7] This parameterization leads to a white noise functional as an integrand in Eq. (2). Its evaluation follows the Hida-Streit white noise path integral method which has been applied to quantum mechanical systems.[3]

3. Winding Probabilities

The path integral Eq. (2), with the parameterization, Eq. (3), can be evaluated using the T-transform in white noise calculus.[7] This results in,[3,8]

$$P(\vartheta_1, \vartheta_0) = \sum_{n=-\infty}^{+\infty} P_n, \tag{4}$$

where P_n is the probability function for n-times winding. The probability that a polymer chain on the plane winds n-times around the z -axis is given by, $W(n, L) = P_n/P$, and has the form,

$$W(n, L) = R\sqrt{\frac{4\pi}{Ll}} \frac{\exp\left[-\frac{R^2}{Ll}\left(2\pi n + \frac{l}{2DR}\int_0^L f \;\; ds\right)^2\right]}{\theta_3\left(\frac{l}{4DR}\int_0^L f \;\; ds\right)}, \tag{5}$$

where, for an arbitrary initial point we have set, $\vartheta_0 = \vartheta_1$. Here, $\theta_3(u)$ is the theta function[9] with $u = (l/4DR)\int f \; ds$.

4. Features of the Model

Physical consequences can be derived from the model and compared with experimental observations. Aside from the Brownian motion inherent in the formalism, we discuss below other features of the model.

4.1. *Chirality*

If we designate a polymer chain which winds counterclockwise ($n > 0$) as left-handed, and the one which winds clockwise ($n < 0$) as right-handed, then the winding probability $W(n, L)$ will depend on the handedness, or chirality of the polymer.[4] From (5), the effect of chirality becomes evident since, $W(-n)/W(n) = \exp\left[(4\pi n/L)\int f \;\; ds\right]$. We note that when the drift coefficient $f\,(s)$ is zero, the winding probability $W(n, L)$, Eq. (5), becomes independent of chirality, or the sign of n.

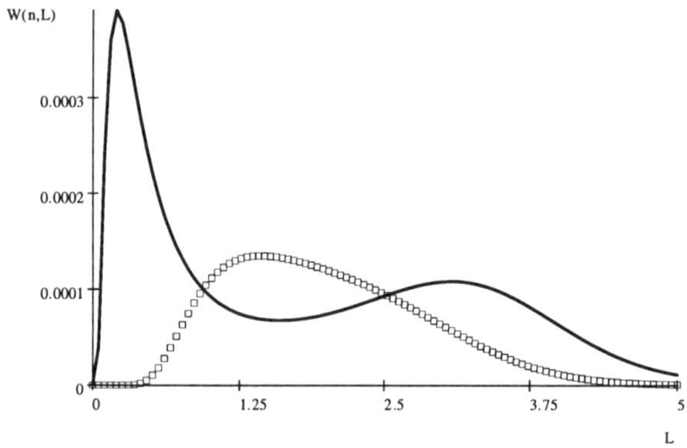

Fig. 1. Solid line is for $n = 300$, open box for $n = 5000$.

4.2. *Conformation arising from linear data*

The behavior of the winding probability $W(n, L)$, Eq. (5), is greatly influenced by the form of the drift coefficient $f(s)$. The $f(s)$, on the other hand, can assume values which differ from monomer to monomer as the variable s $(0 \leq s \leq L)$ runs along the length of the polymer. We therefore have a situation where $f(s)$, which carries linear information encoded throughout the length L of the polymer, modulates the shape of a biopolymer. What is the exact form of the drift coefficient $f(s)$? This would, for example, depend on the sequence of amino acids that link up to form a protein, as well as each monomer's interaction with the aqueous environment and neighboring monomers. We can expand the drift coefficient $f(s)$ in terms of a complete set of orthonormal functions, for example,

$$f(s) = \sum_{m=0}^{\infty} k_m \alpha_m \mathbf{L}_m (\alpha_m s), \qquad (6)$$

where $\mathbf{L}_m$ are the Laguerre polynomials, and k_m and α_m are constants. The constants k_m and α_m may be chosen to best reflect the linear data along the length of the biopolymer.

4.3. *Biopolymers overwind*

Let us consider the first four terms of Eq. (6), i.e., we take $k_m = k$=constant for $m = 0, ..., 3$, and $k_m = 0$ for other values of m. Using $R = 1$ nm, $l = 0.34$ nm, $\alpha_m = 1/\text{m}$, and $(k/D) = 10^5$, Figure 1 shows a graph of the winding

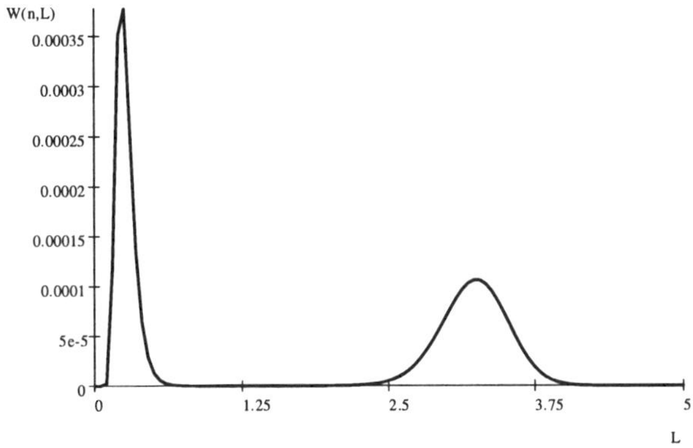

Fig. 2. Two peaks show where a helix is likely to form.

probability $W(n, L)$ vs. L for two values of the winding number: $n = 300$ (solid line) and $n = 5,000$ (open box).

Clearly, for shorter lengths L a lesser winding ($n = 300$) for the polymer dominates. For higher values of L, however, a higher winding is favored ($n = 5,000$) and an overwinding of the polymer is expected. Beyond a critical value of the length L, the lower winding regains dominance and an unwinding should occur.[5] The overwinding of biopolymers when stretched until a critical length is reached has also been shown by experiments.[10,11]

4.4. *Helix-turn-helix structural motif*

A structural motif commonly found in biopolymers, such as proteins, is the helix-turn-helix which can also be demonstrated in the present model. Taking again the first four terms of the expansion in Eq. (6), we look at the graph of $W(n, L)$ vs. L, Figure 2, for $n = 300$ and $(k/D) = 10^{5.5}$.

The two peaks in the graph show the location where a helix is most probable to form. These two peaks are separated by a domain where helices are strongly inhibited. This scenario highly suggests the formation of a helix-turn-helix structural motif.[8]

5. Conclusion

In this paper, we presented a stochastic path integral model which exhibits biomechanical properties of chain-like macromolecules. In particular, experimentally-based features of the model include: (a) a Brownian motion; (b) formation of a helical structure; (c) chirality; (d) information encoded along the length of a polymer; (e) overwinding behavior when stretched; and (f) the formation of a helix-turn-helix structural motif. Future work would focus on other forms of the drift coefficient $f(s)$.

Acknowledgments

The authors wish to thank A. Pelster and W. Janke for their kind invitation to the conference. M. V. C. B. also gratefully acknowledges the support of the Alexander von Humboldt-Stiftung.

References

1. S. F. Edwards, *Proc. Phys. Soc. London* **91**, 513 (1967).
2. S. Prager and H. L. Frisch, *J. Chem. Phys.* **46**, 1475 (1967).
3. C. C. Bernido and M. V. Carpio-Bernido, *J. Phys. A: Math. Gen.* **36**, 4247 (2003).
4. C. C. Bernido, M. V. Carpio-Bernido, and J. B. Bornales, *Phys. Lett. A* **339**, 232 (2005).
5. C. C. Bernido and M. V. Carpio-Bernido, *Phys. Lett. A* **369**, 1 (2007).
6. F. W. Wiegel, *J. Chem. Phys.* **67**, 469 (1977).
7. T. Hida, H. H. Kuo, J. Potthoff, and L. Streit, *White Noise. An Infinite Dimensional Calculus* (Kluwer, Dordrecht, 1993).
8. C. C. Bernido, M. V. Carpio-Bernido, and M. G. O. Escobido, preprint (2007).
9. I. S. Gradshteyn and I. M. Ryzhik, *Table of Integrals, Series, and Products*, 5th Edition (Academic Press, San Diego, 1994).
10. T. Lionnet, S. Joubaud, R. Lavery, D. Bensimon, and V. Croquette, *Phys. Rev. Lett.* **96**, 178102 (2006).
11. J. Gore, Z. Bryant, M. Nöllmann, M. U. Le, N. R. Cozzarelli, and C. Bustamante, *Nature* **442**, 836 (2006).

WHITE NOISE PATH INTEGRALS AND SOME APPLICATIONS

M. V. CARPIO-BERNIDO and C. C. BERNIDO

Research Center for Theoretical Physics, Central Visayan Institute Foundation, Jagna, Bohol 6308, Philippines
E-mail: cbernido@mozcom.com

The Feynman path integral for the quantum mechanical propagator is interpreted as the T-transform (infinite dimensional generalized Fourier transform in the Hida calculus) of a suitable functional in the space of Hida white noise distributions. Essential features of the approach are given and applications in the evaluation of various path integrals are noted.

Keywords: White noise calculus; Propagator.

1. Introduction

The path integral for various quantum mechanical and stochastic systems can be evaluated using the Hida white noise calculus.[1,2] In this mathematically well-founded construction of path integrals introduced by Hida and Streit,[3] there is no analytic continuation to imaginary time, or imaginary mass, and no time-slicing procedure. In particular, the white noise interpretation of the Feynman path integral given by

$$K_V(x,t \mid x_0,t_0) = \int D[x(t)] \exp\left\{\frac{i}{\hbar}\int_{t_0}^{t}\left[\frac{m}{2}\dot{x}^2(t) - V(x)\right]dt\right\}, \qquad (1)$$

particularly highlights the need to shift from classical mathematical methods to modern infinite dimensional analysis. (We note at this point that the Hida calculus differs from the use of white noise in Parisi-Wu stochastic quantization.[4]) This shift is illustrated by the comparison in Table 1.

The propagator in Eq. (1) is directly obtained from the T-transform (infinite dimensional generalized Fourier transform) of a well-defined functional in white noise space.[3,6] The additional appeal of this approach, aside from being mathematically well-defined, is that it could handle various sys-

Table 1. Comparison of finite and infinite dimensional calculus.

	Finite	Infinite
Independent variable	x_j	White Noise: $\omega(\tau)$
Coordinate system	$(x_1, ..., x_n)$	$\{\omega(t);\ t \in \mathbf{R}\}$
Function(al)	$f(x_1, ..., x_n)$	$\Phi\,(\omega(t);\ t \in \mathbf{R})$
Space	$\mathbf{R^n}$	Distributions: S^*
Measure	Lebesgue: dx	Gaussian: $d\mu(\omega)$
Transform	Fourier	T-transform $\int e^{(i<\omega,\xi>)}\Phi\,(\omega)\ d\mu(\omega)$

tems such as those with locally singular potentials like the Dirac delta function,[11,13] the harmonic[3] and time-dependent anharmonic oscillators,[9] the Morse potential,[12] particle in a uniform magnetic field,[10] a particle in a circle,[5,14] the Aharonov-Bohm set-up,[14] particle in a box,[14] and non-smooth and rapidly growing potentials.[7]

The approach appears to be versatile enough as a practical tool for physicists, and investigation has also been done for the non-relativistic problem of a quantum particle on the half-line, $\mathbb{R}_+ = \{x \mid 0 \leq x < \infty\}$, with the general boundary condition,[15] $\left[\alpha\partial_x K^{\beta/\alpha}\,(x,t \mid x_0,t_0) - \beta\ K^{\beta/\alpha}\,(x,t \mid x_0,t_0)\right]_{x=0} = 0$. The ratio, $\gamma = \beta/\alpha$, gives continuous spectra for $\gamma > 0$, and a bound state for $\gamma < 0$. Recently, the white noise path integral has also been applied to relativistic quantum systems[16] as well as in investigations of biopolymer conformations.[17,18]

2. White Noise Path Integrals

In the Hida-Streit approach, the paths are parameterized starting at some initial point x_0 and propagating in Brownian fluctuations $B(t)$ as,

$$x\,(t) = x_0 + (\hbar/m)^{\frac{1}{2}} B(t) = x_0 + (\hbar/m)^{\frac{1}{2}} \int_{t_0}^{t} \omega\,(s)\ ds\ , \tag{2}$$

where $\omega\,(s)$ is a white noise variable. With this parameterization, the Feynman integral for a free particle,[1,3,6] for instance, becomes (let $m = 1$),

$$K\,(x,t \mid x_0,t_0) = \int I_0\,(x,t \mid x_0,t_0)\ \delta\,(x(t) - x)\ d\mu\,(\omega) \tag{3}$$

where the Donsker-Dirac delta function fixes the endpoint of the fluctuating paths at x, the $d\mu(\omega)$ is the Gaussian white noise measure, N a normalization factor, and the Gauss kernel:[1,13]

$$I_0\left(x,t \mid x_0,t_0\right) = N \exp\left(\frac{i+1}{2}\int_{\mathbb{R}} \omega^2\left(\tau\right)\, d\tau\right), \tag{4}$$

is a generalized white noise functional in the Hida white noise space, i.e., $I_0 \in (S)^*$, for the Gel'fand triple, $(S) \subset (L^2) \subset (S)^*$. The integral, Eq. (3), is just a T-transform which yields the free particle propagator,

$$K\left(x,t \mid x_0,t_0\right) = \frac{1}{\sqrt{2\pi i\hbar\left|t-t_0\right|}} \exp\left[\frac{i\left(x-x_0\right)^2}{2\hbar\left|t-t_0\right|}\right]. \tag{5}$$

In the presence of a potential $V\left(x\left(\tau\right)\right)$, the interaction can be incorporated within the framework by considering the white noise functional for the interaction part of the Feynman integrand,

$$I_V = I_0\left(x,t \mid x_0,t_0\right)\ \delta\left(x(t)-x\right) \exp\left(-i\int_{t_0}^{t} V\left(x\left(\tau\right)\right) d\tau\right). \tag{6}$$

By expressing the exponential in a power series, the useful theorem of Khandekar and Streit[11,13] on the generalized Dyson series for the propagator is obtained, i.e., I_V exists as a Hida distribution. In the case of the particle on the half-line with general boundary conditions, the Feynman integrand is given by the white noise functional,

$$I_\delta = I_0\left[\delta\left(x\left(t\right)-x\right) + \delta\left(x\left(t\right)+x\right)\right] e^{-\frac{i}{\hbar}\gamma\int \delta(x)d\tau} \tag{7}$$

with the propagator as the T–transform,[15] $TI_\delta(0) \equiv K_\delta^{\beta/\alpha}(x,t \mid x_0,0)$:

$$TI_\delta(0) = \frac{1}{\sqrt{2\pi i t}}\left(e^{\frac{i}{2t}(x-x_0)^2} + e^{\frac{i}{2t}(x+x_0)^2}\right) - \gamma e^{\gamma(|x|+|x_0|)+\left(i\gamma^2 t/2\right)}$$
$$\times \mathrm{Erfc}\left[\left(\frac{|x|+|x_0|}{\sqrt{2it}}\right) + \gamma\sqrt{\frac{it}{2}}\right]. \tag{8}$$

3. Conclusion

The Feynman integral for the quantum mechanical propagator is the T-transform of well-defined generalized white noise functionals. The mathematically precise formulation with the explicit evaluation of the propagators in closed form allows the possibility of many applications. The extension of the study to general boundary conditions for different geometries and higher dimensions would also be interesting and useful for other applications.

Acknowledgments

The authors are grateful to the conference organizers and the Alexander von Humboldt-Stiftung for supporting our participation in the conference.

References

1. T. Hida, H. H. Kuo, J. Potthoff, and L. Streit, *White Noise. An Infinite Dimensional Calculus* (Kluwer, Dordrecht, 1993).
2. H. H. Kuo, *White Noise Distribution Theory* (CRC, Boca Raton, FL, 1996).
3. L. Streit and T. Hida, *Stoch. Proc. Appl.* **16**, 55 (1983).
4. G. Parisi and Y. S. Wu, *Scientia Sinica* **24**, 483 (1981).
5. A. Lascheck, P. Leukert, L. Streit, and W. Westerkamp, *Soochow J. Math.* **20**, 401 (1994).
6. M. de Faria, J. Potthoff, and L. Streit, *J. Math. Phys.* **32**, 2123 (1991).
7. M. de Faria, M. J. Oliveira, and L. Streit, *J. Math. Phys.* **46**, 063505 (2005).
8. C. Grosche, and F. Steiner, *Handbook of Feynman Integrals* (Springer-Verlag, Berlin, 1998).
9. M. Grothaus, D. C. Khandekar, J. L. Silva, and L. Streit, *J. Math. Phys.* **38**, 3278 (1997).
10. D. de Falco and D. C. Khandekar, *Stoch. Proc. Appl.* **29**, 257 (1988).
11. D. C. Khandekar and L. Streit, *Ann. Phys. (Leipzig)* **1**, 46 (1992).
12. T. Kuna, L. Streit, and W. Westerkamp, *J. Math. Phys.* **39**, 4476 (1998).
13. A. Lascheck, P. Leukert, L. Streit, and W. Westerkamp, *Rep. Math. Phys.* **33**, 221 (1993).
14. C. C. Bernido and M. V. Carpio-Bernido, *J. Math. Phys.* **43**, 1728 (2002).
15. C. C. Bernido, M. V. Carpio-Bernido, and L. Streit, in preparation.
16. C. C. Bernido, J. B. Bornales, and M. V. Carpio-Bernido, *Com. Stoch. Anal.* **1** (2007).
17. C. C. Bernido and M. V. Carpio-Bernido, *J. Phys. A: Math. Gen.* **36,** 4247 (2003); C. C. Bernido and M. V. Carpio-Bernido, *Phys. Lett. A* **369**, 1 (2007).
18. C. C. Bernido, M. V. Carpio-Bernido, and J. B. Bornales, *Phys. Lett. A* **339**, 232 (2005).

BIOPOLYMER CONFORMATIONS AS RANDOM WALKS ON THE EUCLIDEAN GROUP

N. B. BECKER

Laboratoire de Physique, ENS Lyon, 46, Allée d'Italie,
69007 Lyon, France
E-mail: nils.becker@ens-lyon.fr

The function of biopolymers depends to a large part on their shape statistics, on length scales ranging from one to thousands of monomers. I present a continuous model in which the equilibrium ensemble of macromolecular conformations is generated by a random walk with values in the Euclidean group. It includes local bending, twisting, stretching and shearing modes. The model exhibits helical structure on an intermediate length scale, while in the limit of long chains, the well-known worm-like chain behavior is recovered.

Keywords: Random walk; Group space; Biopolymer, DNA.

1. The Rigid Base Pair Model

DNA is arguably the most important, and probably the best investigated biological macromolecule.[a] An appropriate description of its conformations depends on the desired level of detail. On a length scale of a few nm, a particularly useful approximation consists in disregarding all *internal* deformations of the Watson–Crick base-pairs. In the resulting model,[1] each base-pair is considered as rigid body. The DNA molecule thus becomes a rigid base-pair (or rigid body) chain (rbc). While the rbc can describe local elasticity in great detail, its relation to intermediate and large-scale conformations of DNA is not immediate. This article aims at clarifying this point by establishing and parameterizing the corresponding continuous chain model.

In local coordinates, the energy function for weak elastic deformation of a rbc with $n+1$ base-pairs interacting only with nearest neighbors, is a

[a]The following considerations apply to general linear chain molecules with monomers modeled as rigid bodies. However, only for double-stranded DNA are experimental elastic parameter sets available to date, so this example is used exclusively.

sum of quadratic base-pair step energies:[b]

$$E = 1/2 \sum_{i=0}^{n-1} \xi^k_{ii+1} S_{kl} \xi^l_{ii+1}. \tag{1}$$

Here, the six-dimensional coordinate vector ξ_{ii+1} describes the relative orientation and position of the base-pairs i and $i+1$. It is defined such that the relaxed conformation corresponds to all $\xi^k = 0$. The stiffness matrix S is a symmetric, positive definite 6×6 matrix, and in thermal equilibrium $\langle \xi^k_{ii+1} \xi^l_{jj+1} \rangle = \delta_{ij} (S^{-1})^{kl}$. Both the relaxed conformation and S generally depend on the chemical identity of the bases i, $i+1$. Here only the homogeneous case is treated for simplicity. This does include repetitive sequences with short period as shown below.

2. Matrix Representation

To efficiently describe rigid body conformations, let's adopt a representation of the Special Euclidean Group $SE(3)$ which is widely used in mechanical engineering.[2] The conformation in space of base-pair j with respect to base-pair i is given by a rotation matrix R_{ij} and a translation vector p_{ij}, which we arrange into a 4×4 "g-matrix", $g_{ij} = \left[\begin{smallmatrix} R_{ij} & p_{ij} \\ 000 & 1 \end{smallmatrix}\right]$. The unit element will be denoted by $e = g_{ii}$. A suitable choice of local step coordinates is given by[3]

$$g_{ii+1} = g_0 \exp[\xi^k_{ii+1} X_k], \tag{2}$$

where g_0 is the g-matrix of the relaxed conformation, X_k are the infinitesimal matrix generators spanning the Lie algebra $se(3)$, and exp is the matrix exponential[c]. The shorthand $\exp[\xi] = \exp[\xi^i X_i]$ is useful.

3. Combination Formula

Let Ad denote the adjoint representation, so that $(\mathrm{Ad}g)^k_l X_k = g X_l g^{-1}$. One can then see that $\exp[\xi^k X_k] g_0 = g_0 \exp[(\mathrm{Ad}^{-1} g_0\, \xi)^k X_k]$. Using this commutation rule repeatedly on a rbc with steps of the form (2),

$$g_{0n} = g_0^n \exp[\mathrm{Ad}^{1-n} g_0\, \xi_{01}] \exp[\mathrm{Ad}^{2-n} g_0\, \xi_{12}] \cdots \exp[\xi^k_{n-1\,n} X_k]. \tag{3}$$

We write this as a singe effective step of the form $g_{0n} = g_0^n \exp[\xi_{0n}]$. Using a Baker-Campbell-Hausdorff formula on (3), $\xi_{0n} = \sum_{j=1}^n \mathrm{Ad}^{j-n} g_0\, \xi_{j-1j} +$

---

[b]Summation over upper/lower index pairs.
[c]exp is also the exponential map in the Lie group sense.

$O(\xi)^2$, where the higher order terms consist of (nested) commutators of the statistically independent step deformations ξ_{j-1j}. As a result, one obtain combination formulas which are valid up to correction terms of size $O(n\langle\xi^k\xi^l\rangle)^2$:

$$\langle\xi_{0n}\rangle = 0 \text{ and } \langle\xi^k_{0n}\xi^l_{0n}\rangle = \sum_{j=1}^{n}(\mathrm{Ad}^{j-n}g_0)^k_{k'}(S^{-1})^{k'l'}(\mathrm{Ad}^{j-n}g_0)^l_{l'}. \quad (4)$$

4. Continuous Interpolation

We now construct a continuous process $g(s)$ on $SE(3)$ which interpolates a given discrete rbc with energy (1). I.e., whenever s reaches integer values $s = i$, the conformation statistics of $g(i)$ should match with those of g_{0i}. This process will be called continuous rigid base-pair chain (crbc).

We start with an ordinary Brownian motion $\Xi(s) = s\xi_0 + \Phi(s)$ on the vector space $se(3)$, where ξ_0 is a constant drift, $\langle\Phi^k(s)\rangle = 0$, $\langle\Phi^k(s)\Phi^l(s')\rangle = \min(s, s')C^{kl}$, with positive definite 6×6 covariance C. From Ξ, one gets a Brownian motion $g(s)$ on the group space by taking the stochastic exponential,[4–7] which is defined as the solution of

$$\mathrm{d}g(s) = g(s)X_k \circ \mathrm{d}\Xi^k(s),\ g(0) = e. \quad (5)$$

In other words, $g(s)$ is driven by the left translated increments of Ξ.[d]

To relate $g(s)$ to the original discrete rbc, we need an appropriate limiting procedure. So let's define for $m, M \in \mathbb{N}$ the discrete process $g^{(M)}$ by $g^{(M)}(0) = e$ and

$$g^{(M)}(\tfrac{m+1}{M}) = g^{(M)}(\tfrac{m}{M})\exp[\tfrac{1}{M}\xi_0]\exp[\Phi(\tfrac{m+1}{M}) - \Phi(\tfrac{m}{M})]. \quad (6)$$

That is, finite increments of the form (2) and of 'size' $\frac{1}{M}$ are successively multiplied on the right. Using an approximation result[5,7] for the stochastic exponential one can show that $g^{(M)}(\frac{m}{M})$ converges to $g(s)$ in the continuum limit[e].

On the other hand, proceeding as in Section 3 to combine the finite increments of $g^{(M)}(1)$, and then again taking the continuum limit, one obtains $g(1) = \exp[\xi_0]\exp[\xi_{\mathrm{con}}]$ where up to terms of order $O(C^{ij})^2$,

$$\langle\xi_{\mathrm{con}}\rangle = 0,\ \langle\xi^k_{\mathrm{con}}\xi^l_{\mathrm{con}}\rangle = \int_{s=0}^{1}(\mathrm{Ad}^{s-1}\exp[s\xi_0])^k_{k'}C^{k'l'}(\mathrm{Ad}^{s-1}\exp[s\xi_0])^l_{l'}. \quad (7)$$

[d]The stochastic differential equation (5) needs to be interpreted in the Stratonovich sense. Interestingly, the drift correction of the equivalent Ito equation cannot be written as a linear combination of generators X_k!

[e]$M\to\infty$, $\frac{m}{M}\to s$.

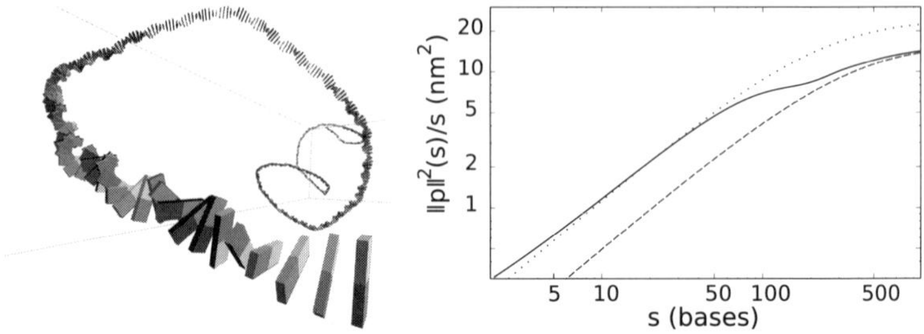

Fig. 1. Left: a thermal conformation of a rbc with repeating sequence CCCCCCTTTAA, showing the superhelical scale $s_{\rm sh}$ ~150 base-pairs. Right: cross-over of the corresponding crbc mean square end-to-end distance (solid) from a local effective worm-like chain (dotted) to a different long-scale effective worm-like chain (dashed) behavior at $s_{\rm sh}$.

We can thus relate the crbc to the discrete rbc by matching the lowest moments of ξ_{ii+1} with those of $\xi_{\rm con}$. That is, the crbc coefficients ξ_0 and C are determined by solving $g_0 = \exp[\xi_0]$, and $\langle \xi^k_{\rm con} \xi^l_{\rm con} \rangle = S^{-1}$, respectively.

An application is shown in Fig. 1. Here, a particular 11-bp sequence repeat was combined into a single compound step using (4). The helical period ~150 bp of the resulting homogeneous rbc is the *superhelical* period of the original DNA. After continuous interpolation using (7), the mean square end-to-end distance $\|p\|^2(s)$ can be computed analytically[8] starting from (5).

References

1. C. Calladine, *J. Mol. Biol.* **161**, 343 (1982).
2. R. Murray, Z. Li, and S. Sastry, *A Mathematical Introduction to Robotic Manipulation* (CRC Press, Boca Raton, 1993).
3. N. B. Becker and R. Everaers, *Phys. Rev. E* **76**, 021923 (2007).
4. H. McKean, *Mem. Col. Sci. Kyoto* **33**, 25 (1960).
5. M. Ibero, *Bull. Soc. Math. France* **100**, 175 (1976).
6. R. Karandikar, *Ann. Proba.* **10**, 1088 (1982).
7. M. Hakim-Dowek and D. Lépingle, *L'exponentielle stochastique des groupes de Lie*, in *Séminaire de Probabilités 20*, Lecture Notes in Mathematics (Springer, Berlin, 1986), p. 352.
8. N. B. Becker, *Sequence Dependent Elasticity of DNA*, PhD thesis, TU Dresden (2007).

TOWARDS A PATH-INTEGRAL FORMULATION OF CONTINUOUS TIME RANDOM WALKS

S. EULE* and R. FRIEDRICH
Institute for Theoretical Physics, Universität Münster,
D-48149 Münster, Germany
** E-mail:seule@ipp.mpg.de*

In recent years the Continuous Time Random Walk (CTRW) has been used to model anomalous diffusion in a variety of complex systems. Since this process is non-Markovian, the knowledge of single time probability distributions is not sufficient to characterize the CTRW. Using the method of subordination we construct an extension of the Wiener path integral for Brownian motion to CTRW processes. This contribution is a step towards a path integral formulation of CTRWs.

Keywords: Continuous time random walk; Anomalous diffusion; Subordination; Wiener path integral.

1. Introduction

Some 40 years ago the Continuous Time Random Walk (CTRW) was introduced by Montroll and Weiss as an extension to the ordinary random walk.[1] Instead of successive jumps with definite jumpwidth, the CTRW is subject to fluctuating jumplengths as well as random waiting times between the jumps. These random variables are drawn from two probability distributions for the jumplength $\psi(\eta)$ and for the waiting times $\omega(\tau)$ respectively.

The CTRW exhibits strong non-Markovian and non-Gaussian behavior and leads in the diffusion limit either to sub- or super-diffusive properties, depending on the jumplength- and waitingtime distributions. The time evolution of the probability distribution (PDF) of a CTRW-process can be described with the help of fractional calculus and leads to fractional generalizations of PDF-evolution equations.[2]

In recent years many physical, chemical, biological and medical systems which exhibit 'Anomalous Diffusion' have been modelled using the mathematical theory of CTRWs.[2,3]

Since it lacks the Markov-property, the knowledge of single-time PDFs

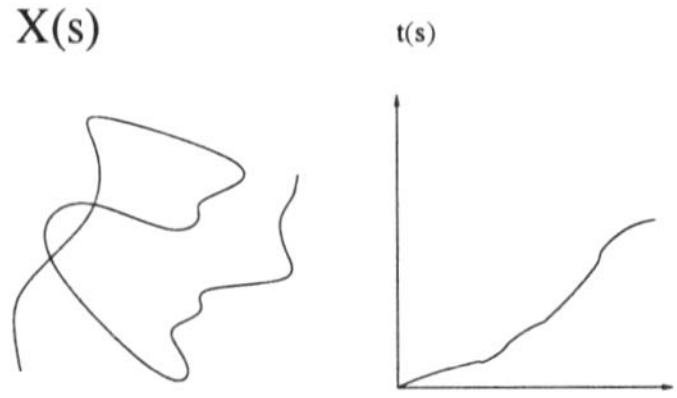

Fig. 1. The processes $X(s)$ and $t(s)$.

is not sufficient to fully characterize a CTRW. Using the method of subordination and applying an idea of Fogedby to describe CTRWs[4] we perform a step towards a path integral formulation of CTRW processes. As a result we obtain an extension of the Wiener measure leading to anomalous behavior of the transition amplitudes.

2. Fogeby's Formulation of CTRWs

In contrast to the classical description of CTRWs based on a Master equation by Montroll and Weiss, Fogedby considered a realization of a CTRW based on a system of coupled Langevin equations[4]

$$\frac{dx(s)}{ds} = F(x) + \eta(s), \quad \frac{dt(s)}{ds} = \tau(s)\,. \tag{1}$$

The random walk is in this case parametrized via an 'internal' variable s. For continuous random walks the variable s can be understood as an 'internal time' The position of the walker is described by the first equation in Eq. (1). The second process assigns to each s a physical 'wall-clock' time t. The coupled process describes the CTRW via the transformation $x(t) = x(s(t))$.

From now on we assume that $\eta(s)$ are Gaussian random numbers, so that the probability density $f_0(x, s)$ is described by a generic Fokker-Planck equation. Due to causality $\tau(s)$ has to be positive. They are usually assumed to be Lévy distributed, where the distribution has to be fully skewed to ensure that $t(s)$ is non-decreasing.[5] The solution to the coupled system (1) $x(t) = x(s(t))$ has been obtained in Ref. 6:

$$f(x,t) = f(x, s(t)) = \int_0^\infty h(s,t) f_0(x,s)\, ds\,. \tag{2}$$

Therefore the distribution $h(s,t)$ of the inverse process $s(t)$ has to be calculated. This can be done analytically and the resulting distribution is called inverse Lévy distribution.[6] Recently this concept has been extended to position-velocity space.[7]

3. Towards a Path-Integral Formulation of the CTRW

It is now straightforward to extend this concept to N-point probability distributions. Corresponding to Eq. (2) we can formulate the two point pdf

$$f(x_2,t_2;x_1,t_1) = \int_0^\infty ds_1 \int_0^\infty ds_2 h(s_2,t_2;s_1,t_1) f_0(x_2,s_2;x_1 s_1)\,. \qquad (3)$$

For detailed results concerning the two point PDFs of CTRWs see Ref. 8. An important point to mention is that the $s(t)$-process is neither stationary nor Markovian. For this reason the multi-point probability distributions $h(s_2,t_2;s_1,t_1)$ do not factorize.

If we assume the noise in Eq. (1) to be Gaussian and neglect the deterministic drift, we are now able to write down the probability to find a path which passes through gates a_i, b_i (i.e. $a_i \le x_i \le b_i$) at time t_i

$$\begin{aligned} P(a_1 \le x(t_1) \le b_1; \ldots a_N \le x(t_N) \le b_N) &= \qquad (4) \\ T(s_i) \int_0^\infty ds_1 \ldots \int_0^\infty ds_N h(t_N - t_{N-1}, s_N - s_{N-1}; \ldots; t_1, s_1) \\ \times \int_{a_1}^{b_1} dx_1 \ldots \int_{a_N}^{b_N} dx_N f(x_N - x_{N-1}, s_N - s_{N-1}; \ldots; x_1, s_1)\,. \end{aligned}$$

Here $T(s_i)$ is a time-ordering operator which ensures that for $i < j$, $s_i \le s_j$. In the discrete case, this is achieved by a multidimensional Heavyside-function. Without this timeordering we would have an overcounting of contributing paths that violate causality. Due to the assumed Gaussianity of the $x(s)$-process the PDF for this process factorizes and due to the stationarity we can rearrange the order of integration. After letting the width of the gates go to 0 and taking the limits $(s_i - s_{i-1}) \to 0$ and $N \to \infty$ a Wiener measure is obtained for this process.

4. A Generalization of the Wiener Path Integral

As has already been pointed out the subordination process $s(t)$ in Fogedby's approach is non-Markovian. For this reason we now leave the lines of Fogedby and assume the PDF $h(t_N - t_{N-1}, s_N - s_{N-1}; \ldots; t_1, s_1)$ to be separable. Clearly this violates Fogedby's description of CTRWs, but nevertheless leads to an interesting generalization of the Wiener path integral:

$$P(a_1 \leq x(t_1) \leq b_1; \ldots a_N \leq x(t_N) \leq b_N) = \tag{5}$$
$$T(s_i) \int_0^\infty ds_1 \ldots \int_0^\infty ds_N h(t_N - t_{N-1}, s_N - s_{N-1}) \ldots h(t_1, s_1) \times$$
$$\times \int_{a_1}^{b_1} dx_1 \ldots \int_{a_N}^{b_N} dx_N f(x_N - x_{N-1}, s_N - s_{N-1}) \ldots f(x_1, s_1)\,.$$

Furthermore, we want to assume the process to be stationary, so we can rearrange the order of integration. If we now narrow the gates to 0 and take the limits $(s_i - s_{i-1}) \to 0$ and $N \to \infty$ we obtain

$$P(x(t)) = \lim_{\Delta s \to 0} \lim_{N \to \infty} \prod_{j=1}^{\infty} \tag{6}$$
$$\times \left[\int_0^{s_{j-1}} h(t_j - t_{j-1}, s_j - s_{j-1}) f(x_j - x_{j-1}, s_j - s_{j-1})\, ds_j \right] dx_j\,.$$

Here we set $s_0 = \infty$ for the first limit of integration. The expression in brackets can be expressed in terms of Fox-functions for an inverse Lévy-type process $s(t)$. For $s(t) = t$ the PDF $h(s, t)$ becomes a delta distribution $h(s, t) = \delta(s - t)$ and Eq. (6) reduces to the Wiener path integral. The limit procedure in Eq. (6) has to be investigated further. Furthermore the form of the time-ordering operator in the continuum-limit is not clear.

5. Conclusions

We have presented an extension of the Wiener path integral. To this end we presented a stochastic process termed Continuous Time Random Walk (CTRW) and an approach to this process via two coupled Langevin equations. We have presented an exact expression for a path of a CTRW. Furthermore we have obtained a generalization of the Wiener path integral based on the idea of subordination.

References

1. E. W. Montroll and G. H. Weiss, *J. Math. Phys.* **6**, 167 (1965).
2. R. Metzler and J. Klafter, *Phys. Rep.* **339**, 1 (2000).
3. G. M. Zaslavsky, *Phys. Rep.* **371**, 461 (2002).
4. H. C. Fogedby, *Phys. Rev. E* **50**, 1657 (1994).
5. W. Feller, *An Introduction to Probability Theory and its Applications* (John Wiley, New York, 1965).
6. E. Barkai, *Phys. Rev. E* **63**, 046118 (2001).
7. R. Friedrich, F. Jenko, A. Baule, and S. Eule, *Phys. Rev. Lett.* **96**, 230601 (2006).
8. A. Baule and R. Friedrich, *Phys. Rev. E* **71**, 026101 (2005).

SELF-AVOIDING WALKS ON FRACTALS: SCALING LAWS

V. BLAVATSKA*,† and W. JANKE*

*Institut für Theoretische Physik, Universität Leipzig,
Postfach 100 920, D-04009 Leipzig, Germany
†Institute for Condensed Matter Physics, National Academy of Sciences of Ukraine,
79011 Lviv, Ukraine
E-mail: blavatska@itp.uni-leipzig.de, Wolfhard.Janke@itp.uni-leipzig.de

The scaling behavior of linear polymers in disordered media, modelled by self-avoiding walks (SAWs) on the backbone of percolation clusters in two, three and four dimensions is studied by numerical simulations. We apply the pruned-enriched Rosenbluth chain-growth method (PERM). Our numerical results bring about the estimates of critical exponents, governing the scaling laws of disorder averages of the configurational properties of SAWs.

Keywords: Self-avoiding walk; Percolation; Quenched disorder; Scaling.

The model of a self-avoiding walk (SAW) on a regular lattice is one of the most successful in describing the universal configurational properties of polymer chains in good solvent.[1] In particular, the average square end-to-end distance $\langle R^2 \rangle$ of a SAW with N steps obeys the scaling law

$$\langle R^2 \rangle \sim N^{2\nu_{\mathrm{SAW}}}, \tag{1}$$

where ν_{SAW} is an universal exponent depending on the space dimension d.

A question of great interest is how linear polymers behave in a disordered porous medium, which in the language of SAWs can be modelled by a structurally disordered lattice. In the present study, we consider the particular case when the underlying lattice is exactly at the percolation threshold. In this regime the scaling law (1) holds with an exponent $\nu_{p_c} \neq \nu_{\mathrm{SAW}}$.[2–8] Note that the percolation cluster itself is a fractal with fractal dimension dependent on the space dimension d.

To date there do not exist many works dedicated to Monte Carlo (MC) simulations of our problem and they do still exhibit some controversies. Early MC studies of SAW statistics on disordered (diluted) lattices at the percolation threshold indicate the rather surprising result that in two dimensions the critical exponent ν_{p_c} is *not* different from the pure lattice value. Later, some numerical uncertainties were corrected and the value for ν_{p_c} found in two dimensions is in a new universality class. This result has been confirmed in a more accurate study by Grassberger.[3] In the case of three and four dimensions, there also exist estimates indicating a new universality class,[2] but no satisfactory numerical values have been obtained so far.

Here we use the pruned-enriched Rosenbluth method (PERM) for growing chains, based on the original Rosenbluth-Rosenbluth (RR) algorithm[9] combined with enrichment strategies. PERM suppresses the weight fluctuations of the growing chain by pruning configurations with too small weights, and by enriching the sample with copies of high-weight configurations.[10] This is performed by choosing thresholds $W_n^<$ and $W_n^>$ depending on the estimate of the partition sums of n-monomer chains. If the current weight W_n of an n-monomer chain is less than $W_n^<$, this low-weight chain is pruned with probability 1/2, otherwise it is kept and its weight is doubled. If W_n exceeds $W_n^>$, the configuration is doubled and the weight of each copy is taken as half the original weight. For updating the threshold values we apply similar rules as in Ref. 11: $W_n^> = C(Z_n/Z_1)(c_n/c_1)^2$ and $W_n^< = 0.2W_n^>$, where c_n denotes the number of created chains having length n, and the parameter C controls the pruning-enrichment statistics.

Studying properties of percolative lattices, one encounters two possible statistical averages. In the first one considers only incipient percolation clusters, whereas the other statistical ensemble includes all the clusters which can be found on a percolative lattice. For the ensemble of all clusters, the SAW can start on any of the clusters, and for a N-step SAW, performed on the ith cluster, we have $\langle R^2 \rangle \sim l_i^2$, where l_i is the average size of the ith cluster. In what follows, we will be interested in the former case, that is, we perform averaging on the backbone of the percolation cluster which is defined as follows. Assume that each bond (or site) of the cluster is a resistor and that an external potential drop is applied at two ends of the cluster. The backbone is the subset of the cluster consisting of all bonds (or sites) through which the current flows, i.e., it is the structure left when all "dangling ends" are eliminated from the cluster.

To extract the backbone of the percolation cluster, we use the algorithm first introduced in Ref. 14. As starting point we choose the "seed" of the cluster, and apply the PERM algorithm, taking into account, that a SAW can have its steps only on the sites belonging to the cluster. Final results are found by averaging over all generated clusters, i.e., over different realizations

Table 1. The exponent ν_{p_c} for a SAW on a percolation cluster. MC: Monte Carlo simulations, EE: exact enumerations, RG: field-theoretic renormalization group. The first line shows $\nu_{\rm SAW}$ for SAWs on the regular lattice (d=2 (Ref. 12), d=3 (Ref. 13)).

	d	2	3	4	5	6
	$\nu_{\rm SAW}$	3/4	0.5882(11)	1/2	1/2	1/2
EE	Ref. 5	0.770(5)	0.660(5)			
	Ref. 6	0.778(15)	0.66(1)			
RG	Ref. 7	0.785	0.678	0.595	0.536	1/2
MC	Ref. 2	0.77(1)				
	Ref. 3	0.783(3)				
	our results	0.782 ± 0.003	0.667 ± 0.003	0.586 ± 0.003		

of disorder. That is, the case of so-called "quenched disorder" is considered, where the average over different realizations of disorder is taken after the configurational average has been performed. We use lattices of size up to $L_{\rm max} = 300, 200, 50$ in $d = 2, 3, 4$, respectively, and performed averages over 1000 clusters in each case.

To estimate critical exponents, a linear extrapolation with lower cutoff for the number of steps $N_{\rm min}$ is used. The results of fitting the data for $\overline{\langle r(N)\rangle}$ with $r \equiv \sqrt{R^2}$ on the backbone of percolation clusters are presented in Table 1. Note, that in percolation theory the upper critical dimension is $d_c = 6$, and thus our result for the exponent ν_{p_c} in $d = 4$ dimensions is non-trivial.

Since we can only construct lattices of a finite size L, it is not possible to perform very long SAWs on it. For each L, the scaling (1) holds only up to some "marginal" number of SAW steps $N_{\rm marg}$, as is shown in Fig. 1(a). Let us assume that $N_{\rm marg} \sim L^{\omega}$, and that for a SAW confined inside a lattice of size L finite-size scaling holds:

$$\langle r\rangle \sim N^{\nu} g\left(\frac{N}{L^{\omega}}\right), \qquad (2)$$

where $g(x) = \text{const}$ when $N \ll L^{\omega}$ so that Eq. (1) is recovered. The crossover occurs at $\langle r\rangle \sim L, N = N_{\rm marg}$, which leads to $\omega = 1/\nu$.

Having estimated values for the critical exponent ν_{p_c}, we can proceed with testing the finite-size scaling assumption (2): the data for different lattice sizes L collapse indeed onto a single curve if we have found the correct values for the critical exponents, as is demonstrated in Fig. 1(b).

To conclude, we obtained numerical estimates for exponents governing the configurational properties of SAWs on diluted lattices at the percolation threshold in $d = 2, 3, 4$. The statistical averaging was performed on the

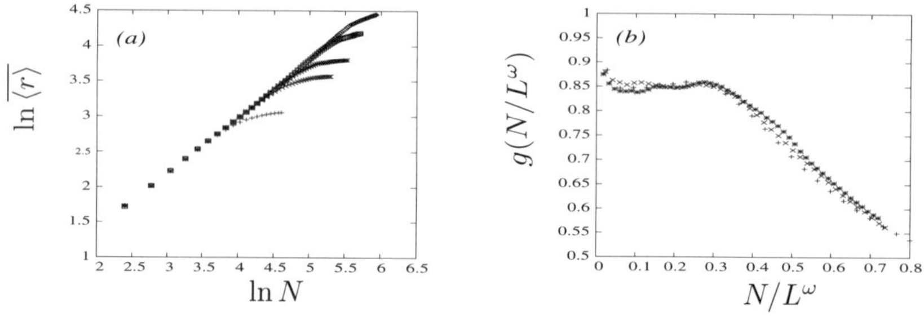

Fig. 1. (a) Mean squared end-to-end distance vs number of steps on a double logarithmic scale for SAWs on the backbone of percolation clusters in $d = 2$. The lattice size L increases from below: $L = 50, 80, 100, 150, 200$. (b) The scaling function $g(N/L^{\omega}) = N^{-\nu}\langle r \rangle$ as a function of its argument at data collapse for three different lattice sizes $L = 100, 150, 200$.

backbone of the incipient percolation cluster. The results obtained describe scaling of SAWs in a new universality class.

Acknowledgment

V.B. is grateful for support through the "Marie Curie International Incoming Fellowship" EU Programme and to the Institut für Theoretische Physik, Universität Leipzig, for hospitality.

References

1. J. des Cloizeaux and G. Jannink, *Polymers in Solution* (Clarendon Press, Oxford, 1990); P.-G. de Gennes, *Scaling Concepts in Polymer Physics* (Cornell University Press, Ithaca and London, 1979).
2. K. Y. Woo and S. B. Lee, *Phys. Rev. A* **44**, 999 (1991).
3. P. Grassberger, *J. Phys. A* **26**, 1023 (1993).
4. S. B. Lee, *J. Korean Phys. Soc.* **29**, 1 (1996).
5. M. D. Rintoul, J. Moon, and H. Nakanishi, *Phys. Rev. E* **49**, 2790 (1994).
6. A. Ordemann, M. Porto, H. E. Roman, S. Havlin, and A. Bunde, *Phys. Rev. E* **61**, 6858 (2000).
7. C. von Ferber, V. Blavatska, R. Folk, and Yu. Holovatch, *Phys. Rev. E* **70**, 035104(R) (2004).
8. H.-K. Janssen and O. Stenull, *Phys. Rev. E* **75**, 020801(R) (2007).
9. M. N. Rosenbluth and A. W. Rosenbluth, *J. Chem. Phys.* **23**, 356 (1955).
10. P. Grassberger, *Phys. Rev. E* **56**, 3682 (1997).
11. M. Bachmann and W. Janke, *Phys. Rev. Lett.* **91**, 208105 (2003); *J. Chem. Phys.* **120**, 6779 (2004).
12. B. Nienhuis, *Phys. Rev. Lett.* **49**, 1062 (1982).
13. R. Guida and J. Zinn Justin, *J. Phys. A* **31**, 8104 (1998).
14. H. J. Herrmann, D.C. Hong, and H. E. Stanley, *J. Phys. A* **17**, L261 (1984).

SMEARING DISTRIBUTIONS AND THEIR USE IN FINANCIAL MARKETS

P. JIZBA* and H. KLEINERT*,†

*Institute for Theoretical Physics, Freie Universität Berlin, Arnimallee 14, D-14195 Berlin, Germany

†Nomura International, Nomura House, 1 St Martin's-le-Grand, London, EC1 4ANP, UK

E-mail: jizba@physik.fu-berlin.de, kleinert@physik.fu-berlin.de

It is shown that superpositions of path integrals with arbitrary Hamiltonians and different scaling parameters v ("variances") obey the Chapman-Kolmogorov relation for Markovian processes if and only if the corresponding smearing distributions for v have a specific functional form. Ensuing "smearing" distributions substantially simplify the coupled system of Fokker-Planck equations for smeared and unsmeared conditional probabilities. Simple application in financial models with stochastic volatility is presented.

Keywords: Chapman-Kolmogorov equation; Path integrals; Heston's model.

1. Introduction

One often encounters in practical applications probabilities formulated as a superposition (or "smearing") of path integrals (PI) of the form

$$P(x_b, t_b|x_a, t_a) = \int_0^\infty \mathrm{d}v\ \omega(v, t_{ba}) \int_{x(t_a)=x_a}^{x(t_b)=x_b} \mathcal{D}x\mathcal{D}p\ e^{\int_{t_a}^{t_b} \mathrm{d}\tau (ip\dot{x} - vH(p,x))}. \quad (1)$$

Here $\omega(v, t_{ba})$ is some positive, continuous and normalizable function on $\mathbb{R}^+ \times \mathbb{R}^+$ with $t_{ba} = t_b - t_a$ being the time difference. Examples of (1) can be found in financial markets,[1,2] in polymer physics,[1,3] in superstatistics,[4,5] etc.

Whenever a smeared PI fulfills the Chapman-Kolmogorov equation (CKE) for continuous Markovian processes, the Feynman-Kac formula guarantees that such a superposition itself can be written as PI, i.e.

$$P(x_b, t_b|x_a, t_a) \ = \ \int_{x(t_a)=x_a}^{x(t_b)=x_b} \mathcal{D}x\mathcal{D}p\ e^{\int_{t_a}^{t_b} \mathrm{d}\tau \left(ip\dot{x} - \bar{H}(p,x)\right)}. \quad (2)$$

The new Hamiltonian $\bar{H}$ typically depends on first few momenta of $\omega(v,t)$.

Under normal circumstances, the smeared PI does not saturate CKE for Markovian processes, in fact, ad hoc choices of smearing distributions typically introduce memory into a dynamics. Our goal here will be to isolate the general class of continuous smearing distributions that conserve CKE. In Ref. 7 we have shown that ensuing distributions have important applications in evaluations of PI or in simplifications of the associated stochastic differential equations. In this note we shall briefly present the latter application. On the way we also mention a simple implication in economical models with stochastic volatility.

2. Most General Class of Smearing Distributions

We look for $\omega(v,t)$ fulfilling CKE for any intermediate time t_c:

$$P(x_b,t_b|x_a,t_a) \;=\; \int_{-\infty}^{\infty} dx\, P(x_b,t_b|x,t_c)P(x,t_c|x_a,t_a)\,. \tag{3}$$

It can be shown[7] that Eq. (3) is fulfilled only when the integral equation

$$\int_0^z dz'\;\omega(z',t)\,a\,\omega\!\left(a(z-z'),\frac{t}{a}\right) = b\,\omega\!\left(bz,\frac{t}{b}\right)\,, \tag{4}$$

($a,b \in \mathbb{R}^+$ and $1+1/a = 1/b$) holds. By defining the Laplace image function $\tilde{\omega}$ as

$$\tilde{\omega}(\xi,t) = \int_0^{\infty} dz\; e^{-\xi z}\omega(z,t), \qquad \mathrm{Re}\;\xi > 0\,, \tag{5}$$

Eq. (4) can be equivalently formulated as the functional equation for $\tilde{\omega}$

$$\tilde{\omega}(\xi,t)\,\tilde{\omega}\!\left(\frac{\xi}{a},\frac{t}{a}\right) = \tilde{\omega}\!\left(\frac{\xi}{b},\frac{t}{b}\right)\,. \tag{6}$$

Assumed normalizability and positivity of $\omega(v,t)$ implies that smearing distributions are always Laplace transformable. After setting $\alpha = 1/a$ we get

$$\tilde{\omega}(\xi,t)\,\tilde{\omega}(\alpha\xi,\alpha t) = \tilde{\omega}(\xi+\alpha\xi,t+\alpha t)\,. \tag{7}$$

This equation can be solved by iterations. An explicit general solution of Eq. (7) was found in Ref. 7 and it reads

$$\tilde{\omega}(\xi,t) \;=\; \begin{cases} [G(\xi/t)]^t, & \text{when } \; t\neq 0\,,\\ \kappa^{\xi}, & \text{when } \; t=0\,. \end{cases} \tag{8}$$

$G(x)$ is an arbitrary continuous function of x. Constant κ is determined through the initial-time value of $\omega(v,t)$. In particular, $t \to 0$ solution (cf., Eq. (8)) gives

$$\lim_{\tau\to 0}\omega(v,\tau) \;=\; \theta(v+\log\kappa)\delta(v+\log\kappa) \;=\; \delta^{+}(v+\log\kappa)\,. \tag{9}$$

Let us also notice that (7) implies $\tilde{\omega}(\zeta,\tau) > 0$ for all τ and ζ which gives $G(x) > 0$ for all x. This allows us to write

$$[G(\zeta/\tau)]^{\tau} = e^{-F(\zeta/\tau)\tau}\,. \tag{10}$$

Here $F(x)$ is some continuous function of x. Final $\omega(v,t)$ can be obtained via real Laplace's inverse transformation known as Post's inversion formula:[6]

$$\omega(v,t) = \lim_{k\to\infty}\frac{(-1)^k}{k!}\left(\frac{k}{v}\right)^{k+1}\left.\frac{\partial^k\tilde{\omega}(x,t)}{\partial x^k}\right|_{x=k/v}\,. \tag{11}$$

Real inverse transform (11) is essential because the solution of the functional equation (7) was found only for real variables in $\tilde{\omega}(\xi,t)$. In fact, complex functional equations are notoriously difficult to solve.

We finally point out that the result (8) is true also in the case when $vH(p,x) = vH_1(p,x) + H_2(p,x)$, such that $[H_1,H_2] = 0$.

3. Explicit Form of $\bar{H}$

To find $\bar{H}$ we use Post's formula (11) which directly gives

$$\begin{aligned}
&P(x_b,t_b;x_a,t_a)\\
&= -\lim_{k\to\infty}\int_0^{\infty}\mathrm{d}x\,\frac{(-x)^{k-1}}{(k-1)!}\frac{\partial^k\tilde{\omega}(x,t)}{\partial x^k}\int_{x(t_a)=x_a}^{x(t_b)=x_b}\mathcal{D}x\mathcal{D}p\;e^{\int_{t_a}^{t_b}\mathrm{d}\tau\,(ip\dot{x}-kH/x)}\\
&= \int_{x(t_a)=x_a}^{x(t_b)=x_b}\mathcal{D}x\mathcal{D}p\left[\int_0^{\infty}\mathrm{d}y\;\omega(y,t)\;e^{-y\int_{t_a}^{t_b}\mathrm{d}\tau H}\right]e^{i\int_{t_a}^{t_b}\mathrm{d}\tau\,p\dot{x}}\\
&= \int_{x(t_a)=x_a}^{x(t_b)=x_b}\mathcal{D}x\mathcal{D}p\;e^{\int_{t_a}^{t_b}\mathrm{d}\tau\,(ip\dot{x}-F(H))}\,.
\end{aligned} \tag{12}$$

In passing from second to third line we have used the asymptotic expansion of the modified Bessel function $K_k(\sqrt{k}x)$ for large k. On the last line we have utilized the definition of the Laplace image. Result (12) thus allows to identify $\bar{H}$ with $F(H)$. Let us finally mention that the normalization

$$1 = \int_0^{\infty}\mathrm{d}v\;\omega(v,t) = \tilde{\omega}(0,t) \tag{13}$$

implies that $F(0) = 0$.

4. Kramers-Moyal Expansion for $\omega(v,t)$

Let us now mention a simple application in stochastic processes. To this end we notice that both $P(x_b,t_b|x_a,t_a)$ and

$$P_v(x_b,t_b|x_a,t_a) \equiv \int_{x(t_a)=x_a}^{x(t_b)=x_b} \mathcal{D}x\mathcal{D}p\; e^{\int_{t_a}^{t_b} \mathrm{d}\tau (ip\dot{x} - vH)}, \tag{14}$$

fulfil CKE, so they can be alternatively evaluated by solving Kramers-Moyal's (KM) equations:[1,8]

$$\frac{\partial}{\partial t_b} P(x_b,t_b|x_a,t_a) = \mathbb{L}_{KM}\, P(x_b,t_b|x_a,t_a)\,, \tag{15}$$

$$\frac{\partial}{\partial t_b} P_v(x_b,t_b|x_a,t_a) = \mathbb{L}_{KM}^{v}\, P_v(x_b,t_b|x_a,t_a)\,, \tag{16}$$

where the KM operator $\mathbb{L}_{KM}$ is

$$\mathbb{L}_{KM} = \sum_{n=1}^{\infty}\left(-\frac{\partial}{\partial x_b}\right)^n D^{(n)}(x_b,t_b)\,, \tag{17}$$

and similarly for $\mathbb{L}_{KM}^{v}$. The coefficients $D^{(n)}$ and $D_v^{(n)}$ are defined through the corresponding short-time transitional probabilities, so e.g., $D_v^{(n)}(x,t)$ is

$$D_v^{(n)}(x,t) \;=\; \frac{1}{n!}\lim_{\tau\to 0}\frac{1}{\tau}\int_{-\infty}^{\infty} \mathrm{d}y\,(y-x)^n P_v(y,t+\tau|x,t)\,. \tag{18}$$

Equation (16) can be cast into an equivalent (and more convenient) system of equations. For this we rewrite (4) as

$$\omega(z,t) = \int_0^{\infty} \mathrm{d}z'\,\omega(z',t')P^{\omega}(z,t|z',t')\,, \tag{19}$$

with the conditional probability

$$P^{\omega}(z,t|z',t') \;=\frac{t}{t-t'}\,\theta(tz-t'z')\,\omega\!\left(\frac{tz-t'z'}{t-t'}, t-t'\right),$$

$$\int_0^{\infty}\mathrm{d}z\,P^{\omega}(z,t|z',t') = 1\,, \qquad \lim_{\tau\to 0} P^{\omega}(z,t+\tau|z',t) = \delta^{+}(z-z')\,. \tag{20}$$

Equations (19) and (20) ensure that the transition probability $P^{\omega}(z,t|z',t')$ obeys CKE for a Markovian process. Since the process is Markovian, we can define KM coefficients $K^{(n)}$ in the usual way as

$$K^{(n)}(v,t) \;=\; \lim_{\tau\to 0}\frac{1}{n!\tau}\int_{-\infty}^{\infty}\mathrm{d}x\,(x-v)^n P^{\omega}(x,t+\tau|v,t)\,. \tag{21}$$

From ensuing CKE for a short-time transition probability one may directly write down the KM equation for $\omega(v, t_{ba})$

$$\frac{\partial}{\partial t_{ba}}\,\omega(v,t_{ba}) = \mathbb{L}_{KM}^{(\omega)}\,\omega(v,t_{ba})\,, \quad \mathbb{L}_{KM}^{(\omega)} = \sum_{n=1}^{\infty}\left(-\frac{\partial}{\partial v}\right)^{n} K^{(n)}(v,t_{ba})\,. \tag{22}$$

In cases when both (16) and (22) are naturally or artificially truncated at $n = 2$ one gets two coupled Fokker-Planck equations

$$\frac{\partial}{\partial t}\,\omega(v,t) = \mathbb{L}_{FP}^{(\omega)}\,\omega(v,t), \quad \frac{\partial}{\partial t_b} P_v(x_b,t_b|x_a,t_a) = \mathbb{L}_{FP}^{v}\,P_v(x_b,t_b|x_a,t_a), \tag{23}$$
$$\mathbb{L}_{FP}^{(\omega)} = \mathbb{L}_{KM}^{(\omega)}(n=1,2)\,, \quad \mathbb{L}_{FP}^{v} = \mathbb{L}_{KM}^{v}(n=1,2)\,.$$

On the level of sample paths the system (23) is represented by two coupled Itō's stochastic differential equations

$$\begin{aligned} \mathrm{d}x_b &= D_v^{(1)}(x_b,t_b)\,\mathrm{d}t_b + \sqrt{2D_v^{(2)}(x_b,t_b)}\,\mathrm{d}W_1\,, \\ \mathrm{d}v &= K^{(1)}(v,t_{ba})\,\mathrm{d}t_{ba} + \sqrt{2K^{(2)}(v,t_{ba})}\,\mathrm{d}W_2\,. \end{aligned} \tag{24}$$

Here $W_1(t_b)$ and $W_2(t_{ba})$ are respective Wiener processes.

5. Economical Models with Stochastic Volatility

Note that in (24) the dynamics of the variance v is explicitly separated from the dynamics of x_b. This is a desirable starting point, for instance, in option pricing models.[1] As a simple illustration we discuss the stochastic volatility model presented in Ref. 2. To this end take G to be

$$G(x) = \left(\frac{b}{x+b}\right)^{c}, \quad b \in \mathbb{R}^{+};\ c \in \mathbb{R}_0^{+}\,. \tag{25}$$

This gives

$$\tilde{\omega}(\zeta,t) = \left(\frac{bt}{\zeta+bt}\right)^{ct} \quad \Rightarrow \quad \omega(v,t) = \frac{(bt)^{ct} v^{ct-1}}{\Gamma(ct)}\, e^{-btv}\,. \tag{26}$$

$F(0) = 0$ as it should. Distribution (26) corresponds to the Gamma distribution[2,9] $f_{bt,ct}(v)$. The Hamiltonian $\bar{H}$ associated with (26) reads

$$\bar{H}(p,x) = \bar{v}b\,\log\left(\frac{H(p,x)}{b} + 1\right), \tag{27}$$

where $\bar{v} = c/b$ is the mean of $\omega(v,t)$. As H we use the Hamiltonian from Refs. 1,2 which has the form $\mathbf{p}^2/2 + i\mathbf{p}(r/v - 1/2)$, r is a constant. This choice ensures that $P_v(x_b, t_b|x_a, t_a)$ represents a riskfree martingale distribution.[1] Full discussion of this model without truncation is presented in Ref. 2.

We consider here the truncated-level description that is epitomized by Itō's stochastic equations (24). The corresponding drift and diffusion coefficients $D_v^{(1)}$ and $D_v^{(2)}$ are then (cf. Eq. (18))

$$D_v^{(1)}(x, t_b) = \left(r - \frac{v}{2}\right), \qquad D_v^{(2)}(x, t_b) = \frac{v}{2}. \tag{28}$$

For the coefficients $K^{(n)}$ an explicit computation gives

$$K^{(1)}(v, t_{ba}) = \frac{1}{t_{ba}}(\bar{v} - v), \quad K^{(2)}(v, t_{ba}) = \frac{1}{t_{ba}^2}\frac{c}{2b^2}. \tag{29}$$

Consequently, the Itō's system takes the form

$$dx_b = \left(r - \frac{v}{2}\right) dt_b + \sqrt{v}\, dW_1, \quad dv = \frac{1}{t_{ba}}(\bar{v} - v)\, dt_{ba} + \frac{1}{t_{ba}}\sqrt{\frac{\bar{v}}{b}}\, dW_2. \tag{30}$$

Let us now view x_b as a logarithm of a stock price S, and v and r as the corresponding variance and drift. If, in addition, we replace for large t_{ab} the quantity $\sqrt{\bar{v}}$ with $\sqrt{v}$, the systems (30) reduce to

$$dS = rS\, dt_b + \sqrt{v}S\, dW_1, \quad dv = \gamma(\bar{v} - v)\, dt_{ba} + \varepsilon\sqrt{v}\, dW_2. \tag{31}$$

The system of equations (30) corresponds to Heston's stochastic volatility model,[1,10] with mean-reversion speed $\gamma = 1/t_{ba}$ and volatility $\varepsilon = 1/(t_{ba}\sqrt{b})$.

Acknowledgments

This work has been partially supported by the Ministry of Education of the Czech Republic under grant MSM 6840770039, and by the Deutsche Forschungsgemeinschaft under grant Kl256/47.

References

1. H. Kleinert, *Path Integrals in Quantum Mechanics, Statistics, Polymer Physics and Financial Markets*, 4th edition (World Scientific, Singapore, 2006).
2. P. Jizba, H. Kleinert, and P. Haener, arXiv:0708.3012 (physics.soc-ph).
3. A. Kholodenko, *Ann. Phys.* **202**, 186 (1990).
4. F. Sattin, *Physica A* **338**, 437 (2004).
5. C. Beck and E. Cohen, *Physica A* **322**, 267 (2003).
6. E. Post, *Trans. Amer. Math. Soc.* **32**, 723 (1930).
7. P. Jizba and H. Kleinert, FU Berlin preprint, to appear shortly.
8. N. V. Kampen, *Stochastic Processes in Physics and Chemistry*, 2nd edition (North Holland, Amsterdam, 1993).
9. W. Feller, *An Introduction to Probability Theory and Its Applications*, Vol. II, 2nd edition (John Wiley, New York, 1966).
10. S. Heston, *Trans. Amer. Math. Soc.* **6**, 327 (1993).

LIST OF PARTICIPANTS

Ciprian Sorin Acatrinei
Institut Fizyki
Uniwersytet Jagielloński w Krakowie
Kraków, Poland
acatrine@th.if.uj.edu.pl

Francisco Ednilson Alves dos Santos
Institut für Theoretische Physik
Freie Universität Berlin
Berlin, Germany
santos@physik.fu-berlin.de

Jan Ambjorn
Niels Bohr Institutet
Kobenhavns Universitet
Copenhagen, Denmark
ambjorn@nbi.dk

Joachim Ankerhold
Institut für Theoretische Physik
Universität Ulm
Ulm, Germany
joachim.ankerhold@uni-ulm.de

Michael Bachmann
Institut für Theoretische Physik
Universität Leipzig
Leipzig, Germany
michael.bachmann@itp.uni-leipzig.de

Antun Balaž
Institute of Physics
Scientific Computing Laboratory
Belgrade, Serbia
antun@phy.bg.ac.yu

Nils Becker
Laboratoire de Physique
École Normale Supérieure
Lyon, France
nils.becker@ens-lyon.fr

Christopher C. Bernido
Research Center for Theoretical Physics
Central Visayan Institute Foundation
Jagna, Bohol, Philippines
cbernido@mozcom.com

Rainer Bischof
Institut für Theoretische Physik
Universität Leipzig
Leipzig, Germany
rainer.bischof@itp.uni-leipzig.de

Elmar Bittner
Institut für Theoretische Physik
Universität Leipzig
Leipzig, Germany
elmar.bittner@itp.uni-leipzig.de

Viktoria Blavatska
Institut für Theoretische Physik
Universität Leipzig
Leipzig, Germany
viktoria.blavatska@itp.uni-leipzig.de

Igor Bogolubsky
Laboratory of Information Technology
Joint Institute for Nuclear Research
Dubna, Russia
bogolubs@lxpub04.jinr.ru

Juraj Boháčik
Fyzikálny Ùstav
Slovenská Akadémia vied Bratislava
Bratislava, Slovakia
bohacik@savba.sk

Jürgen Bosse
Institut für Theoretische Physik
Freie Universität Berlin
Berlin, Germany
juergen.bosse@physik.fu-berlin.de

Petr Braun
Fachbereich Physik
Universität Duisburg-Essen
Duisburg, Germany
petr.braun@uni-due.de

Fons Brosens
Department Fysica
Universiteit Antwerpen
Antwerpen, Belgium
fons.brosens@ua.ac.be

Zdzislaw Burda
Institut Fizyki
Universytet Jagielloński w Krakowie
Kraków, Poland
burda@th.if.edu.pl

Maria Victoria Carpio-Bernido
Research Center for Theoretical Physics
Central Visayan Institute Foundation
Jagna, Bohol, Philippines
cbernido@mozcom.com

David M. Ceperley
National Center
for Supercomputing Applications
University of Illinois Urbana-Champaign
Urbana, USA
ceperley@uiuc.edu

David Chandler
Department of Chemistry
University of California
Berkeley, USA
chandler@cchem.berkeley.edu

Alexander Chervyakov
Institut für Theoretische Physik
Freie Universität Berlin
Berlin, Germany
chervyak@physik.fu-berlin.de

Diego Julio Cirilo-Lombardo
Bogoliubov Laboratory of Theoretical Physics
Joint Institute for Nuclear Research
Dubna, Russia
diego@theor.jinr.ru

Daniel Cremer
Institut für Theoretische Physik
Rheinisch-Westfälische Technische Hochschule Aachen
Aachen, Germany
dcremer@physik.rwth-aachen.de

Alessandro Cuccoli
Dipartimento di Fisica
Università di Firenze
Sesto Fiorentino, Italy
cuccoli@fi.infn.it

Christoph Dehne
Institut für Theoretische Physik
Universität Leipzig
Leipzig, Germany
christoph.dehne@itp.uni-leipzig.de

Jan Dely
Ùstav Fyzikàlnych Vied
Univerzita Pavla Jozefa Šafàrika
Košice, Slovakia
jan.dely@upjs.sk

Jozef T. Devreese
TFVS - Solid State Physics
Universiteit Antwerpen
Antwerpen, Belgium
jozef.devreese@ua.ac.be

Cécile DeWitt-Morette
Department of Physics
University of Texas at Austin
Austin, USA
cdewitt@physics.utexas.edu

Jürgen Dietel
Institut für Theoretische Physik
Freie Universität Berlin
Berlin, Germany
dietel@physik.fu-berlin.de

Alain Moise Dikandé
Faculté des Sciences
Université de Buea
Buea, Cameroon
bithadel@yahoo.com

Volker Dohm
Institut für Theoretische Physik
Rheinisch-Westfälische Technische Hochschule Aachen
Aachen, Germany
vdohm@physik.rwth-aachen.de

Garii V. Efimov
Bogoliubov Laboratory of Theoretical Physics
Joint Institute for Nuclear Research
Dubna, Russia
efimovg@thsun1.jinr.ru

Stephan Eule
Institut für Theoretische Physik
Westfälische Wilhelms-Universität Münster
Münster, Germany
yeahhart@uni-muenster.de

Andrei A. Fedorenko
Laboratoire de Physique Théorique
École Normale Supérieure
Paris, France
andrei.fedorenko@lpt.ens.fr

Franco Ferrari
Instytut Fizyki
Uniwersytet Szczeciǹski
Szczecin, Poland
ferrari@fermi.fiz.univ.szczecin.pl

Oleksandr Fialko
Institut für Physik
Universität Augsburg
Augsburg, Germany
oleksandr.fialko@physik.uni-augsburg.de

Ignat V. Fialkovsky
High Energy & Particle Physics Department
St. Petersburg State University
St. Petersburg, Russia
ifialk@gmail.com

Reinhard Folk
Institut für Theoretische Physik
Universität Linz
Linz, Austria
reinhard.folk@jku.at

Erwin Frey
Arnold-Sommerfeld-Zentrum
für Theoretische Physik
Ludwig-Maximilians-Universität München
München, Germany
frey@lmu.de

Kazuo Fujikawa
Institute of Quantum Science
Nihon University
Tokyo, Japan
fujikawa@phys.cst.nihon-u.ac.jp

Gurjav Ganbold
Bogoliubov Laboratory of Theoretical
Physics
Joint Institute for Nuclear Research
Dubna, Russia
ganbold@theor.jinr.ru

Remo Garattini
Dipartimento di Progettazione e Tecnologie
Università degli studi di Bergamo
Dalmine, Italy
remo.garattini@unibg.it

Konstantin Glaum
Institut für Theoretische Physik
Freie Universität Berlin
Berlin, Germany
glaum@physik.fu-berlin.de

John Gracey
Department of Physics
University of Liverpool
Liverpool, United Kingdom
jag@amtp.liv.ac.uk

Robert Graham
Fachbereich Physik
Universität Duisburg-Essen
Duisburg, Germany
robert.graham@uni-duisburg-essen.de

Daniel Grumiller
Center for Theoretical Physics
Massachusetts Institute of Technology
Cambridge, USA
grumi@lns.mit.edu

Martin C. Gutzwiller
Department of Physics
Yale University
New York, USA
moongutz@aol.com

Fritz Haake
Fachbereich Physik
Universität Duisburg-Essen
Duisburg, Germany
fritz.haake@uni-duisburg-essen.de

Zbigniew Haba
Wydzial Fizyki i Astronomii
Universytet Wroclawski
Wroclaw, Poland
yhab@itf.uni.wroc.pl

Christian Hagendorf
Laboratoire de Physique Théorique
École Normale Supérieure
Paris, France
hagendor@lpt.ens.fr

Sigmund Heller
Physikalisches Institut
Universität Freiburg
Freiburg, Germany
sigmund.heller@physik.uni-freiburg.de

Malte Henkel
Laboratoire de Physique des Matériaux
Université Henri Poincaré Nancy I
Vandoeuvre les Nancy, Cedex, France
henkel@lpm.u-nancy.fr

Yurij Holovatch
Institute for Condensed Matter Physics
National Academy of Science of Ukraine
Lviv, Ukraine
hol@icmp.lviv.ua

Dirk Homeier
Forschungsgruppe Elementarteilchenphysik
DESY
Zeuthen, Germany
jonharker@web.de

Akira Inomata
Department of Physics
State University of New York at Albany
Albany, USA
inomata@albany.edu

Roman R. Jackiw
Center for Theoretical Physics
Massachusetts Institute of Technology
Cambridge, USA
jackiw@lns.mit.edu

Wolfhard Janke
Institut für Theoretische Physik
Universität Leipzig
Leipzig, Germany
wolfhard.janke@itp.uni-leipzig.de

Karl Jansen
Forschungsgruppe Elementarteilchenphysik
DESY
Zeuthen, Germany
karl.jansen@desy.de

Ahmed Jellal
Faculté des Sciences
Université Chouaib Doukkali
El Jadida, Marocco
ajellal@ictp.it

Petr Jizba
Institut für Theoretische Physik
Freie Universität Berlin
Berlin, Germany
jizba@physic.fu-berlin.de

Boris Kastening
Institut für Theoretische Physik
Rheinisch-Westfälische Technische
Hochschule Aachen
Aachen, Germany
kastening@physik.rwth-aachen.de

Claus Kiefer
Institut für Theoretische Physik
Universität zu Köln
Köln, Germany
kiefer@thp.uni-koeln.de

John R. Klauder
Department of Physics
University of Florida
Gainesville, USA
klauder@phys.ufl.edu

Hagen Kleinert
Institut für Theoretische Physik
Freie Universität Berlin
Berlin, Germany
kleinert@physik.fu-berlin.de

Ben Klünder
Fachbereich Physik
Universität Duisburg-Essen
Duisburg, Germany
benkluender@gmx.de

Hans Konrad Knörr
Institut für Theoretische Physik
Universität Erlangen-Nürnberg
Erlangen, Germany
hansknoerr@gmx.net

Evgeny Kochetov
Centro Internacional de Física
da Matéria Condensada
Universidade do Brasília
Brasilia, Brazil
kochetov@unb.br

Sigmund Kohler
Institut für Physik
Universität Augsburg
Augsburg, Germany
sigmund.kohler@physik.uni-augsburg.de

Klaus Kroy
Institut für Theoretische Physik
Universität Leipzig
Leipzig, Germany
klaus.kroy@itp.uni-leipzig.de

Naoto Kumano-go
Department of Mathematics
Kogakuin University
Tokyo, Japan
ft24343@ns.kogakuin.ac.jp

Frank Lee
Physics Department
George Washington University
Washington DC, USA
fxlee@gwu.edu

Hajo Leschke
Institut für Theoretische Physik
Universität Erlangen-Nürnberg
Erlangen, Germany
hajo.leschke@physik.uni-erlangen.de

Aristeu Lima
Institut für Theoretische Physik
Freie Universität Berlin
Berlin, Germany
lima@physik.fu-berlin.de

József Lőrinczi
School of Mathematics
Loughborough University
Loughborough, United Kingdom
lorinczi@ma.tum.de

Marcos Gomes Eleuterio da Luz
Departamento de Física
Universidade Federal do Paraná
Curitiba, Brazil
luz@fisica.ufpr.br

Cyril Malyshev
Steklov Institute of Mathematics
Russian Academy of Sciences
St. Petersburg, Russia
malyshev@pdmi.ras.ru

Victor Martin-Mayor
Facultad de Ciencias Físicas
Universidad Complutense de Madrid
Madrid, Spain
victor@lattice.fis.ucm.es

Edmund Menge
Institut für Theoretische Physik
Universität Erlangen-Nürnberg
Erlangen, Germany
edmund.menge@googlemail.com

Sebastian Müller
Cavendish Laboratory
University of Cambridge
Cambridge, United Kingdom
sm645@cam.ac.uk

David R. Nelson
Lyman Laboratory
Harvard University
Cambridge, USA
nelson@physics.harvard.edu

Vladimir V. Nesterenko
Bogoliubov Laboratory of Theoretical Physics
Joint Institute for Nuclear Research
Dubna, Russia
nestr@theor.jinr.ru

Thomas Neuhaus
John von Neumann Institute for Computing
Forschungszentrum Jülich
Jülich, Germany
t.neuhaus@fz-juelich.de

Giuliano Niccoli
Laboratoire de Physique et Modélisation
Université de Cergy-Pontoise
Cergy-Pontoise, France
giuliano.niccoli@ens-lyon.fr

Peter Nielaba
Abteilung Physik
Universität Konstanz
Konstanz, Germany
peter.nielaba@uni-konstanz.de

Flavio S. Nogueira
Institut für Theoretische Physik
Freie Universität Berlin
Berlin, Germany
nogueira@physik.fu-berlin.de

Andreas Nußbaumer
Institut für Theoretische Physik
Universität Leipzig
Leipzig, Germany
andreas.nussbaumer@itp.uni-leipzig.de

Jaroslaw Paturej
Institute of Physics
University of Szczecin
Szczecin, Poland
jpaturej@univ.szczecin.pl

Axel Pelster
Fachbereich Physik
Universität Duisburg-Essen
Duisburg, Germany
axel.pelster@uni-duisburg-essen.de

Vladimir N. Plechko
Bogoliubov Laboratory of Theoretical Physics
Joint Institute for Nuclear Research
Dubna, Russia
plechko@thsun1.jinr.ru

Lev Plimak
Institut für Theoretische Physik
Universität Ulm
Ulm, Germany
lev.plimak@uni-ulm.de

Lev V. Prokhorov
V. A. Fock Institute of Physics
St. Petersburg State University
St. Petersburg, Russia
lev.prokhorov@pobox.spbu.ru

Heiko Rieger
Institut für Theoretische Physik
Universität des Saarlandes
Saarbrücken, Germany
h.rieger@mx.uni-saarland.de

Ray Rivers
Physics Department
Imperial College London
London, United Kingdom
r.rivers@imperial.ac.uk

Gerd Röpke
Institut für Physik
Universität Rostock
Rostock, Germany
gerd.roepke@uni-rostock.de

Roland Rosenfelder
TEM/LTP
Paul-Scherrer Institut
Villigen, Switzerland
roland.rosenfelder@psi.ch

Steffen Röthel
Institut für Theoretische Physik
Westfälische Wilhelms-Universität Münster
Münster, Germany
steffen.roethel@uni-muenster.de

Remo Ruffini
International Center for Relativistic Astrophysics
Università degli Studi di Roma "La Sapienza"
Rome, Italy
ruffini@icra.it

Tilman Sauer
Einstein Papers Project
California Institute of Technology MC 20-7
Pasadena, USA
tilman@einstein.caltech.edu

Virulh Sa-yakanit
Center of Excellence in Forum
for Theoretical Science
Chulalongkorn University
Bangkok, Thailand
svirulh@chula.ac.th

Adriaan M. J. Schakel
Institut für Theoretische Physik
Freie Universität Berlin
Berlin, Germany
schakel@physik.fu-berlin.de

Lawrence S. Schulman
Department of Physics
Clarkson University
Potsdam, NY, USA
schulman@clarkson.edu

Moritz Schütte
Institut für Theoretische Physik
Freie Universität Berlin
Berlin, Germany
schuett@physik.fu-berlin.de

Boris N. Shalaev
A. F. Ioffe Physical and Technical Institute
Russian Academy of Sciences
St. Petersburg, Russia
shalaev@mail.ioffe.ru

Andreas Sinner
Institut für Theoretische Physik
Universität Frankfurt/Main
Frankfurt am Main, Germany
sinner@itp.uni-frankfurt.de

Andrey A. Slavnov
Steklov Mathematical Institute
Russian Academy of Sciences
Moscow, Russia
slavnov@mi.ras.ru

Semjon Stepanow
Institut für Physik
Martin Luther-Universität Halle-Wittenberg
Halle, Germany
semjon.stepanow@physik.uni-halle.de

Ludwig Streit
Forschungszentrum BiBoS
Universität Bielefeld
Bielefeld, Germany
streit@physik.uni-bielefeld.de

Zoryana Usatenko
Institut für Theoretische Physik
Leibnitz Institut für Polymerforschung
Dresden, Germany
usatenko@ipfdd.de

Alexei Vagov
Physics Department
Lancaster University
Lancaster, United Kingdom
a.vagov@lancaster.ac.uk

Ruggero Vaia
Instituto dei Sistemi Complessi
Sezione di Firenze
Sesto Fiorentino, Italy
ruggero.vaia@isc.cnr.it

Ivana Vidanović
Institute of Physics
Scientific Computing Laboratory
Belgrade, Serbia
ivanavi@phy.bg.ec.yu

Bartlomiej Waclaw
Institut für Theoretische Physik
Universität Leipzig
Leipzig, Germany
bwaclaw@googlemail.com

Martin Weigel
Institut für Physik
Johannes Gutenberg-Universität Mainz
Mainz, Germany
weigel@uni-mainz.de

Ulrich Weiß
2. Institut für Theoretische Physik
Universität Stuttgart
Stuttgart, Germany
weiss@theo2.physik.uni-stuttgart.de

Fritz Wiegel
Universteit van Amsterdam
Amsterdam, The Netherlands
ruby.p@zonnet.nl

Kay Jörg Wiese
Laboratoire de Physique Théorique
École Normale Supérieure
Paris, France
wiese@lpt.ens.fr

Günter Wunner
1. Institut für Theoretische Physik
Universität Stuttgart
Stuttgart, Germany
wunner@itp1.uni-stuttgart.de

Klaus Ziegler
Institut für Physik
Universität Augsburg
Augsburg, Germany
klaus.ziegler@physik.uni-augsburg.de

Jean Zinn-Justin
Service de Physique Theorique
CEA-Saclay
Gif-sur-Yvette, France
jean.zinn-justin@cea.fr

Martin Zirnbauer
Institut für Theoretische Physik
Universität zu Köln
Köln, Germany
zirn@thp.uni-koeln.de

AUTHOR INDEX

KEYWORD INDEX